ADVANCED THERMODYNAMICS FOR ENGINEERS

McGraw-Hill Series in Mechanical Engineering

Consulting Editors

Jack P. Holman, *Southern Methodist University*
John R. Lloyd, *Michigan State University*

Anderson: *Modern Compressible Flow: With Historical Perspective*
Arora: *Introduction to Optimum Design*
Bray and Stanley: *Nondestructive Evaluation: A Tool for Design, Manufacturing, and Service*
Burton: *Introduction to Dynamic Systems Analysis*
Culp: *Principles of Energy Conversion*
Dally: *Packaging of Electronic Systems: A Mechanical Engineering Approach*
Dieter: *Engineering Design: A Materials and Processing Approach*
Driels: *Linear Control Systems Engineering*
Eckert and Drake: *Analysis of Heat and Mass Transfer*
Edwards and McKee: *Fundamentals of Mechanical Component Design*
Gebhart: *Heat Conduction and Mass Diffusion*
Gibson: *Principles of Composite Material Mechanics*
Hamrock: *Fundamentals of Fluid Film Lubrication*
Heywood: *Internal Combustion Engine Fundamentals*
Hinze: *Turbulence*
Holman: *Experimental Methods for Engineers*
Howell and Buckius: *Fundamentals of Engineering Thermodynamics*
Hutton: *Applied Mechanical Vibrations*
Juvinall: *Engineering Considerations of Stress, Strain, and Strength*
Kane and Levinson: *Dynamics: Theory and Applications*
Kays and Crawford: *Convective Heat and Mass Transfer*
Kelly: *Fundamentals of Mechanical Vibrations*
Kimbrell: *Kinematics Analysis and Synthesis*
Krieder and Rabl: *Heating and Cooling of Buildings*
Martin: *Kinematics and Dynamics of Machines*
Modest: *Radiative Heat Transfer*
Norton: *Design of Machinery*
Phelan: *Fundamentals of Mechanical Design*
Raven: *Automatic Control Engineering*
Reddy: *An Introduction to the Finite Element Method*
Rosenberg and Karnopp: *Introduction to Physical Systems Dynamics*
Schlichting: *Boundary-Layer Theory*
Shames: *Mechanics of Fluids*
Sherman: *Viscous Flow*
Shigley: *Kinematic Analysis of Mechanisms*
Shigley and Mischke: *Mechanical Engineering Design*
Shigley and Uicker: *Theory of Machines and Mechanisms*
Stiffler: *Design with Microprocessors for Mechanical Engineers*
Stoecker and Jones: *Refrigeration and Air Conditioning*
Ullman: *The Mechanical Design Process*
Vanderplaats: *Numerical Optimization: Techniques for Engineering Design, with Applications*
Wark: *Advanced Thermodynamics for Engineers*
White: *Viscous Fluid Flow*
Zeid: *CAD/CAM Theory and Practice*

ADVANCED THERMODYNAMICS FOR ENGINEERS

Kenneth Wark, Jr.
Associate Professor of Mechanical Engineering
Purdue University

McGraw-Hill, Inc.

New York St. Louis San Francisco Auckland Bogotá
Caracas Lisbon London Madrid Mexico City Milan
Montreal New Delhi San Juan Singapore Sydney Tokyo Toronto

This book was set in Times Roman by Publications Services, Inc.
The editors were John J. Corrigan and Eleanor Castellano;
the production supervisor was Leroy A. Young.
The cover was designed by Joseph Gillians.
R. R. Donnelley & Sons Company was printer and binder.

ADVANCED THERMODYNAMICS FOR ENGINEERING

This book is printed on recycled, acid-free paper containing 10% postconsumer waste.

1 2 3 4 5 6 7 8 9 0 DOC DOC 9 0 9 8 7 6 5 4

ISBN 0-07-068292-5

Library of Congress Cataloging-in-Publication Data

Wark, Kenneth, (date).
Advanced thermodynamics for engineers / Kenneth Wark, Jr.
p. cm. — (McGraw-Hill series in mechanical engineering)
Includes bibliographical references.
ISBN 0-07-068292-5
1. Thermodynamics. I. Title. II. Series.
TJ265.W35 1995
621.402'1—dc20 94-13915

ABOUT THE AUTHOR

Kenneth Wark, Jr., received his B.S. and M.S. degrees in chemical engineering from Purdue University and the University of Illinois, respectively. He joined the faculty of mechanical engineering at Purdue after receiving his Ph.D. degree from that school. In addition to his primary teaching responsibilities in undergraduate and graduate courses in thermodynamics, he has been involved in courses in heat transfer, fluid mechanics, combustion, and design. He was one of the first recipients of the best teacher awards from the Purdue chapter of Tau Beta Pi. His experience outside of Purdue has involved work with the General Electric Company, Boeing Aircraft, Carrier Corporation, U.S. Steel (USX), Allison Division of General Motors, Atlantic-Richfield, Standard Oil of Indiana, NASA-Lewis Field, and the Dow Chemical Company. In addition to this text, Dr. Wark is the author of an undergraduate text *Thermodynamics,* Fifth Edition, McGraw-Hill, Inc., New York, 1988, and is a coauthor (with Dr. Cecil Warner) of *Air Pollution—Its Origin and Control,* Second Edition, Harper/Collins, New York, 1981.

CONTENTS

PREFACE

This textbook is the outgrowth of teaching a senior-graduate-level course in thermodynamics at Purdue University over the past 25 years. Other instructors have shared this particular teaching responsibility, and I am indebted to them for the numerous discussions on the theory and practice of thermodynamics on all levels. Over this time period we have used many different texts, few of which are still in print. As a result, the need for hand-out notes for the course increased over recent years. The text evolved from these notes. In their final form all chapters have been class tested several times, while some have been in use over the past three to four years. Student feedback has been invaluable.

Engineering graduate students have varied backgrounds, including some with nonengineering undergraduate degrees. Hence the first two chapters are used to review and extend basic first- and second-law concepts. Special attention is given to the role of entropy production in nonreactive processes. With national attention on energy usage, the third and fourth chapters discuss the theory and application of availability (or exergy) concepts to nonreactive processes and cycles. Chapters 5 and 6 present the theory and experimental correlations for the important properties of simple compressible substances. Emphasis is directed toward the use of PvT equations of state in the evaluation of other properties in the single- and two-phase regions.

Chapter 7 is a short discourse on the third law, modeling of specific-heat data, and experimental work in this area. Chapter 8 covers the general theory of multicomponent, single-phase systems, including the concepts of a partial molal property and the chemical potential. Chapter 9 extends this discussion to multiphase, multicomponent systems. General coverage of the first- and second-law applications to reacting systems is presented in Chapter 10. The affinity of a system is introduced as a means for determining the direction of change for a chemical reaction. In addition, a comparison of real- versus ideal-gas equilibrium is presented. A major topic in this chapter is the availability analysis of reactive systems with respect to a restricted dead state. Chapter 11 introduces chemical availability and the environmental dead state for reactive systems, and Chapter 12 deals with the chemical availability of fuels and the concept of the standard chemical availability. Chapter 13 provides insight into the microscopic or molecular origins of the entropy function, heat and work, and the second law of thermodynamics. This chapter could easily be assigned in conjunction with Chapter 2.

As an aid in course planning, some comments on time allotment to various chapters may be helpful. It is difficult to predict the class time to spend on the first two chapters. Instructors, however, may feel it prudent to spend considerable time in the review and expansion of basic concepts, especially with respect to the second law. For the remaining chapters, based on class experience, the author suggests that the following minimum class period allotment may be appropriate:

Chapter	3	4	5	6	7	8	9	10	11	12
Periods	4–5	3	2	6	3	5–6	6–7	6	1	3

The time to spend on microscopic aspects of entropy and the second law is also one of personal judgment.

The course for which this text was written has always been directed toward engineering applications and problem-solving techniques. To this end the text contains nearly 400 problems which cover all major subject matter. Most of these problems deal with the engineering aspects of process and cycle analysis. All the problems are expressed in SI units. The only exception is the use of either 1 atm or 1 bar as a standard reference pressure. The use of a personal computer is appropriate for problem solving in a number of cases. Polynomial curve fitting of data and the solution of cubic equations are two examples, as well as the use of a data plotter. A software package which contains both files of thermodynamic data and equation-solving techniques also is quite useful. Some problems in Chapter 10 require data from the JANAF tables. Examples of these tables are not provided in the Appendix, but the unabridged tables are readily available in engineering libraries. Instructors may wish to provide tables for a few basic substances to their classes. A solution manual, which is detailed and complete, is available. In addition to the helpful comments from students, I am indebted to Janice Napier and Kathy Taylor, who expertly typed a large portion of the original manuscript. I would also like to thank the following reviewers for their many helpful comments and suggestions: Robert Balmer, University of Wisconsin; Eugene Brown, Virginia Tech University; Baki M. Cetegen, University of Connecticut; Creighton Depew, University of Washington; John R. Lloyd, Michigan State University; Mohamad Metghalehi, Northeastern University; Michael Moran, Ohio State University; Henry Perkins, University of Arizona; G. P. Peterson, Texas A&M University; Ramendra Roy, Arizona State University; J. D. A. Walker, Lehigh University; and James Welty, Oregon State University.

Kenneth Wark, Jr.

ADVANCED THERMODYNAMICS FOR ENGINEERS

CHAPTER 1

BASIC RELATIONS AND THE FIRST LAW

This chapter reviews the basic concepts of thermodynamics and discusses the application of the general energy balance to transient systems.

1-1 REVIEW OF PROPERTY EVALUATION AND BASIC LAWS

This section begins with a review of basic definitions, followed by a convenient summary of equations and procedures for evaluation of thermodynamic properties for simple compressible substances (ideal gases and incompressible liquids) and, where tables are available, for materials of more complex behavior. In addition, some basic relationships for the first and second laws are reviewed. However, this section does not develop or examine theory in detail. For more information see any basic engineering thermodynamic text.

1-1-1 Basic Terminology

Any thermodynamic analysis begins with the precise specification of the matter or space under investigation. A thermodynamic *system* is a three-dimensional region of space which is bounded by an arbitrary surface. The *control surface* or *boundary* may be real or imaginary, may be at rest or in motion, and may change its size or shape. Difficulties frequently arise when systems are not adequately defined before an analysis is begun. A system could be as simple as the matter within a piston-cylinder device, or as complex as portions of an oil refinery. All physical space which lies outside the arbitrarily selected boundary of a system is called the *surroundings* or *environment*. An important characteristic of a boundary is that it neither contains matter nor occupies a volume in space. Hence the thickness of a boundary is mathematically zero and a property value at a point on the boundary is shared by both the system and the local environment.

The analysis of thermodynamic processes includes the study of mass and energy transfer across the boundaries. A *closed system* or *control-mass* analysis involves no mass transfer across the boundaries. However, energy transfer is allowed, as well as a change in chemical composition within the system. Conversely, mass and energy may cross the boundary of an *open (flow) system,* which is also called a *control volume*.

The system itself is described by its properties. These are macroscopic characteristics to which a numerical value can be assigned without knowledge of the history of the system. The origins of properties include those (1) directly measurable, (2) defined by laws of thermodynamics, and (3) defined by mathematical combinations of other properties. The description of the condition of a system at a given instant is known as the *state* of the system. A state can be completely identified from a small subset of (independent) property values. The values of all remaining (dependent) properties can be determined in terms of the subset. Properties are defined only when a system is in equilibrium. Any transformation of a system from one equilibrium state to another (the end states) is called a *process*. The *path* of a process is the series of states through which the system passes. When a system is infinitesimally close to equilibrium at all times during a process, the process is called *quasistatic*. A complete description of a process requires specification of the end states, the path, and the interactions across its boundaries.

When a system undergoes a change in state, the change in the value of a property does not depend on the process connecting the two end states. Hence the change in value of a property is fixed only by the end states. Conversely, if the value of a quantity is independent of the process between two end states, then that quantity is the change in a property. If the value of a quantity depends on the nature of the process between two end states, the quantity is not a property. Examples of this latter quantity are heat, work, and mass interactions, as well as entropy production and irreversibility. (These latter two items are discussed in detail in the next two chapters.) Properties are classified as either extensive or intensive. *Extensive* properties depend on the size of the system. Examples include mass, volume, and energy. The values of *intensive* properties are independent of the size of the system, such as pressure and temperature. However, they may vary in value with both position and time. When an extensive property value is divided by the mass of the system, the resulting property value per unit mass is called a *specific* property. A specific property is an intensive property. A *phase* of a system is the collection of all regions of the system that have the same intensive state and chemical composition.

A further classification of properties frequently is made in terms of intrinsic and extrinsic characteristics. In general, for any system:

1. An *extrinsic quantity* is a quantity whose value is *independent* of the nature of the substance within the system boundaries.
2. An *intrinsic quantity* is a quantity whose value is dependent on the nature of the substance composing the system.

These terms, for example, can be used with respect to properties of a system. Examples of extrinsic properties include translational velocity, rotational energy, and electric field strength. Intrinsic properties include temperature, pressure, and internal energy.

1-1-2 Conservation of Energy Principles

The conservation of energy principle for a closed system is

$$Q + W = \Delta E \tag{1-1}$$

where Q and W represent heat and work interactions, respectively, and E represents all forms of energy. Consider a closed system which undergoes a cycle, for which ΔE is zero. This system is a perpetual motion machine of the first kind (a PMM1) if no other external effect occurs other than net work output, or if the net work output is greater than the net heat input. A *simple compressible* substance is one for which the only relevant reversible (quasistatic) work mode is boundary work. For a simple compressible system which is stationary the energy balance for the *control mass* becomes

$$Q + W = \Delta U \tag{1-2}$$

where U is the intrinsic internal energy of the substance. The quantity W may represent several forms of work which may be present during a specified process. On a unit-mass basis the above equation becomes

$$q + w = \Delta u \tag{1-3}$$

In the above equations the heat Q (or q) and the work W (or w) are taken as positive when they occur into the system.

For the steady-state flow of a simple compressible fluid through a control volume (CV) the conservation of energy is frequently written in two formats. On a rate basis

$$\dot{Q} + \dot{W} = \sum_{\text{out}} \dot{m}_e \left(h + \frac{V^2}{2} + gz\right)_e - \sum_{\text{in}} \dot{m}_i \left(h + \frac{V^2}{2} + gz\right)_i \tag{1-4}$$

and on a unit-mass basis for a control volume with one inlet (1) and one outlet (2),

$$q + w = (h_2 - h_1) + \left(\frac{V_2^2}{2} - \frac{V_1^2}{2}\right) + g(z_2 - z_1) \tag{1-5}$$

where w usually represents any shaft work present, the specific enthalpy $h \equiv u + Pv$, and $\dot{m}(= \rho AV)$ is the mass flow rate at a given position.

1-1-3 Second-Law Relations

A fundamental relationship relating the entropy change of a control mass (closed system) to the heat interactions that occur is

$$dS \geq \frac{\delta Q}{T} \tag{1-6}$$

where the inequality applies to internally irreversible processes and the equality to internally reversible processes. The equation assumes a uniform temperature at the system boundary. The above inequality can be written as an equality by including a term σ, which represents the contribution to the entropy change due to the presence of internal irreversibilities. Thus the above equation is frequently written as

$$dS = \frac{\delta Q}{T} + \sigma \tag{1-7}$$

where $\sigma \geq 0$. For internally reversible processes $\delta Q/T$ is directly related to the entropy change; that is,

$$\Delta S = \int_1^2 \left(\frac{\delta Q}{T}\right)_{\text{int rev}} \tag{1-8}$$

Heat reservoirs are closed systems which are defined as internally reversible and of constant temperature. For such systems the above equation reduces to $\Delta S = Q/T$.

The relationship for dS leads to two well-known forms of the increase in entropy principle,

$$dS_{\text{adia}} \geq 0 \qquad \text{and} \qquad dS_{\text{isol}} \geq 0 \tag{1-9}$$

Thus for adiabatic or isolated systems the entropy always increases in the presence of irreversibilities, or in the limit of reversible processes the entropy change is zero.

1-1-4 Property Considerations of a Simple Compressible Substance

As noted earlier, a *simple compressible* substance is one for which the only relevant, intrinsic, reversible work mode is boundary ($P\,dV$) work. On the basis of the *state postulate* the number of independent, intensive, intrinsic properties required to fix the state of such a substance is two. In general, any two may be chosen as long as they are independent of one another. (In certain cases one property may be a function of only one other property and these two cannot be chosen as the independent variables.)

In addition to P, T, and v, we shall be concerned with the properties u (internal energy), $h = u + Pv$ (enthalpy), s (entropy), $a = u - Ts$ (Helmholtz function), $g = h - Ts$ (Gibbs function), and two specific heats:

$$c_p = \left(\frac{\partial h}{\partial T}\right)_p \qquad \text{and} \qquad c_v = \left(\frac{\partial u}{\partial T}\right)_v \tag{1-10}$$

A combination of the first law in the form $\delta q + \delta w = du$, the restriction that $\delta w = -P\,dv$, and the defining equation $ds = \delta q/T$ leads to the $T\,ds$ equations. The $T\,ds$ equations provide the basis for property relations. Two important forms are

$$T\,ds = du + P\,dv \qquad \text{and} \qquad T\,ds = dh - v\,dp \tag{1-11}$$

1-1-5 Ideal Gases

The basic equations applying to all ideal gases are

$$Pv = RT \qquad c_p - c_v = R \qquad k = \frac{c_p}{c_v} \tag{1-12}$$

$$dh = c_p\,dT \qquad du = c_v\,dT \tag{1-13}$$

$$ds = c_p\frac{dT}{T} - R\frac{dP}{p} \qquad ds = c_v\frac{dT}{T} + R\frac{dv}{v} \tag{1-14}$$

Note that h and u are functions of T *only* but that s is a function of *two* variables.

1-1-6 Ideal Gases with Constant Specific Heats

In general, c_p and c_v are functions of T only for an ideal gas. For some gases (such as the monatomic gases He and Ar) the specific heats are constant over large temperature ranges. For many other gases c_p and c_v change slowly and often can be taken to be constant over limited temperature ranges. If c_p and c_v are taken to be constant, the expressions for dh, du, and ds can be easily integrated. The results are

$$h_2 - h_1 = c_p(T_2 - T_1) \qquad \text{and} \qquad u_2 - u_1 = c_v(T_2 - T_1) \tag{1-15}$$

$$s_2 - s_1 = c_p \ln\frac{T_2}{T_1} - R\ln\frac{P_2}{P_1} = c_v\ln\frac{T_2}{T_1} + R\ln\frac{v_2}{v_1} \tag{1-16}$$

The *isentropic* process, where s is a constant, is an important one. When $s_2 = s_1$, the two expressions above can be manipulated to give

$$\left(\frac{T_2}{T_1}\right)_s = \left(\frac{P_2}{P_1}\right)^{(k-1)/k} = \left(\frac{v_1}{v_2}\right)^{(k-1)} \quad \text{and} \quad \left(\frac{P_2}{P_1}\right)_s = \left(\frac{v_1}{v_2}\right)^k \tag{1-17}$$

1-1-7 Ideal Gases with Variable Specific Heats and the Gas Tables

The gas tables (such as those by Keenan, Chao, and Kaye) enable one to account for variable specific heats by tabulating those integrals which are tedious or difficult to evaluate. Enthalpy and internal energy are calculated from

$$h - h_{\text{ref}} = \int_{T_{\text{ref}}}^{T} c_p \, dT \quad \text{and} \quad u - u_{\text{ref}} = \int_{T_{\text{ref}}}^{T} c_v \, dT$$

where equations for c_p and c_v as a function of temperature must be known. (Note that in these tables h_{ref} and u_{ref} are arbitrarily assigned the value zero at $T_{\text{ref}} = 0$.) Thus Δh and Δu are found directly from the table as the difference between two values at the given two temperatures. The entropy evaluation is written as

$$s_2 - s_1 = s_2^o - s_1^o - R \ln \frac{P_2}{P_1} \quad \text{and} \quad s^o = \int_0^T \frac{c_p}{T} \, dT \tag{1-18}$$

The s^o values are tabulated solely as a function of temperature. In a constant entropy process

$$s_2^o - s_1^o = R \ln \frac{P_2}{P_1} \quad \text{(isentropic)} \tag{1-19}$$

The gas tables sometimes also provide a somewhat more convenient scheme for isentropic processes. A quantity p_r is defined as $s^o \equiv R \ln p_r$ (so p_r is a function of T only). When $s_2 = s_1$, then

$$\frac{p_{r2}}{p_{r1}} = \frac{P_2}{P_1} \quad \text{and} \quad \frac{v_{r2}}{v_{r1}} = \frac{v_2}{v_1} \quad \text{(isentropic)} \tag{1-20}$$

The values of p_r and v_r are tabulated along with h, u, and s^o as a function of temperature in some tables.

1-1-8 Incompressible Materials

Many liquids and solids, at least over limited temperature ranges, are quite insensitive to pressure. Their behavior can be approximated by v = constant. The consequences are that

$$c_p = c_v = c, \quad du = c \, dT, \quad dh = c \, dT + v \, dp, \quad \text{and} \quad ds = \frac{c}{T} \, dT \tag{1-21}$$

If c is assumed constant

$$u_2 - u_1 = c(T_2 - T_1) \qquad h_2 - h_1 = c(T_2 - T_1) + v(P_2 - P_1) \tag{1-22}$$

and

$$s_2 - s_1 = c \ln \frac{T_2}{T_1} \tag{1-23}$$

1-1-9 Standard Tables of Properties

These tables are used particularly if the substance in question may exist as either liquid or vapor in the region of interest. Remember that for *mixtures* of liquid and vapor T depends *only* on P. The saturated-liquid and saturated-vapor properties are functions of P *or* T. Properties of mixtures are weighted average values. The quality x is defined as

$$x = \frac{\text{mass of vapor}}{\text{mass of mixture}}$$

Therefore

$$h = h_f + xh_{fg} \qquad u = u_f + xu_{fg} \qquad s = s_f + xs_{fg} \qquad v = v_f + xv_{fg}$$

where $y_{fg} = y_g - y_f$. In some tables no internal energy values are given. Use $u = h - Pv$.

If the substance is superheated vapor, the properties v, h, s (and sometimes u) are tabulated as functions of T and P. Only a few tables (notably H_2O) include the compressed liquid region, where the pressure is greater than the saturation pressure at the given temperature. Properties are presented in the same way as for superheated vapor. Since most tables do not include compressed liquid data, it may be necessary to estimate these properties. This can be done by treating the compressed liquid as an incompressible material so far as pressure is concerned. For such a liquid u, v, and s are *independent* of pressure so that saturated-liquid values (at the same temperature) may be used. To evaluate the enthalpy, one uses $dh = c\,dT + v\,dP$. Let

$$\begin{aligned} h &= \text{enthalpy of compressed liquid at } P \text{ and } T \\ h_f &= \text{enthalpy of saturated liquid at } P_{\text{sat}} \text{ and } T \\ v_f &= \text{specific volume of saturated liquid at } P_{\text{sat}} \text{ and } T \end{aligned}$$

Then $h = h_f + v_f(P - P_{\text{sat}})$.

1-1-10 Adiabatic (First-Law) Efficiencies

The actual flow through many engineering devices is approximately adiabatic. One method of estimating the irreversible performance of an adiabatic device is to compare it to the isentropic performance. This type of comparison is expressed in a parameter known as the *adiabatic, isentropic,* or *first-law efficiency* of the device. The comparison is made for the *same initial state* and the *same final pressure*.

Such efficiencies are applicable to systems such as turbines, compressors, pumps, and nozzles. Use of the subscripts "a" and "s" to represent the actual and isentropic flows, and 1 and 2 to represent the inlet and outlet states, respectively, leads to the definitions of the various efficiencies noted below.

TURBINE EFFICIENCY. The turbine efficiency η_t is the ratio of the actual work (or power) to the isentropic work; that is,

$$\eta_t = \frac{w_a}{w_s} = \frac{\dot{w}_a}{\dot{w}_s} \approx \frac{h_1 - h_{2a}}{h_1 - h_{2s}} \qquad \text{(turbine)} \tag{1-24}$$

where the approximation sign applies when the kinetic-energy change across the turbine is negligible. Turbine efficiencies generally range from 80 to 90 percent.

NOZZLE EFFICIENCY. The nozzle efficiency η_n is the ratio of the actual kinetic-energy change to the isentropic kinetic-energy change. In many cases one can reasonably neglect the inlet kinetic energy in both cases. Thus for a nozzle where $\Delta \text{ke} = \Delta h$

$$\eta_n = \frac{V_{2a}^2/2 - V_1^2/2}{V_{2s}^2/2 - V_1^2/2} = \frac{h_1 - h_{2a}}{h_1 - h_{2s}} \approx \frac{V_{2a}^2/2}{V_{2s}^2/2} \qquad \text{(nozzle)} \tag{1-25}$$

Nozzle efficiencies usually range upward from 90 percent.

COMPRESSOR EFFICIENCY. The compressor efficiency η_c is defined as the ratio of the isentropic work (or power) to the actual work. That is,

$$\eta_c = \frac{w_s}{w_a} = \frac{\dot{w}_s}{\dot{w}_a} \approx \frac{h_{2s} - h_1}{h_{2a} - h_1} \qquad \text{(compressor)} \tag{1-26}$$

where the approximation is valid when kinetic-energy changes may be neglected. Compressor efficiencies typically range from 70 to 85 percent.

PUMP EFFICIENCY. The pump efficiency η_p is defined similarly to that of a compressor.

$$\eta_p = \frac{w_s}{w_a} = \frac{\dot{w}_s}{\dot{w}_a} \approx \frac{h_{2s} - h_1}{h_{2a} - h_1} \qquad \text{(pump)} \tag{1-27}$$

where the approximation holds when Δke is negligible. Pump efficiencies range from 50 to 90 percent.

1-2 FIRST LAW FOR CLOSED SYSTEMS

The conservation of energy principle enunciated in Sec. 1-1-2 for closed systems states that

$$Q + W = \Delta E \tag{1-1}$$

Historically the above equation has been arrived at by two different starting points. These two viewpoints can be traced to Poincaré [1] and Carathéodory [2]. A lengthy discussion of the historical development of the first and second laws appears in Bejan [3]. These two approaches are summarized below. The methods differ in the primary concepts and experimental evidence used to build the theory.

1-2-1 Carathéodory's Approach

The primary concepts in this approach are work transfer and the adiabatic boundary. The following thermodynamic definition of work is frequently presented:

> *Work* is done by a system on its surroundings if the sole effect external to the system could have been the raising of a mass in a gravitational field.

This operational definition could also be expressed, for example, in terms of energy storage in a translational spring [4] or some other conservative mechanical system. Historically, however, engineers since the early steam engine era have measured the capacity of engines in terms of the maximum height a given weight could be raised. This convention of standardizing work measurements in terms of gravitational potential-energy changes remains in practice. In addition, an *adiabatic surface* is defined as a boundary of a system across which only work interactions can occur. Hence an adiabatic process is one during which all interactions can be classified as work interactions.

On the basis of these two primary concepts—work and the adiabatic boundary—a statement of the first law of thermodynamics emerges based on some indirect experimental results. To paraphase Zemansky [5]:

> If a closed system undergoes an adiabatic change of state, the work transfer experienced by the system is the same for all adiabatic paths connecting the same initial and final states.

That is, the total value of the work interactions causing the change of state depends only on the end states and not on the details of the process. However, as discussed in Sec. 1-1-1, if the value of a quantity is independent of the process between two end states, then that quantity is the measure of the change in a property. In this case work is used to define the change in energy E of a closed system. Thus

$$\delta W \equiv dE \qquad \text{(adiabatic process)} \tag{1-28a}$$

and for a finite process

$$W = \Delta E = E_2 - E_1 \qquad \text{(adiabatic process)} \tag{1-28b}$$

where the sign of W is such that, if the system does work, its energy decreases.

As a consequence of this approach, heat and temperature become derived concepts. During a nonadiabatic process another interaction called *heat* is present. When a closed system undergoes a nonadiabatic change of state, the measured work does not equal $E_2 - E_1$. By definition, the value of the heat interaction during such a change of state is given by

$$\delta Q \equiv dE - \delta W = \delta W_{\text{adia}} - \delta W \tag{1-29}$$

For a finite change of state this equation takes the expected form of Eq. (1-1). In addition, when heat transfer is zero between two systems separated by a nonadiabatic wall, the two systems are defined to be of equal temperature. Heat is energy in transition which occurs by virtue of a temperature difference.

1-2-2 Poincaré's Approach

A second approach to a first-law statement and a mathematical formulation of a conservation of energy principle is to select heat, work, and temperature as primary concepts. This requires that the concepts of heat and work be defined unambiguously. The experimental foundation for the first law in this case is the "mechanical equivalent of heat" measurements by Joule in the 1840s, and by others. A system is altered from a given initial state to a specified final state either solely by a work effect or a heat interaction. It is found that the quantity of heat required is proportional to the work required. (This proportionality, rather than equality, was due to Q and W being measured in different units in the midnineteenth century.) When the state is altered solely by a work effect and a heat effect is used to return the system to its initial state, the same proportionality between Q and W is observed. This type of data leads to the conclusion that when a system undergoes a cyclic process, the sum of the work transfer and the heat transfer for the complete cycle is zero. The mathematical formulation is

$$\oint \delta Q + \oint \delta W = 0 \tag{1-30}$$

This is the Poincaré formulation of the first law. As a final step toward a conservation of energy principle, the above relation is applied to two cyclic processes between states 1 and 2 of a closed system. The first cycle consists of path A from 1 to 2 and path C from 2 to 1. A second cycle consists of a different path B from 1 to 2 but the same path C from 2 to 1. For those two cycles Eq. (1-30) becomes

$$\int_1^2 \delta Q_A + \int_2^1 \delta Q_C + \int_1^2 \delta W_A + \int_2^1 \delta W_C = 0$$

$$\int_1^2 \delta Q_B + \int_2^1 \delta Q_C + \int_1^2 \delta W_B + \int_2^1 \delta W_C = 0$$

When the second equation is subtracted from the first equation and the result is rearranged, we find that

$$\int_1^2 (\delta Q + \delta W)_A = \int_1^2 (\delta Q + \delta W)_B$$

Because processes A and B between end states 1 and 2 are completely arbitrary, the quantity $(\delta Q + \delta W)$ must be the same for *all* paths between states 1 and 2. Therefore the value of this quantity is independent of the path and depends only on the initial and final states. Thus the quantity $(\delta Q + \delta W)$ is a measure of the differential change in a property. By definition, then,

$$\delta Q + \delta W \equiv dE$$

and we have arrived at the same conservation of energy principle for closed systems as developed by the Carathéodory approach.

Thus both the Carathéodory and Poincaré approaches, although based on different empirical data, contain two related results: (1) the existence of an energy function and (2) the mathematical formulation of a conservation of energy principle for a closed system. Either approach has its adherents. The development of Carathéodory's scheme has the advantage that heat transfer, and the standardization of its measurement, is not a primary concept.

1-3 ALLOWED STATES

One method of classifying interactions as either heat or work is based on the concept of allowed states, and is due to Hatsopoulos and Keenan [6]. By their definition, the *allowed states* of a system are all the states which the system can assume consistent with (1) the laws of matter, (2) passive resistances, and (3) constraints. The *laws of matter* include the:

1. Law of continuity
2. Law of conservation of charge
3. Law of conservation of chemical species

Passive resistances are those internal constraints which prevent the attainment of certain equilibrium states which otherwise might be expected. A short list of examples might include:

1. Impermeable walls, which prevent the passage of matter from one region of the system to another region
2. Adiabatic walls, which prevent the exchange of heat from one part of the system to another part
3. Solid friction, which might prevent the movement of one part of a system relative to another part
4. Anticatalysts, which prevent a change in chemical composition

The removal of a passive resistance constitutes a change in the nature of a system and alters the total number of states possible. Finally, *constraints* are those restrictions imposed by the environment on the states that a system might assume. They include one or more of the following:

1. An upper limit of volume
2. A conservative force field established by bodies in the environment
3. An electrically insulating wall, which prevents the flow of electricity between the system and its environment

The above conditions establish the states in which a system might be found, that is, the "allowed states."

Consider two systems initially in equilibrium. A general characteristic of all work interactions is that at least one of the two interacting systems is *not* in an equilibrium state. To place a system into nonequilibrium requires that a passive resistance or constraint must be altered, which in turn alters the number of allowed states of that system. Hence:

Any work interaction requires an alteration in the allowed states.

On the other hand, a heat interaction only requires that two systems initially in equilibrium be brought into thermal communication:

Any heat interaction does not require an alteration in the allowed states.

Thus there is a significant difference between heat and work interactions when expressed in terms of the allowed states of the interacting systems. In Chap. 2 another differentiation between heat and work will be noted based on entropy transfer across a boundary.

1-4 UNCOUPLED AND COUPLED SYSTEMS

A major thrust of engineering thermodynamics is toward the study of energy conversion devices. These devices generally fall into two categories: (1) work input to work output and (2) heat input to work output. In either case the energy conversion device falls into a class called *coupled systems*. Elements of systems which are uncoupled do not lend themselves to input-output conversion devices. Thus the differentiation between coupled and uncoupled interactions is of some importance.

1-4-1 Uncoupled Systems

In an *uncoupled* system each type of interaction (work or heat) leads to the change in a form of energy storage unique to that interaction. An uncoupled energy system is easily explained by writing the conservation of energy principle for a closed system in the following format:

$$Q + W_1 + W_2 + \cdots = \Delta U + \Delta E_1 + \Delta E_2 + \cdots$$

For an uncoupled system the value of Q equals ΔU, W_1 equals the value of ΔE_1, W_2 equals the value of ΔE_2, and so forth. Thus not only are the interactions separate, but the energy storage modes they affect are separate. As a simple physical example, consider an incompressible block of mass m and specific heat c initially at position z_1, velocity V_1, and temperature T_1. By heat and nondissipative work interactions its final state is (z_2, V_2, T_2). For the overall process

$$Q = mc(T_2 - T_1) \qquad W_1 = \Delta E_1 = m(V_2^2 - V_1^2)/2 \qquad W_2 = \Delta E_2 = mg(z_2 - z_1)$$

These are familiar concepts and equations. The major point is that a separate system behavior (in terms of energy storage) is associated with each type of interaction. The heat interaction does not affect V or z. That is, we do not expect a heat addition to result in either an increase in translational kinetic or gravitational potential energy. Table 1-1 lists some examples of uncoupled forms of work.

In the ideal situation the velocity and elevation changes discussed above could be reversed. The work interactions for the reversed progress would be equal in magnitude but opposite in directions compared to the original process. When work is completely recoverable during a cycle, the system is known as a *conservative system*. Mathematically this requires that

$$\oint \delta W = 0 \qquad \text{and} \qquad \oint \delta Q = 0$$

for uncoupled systems. [The cyclic heat relation above is required by Eq. (1-30).]

In actual systems we know that dissipative (frictional) effects of various kinds are present. As a result, work is not recoverable during such a cyclic process, and the system is said to be *nonconservative*. A major characteristic of systems undergoing a cyclic process with dissipative effects present is that the net work is always positive (into the system). On the basis of the conservation of energy, then, the net heat transfer must be negative (out of the system) for the cycle.

TABLE 1-1
Some common examples of uncoupled forms of work and energy storage

Energy storage form	Force displacement relation	Work transfer, δW	Macroscopic energy storage
Kinetic, translational	$F = m\frac{dV}{dt}$	$F\,ds$	$\frac{m}{2}(V_2^2 - V_1^2)$
Kinetic, rotational	$\tau = I\frac{d\omega}{dt}$	$-\tau\,d\theta$	$\frac{I}{2}(\omega_2^2 - \omega_1^2)$
Potential, gravitational	$F = mg$	$F\,dz$	$mg(z_2 - z_1)$
Electrical, capacitance	$\mathscr{E} = \frac{Q_c}{C}$	$\mathscr{E}\,dQ_c$	$\frac{1}{2C}(Q_{c2}^2 - Q_{c1}^2)$
Electrical, inductance	$\mathscr{E} = L\frac{di}{dt}$	$\mathscr{E}\,dQ_c$	$\frac{L}{2}(i_2^2 - i_1^2)$
Potential, electrostatic			$\frac{V\epsilon_0}{2}(E_2^2 - E_1^2)$
Potential, magnetostatic			$\frac{V\mu_0}{2}(H_2^2 - H_1^2)$

1-4-2 Coupled Systems

Unlike uncoupled systems, the *coupling* of energy interactions is said to occur when two or more interactions affect the *same* mode of energy storage. Of primary importance for engineering applications are coupled systems in which the internal energy U is affected by both heat and work interactions. In such cases separate energy storage modes are not attributed to the separate thermal and work effects. As a result, both Q and W affect the same thermodynamic properties of the system. In fact, a heat effect or work effect *alone* will change *both* thermal and mechanical properties of a coupled system. Thus the thermal and mechanical properties themselves are coupled. Table 1-2 lists some examples of coupled, quasistatic forms of work.

All energy conversion systems are examples of coupled behavior. Recall for uncoupled systems that

$$\oint \delta W = 0 \qquad \text{and} \qquad \oint \delta Q = 0$$

In contrast, for heat-to-work converters it is desired to produce net work output by the cyclic device. Thus for such a coupled thermodynamic system undergoing a cyclic process

$$\oint \delta W \neq 0 \qquad \text{and} \qquad \oint \delta Q \neq 0 \tag{1-31}$$

However, in order to satisfy the conservation of energy principle, the sum of the two cyclic integrals above must equal zero [see Eq. (1-30)]. Thus the test for the

TABLE 1-2
Examples of intrinsic, quasistatic, coupled-work interactions

Type of work	Generalized force, F_k	Generalized displacement, X_k	Work equations
Boundary	P	V	$-P\,dV$
Spring	F_s	$x\ (= L - L_0)$	$F_s\,dx$
Solid	σ	ϵ	$V\sigma\,d\epsilon$
Capacitor or chemical cell	$\mathscr{E}$	Q_c	$\mathscr{E}\,dQ_c$
Surface	γ	A	$\gamma\,dA$
Torsion	τ	θ	$\tau\,d\theta$
Electric polarization	E	P	$VE \cdot dP$
Magnetic polarization	H	M	$V\mu_0 H \cdot dM$

coupling of two different energy interactions and their thermal and mechanical properties is:

> For at least one cycle executed by the system, *in the absence of any dissipative effects*, the net value for *each* mode of energy transfer is nonzero.

This generalization is exemplified by Eq. (1-31) for heat-to-work conversion devices. In heat engine design one should select a working fluid whose mechanical and thermal properties are as strongly coupled as possible. Fluids in the gas phase fulfill this requirement quite well. For example, at low pressure $Pv = RT$. The thermal property (T) is strongly linked (coupled) to the mechanical properties (P and v). For this reason gases (ideal or real) are widely used in heat engines, as exemplified by gas turbine (GT), spark ignition (SI), and compression ignition (CI) engines.

Work interactions may be coupled to produce work-to-work conversion devices. An example is a parallel plate capacitor which undergoes both a change in plate spacing and a change in charge. For this adiabatic, nondissipative device the energy equation becomes

$$E_2 - E_1 = \int_{Q_{c,1}}^{Q_{c,2}} \mathscr{E}\, dQ_c + \int_{x_1}^{x_2} F\, dx \tag{1-32}$$

The property (constituitive) relations for this system are given by

$$\mathscr{E} = \frac{Q_c x}{\epsilon_0 A} \qquad \text{and} \qquad F = \frac{\mathscr{E} Q_c}{2x} = \frac{Q_c^2}{2\epsilon_0 A}$$

where ϵ_0 is the permittivity of free space, A is the area of a plate, and F is the external force required to hold one plate in a fixed position. Note that the constitutive relations contain common terms, and hence are coupled. Thus the two work interactions are likewise coupled. A more complete discussion of uncoupled and coupled systems appears in Ref. [4].

1-4-3 Intrinsic and Extrinsic Energies

Another viewpoint of uncoupled versus coupled behavior is through the concepts of intrinsic versus extrinsic energy effects. (The general definitions of intrinsic and extrinsic appear in Sec. 1-1.) The conservation of energy for a closed system, Eq. (1-1), can be expanded into the form

$$\delta Q + \delta W^{\text{ext}} + \delta W^{\text{int}} = dE^{\text{ext}} + dE^{\text{int}} \tag{1-33}$$

where the superscripts "ext" and "int" represent extrinsic and intrinsic forms, respectively, of work and energy. It is important to note that heat transfer affects *only* the intrinsic state of a substance. The total kinetic forms of energy include the macroscopic translational and rotational forms relative to the center of mass, as well as the kinetic energy of the particles within the system. They are represented by

$$E_{\mathrm{KE,tot}} = E_{\mathrm{trans}}^{\mathrm{ext}} + E_{\mathrm{rot}}^{\mathrm{ext}} + U_{\mathrm{KE}}^{\mathrm{int}}$$

where $E = U$ for intrinsic stored energies.

The total potential energy includes the gravitational form, as well as those due to stationary and moving charges (electrostatic and magnetostatic). A fourth form is the intrinsic potential energy of its particles within the system. These may be represented as

$$E_{\mathrm{PE,tot}} = E_{\mathrm{grav}}^{\mathrm{ext}} + E_{\mathrm{elec}}^{\mathrm{ext}} + E_{\mathrm{mag}}^{\mathrm{ext}} + U_{\mathrm{PE}}^{\mathrm{int}}$$

Five of the seven listed forms of kinetic and potential energies are extrinsic and measurable. The sum of the two remaining intrinsic terms is defined as the internal energy U of the substance. It is not directly measurable but must be evaluated in terms of measured interactions or from expressions relating it to other measurable properties. On the basis of the preceding discussion Eq. (1-33) becomes

$$\delta Q + \delta W^{\mathrm{ext}} + \delta W^{\mathrm{int}} = dU^{\mathrm{int}} + dE_{\mathrm{trans}}^{\mathrm{ext}} + dE_{\mathrm{rot}}^{\mathrm{ext}} + dE_{\mathrm{grav}}^{\mathrm{ext}} + dE_{\mathrm{elec}}^{\mathrm{ext}} + dE_{\mathrm{mag}}^{\mathrm{ext}}$$

However, all the extrinsic forms of work (the quantity δW^{ext} is a sum of terms) are uncoupled on a one-to-one basis with their corresponding dE^{ext} terms. Dropping out the uncoupled terms, we find that

$$\delta Q + \delta W^{\mathrm{int}} = dU \tag{1-34}$$

This is an *intrinsic* or *coupled energy balance*. Heat transfer and various forms of intrinsic work interactions together bring about a change in the intrinsic (molecular) energy U of the system. Thus the internal energy is defined, through the first law, in terms of coupled properties of the system. For example, for a simple compressible system the internal energy is usually expressed by a relationship of the type $u = f(T, v)$. The temperature T and specific volume v are thermal and mechanical properties, respectively.

There are situations where systems are influenced by both uncoupled and coupled behavior. For example, consider a coupled system which undergoes a change in elevation while also experiencing heat and intrinsic work effects. In this case the conservation of energy equation becomes

$$Q + W = \Delta U + \Delta E_{\mathrm{grav}}$$

where W includes the uncoupled work associated with the elevation change and the coupled intrinsic work.

1-5 GENERAL CONSERVATION OF ENERGY PRINCIPLE FOR A CONTROL VOLUME

The conservation of energy principle may be applied to a control-volume analysis by properly extending the energy equation developed in Sec. 1-2 for a control mass.

1-5-1 General Considerations

We shall consider the control volume shown enclosed within the dashed line in Fig. 1-1(a). The control surface at sections 1 and 2 is open to the transfer of mass in or out of the control volume. At some initial time t we focus our attention on the control mass, which is the sum of the mass within the control volume at that instant and the mass m_{in} within the volume element marked A, which lies adjacent to section 1. At time $t + \Delta t$ this control mass has moved so that all the mass originally in region A is now just inside the control volume. In the same time interval part of the control mass has been pushed out of the control volume into the region marked B, which is adjacent to section 2 [Fig. 1-1(b)]. The mass and volume of regions A and B are taken to be small compared to those of the control volume. During the time interval Δt heat and work interactions may have occurred to the control mass. An energy balance on the control mass is

$$Q + W = \Delta E_{\text{CM}} = E_{\text{CM},t+\Delta t} - E_{\text{CM},t} \tag{1-35}$$

Since the masses associated with regions A and B are small, compared to the control volume, the heat and work effects on the control mass are essentially the same as those which cross the boundary of the control volume.

The right-hand side of Eq. (1-35) may be expressed in terms of the control volume in the following manner: The energy E of the control mass at time t is the sum of the energies of the mass within the control volume at that instant plus the energy of the mass in region A. This may be expressed as

$$E_{\text{CM},t} = E_{\text{CV},t} + \Delta E_{\text{in}} \tag{a}$$

where the subscripts again indicate the physical significance of the terms and ΔE_{in} is the increment of energy associated with the mass which passes from region A into the control volume. In a similar fashion, the energy of the control mass at time $t + \Delta t$ would be

$$E_{\text{CM},t+\Delta t} = E_{\text{CV},t+\Delta t} + \Delta E_{\text{out}} \tag{b}$$

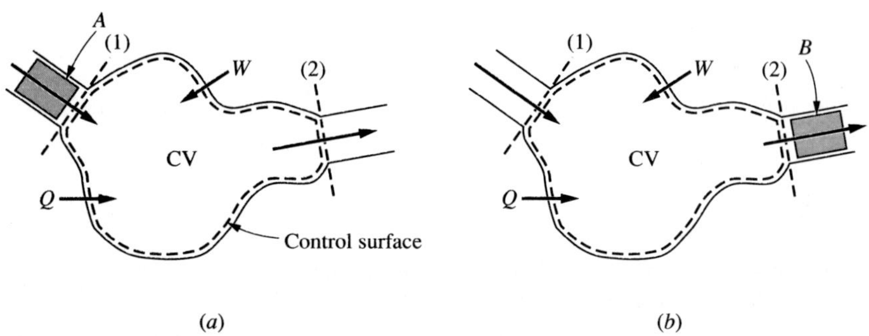

FIGURE 1-1
The development of conservation principles for a control-volume analysis. (a) Control mass at time t; (b) control mass at time $t + \Delta t$.

The quantity ΔE_{out} is the increment of energy associated with the mass which passes from the control volume into region B. Subtracting Eq. (b) from Eq. (a), we find that

$$E_{\text{CM},t+\Delta t} - E_{\text{CM},t} = (E_{\text{CV},t+\Delta t} - E_{\text{CV},t}) - \Delta E_{\text{in}} + \Delta E_{\text{out}} \tag{c}$$

The left-hand side of Eq. (c) is equal to the right-hand side of Eq. (1-35). Substitution of Eq. (c) into Eq. (1-35) yields

$$Q + W = E_{\text{CV},t+\Delta t} - E_{\text{CV},t} - \Delta E_{\text{in}} + \Delta E_{\text{out}} \tag{d}$$

The conservation of energy principle for a control volume is now placed on a rate basis by dividing Eq. (d) by the time interval Δt. Hence

$$\frac{Q}{\Delta t} + \frac{W}{\Delta t} = \frac{E_{\text{CV},t+\Delta t} - E_{\text{CV},t}}{\Delta t} - \frac{\Delta E_{\text{in}}}{\Delta t} + \frac{\Delta E_{\text{out}}}{\Delta t} \tag{e}$$

The limit is now taken as Δt approaches zero so that each term may be expressed as a derivative. As a result,

$$\dot{Q} + \dot{W} = \frac{dE_{\text{CV}}}{dt} - \frac{dE_{\text{in}}}{dt} + \frac{dE_{\text{out}}}{dt} \tag{1-36}$$

where $\delta Q/dt \equiv \dot{Q}$ and $\delta W/dt \equiv \dot{W}$. $\dot{Q}$ is the instantaneous rate of heat transfer to the control volume, and $\dot{W}$ is the rate of work done on the system, or the power.

At this point it is desirable to change the last two quantities in Eq. (1-36) to terms identifiable with the mass crossing the control surface. Because the mass in region A would be differentially small for a time interval dt, this small mass which lies adjacent to section 1 would have the thermodynamic properties associated with that particular section if the condition of local equilibrium were valid. Consequently, $E_{\text{in}} = e_1 m_1$, or $dE_{\text{in}}/dt = e_1(dm_1/dt)$. A similar expression holds for the transport of energy at section 2. Further, dm/dt is defined as the mass flow rate $\dot{m}$. Thus $dE_{\text{in}}/dt = e_{\text{in}}\dot{m}_{\text{in}}$ and $dE_{\text{out}}/dt = e_{\text{out}}\dot{m}_{\text{out}}$. Substitution of these two expressions into Eq. (1-36) leads to

$$\dot{Q} + \dot{W} = \frac{dE_{\text{CV}}}{dt} - e_{\text{in}}\dot{m}_{\text{in}} + e_{\text{out}}\dot{m}_{\text{out}} \tag{1-37}$$

The energy e represents all forms of conceivable energy that might be associated with a unit of mass.

Finally, the work term in Eq. (1-37) must be separated into three parts. The first of these is the flow or displacement work associated with mass crossing a control surface. The work δW necessary to push a differential mass δm into a control volume is the product of the force PA and the distance $V\,dt$ under uniform flow, or $PAV\,dt$. Recalling that $\dot{m} = \rho AV$, we find that $\delta W_{\text{flow}} = Pv\dot{m}\,dt$, or on a rate basis, $\dot{W}_{\text{flow}} = Pv\dot{m}$. Secondly, we must consider shaft work which is measured by devices external to the control volume, such as a dynamometer. Thirdly, other forms of work may be present in special situations. For example, if the boundary of the control volume is not rigid, boundary or $P\,dV$ work may occur. The second and third forms of work discussed above will be lumped together as $\dot{W}_{\text{net}}$ in the conservation of energy principle. On the basis of the preceding discussion Eq. (1-37) may be written as

$$\dot{Q} + \dot{W}_{\text{net}} + (Pv)_{\text{in}}\dot{m}_{\text{in}} - (Pv)_{\text{out}}\dot{m}_{\text{out}} = \frac{dE_{\text{CV}}}{dt} - e_{\text{in}}\dot{m}_{\text{in}} + e_{\text{out}}\dot{m}_{\text{out}} \tag{1-38}$$

where the term $\dot{W}_{\text{net}}$ represents shaft work and any other forms of work associated with the control volume.

In applying Eq. (1-38), we commonly restrict the types of energy associated with flow of mass across the system boundary to internal, linear kinetic, and gravitational forms. That is,

$$e = u + \frac{V^2}{2} + gz \tag{1-39}$$

In addition, $h \equiv u + Pv$. By collecting common terms in Eq. (1-38), and by making use of the preceding expressions for e and h, the energy balance on a rate basis for a control volume with multiple inlets and outlets is written as

$$\dot{Q} + \dot{W}_{\text{net}} + \sum_{\text{in}} \dot{m}_i \left(h + \frac{V^2}{2} + gz\right)_i - \sum_{\text{out}} \dot{m}_e \left(h + \frac{V^2}{2} + gz\right)_e = \frac{dE_{\text{CV}}}{dt} \tag{1-40}$$

All terms on the left of Eq. (1-40) represent interactions which add energy to or remove energy from the control volume. The term on the right is the resultant change of energy of the control volume due to these interactions. By an analysis similar to that used above a *conservation of mass* (or *continuity of flow*) *principle* on a rate basis for a control volume takes the form

$$\sum_{\text{in}} \dot{m}_i - \sum_{\text{out}} \dot{m}_e = \frac{dm_{\text{CV}}}{dt} \tag{1-41}$$

where the subscripts i and e represent inlet and exit states. This is a general expression for the conservation of mass of a control volume.

1-5-2 Steady State

Many engineering systems can be modeled as one of steady state. In a *steady-state* control-volume properties at a given position do not change with time. On the basis of the definition of steady state the total energy within the control volume under steady state remains constant with time. This requires that $dE_{\text{CV}}/dt = 0$. As a result, Eq. (1-40) reduces to

$$\dot{Q} + \dot{W}_{\text{shaft}} = \sum_{\text{out}} \dot{m}_e \left(h + \frac{V^2}{2} + gz\right)_e - \sum_{\text{in}} \dot{m}_i \left(h + \frac{V^2}{2} + gz\right)_i \tag{1-42}$$

The mass flow rates $\dot{m}$ are not equal at each open boundary for control volumes with multiple inlets and outlets. However, the values are related by the continuity of flow equation. In steady state $dm_{CV}/dt = 0$. Therefore Eq. (1-41) reduces to

$$\sum_{\text{in}} \dot{m}_i = \sum_{\text{out}} \dot{m}_e \tag{1-43}$$

where $\dot{m} = \rho AV = AV/v$.

There are numerous applications of steady-state processes that involve only one inlet (position 1) and one outlet (position 2). In this circumstance $\dot{m}_i = \dot{m}_e = \dot{m}$ and Eq. (1-42) reduces to

$$\begin{aligned}\dot{Q} + \dot{W}_{\text{shaft}} &= \dot{m}\left[\left(h + \frac{V^2}{2} + gz\right)_2 - \left(h + \frac{V^2}{2} + gz\right)_1\right] \\ &= \dot{m}\left[(h_2 - h_1) + \frac{V_2^2 - V_1^2}{2} + g(z_2 - z_1)\right]\end{aligned} \tag{1-44}$$

The above conservation of energy equation for a control volume with one inlet and one outlet can be written on a unit-mass basis in the form

$$\begin{aligned}q + w_{\text{shaft}} &= \left(h + \frac{V^2}{2} + gz\right)_2 - \left(h + \frac{V^2}{2} + gz\right)_1 \\ &= (h_2 - h_1) + \frac{V_2^2 - V_1^2}{2} + g(z_2 - z_1)\end{aligned} \tag{1-45}$$

Equations (1-42), (1-44), and (1-45) are important statements of the conservation of energy principle for a control volume under steady-state conditions.

1-6 TRANSIENT FLOW ANALYSIS

The analysis of nonsteady-flow devices serves to strengthen one's understanding of and one's ability to use the general conservation principles related to mass and energy. The importance of the proper selection of a control volume often becomes apparent in such analyses. Also, there are useful applications of devices which operate under nonsteady flow. For example, some wind tunnels operate on the discharge of a gas which has been stored under high pressure in a large tank. Compressed air stored in a cylinder might be used to drive a gas turbine coupled with a generator as an auxiliary electric power source. In steady-state analyses we usually consider the control volume fixed in space, size, and shape. We now wish to relinquish two of these contraints and to allow the size and shape to vary, if required. However, the control volume will still be considered to be fixed in location.

Transient flow problems arise when the control volume is not in steady state. Heat, work, and mass transfer rates, as well as properties at an open boundary, may vary with time. As a result, properties within the control volume vary. For example, the energy within the control volume varies so that $dE_{CV}/dt \neq 0$. Hence we must return to Eq. (1-40) as a starting point.

$$\dot{Q} + \dot{W}_{net} - \sum_{in} \dot{m}_i \left(h + \frac{V^2}{2} + gz\right)_i - \sum_{out} \dot{m}_e \left(h + \frac{V^2}{2} + gz\right)_e = \frac{dE_{CV}}{dt} \quad (1\text{-}40)$$

This equation can be used to develop other useful forms of the transient flow energy equation.

Since some of the terms in the above equation vary with time, the equation must be integrated over a time t to give the total energy change of the control volume during this interval. As a result, for a control volume which remains fixed in space so that $E_{CV} = U_{CV}$

$$\int_0^t \dot{Q}\,dt + \int_0^t \dot{W}_{net}\,dt + \sum_{in} \int_0^t \left(h + \frac{V^2}{2} + gz\right)_i \dot{m}_i\,dt - \sum_{out} \int_0^t \left(h + \frac{V^2}{2} + gz\right)_e \dot{m}_e\,dt = \int_0^t \frac{dU_{CV}}{dt}\,dt \quad (1\text{-}46)$$

To carry out these integrations, we need to know how the various properties and rates vary with time.

The integration of the control-volume energy term on the right-hand side may take one of several formats, depending on the type of process under investigation. Among the possibilities are

$$\int \frac{dU_{CV}}{dt}\,dt = \int dU_{CV} = \int d(mu) = m_2u_2 - m_1u_1 = \Delta U_{CV}$$

and

$$\int dU_{CV} = \int d(mu) = \int (m\,du + u\,dm)$$

Both these choices are used in the following section. Another useful formulation of Eq. (1-40) is obtained by multiplying the equations by dt. The resulting relationship is valid for a differential change of state for a control volume over a time dt. Thus

$$\delta Q + \delta W_{net} + \sum_{in} \left(h + \frac{V^2}{2} + gz\right)_i dm_i - \sum_{out} \left(h + \frac{V^2}{2} + gz\right)_e dm_e = dU_{CV} \quad (1\text{-}47)$$

Integration of Eq. (1-47) over a time period of interest yields

$$Q + W_{net} + \int \left(h + \frac{V^2}{2} + gz\right)_i dm_i - \int \left(h + \frac{V^2}{2} + gz\right)_e dm_e = \Delta U_{CV} \quad (1\text{-}48)$$

where only one inlet term and one outlet term are shown.

In the analysis of transient flow systems several models frequently are applicable to real situations. These include the models of the *uniform state* and of *uniform flow*. These are defined as follows:

1. A *uniform state* requires that the state within the control volume at any instant be uniform through the control volume (or over several different regions of the control volume). However, the state within the control volume may change with time.
2. A condition of *uniform flow* requires that the state of the mass crossing a control surface be invariant with time. However, the mass flow rate across that particular control surface may vary with time.

The condition of uniform state implies rapid or instantaneous approach to equilibrium at all times for the mass within the control volume. This is met by many transient fluid systems in practice. The condition of uniform flow is frequently met when the flow into a transient system is supplied from a very large reservoir of matter.

In a transient flow system a net accumulation or depletion of mass within the control volume may occur. As a result, the sum of the rates of mass entering and leaving the control volume is not zero but must equal the time rate of change of the mass within the system. Symbolically this was given earlier by Eq. (1-41),

$$\sum_{\text{in}} \dot{m}_i - \sum_{\text{out}} \dot{m}_e = \frac{dm_{\text{CV}}}{dt} \tag{1-41}$$

In some transient flow problems the mass rate of flow varies with time. In this case the above equation must be integrated over a period t to give the total mass change of the control volume during this interval. Therefore

$$\sum_{\text{in}} \int_0^t \dot{m}_i \, dt - \sum_{\text{out}} \int_0^t \dot{m}_e \, dt = \Delta m_{\text{CV}} \tag{1-49}$$

To carry out these integrations, one needs to know how the mass flow rates at various portions of the control surface vary with time.

1-7 CHARGING AND DISCHARGING RIGID VESSELS

This section deals with two common unsteady-state problems. One of these is the charging of a tank with a fluid from a pressurized line. The other is the discharging of a fluid from a pressurized tank into a region of lower pressure.

1-7-1 Charging Processes

A common and typical example of an unsteady-state flow problem is the charging of a rigid vessel by supplying a fluid from some outside high-pressure source. The rigid vessel may initially be evacuated, or may contain a finite amount of matter within it before the charging process begins. Generally, the mass enters only at one section of the control surface and no efflux of matter occurs.

As a more definite illustration of the problem, consider an evacuated rigid tank connected through a closed valve to a high-pressure line. This situation is shown in Fig. 1-2(a). A control volume has been selected, as indicated by the dashed line in the figure, which includes essentially the volume inside the rigid tank. We shall assume that the volume in the region of the valve and entrance pipe is negligible. Application of Eq. (1-48) to this situation reveals that

$$Q + W_{\text{net}} + \int \left(h + \frac{V^2}{2} + zg \right) dm = \Delta U_{\text{CV}}$$

The net work term is zero in value on the basis of the statement of the process. Usually the kinetic and potential energies of any unit mass in the line may be neglected. Thus the conservation of energy equation for the selected control volume reduces to

$$Q + \int h \, dm = \Delta U_{\text{CV}}$$

At this point one must make a decision about the quantity Q. If the tank were heavily insulated or if the tank were allowed to fill rapidly, the amount of heat transferred would be quite small even if an appreciable temperature gradient existed between the contents of the tank and the environment. Therefore we shall analyze the process on the basis of negligible heat transfer. The overall internal

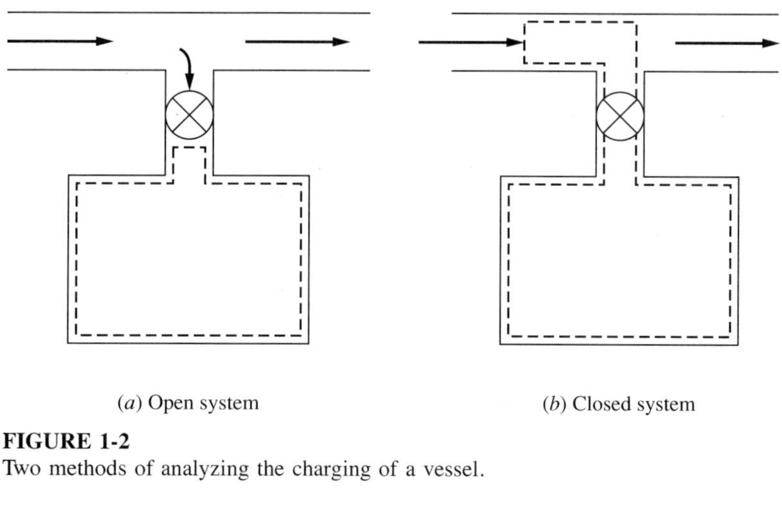

(a) Open system (b) Closed system

FIGURE 1-2
Two methods of analyzing the charging of a vessel.

energy change of the control volume can be written as $\Delta(mu)$. Thus the above energy equation becomes

$$\int h\,dm = m_f u_f - m_i u_i \tag{1-50}$$

where i and f represent the initial and final states within the control volume. The integration is carried out from zero mass to the final mass which enters from the line.

Frequently the integral of $h\,dm$ is difficult to evaluate because the enthalpy of the mass crossing the control surface varies with respect to time. However, in our particular case we can assume that the quantity of fluid bled from the line is small compared to the quantity flowing through the line. Consequently, the properties of the fluid at the entrance to the valve remain constant, so we have a case of *uniform flow*. If we designate the line conditions by the subscript L, then integration of the above equation leads to

$$h_L(m_L - 0) = m_f u_f - m_i u_i$$

where the terms on the right refer solely to the control volume. For this particular problem m_i is zero (the tank initially was evacuated), and thus m_f equals m_L. Hence

$$h_L m_L = m_L u_f$$

or

$$h_L = u_f$$

This relationship fixes the final state within the control volume. Under the conditions specified the enthalpy of the fluid in the high-pressure line is numerically equal to the internal energy of the fluid within the control volume if a *uniform state* exists within at any instant.

The above result is completely general in the sense that the nature of the fluid has never been specified. Although equally applicable to any real gas, the equation above is of special interest when employed for the flow of an ideal gas into the tank. Since $h = c_pT$ and $u = c_vT$ for an ideal gas (if the values of the two specific heats are assumed constant over the required temperature range), then

$$c_pT_L = c_vT_f$$

or

$$T_f = kT_L$$

where k is the specific-heat ratio. This important result shows that T_f is always greater than T_L for the flow of an ideal gas into a rigid tank under the following major idealizations:

1. Adiabatic, rigid control volume
2. Uniform state within the tank at any instant
3. Negligible kinetic energy of the inflowing gas
4. Uniform inlet flow

Since k for many gases ranges from 1.25 to 1.40, the absolute temperature rise may be as much as 40 percent. For real gases one will have to rely on tabular data in order to ascertain the final state of the fluid after it reaches a certain pressure within the tank.

The problem of charging a tank from a steady-flow line is amenable to analysis also on the basis of a closed system. The closed system selected for analysis is shown in Fig. 1-2(b). The conservation of energy equation in this case is

$$Q + W = \Delta E$$

Again, we shall neglect kinetic- and potential-energy terms and assume adiabatic conditions. The energy equation becomes

$$W = \Delta U = m_f u_f - m_i u_i$$

Unlike the control-volume approach, however, neither the work term nor the initial mass term m_i is zero. The work done is the boundary work related to pushing the mass contained within the system, but outside the valve, into the tank. Consequently, the energy equation becomes

$$-P(V_f - V_i) = m_f u_f - m_i u_i$$

or

$$-P(0 - m_L v_L) = m_f u_f - m_L u_L$$

Finally, combining the second term on the right with the term on the left, we find that

$$m_L(P_L v_L + u_L) = m_f u_f$$

and

$$h_L = u_f$$

This result is identical to that obtained by a control-volume analysis. Both were based on comparable idealizations. However, it should be noted that the terms included in the two analyses are quite different.

1-7-2 Discharging Processes

From a physical viewpoint the discharge of a fluid from a pressurized vessel is similar to the preceding case of the charging of a vessel. However, from a thermodynamic viewpoint there is little similarity in the relations which result from applying the conservation of energy principle to the two processes. We shall again neglect the kinetic and potential energies associated with either the control volume or the mass leaving the vessel through a suitable valve. Thus $e = u$. The absence of shaft work again is noted. In the absence of specific information on heat transfer we shall assume that the effect is negligible. If Eq. (1-47) is now applied to the control volume shown in Fig. 1-3(a) and the above comments are taken into consideration, we find that

$$(u + Pv)\,dm = dU_{\text{CV}}$$

or

$$h\,dm = d(mu) = m\,du + u\,dm$$

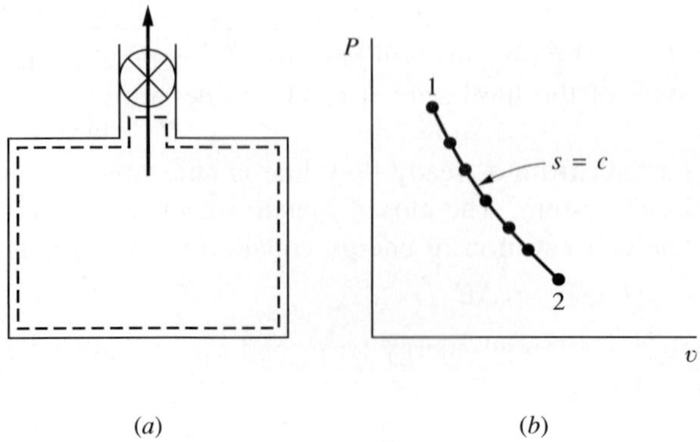

FIGURE 1-3
(*a*) Schematic diagram, (*b*) Pv process diagram for flow from a pressurized vessel.

This problem is unlike the charging of a vessel from a pressurized line in that the properties of the fluid crossing the control surface are continually changing. For this reason we have elected to write the conservation equation in its differential form. The above equation can be arranged into the following forms:

$$(h - u)\,dm = m\,du$$

and

$$\frac{du}{h - u} = \frac{dm}{m}$$

It should be noted that the specification of property values, such as h and u, for the mass within the control volume requires that the fluid be essentially in equilibrium at all times. Hence the process of discharging the gas is quasistatic and one of uniform state.

The equation above is completely general. That is, we have not specified the nature of the fluid leaving the pressurized tank. We may introduce two further relationships without losing this generality. First of all, from the definition of the enthalpy function $h - u = Pv$. Second, the volume of the control volume remains constant with time. Then from the basic definition $V = mv$ we find that $dV = m\,dv + v\,dm = 0$. Therefore

$$\frac{dm}{m} = -\frac{dv}{v}$$

Substitution of these two basic relations into the conservation of energy equation leads to

$$\frac{du}{Pv} = -\frac{dv}{v}$$

or upon rearrangement

$$du + P\,dv = 0$$

For a closed system of fixed composition the first $T\,ds$ equation [Eq. (1-11)] requires that

$$T\,ds = du + P\,dv \tag{1-11}$$

for a simple compressible substance, regardless of the nature of the process. Comparison of these last two equations forces one to conclude that, for the process,

$$ds = 0 \qquad \text{(flow from a pressurized tank)}$$

Therefore for any fluid which flows from a pressurized vessel under the stated assumptions or idealizations the *specific entropy* within the control volume remains constant. The total entropy of the control volume decreases, however, since the amount of mass decreases. The overall process is highly irreversible so that the net entropy change of the control volume plus its environment must be positive.

Figure 1-3(b) illustrates the process path for the discharging of a fluid from a pressurized tank. The path is for the fluid which remains inside the tank. Note that such a process must fulfill the following idealizations:

1. Adiabatic, rigid control volume
2. Equilibrium within the tank at any instant (uniform state)
3. Negligible kinetic- and potential-energy changes

The fact that the process is isentropic is not surprising, since it is adiabatic and quasistatic and all internal dissipative effects and finite gradients have been ruled out. That is, the process within the tank is adiabatic and internally reversible.

As a specific example, consider the flow of an ideal gas from a pressurized tank. The approximate validity of the ideal-gas equation will depend on the pressure range involved. Since the process is isentropic, the isentropic relations presented in Sec. 1-1-4 for ideal gases are valid for relating pressure, temperature, and specific volume. A relationship for the mass remaining in the tank at any instant can also be obtained for an ideal gas. We noted earlier for a constant volume process that

$$\frac{dm}{m} = -\frac{dv}{v}$$

In addition, if one assumes constant specific heats, the isentropic relation $T(v)^{k-1}$ = constant is valid [see Eq. (1-17)]. Differentiation of this relation yields

$$\frac{dT}{T} = (1-k)\,\frac{dv}{v}$$

Therefore equating dv/v in the last two equations leads to

$$\frac{1}{k-1}\frac{dT}{T} = \frac{dm}{m}$$

Finally, integration of the above equation yields

$$\frac{m_2}{m_1} = \left(\frac{T_2}{T_1}\right)^{1/(k-1)}$$

If the pressure ratio is known, the comparable equation is

$$\frac{m_2}{m_1} = \left(\frac{P_2}{P_1}\right)^{1/k}$$

A third equation can be derived for the mass as a function of the specific volume, or density.

Example 1-1. A tank with a volume of 1.5 m^3 is filled with air at a pressure of 7 bars and a temperature of 250°C. Determine (*a*) the final temperature, (*b*) the percent of the mass left in the tank, and (*c*) the quantity of mass, in kilograms (kg), that left the tank if the gas is permitted to leave the tank under adiabatic conditions until the pressure drops to 1.0 bar.

Solution

(*a*) Under the conditions of the problem it is realistic to assume ideal-gas behavior. If the mass within the tank changes its state isentropically, then, since k is roughly equal to 1.4,

$$T_2 = T_1\left(\frac{P_2}{P_1}\right)^{(k-1)/k} = 523\left(\frac{1}{7}\right)^{0.285} = 300 \text{ K} = 27°\text{C}$$

(*b*) The percent of the mass left in the tank can be found from the relation

$$\frac{m_2}{m_1} = \left(\frac{P_2}{P_1}\right)^{1/k} = \left(\frac{1}{7}\right)^{0.715} = 0.249$$

Therefore only 25 percent of the mass is left when a pressure of 1 bar is reached, and the temperature at that instant will be 27°C.

(*c*) The mass initially in the tank is given by the ideal-gas relation

$$m = \frac{PV}{RT} = \frac{7.0(1.5)(28.97)}{0.08314(523)} = 7.00 \text{ kg}$$

Hence the mass bled from the tank is 0.751(7.00) = 5.26 kg.

There are many other types of transient flow problems of theoretical, as well as of practical interest. The solution of each type depends on the restrictions and idealizations placed on the process. In every case it is best to begin with basic or fundamental equations and to proceed logically from that point.

1-8 TRANSIENT ANALYSIS WITH BOUNDARY WORK

Consider the situation shown in Fig. 1-4. A weighted piston-cylinder assembly initially containing a known quantity of a gas at a known equilibrium state is connected through a valve to a high-pressure line. To simplify the problem, the fluid in the line is the same as that in the piston cylinder. We are interested in determining the temperature and the quantity of mass within the cylinder after the volume has changed by a known amount. Since the upper boundary will move, the control volume shown in Fig. 1-4 changes size and boundary work occurs. It is necessary to place a restriction on the process if the boundary work is to be evaluated. Even though the pressure in the line may be substantially above the pressure within the cylinder (as maintained by the weighted piston), we must assume that the entering gas quickly equilibrates with the gas already present in the control volume. Thus a uniform and constant pressure P is maintained by the piston on the gas. Then the work done by the gas on the surroundings is simply $-P\Delta V$. We shall assume that the process is adiabatic and that the entering velocity is negligible. Although the center of mass of the control volume changes in elevation, this potential-energy effect can be neglected. As a result, $e = u$.

The basic conservation of energy equation for the control volume in this case, based on Eq. (1-48), is

$$Q + W_{\text{shaft}} + W_{\text{boundary}} + \int (u + Pv)\, dm = \Delta U_{\text{CV}}$$

For this specific situation the above equation reduces to

$$W_{\text{boundary}} + \int h\, dm = \Delta U_{\text{CV}}$$

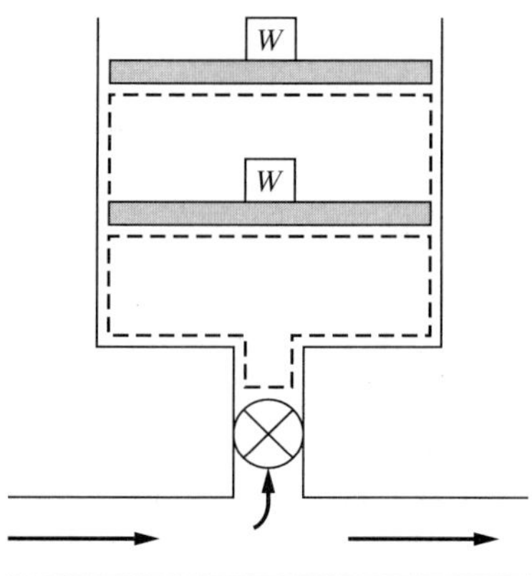

FIGURE 1-4
A control volume with a moving boundary.

In terms of the properties for the initial and final states we find that

$$-P(V_f - V_i) + \int h\,dm = m_f u_f - m_i u_i$$

As a final step, since the flow through the line is steady, we can integrate the quantity $h\,dm$ directly and obtain the equation

$$-P(V_f - V_i) + h_L m_L = m_f u_f - m_i u_i \tag{1-51}$$

where the line condition is designated by the subscript "L." The unknowns in this equation are m_L, m_f, and u_f. However, the mass quantities are related by

$$m_L = m_f - m_i$$

Thus we actually have one equation with two unknowns, m_f and u_f.

The second required relationship in this case is found from a knowledge of the final volume. Note that $V_f = m_f v_f$. Since V_f is given, the unknowns in this relationship are m_f and v_f. Hence another unknown, v_f, is introduced. However, u_f and v_f are not independent variables. On the basis of the state principle for a given pressure within the control volume only u_f or v_f can be varied independently, but not both. Thus a suitable equation of state which uniquely relates u_f and v_f for a given P is the desired third relation which permits the two equations already presented to be satisfied.

Example 1-2. A piston-cylinder assembly is attached through a valve to a source of air which is flowing in steady state through a pipe. Initially, inside the cylinder the volume is 0.01 m^3, the temperature of the air inside is 40°C, and the pressure is 1 bar. The valve is then opened to the air-supply line which is maintained at 6 bars and 100°C. If the cylinder pressure remains constant, what will be the final equilibrium temperature, in degrees Celsius, and how much mass will have entered through the valve, in grams, when the cylinder volume reaches 0.02 m^3?

Solution. The control volume is selected to lie just inside the solid boundaries of the piston-cylinder assembly (see Fig. 1-4) and the gas is assumed to behave ideally and with constant specific heats. Then Eq. (1-51) developed above for the conservation of energy takes on the form

$$\begin{aligned}-P(V_f - V_i) + h_L(m_f - m_i) &= m_f c_v T_f - m_i c_v T_i \\ &= m_f c_v \frac{PV_f}{m_f R} - m_i c_v T_i \\ &= \frac{PV_f}{k-1} - m_i c_v T_i\end{aligned}$$

Hence by introducing the equation of state for an ideal gas and the relation $c_v = R/(k-1)$, we find that the resulting equation contains only the single unknown m_f. Before substituting numerical values, it is convenient to calculate first the initial mass within the piston-cylinder assembly.

$$m_i = \frac{PV}{RT} = \frac{1 \text{ bar} \times 0.01 \text{ m}^3}{0.08314 \text{ (bar} \cdot \text{m}^3/\text{kmol} \cdot \text{K)} \times 313 \text{ K}} \times \frac{29 \text{ kg}}{\text{kmole}}$$
$$= 0.0111 \text{ kg} = 11.1 \text{ g}$$

Note also that $h_L = c_p T_L$, and that c_p, c_v, and k values for air are found in the Appendix. Substitution of known values into the energy equation then yields [expressing each term in newton-meters (N · m) or joules (J), and m_f in grams],

$$-1(0.02 - 0.01)10^5 + 1.01(m_f - 11.1)(373) = \frac{1(0.02)(10^5)}{0.4} - 11.1(0.72)(313)$$
$$-1000 + 377m_f - 4180 = 5000 - 2500$$
$$m_f = 7680/377 = 20.4 \text{ g}$$

Thus the mass entering the assembly during the filling process is

$$m_L = m_f - m_i = 20.4 - 11.1 = 9.3 \text{ g}$$

The final equilibrium temperature is found from the ideal-gas relation,

$$T_f = T_i \frac{V_f}{V_i} \frac{m_i}{m_f} = 313 \times \frac{0.02}{0.01} \times \frac{11.1}{20.4} = 341 \text{ K} = 68°\text{C}$$

Thus the temperature of the resulting mixture in the assembly increases nearly 30°C above the initial temperature.

If the fluid had been a real gas, the solution would not have been quite as straightforward. In this latter case the property relations are usually not known in an explicit algebraic form. Consequently, an iterative process often is required during the numerical evaluation.

PROBLEMS

1-1. A vertical piston-cylinder assembly contains 0.050 m^3 of a gas at 1.0 bar. The piston is restrained by a linear spring as well as the atmosphere. Initially, the atmospheric pressure of 1.0 bar just balances the gas pressure, and the spring exerts no force on the piston. (Neglect the weight of the piston.) The gas is now heated quasistatically until its volume and pressure are 0.080 m^3 and 1.9 bars, respectively.

a. Develop an equation for the relationship between the pressure and the volume of the gas, with the pressure in N/m^2 and the volume in cubic meters.

b. Determine the work done by the gas, in kilojoules (kJ).

c. Of the total work done by the gas, determine the percent done on the spring.

1-2. A gas is contained in a piston-cylinder apparatus at an initial state of 1 bar and 300 K. The cylinder diameter is 0.10 m, and the face of the piston initially is 0.40 m from the base of the vertical cylinder. A spring is located between the piston face and cylinder base, and in its initial position the elastic spring is just touching the piston face (unconstrained state). The spring constant is 50 kN/m, and the mass of the piston is negligible. The atmosphere outside the cylinder is at 1 bar. For a work input of 1.06 kJ along the external shaft connected to the piston, determine (a) the distance that the piston is displaced, in meters, and (b) the heat transfer, in kilojoules, if the process is isothermal.

1-3. Two tanks of the same volume are connected by a pipe containing a valve. Initially, the valve is closed, one tank is evacuated, and the other tank contains 150 g of saturated water vapor at 40 bars. The valve is opened and the steam flows into the evacuated tank until equilibrium of P and T prevails. If the steam pressure at the end of the process is 20 bars, determine (a) the final temperature, in degrees Celsius, (b) the heat transfer, in kilojoules, and (c) the entropy change of the steam, in kJ/K. If the temperature of the surroundings is 27°C, (d) determine the total entropy change for the process, in kJ/K.

1-4. Steam at 200°C and 7.0 bars enters a mixing tank at a rate of 3.0 kg/min through a pipe with a cross-sectional area of 1.20 cm^2. Liquid water at 30°C and 7.0 bars enters the tank through another pipe at a rate of 25.0 kg/min. The mixing process leads to a stream of heated water which leaves the tank at 7.0 bars. Determine the final temperature of the fluid leaving the tank, in degrees Celsius, if the kinetic energy of any liquid stream is neglected.

1-5. Steam at 7 bars and temperature T enters an adiabatic mixing tank. Compressed liquid water at 40°C and 7 bars enters at another opening, and the mixture leaves as a saturated vapor at 7 bars. Determine the temperature of the entering steam, in degrees Celsius, if the ratio of the mass flow rate of compressed liquid to superheated vapor is 0.130/1. Neglect kinetic-energy effects.

1-6. A vertical cylinder, closed at the bottom and fitted on top with a piston, contains air initially at 27°C with a volume of 320 cm^3. The piston is held in place externally by an elastic spring, the atmosphere at 1.00 bar, and the weight of the piston of 96.0 N. The initial spring force is 448 N, the spring constant is 160 N/cm, and the piston area is 64.0 cm^2. During a process heat is added until the final air pressure is 2.75 bars. Determine

a. the work done on the piston, in joules
b. the net work on the piston by the gas, in joules
c. the temperature change of the air, in degrees Celsius
d. the heat transfer, in joules

1-7. One kilogram of air in a piston-cylinder device undergoes a thermodynamic cycle composed of the following reversible processes: (1-2) air at 2.4 bars and 250 K is expanded isothermally to 1.2 bars; (2-3) air is then compressed adiabatically back to its initial volume: (3-1) air is finally cooled at constant volume back to its initial state. Account for variable specific heats.

a. Sketch the Ts and Pv diagrams for the cycle.
b. Calculate the work, heat transfer, and entropy change for process (1-2).
c. Calculate the work, heat transfer, and entropy change for process (2-3).
d. Calculate the work, heat transfer, and entropy change for process (3-1).
e. If the machine operates as a heat pump, determine the coefficient of performance (COP) of the machine.

1-8. Sealed inside a rigid, poorly insulated vessel is 3.0 m^3 of diatomic nitrogen. The system is at 1.0 bar, and the saturated nitrogen is 30.0 percent vapor by mass. The vessel will rupture if the pressure reaches 10.0 bars. Heat transfer from the surrounding atmosphere enters at an average rate of 70.0 kJ/h. Determine the time interval before the vessel ruptures, in hours.

1-9. A well-insulated tank contains 400 g of a liquid at 20°C. A paddle wheel within the liquid is driven by a pulley-weight system. When the weight descends, the paddle wheel turns. Neglect friction in the mechanical pulley-weight system, and neglect any energy storage in the mass of the paddle wheel and tank walls. Assume the liquid is incompressible with a specific heat of 1.80 kJ/kg·K. The gravitational force on the weight during a specific experiment is 18,000 N, and the weight descends 3.0 m.

a. For this experiment, determine the final temperature of the fluid, in degrees Celsius, the entropy change of the liquid in J/K, and the entropy change of the total process, in J/K. Is the process reversible? Explain.
b. The experiment is now repeated, except that the insulation is removed from the tank. The liquid exchanges heat with the surroundings, which are at 20°C, until the liquid and the surroundings are in thermal equilibrium. Determine the entropy change of the liquid, and for the overall process, in J/K.
c. Compare the irreversibility of parts (a) and (b). Explain.

1-10. Two tanks of the same volume are connected by a pipe containing a valve. Initially, the valve is closed, one tank is evacuated, and the other tank contains 30 g of saturated water vapor at 10 bars. The valve is opened and the steam flows into the evacuated tank until pressure equilibrium prevails. If the steam temperature at the end of the process is 200°C, determine (a) the final pressure, in bars, (b) the heat transfer, in kilojoules, and (c) the total entropy change, in kJ/K, if the heat is supplied from a reservoir at 200°C.

1-11. Carbon dioxide (CO_2) flows steadily at a rate of 2.0 kg/min through an uninsulated device from a state of 3 bars and 700 K to a state of 1 bar and 650 K. Power output from the fluid is 16.0 kJ/min. Assume CO_2 is an ideal gas. Determine whether the process is reversible, irreversible, or impossible. Justify your answer. The atmospheric temperature is 27°C.

1-12. A horizontal cylinder is insulated around its side and one end and is separated initially into two 0.040 m^3 compartments by an adiabatic, frictionless piston, as shown in Fig. P1-12. One side contains air and the other side contains argon, both initially at 27°C and 100 kPa. The argon end is completely insulated. Heat is added slowly to the air side from a reservoir at 600 K until the pressure of the argon is 135 kPa. Determine

a. the internal energy change of the argon, in kilojoules
b. the final volume of the air, in cubic meters
c. the heat transferred into the air, in kilojoules
d. the entropy change of the air and of the overall process, in kJ/K

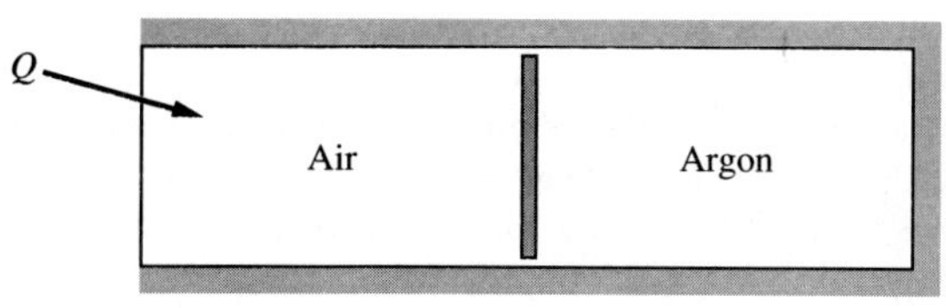

FIGURE P1-12

1-13. An adiabatic steam turbine drives an adiabatic water pump. Steam enters the turbine at 10 bars and 240°C, and exhausts at 1.5 bars. Water enters the pump at 1 bar and 15°C and leaves at 60 bars.

a. What is the ratio of water flow rate to steam flow rate if the turbine and pump are internally reversible?

b. What is this ratio if the adiabatic efficiencies of the turbine and pump are 85 and 75 percent, respectively, for the same inlet conditions and outlet pressures?

c. What is the percent decrease in pumping capability when real performance is considered, compared to ideal performance?

1-14. A water pump is driven by a steam turbine. The steam at the inlet is at 10 bars and 280°C, and leaves the turbine at 1.5 bars. Saturated liquid water enters the pump at 20°C at 100,000 kg/h, and the water exits at 100 bars. Consider both the pumping and expansion processes to be adiabatic and neglect kinetic-energy changes.

a. Determine the flow rate of steam required to drive the turbine if both processes are reversible, in kg/h.

b. If the turbine is replaced with an electric motor, what capacity motor is required, in kilowatts.

c. Determine the flow rate of steam required to drive the turbine if the turbine is 80.0 percent efficient and the pump is 60.0 percent efficient, in kg/h.

1-15. Steam is condensed on the outside of some heat exchanger tubes by passing air through the inside of the tubes. The air enters at 1.30 bars, 20°C, and 15 m/s, and exits at 90°C. The steam enters at 3 bars and 200°C with a mass flow rate of 7.0 kg/min, and leaves as a saturated liquid at 3 bars. Find (a) the mass flow rate of air required, in kg/min, and (b) the number of tubes required if the internal diameter of each tube is 1.90 cm.

1-16. A well-insulated tank contains 10.0 kg of a liquid with a specific heat of 1.80 kJ/kg·°C and an initial temperature of 20.0°C. A paddle wheel within the liquid is driven by a pulley-weight system. The paddle wheel has a mass of 1.3 kg and a specific heat of 0.80 kJ/kg·°C. During an experiment the gravitational force on the weight is 150,000 N and the weight drops 5.0 m as it turns the paddle wheel. Neglect thermal storage in the tank walls.

a. Determine the final temperature, in degrees Celsius, and the entropy change for the overall process, in kJ/K.

b. Is the process reversible or irreversible?

c. The experiment is now repeated, except that the insulation is removed. The contents of the tank exchange heat with the surroundings, which are at 20°C, until the contents and surroundings are at thermal equilibrium. Determine the entropy change of the contents of the tank and the total entropy change for the process, in kJ/K.

d. Compare the irreversibility of process (a) versus (c). Explain.

1-17. An electrical resistor has a mass of 10 g and a resistance of 32 ohms. Assume the material is incompressible with a constant specific heat of 0.80 J/g·K. Initially, the resistor is wrapped in a very good thermal insulation and a current of 1.0 A is allowed to flow for 20.0 s. If its initial temperature is 0°C, determine (a) the final temperature, in degrees Celsius, (b) the value of the integral of $\delta Q/T$ for the resistor, in J/K, and (c) the total entropy change for the process, in J/K. (d) Is this a reversible process? State your reasons. Now, the same resistor is immersed in a water-ice bath which is well insulated from its surroundings. However, the bath and the resistor stay at a constant 0°C even when a current of 1.0 A flows through the resistor for 20.0 s. For this process, find (e) the value of the integral of $\delta Q/T$ for the resistor, (f) ΔS for the resistor, in J/K, (g) ΔS for the bath, in J/K, and (h) ΔS for the overall process, in J/K. (i) Is the first or second process more irreversible? State your reasons.

1-18. Water and steam enter an insulated mixing chamber in steady flow. The liquid water enters with a rate of 100 kg/min at 20°C and 3 bars. The steam enters at 320°C and 3 bars, and the mixture leaves the mixing chamber at (a) 130°C and 3 bars, and (b) at 3 bars as a saturated vapor. Determine (1) the amount of steam needed, in kg/min, and (2) the exit velocity in m/s if the exit area is 350 cm^2.

1-19. A cylinder is insulated around its side and on one end and is separated initially into two 0.10-m^3 compartments by an adiabatic, frictionless piston, as shown in Fig. P1-19. One side contains air and the other side contains helium, both initially at 27°C and 1.0 bar. The helium end is completely insulated, but heat is added slowly to the air side until the pressure of both gases reaches 2.0 bars. Find (a) the internal energy change of the helium, in kilojoules, (b) the final temperature of the air, in kelvins, (c) the heat transferred into the air, in kilojoules, and (d) the total entropy change of the air, helium, and reservoir, in kJ/K, if the heat transferred originated from a heat reservoir at 1000 K.

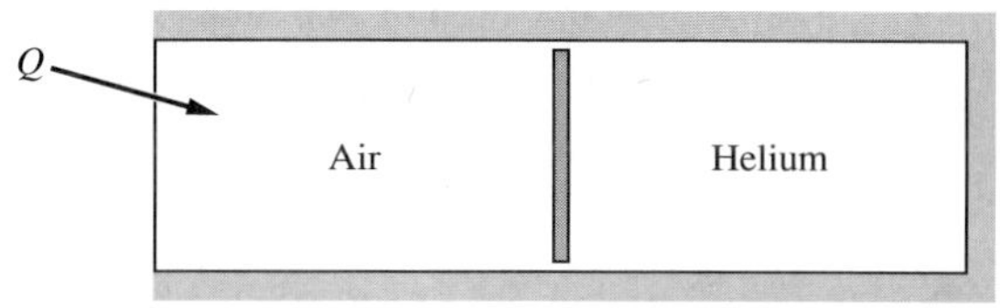

FIGURE P1-19

1-20. A kilogram of a substance undergoes an internally reversible process from state 1 to state 2. For this process it is determined that the heat transfer is 2000 J to the substance, and that the change in entropy is 4.0 J/K. Other processes for this same system will be described below. Some of the questions then asked may be answered by definite numbers, some only by "greater than" or "less than" certain numbers, and some do not permit any definitive answer. Explain your answers.

a. The system undergoes a different internally reversible process from 1 to 2. Find the values of (1) Q, (2) ΔS, and (3) $\int \delta Q/T$.

b. The system undergoes an irreversible process from 1 to 2. Find the values of (1) Q, (2) ΔS, and (3) $\int \delta Q/T$.

c. The system undergoes an irreversible cycle, starting and ending at 1. Find the values of (1) Q, (2) ΔS, and (3) $\int \delta Q/T$.

1-21. The pressure ratio of an air-standard Brayton gas-turbine cycle is 6.0/1, and the inlet conditions to the compressor are 100 kPa and 27°C. The turbine is limited to a temperature of 1100 K, and the mass flow rate is 5.0 kg/s. The compressor and turbine adiabatic efficiencies are 80 and 85 percent, respectively. Determine (a) the actual compressor and turbine work, in kJ/kg, (b) the thermal efficiency, (c) the net power output, in kilowatts, and (d) the volume flow rate at the compressor inlet, in m^3/min. If the heat addition is accomplished by the complete combustion of a fuel with a heating value of 43,000 kJ/kg fuel, (e) estimate the air-fuel ratio used in the combustor, in kg/kg. Use the air table in the Appendix for all property data.

1-22. Refrigerant-134a (R-134a) is compressed adiabatically at a rate of 1.0 m^3/min (inlet conditions) in a steady-flow compressor from 4.0 to 16.0 bars. The temperature of the gas entering the compressor is 20°C and the compressor adiabatic efficiency is 75 percent. Determine (1) the mass flow rate, in kg/min and (2) the shaft power required, in kilowatts, if (a) the gas were an ideal gas with constant specific-heat values of $c_v = 0.78$ and $c_p = 0.86$ kJ/kg·K and (b) the gas data are obtained from the superheat table.

1-23. Two tanks of the same volume are connected by a pipe containing a valve. Initially, the valve is closed, one tank is evacuated, and the other tank contains 40 g of steam at 15 bars and 280°C. The valve is opened and the steam flows into the evacuated tank until pressure equilibrium is reached. If the steam temperature at the end of the process is 440°C in both tanks, determine

a. the final equilibrium pressure, in bars
b. the heat transfer, in kilojoules
c. the entropy change, in kJ/K, of the steam
d. the total entropy change, in kJ/K

1-24. An insulated tank with a volume of 0.50 m^3 contains air at 1.0 bar and 25°C. The tank is connected through a valve to a line which carries air continuously at 7.0 bars and 177°C. If the valve is opened and air is allowed to flow into the tank until the pressure reaches 5.0 bars, determine (a) the amount of mass which has entered, in kg, and (b) the final temperature of the air in the tank, in degrees Celsius. (Use constant specific-heat data in your solution.)

1-25. A tank contains R-134a at 14.0 bars and 60°C and has a volume of 0.20 m^3. The contents of the tank are heated, but a relief valve allows fluid to escape so that the pressure remains constant. If the final temperature is 130°C, find by a least square fit of data and numerical integration or by graphical integration the quantity of heat added, in kilojoules.

1-26. A very large reservoir contains steam at 30 bars and an unknown temperature T_R. Steam is allowed to flow from this reservoir into a small insulated rigid tank with a volume of 5.0 m^3. The tank initially contains saturated water vapor at 10 bars. The flow of steam from the reservoir continues until the pressure in the tank builds up to 20 bars. At this instant the measured temperature in the tank is 320°C. Determine the temperature T_R of the reservoir, in degrees Celsius.

1-27. A very large reservoir contains air at 10 bars and an unknown temperature T_R. Air is allowed to flow from the reservoir into a small insulated tank with a volume of 0.150 m^3. The tank initially contains air at 2.0 bars and 27°C. The flow from the reservoir to the tank continues until the pressure in the tank builds up to 4.0 bars. At this instant the measured temperature in the tank is 87°C. Determine the temperature T_R of the reservoir, in degrees Celsius.

1-28. Air, an ideal gas with constant specific heats, flows slowly from the atmosphere at T_a and P_a into an insulated cylinder fitted with a spring-loaded piston. The spring is linear; that is, the force-displacement relation is a straight line. Initially, the volume of air in the cylinder is zero. The spring is preloaded so that the pressure at the piston face must rise to P_0 before the piston will move. When the cylinder pressure reaches P_a, the volume is V_f. Assume constant specific heats. (a) Show that the temperature of the air in the cylinder at this point is given by

$$T_f = \frac{T_a}{1 - \frac{R}{2c_p}\left(1 - \frac{P_0}{P_a}\right)}$$

In addition, determine the value of T_f under the special conditions that (b) $P_0 = 0$ and (c) $P_0 = P_a$.

1-29. A piston-cylinder assembly maintained at a constant pressure of 1.00 bar (weightless piston) initially contains 0.0010 kmol of air at 27°C. The assembly is connected to an air line maintained at 7.0 bars and 400 K. Air is allowed to flow from the line to the assembly until the volume is doubled. During this process heat enters the cylinder in the amount of 1000 kJ/kmol of gas finally in the tank. For air, let $u = 20.9T$ and $h = 29.2T$, both expressed in kJ/kmol and with T in kelvins.

a. Use a control-volume energy analysis to derive an equation in the form $C_1 N_f T_f + C_2 N_f + C_3 = 0$, where C_1, C_2, and C_3 are numerical values and the subscript f stands for the final state in the cylinder.

b. Derive another independent equation which shows that $N_f T_f = C_4$, where C_4 is some numerical value.

c. Determine the values of the mass which has entered the assembly, in kilomoles, and the final temperature in the cylinder, in kelvins.

1-30. A tank containing R-134a at 12 bars and 50°C has a volume of 0.10 m^3. The contents of the tank are heated, but a relief valve allows fluid to escape so that the pressure in the tank remains constant. If the final temperature is 120°C, determine by a least square fit of the data and numerical integration (or by graphical integration) the quantity of heat added, in kilojoules.

1-31. The upper tank in Fig. P1-31 contains 0.060 m^3 of steam initially at 80 bars and 400°C. The lower tank contains a liquid which is to be forced out by admitting steam from the upper tank. The liquid initially completely fills the lower tank, and

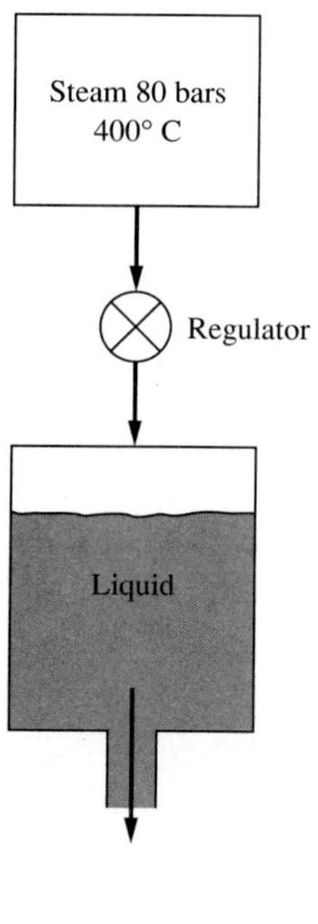

FIGURE P1-31

the steam and liquid are not miscible. The pressure in the lower tank is kept constant at 30 bars by means of a valve between the two tanks. The steam continues to push liquid from the lower tank until the pressure in the upper tank has fallen to 30 bars. There is negligible heat transfer between the tanks, between the tanks and the surroundings, and between the liquid and the steam. Determine

a. the final temperature of the steam in the upper tank, in degrees Celsius
b. the final temperature of the steam in the lower tank, in degrees Celsius
c. the volume of the liquid displaced, in cubic meters

(*Suggestion:* Leave your analysis in terms of symbols as long as possible. Some analyses require a least square fit and numerical integration.)

1-32. A pressurized tank contains 0.1 kg of water vapor at 40 bars and 280°C. Mass is allowed to flow from the tank until the pressure reaches 10 bars. However, during the process heat is added to the fluid within the tank to keep it at a constant temperature. Determine the amount of heat added, in kilojoules, by

a. a solution which requires approximating the enthalpy of the steam leaving the tank
b. a solution which requires a least square curve fit of data and numerical integration.

1-33. Rework Prob. 1-30 under the following conditions. The tank of volume V initially contains an ideal gas at P_1 and T_1. Heat is added until the temperature rises to T_2, the pressure remaining constant. Derive an expression for the total heat transferred during the process, Q, in terms of the independent variables, P, V, T_1, T_2, and the specific-heat ratio k.

1-34. A 1.0-m^3 tank contains 0.5 m^3 of R-134a saturated liquid and 0.5 m^3 of saturated vapor at 9.0 bars. The tank is fitted with a pressure relief valve that allows only saturated vapor to escape and that keeps the pressure inside the tank constant. Determine the heat transfer required, in kilojoules, to evaporate one-half the mass of liquid R-134a that was originally present.

1-35. Saturated steam in the amount of 0.10 kg initially is maintained at 1.0 bar within a piston-cylinder assembly. The pressure within the cylinder is maintained by a spring on the other face of the piston. A steam line containing saturated steam at 9.0 bars is connected to the assembly through a valve. The valve is opened until the pressure in the cylinder reaches 5.0 bars. During the period of time that the valve is opened heat transfer to the cylinder contents is 32.0 kJ and 0.20 kg of steam flows into the cylinder. The work done on the spring is 40.0 kJ. Determine the final quality (if saturated) or the final temperature in degrees Celsius (if superheated) of the steam within the cylinder.

REFERENCES

1. Poincaré, H., *Thermodynamique,* Georges Carre, Paris, 1892.
2. Carathéodory, C., *Math. Ann.* (Berlin), **67**:355–386, 1909. [English translation, J. Kestin, ed., *The Second Law of Thermodynamics,* Part I, Dowden, Hutchinson, and Ross, Stroudsburg, PA, 1976].
3. Bejan, A., *Advanced Engineering Thermodynamics,* Chap. 1, John Wiley & Sons, New York, 1988.
4. Cravalho, E. G., and J. L. Smith, Jr., *Engineering Thermodynamics,* Pitman, Boston, MA, 1981.
5. Zemansky, M., *Heat and Thermodynamics,* 5th ed., McGraw-Hill, Inc., New York, 1968.
6. Hatsopoulos, G. N., and J. H. Keenan, *Principles of General Thermodynamics,* John Wiley & Sons, New York, 1965.

CHAPTER 2

THE SECOND LAW OF THERMODYNAMICS

In an introductory course in thermodynamics attention is focused on the second law and related theorems and corollaries. The concepts of reversibility (and irreversibility), entropy, and entropy production are important in the analysis of energy processes and systems. Certain types of reversible cyclic equipment, called *Carnot devices,* are found to be useful in the study of simplified heat engines, refrigerators, and heat pumps. These and other related topics will be reviewed and extended in this chapter.

2-1 THE SECOND LAW FOR CLOSED SYSTEMS

The early development of the steam engine occurred in the latter half of the eighteenth century. The conversion efficiency of heat to work of these early devices was usually less than 5 percent. An important first step in the theoretical analysis of heat engines was a memoir by Sadi Carnot, a French engineer, in 1824 [1]. This work was an early step historically in the development of the second law. Although developed in the era of the caloric theory, it correctly presented some important principles which still bear his name (and will be reviewed shortly).

The two statements of the second law most frequently cited in introductory textbooks are those due to Clausius and to Kelvin and Planck.

Both originated in the middle of the nineteenth century.

Both are based on observations of specific irreversible processes.

Both state that the reverse of these observed processes cannot occur naturally.

It is observed that heat always flows spontaneously (of its own accord) from a region of higher temperature to one of lower temperature. In 1850 Clausius postulated that the reverse process is impossible. That is:

> Heat cannot pass spontaneously (unaided) from a region of lower temperature to a region of higher temperature.

The inclusion of the word "spontaneously" in the Clausius statement of the second law is critical. Heat will pass from a lower to a higher temperature if an appropriate device (a refrigerator or heat pump) is used as an aid. But the absence of such a device precludes the passage of heat spontaneously.

The Kelvin-Planck statement focuses on the heat-to-work conversion process. The first law places no restriction on the conversion of heat into work or work into heat. The latter process is observable, for example, through a frictional process. However, it was also observed that a heat engine (closed, cyclic heat-to-work converter) would not operate if it received heat from a single source. Hence a major restriction existed in the design and operation of heat engines. This was first recognized formally in 1851 by William Thomson (Lord Kelvin) [2], who stressed the impossibility of producing work by some device by means of cooling matter below the temperature of the surroundings. Another version of this approach, using the concepts of a cyclic device and a heat reservoir, was presented in 1897 by Planck [3]. In honor of both scientists the Kelvin-Planck statement of the second law may be paraphrased in the following manner:

> It is impossible to construct a heat engine which produces no other effects than the extraction of heat from a single source and the production of an equivalent amount of work.

A device which violates the above statement is known as a *perpetual-motion machine of the second kind* (PMM2). It produces work solely by the cooling of a body. Hence we cannot directly use the vast quantities of energy in the oceans and the atmosphere on earth to produce work without some additional changes taking place in the interacting systems.

The preceding second-law statements rely heavily on observations made with respect to cyclic devices such as refrigerators and heat engines. Since the second law has general applicability to a wide range of phenomena, it should be possible to provide other formulations which avoid reference to specific applications. One of these is due to Hatsopoulos and Keenan [4], who define a stable equilibrium state of a system such that a finite change of state cannot occur without a corresponding finite change occurring in the state of the environment. Making use of this definition, Hatsopoulous and Keenan proposed the following second-law statement in 1965:

> A system having specified allowed states and an upper bound in volume can reach from any given state a stable state and leave no net effect on the environment.

This statement, which is not deducible from the definition of a stable state, simply predicts the existence in nature of stable equilibrium states. In 1909 Carathéodory [5] presented a highly mathematical approach to the second law. As reworded by Zemansky [6]:

> In the neighborhood (however close) of any equilibrium state of a system of any number of thermodynamic coordinates, there exist states that cannot be reached (are inaccessible) by reversible adiabatic processes.

He showed how to derive from this statement all the consequences usually associated with the methods of Kelvin-Planck and Clausius. In turn, Landsberg [7] has shown that Carathéodory's principle can be deduced from Kelvin's principle.

The work of Landsberg cited above is an example of the equivalency of various second-law statements. All four of the second-law statements discussed in this section are entirely equivalent. This means that all the theorems and corollaries derivable from one statement are inherent in the other statements. The equivalency of two statements is determined through the principles of mathematical logic. The equivalency of two statements, X and Y, is written symbolically as

$$X \equiv Y \tag{2-1}$$

Equivalency occurs if, and only if, the negation of one statement implies the negation of the other statement, and vice versa. The symbol used for "implies" is $\rightarrow$. Hence the validity of Eq. (2-1) requires symbolically that

$$-X \rightarrow -Y \qquad \text{and} \qquad -Y \rightarrow -X \tag{2-2}$$

This technique may be used to verify the complete equivalence of the four alternative statements of the second law presented above.

The early development of topics such as mechanics, electromagnetic theory, and thermodynamics was based in each case on inductive analysis of macroscopic experimental observations. In thermodynamics the observations of certain processes lead to the concept of state functions such as energy and entropy. Later in the history of each of these fields a postulational approach was shown to be more concise and internally consistent. An important postulational approach to thermodynamics is found in the text by Callen [8]. In this approach the concepts of state functions, energy, and entropy are presented as fundamental concepts. The postulates presented by Callen in his text are as follows:

Postulate I. "There exist particular states (called equilibrium states) of simple systems that, macroscopically, are characterized completely by the internal energy U, the volume V, and the mole numbers $N_1, N_2, \ldots, N_r$ of the chemical components."

Postulate II. "There exists a function (called the entropy S) of the extensive parameters of any composite system, defined for all equilibrium states and having the following property. The values assumed by the extensive parameters in the absence of an internal constraint are those that maximize the entropy over the manifold of constrained equilibrium states."

Postulate III. "The entropy of a composite system is additive over the constituent subsystems. The entropy is continuous and differentiable and is a monotonically increasing function of the energy."

Postulate IV. "The entropy of any system vanishes in the state for which $(\partial U/\partial S)_{V,N} = 0$, that is, at the zero of temperature."

Note that the internal energy and entropy play a dominant role in these postulates. Contained within these postulates are the zeroth, first, second, and third laws, and the state postulate found in conventional thermodynamic textbooks based on inductive logic.

2-2 REVERSIBILITY

Two of the major goals of second-law analysis are: (1) determining the "ideal" or "theoretically best" performance of energy devices and (2) evaluating quantitatively the factors that diminish performance in actual practice. Such analysis requires an understanding of the concept of reversibility and the reversible process. In general, a *reversible process* is defined as follows:

> A process commencing from an initial equilibrium state is called reversible (or totally reversible) if at any time during the process both the system and the environment with which it interacts can be returned to their initial states.

The concept of reversibility by its definition requires *restorability.* But the requirement of restorability is quite strong, applying to both the system and the surroundings with which it interacts. Normally interest is centered during a given process solely on the system. However, a reversible process requires something of the environment. As a consequence of the definition of a reversible process, *it is the nature of a reversible process that all heat and work interactions which occur during the original (forward) process are equal in magnitude but opposite in direction during the reversed process*. Thus no net history is left in the surroundings when the system regains its initial state.

The reversible process is an *idealization*. It is a concept which can be approximated very closely at times by actual devices but is never matched. There are two fundamental reasons that actual processes are never truly reversible. First, we require the absence of solid friction at the bearing surfaces of pistons, flywheels, and other mechanical devices. A capacitive-inductive circuit should not contain a resistive element, and a spring must be made of an ideal elastic material. In many actual cases the effects of friction, electric resistance, and inelasticity can be substantially reduced, but their complete elimination is not possible. These physical phenomena are called *dissipative effects,* because in all cases a portion of the energy in the system is converted or dissipated to a less useful form. Only in the absence of dissipative effects can a process within an isolated system be considered (totally) reversible.

A second criterion must be met if a process involving a system and its local environment is to be reversible. The requirement that a process be *quasistatic* in order to be reversible is a general one. Thus only infinitesimal unbalanced forces are present during a reversible process. For example, heat transfer is not reversible if its occurrence is due to a finite temperature difference ΔT between the system and its surroundings. The transfer of heat is made reversible if the temperature difference between the two interacting systems is made infinitesimally small, that is, of magnitude dT.

An irreversible process is such that either the system or its surroundings cannot be returned to their initial states. *Any system which is returned to its initial state after experiencing an irreversible process will leave a history in the surroundings due to the irreversibilities.* Irreversibilities arise from the two sources outlined above:

1. Presence of inherent dissipative effects
2. Presence of a nonquasistatic process

The presence of either class of effects is sufficient to make a process irreversible. Since all actual processes include such effects, the reversible process is a limiting process toward which all actual processes may approach in performance but may never succeed in realization. Nevertheless, a reversible process is an appropriate starting place on which to base thermodynamic calculations.

The absence of inherent dissipative effects and nonequilibrium effects in a system and its environment leads to the concept of a *totally reversible process*. Frequently it is advantageous to consider processes for which irreversibilities are absent within the system but are not necessarily absent from the surroundings. Such processes are called *internally reversible*. *Only quasistatic work interactions may occur during internally reversible processes.* Work effects associated with paddle wheels or electric resistors within the system are not permitted, since these effects by nature are irreversible. Finally, an *externally reversible process* is one for which irreversibilities may be present within the boundaries of the system of interest, but the surroundings which interact with the system must undergo only reversible changes.

In the development of second-law concepts, it is useful to imagine systems which can supply or receive energy in the form of heat or work in an internally reversible manner. These two idealized systems are known as heat and work reservoirs. A *heat reservoir* is defined as a closed system with the following characteristics:

1. The only interactions of interest across its boundaries are heat interactions.
2. Changes within the reservoir are internally reversible.
3. Its temperature remains uniform and constant during a process.

The only significant property of a heat reservoir is its *temperature,* which must remain constant. Internal reversibility implies that there are no dissipative effects within the reservoir during a change in its state. A heat reservoir simply acts as a *heat source* or *heat sink* during a process, depending on whether the direction of the heat transfer is from or to the reservoir. It is an idealization which in practice is closely approximated, for example, by portions of the environment. Because the name is a misnomer, this type of reservoir frequently is called a *thermal energy reservoir,* or simply a thermal reservoir. Heat transfer to or from a thermal energy reservoir results in an increase or decrease in the internal energy of the reservoir.

Another useful concept in thermodynamics is that of a work reservoir. Analogous to a heat reservoir, a *work reservoir* acts solely as a source or sink for the exchange of work with another system. Changes of state of a work reservoir occur in an internally reversible manner. Any idealized mechanical, electric, magnetic, or elastic system could operate as a work reservoir. Examples include frictionless pulley-weight systems and rotating flywheels, and elastic springs, among others.

2-3 CYCLES WITH TWO HEAT RESERVOIRS

The Kelvin-Planck statement of the second law precludes a heat engine producing work while receiving heat from a *single* thermal reservoir. That is, the *net* work for the cycle cannot be negative (output). However, there is no restriction on the possibility of net work input or zero work for the closed-system cycle. Thus an analytical expression for the Kelvin-Planck statement of the second law is

$$\sum_{\text{cycle}} W_{\text{closed}} \geq 0 \qquad \text{(single thermal reservoir)} \tag{2-3a}$$

or

$$\oint \delta W_{\text{closed}} \geq 0 \qquad \text{(single thermal reservoir)} \tag{2-3b}$$

(Recall that work input is defined as positive in this text.) In addition, for a cyclic process of a closed system $\sum W = -\sum Q$. As a result, another important analytical form of the Kelvin-Planck statement is

$$\sum_{\text{cycle}} Q_{\text{closed}} \leq 0 \qquad \text{(single thermal reservoir)} \tag{2-4a}$$

or

$$\oint \delta Q \leq 0 \qquad \text{(single thermal reservoir)} \tag{2-4b}$$

Thus a sign restriction exists on W and Q for any closed system which undergoes an integral number of cycles while in communication with a single thermal reservoir. In terms of the concept of reversibility the equality sign in the above equations applies when the process is totally reversible. The inequality sign applies when irreversibilities are present during the cycle. The proof appears in a number of introductory texts of thermodynamics.

The next logical step is to determine whether a heat engine is possible which operates solely between *two different* thermal reservoirs and whether there are sign restrictions on the heat terms. A schematic of such a heat engine, labeled A,

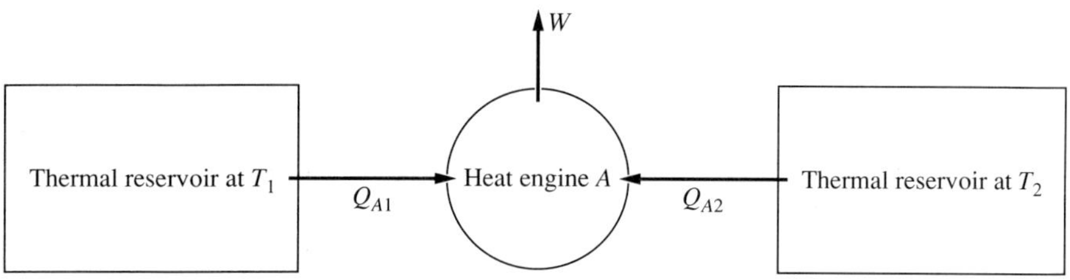

FIGURE 2-1
Schematic of a heat engine operating between two different thermal reservoirs.

which operates between temperatures T_1 and T_2 is shown in Fig. 2-1. The arrows for Q_{A1} and Q_{A2} are arbitrarily shown in a positive direction relative to the heat engine. However, in general, the heat terms might be negative. The limitations on the cyclic engine A are determined by first- and second-law considerations. The first law requires that

$$Q_{A1} + Q_{A2} + W_A = 0$$

where the work output W_A is a negative value for the heat engine. With the sign on W_A fixed, there are three possibilities or options for the signs on the heat terms. One option can be ruled out immediately. Both Q_{A1} and Q_{A2} cannot be negative, based on the first-law statement above. However, the first law permits two other options:

1. Both Q_{A1} and Q_{A2} are positive in value.
2. Q_{A1} and Q_{A2} are of opposite signs such that the product $Q_{A1}Q_{A2} < 0$ and the absolute sum $|Q_{A1} + Q_{A2}| > 0$.

The possible validity of these two remaining options depends on the application of the second law in the form of the Kelvin-Planck statement.

Figure 2-2 shows heat engine A operating according to option 1, where both Q_{A1} and Q_{A2} are taken into the heat engine from thermal reservoirs T_1 and T_2, respectively. Also, operating in conjunction with reservoir T_2 is cyclic device B. Since only one reservoir is in contact with B, the Kelvin-Planck statement requires that $Q_B < 0$. Now A and B each undergoes an integral number of cycles so that $Q_{A2} = -Q_{B2}$. That is, the heat rejected by system B to thermal reservoir T_2 equals the heat supplied by thermal reservoir T_2 to heat engine A. Thus reservoir T_2 also undergoes a cycle. As a result, the composite system of A, B, and T_2, shown by the dashed line in Fig. 2-2, undergoes a cycle and produces work while exchanging heat solely with a single reservoir at T_1. This violates the Kelvin-Planck statement. The only assumption made in this proof is that both Q_{A1} and Q_{A2} are inputs to the engine. Therefore this assumption cannot be correct, and option 1 is invalid.

In conclusion, the heat engine shown in Fig. 2-2 cannot operate with both heat inputs or both heat outputs. Only option 2 is a viable alternative such that

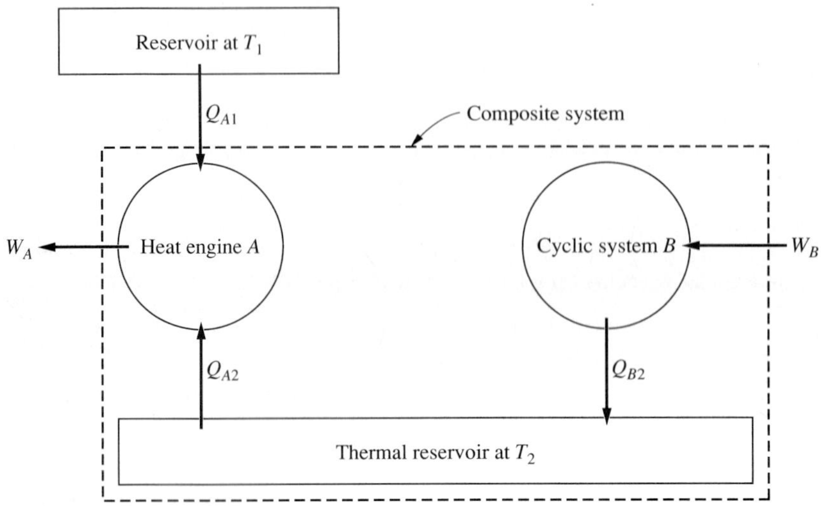

FIGURE 2-2
Schematic for resolution of the validity of option 1 for a two-thermal-reservoir heat engine.

Q_{A1} and Q_{A2} are of opposite signs and their net value is positive. (This alternative does not violate the second law, as noted below.) Hence all heat engines operating between two thermal reservoirs of different temperature must have heat supplied from one reservoir and rejected to another reservoir. However, one point remains unresolved. We know that $T_1 \neq T_2$, but the sign on $(T_1 - T_2)$ is not known. The question is, does the larger of T_1 and T_2 match up with the heat-supply reservoir or the heat-rejection reservoir?

Some other information on the ratio Q_2/Q_1 must be ascertained before the preceding question is resolved. Toward this end we shall define a *Carnot-type* heat engine as one which operates in a totally reversible manner while exchanging heat solely with two different thermal reservoirs. The nature of the actual device is unimportant. The definition does require that the engine itself be internally reversible and that the two heat transfers Q_1 and Q_2 between the thermal reservoirs and the engine occur across a differential temperature dT. Now, consider Fig. 2-3, which shows two heat engines, B and C, operating between the same thermal reservoirs with temperatures T_1 and T_2, where $T_2 \neq T_1$. Engine B operates under any arbitrary cycle, whereas engine C operates under a reversible Carnot-type cycle. Since engine C is reversible, it may operate as a heat pump (or refrigerator), as well as a heat engine. The values of all interactions will be of equal magnitude but opposite in sign when reversed. For the purpose of the following proof cyclic device C is shown operating in the reverse direction from that of engine B. We also shall arbitrarily set $Q_{B1} > 0$ and $Q_{B2} < 0$ with respect to engine B. Finally, to take advantage of the analytical form of the Kelvin-Planck statement as given by Eq. (2-4b),

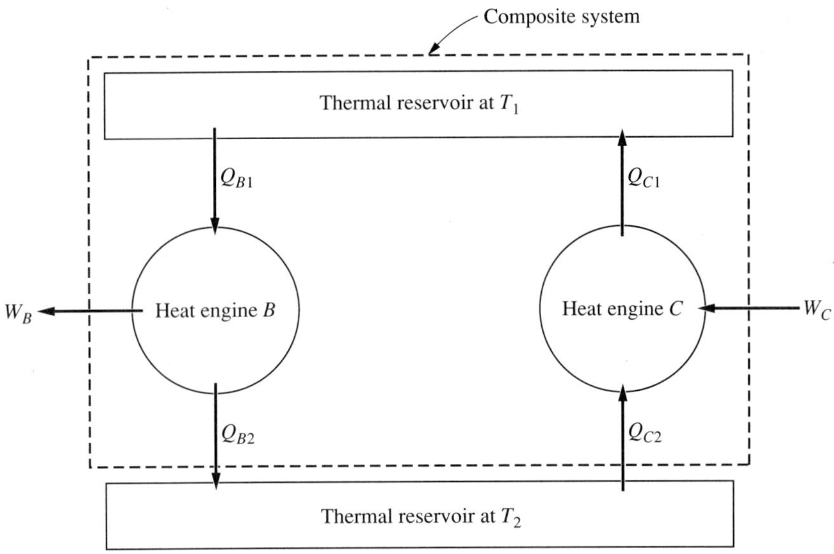

FIGURE 2-3
Schematic of an arbitrary heat engine B and a reversed Carnot-type device C operating between the same two thermal reservoirs.

$$\oint \delta Q \leq 0 \tag{2-4b}$$

the size and integral number of cycles of each engine can be adjusted so that $(Q_{B1} + Q_{C1}) = 0$. That is, the heat rejected by engine C to reservoir T_1 equals the heat supplied to engine B from reservoir T_1. As a result, reservoir T_1, as well as engine B and reversed Carnot engine C, operate in a cyclic manner. A composite system made up of these three subsystems, shown by the dashed line in Fig. 2-3, thus also executes a cycle while exchanging heat with a single reservoir T_2. For the composite system the Kelvin-Planck analytical statement above requires that

$$Q_{B2} + Q_{C2} \leq 0$$

Recall that $Q_{B1} = -Q_{C1}$, where both sides of this equation are positive. Hence the above equation can be written as

$$\frac{Q_{B2}}{Q_{B1}} - \frac{Q_{C2}}{Q_{C1}} \leq 0$$

or, after rearranging and multiplying both sides by minus 1, we obtain

$$\frac{-Q_{B2}}{Q_{B1}} \geq \frac{-Q_{C2}}{Q_{C1}} \tag{2-5}$$

where both sides are still positive in value because $Q_{B2} < 0$ and $Q_{B1} > 0$. Because cycle C is totally reversible, the value of $-Q_{C2}/Q_{C1}$ remains the same whether cycle C operates as a heat engine or a heat pump. In addition, engine B is any (unspecified) heat engine. Hence Eq. (2-5) can be written more generally for *any* heat engine as

$$\frac{-Q_2}{Q_1} \geq \left(\frac{-Q_2}{Q_1}\right)_{\text{rev}} \tag{2-6}$$

where the inequality applies to any irreversible heat engine. Thus Eq. (2-6) states that the absolute value of the ratio of heat rejected at T_2 to heat supplied at T_1 for any actual (irreversible) heat engine is always equal to or greater than the value for a totally reversible (Carnot-type) heat engine in contact with the same two thermal reservoirs. (A more detailed analysis can be found in Bejan [9].)

The thermal efficiency η_{th} of any heat engine is defined as

$$\eta_{\text{th}} \equiv \frac{|W_{\text{net}}|}{Q_{\text{in}}} = \frac{Q_1 - |Q_2|}{Q_1} = 1 - \left|\frac{Q_2}{Q_1}\right|$$

On the basis of Eq. (2-6) and the above relation we find that

$$\eta_{\text{th,act}} \leq \eta_{\text{th,rev}} \tag{2-7}$$

Not unexpectedly, we have shown that the presence of irreversibilities during the operation of a two-reservoir heat engine reduces its thermal efficiency in comparison to a totally reversible heat engine operating between the same reservoirs. As a consequence, a second important question arises from Eqs. (2-6) and (2-7): What is the limiting (minimum) value of $(-Q_2/Q_1)$ or the limiting (maximum) value of $\eta_{\text{th,rev}}$ as an actual engine approaches the performance of a totally reversible heat engine? The answers to the questions posed in this section are found in the corollaries developed below.

2-3-1 Carnot's Principle

The inequality expressed by Eq. (2-7) is the first of three corollaries to the second law known collectively as *Carnot's principle*. The three corollaries are:

1. The thermal efficiency of any irreversible heat engine is always less than that of a totally reversible heat engine operating between the same two heat reservoirs.
2. The thermal efficiencies of all totally reversible heat engines operating between the same two heat reservoirs are equal.
3. An absolute temperature scale may be defined which is independent of the nature of the measuring substance.

Statement 1 follows directly from Eq. (2-7). The proof of statement 2 is as follows. Reconsider Fig. 2-3, only both heat engines B and C are totally reversible. If the thermal efficiency of either one is assumed greater than the other, a violation of the Kelvin-Planck statement can be established. Thus the assumption is incorrect and statement 2 holds.

2-3-2 The Thermodynamic Temperature Scale

On the basis of statement 2 of Carnot's principle the thermal efficiency of totally reversible heat engines is independent of the construction, cycle design, and working fluid of the engine. As a result, the limiting values of $(-Q_2/Q_1)$ and $\eta_{\text{th,rev}}$ can only be a function of T_1 and T_2. They are the only variables left in the design of a totally reversible engine. In the process of seeking the appropriate temperature function another fundamental question arises: What is the nature of the thermometer to be used for measuring T_1 and T_2? It is well known that thermometers which rely on different thermometric substances frequently are not in good agreement, even over reasonably small temperature ranges.

Mathematically the dependence of $(-Q_2/Q_1)$ on T_1 and T_2 can be expressed by

$$\frac{Q_2}{Q_1} = f(T_2, T_1) \tag{2-8}$$

where $f(T_2, T_1)$ is some as yet unknown function and both Q values are considered positive in this discussion. To ascertain the nature of the function, consider the three Carnot-type heat engines X, Y, and Z shown in Fig. 2-4. They operate between pairs of heat reservoirs (T_1, T_2), (T_1, T_3), and (T_2, T_3). Application of Eq. (2-8) to the three heat engines in Fig. 2-4 yields

$$\frac{Q_2}{Q_1} = f(T_2, T_1) \qquad \frac{Q_3}{Q_2} = f(T_3, T_2) \qquad \frac{Q_3}{Q_1} = f(T_3, T_1)$$

Since $(Q_2/Q_1) = (Q_3/Q_1)/(Q_3/Q_2)$, then

$$f(T_2, T_1) = \frac{f(T_3, T_1)}{f(T_3, T_2)}$$

The left-hand side of the above equation does not depend on T_3, and T_1, T_2, and T_3 are independent variables. The above equation and these restrictions are satisfied only when the function is written as

$$f(T_2, T_1) = \frac{\phi(T_2)}{\phi(T_1)}$$

where ϕ is another function. Substitution of this required format into Eq. (2-8) leads to the general relationship,

$$\frac{Q_2}{Q_1} = \frac{\phi(T_2)}{\phi(T_1)} \tag{2-9}$$

This is as far as thermodynamic theory goes. At this point the choice of the mathematical form of $\phi(T)$ is completely arbitrary. Any monotonic function can be used to define a *thermodynamic temperature scale* which is independent of the physical (thermometric) properties of any substance. The *Kelvin scale* is defined by selecting $\phi(T) \equiv T$. If a Carnot-type heat engine operates between T_1 and T_2

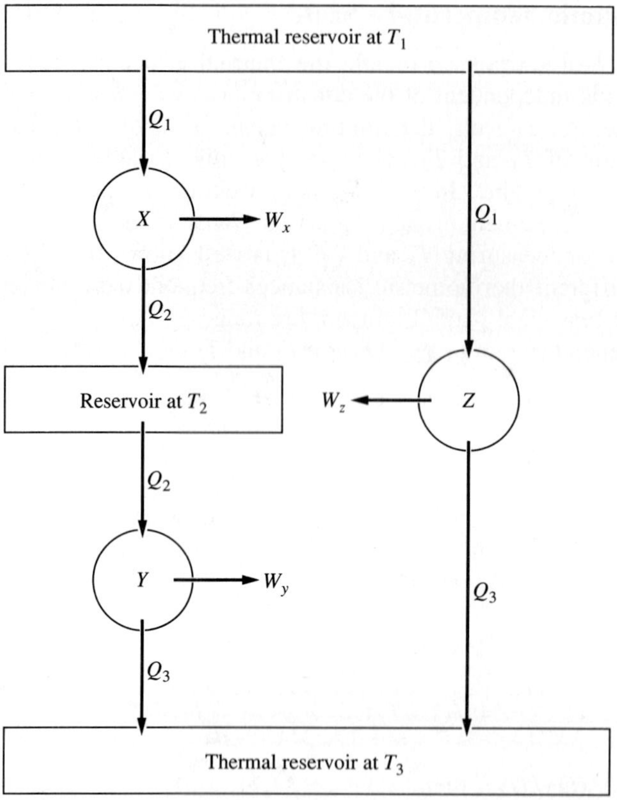

FIGURE 2-4
Three Carnot-type heat engines operating between selected pairs of three thermal reservoirs.

on the Kelvin scale, then (recognizing that Q_1 and Q_2 in practice are of opposite signs)

$$\left|\frac{Q_2}{Q_1}\right| = \frac{T_2}{T_1} \tag{2-10}$$

Values on this thermodynamic temperature scale are established by operating a Carnot-type heat engine between any reservoir at T and a reference reservoir at T_{ref}. An international committee has set the reference temperature at 273.16 (exactly) on the *Kelvin* scale for the triple state of water (491.69 on the Rankine scale). The reason for selecting this value is historical in nature. Thus, in general,

$$T \text{ (kelvins)} = 273.16\left|\frac{Q}{Q_{\text{tp}}}\right| \tag{2-11}$$

where Q and Q_{tp} are the quantities of heat exchanged by a reversible heat engine between kelvin temperatures T and 273.16 for a reservoir at the triple state of water. A Carnot-type heat engine has another important function besides establishing an absolute temperature scale. The device itself can be used to transfer heat reversibly when a *finite* temperature difference exists between two systems, and Eq. (2-10) is valid for that reversible exchange. The only other way to transfer heat reversibly is to maintain a *differential* temperature difference dT between the two systems.

The major advantage of the thermodynamic temperature scale is its universality. The nature of the working fluid is not a factor. Equation (2-10) also provides answers to the two questions posed earlier. First, since we choose $Q_1 > 0$ and $Q_2 < 0$ in this development (recall Fig. 2-3), and $W < 0$, then $Q_2 < Q_1$ in an absolute sense. Thus Eq. (2-11) requires that $T_2 < T_1$, and heat Q_1 is necessarily supplied from the higher temperature reservoir. Second, the limiting (minimum) value of $(-Q_2/Q_1)$ is T_2/T_1 on the absolute Kelvin scale. Likewise, the limiting (maximum) value for the thermal efficiency, called the *Carnot efficiency,* is

$$\eta_{\text{th,rev}} \equiv \eta_{\text{Carnot}} = 1 - \frac{T_2}{T_1} \tag{2-12}$$

where T_1 and T_2 are again thermodynamic temperatures. Finally, a combination of Eqs. (2-6) and (2-10) reveals that for *any* heat engine operating between two reservoirs (a $2T$ engine)

$$-\frac{Q_2}{Q_1} \geq \left(\frac{-Q_2}{Q_1}\right)_{\text{rev}} = \frac{T_2}{T_1}$$

or upon rearrangement and multiplication by minus 1

$$\frac{Q_1}{T_1} + \frac{Q_2}{T_2} \leq 0 \qquad (2T\text{-heat engine}) \tag{2-13}$$

This is an important *alternative* second-law statement. Although its derivation is based on $W < 0$, the result is independent of the sign convention on W, and therefore is general for any cyclic closed system exchanging heat with two thermal reservoirs [see Bejan [9]].

If a Carnot-type heat engine operates with an ideal gas as the working fluid, specific calculations show that

$$\left|\frac{Q_2}{Q_1}\right| = \frac{T_{2,\text{ideal}}}{T_{1,\text{ideal}}}$$

where T_{ideal} is a temperature measured by an ideal-gas thermometer. In comparison to Eq. (2-10) the above result shows that T_{ideal} equals the absolute temperature. This equality is a fortuitous coincidence, since the two scales stem from two completely different concepts. It does indicate that a properly designed "ideal-gas thermometer" will indicate "thermodynamic" temperatures within the accuracy and range of validity of the instrument.

The measurement of temperatures on a thermodynamic scale is complex. As a result, a practical temperature scale has been adopted among the various national laboratories (like NIST, the National Institute for Standards and Technology in the United States) operating through the international General Conferences on Weights and Measures. Reproducible melting or boiling points of chosen pure substances are fixed by the General Conferences.

2-4 THE ENTROPY FUNCTION

The preceding sections have focused attention on the behavior of heat engines operating between a single thermal reservoir or between two different thermal reservoirs. Several of the important relationships previously developed will now be employed in a different setting. We shall continue to examine a cyclic closed system for the moment. However, the possible presence of irreversibilities is now restricted to the system itself. Heat transfer to or from the system must be reversible so that any process under consideration is externally reversible. This latter requirement can be met by the process schematic shown in Fig. 2-5. A closed system, internally reversible or irreversible, undergoes a cyclic process during which interactions δQ and δW may occur for each infinitesimal change of state. To ensure reversible heat transfer, a Carnot-type heat engine C is placed between a single thermal reservoir at T_R and the system at T. Since T may change as the cycle progresses, a number of different heat engines may be required, all of which exchange heat with the same reservoir at T_R. (Only one engine is shown in the figure.) A subscript C indicates quantities relative to any engine. For each of these reversible heat engines Eq. (2-10) requires that

$$-\frac{\delta Q_{R,C}}{T_R} = \frac{\delta Q_C}{T} = -\frac{\delta Q}{T_b} \tag{2-14}$$

where, by sign convention, $\delta Q_C = -\delta Q$, and δQ is the heat transfer relative to the system and can be positive or negative. The subscript b emphasizes that T_b is the absolute temperature at that part of the boundary where the heat transfer occurs.

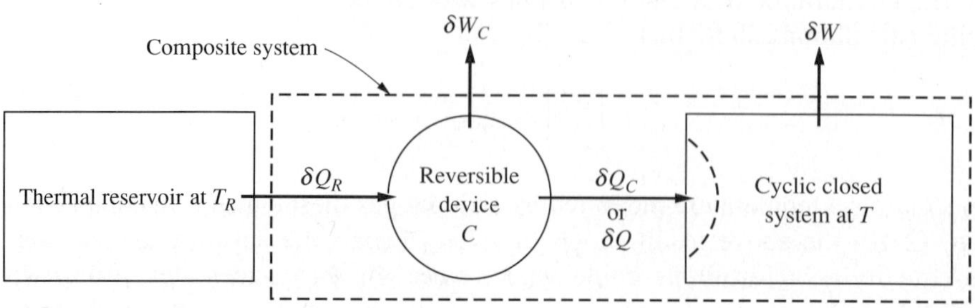

FIGURE 2-5
Schematic of an arbitrary closed system undergoing a cycle while exchanging heat reversibly with a single thermal reservoir at T_R.

To make use of a previously developed relation, we now consider the composite system shown by the dashed line in Fig. 2-5. The composite cyclic system exchanges heat with a single reservoir. Thus Eq. (2-4b) applies, that is,

$$\oint \delta Q_{R,C} \leq 0 \tag{2-4b}$$

where the cyclic integral indicates a summation process of $\delta Q_{R,C}$ for all the necessary reversible heat engines employed as the system undergoes a cycle. The substitution of Eq. (2-14) into Eq. (2-4b) yields

$$T_R \oint \frac{\delta Q}{T} \leq 0$$

or since T_R is a positive number,

$$\oint \frac{\delta Q}{T} \leq 0 \tag{2-15}$$

This result is known as the *Clausius inequality.* The quantity $\delta Q/T$ is the ratio of the heat δQ exchanged by the closed system to the temperature T *at the boundary* where δQ occurs. Following previous notation, the equality sign applies to internally reversible cycles and the inequality applies to internally irreversible process.

The internally reversible (int rev) format of Eq. (2-15)

$$\oint \left(\frac{\delta Q}{T}\right)_{\text{int rev}} = 0 \tag{2-16}$$

leads to an important result. A corollary to the definition of a property is that any quantity whose change during a cycle is always zero is a property of the system. Therefore $\delta Q/T$ represents the change in a thermodynamic property. The property is denoted by S and was named *entropy* by Clausius in 1865. Thus

$$dS \equiv \frac{\delta Q}{T} \qquad \text{(internally reversible)} \tag{2-17}$$

or for a finite, internally reversible path

$$\Delta S = S_2 - S_1 = \int_1^2 \left(\frac{\delta Q}{T}\right)_{\text{int rev}} \qquad \text{(closed system)} \tag{2-18}$$

Thus the first and second laws are similar in that both lead to the definition of a change in a property (E and S) in terms of interactions at the boundary of a closed system. For an internally reversible process the change in entropy ($S_2 - S_1$) is due solely to the transfer of heat. Thus the phenomenon of heat transfer across a boundary has associated with it the transfer of entropy as well. The term on the right-hand side of Eqs. (2-17) and (2-18) is a measure of the *entropy transfer* or *entropy flux*.

The relationship between dS and $\delta Q/T$ for internally *irreversible* processes is determined from Eq. (2-15). Consider a cyclic process composed of a reversible process R from state 1 to state 2 and an irreversible process I from 2 to 1. On the basis of Eq. (2-15) we write

$$\oint \frac{\delta Q}{T} = \int_1^2 \left(\frac{\delta Q}{T}\right)_R + \int_2^1 \left(\frac{\delta Q}{T}\right)_I < 0$$

Only the inequality applies, because part of the overall cycle is irreversible. Substitution of Eq. (2-18) into the above relation yields

$$\Delta S > \int_1^2 \left(\frac{\delta Q}{T}\right)_I \qquad \text{(irreversible process)}$$

A combination of this equation with Eq. (2-18) leads to the *generalization* that

$$\Delta S \geq \int_1^2 \frac{\delta Q}{T} \tag{2-19}$$

where the equality and inequality signs apply to internally reversible and irreversible processes, respectively. As a consequence, the entropy transfer *never* exceeds, algebraically, the entropy change.

Equation (2-19) leads directly to one form of the *increase in entropy principle*. In the absence of heat transfer any closed system must fulfill the condition that

$$dS_{\text{adia}} \geq 0 \qquad \text{(closed system)} \tag{2-20}$$

The entropy function always increases due to internal irreversibilities during an adiabatic closed process. In the limiting case of an internally reversible adiabatic process the entropy will remain constant. For an isolated system $\delta Q = 0$ by definition. Therefore

$$dS_{\text{isol}} \geq 0 \tag{2-21}$$

This second form of the increase in entropy principle is extremely general in its application, because an isolated system analysis includes changes to all subsystems participating in a process.

2-5 CLOSED-SYSTEM ENTROPY GENERATION

In the preceding section the fundamental relationship $dS \geq \delta Q/T$ was developed for any closed system. This equation indicates that $\delta Q/T$ is a measure of dS when evaluated during an internally reversible process, and the inequality holds when internal irreversibilities are present. This inequality can be transformed into an equality by the introduction of the concept of entropy generation.

2-5-1 Entropy Generation

Due to the inequality in the relationship $dS \geq \delta Q/T$, it is apparent that $\delta Q/T$ is not an adequate measure of dS for an internally irreversible process; in fact, its contribution is always too small. But the only difference between internally reversible and irreversible processes is the physical presence of the irreversibilities in the latter case. This implies that internal irreversibilities themselves contribute to the change in S. This contribution, noted by σ on an extensive basis and called the *entropy production* or *entropy generation,* is always positive. Thus σ is defined as

$$\sigma \equiv S_2 - S_1 - \int \frac{\delta Q}{T} \geq 0 \qquad (2\text{-}22)$$

or for any closed system or control mass (CM) of uniform temperature

$$dS_{\text{CM}} = \frac{\delta Q_{\text{act}}}{T} + \delta\sigma \qquad \text{(any closed system)} \qquad (2\text{-}23)$$

Hence the inequality in Eq. (2-19) can be made an equality if a term is added to account for the presence of internal irreversibilities. The symbol σ_m (or σ/m) will be used for intensive values of the entropy production. This equation retains the concept that heat transfer always contributes to the entropy change of a closed system, whether the process is reversible or irreversible. This contribution can be positive, negative, or zero, depending on the direction of the actual heat transfer. For a system of uniform temperature Eq. (2-23) on an intensive basis becomes

$$ds = \frac{\delta q}{T} + \delta\sigma_m \qquad \text{(any closed process)} \qquad (2\text{-}24)$$

Note that the differential symbol δ is used before σ (instead of the symbol d) in Eqs. (2-23) and (2-24). This occurs because entropy production, like heat and work, is a function of the path of the process. Entropy production depends on the magnitude of the internal irreversibilities present during the process.

In some situations heat transfer may occur at several positions on the boundary. If the temperature at these positions is different, then a more general expression for Eq. (2-23) is

$$dS_{\text{CM}} = \sum_{i=1}^{n} \frac{\delta Q_i}{T_i} + \delta\sigma \qquad (2\text{-}25)$$

where T_i is the temperature at the point where δQ_i enters. On a unit-mass basis

$$ds_{\text{CM}} = \sum_{i=1}^{n} \frac{\delta q_i}{T_i} + \delta\sigma_m \qquad (2\text{-}26)$$

where the summation in both equations is over n different locations on the boundary. Equations (2-23) and (2-25) also may be expressed on a rate basis. For example, Eq. (2-25) becomes

$$\frac{dS_{\mathrm{CM}}}{dt} = \sum_{i=1}^{n} \frac{\dot{Q}_i}{T_i} + \dot{\sigma} \tag{2-27}$$

where $\dot{\sigma} = \delta\sigma/dt$, the time rate of entropy production due to irreversibilities within the system.

The first term on the right-hand side of Eqs. (2-23) through (2-27) is called the *entropy transfer* or *entropy flux* term, as noted in Sec. 2-4. As a result, these equations may be written as

$$\text{Entropy change} = \text{entropy transfer} + \text{entropy production} \tag{2-28}$$

$$\begin{pmatrix} > \\ = 0 \\ < \end{pmatrix} \qquad \begin{pmatrix} > \\ = 0 \\ < \end{pmatrix} \qquad (\geq 0)$$

The terms in the parentheses below Eq. (2-28) indicate the possible values of the quantities in the equation.

In summary, Eqs. (2-23) through (2-27) indicate that the change in entropy of any closed system is due to only two separate physical effects:

1. One effect is the transfer of heat to or from the system, as measured by $\delta Q_{\mathrm{act}}/T$. This contribution to dS is called the *entropy flux* or *entropy transfer* term. Since the actual heat transfer can be positive, negative, or zero, the entropy transfer term can take on any value.
2. The second effect is the presence of irreversibilities within the system. This contribution is called the *entropy production* or *entropy generation* term. This contribution to dS is either equal to or greater than zero.

As a consequence, the only way to decrease the entropy of any closed system is to transfer heat *from* it. However, in this case the heat-transfer contribution must be more negative than the positive contribution of any internal irreversibilities. Finally, when Eq. (2-25) is applied to an isolated system, we find that

$$dS_{\mathrm{isol}} = dS_{\mathrm{sys}} + dS_{\mathrm{surr}} = \delta\sigma_{\mathrm{tot}} \geq 0 \tag{2-29}$$

As expected, ΔS of an isolated system is due solely to the various entropy generation terms within the system.

Example 2-1. Oxygen initially at 300 K and 100 kPa is contained in a rigid tank. Two alternative processes are to be considered to increase its temperature to 500 K. The gas receives energy (1) adiabatically by means of a paddle wheel driven by an external source and (2) by means of an external heat reservoir at 600 K. In the second case the temperature at the boundary is taken as the gas temperature throughout the process. Find (*a*) the entropy production within the system in both cases and (*b*) the total entropy production in both cases, in kJ/kmol · K. (*c*) Then compare the results.

Solution. The specific entropy generation for a closed system is given by Eq. (2-26), namely,

$$\frac{\sigma}{m} = s_2 - s_1 - \sum_i \frac{q_i}{T_i}$$

(*a*) (1) For the paddle-wheel process there is no heat exchanged. Therefore the term $\sum q_i/T_i = 0$. The specific entropy generation, then, is equal to the specific entropy change, $s_2 - s_1$. Using data from the ideal-gas oxygen table, we find that

$$\begin{aligned}\sigma_m &= \Delta s = s_2^0 - s_1^0 - R \ln \frac{P_2}{P_1} = s_2^0 - s_1^0 - R \ln \frac{T_2}{T_1} \\ &= 220.589 - 205.213 - 8.314 \ln \frac{500}{300} = 11.13 \text{ kJ/kmol} \cdot \text{K}\end{aligned}$$

where $P_2/P_1 = T_2/T_1$ from the ideal-gas equation at constant volume.

(2) When the temperature is increased by means of heat addition, there are no apparent sources of irreversibility within the oxygen. Hence σ_m for this process is zero.

(*b*) (1) If we assume that the work used to drive the paddle wheel came from a work reservoir, then σ external to the tank is zero. Thus the total entropy production also equals 11.13 kJ/kmol · K.

(2) For the heat-addition process the system boundary is drawn around both the tank and the heat reservoir. Thus there is no heat transfer across this boundary and the total entropy production equals the total entropy change of the overall system. That is,

$$\sigma_{m,\text{tot}} = \Delta s_{\text{oxygen}} + \Delta s_{\text{reservoir}} = (s_2 - s_1)_{\text{oxygen}} + \frac{q_R}{T_R}$$

The value of q_R is the negative of the heat transfer to the oxygen. This latter quantity is found from an energy balance. Since $w = 0$ for the constant-volume process,

$$q = \Delta u - w = u_2 - u_1 = 10{,}614 - 6242 = 4372 \text{ kJ/kmol}$$

where the u data come from the oxygen ideal-gas table. Hence for a reservoir temperature of 600 K

$$\sigma_{m,\text{tot}} = 11.13 + -\frac{4372}{600} = 3.84 \text{ kJ/kmol} \cdot \text{K}$$

Because σ_m for the oxygen in this case is zero, the value of $\sigma_{m,\text{tot}}$ is also σ_Q, that due to irreversible heat transfer.

(*c*) The values of the total entropy production are quite different, 11.13 versus 3.84 kJ/kmol·K. Thus the paddle-wheel procedure is far more irreversible than the heat-addition process to achieve the same change of state for the oxygen. There are two factors which could alter this comparison. First, any irreversibility in the work mechanism external to the tank would make the use of the paddle wheel even more unfavorable. Second, an increase in the reservoir temperature T_R would diminish the difference between the σ_{tot} values for the two processes. For example, a T_R value of 1000 K would increase σ_{tot} for the heat-transfer process to 6.76 kJ/kmol · K. Thus T_R should be kept as low as possible, consistent with any heat-transfer rate requirement.

2-5-2 Comparison of Reversible and Irreversible Work Interactions

The presence of irreversibilities within a system degrades performance. This is especially important in work-producing and work-absorbing devices. For any internally reversible (rev) process of a closed system the $T\,ds$ equation from Chap. 1 can be written on an extensive basis as

$$T\,dS = dU - \delta W_{\text{rev}} \tag{2-30}$$

where δW_{rev} would be represented by $-P\,dV$ for a simple compressible substance. Also, for any process involving a closed system the conservation of energy principle requires that

$$\delta Q_{\text{act}} + \delta W_{\text{act}} = dU \tag{2-31}$$

The subscript "act" signifies that the δQ and δW terms are actual values and apply to either internally irreversible or reversible changes. When Eqs. (2-30) and (2-31) are applied to a process with specified end states, dU in both equations represents the same quantity, since U is a property. If we use Eq. (2-31) to eliminate dU from Eq. (2-30), then

$$T\,dS = \delta Q_{\text{act}} - \delta W_{\text{rev}} + \delta W_{\text{act}}$$

or upon rearrangement

$$dS_{\text{closed}} = \frac{\delta Q_{\text{act}}}{T} + \frac{1}{T}(\delta W_{\text{act}} - \delta W_{\text{rev}}) \tag{2-32}$$

However, for any closed system Eq. (2-23) requires that

$$dS_{\text{closed}} = \frac{\delta Q_{\text{act}}}{T} + \delta\sigma \tag{2-23}$$

A comparison of the last two equations shows that the entropy production $\delta\sigma$ of a closed system is related to work interactions by

$$\delta\sigma = \frac{(\delta W_{\text{act}} - \delta W_{\text{rev}})}{T} \geq 0 \tag{2-33}$$

Thus one of the manifestations of entropy generation is that the reversible and actual work for a process between given states are not equal. In fact, since σ is always equal to or greater than zero [see Eq. (2-28)], then

$$\delta W_{\text{act}} \geq \delta W_{\text{rev}} \tag{2-34}$$

This important relation is valid, regardless of the direction of change of the closed system. Hence, taking into account our sign convention for work interactions, we may write

$$W_{\text{in,act}} \geq W_{\text{in,rev}} \qquad \text{and} \qquad W_{\text{out,act}} \leq W_{\text{out,rev}} \tag{2-35}$$

Noting that the inequality sign on Eq. (2-34) denotes the presence of irreversibilities, we have shown that irreversibilities within a closed system reduce the work output and increase the work input for a specified change of state of a substance.

2-5-3 Entropy Generation and the Heat-Transfer Process

Heat-transfer processes are generally not reversible due to the presence of a finite temperature difference. Hence entropy production is associated with a heat-transfer process. The entropy production is associated with the space where the temperature gradient exists. Consider the region shown by the dashed line in Fig. 2-6 between systems A and B. To simplify the analysis, we shall assume that the source and sink for heat transfer are heat reservoirs. That is, the heat source and heat sink have constant temperatures T_A and T_B, respectively, and are internally reversible. If the heat transfer takes place at boundaries of the system where the temperature does not change during the heat exchange, Eq. (2-25) may be integrated to the form

$$\Delta S_{\text{CM}} = \sum_{i=1}^{n} \frac{Q_i}{T_i} + \sigma \tag{2-36}$$

and on a unit-mass basis

$$\Delta s_{\text{CM}} = \sum_{i=1}^{n} \frac{q_i}{T_i} + \sigma_m \tag{2-37}$$

For the passage of heat from body A to body B the properties of the dashed region in Fig. 2-6 remain fixed during the heat-transfer process. Hence the entropy change ΔS_{CM} for the region is zero in Eq. (2-36). However, there are entropy transfers Q/T_A entering and Q/T_B leaving the region. Applied specifically to the dashed line region of Fig. 2-6, Eq. (2-36) shows that the entropy production σ_Q is given by

$$\begin{aligned}\sigma_Q &= -\sum_i \frac{Q_i}{T_i} = -\left(\frac{Q}{T_A} + \frac{-Q}{T_B}\right)\\ &= Q\left(\frac{1}{T_B} - \frac{1}{T_A}\right) = \frac{Q}{T_A T_B}(T_A - T_B)\end{aligned} \tag{2-38}$$

where σ_Q is a special symbol for *entropy production due to heat transfer* between a finite temperature difference, Q is taken as *positive* in value, and $T_A > T_B$.

Equation (2-38) applies to the situation where T_A and T_B are constant. In some cases either system A or B, or both, may be relatively small. As a result,

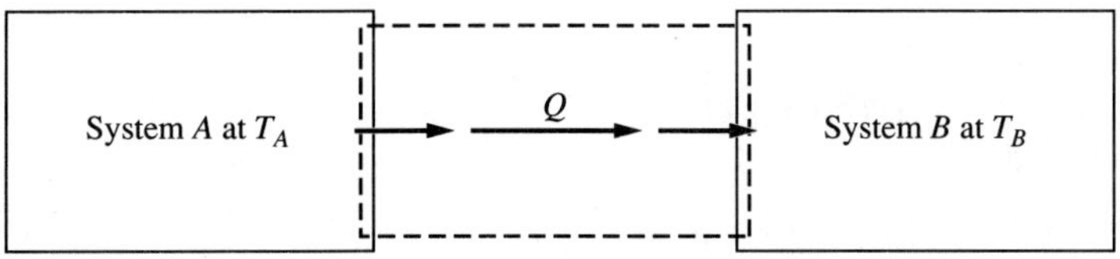

FIGURE 2-6
Illustration for the evaluation of entropy generation during irreversible heat transfer.

the temperature of such systems might change significantly due to heat transfer. For this more general situation we shall consider an isolated system comprised of systems A and B and the region of space between them (recall Fig. 2-6). Because the term $\sum Q_i/T_i$ is zero for the isolated system, Eq. (2-25) for a finite process reduces to $\sigma_{\text{tot}} = \Delta S_{\text{tot}}$, or for the three regions

$$\sigma_A + \sigma_B + \sigma_Q = \Delta S_A + \Delta S_B + \Delta S_Q$$

But ΔS_Q for the heat-transfer region is zero by concept. In addition, regions A and B are taken to be *internally reversible* (variable temperature heat source or sink only). As a result, $\sigma_A = \sigma_B = 0$. Thus the preceding equation reduces to

$$\sigma_Q = \Delta S_A + \Delta S_B \tag{2-39}$$

This is the general solution for a heat-transfer process between two internally reversible systems, for which Eq. (2-38) is a special case. [The method described by Eq. (2-39) was used in Example 2-1, part (b)(2).] From another viewpoint Eq. (2-39) can be physically interpreted in the following manner. The entropy flux from system A plus the entropy production in the region between A and B equal the entropy flux into system B.

Example 2-2. Heat transfer occurs between two bodies with temperatures of 300 and 1000 K. Determine the entropy production associated with a heat transfer of 100 kJ, in kJ/K.

Model. Both bodies are of fixed temperature.

Solution. The entropy production associated with heat transfer between two reservoirs is given by Eq. (2-38). In this case

$$\sigma_Q = Q\left(\frac{1}{T_B} - \frac{1}{T_A}\right) = 100\left(\frac{1}{300} - \frac{1}{1000}\right) = 0.233 \text{ kJ/K}$$

It is easily seen that as T_A is lowered in value toward 300 K, the entropy production approaches zero. Note also that the answer is equal to the sum of the changes in entropy of the two heat reservoirs. That is, on the basis of Eq. (2-39)

$$\sigma_Q = \sum \Delta S_{\text{res}} = \sum \frac{Q}{T} = \frac{100}{300} + \frac{-100}{1000} = 0.233 \text{ kJ/K}$$

From another viewpoint the entropy flux from A of 0.100 kJ/K, plus the entropy production in the region between A and B of 0.233 kJ/K, equal the entropy flux into B of 0.333 kJ/K.

2-6 SECOND-LAW ANALYSIS FOR A CONTROL VOLUME

The entropy balance for any closed system has been shown to be

$$\Delta S_{\text{CM}} = \sum_{i=1}^{n} \frac{Q_i}{T_i} + \sigma \tag{2-36}$$

or on a time basis

$$\frac{dS_{\mathrm{CM}}}{dt} = \sum_{i=1}^{n} \frac{\dot{Q}_i}{T_i} + \dot{\sigma} \tag{2-27}$$

The extension of this latter equation to a control volume may be made by an approach similar to that used to develop the general energy equation for a control volume. In Fig. 2-7 a control mass at time t lies within the small region A plus region CV. At time $t + \Delta t$ the mass of interest lies in region CV plus the small region B. At these two times the entropies of the control mass are

$$S_{\mathrm{CM},t} = S_A + S_{\mathrm{CV},t}$$

and

$$S_{\mathrm{CM},t+\Delta t} = S_{\mathrm{CV},t+\Delta t} + S_B$$

Subtraction of the first equation from the second equation yields the entropy change during the time interval Δt, namely,

$$S_{\mathrm{CM},t+\Delta t} - S_{\mathrm{CM},t} = S_{\mathrm{CV},t+\Delta t} - S_{\mathrm{CV},t} + S_B - S_A$$

To express this equation on a time-rate basis, we first divide by Δt. Noting that $S_A = m_1 s_1$ and $S_B = m_2 s_2$, we obtain

$$\frac{S_{\mathrm{CM},t+\Delta t} - S_{\mathrm{CM},t}}{\Delta t} = \frac{S_{\mathrm{CV},t+\Delta t} - S_{\mathrm{CV},t}}{\Delta t} + \frac{m_2 s_2}{\Delta t} - \frac{m_1 s_1}{\Delta t}$$

In the limit as Δt approaches zero, the boundaries of the control mass and control volume coincide, and the above equation becomes

$$\frac{dS_{\mathrm{CM}}}{dt} = \frac{dS_{\mathrm{CV}}}{dt} + \dot{m}_2 s_2 - \dot{m}_1 s_1$$

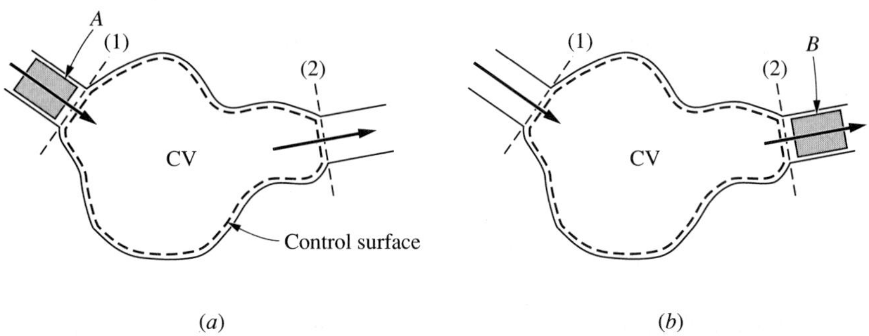

FIGURE 2-7
Development of an entropy relation for a control volume. (a) Control mass at time t; (b) control mass at time $t + \Delta t$.

For a control volume with multiple outlets and inlets the above equation is written as

$$\frac{dS_{\mathrm{CM}}}{dt} = \frac{dS_{\mathrm{CV}}}{dt} + \sum_{\mathrm{out}} \dot{m}s - \sum_{\mathrm{in}} \dot{m}s \tag{2-40}$$

The substitution of Eq. (2-27) shown above for dS_{CM}/dt into Eq. (2-40) yields an important form of the entropy balance for a control volume on a time-rate basis. That is,

$$\frac{dS_{\mathrm{CV}}}{dt} = \sum_{i=1}^{n} \frac{\dot{Q}_i}{T_i} + \sum_{\mathrm{in}} \dot{m}s - \sum_{\mathrm{out}} \dot{m}s + \dot{\sigma}_{\mathrm{CV}} \tag{2-41}$$

The quantity $\dot{Q}_i/T_i$ is the rate of entropy transferred through the portion of the control surface where the instantaneous absolute temperature is T_i. The value of the summation can be positive, negative, or zero. The next two terms account for rates of entropy transfer due to mass transfer. The final term represents the rate of entropy generation due to internal irreversibilities within the control volume, and can only be positive or zero.

It is useful to rearrange Eq. (2-41) to find an expression for the entropy generation within a control volume. Thus

$$\dot{\sigma}_{\mathrm{CV}} = \frac{dS_{\mathrm{CV}}}{dt} + \sum_{\mathrm{out}} \dot{m}s - \sum_{\mathrm{in}} \dot{m}s - \sum_{i=1}^{n} \frac{\dot{Q}_i}{T_i} \geq 0 \tag{2-42}$$

The vast majority of engineering analyses are directed toward control volumes operating in steady state. This requires that $dS_{\mathrm{CV}}\, dt = 0$. Therefore Eq. (2-42) reduces to

$$\dot{\sigma}_{\mathrm{CV}} = \sum_{\mathrm{out}} \dot{m}s - \sum_{\mathrm{in}} \dot{m}s - \sum_{i=1}^{n} \frac{\dot{Q}_i}{T_i} \geq 0 \qquad \text{(steady state)} \tag{2-43}$$

This equation is another illustration that in real processes entropy is not conserved. In steady state mass rates of flow in and out are equal and rates of energy transfer in and out are equal. However, Eq. (2-43) dictates that the *net* rate of transfer of entropy out of a control volume in steady state must be *positive* for *irreversible* processes. For those many steady-flow processes which involve only one inlet and one exit the equation for entropy generation becomes

$$\dot{\sigma}_{\mathrm{CV}} = \dot{m}(s_2 - s_1) - \sum_{i=1}^{n} \frac{\dot{Q}_i}{T_i} \geq 0 \qquad \text{(steady state)} \tag{2-44}$$

where subscripts 1 and 2 again represent inlet and exit positions. Upon rearrangement the specific entropy change for a steady-flow process with one inlet and one outlet becomes

$$(s_2 - s_1)_{CV} = \frac{1}{\dot{m}}\left(\sum_{i=1}^{n} \frac{\dot{Q}_i}{T_i}\right) + \sigma_{m,CV} \qquad \text{(steady state)} \qquad (2\text{-}45)$$

where the specific entropy production within the control volume $\sigma_{m,CV} = \dot{\sigma}_{CV}/\dot{m}$. In comparison, recall that the specific entropy change of a closed system is given by

$$(s_2 - s_1)_{CM} = \sum_{i=1}^{n} \frac{q_i}{T_i} + \sigma_{m,CM} \qquad \text{(closed system)} \qquad (2\text{-}37)$$

The similarity of the terms on the right-hand side of these last two equations is not unexpected.

The conclusions drawn earlier from Eq. (2-26) regarding the magnitude and sign on Δs for a closed system are equally valid for a unit mass passing through a steady-state control volume as represented by Eq. (2-45). These are:

1. The specific entropy change from inlet to exit can increase, decrease, or remain constant.
2. Because $\dot{\sigma}_{CV}$ can never be negative, the only way that the specific entropy change can decrease is for the entropy transfer out accompanying heat transfer out of the control volume to be greater than the entropy produced internally by irreversibilities.
3. When internal irreversibilities are present and heat transfer is absent, the specific entropy will always increase from inlet to exit.
4. In the very special case of no heat transfer and no internal irreversibilities the specific entropy will not change; that is, the flow is isentropic.

It should be recognized that the summation term Q_i/T_i in Eq. (2-42) might be difficult to evaluate numerically, because the values of Q_i and T_i are not known at every position on the control surface. However, the calculation of $\sum Q_i/T_i$ can be avoided by choosing a system which includes the control volume and any heat reservoirs which might supply or receive heat during a process, including the atmosphere. Figure 2-8(a) shows a schematic of a control volume which exchanges heat solely with the environment at T_0 and (arbitrarily) only one other reservoir at T_R.

At any arbitrary position along the control surface we consider a transfer of heat Q_i at a boundary temperature of T_i. Consider that Q_i is transferred irreversibly (due to a finite temperature difference) from T_R to T_i, as shown in Fig. 2-8(b). The entropy production associated with irreversible heat transfer was shown earlier to be given by the sum of the $-Q/T$ terms for the region of interest. For the small heat-transfer region associated with T_i, T_R, and Q_i the entropy production is

$$\sigma_Q = -\left(\frac{Q}{T}\right)_{\text{in}} + \left(\frac{Q}{T}\right)_{\text{out}} = -\left(\frac{Q_i}{T_R} - \frac{Q_i}{T_i}\right)$$

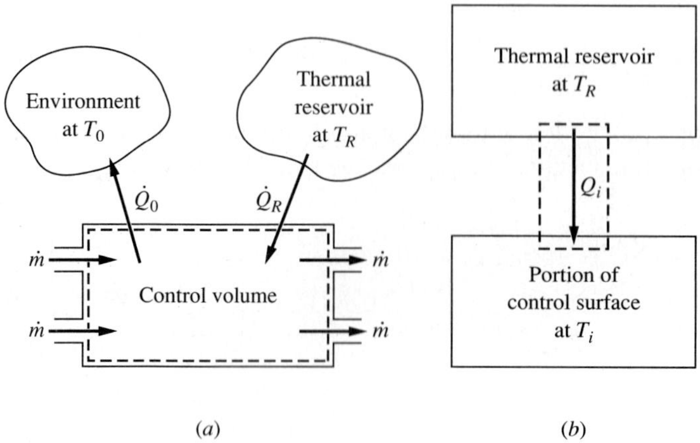

FIGURE 2-8
(a) Control volume exchanging heat solely with the environment at T_0 and one other reservoir at T_R; (b) heat transfer between a reservoir at T_R and a small region of the control volume at T_i.

where Q_i is measured relative to the control volume. If we now sum over the entire control surface where heat transfer occurs between the control volume and T_R, then for n heat-transfer regions along the control surface

$$\sigma_Q = -\sum_{i=1}^{n} \frac{Q_i}{T_R} + \sum_{i=1}^{n} \frac{Q_i}{T_i}$$

Noting that T_R is a constant, and that $\sum Q_i = Q_R$, we find that the summation in the first term on the right reduces to Q_R/T_R. If we also include those heat-transfer regions between the control volume and the reservoir T_0, then a similar term $-Q_0/T_0$ must be added. (Note that signwise Q_R and Q_0 are to be measured relative to the control volume.) Consequently, the above equation becomes

$$\sigma_Q = -\frac{Q_R}{T_R} - \frac{Q_0}{T_0} + \sum_{i=1}^{n} \frac{Q_i}{T_i}$$

Upon rearrangement

$$\sum_{i=1}^{n} \frac{Q_i}{T_i} = \sigma_Q + \frac{Q_0}{T_0} + \frac{Q_R}{T_R} \tag{2-46}$$

A more general statement of Eq. (2-46) would be

$$\sum_{i=1}^{n} \frac{Q_i}{T_i} = \sigma_Q + \frac{Q_0}{T_0} + \sum_{j} \frac{Q_j}{T_j}$$

where the summation on the right-hand side allows for several (j) heat reservoirs other than the atmosphere. The above equation becomes, on a rate basis,

$$\sum_{i=1}^{n} \frac{\dot{Q}_i}{T_i} = \dot{\sigma}_Q + \frac{\dot{Q}_0}{T_0} + \sum_j \frac{\dot{Q}_j}{T_j} \tag{2-47}$$

This is the desired result which will be used to replace $\sum \dot{Q}_i/T_i$ in Eq. (2-42). The replacement expression now contains $\dot{\sigma}_Q$, the entropy generation term for the irreversible heat transfer to or from the T_0 and T_j reservoirs.

The substitution of Eq. (2-47) into Eq. (2-42) yields

$$\dot{\sigma}_{\text{tot}} = \dot{\sigma}_{\text{CV}} + \dot{\sigma}_Q = \frac{dS_{\text{CV}}}{dt} + \sum_{\text{out}} \dot{m}s - \sum_{\text{in}} \dot{m}s - \frac{\dot{Q}_0}{T_0} - \sum_j \frac{\dot{Q}_j}{T_j} \geq 0 \tag{2-48}$$

where Q_0 and Q_j are measured relative to the control-volume boundary. Compared to Eq. (2-42), this equation has the advantage that normally Q_0, Q_j, T_0, and T_j are known quantities in an analysis. In steady state $dS_{\text{CV}}/dt = 0$, and Eq. (2-48) becomes

$$\dot{\sigma}_{\text{tot}} = \sum_{\text{out}} \dot{m}s - \sum_{\text{in}} \dot{m}s - \frac{\dot{Q}_0}{T_0} - \sum_j \frac{\dot{Q}_j}{T_j} \geq 0 \qquad \text{(steady state)} \tag{2-49}$$

Recall Eq. (2-43), which applies solely to any control volume with Q_i measured relative to the control volume at surface temperature T_i, namely,

$$\dot{\sigma}_{\text{CV}} = \sum_{\text{out}} \dot{m}s - \sum_{\text{in}} \dot{m}s - \sum_{i=1}^{n} \frac{\dot{Q}_i}{T_i} \qquad \text{(steady state)} \tag{2-43}$$

Although these latter two equations have a similar format, their usage is quite different.

Finally, consider a control volume with one inlet (state 1) and one outlet (state 2) and heat is exchanged with a single-reservoir at T_R (which could be T_0). Equation (2-49) becomes, after rearrangement,

$$s_2 - s_1 = \frac{q_R}{T_R} + \sigma_{\text{tot}} \tag{2-50}$$

where σ_{tot} is the sum of the entropy generation due to internal irreversibilities and due to irreversible heat transfer. In addition, $s_2 - s_1$ is the specific entropy change of the mass passing through the control volume, and subscript "R" represents the single-reservoir condition.

Example 2-3. Steam enters a turbine at 40 bars, 500°C, and 140 m/s and leaves as a saturated vapor at 100°C and 80 m/s. The measured work output is 746.0 kJ/kg. The average temperature T_b at the outer surface of the turbine where heat transfer occurs to the environment is taken as the average of the inlet and exit steam temperatures. The potential-energy change is negligible. (*a*) Determine the specific entropy production within the turbine, in kJ/kg · K. (*b*) Now, enlarge the control volume so that the local environment at 25°C is included. Find the specific entropy production in this case. (*c*) Discuss why these answers are different.

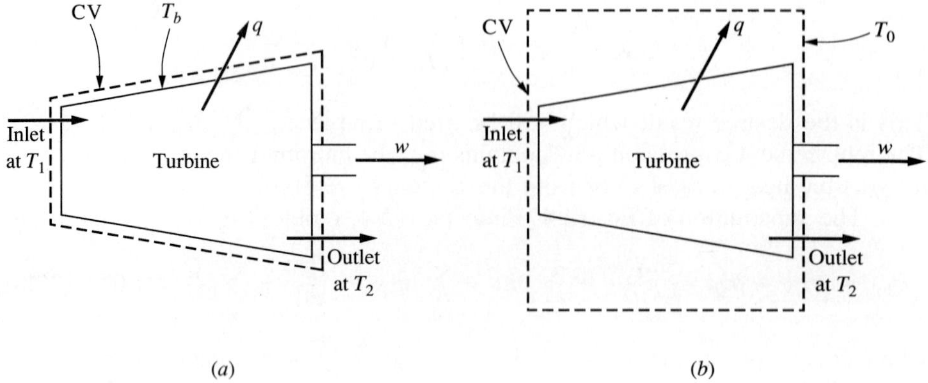

FIGURE 2-9
Figure for Example 2-3.

Solution
(*a*) The specific entropy production $\sigma_{m,\text{CV}}$ for the control volume with one inlet and one exit is given by Eq. (2-45). On the basis of unit mass entering and leaving the control volume this equation becomes

$$\sigma_{m,\text{CV}} = s_2 - s_1 - \sum \frac{q_i}{T_i}$$

The entropy flux at the control surface shown in Fig. 2-9(*a*) is given simply by q/T_b, because the surface temperature T_i is represented by a single boundary temperature T_b. Thus the preceding general relation is approximated in this case by

$$\sigma_{m,\text{CV}} = s_2 - s_1 - \frac{q}{T_b}$$

The only unknown in this equation is q, which is found from a steady-flow energy balance.

$$q = -w + (h_2 - h_1) + \frac{V_2^2 - V_1^2}{2}$$

From the steam tables we find that h_1 and h_2 are 3445.3 and 2676.1 kJ/kg, respectively. Therefore

$$q = -(-746.0) + (2676.1 - 3445.3) + \frac{(80)^2 - (140)^2}{2(1000)} = -29.8 \text{ kJ/kg}$$

In addition, the specific entropy values at the inlet and exit are 7.0901 and 7.3549 kJ/kg · K, respectively. On this basis the specific entropy production is

$$\sigma_{m,\text{CV}} = (7.3549 - 7.0901) - \frac{-29.8}{573} = 0.3168 \text{ kJ/kg} \cdot \text{K}$$

(*b*) When the control volume is moved sufficiently away from the turbine surface, as shown in Fig. 2-9(*b*), the value of T_b becomes T_0. For this situation the specific entropy production is found to be

$$\sigma_{m,\text{tot}} = (7.3549 - 7.0901) - \frac{-29.8}{298} = 0.3648 \text{ kJ/kg} \cdot \text{K}$$

(*c*) The evaluation indicates that the entropy production increases when the larger control volume is considered. In part (*a*) only irreversibilities within the turbine are evaluated. In part (*b*) irreversible heat transfer between the turbine cover and the atmosphere is also included. Thus the entropy production due to heat transfer is $0.3648 - 0.3168 = 0.0480$ kJ/kg · K. This same value is found by employing Eq. (2-38). That is,

$$\sigma_Q = Q\left(\frac{1}{T_B} - \frac{1}{T_A}\right) = 29.8\left(\frac{1}{298} - \frac{1}{573}\right) = 0.0480 \text{ kJ/kg} \cdot \text{K}$$

The entropy production due to heat transfer across the finite temperature difference is only 15 percent of that produced within the turbine. This is not surprising, since the heat loss is only 4 percent of the actual work delivered and the external temperature gradient is relatively small.

By an approach similar to that used in Sec. 2-5-2, it may be shown that both Eqs. (2-33) and (2-34) apply to a steady-flow (sf) control-volume analysis as well. A combination of the second $T\,ds$ equation and the basic entropy generation equation for a unit mass, namely,

$$T\,ds = dh - v\,dP \qquad \text{and} \qquad T\,ds = \delta q_{\text{act}} + T\delta\sigma$$

yields

$$dh = \delta q_{\text{act}} + T\delta\sigma + v\,dP$$

In addition, for a control volume with one inlet and one outlet the energy equation for steady flow can be written as

$$\delta q_{\text{act}} + \delta w_{sf,\text{act}} = dh + d\,\text{ke} + d\,\text{pe}$$

Since dh represents the same quantity in the two preceding equations, on equating these terms we find that

$$\delta w_{sf,\text{act}} = T\delta\sigma + v\,dP + d\,\text{ke} + d\,\text{pe} \tag{2-51}$$

or noting that $\delta w_{sf,\text{rev}} = v\,dP + d\,\text{ke} + d\,\text{pe}$ from mechanics, we have

$$\delta w_{sf,\text{act}} = T\delta\sigma + \delta w_{sf,\text{rev}}$$

Rearrangement of the latter equation shows that on a rate basis for steady flow

$$\delta\dot{\sigma}_{\text{CV}} = \frac{\delta\dot{W}_{sf,\text{act}} - \delta\dot{W}_{sf,\text{rev}}}{T} \geq 0 \tag{2-52}$$

and

$$\delta w_{sf,\text{act}} \geq \delta w_{sf,\text{rev}} \tag{2-53}$$

These latter two expressions are equivalent to Eqs. (2-33) and (2-34) developed for a closed system. Thus in the analysis of compressors, pumps, and fans, with *work input*

$$w_{sf,\text{in,act}} \geq w_{sf,\text{in,rev}} \qquad \text{(compressor, pump, fan)} \tag{2-54}$$

and for turbines with *work output*

$$w_{sf,\text{out,act}} \leq w_{sf,\text{out,rev}} \qquad \text{(turbine)} \tag{2-55}$$

Thus to minimize work input or maximize work output for closed or open systems, we must attempt to minimize internal irreversibilities. With the general equations developed above, we are now in a position to examine the role of entropy production in steady-flow processes.

2-7 ENTROPY PRODUCTION IN SIMPLE CYCLIC DEVICES

A Carnot heat engine, refrigerator, or heat pump is totally reversible. Thus the cyclic system itself, the reservoirs used, and the heat-transfer processes must all be reversible. As a result, there is no entropy production associated with these idealized devices. We now wish to consider heat engines, refrigerators, and heat pumps which are irreversible, and hence have entropy generation during their operation. However, to simplify the analysis, we shall assume that the source and sink for heat transfer are heat reservoirs. That is, the heat source and heat sink have constant temperatures and are internally reversible. Recall that for heat transfer between two systems with temperatures T_A and T_B, the entropy production σ_Q is given by Eq. (2-38), namely,

$$\sigma_Q = -\sum_i \frac{Q_i}{T_i} = Q\left(\frac{1}{T_B} - \frac{1}{T_A}\right) \tag{2-38}$$

where Q is taken as positive in value and $T_A > T_B$. This irreversible heat-transfer process is illustrated in Fig. 2-10(a). We shall now apply this information on entropy production due to irreversible heat transfer to the study of irreversible heat engines, refrigerators, and heat pumps operating between two heat reservoirs.

First, consider the situation where the heat transfers to and from a cyclic heat engine are reversible but the cyclic system itself is irreversible. A heat engine schematic is shown in Fig. 2-10(b). The heat transfers Q_H and Q_L are shown to occur across differential temperature differences. However, because the engine itself operates irreversibly, the actual efficiency must be less than that of a Carnot engine operating between the same two temperatures. For the engine we may write

$$\Delta S_{\text{engine}} = \sum_{i=1}^{n} \frac{Q_i}{T_i} + \sigma_{\text{engine}}$$

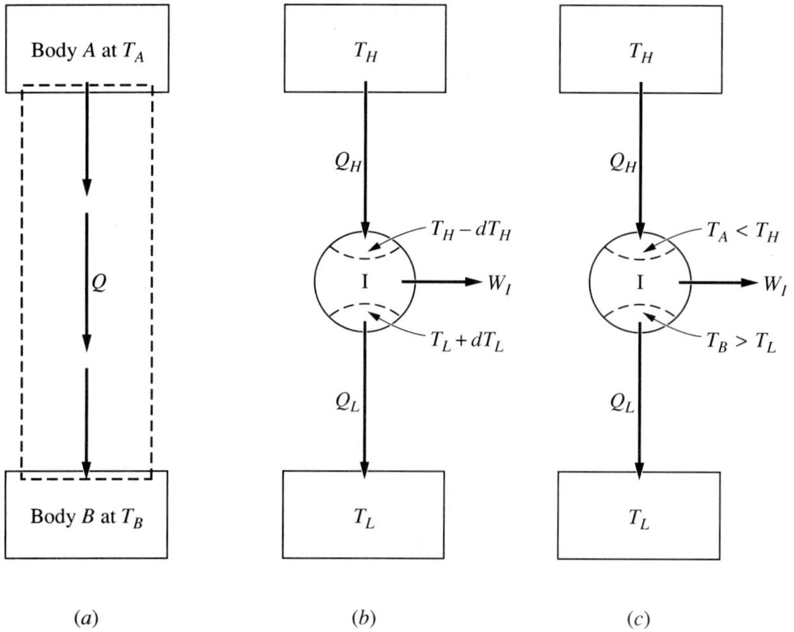

FIGURE 2-10
(a) Irreversible heat transfer from T_A to T_B; (b) heat engine with reversible heat transfer and irreversible engine operation; (c) heat engine with irreversible heat transfer and engine operation.

where T_i is measured at the boundary of the engine. The engine is a cyclic device. Hence for an integral number of cycles $\Delta S_{\text{engine}} = 0$. Thus

$$\sigma_{\text{engine}} = -\sum_{i=1}^{n} \frac{Q_i}{T_i} = -\left(\frac{Q_H}{T_H} + \frac{Q_L}{T_L}\right) \tag{2-56}$$

In addition, $Q_H + Q_L + W_{\text{net}} = 0$ and $\eta_{\text{th}} = -W_{\text{net}}/Q_H$. In all three of the preceding equations the sign on Q_H, Q_L, and W_{net} must be supplied as well as their magnitudes.

Example 2-4. A heat engine operates between two heat reservoirs, the temperatures of which are 1000 and 300 K. Consider the situation where the cyclic system is also at 1000 and 300 K while receiving or rejecting heat [recall Fig. 2-10(b)]. In addition, the measured thermal efficiency of the engine is 50 percent. If the heat supplied to the engine during one cycle of operation is 100 kJ, determine the entropy production, in kJ/K, for the cycle.

Model. The system consists of an irreversible heat engine and two heat reservoirs exchanging heat reversibly.

Solution. There is no temperature difference between the engine and either reservoir when heat is transferred. Hence the heat-transfer processes are reversible and the entropy production is zero for these processes. However, the actual engine efficiency is 50 percent, whereas the Carnot efficiency, given by $1 - T_L/T_H$, is 70 percent. Hence the engine operation is internally irreversible. The entropy production associated with the engine operation is found from Eq. (2-56), namely,

$$\sigma_{\text{engine}} = -\sum_{i=1}^{n} \frac{Q_i}{T_i}$$

For $Q_H = 100$ kJ and $\eta_{\text{th,act}} = 50$ percent $= |W_{\text{net}}|/Q_H$, then $W_{\text{net}} = 50$ kJ. From an energy balance on the engine, $Q_L = -50$ kJ. Hence the entropy production for the actual heat engine is

$$\sigma_{\text{engine}} = -\left[\frac{100}{1000} + \frac{-50}{300}\right] = 0.0667 \text{ kJ/K}$$

Figure 2-10(c) illustrates the situation where the heat transfers are irreversible and the heat engine may be reversible or irreversible. Heat transfer Q_H occurs between T_H and T_A, and heat transfer Q_L occurs between T_B and T_L. Hence the heat engine effectively operates only between T_A and T_B, and not between T_H and T_L. Consequently, the sum of the entropy generation terms for the two heat-transfer processes and the engine process is

$$\begin{aligned}\sigma_{\text{tot}} &= \sigma_{Q,\text{high}} + \sigma_{\text{engine}} + \sigma_{Q,\text{low}} \\ &= \left(\frac{Q_H}{T_H} + \frac{Q_H}{T_A}\right) - \left(\frac{Q_H}{T_A} + \frac{Q_L}{T_B}\right) - \left(\frac{Q_L}{T_B} + \frac{Q_L}{T_L}\right) \qquad \text{(2-57)}\end{aligned}$$

Several terms in the above equation will cancel out. However, if left in its present format, Eq. (2-57) will yield the values of the separate entropy production quantities. A comparison of these values will show which process contains the largest (or smallest) losses. In practice, this would help the engineer to decide which portion of the overall process needs improvement.

It is obvious that if a boundary is drawn around the entire heat-engine process, then

$$\sigma_{\text{tot}} = \frac{Q_H}{T_H} + \frac{Q_L}{T_L}$$

This format would be useful in comparing several different heat engines overall, but it gives no information on the contributions of processes associated with the cycle. A similar analysis can be made for refrigerators and heat pumps operating between two reservoirs.

Example 2-5. Consider the same physical situation as in Example 2-4, except that the engine temperature during heat addition is 900 K and is 350 K during heat rejection [recall Fig. 2-9(c)]. The heat engine is considered to be reversible. Determine the entropy production for all the irreversibilities present in the overall process, in kJ/K.

Model. Reversible heat engine and irreversible heat transfers

Solution. The thermal efficiency of the reversible heat engine is now based on temperatures of 900 and 350 K and is 61.1 percent. For the same heat input of 100 kJ as in Example 2-4 the net work output becomes 61.1 kJ and the heat rejection is 38.9 kJ. The entropy production within the reversible heat engine is zero. However, the heat-transfer processes themselves are irreversible, because a finite temperature difference exists in each case. For the heat-transfer process to the engine

$$\sigma_{Q,\text{high}} = -\left[\frac{100}{1000} + \frac{-100}{900}\right] = 0.0111 \text{ kJ/K}$$

In addition, for the engine heat-rejection process

$$\sigma_{Q,\text{low}} = -\left[\frac{38.9}{350} + \frac{-38.9}{300}\right] = 0.0185 \text{ kJ/K}$$

Hence for the overall process, $\sigma_{\text{tot}} = 0.0111 + 0.0185 = 0.0296$ kJ/K. The low-temperature heat-transfer process makes a sizable contribution (62 percent) to the overall entropy production.

Example 2-6. Consider the same physical situation as in Example 2-5, except that the engine is now irreversible and has a thermal efficiency of 40 percent. Determine the entropy production for all the irreversibilities present in the overall process, in kJ/K.

Model. Irreversible heat engine and heat transfers

Solution. The thermal efficiency is now 40 percent. For the same heat input of 100 kJ the net work output becomes 40 kJ and the heat rejection is 60 kJ. In this situation the entropy production for the engine is

$$\sigma_{\text{engine}} = -\left[\frac{100}{900} + \frac{-60}{350}\right] = 0.0603 \text{ kJ/K}$$

The entropy production for the heat-transfer process to the engine remains the same as in Example 2-5, namely, 0.0111 kJ/K. In addition, for the engine heat-rejection process

$$\sigma_{Q,\text{low}} = -\left[\frac{60}{350} + \frac{-60}{300}\right] = 0.0286 \text{ kJ/K}$$

Hence for the overall process $\sigma_{\text{tot}} = 0.0603 + 0.0111 + 0.0286 = 0.1000$ kJ/K. The engine operation in this case is the major contributor (60 percent) to the overall entropy production.

The analysis of entropy generation in simple refrigeration and heat pump cycles would be similar to that shown above for a heat-engine cycle.

2-8 ENTROPY PRODUCTION AND FIRST-LAW EFFICIENCIES

The first-law efficiencies of several devices were reviewed in Sec. 1-1-10. These efficiencies can be written in terms of entropy production within the device. This is not surprising, because both first-law efficiencies and entropy production relate reversible and irreversible performances. Recall Eq. (2-51), which applies to a steady-state control volume with one inlet and one outlet:

$$\begin{aligned} \delta w_{sf,\mathrm{act}} &= T\delta\sigma + v\,dP + d\mathrm{ke} + d\mathrm{pe} \\ &= T\,\delta\sigma + \delta w_{sf,\mathrm{rev}} \end{aligned} \tag{2-51}$$

This equation, then, is combined with the first-law efficiency for various work devices. As an example, consider a steady-flow turbine for which the adiabatic (first-law) efficiency η_t is given by

$$\eta_t = \frac{\delta w_{\mathrm{act}}}{\delta w_{\mathrm{isen}}}$$

when written for a differential change of state. In addition, in the absence of kinetic- and potential-energy changes, $\delta w_{\mathrm{rev},sf} = v\,dP$. When Eq. (2-51) is substituted into the above relation we find that

$$\eta_t = \frac{T\,\delta\sigma_{\mathrm{CV}} + \delta w_{\mathrm{rev}}}{\delta w_{\mathrm{rev}}} = \frac{T\delta\sigma_{\mathrm{CV}} + v\,dP}{v\,dP}$$

For a turbine dP is negative and for any device $\delta\sigma_{\mathrm{CV}}$ is positive. Hence greater internal irreversibilities lead to larger σ values and smaller turbine first-law efficiencies. Thus turbine performance is directly affected by entropy generation. A similar analysis would hold for compressors, fans, and pumps.

In nonwork devices such as duct flow, nozzles, and diffusers Eq. (2-51) still applies. In the case where shaft work and potential-energy change are zero, this equation can be written as

$$v\,dP = -T\delta\sigma - d\,\mathrm{ke}$$

In the case of internally reversible flow $\delta\sigma$ is zero and $v\,dP + d\,\mathrm{ke} = 0$. For example, the drop in pressure in an adiabatic nozzle is responsible for the acceleration of the flow. For internally irreversible flow through an adiabatic nozzle $\delta\sigma > 0$. Thus

$$d\,\mathrm{ke}_{\mathrm{act}} = -T\,\delta\sigma - v\,dP \qquad \text{(irreversible nozzle)}$$

For a specified value of dP (which is negative) the change in kinetic energy is decreased by the presence of entropy production. In terms of the adiabatic nozzle efficiency η_n we find that

$$\eta_n = \frac{d\,\mathrm{ke}_{\mathrm{act}}}{d\,\mathrm{ke}_{\mathrm{rev}}} = \frac{-v\,dP - T\delta\sigma}{-v\,dP}$$

Like a turbine, dP is negative and $\delta\sigma$ is positive. Thus the presence of entropy production due to internal irreversibilities degrades performance.

An interesting application of entropy production is to the behavior of a fluid passing through a horizontal, insulated pipe or duct under steady state. Equation (2-51) dictates in this case that

$$V\,dV + T\,\delta\sigma + v\,dP = 0 \tag{2-58}$$

However, in courses in fluid mechanics the concept of "head loss" h_F is introduced in terms of a "friction factor" f. The format usually presented is

$$-v\,dP - V\,dV = \delta h_F = \frac{fV^2\,dx}{2D}$$

or

$$V\,dV + \frac{fV^2\,dx}{2D} + v\,dP = 0$$

A term-by-term comparison of the above equation with Eq. (2-58) shows that

$$T\,\delta\sigma_{\text{CV}} = \frac{fV^2\,dx}{2D} = \delta h_F \tag{2-59}$$

Hence the term $T\,\delta\sigma_{\text{CV}}$ is a measure of the head loss due to internal friction. In the situation where velocity changes are negligible Eq. (2-59) indicates that σ_{CV} is directly related to the pressure drop in the duct. The rate of entropy production in this case would be given by

$$T\,\delta\sigma_{\text{CV}} = -v\,dP \qquad \text{or} \qquad \dot{\sigma}_{\text{CV}} = -\dot{m}\int_{\text{in}}^{\text{out}}\left(\frac{v}{T}\right)dP$$

This result indicates that the rate of entropy production due to frictional shear stress is proportional to the mass flow rate and the pressure drop.

This relationship between $\dot{\sigma}$ and ΔP is clearly shown by employing the ideal-gas and incompressible models for fluid flow through ducts. The steady-state entropy balance for adiabatic flow through a duct, based on Eq. (2-44), is $\dot{\sigma}_{\text{CV}} = \dot{m}(s_2 - s_1)$, where state 1 is the inlet. The steady-state energy equation for adiabatic flow through a duct with negligible changes in kinetic and potential energies reduces to $h_1 = h_2$. For an ideal gas this requires that $T_1 = T_2$ and that $s_2 - s_1 = -R\ln P_2/P_1$. Substitution of this result into the steady-state entropy balance shows that

$$\dot{\sigma}_{\text{CV}} = \dot{m}R\ln\frac{P_1}{P_2} = -\dot{m}R\ln\left(1 + \frac{\Delta P}{P_1}\right) \qquad \text{(ideal gas)}$$

When $(P_1 - P_2) \ll P_1$, the relation $\ln(1 + x) \approx x$ may be used to transform the above equation into

$$\dot{\sigma}_{\text{CV}} = -\dot{m}R\frac{\Delta P}{P} = \dot{m}R\left(\frac{P_1 - P_2}{P_1}\right) \qquad \text{(ideal gas)} \tag{2-60}$$

For an incompressible fluid $\Delta s = c \ln(T_2/T_1)$, where the specific heat c is assumed to be constant. Also, since $h_1 = h_2$, for an incompressible fluid $c\,\Delta T + v\,\Delta P = 0$, or $\Delta T/T_1 = -v\,\Delta P/cT_1$. Consequently,

$$\dot{\sigma}_{CV} = \dot{m}c \ln \frac{T_2}{T_1} = \dot{m}c \ln\left(1 + \frac{\Delta T}{T_1}\right) = \dot{m}c \ln\left(1 - \frac{v\Delta P}{cT_1}\right) \qquad \text{(incompressible)}$$

When $v\,\Delta P \ll cT_1$, then the relation $\ln(1 - x) \approx -x$ allows the above equation to be written as

$$\dot{\sigma}_{CV} = -\frac{\dot{m}v\,\Delta P}{T_1} = \frac{\dot{m}v(P_1 - P_2)}{T_1} \qquad \text{(incompressible)} \qquad (2\text{-}61)$$

Both Eqs. (2-60) and (2-61) show that σ_{CV} is a direct function of $\dot{m}$ and ΔP for these models.

PROBLEMS

2-1. Present a proof that the following process is irreversible. A piece of copper is cooled from 100 to 20°C by submersion into a very large body of water.

2-2. Present a proof that the following process is irreversible. A saturated liquid confined to a piston-cylinder device of constant pressure is stirred until one-half of it evaporates.

2-3. Present a proof that the following process is irreversible. A gas expands slowly through a small hole from one chamber into another chamber where the pressure initially is considerably lower.

2-4. A vessel initially contains 5.0 kg of liquid water and 2.0 kg of ice at 0°C. Energy is added until the ice has just melted. The temperature at the boundary where heat transfer occurs is taken to be the system temperature during the process. The enthalpy of melting is 333.5 kJ/kg. Consider the following processes used to melt the ice.

- **a.** Heat is added from the environment at 20°C. Determine the entropy flux and the total entropy generation, both in kJ/K.
- **b.** Heat is added from a reservoir at 727°C. Determine the same quantities as in part (a).
- **c.** Paddle-wheel work is used to change the state. Determine the total entropy production for the process, in kJ/K.
- **d.** Comment on the relative degree of irreversibility for the three processes.

2-5. A piston-cylinder contains 0.10 kg of saturated liquid water at 10 bars. Energy is added until the liquid has just evaporated. The boundary temperature where heat transfer occurs is taken as the saturation temperature of water at 10 bars. Consider the following three processes.

- **a.** First, heat is added from a reservoir at 227°C. Determine the entropy flux for the fluid and the total entropy production, both in kJ/K.
- **b.** Second, heat is added from a reservoir at 2000 K. Find the same quantities requested in part (a).

c. Third, the energy is added by electrical work on a resistor within the cylinder. Determine the entropy generation within the fluid and the total entropy production for the process, both in kJ/K.

d. Fourth, one-half the energy required comes from heat addition at 227°C, the other half comes from electrical work input. Find the entropy production due to the electrical work and the total entropy production for the overall process, both in kJ/K.

e. Compare the relative irreversibility of the three processes.

2-6. A piston-cylinder contains 1 L (liter) of saturated liquid R-134a maintained at 6.0 bars. Energy is added until one-half the liquid has evaporated. The boundary temperature where heat transfer occurs is taken as the saturation temperature of the R-134a at 6.0 bars. Consider the following three processes.

a. First, heat is added from a reservoir at 127°C until the required liquid has evaporated. Determine the entropy flux for the R-134a and the entropy production for the heat-transfer process, both in J/K.

b. Second, the heat added is from a reservoir at 600 K. Find the same quantities requested in part (a).

c. Third, one-half the energy required is added by heat transfer from the 127°C reservoir, whereas the other half is added by electrical work on a resistor within the cylinder. Determine the entropy production within the R-134a and for the heat-transfer process, both in J/K.

d. Compare the total entropy production for the three processes.

2-7. A closed rigid tank contains 0.10 kg of air initially at 1 bar and 350 K. Energy is added until the temperature reaches 400 K. Consider the following methods of energy addition.

a. Heat is added from a reservoir at 500 K, and the system temperature where heat transfer occurs is 500 K. Find the entropy production for the air, in J/K.

b. The same situation as part (a), except that the temperature at the boundary where heat transfer occurs is the same as the air temperature within the system throughout the process. Find the entropy generation for both the air and the overall process, in J/K.

c. The energy is added in the form of paddle-wheel work. Find the total entropy generation for the process, in J/K.

2-8. Carbon dioxide (CO_2) at 110 kPa and 290 K is contained within a constant-pressure piston-cylinder assembly with an initial volume of 0.020 m^3. Energy is added until the temperature reaches 350 K. Consider the following processes.

a. Heat is added from a reservoir at 500 K, and the system boundary temperature is the same as the gas temperature throughout the process. Find the entropy change and the entropy flux for the gas, both in J/K.

b. Calculate the entropy production for an enlarged system which includes the piston-cylinder device as well as the reservoir, in J/K.

c. Calculate the entropy production of the gas and the total value if the system boundary temperature throughout the process is 400 K, in J/K.

d. The process is repeated by using electrical resistor work instead of heat addition. Find the total entropy production, in J/K.

e. Compare the relative irreversibility of the processes.

2-9. A closed rigid tank with a volume of 2 m^3 contains hydrogen gas initially at 320 K and 180 kPa. Heat transfer from a reservoir at 500 K takes place until the gas temperature reaches 400 K.

a. Calculate the entropy change and entropy production for the hydrogen gas during the process, both in kJ/K, if the boundary temperature for the gas is the same as the gas temperature throughout the process.
b. Determine the entropy production, in kJ/K, for an enlarged system which includes the tank and the reservoir.
c. Explain why the entropy production values differ for parts (a) and (b).
d. Find the entropy production for the gas and the total value, in kJ/K, if the system boundary temperature is 450 K throughout the process.
e. The hydrogen gas process is repeated, but paddle-wheel work is used instead of heat transfer. Calculate the entropy production in this case.
f. Compare relative irreversibility of processes (b), (d), or (e).

2-10. A 1.5-kg block of copper at 100°C and a 1.0-kg block of aluminum at 27°C initially are isolated from each other and from the local environment. They are then placed in thermal contact with each other but are still isolated from the environment. Determine

a. the equilibrium temperature in degrees Celsius
b. the entropy generation for the process, in kJ/K.

2-11. A rigid insulated tank is divided into two compartments by a noninsulated partition. Initially, 0.40 kmol of CO_2 is rapidly introduced into one compartment at 4.0 bars and 300 K. At the same time 0.10 kmol of CO is introduced into the other compartment at 1.0 bar and 500 K. Determine (a) the equilibrium temperature, in kelvins, and (b) the entropy production for the overall process, in kJ/K. Use CO and CO_2 tabular data.

2-12. A piston-cylinder device maintained at a constant pressure of 0.10 MPa contains nitrogen gas. The gas is cooled from 420 to 310 K. Compute (a) the heat removed, in kJ/kg, and (b) the entropy production for the heat-transfer process, in kJ/kg · K, if the temperature of the environment is 22°C. Use the nitrogen ideal-gas table for data.

2-13. Nitrogen flows steadily and isothermally through a duct at 400 K such that the heat transfer into the fluid is 5.56 kJ/kg. The initial and final pressures are 2.40 and 1.80 bars, respectively, and the diameter ratio D_1/D_2 for the duct is 1.40. Determine (a) the entropy generation for the fluid, in kJ/kg · K, and (b) the inlet and exit velocities of the fluid, in m/s, if the potential-energy change is negligible.

2-14. Air flows isothermally at 60°C through a constant-area duct such that the entropy increase due solely to internal irreversibilities is 0.0544 kJ/kg · K. Determine

a. the heat added or removed, in kJ/kg, if the initial and final pressures are 400 and 320 kPa, respectively
b. the inlet and exit velocities, in m/s

2-15. Oxygen flows steadily and isothermally at 300 K through a constant-area duct such that the entropy production due to internal irreversibilities is 0.0495 kJ/kg · K. Find

a. the heat added or removed, in kJ/kg, if the initial and final pressures are 3.0 and 2.4 bars, respectively
b. the exit and inlet velocities, in m/s

2-16. The entropy production is 0.1406 kJ/kg · K when CO_2 is throttled adiabatically from 8 bars. Determine (a) the final pressure in bars and (b) the ratio of inlet to outlet area if it is assumed that the inlet and exit velocities are identical.

2-17. The entropy production σ for nitrogen, an ideal gas, is 0.360 kJ/kg · K when it is throttled from an initial pressure of 400 kPa. Determine (a) the final pressure, in kilopascals, to which the gas is throttled and (b) the ratio of the inlet to outlet area if the inlet and exit velocities are identical.

2-18. Oxygen, an ideal gas, is throttled from 6.0 to 3.0 bars. Determine (a) the internal entropy production for the fluid, in kJ/kg · K, and (b) the ratio of the inlet to outlet area if the velocities are small, but $V_2/V_1 = 2/1$.

2-19. Refrigerant-134a enters a compressor as a saturated vapor at 2.4 bars. At the exit the pressure is 7.0 bars, and the entropy has increased by 0.0645 kJ/kg · K under adiabatic conditions. The changes in kinetic and potential energies may be neglected. (a) Determine the work input required, in kJ/kg. (b) For the same initial conditions and the same final pressure, the compressor is run without insulation. The entropy production is assumed to remain the same as in part (a). However, the environment at 22°C experiences an entropy increase of 0.0190 kJ/K per kilogram of R-134a. If the work input required in this case is found to be 41.77 kJ/kg, determine (b) the heat transfer, in kJ/kg, (c) the final temperature, in degrees Celsius, and (d) the entropy transfer due to heat transfer for R-134a, in kJ/kg · K.

2-20. Steam enters a turbine at 120 bars and 440°C and expands to 1 bar. At the exit the entropy is 0.650 kJ/kg · K greater than at the inlet. The process is adiabatic and changes in kinetic and potential energies may be neglected.

a. Determine the work done by the steam, in kJ/kg.

b. Now, for the same initial conditions and the same final pressure the turbine is run without insulation. It is found that the entropy increase of the steam is now 0.600 kJ/kg · K, and that the heat transfer from the steam occurs to the environment, which is at 300 K. The entropy change of the environment is 0.0850 kJ/K per kilogram of steam. Determine the heat transfer and the work output for the steam turbine under these new conditions, in kJ/kg.

2-21. Steam enters a turbine at 100 bars and 400°C. At the exit the pressure is 1.0 bar and the entropy has increased by 0.600 kJ/kg · K under adiabatic conditions. The changes in kinetic and potential energies may be neglected. (a) Determine the work, in kJ/kg. For the same initial conditions and final pressure the turbine is now run without insulation. The entropy production due to internal irreversibilities is assumed to be the same as in part (a). However, the environment at 300 K is found to have an entropy change of +0.0800 kJ/(kg steam) · K. Also, the entropy change of the steam due to heat exchange with the environment has an absolute value of 0.0460 kJ/kg · K. (b) Determine the quality of the steam at the turbine outlet, and compare to the answer from part (a). Also, (c) determine the heat transfer to the environment, in kJ/kg steam, and (d) find the amount of work output of the turbine, in kJ/kg, and compare to part (a).

2-22. Consider Prob. 1-4. Determine the value of the entropy production for the process, in kJ/K · min.

2-23. Water and steam enter an insulated mixing chamber in steady flow. The liquid water enters at 20°C and 3 bars with a rate of 100 kg/min. The steam enters at 320°C and 3 bars, and the mixture leaves the chamber at (a) 130°C and 3 bars and (b) at 3 bars as a saturated vapor. Determine the value of the entropy production for the processes, in kJ/K · min (see Prob. 1-18).

2-24. Consider Prob. 1-5. Determine the value of the entropy production for the process, in kJ/K · min, if the steam enters at 200 kg/min.

2-25. A heat pump unit operates between two heat reservoirs with temperatures of 270 and 330 K. The heat pump itself has a temperature of 255 K during heat addition from the 270 K reservoir and a temperature of 350 K during heat rejection to the 330 K reservoir. The COP of the actual unit is 3.40, and the heat-transfer rate from the 270 K reservoir is 2000 kJ/min. Determine the entropy production rate, in kJ/min · K, for

a. the heat pump unit itself
b. the low-temperature heat-transfer process
c. the high-temperature heat-transfer process
d. Which process has the largest irreversibility?

Now, for the same temperatures and a heat-transfer rate of 2000 kJ/min, reduce the actual COP to 3.10. Again, determine the entropy production rate for

e. the refrigeration unit itself
f. the low-temperature heat-transfer process
g. the high-temperature heat-transfer process
h. Which process now has the largest irreversibility?

2-26. A refrigeration unit operates between two heat reservoirs with temperatures of 240 and 320 K. The refrigeration unit itself has a temperature of 230 K during heat addition from the 240 K reservoir and a temperature of 340 K during heat rejection to the 320 K reservoir. The COP of the actual unit is 2.00, and the heat-transfer rate from the 240 K reservoir is 1000 kJ/min. Determine the entropy production rate, in kJ/min · K, for

a. the refrigeration unit itself
b. the low-temperature heat-transfer process
c. the high-temperature heat-transfer process
d. Which process has the largest irreversibility?

Now, for the same temperatures and a heat-transfer rate of 1000 kJ/min, reduce the actual COP to 1.50. Again, determine the entropy production rate in kJ/min · K for

e. the refrigeration unit itself
f. the low-temperature heat-transfer process
g. the high-temperature heat-transfer process
h. Which process now has the largest irreversibility?

2-27. A heat engine operates between two heat reservoirs with temperatures of 560 and 280 K.

a. The heat engine itself has a temperature of 560 K during heat addition from the 560 K reservoir and a temperature of 280 K during heat rejection to the 280 K reservoir. The thermal efficiency of the actual unit is 0.40, and the heat-transfer rate from the 560 K reservoir is 1000 kJ/min. Determine the entropy production rate, in kJ/min · K, for (1) the heat engine itself, (2) the low-temperature heat-transfer process, and (3) the high-temperature heat-transfer process.
b. Now, for the same reservoir temperatures, heat-transfer rate, and engine thermal efficiency the engine itself receives heat at 540 K and rejects heat at 300 K. As before, determine the entropy production rate for (1) the heat engine itself, (2) the low-temperature heat-transfer process, and (3) the high-temperature heat-transfer process. (4) Which process has the largest irreversibility?

c. Finally, for the same reservoir temperatures, heat-transfer rate, and engine temperatures cited in part (b) the thermal efficiency is increased to 0.42. For these conditions, determine the entropy production rate for (1) the heat engine itself, (2) the low-temperature heat-transfer process, and (3) the high-temperature heat-transfer process. (4) Which process now has the largest irreversibility?

2-28. A supply line is attached to the bottom of a vertical piston-cylinder apparatus which initially has the piston touching the bottom of the cylinder. A compressed ideal gas at reservoir conditions of (T_0, P_R) is slowly admitted to the cylinder through a valve, which lowers the pressure to P_L. The pressure P_L is maintained constant within the cylinder throughout the process by the weight of the piston and the atmosphere acting on the outside of the piston. The cylinder is not insulated so that the temperature within the apparatus remains at T_0. For a control volume consisting of the gas within the cylinder and the supply line containing the valve, determine (a) the total heat interaction between the system and the environment, in J, and (b) the entropy generation, in J/K, for the following conditions: $P_R = 5$ bars, $P_L = 2$ bars, $V_{\text{final}} = 1.0$ L, and $T_0 = 300$ K.

REFERENCES

1. Carnot, S., *Reflections on the Motive Power of Heat and on Machines Fitted to Develop This Power,* Bachelier, Paris, 1924; also *Reflections on the Motive Power of Fire and Other Papers,* Dover, New York, 1960.
2. Thomson, W., "On the Dynamical Theory of Heat, with Numerical Results Deduced from Mr. Joule's Equivalent of a Thermal Unit, and M. Regnault's Observation on Steam," *Trans. R. Soc.,* Edinburg, 1851.
3. Planck, M., *Treatise on Thermodynamics,* 3rd ed., translated by A. Ogg, Dover, New York, 1945 (first German ed. in 1897).
4. Hatsopoulos, G. N., and J. H. Keenan, *Principles of General Thermodynamics,* John Wiley & Sons, New York, 1965.
5. Carathéodory, C., "Grundlagen der Thermodynamik," *Math. Ann.,* **67**:355, 1909.
6. Zemansky, M. W., *Heat and Thermodynamics,* 5th ed., McGraw-Hill, Inc., New York, 1968.
7. Landsberg, P. T., "A deduction of Carathéodory's principle from Kelvin's principle," *Nature* (London), **201**:485, 1964.
8. Callen, Herbert B., *Thermodynamics,* John Wiley & Sons, New York, 1960.
9. Bejan, A., *Advanced Engineering Thermodynamics,* Chap 2., John Wiley & Sons, New York, 1988.

CHAPTER 3

AVAILABILITY ANALYSIS

This chapter is the first of four chapters in this text which are devoted to the study of availability or exergy concepts. In this chapter we shall define and apply the concepts to processes involving closed and open steady-state systems.

3-1 INTRODUCTION

The first law of thermodynamics, when expressed as a conservation of energy principle, is concerned with the quantity of various forms of energy. From an engineering viewpoint a quantity of energy also has quality. Quality is used in the sense of "degree of usefulness" to society. Changes in systems are brought about by heat and work interactions. Observations of past engineering achievements indicate that work interactions are far more important than heat interactions in bringing about changes of state. Since work interactions have a higher degree of "usefulness," it may be said that such energy forms are of a higher quality than heat interactions. Even the second law places a higher standard on work than on heat. Work is completely convertible into heat, but the conversion of heat into work by a cyclic device is highly restricted. Thus not only is work more useful than heat, but it is also more difficult to obtain in many instances.

This measure of "usefulness of energy" may be applied to forms of energy in a given state, as well as to energy transformations. The capacity of a given quantity of energy to produce work is accepted as a meaningful measure of the quality of that energy. In a broad sense, then:

> The *quality* of energy is the *potential* of that energy to produce *useful work*.

The useful work associated with a quantity of energy, called the *work potential* of that energy, requires a datum for its measurement. This datum is usually taken to be the local environment. As a result:

> The *work potential* of a given quantity of energy is defined as the *maximum possible useful work* that can be obtained from that energy in a *given environment*.

Whenever the work potential of a given quantity of energy is reduced during a process, the energy is said to have been *degraded*. Thus the first and second laws become statements of the conservation and degradation of energy and take on the following format. Every time energy changes form or is transferred from one system to another:

> The total amount of energy is constant (first law).
>
> The potential for producing useful work is reduced forever (second law).

Thus during energy transformations and transfer, energy is both conserved and degraded. It is this *potential* to do *useful work* for society that determines the quality of energy. Energy conservation typically means reduction in energy usage. Of equal importance is the need to reduce the degradation of energy during its use.

Note that the first and second laws cited above are both expressed in terms of energy. This type of formulation is deliberate. If we wish to have a quantitative measure of the work potential of a system in a given state or the change in work potential during a process, we must simultaneously apply the two laws. This is quite different from many introductory discussions of thermodynamics, where the two laws are commonly used as separate steps in an analysis. The concepts in this chapter will help tie together many apparently unrelated ideas. Engineers use thermodynamics to develop energy processes. Many of these processes require the use of work interactions. Among the major goals of engineering thermodynamics are:

> To optimize methods for converting various forms of energy into work
>
> To optimize the use of work interactions to bring about desired results
>
> To minimize the loss in work potential of a system even in the absence of work interactions

The purpose of this chapter is to quantify the general techniques for optimizing energy processes. As we combine these laws into a single relationship, we shall derive a general solution and then proceed to illustrate specific applications to control-mass and control-volume analyses.

3-2 REVERSIBLE WORK, AVAILABILITY, IRREVERSIBILITY, AND SECOND-LAW EFFICIENCY

In this section we begin by developing a general equation for evaluating work interactions in the presence of irreversibilities. This will enable us to quantify the concept of work potential for closed and open systems. The first law contains a term for work, but there is no energy term which accounts for irreversibilities. The second law contains a term for irreversibilities (the entropy production or generation term), but no term is specifically written in terms of work. In order to develop an expression directly relating work to entropy generation (or irreversibilities), we combine the first and second laws through common terms. Figure 3-1 illustrates a general control volume with multiple inlets and outlets for mass flow. In addition, heat $\dot{Q}_0$ may be exchanged with the atmosphere at T_0, as well as heat quantities $\dot{Q}_j$ with j additional reservoirs at temperatures $T_1, \ldots, T_j$. The total heat-transfer rate $\dot{Q}$ is given by $\dot{Q} = \dot{Q}_0 + \sum \dot{Q}_j$. Finally, work interactions may also occur across the boundary. The term $\dot{W}_{\text{net}}$ represents the sum of all possible work effects, such as those due to a volume change, shear, electrical effects, and others. As shown in Fig. 3-1, part of the volume change may be done against the atmosphere at P_0.

The general conservation of energy equation for the control volume in Fig. 3-1 is

$$\dot{Q} + \dot{W}_{\text{net}} + \sum_{\text{in}} \left(h + \frac{V^2}{2} + gz\right)\dot{m} - \sum_{\text{out}} \left(h + \frac{V^2}{2} + gz\right)\dot{m} = \frac{dE_{\text{CV}}}{dt} \tag{1-40}$$

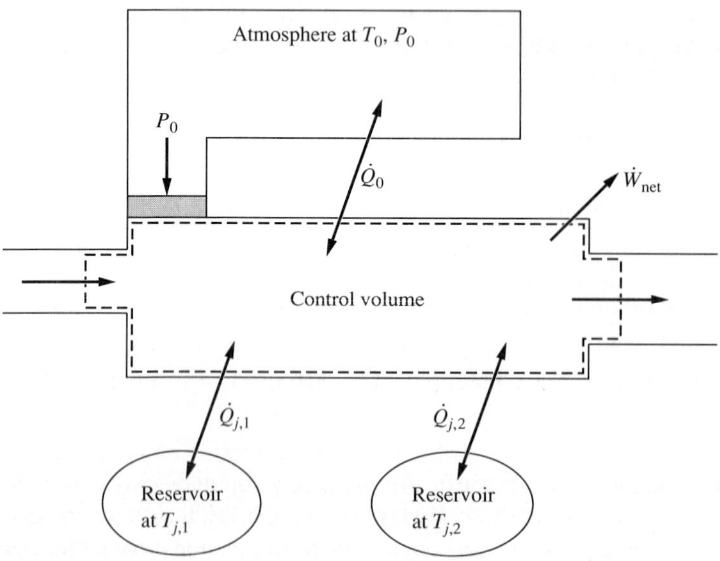

FIGURE 3-1
Schematic for the development of the work potential for a process.

By rewriting $\dot{Q}$ as a sum of terms to the various reservoirs, we can write Eq. (1-40) as

$$\dot{Q}_0 + \sum_j \dot{Q}_j + \dot{W}_{\text{net}} + \sum_{\text{in}} \left(h + \frac{V^2}{2} + gz\right)\dot{m} - \sum_{\text{out}} \left(h + \frac{V^2}{2} + gz\right)\dot{m} = \frac{dE_{\text{CV}}}{dt} \tag{3-1}$$

The data points for the velocities and the elevations and the reference state for the enthalpy (or the internal energy) must be the same for the inlets and outlets. Since the work potential is measured in terms of *useful* work, $\dot{W}_{\text{net},u}$, we must correct $\dot{W}_{\text{net}}$ by the work done against the atmosphere, namely, by $P_0 d(\text{Vol})/dt$. That is,

$$\dot{W}_{\text{net},u} = \dot{W}_{\text{net}} + P_0 \left[\frac{d(\text{Vol})}{dt}\right]_{\text{CV}} \tag{3-2}$$

In addition, the form of the entropy balance appropriate here is Eq. (2-48), namely,

$$\dot{\sigma}_{\text{tot}} = \dot{\sigma}_{\text{CV}} + \dot{\sigma}_Q = \frac{dS_{\text{CV}}}{dt} + \sum_{\text{out}} \dot{m}s - \sum_{\text{in}} \dot{m}s - \frac{\dot{Q}_0}{T_0} - \sum_j \frac{\dot{Q}_j}{T_j} \tag{2-48}$$

We shall now use the above three equations to develop relations for $\dot{W}_{\text{net},u}$ under actual and reversible conditions, in terms of specified property values, mass flow rates, reservoir heat-transfer rates, and the environment temperature T_0 and pressure P_0.

3-2-1 Reversible Work

In order to combine the first- and second-law statements, we must recognize an important engineering concept. Although the physical quantities cited above may be fixed by the designer, the heat-transfer rate $\dot{Q}_0$ to or from the atmosphere is arbitrary. It is not under the control of the design engineer but varies as $\dot{W}$ changes to satisfy the first law. Hence it is justifiable to eliminate $\dot{Q}_0$ between Eqs. (3-1) and (2-48) while also substituting Eq. (3-2). The important result is

$$\begin{aligned}\dot{W}_{\text{act},u} = {} & \frac{d(E + P_0V - T_0S)_{\text{CV}}}{dt} + \sum_{\text{out}} \dot{m}\left(h + \frac{V^2}{2} + gz - T_0 s\right) \\ & - \sum_{\text{in}} \dot{m}\left(h + \frac{V^2}{2} + gz - T_0 s\right) - \sum_j \dot{Q}_j\left(1 - \frac{T_0}{T_j}\right) + T_0\dot{\sigma}_{\text{tot}}\end{aligned} \tag{3-3}$$

where the direction of the Q_j terms again is measured relative to the control surface. This is a general equation for the rate of actual useful work (power) associated with a control volume and the heat reservoirs with which it interacts. Although $\dot{Q}_0$ does not appear explicitly in the equation, heat transfer to or from the environment at T_0 is automatically included. Again, the summation sign in front of $\dot{Q}_j$ allows for j possible heat reservoirs at T_j. The term $\dot{\sigma}_{\text{tot}}$ includes entropy generation within the control volume as well as that associated with irreversible heat transfer.

Similar to the development in Chap. 2 of relationships between $\dot{\sigma}_{\text{CV}}$ and dS_{CV}/dt, Eq. (3-3) can also be written solely in terms of interactions at the control surface. To obtain this format, we write Eq. (1-40) for the conservation of energy as

$$\sum_{i=1}^{n} \dot{Q}_i + \dot{W}_{\text{net}} + \sum_{\text{in}} \left(h + \frac{V^2}{2} + gz\right)\dot{m} - \sum_{\text{out}} \left(h + \frac{V^2}{2} + gz\right)\dot{m} = \frac{dE_{\text{CV}}}{dt} \tag{3-4}$$

where Q is expressed as a sum of heat interactions $\sum Q_i$ at the control surface, rather than in terms of heat interactions to or from reservoirs at T_0 and T_j. The appropriate entropy balance solely for the control volume is Eq. (2-42), namely,

$$\dot{\sigma}_{\text{CV}} = \frac{dS_{\text{CV}}}{dt} + \sum_{\text{out}} \dot{m}s - \sum_{\text{in}} \dot{m}s - \sum_{i=1}^{n} \frac{\dot{Q}_i}{T_i} \tag{2-42}$$

Since Q_0 does not appear in either of the above two equations, a different strategy must be used to combine the equations. The simplest procedure is to multiply Eq. (2-42) by T_0 and to subtract the resulting equation from Eq. (3-4). The desired result is

$$\begin{aligned} \dot{W}_{\text{act},u} &= \frac{d(E + P_0V - T_0S)_{\text{CV}}}{dt} + \sum_{\text{out}} \dot{m}\left(h + \frac{V^2}{2} + gz - T_0s\right) \\ &\quad - \sum_{\text{in}} \dot{m}\left(h + \frac{V^2}{2} + gz - T_0s\right) - \sum_{i=1}^{n} \dot{Q}_i\left(1 - \frac{T_0}{T_i}\right) + T_0\dot{\sigma}_{\text{CV}} \end{aligned} \tag{3-5}$$

Equation (3-5) expresses $\dot{Q}$ in terms of $\dot{Q}_i$ at sections of the control surface at T_i, rather than the parameters $\dot{Q}_0$ and $\sum \dot{Q}_j$ used to develop Eq. (3-3). These latter quantities represent the overall heat-transfer rates associated with the environment at T_0 and those reservoirs at T_j.

Recall from Sec. 2-6 that the relation between entropy generation and work effects for a control volume on a time basis is given by

$$\delta\dot{\sigma}_{\text{CV}} = \frac{\delta\dot{W}_{sf,\text{act}} - \delta\dot{W}_{sf,\text{rev}}}{T} \geq 0 \tag{2-52}$$

When $\dot{\sigma}_{\text{CV}}$ is zero, the process is internally reversible and $\dot{W}_{sf,\text{act}} = \dot{W}_{sf,\text{rev}}$. Thus the optimum or reversible value for the rate of useful work, in terms of Eqs. (3-3) and (3-5), is given either by

$$\begin{aligned} \dot{W}_{\text{rev},u} &= \frac{d(E + P_0V - T_0S)_{\text{CV}}}{dt} + \sum_{\text{out}} \dot{m}\left(h + \frac{V^2}{2} + gz - T_0s\right) \\ &\quad - \sum_{\text{in}} \dot{m}\left(h + \frac{V^2}{2} + gz - T_0s\right) - \sum_{j} \dot{Q}_j\left(1 - \frac{T_0}{T_j}\right) \end{aligned} \tag{3-6}$$

or,
$$\dot{W}_{\text{rev},u} = \frac{d(E + P_0V - T_0S)_{\text{CV}}}{dt} + \sum_{\text{out}} \dot{m}\left(h + \frac{V^2}{2} + gz - T_0 s\right)$$
$$- \sum_{\text{in}} \dot{m}\left(h + \frac{V^2}{2} + gz - T_0 s\right) - \sum_{i=1}^{n} \dot{Q}_i\left(1 - \frac{T_0}{T_i}\right) \qquad (3\text{-}7)$$

Equation (3-6) applies to a control volume and the reservoirs with which it interacts. On the other hand, Eq. (3-7) applies solely to the control volume of interest. The last term in Eq. (3-7) may be difficult to evaluate accurately, because it requires knowledge of Q_i at a portion of the control surface at temperature T_i. Nevertheless, it is an important result, since the reversible work is expressed solely in terms of changes within or at the boundaries of the system. A major goal of the next three sections is to replace the first four terms in Eqs. (3-3), (3-5), (3-6), and (3-7) with some form of an *availability* function.

The numerical evaluation of reversible work depends on the values chosen for T_0 and P_0. A logical choice for (T_0, P_0) are those values traditionally associated with the *standard atmosphere*, namely, 25°C and 1 atm. For comparative purposes it is recommended that

$$T_0 = 298.15 \text{ K} \qquad \text{and} \qquad P_0 = 1.01325 \text{ bars} = 0.101325 \text{ MPa} \qquad (3\text{-}8)$$

be regarded as the *standard environmental state*. Values of T_0 slightly different from 25°C are frequently used in practice.

3-2-2 Availability or Exergy

In the preceding section equations were developed for the reversible work associated with systems undergoing specified changes of state and exchanging heat with the environment or other heat reservoirs. It is useful to reformulate these equations in terms of the *availability functions* for closed and open systems. To this end it is necessary first to determine the work potential of a system at a given state as it proceeds toward a state of equilibrium with the environment while exchanging heat solely with the environment. Once the system and environment are in equilibrium, no further change of state of the system can occur spontaneously, and hence no further work is performed. Hence the above described process leads to the *maximum reversible work* or *work potential* associated with the state of a system. When a system and its environment are in equilibrium with each other, the system is said to be in its *dead state*. Specifically, a system in a dead state is in thermal and mechanical equilibrium with the environment at T_0 and P_0. The numerical values of (T_0, P_0) recommended for the dead state are those of the *standard environmental state* discussed earlier, namely, 298.15 K and 1.01325 bars (1 atm).

Additional requirements for the dead state are that the velocity of a closed system or fluid stream be zero and that the gravitational potential energy be zero. This latter requirement is fulfilled by arbitrarily setting some elevation on earth, such as sea level or ground level, as zero. These restrictions of temperature, pressure, velocity, and elevation characterize a *restricted dead state* associated with thermomechanical equilibrium with the atmosphere. It is restricted in the sense that chemical equilibrium with the environment is not considered. That is, the control mass is not allowed to pass into or react chemically with the environment. The work potential of a system relative to its dead state, which exchanges heat solely with the environment, is called the thermomechanical *availability* or *exergy* of that state.

The methods for evaluating the availability of closed and open steady-state systems, as well as for heat-transfer processes, are presented in several of the following sections. The availability transfer associated with a work interaction is immediately obvious. Work interactions by concept are reversible at the point where they occur at the boundary. Consequently, the availability transfer associated with the transfer of work (excluding work against the environment) *equals* the value of the useful work itself.

3-2-3 Irreversibility

Of major concern to engineers is the loss in capability of producing work (or the increase in required work input) when irreversibilities are present. One method of measuring such losses is through the concept of "adiabatic efficiencies" for work devices, which frequently are termed "first-law" efficiencies. A more fundamental measure of the loss in work capability of a system (relative to its local environment) as a result of a process with *specified end states* is the difference between the *actual work* and the *reversible work*. On the basis of Eqs. (3-5) and (3-7) this difference, $W_{\text{act}} - W_{\text{rev}}$, for a control volume on a rate basis is given by

$$\dot{W}_{\text{act},u} - \dot{W}_{\text{rev},u} = \dot{W}_{\text{act}} - \dot{W}_{\text{rev}} = T_0\dot{\sigma}_{\text{CV}} \geq 0 \tag{3-9}$$

This equation indicates that when any system changes state irreversibly, the work capability of the system is reduced (destroyed) at a rate proportional to the rate of entropy generation within the system. (This proportionality historically is known as the Gouy-Stodola theorem [1, 2].) Equation (3-9) may be rearranged into the format

$$\dot{W}_{\text{act}} = \dot{W}_{\text{rev}} + T_0\dot{\sigma}_{\text{CV}}$$

Since $\dot{\sigma}_{\text{CV}} \geq 0$, we can restate the inequality between W_{act} and W_{rev}, presented earlier as Eqs. (2-35) and (2-54), in the following ways:

$$\dot{W}_{\text{act}} \geq \dot{W}_{\text{rev}}, \qquad W_{\text{act}} \geq W_{\text{rev}}, \qquad \text{and} \qquad W_{\text{act},u} \geq W_{\text{rev},u} \tag{3-10}$$

Hence the presence of irreversibilities leads to a reduction in output or an increase in input when an actual process is compared to a totally reversible process between the same end states.

The quantity of work capacity of a system destroyed or lost during a process is defined as the *irreversibility I* of the process. It is a measure of the availability or exergy loss due to internal irreversibilities. With references to Eq. (3-9), the irreversibility $\dot{I}$ on a time-rate basis for a control volume is

$$\dot{I}_{\text{CV}} \equiv \dot{W}_{\text{act}} - \dot{W}_{\text{rev}} = \dot{W}_{\text{act},u} - \dot{W}_{\text{rev},u} = T_0\dot{\sigma}_{\text{CV}} \geq 0 \tag{3-11}$$

For a vast majority of control-volume analyses there is no change in volume. Hence the use of "useful" as a subscript on $\dot{W}$ is redundant, and it is usually omitted. For a closed system on an extensive basis

$$I_{\text{CM}} \equiv W_{\text{act}} - W_{\text{rev}} = W_{\text{act},u} - W_{\text{rev},u} = T_0\sigma_{\text{CM}} \geq 0 \tag{3-12}$$

and on a unit-mass basis, where $i \equiv \text{I}/m$,

$$i_{\text{CM}} = w_{\text{act}} - w_{\text{rev}} = w_{\text{act},u} - w_{\text{rev},u} = T_0\sigma_{\text{CM}} \geq 0 \tag{3-13}$$

The parameter I is also known as the *lost work*, W_{lost}. The irreversibility of a process, like heat and work, is a function of the path of the process, that is, the design and operation of the system. The parameter I is shown schematically in Fig. 3-2 on the horizontal work-transfer rate axis. Note that $\dot{W}_{\text{act}}$ always lies to the right of $\dot{W}_{\text{rev}}$ so that I is always *positive* as required by Eq. (3-9). The major purpose of an availability analysis is to detect and evaluate quantitatively the irreversibilities of a process. Such an analysis could then indicate areas for thermodynamic improvement of the process.

If we include external reservoirs and the environment in our system of interest, then Eq. (3-11) takes on the format

$$\dot{I}_{\text{tot}} = \dot{W}_{\text{act}} - \dot{W}_{\text{rev}} = T_0\dot{\sigma}_{\text{tot}} = T_0(\dot{\sigma}_{\text{CV}} + \dot{\sigma}_Q) \geq 0$$

Similar to the control-volume analysis above, I and σ in Eqs. (3-12) and (3-13) may refer to the control mass alone, or could include the heat-transfer processes as well. Thus the concept of irreversibility can be applied to a control volume or control mass alone, or can be extended to include other interacting systems.

3-2-4 Second-Law Efficiency or Effectiveness

The first-law efficiencies η of some common devices were reviewed in Chap. 1. These expressions are ratios of selected energy quantities. In the context of this

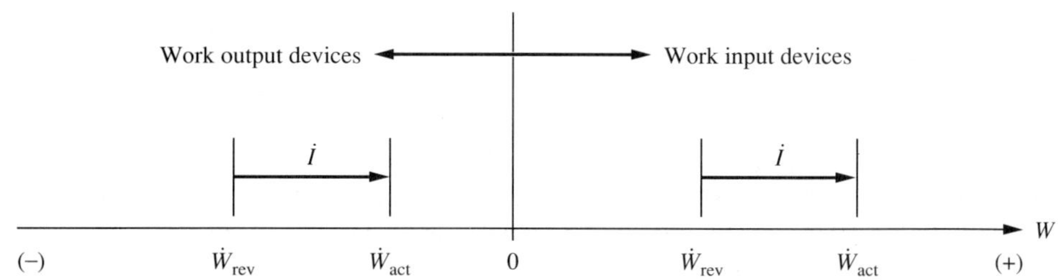

FIGURE 3-2
Illustration of the irreversibility I in terms of reversible and actual work values for work-producing and absorbing devices.

chapter the usefulness of energy is more appropriately described by its exergy or availability. Since availability has its origin in the second law, a performance parameter for a process based on availability concepts is known as a *second-law* or *exergetic efficiency* η_{II}, or as a *second-law effectiveness*, or simply as an *effectiveness* ϵ. (Both symbols are in current use, among others.) A first-law efficiency gauges how well energy is used, whereas an effectiveness indicates how well exergy or availability is used. Both concepts are important in thermodynamic analysis.

First- and second-law efficiencies are different in one other important respect. The first law is a conservation principle, and first-law efficiencies tend to fall into two general categories. Equipment first-law efficiencies compare actual energy changes to theoretical energy changes under specified restrictions. Examples include turbines, compressors, nozzles, and pumps. Cycle first-law efficiencies compare desired energy output to required (costly) energy input. Thermal efficiency and coefficient of performance (COP) are typical examples. On the other hand, entropy and availability from a second-law viewpoint are nonconserved properties. In the presence of irreversibilities the total entropy increases and the total availability decreases. The former effect is measured by the entropy production σ, and the latter effect is measured by the irreversibility I. Hence second-law efficiencies measure losses in availability during a process. One typical definition of the second-law effectiveness ϵ is

$$\epsilon \equiv \frac{\text{useful availability out}}{\text{availability in}} = 1 - \frac{\text{availability destruction and losses}}{\text{availability input}}$$

where losses imply *nonuseful* transfers across the boundary. A second approach, especially useful for steady-state devices, is

$$\epsilon = \frac{\text{rate of availability (exergy) output}}{\text{rate of availability (exergy) input}}$$

The second law stresses the fact that two forms of the same quantity of energy may have quite different availabilities. Thus energy is "weighted" according to its availability. Unlike a first-law efficiency, an effectiveness accounts for losses in work capability during a process. Note, however, that the definitions of ϵ given above are quite general. In specific applications one must still decide what is the desired output, what is regarded as input, and what is counted as a loss. In later examples we shall demonstrate that different decisions about these items lead to different expressions for the effectiveness of a given process. Hence different ϵ values may exist for the same data. It is not a question of which value is better, or correct. It is a matter of consistency when comparing sets of data. Fortunately, the availability balances for closed and open systems are good guidelines for establishing appropriate second-law efficiencies. In subsequent sections discussions and examples will illustrate the evaluation of the effectiveness ϵ or η_{II} for a variety of systems.

3-3 AVAILABILITY TRANSFER ASSOCIATED WITH HEAT TRANSFER

Entropy transfer associated with heat transfer Q_i across a system boundary at T_i is represented by the quantity Q_i/T_i. A transfer of *availability* is also associated with heat transfer. To derive this relation, consider Fig. 3-3, which shows heat transfer δQ_i occurring for $T_i > T_0$, where T_i is the temperature at the boundary of a system. The availability associated with δQ_i is the reversible work obtainable as the heat is transferred to the dead state. To produce work, δQ_i passes through a reversible heat engine which rejects to T_0. Using an overbar to designate quantities relative to the engine, we find the first and second laws to require that

$$\delta \bar{W}_{\text{eng,rev}} = -\delta \bar{Q}_i - \delta \bar{Q}_0 \qquad \text{and} \qquad \frac{\delta \bar{Q}_i}{T_i} = \frac{-\delta \bar{Q}_0}{T_0}$$

Sign convention requires that $\delta Q_i = -\delta \bar{Q}_i$, where δQ_i is measured relative to the system at T_i. A combination of these relations yields

$$\delta \bar{W}_{\text{eng,rev}} = \delta Q_i \left(1 - \frac{T_0}{T_i}\right) \tag{3-14}$$

The right-hand side of this equation, then, is the availability transfer associated with heat transfer δQ_i as it crosses the system boundary at T_i. This same result is obtained if the directions of δQ_i and δQ_0 are reversed and $T_i < T_0$.

For a finite heat transfer Q_i occurring at boundary temperature T_i the above equation becomes

$$\bar{W}_{\text{eng,rev}} = Q_i \left(1 - \frac{T_0}{T_i}\right) \equiv \Phi_{Q,i} \tag{3-15}$$

where $\Phi_{Q,i}$ defines the *availability transfer associated with heat transfer* Q_i at boundary temperature T_i. (Φ will be the general symbol for the availability of an extensive closed system; see Sec. 3-4.) For the more general situation where both

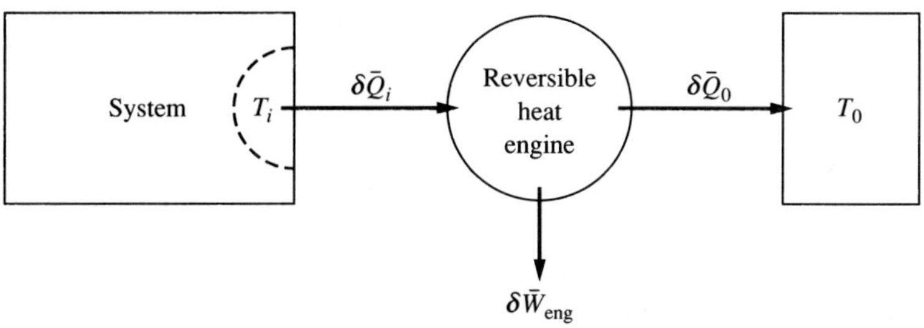

FIGURE 3-3
Schematic for the development of the expression for the work potential Φ_Q associated with heat transfer.

Q_i and T_i vary along the boundary of the system (recall Fig. 3-1) a summation of terms must be considered. In this situation Eq. (3-15) becomes

$$\Phi_Q \equiv \sum_{i=1}^{n} Q_i \left(1 - \frac{T_0}{T_i}\right) \tag{3-16}$$

where Φ_Q is defined as the total availability transfer associated with the heat-transfer quantities Q_i which occur at boundary temperatures T_i. The rate of availability transfer associated with heat transfer at a boundary is given simply by

$$\dot{\Phi}_Q \equiv \sum_{i=1}^{n} \dot{Q}_i \left(1 - \frac{T_0}{T_i}\right) \tag{3-17}$$

The term on the right of Eq. (3-17) has appeared previously in Eqs. (3-5) and (3-7) for $\dot{W}_{\text{act},u}$ and $\dot{W}_{\text{rev},u}$ in the preceding section.

A special application of Eq. (3-15) is to the availability transfer to or from a single thermal reservoir at $T_i = T_R$, where T_R is constant and uniform. Since $Q_i \equiv Q_R$ in this case, the equation can be written as

$$\Phi_{Q,R} \equiv Q_R \left(1 - \frac{T_0}{T_R}\right) \qquad \text{(thermal reservoir)} \tag{3-18}$$

where $\Phi_{Q,R}$ is defined as the availability of heat transfer Q_R to or from a reservoir at T_R. Note that one does not associate availability transfer due to heat transfer at the boundary of a reservoir at T_0. If j reservoirs are under consideration, then Eq. (3-18) can be expanded to mass and rate formats such as

$$\sum_j \Phi_{Q,j} \equiv \sum_j Q_j \left(1 - \frac{T_0}{T_j}\right) \qquad \text{or} \qquad \sum_j \dot{\Phi}_{Q,j} \equiv \sum_j \dot{Q}_j \left(1 - \frac{T_0}{T_j}\right) \tag{3-19}$$

which define the parameter $\sum \Phi_{Q,j}$. The last term of Eq. (3-19) on a rate basis has previously appeared as a term in Eqs. (3-3) and (3-6) in the preceding section.

The format of Eqs. (3-16) and (3-18) is not surprising. Both equations represent the work of a Carnot heat engine rejecting to the atmosphere at T_0. (Note that the irreversibility of any thermal reservoir is zero, because by definition such a reservoir is internally reversible.) An important point must be noted with respect to Eqs. (3-16) and (3-18): If the system temperature is *greater* than T_0, then the system *gains* availability when heat is added, and vice versa. However, if the system temperature is *less* than T_0, then the system *loses* availability when heat is added and *gains* availability when heat transfer is out. That is, the energy flow and availability flow are in *opposite* directions in this latter case. This effect is important in the study of refrigeration processes.

In an internally irreversible process $dS \neq \delta Q/T$. In addition, when an irreversibility exists within a system, the temperature T of the system may be undefined. In this case, for either closed or open systems, Eq. (3-16) for the availability transfer associated with heat transfer may have to be modified to an approximation format

$$\Phi_Q = Q\left(1 - \frac{T_0}{T_s}\right) \tag{3-20}$$

where T has been replaced by an average surface temperature T_s at the solid physical boundary where heat transfer occurs.

The irreversibility I_Q of a heat-transfer process is determined by employing Eq. (3-5). For a steady-state process without mass transfer or work interactions this equation reduces to

$$I_Q = \Phi_Q = \sum \Phi_{Q,i} \tag{3-21a}$$

That is, I_Q is measured by the sum of the availability transfers in and out of the heat-transfer region. Substitution of Eq. (3-16) into the above relation yields, for a supply reservoir at T_A and a sink reservoir at T_B,

$$I_Q = Q\left(1 - \frac{T_0}{T_A}\right) - Q\left(1 - \frac{T_0}{T_B}\right) = T_0 Q\left(\frac{1}{T_B} - \frac{1}{T_A}\right) \tag{3-21b}$$

where Q is a positive value and $T_A > T_B$. Equation (3-21) is equally valid on a rate basis.

Equation (3-21a) for I_Q has a graphical interpretation in terms of a TS diagram. The heat quantity $Q = -Q_A = Q_B$. For the internally reversible reservoirs at T_A and T_B the heat quantities are given by $T_A \Delta S_A$ and $T_B \Delta S_B$. But $T\,\Delta S$ is a rectangular area on a TS plot. The area representations of Q_A and Q_B are shown in Fig. 3-4. The areas under lines 1-2 and 3-4 must be equal in magnitude and have a common left border for convenience in comparing areas. The gray shaded area beneath line 5-6 measures the irreversibility, since $I_Q = T_0(\Delta S_A + \Delta S_B)$. Note that the gray shaded area approaches zero in magnitude as $T_B \to T_A$.

In the preceding development the availability transfer associated with heat transfer was restricted to a fixed and uniform boundary temperature or a sum of terms which account for different T_i values at various parts of the boundary. This terminology is especially appropriate for steady-state control-volume analysis. In many closed-system processes, however, the boundary temperature is uniform but varies during the process. In this case the equations for Φ_Q listed above must be altered. In this new situation we begin with Eq. (3-14) in the form

$$\delta W_{\text{eng,rev}} = \left(1 - \frac{T_0}{T}\right)\delta Q \tag{3-22}$$

and for a finite *internally reversible* process of a closed system with variable temperature

$$\Phi_Q = W_{\text{eng,rev}} = \int_1^2 \left(1 - \frac{T_0}{T}\right)\delta Q = \int_1^2 \delta Q - T_0 \int_1^2 \frac{\delta Q}{T}$$

$$= (U_2 - U_1) + \int_1^2 P\,dV - T_0(S_2 - S_1) \tag{3-23}$$

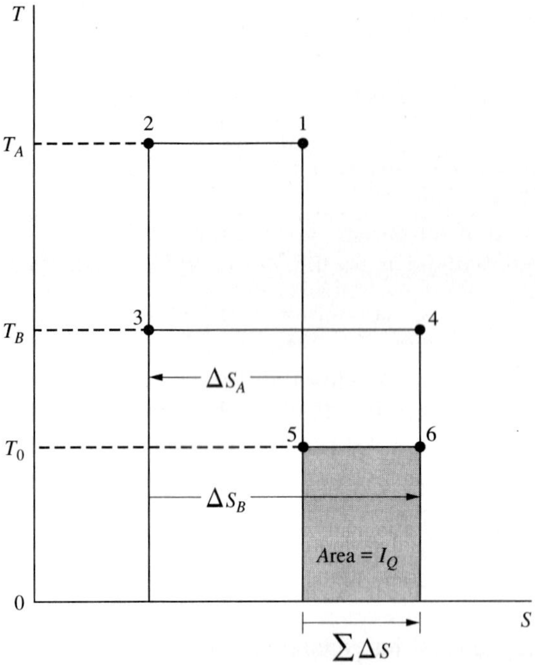

FIGURE 3-4
Graphical representation of irreversibility I_Q on a TS diagram when heat transfer Q occurs from thermal reservoir A to reservoir B.

where δQ has been replaced by $(dU + P\,dV)$ and $\delta Q/T$ by dS. In this restricted case Φ_Q may be evaluated in terms of property changes within the closed system.

Example 3-1. In Example 2-2 a heat transfer of 100 kJ occurred between two thermal reservoirs with temperatures of 300 and 1000 K. For a T_0 value of 300 K, determine the availability transfers and the irreversibility of the heat exchange process, in kilojoules.

Solution. The availability transfer is given by Eq. (3-18), namely,

$$\Phi_{Q,R} = Q_R\left(1 - \frac{T_0}{T_R}\right)$$

For the 1000 K reservoir

$$\Phi_{Q,1000} = -100\left(1 - \frac{300}{1000}\right) = -70 \text{ kJ}$$

The availability transfer to the 300 K reservoir is zero since T_0 is 300 K. The irreversibility of the process is given by Eq. (3-21).

$$I_Q = T_0 Q\left(\frac{1}{T_B} - \frac{1}{T_A}\right) = 300(100)\left(\frac{1}{300} - \frac{1}{1000}\right) = 70 \text{ kJ}$$

Note that the irreversibility or loss in work potential equals the net transfer of availability.

A simple but industrially important example of second-law efficiencies is in the use of energy from a high-temperature combustion source to provide thermal energy to processes occurring at lower temperatures. Process temperatures typically might range from barely above the ambient temperature to 700–800 K. Combustion source temperatures might range from 1000 to 2000 K. In general, a heat transfer $\dot{Q}_S$ from the combustion source appears as either a useful process heat quantity $\dot{Q}_U$ or heat losses $\dot{Q}_i$ to the environment. To simplify our analysis, we assume the source and use temperatures to be constant at T_S and T_U, and any heat losses $\dot{Q}_i$ occur at boundary temperatures T_i. For a steady-state process the energy and availability balances for the closed system restricted to heat transfers across its boundary are

$$0 = \dot{Q}_S - \dot{Q}_U - \sum_i \dot{Q}_i$$

$$0 = \dot{Q}_S\left(1 - \frac{T_0}{T_S}\right) - \dot{Q}_U\left(1 - \frac{T_0}{T_U}\right) - \sum_i \dot{Q}_i\left(1 - \frac{T_0}{T_i}\right) - \dot{I}$$

In the development of an effectiveness for the heat-transfer process it is useful to rearrange the availability balance so that the availability input is on the left-hand side of the equation. That is,

$$\dot{Q}_S\left(1 - \frac{T_0}{T_S}\right) = \dot{Q}_U\left(1 - \frac{T_0}{T_U}\right) + \sum_i \dot{Q}_i\left(1 - \frac{T_0}{T_i}\right) + \dot{I}$$

or

$$\dot{\Phi}_{Q,S} = \dot{\Phi}_{Q,U} + \dot{\Phi}_{Q,i} + \dot{I}$$

This equation indicates that the availability $\dot{\Phi}_{Q,S}$ carried in by the heat transfer $\dot{Q}_S$ appears eventually as availability transfers $\dot{\Phi}_{Q,U}$ and $\dot{\Phi}_{Q,i}$, or is destroyed by irreversibilities $\dot{I}$ within the system.

Of the three products of the process, only one is useful. If the effectiveness is defined as *useful* product over required input, then

$$\epsilon = \frac{\dot{Q}_U(1 - T_0/T_U)}{\dot{Q}_S(1 - T_0/T_S)} = \frac{\dot{\Phi}_{Q,U}}{\dot{\Phi}_{Q,S}} \tag{3-24}$$

where both Q_U and Q_S are taken to be positive. Each term in this case represents the work obtainable from a reversible engine which receives heat Q at either T_S or T_U and rejects heat at T_0. The above equation indicates that two factors control the effectiveness. First, the ratio $\dot{Q}_U/\dot{Q}_S$ should be as close to unity as possible (heat losses should be minimized). Second, the value of T_U should approach that of T_S. Since T_U is normally a required value of the end use, then T_S should be

TABLE 3-1
Summary of data for Example 3-2

Use*	T_U, K	Φ_U, kJ	I_Q, kJ	ϵ
Atmospheric sink	300	0	70.0	0.00
Space heating	313	4.2	65.8	0.06
Process steam generation	473	36.6	33.4	0.52
Furnace operation	693	56.7	13.3	0.81

* Based on $Q = 100$ kJ, $T_H = 1000$ K, and $T_0 = 300$ K

lowered, if possible. If a good match between T_S and T_U is not possible, a poor effectiveness will result. The example below illustrates this point for a fixed value of T_S and several possible T_U values required in industry.

Example 3-2. In Example 3-1 a heat transfer of 100 kJ occurs between reservoirs of 1000 and 300 K, and $T_0 = 300$ K. For (*a*) the situation described above, and for (*b*) additional higher use temperatures of 40, 200, and 420°C, determine the effectiveness of the heat-transfer process. Neglect heat losses.

Solution. The second-law effectiveness in all these cases is based on Eq. (3-24),

$$\epsilon = \frac{\Phi_{Q,U}}{\Phi_{Q,S}} = \frac{Q_U(1 - T_0/T_U)}{Q_S(1 - T_0/T_S)}$$

The availability of the heat supplied for all T_U values is

$$\Phi_{Q,S} = Q_S\left(1 - \frac{T_0}{T_S}\right) = 100\left(1 - \frac{300}{1000}\right) = 70 \text{ kJ}$$

(*a*) In the original case $T_U = T_0$. Therefore the availability supplied to the reservoir at T_0 is zero and $\epsilon = 0$. All work capability of the energy has been lost.

(*b*) The sink temperatures of 40, 200, and 420°C might be chosen to represent, for example, some operating temperatures commonly found in industrial practice, such as space heating, process steam generation, and furnace operation. The values of T_U, Φ_U, and ϵ for the four different temperatures are summarized in Table 3-1. The results are not surprising: the larger the ΔT during heat transfer, the greater the value of I_Q and the smaller the value of ϵ. Thus the smaller the difference between the source and use temperatures, the more effective is the use of thermal energy.

3-4 REVERSIBLE WORK, AVAILABILITY, AND IRREVERSIBILITY FOR A CLOSED SYSTEM

The general equations for actual and reversible work developed in Sec. 3-2 will now be applied to closed systems.

3-4-1 Actual and Reversible Work for a Closed System

A closed system by definition is one for which mass flow across its boundaries is excluded. Hence Eq. (3-5) reduces to

$$\dot{W}_{act,u} = \frac{d(E + P_0V - T_0S)}{dt} - \sum_{i=1}^{n} \dot{Q}_i \left(1 - \frac{T_0}{T_i}\right) + T_0\dot{\sigma}_{CM} \qquad \text{(closed)} \qquad (3\text{-}25)$$

and in the format involving j thermal reservoirs Eq. (3-3) becomes

$$\dot{W}_{act,u} = \frac{d(E + P_0V - T_0S)}{dt} - \sum_{j} \dot{Q}_j \left(1 - \frac{T_0}{T_j}\right) + T_0\dot{\sigma}_{tot} \qquad \text{(closed)} \qquad (3\text{-}26)$$

where $E = U + \text{KE} + \text{PE}$. On an extensive basis rather than on a rate basis, and making use of Eqs. (3-16) and (3-18), the above two equations become

$$\begin{aligned} W_{act,u} &= \Delta(E + P_0V - T_0S) - \sum_{i=l}^{n} Q_i \left(1 - \frac{T_0}{T_i}\right) + T_0\sigma_{CM} \\ &= \Delta(E + P_0V - T_0S) - \Phi_Q + T_0\sigma_{CM} \qquad \text{(closed)} \qquad (3\text{-}27) \end{aligned}$$

and

$$\begin{aligned} W_{act,u} &= \Delta(E + P_0V - T_0S) - \sum_{j} Q_j \left(1 - \frac{T_0}{T_j}\right) + T_0\sigma_{tot} \\ &= \Delta(E + P_0V - T_0S) - \sum_{j} \Phi_{Q,j} + T_0\sigma_{tot} \qquad \text{(closed)} \qquad (3\text{-}28) \end{aligned}$$

where the direction of both Q_i and Q_j are measured relative to the closed-system boundary. Both Eqs. (3-27) and (3-28) may be written on an intensive basis for the actual useful work by replacing E, V, S, and Q by e, v, s, and q. In addition, the $T_0\sigma$ terms may be replaced by the irreversibility I. Although only the heat transfer Q_j appears explicitly in Eq. (3-28), both equations above implicitly include any necessary heat transfer Q_0 with the environment at T_0. On either an extensive or intensive basis the reversible work is found by setting $\sigma_{CM} = \sigma_{tot} = 0$. That is,

$$W_{rev,u} = \Delta E + P_0\,\Delta V - T_0\,\Delta S - \Phi_Q \qquad \text{(closed)} \qquad (3\text{-}29)$$

and

$$W_{rev,u} = \Delta E + P_0\,\Delta V - T_0\,\Delta S - \Phi_{Q,R} \qquad \text{(closed)} \qquad (3\text{-}30)$$

where the last equation has been written for a single reservoir at T_R. Now, consider a *stationary* closed system undergoing a process where no reservoirs at T_j are present. The specific useful work transfer for a process between two end states 1 and 2, while exchanging heat solely with the environment and under reversible conditions, is given by

$$w_{rev,u} = (u_2 - u_1) + P_0(v_2 - v_1) - T_0(s_2 - s_1) \qquad (\text{closed, } q_R = 0) \qquad (3\text{-}31)$$

Under these restrictive conditions the only data required are u, v, and s at the end states and the environmental pressure and temperature.

3-4-2 Closed-System Availability

A formal definition of *closed-system availability* is as follows:

> The availability of a closed system in a given state is the maximum useful work output that may be obtained from a system-atmosphere combination as the system proceeds from a specified equilibrium state to the dead state, while exchanging heat solely with the environment.

A mathematical expression for the closed-system availability on an extensive basis follows directly from Eq. (3-30). The definition of availability excludes any heat transfer Q_j to or from a reservoir at T_j. Hence the last term in Eq. (3-30) is zero. Thus for a change of state to the dead state the general equation for reversible useful work reduces to

$$W_{\text{rev},u} = E_0 - E + P_0(V_0 - V) - T_0(S_0 - S)$$

where $E = (U + V^2/2 + gz)$ and the subscript "0" represents properties at the dead state. This equation fulfills the definition of closed-system availability, except for a traditional sign convention. The value of W obtained by using the above equation will be negative, but the availability of a closed system historically is always taken to be positive. To satisfy this convention, we define the extensive *closed-system availability* Φ as

$$\Phi \equiv E - U_0 + P_0(V - V_0) - T_0(S - S_0) \tag{3-32}$$

where in the dead state $E_0 = U_0$. The function Φ is also known as the *nonflow exergy* [3] associated with the restricted dead state (T_0, P_0). It is always positive in value whether T is greater than or less than T_0 and whether P is greater than or less than P_0. The greater the difference $(T - T_0)$ or the difference $(P - P_0)$, the greater is the value for the closed-system availability. Progress toward the dead state requires not only that the process be internally reversible, but also that any heat interaction with the environment be reversible. Reversible heat transfer requires that either the heat interaction occurs through a reversible cycle operating between T and T_0 or the part of the system where heat transfer occurs has a temperature only differentially different from T_0.

Similar to energy analyses of closed systems, availability analyses are frequently carried out on a unit-mass or molar basis. The specific closed-system availability $\phi \equiv \Phi/m$ and is given by

$$\phi \equiv (u - u_0) + P_0(v - v_0) - T_0(s - s_0) + V^2/2 + gz \tag{3-33}$$

where the symbols without a subscript refer to the actual state of the closed system. Application of Eq. (3-33) to a closed-system process with initial and final states 1 and 2 yields

$$\Delta\phi = \phi_2 - \phi_1 = (u_2 - u_1) + P_0(v_2 - v_1) - T_0(s_2 - s_1) + \Delta\text{ke} + \Delta\text{pe} \tag{3-34}$$

where ke and pe are the specific kinetic and gravitational potential-energy terms. The latter two terms in the equation are frequently zero for closed-system analysis. (It is important to note that sometimes the kinetic- and potential-energy terms are excluded from the definition of closed-system availability. This is a matter of choice.)

Finally, the use of Eq. (3-34) on an extensive basis in conjunction with Eqs. (3-27) and (3-28) for the actual, useful work leads to

$$W_{\text{act},u} = \Phi_2 - \Phi_1 - \sum_j Q_j\left(1 - \frac{T_0}{T_j}\right) + I_{\text{tot}}$$

$$= \Phi_2 - \Phi_1 - \sum_j \Phi_{Q,j} + I_{\text{tot}} \qquad \text{(closed system)} \qquad (3\text{-}35)$$

and

$$W_{\text{act},u} = \Phi_2 - \Phi_1 - \sum_{i=1}^{n} Q_i\left(1 - \frac{T_0}{T_i}\right) + I_{\text{CM}}$$

$$= \Phi_2 - \Phi_1 - \Phi_Q + I_{\text{CM}} \qquad \text{(closed system)} \qquad (3\text{-}36)$$

In the first equation above $W_{\text{act},u}$ is expressed in terms of heat exchange with one or more reservoirs, as well as with the local environment at T_0, whereas in the second equation it is expressed in terms of heat exchange along the closed-system boundary. Both Eqs. (3-35) and (3-36) predict the reversible useful work when the irreversibility terms are zero. In this latter form they are equivalent to Eqs. (3-29) and (3-30).

By rearrangement Eq. (3-36) takes on the form of an *availability balance* for a control mass, namely,

$$\Phi_2 - \Phi_1 = \sum_{i=1}^{n} Q_i\left(1 - \frac{T_0}{T_i}\right) + (W_{\text{act}} + P_0\,\Delta V) - I_{\text{CM}}$$

$$= \Phi_Q + (W_{\text{act}} + P_0\,\Delta V) - I_{\text{CM}} \qquad (3\text{-}37\text{a})$$

Thus the change in Φ for a closed system is brought about by three effects, as shown by the three terms on the right of Eq. (3-37a): (1) heat transfer, (2) work interactions, and (3) internal irreversibilities. The first term on the right is a measure of the availability transfer associated with heat transfer. (This term was discussed in detail in Sec. 3-3.) Note that Φ can only be increased by heat and/or useful work transfer *into* the system. The presence of I_{CM} accounts for the destruction of availability due to irreversibilities within the control mass.

3-4-3 Irreversibility of a Closed System

An expression for the internal irreversibility of any closed system is found directly by rearrangement of Eq. (3-37a). That is,

$$I_{CM} = T_0\sigma_{CM} = -(\Phi_2 - \Phi_1) + \sum_{i=1}^{n} Q_i\left(1 - \frac{T_0}{T_i}\right) + (W_{act} - P_0\,\Delta V) \geq 0 \quad \text{(3-37b)}$$

and on a unit-mass basis,

$$i_{CM} = T_0\sigma_m = -(\phi_2 - \phi_1) + \phi_Q + w_{act,u} \quad \text{(3-38)}$$

The above equations apply solely to the closed system of interest. From a broader viewpoint, a closed system frequently interacts with other systems during a process. Because both I and ΔS can be interpreted as measuring losses during a process, a relationship should exist between them even when a composite of several interacting systems is analyzed. As an example, we shall develop an expression for the total irreversibility I_{tot} for a stationary closed system and its interacting heat reservoirs in terms of ΔS_{tot}. We begin by using Eq. (3-12), namely,

$$I_{tot} = W_{act,u} - W_{rev,u} \quad \text{(3-12)}$$

The first term on the right may be replaced by $W_{act,u} = W_{act,tot} + P_0\,\Delta V$, and the second term by the reversible form of Eq. (3-27) where $E = U$. The result is

$$\begin{aligned} I_{tot} &= (W_{act,tot} + P_0\,\Delta V)_{CM} - (\Delta U + P_0\,\Delta V - T_0\,\Delta S)_{CM} + \sum_j Q_j\left(1 - \frac{T_0}{T_j}\right) \\ &= W_{act,tot} - \Delta U_{CM} + T_0\,\Delta S_{CM} + \sum_j Q_j - \sum_j Q_j\frac{T_0}{T_j} \end{aligned}$$

In addition, $W_{act,tot} - \Delta U_{CM} = -Q$ from the conservation of energy principle for a control mass, where Q is the sum of all heat interactions. As noted earlier, this latter term is given by $Q = Q_0 + \sum Q_j$. Use of these latter two relations in the equation for I_{tot} leads to

$$I_{tot} = -Q_0 + T_0\,\Delta S_{CM} - \sum_j Q_j\frac{T_0}{T_j}$$

where Q_0 and Q_j are measured signwise relative to the closed-system boundary. As a result, $-Q_0 = T_0\,\Delta S_0$ and $-Q_j = T_j\,\Delta S_j$ for the heat reservoirs. Thus the equation for the total irreversibility of a closed-system process which involves heat-transfer processes becomes

$$\begin{aligned} I_{tot} &= T_0\left(\Delta S_0 + \Delta S_{CM} + \sum_j \Delta S_{R,j}\right) \\ &= T_0\,\Delta S_{tot} = T_0\,\Delta S_{isol} \end{aligned} \quad \text{(3-39)}$$

On a unit-mass basis $i_{tot} = T_0\,\Delta s_{tot}$. Hence for a closed system interacting with heat reservoirs the total irreversibility is directly proportional to the total entropy change of the composite (isolated) system.

3-5 CLOSED-SYSTEM AVAILABILITY EXAMPLES

In this section a few examples for the availability analysis of closed systems are presented. The general availability balance has the form

$$\Phi_2 - \Phi_1 = \Phi_Q + (W_a + P_0\,\Delta V) - I_{\text{CM}} \tag{3-37}$$

while the irreversibility of the control mass is given by

$$I_{\text{CM}} = W_{\text{act},u} - W_{\text{rev},u}$$

The second-law efficiency or effectiveness ϵ follows the general formats presented in Sec. 3-2-2.

3-5-1 Energy Addition to a Closed System

The preceding example in Sec. 3-3 illustrated the effect of the variation of use temperature T_U with fixed source temperature T_S on the second-law efficiency of the heat-transfer process. A closely connected problem is to determine the effectiveness of different types of energy transfer used to achieve a given change of state for a system. For example, heat transfer, electrical work, or paddle-wheel work could be used separately to alter the state of a closed rigid system. The energy equation for such processes would be either $q = \Delta u$ or $w = \Delta u$, where w represents either of the two work forms. The general availability balance would be given in this case by Eq. (3-37),

$$\Delta\Phi = \Phi_Q + W_{\text{act},u} - I_{\text{CM}}$$

or rearranging so that the inputs are on the left, we obtain

$$\Phi_Q + W_{\text{act},u} = \Delta\Phi + I_{\text{CM}}$$

The effectiveness of the heat-transfer process ϵ_Q and work processes ϵ_W, defined in terms of useful outcome over input, would be

$$\epsilon_Q = \frac{\Delta\Phi_{\text{CM}}}{\Phi_Q} \qquad \text{and} \qquad \epsilon_W = \frac{\Delta\Phi_{\text{CM}}}{W_{\text{act},u}} \tag{3-40}$$

The differences in these effectivenesses lie in the relative values of Φ_Q and $W_{\text{act},u}$ for a given change of state. The example below illustrates the situation.

Example 3-3. With reference to Example 2-1, oxygen initially at 300 K and 100 kPa is contained in a rigid tank. Two alternative processes are to be considered to increase its temperature to 500 K. The gas receives energy (*a*) adiabatically by means of a paddle wheel driven by an external source and (*b*) by means of an external heat reservoir at 600 K. In the second case the temperature at the boundary is taken as the gas temperature throughout the process. Determine (1) the availability change and (2) the irreversibility, in kJ/kmol, and (3) the effectiveness of the overall process for the two parts of the problem. The value of T_0 is 300 K.

Solution

(*a*) (1) The availability change of the oxygen is given by Eq. (3-34). Using Δu and Δs data from Prob. 2-1, we find that

$$\Delta\phi = \Delta u + P_0\,\Delta v - T_0\,\Delta s = 4372 + 0 - 300(11.13) = 1033 \text{ kJ/kmol}$$

If we considered the work source to be reversible, then its availability change is equal to the paddle-wheel work it provides, or 4372 kJ/kmol.

(2) The irreversibility of the overall process, which is also the irreversibility within the tank, is given by Eq. (3-37). On a molar basis

$$i = -(\phi_2 - \phi_1) + \phi_Q + w_{\text{net}} = -1033 + 0 + 4372 = 3339 \text{ kJ/kmol}$$

This value can also be obtained from $i = T_0\sigma_{\text{tot}}$. From Prob. 2-1, $\sigma_{\text{tot}} =$ 11.13 kJ/kmol · K. So $i = 300(11.13) = 3339$ kJ/kmol.

(3) The effectiveness of either process can be determined from the ratio of outcome to input. The outcome in both cases is the increase in availability of the oxygen. This was found above to be 1033 kJ/kmol. For this first process the input of availability is the value of the paddle-wheel work itself. This is 4372 kJ/kmol. Hence for process (*a*) on a unit-mass basis

$$\epsilon_a = \epsilon_W = \frac{\text{outcome}}{\text{input}} = \frac{\Delta\phi}{w_{\text{pad}}} = \frac{1033}{4372} = 0.236$$

Hence roughly 76 percent of the work input has been lost in terms of the work potential of the energy exchanged.

(*b*) (1) The availability change of the gas remains the same as in part (*a*), namely, 1033 kJ/kmol. The availability change of the reservoir is given by

$$\phi_{Q,R} = q_R\left(1 - \frac{T_0}{T_R}\right) = -4372\left(1 - \frac{300}{600}\right) = -2186 \text{ kJ/kmol}$$

The net change in the availability of the overall process is $1033 - 2186 = -1153$ kJ/kmol.

(2) The total irreversibility in this case for the composite closed system consisting of the tank and reservoir is

$$\begin{aligned} i &= -\sum(\Delta\phi) + \phi_Q + w_{\text{net}} \\ &= -(1033 - 2186) + 0 + 0 = 1153 \text{ kJ/kmol} \end{aligned}$$

This value could also be found from $T_0\sigma_{\text{tot}}$. From Prob. 2-1, $\sigma_{\text{tot}} = 3.84$ kJ/kmol·K. Hence $i = 300(3.84) = 1152$ kJ/kmol, which is due solely to the irreversible heat-transfer process.

(3) For the second process the input is the availability ϕ_R associated with the heat transfer q_R leaving the reservoir at T_R. Since the energy added as heat or work is the same value, then

$$\phi_R = q_R\left(1 - \frac{T_0}{T_R}\right) = 4372\left(1 - \frac{300}{600}\right) = 2186 \text{ kJ/kmol}$$

The effectiveness of process b is

$$\epsilon_b = \epsilon_Q = \frac{\Delta\phi}{\phi_R} = \frac{1033}{2186} = 0.473$$

In comparison, for parts (a) and (b), the lost work potential when using paddle-wheel work is 3339 kJ/kmol versus 1153 kJ/kmol for the heat-addition process. Hence work addition to achieve a temperature rise of a fluid is a poor choice over a simple heat-addition process. However, the value of T_R for heat addition must be kept reasonably close to the system temperature. Although the energy added in each case was the same, the heat-addition process made more effective use of the availability supplied.

3-5-2 Adiabatic Compressor Work

For the adiabatic compression of a gas in a piston-cylinder device the energy and availability balances and the irreversibility are

$$\begin{aligned} w_{\text{act}} &= w_{\text{act},u} - P_0\,\Delta v = \Delta u \\ \phi_2 - \phi_1 &= w_{\text{act}} + P_0\,\Delta v - i_{\text{CM}} \\ i_{\text{CM}} &= w_{\text{act},u} - w_{\text{rev},u} = T_0\,\Delta s_{\text{CM}} \end{aligned}$$

The effectiveness ϵ_c in this case is a measure of the increase in the specific availability of the fluid per unit of actual work input. That is,

$$\epsilon_c \equiv \frac{\Delta\phi}{w_a} \tag{3-41}$$

The example below illustrates this calculation for both an ideal gas and a real gas.

Example 3-4. Carbon dioxide (CO_2) is compressed adiabatically in a piston cylinder from 10 bars and 300 K to 50 bars and 450 K. The fluid is modeled as (a) an ideal gas and (b) a real gas. For the two models, determine (1) the availability change and (2) the irreversibility, in kJ/kg, and (3) the second-law effectiveness.

Solution

(a) For an ideal gas the change in specific availability for a closed system is

$$\begin{aligned} \Delta\phi &= (u_2 - u_1) + P_0(v_2 - v_1) - T_0(s_2 - s_1) \\ &= (u_2 - u_1) + RP_0\left(\frac{T_2}{P_2} - \frac{T_1}{P_1}\right) - T_0\left(s_2^o - s_1^o - R\ln\frac{P_2}{P_1}\right) \end{aligned}$$

(1) Substitution of tabular data from the CO_2 ideal-gas table leads to

$$\begin{aligned} \phi_2 - \phi_1 &= (11{,}742 - 6939) + 8.3145(1)\left(\frac{450}{50} - \frac{300}{10}\right) \\ &\quad - 298.15(230.194 - 213.915 - 8.314\ln 5) \\ &= 4803 - 175 - 864 = 3764\ \text{kJ/kmol} = 85.53\ \text{kJ/kg} \end{aligned}$$

(2) The actual useful work $w_{\text{act},u} = \Delta u + P_0\,\Delta v$. These two terms are given by the first two terms above in the $\Delta\phi$ calculation. Therefore $w_{\text{act},u} = 4628$ kJ/kmol = 105.16 kJ/kg. As a result,

$$i_{\text{CM}} = w_{\text{act},u} - w_{\text{rev},u} = 105.16 - 85.53 = 19.63 \text{ kJ/kg}$$

This value is also given by $T_0\,\Delta s$, which is the last term in the $\Delta\phi$ calculation. Hence $i_{\text{CM}} = 864$ kJ/kmol = 19.63 kJ/kg.

(3) The effectiveness $\epsilon = \Delta\phi/w_{\text{act}}$. The actual work $w = \Delta u = 4803$ kJ/kmol = 109.13 kJ/kg. Thus

$$\epsilon = \frac{85.53}{109.13} = 0.784$$

The fluid has gained 78 percent of the work potential added during the process.

(*b*) (1) Using data from the CO_2 superheat table in the Appendix, we find that

$$\begin{aligned}\Delta\phi &= (461.15 - 361.39) + (0.01605 - 0.0538)(10^{-2}) - 298.15(1.8019 - 1.7514)\\ &= 99.76 - 3.78 - 15.06 = 80.92 \text{ kJ/kg}\end{aligned}$$

This compares to 85.53 kJ/kg for the ideal-gas model, which is 5 percent too low.

(2) The actual useful work again is the sum of the first two terms above, or 95.98 kJ/kg. Hence the irreversibility is

$$i = w_{\text{act},u} - w_{\text{rev},u} = 95.98 - 80.92 = 15.06 \text{ kJ/kg}$$

This agrees with the third term in the $\Delta\phi$ calculation, which is the $T_0\,\Delta s$ term. This real-gas i value compares to 19.63 kJ/kg for the ideal-gas model.

(3) Finally, the effectiveness $\epsilon = \Delta\phi/w_{\text{act}} = 80.92/99.76 = 0.81$. This compares to 0.784 for the ideal-gas model. The differences in the three answers reflect the error in assuming an ideal-gas model.

3-5-3 Energy Exchange between Incompressible Substances

There is a considerable loss in availability when two incompressible substances of different temperatures are brought into thermal contact. When no heat transfer occurs to the environment, the closed-system energy equation reduces to $\Delta U = 0$. Since the internal energy change of an incompressible substance is $\Delta U = mc\,\Delta T$, the energy equation for substances A and B becomes

$$[mc(T_f - T_i)]_A + [mc(T_f - T_i)]_B = 0$$

where i and f represent the initial and final states, respectively. Because $\Delta s = mc \ln T_f/T_i$, the availability change $\Delta\Phi$ for an incompressible substance is given by

$$\Delta\Phi = \Delta U + P_0\Delta V - T_0\Delta S = mc(T_f - T_i) - mc\,T_0 \ln\frac{T_f}{T_i}$$

The effectiveness of the process is best described by the gain in availability by one substance divided by the negative loss in availability of the other substance. If T_A is initially greater than T_B, then

$$\epsilon \equiv \frac{\Delta\Phi_B}{-\Delta\Phi_A}$$

The following example illustrates a typical calculation.

Example 3-5. An 18-kg block of lead initially at 200°C is quenched in a tank containing 0.030 m^3 of water initially at 25°C. Assume no heat loss to the surroundings and that $T_0 = 25$°C. Determine (*a*) the total change in availability, in kilojoules, and (*b*) the second-law effectiveness when temperature equilibrium is reached.

Solution. The first step is to determine the equilibrium temperature from an energy analysis. Assuming the substances to be incompressible, the work is zero and the energy equation reduces to $\Delta U = 0$. In terms of the individual substances this becomes

$$(mc\,\Delta T)_{\text{lead}} + (mc\,\Delta T)_{\text{water}} = 0$$

The final temperature depends on the choice of specific-heat data. The value for water is close to 4.18 kJ/kg · K, but c for lead varies between 0.129 kJ/kg · K at 27°C and 0.136 kJ/kg · K at 200°C. If we choose 0.133 for lead, then

$$(mc)_{\text{water}} = \frac{Vc}{v} = \frac{0.030(4.18)(10^3)}{1.003} = 125 \text{ kJ/K}$$

$$(mc)_{\text{lead}} = 18(0.133) = 2.4 \text{ kJ/K}$$

Hence $(mc)_{\text{water}}/(mc)_{\text{lead}} = 50/1$. The energy equation then predicts $\Delta T(\text{water}) \ll \Delta T(\text{lead})$. That is, the final equilibrium temperature is slightly above 25°C. Therefore the chosen c value for lead of around 0.133 kJ/kg · K is appropriate for the process. The energy equation becomes

$$125(T_f - 25) + 2.4(T_f - 200) = 0$$

from which $T_f = 28.30$°C $= 301.45$ K.

(*a*) For an incompressible substance

$$\Delta\Phi = m(\Delta u + P_0\,\Delta v - T_0\,\Delta s) = mc\left(\Delta T - T_0 \ln \frac{T_2}{T_1}\right)$$

For lead and water

$$\Delta\Phi_{\text{lead}} = 2.4\left[(28.3 - 200) - 298.15 \ln \frac{301.45}{473.15}\right] = -89.3 \text{ kJ}$$

$$\Delta\Phi_{\text{water}} = 125.0\left[(28.3 - 25.0) - 298.15 \ln \frac{301.45}{298.15}\right] = 2.50 \text{ kJ}$$

Therefore the total availability change is -86.8 kJ.

(*b*) The effectiveness is

$$\epsilon = \frac{\Delta\Phi_{\text{water}}}{-\Delta\Phi_{\text{lead}}} = \frac{2.50}{-(-89.3)} = 0.028$$

Because the final temperature is close to the initial temperature of the cold substance (water), the retention of the overall availability is quite poor.

3-6 REVERSIBLE WORK, AVAILABILITY, AND IRREVERSIBILITY FOR A STEADY-STATE CONTROL VOLUME

General expressions for the reversible work associated with a control volume are given by Eqs. (3-6) and (3-7). We now address the important class of engineering problems known as steady-state, steady flow.

3-6-1 Reversible Work in Steady State

Because properties are invariant with time under steady-state conditions, the quantity $d(E + P_0V - T_0S)_{\text{CV}}/dt = 0$ for a control-volume analysis. As a result, Eqs. (3-3) and (3-5) reduce under *steady-state* conditions to

$$\dot{W}_{\text{act},sf} = \sum_{\text{out}} \dot{m}_e(h^* - T_0 s)_e - \sum_{\text{in}} \dot{m}_i(h^* - T_0 s)_i - \sum_j \dot{Q}_j\left(1 - \frac{T_0}{T_j}\right) + T_0\dot{\sigma}_{\text{tot}} \tag{3-42}$$

and

$$\dot{W}_{\text{act},sf} = \sum_{\text{out}} \dot{m}_e\left(h^* - T_0 s\right)_e - \sum_{\text{in}} \dot{m}_i(h^* - T_0 s)_i - \sum_{i=1}^{n} \dot{Q}_i\left(1 - \frac{T_0}{T_i}\right) + T_0\dot{\sigma}_{\text{CV}} \tag{3-43}$$

In these two equations $h^* \equiv h + V^2/2 + gz$. The reversible steady-flow work is found by deleting the irreversibility terms $T_0\dot{\sigma}_{\text{tot}}$ and $T_0\dot{\sigma}_{\text{CV}}$. The above equations can be shortened further symbolically by introducing the stream availability parameter b into control-volume analysis, where

$$b \equiv h^* - T_0 s = h + \text{ke} + \text{pe} - T_0 s \tag{3-44}$$

As a result, Eqs. (3-42) and (3-43) become

$$\dot{W}_{\text{act},u} = \sum_{\text{out}} \dot{m}b - \sum_{\text{in}} \dot{m}b - \sum_j Q_j\left(1 - \frac{T_0}{T_j}\right) + \dot{I}_{\text{tot}} \tag{3-45}$$

and

$$\dot{W}_{\text{act},u} = \sum_{\text{out}} \dot{m}b - \sum_{\text{in}} \dot{m}b - \sum_{i=1}^{n} \dot{Q}_i\left(1 - \frac{T_0}{T_i}\right) + \dot{I}_{\text{CV}} \tag{3-46}$$

For a control volume with one inlet (1) and one outlet (2), with unit mass entering and leaving, and a reversible process, Eqs. (3-42) and (3-43) become

$$w_{\text{rev},sf} = (h + \text{ke} + \text{pe} - T_0 s)_2 - (h + \text{ke} + \text{pe} - T_0 s)_1 - \sum_j q_j \left(1 - \frac{T_0}{T_j}\right) \tag{3-47}$$

and

$$w_{\text{rev},sf} = (h + \text{ke} + \text{pe} - T_0 s)_2 - (h + \text{ke} + \text{pe} - T_0 s)_1 - \sum_{i=1}^{n} q_i \left(1 - \frac{T_0}{T_i}\right) \tag{3-48}$$

As in preceding sets of equations, Eq. (3-47) applies to the control volume and all interacting reservoirs, whereas Eq. (3-48) applies solely to effects occurring at the control surface.

3-6-2 Stream or Flow Availability

For flow through a control volume the availability associated with a fluid crossing the control surface is known as the stream or flow availability. By definition:

> The stream availability of a fluid is defined as the maximum work output which can be obtained as the fluid is changed reversibly from the given state to a dead state while exchanging heat solely with the environment.

The equation for evaluating stream availability is most easily developed by applying Eq. (3-47) to the situation where q_j is zero. With this restriction, Eq. (3-47) reduces to

$$w_{\text{rev},sf} = (h + \text{ke} + \text{pe} - T_0 s)_2 - (h + \text{ke} + \text{pe} - T_0 s)_1$$

For a change of state to a dead state this becomes

$$w_{\text{rev},sf} = (h_0 - T_0 s_0) - \left(h + \frac{V^2}{2} + gz - T_0 s\right)$$

Work output in this text is negative by convention, but the availability function is considered to measure positive work output. Hence the *stream* or *flow availability* ψ on a unit-mass or molar basis is defined by the negative of the above equation; that is,

$$\psi \equiv \left(h + \frac{V_2}{2} + gz - T_0 s\right) - (h_0 - T_0 s_0)$$

$$= (h - h_0) - T_0(s - s_0) + \frac{V^2}{2} + gz \tag{3-49}$$

where z for the actual state is measured relative to the dead state datum. The quantity ψ is also known as the *stream* or *flow exergy* associated with the restricted dead state. Thus ψ is the *thermomechanical stream availability* and does not account for chemical equilibrium with the environment. Application of Eq. (3-49) to a steady-flow process with inlet and exit states denoted by 1 and 2 yields

$$\psi_2 - \psi_1 = (h_2 - h_1) - T_0(s_2 - s_1) + (\text{ke}_2 - \text{ke}_1) + (\text{pe}_2 - \text{pe}_1) \qquad (3\text{-}50)$$

Unlike the nonflow availability ϕ, the value of ψ can take on negative values. This can occur when the pressure is less than the environment pressure P_0. This point is illustrated by data for the stream availability of an ideal gas. When ideal-gas relations are substituted into Eq. (3-49), the kinetic- and potential-energy terms are neglected, and the equation is rearranged into a dimensionless format, then

$$\frac{\psi}{c_p T_0} = \frac{T}{T_0} - \ln\frac{T}{T_0} + \ln\left(\frac{P}{P_0}\right)^{(k-1)/k}$$

This equation has been plotted in Fig. 3-5 for a k value of 1.40, which is typical for diatomic gases around room temperature. Note that ψ is negative for a range of values of T/T_0 when $P/P_0 < 0$.

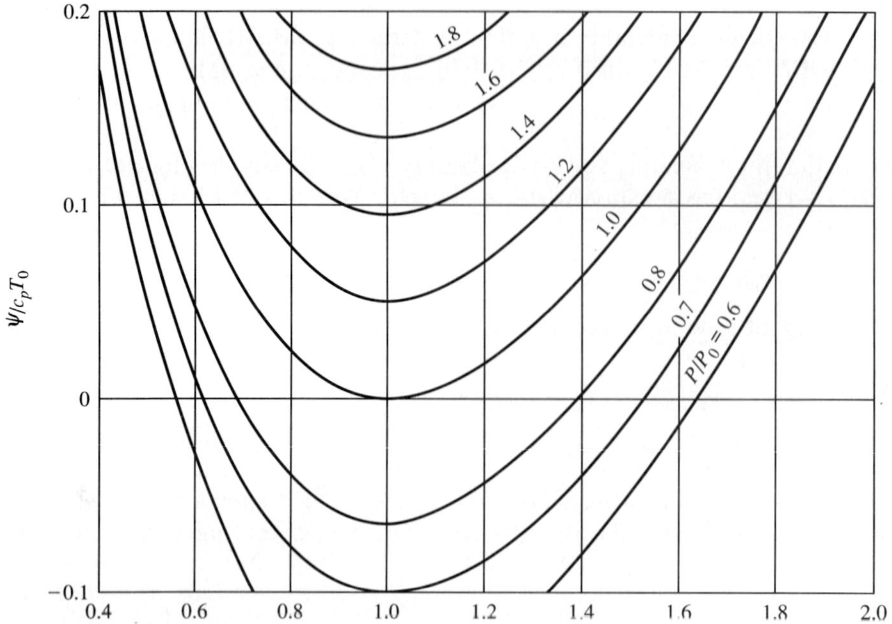

FIGURE 3-5
Dimensionless steady-flow availability ψ/c_pT_0 of an ideal gas as a function of T/T_0 for selected values of P/P_0 and $k = 1.4$.

A graphical interpretation of Eq. (3-49) for ψ_1 at state (T_1, P_1) is shown on the *hs* diagram of Fig. 3-6. The sum of the terms $(h_1 - h_0)$ and $-T_0(s_1 - s_0)$ can be represented as a line length on the plot. (Extrinsic effects such as ke and pe are neglected.) On the diagram the intersection of the isobars P and P_0 and the isotherms T and T_0 locate the two states of interest, namely, (T_1, P_1) and (T_0, P_0). The vertical length between states 1 and B directly represents $(h_1 - h_0)$. To establish the second term, we draw a reference line through state 0 with a slope numerically equal to T_0 so that $(\partial h/\partial s)_p = T_0$ at state 0. This reference line is extended until it crosses line s_1 at state A. From the geometry, $h_B - h_A = T_0(s_0 - s_A) = -T_0(s_1 - s_0)$. As a result,

$$h_1 - h_A = (h_1 - h_B) + (h_B - h_A) = (h_1 - h_0) - T_0(s_1 - s_0) \equiv \psi_1$$

Hence the vertical line length $(h_1 - h_A)$ represents the steady-flow availability or exergy ψ_1 for any state 1 on the *hs* diagram.

The change in ψ given by Eq. (3-50) can be represented by areas on a *Ts* diagram (neglecting Δke and Δpe terms). On the *Ts* plot of Fig. 3-7 states 1 and 2 are located on pressure lines P_1 and P_2. The constant enthalpy lines h_1 and h_2 are also shown, as well as the dead state temperature line T_0. The term $T_0(s_2 - s_1)$ is seen as the shaded area below the line T_0 and between states A and B. Now, consider the constant-pressure process 1-C, where state C also lies on the h_2 line. If process 1-C is internally reversible, then the area under curve 1-C represents q for the process. Also, since $w_{\text{rev}} = \int v\,dp = 0$ for the steady-flow process, the energy equation dictates that $q = h_C - h_1$. Hence the area under process 1-C represents $h_C - h_1$. The sum of the shaded areas under lines C-1 and A-B equals $(h_1 - h_2) + T_0(s_2 - s_1)$, or

$$\text{Area } C\text{-1} + \text{area } A\text{-}B = h_1 - h_2 - T_0(s_1 - s_2) = \psi_1 - \psi_2$$

or

$$\psi_2 - \psi_1 = -(\text{area } C\text{-1} + \text{area } A\text{-}B)$$

Thus the sum of the two areas represents the magnitude of the stream availability change between the two states.

Restructuring Eq. (3-5) in terms of availability concepts and symbols leads to a general *availability balance* for a control volume.

$$\frac{d\Phi_{\text{CV}}}{dt} = \sum \dot{Q}_i\left(1 - \frac{T_0}{T_i}\right) + \sum_{\text{in}} \dot{m}b - \sum_{\text{out}} \dot{m}b + \left(\dot{W}_{\text{act}} + P_0\frac{dV_{\text{CV}}}{dt}\right) - \dot{I}_{\text{CV}} \quad (3\text{-}51)$$

where $\sum \dot{Q}_i(1 - T_0/T_i) = \dot{\Phi}_Q$. The time rate of change of availability within a control volume depends on the availability transfer associated with the heat, mass, and work transfer at the control surface, as well as on the destruction due to internal irreversibilities. A schematic for the availability or exergy accounting of an open system is shown in Fig. 3-8. For a control volume in steady state the term $d\Phi_{\text{CV}}/dt$ is set equal to zero. If only one inlet flow and one outlet flow exist, then

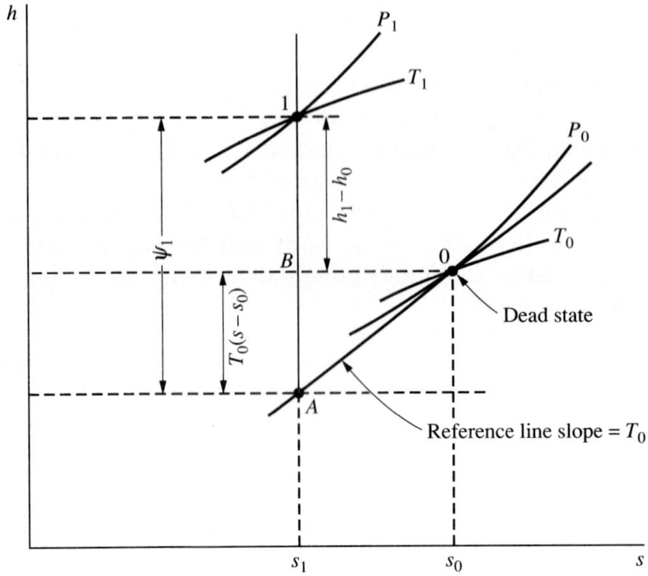

FIGURE 3-6
Graphical interpretation of steady-flow availability ψ on an *hs* diagram.

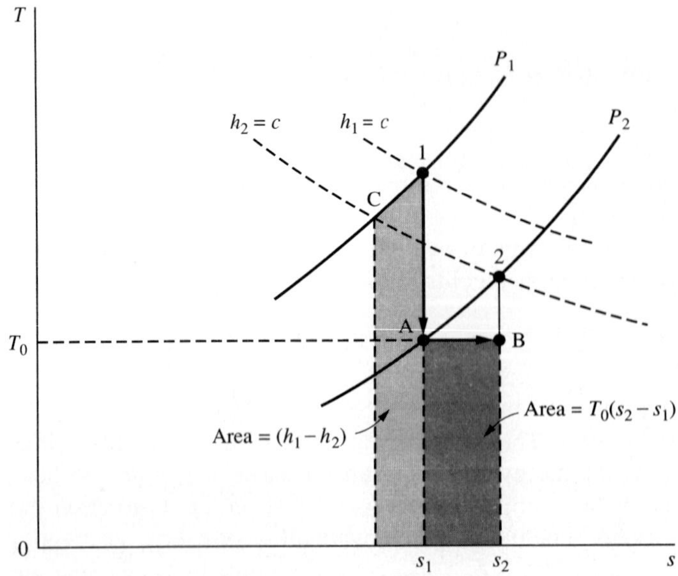

FIGURE 3-7
Graphical interpretation of $\Delta\psi$ on a *Ts* diagram.

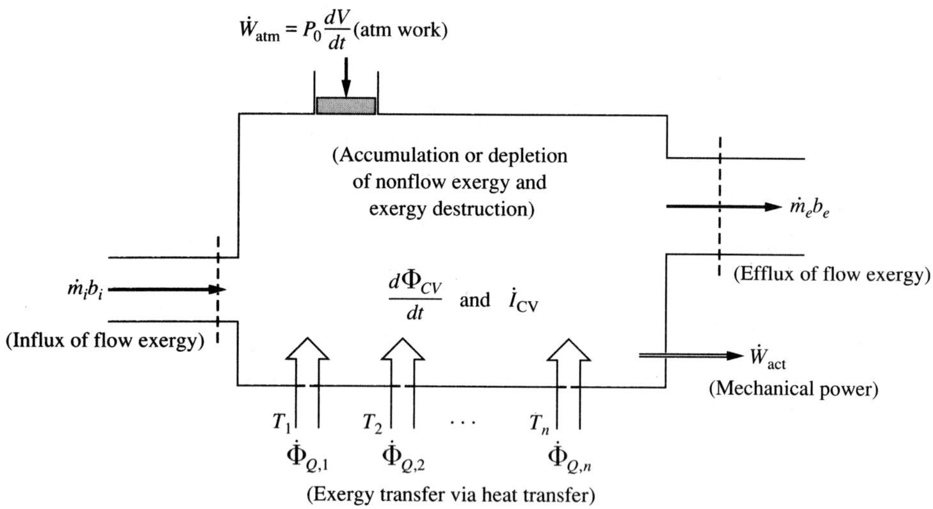

FIGURE 3-8
Schematic for general availability or exergy accounting for an open system. ($b = h + \text{ke} + \text{pe} - T_0 s$)

$$\begin{aligned} 0 &= \dot{\Phi}_Q - \dot{m}\,\Delta b + \dot{W}_{act,sf} - \dot{I}_{CV} \\ &= \dot{\Phi}_Q - \dot{m}\,\Delta\psi + \dot{W}_{act,sf} - \dot{I}_{CV} \qquad \text{(steady flow)} \end{aligned} \tag{3-52}$$

For a steady-state control volume the rate of energy entering equals the rate of energy leaving. However, Eq. (3-52) shows that in steady state the total rate of availability entering must *exceed* the total rate leaving when internal irreversibilities are present.

3-6-3 Irreversibility for Steady Flow

In steady state the rearrangement of Eq. (3-51) yields

$$\dot{I}_{CV} = T_0\dot{\sigma}_{CV} = \sum \dot{Q}_i\left(1 - \frac{T_0}{T_i}\right) + \sum_{in} \dot{m}b - \sum_{out} \dot{m}b + \dot{W}_{act}$$

On a unit-mass basis for one inlet and one outlet the above equation becomes

$$i_{CV} = T_0\frac{\dot{\sigma}_{CV}}{\dot{m}} = q_i\left(1 - \frac{T_0}{T_i}\right) - (b_2 - b_1) + w_{act} \qquad \text{(steady flow)} \tag{3-53}$$

The above equation is equivalent to Eq. (3-38) for a closed system.

An expression for the irreversibility I in terms of ΔS_{tot}, equivalent to Eq. (3-39) for a stationary closed system, can be developed for a steady-state control volume. We begin by using Eq. (3-11), namely,

$$\dot{I} = \dot{W}_{act} - \dot{W}_{rev} \tag{3-11}$$

For a steady-state control volume with multiple inlets and outlets and heat transfer with j reservoirs Eq. (3-42) requires that

$$\dot{W}_{\mathrm{rev},sf} = \sum_{\mathrm{out}} \dot{m}(h^* - T_0 s) - \sum_{\mathrm{in}} \dot{m}(h^* - T_0 s) - \sum_j \dot{Q}_j \left(1 - \frac{T_0}{T_j}\right) \tag{3-54}$$

In addition, the steady-state energy equation for a control volume is

$$\dot{W}_{\mathrm{act},sf} = \sum_{\mathrm{out}} \dot{m}h^* - \sum_{\mathrm{in}} \dot{m}h^* - \dot{Q}$$

Substitution of these two equations in Eq. (3-11) for $\dot{I}$ yields

$$\dot{I}_{sf} = -\dot{Q} + T_0\left(\sum_{\mathrm{out}} \dot{m}s - \sum_{\mathrm{in}} \dot{m}s\right) + \sum_j \dot{Q}_j - \sum_j \dot{Q}_j \frac{T_0}{T_j}$$

However, $\dot{Q} = \dot{Q}_0 + \sum \dot{Q}_j$, where $\dot{Q}_0$ and $\sum \dot{Q}_j$ are measured signwise relative to the control volume. Thus the above equation becomes

$$\dot{I}_{sf} = -T_0 \frac{\dot{Q}_0}{T_0} + T_0\left(\sum_{\mathrm{out}} \dot{m}s - \sum_{\mathrm{in}} \dot{m}s\right) - T_0 \sum_j \frac{\dot{Q}_j}{T_j}$$

Due to the sign convention on Q_0 and Q_j, the entropy changes of reservoirs at T_0 and T_j are given by $-\dot{Q}_0/T_0 = \Delta\dot{S}_0$ and $-\dot{Q}_j/T_j = \Delta\dot{S}_j$. Hence the above equation is recast as

$$\dot{I}_{sf} = T_0\left(\Delta\dot{S}_0 + \sum_{\mathrm{out}} \dot{m}s - \sum_{\mathrm{in}} \dot{m}s + \sum_j \Delta\dot{S}_j\right) = T_0 \frac{dS_{\mathrm{tot}}}{dt} \tag{3-55}$$

On a unit-mass basis for a control volume with one inlet (1) and one outlet (2) and one reservoir (in addition to the environment)

$$i_{sf} = T_0(\Delta s_0 + s_2 - s_1 + \Delta s_R) = T_0\,\Delta s_{\mathrm{tot}} \tag{3-56}$$

Thus, as noted earlier for a closed system, a direct proportionality exists between the total irreversibility I and the total entropy change associated with the process.

3-7 CONTROL-VOLUME AVAILABILITY EXAMPLES

In this section we shall consider steady-flow processes involving turbines, compressors, nozzles, diffusers, pumps, throttling devices, and heat exchangers. For steady flow through those devices with one inlet (1) and one outlet (2) the availability balance on a unit-mass basis is

$$w_{\mathrm{act},sf} = -\sum_i Q_i \left(1 - \frac{T_0}{T}\right) + \psi_2 - \psi_1 + T_0 \sigma_{\mathrm{CV}} \tag{3-57}$$

where $T_0\sigma_{CV} = i_{CV}$. Methods for determining the second-law performance of a number of important steady-state devices are presented below.

3-7-1 Turbines

In steady state the energy balance for a gas turbine on a unit-mass basis with a negligible potential-energy change is

$$q + w = h_e - h_i + \text{ke}_e - \text{ke}_i$$

It is quite common as a first approximation to assume the expansion process is adiabatic and the kinetic-energy change is negligible. As a result, for a process proceeding reversibly from state 1 to state 2, the work is represented by $(h_{2s} - h_1)$. A graphical depiction of isentropic expansion work is shown on a *Ts* diagram in Fig. 3-9(*a*). The constant-enthalpy line h_{2s} has been extended from state 2*s* until it crosses the pressure line P_1 at state *A*. Similar to the discussion for the graphical interpretation of $\Delta\psi$ using Fig. 3-7(*a*), the shaded area shown below the pressure line P_1 and between states 1 and *A* equals $h_A - h_1 = h_{2s} - h_1$. Thus the designated area is a measure of the isentropic work for process 1-2*s*. This area representation of isentropic work is general; it applies to ideal gases, real gases, and incompressible liquids.

Example 3-6. Show mathematically for an ideal gas that the area under curve *A*-1 in Fig. 3-9(*a*) represents the isentropic work for an expansion from state 1 to state 2*s*.

Solution. The area under the constant-pressure line 1-*A* for an internally reversible process involving an ideal gas is

$$\text{Area} = \int_1^A T\,ds = \int_1^A T\left(\frac{c_p\,dT}{T} - \frac{R\,dP}{P}\right) = \int_1^A c_p\,dT \equiv h_A - h_1$$

But $h_A = h_{2s}$. Therefore the area equals $h_{2s} - h_1$, which equals w_s. The area is negative in value, since the integration is in a negative direction with respect to *s*.

The internally irreversible expansion between states 1 and 2*a* shown in Fig. 3-9(*a*) is usually evaluated in terms of the first-law adiabatic efficiency η_t, where

$$\eta_t \equiv w_a/w_s = (h_{2a} - h_1)/(h_{2s} - h_1)$$

The loss of work output (defined as a positive value) due to frictional irreversibilities is $w_a - w_s$, or $h_{2a} - h_{2s}$. This quantity, frequently called the "frictional heating" q_F, also has a graphical interpretation in Fig. 3-9(*a*). Like the preceding discussion on the area representation of isentropic work, if process 2*s*-2*a* at constant pressure P_2 is internally reversible, then the area beneath the curve is a measure of $(h_{2a} - h_{2s})$, or the loss in adiabatic work output for the irreversible process compared to the reversible case. The irreversibility *i* of process 1-2*a* is measured by $T_0\sigma_{CV} = T_0(s_2 - s_1)$. This is shown by the dark shaded area below

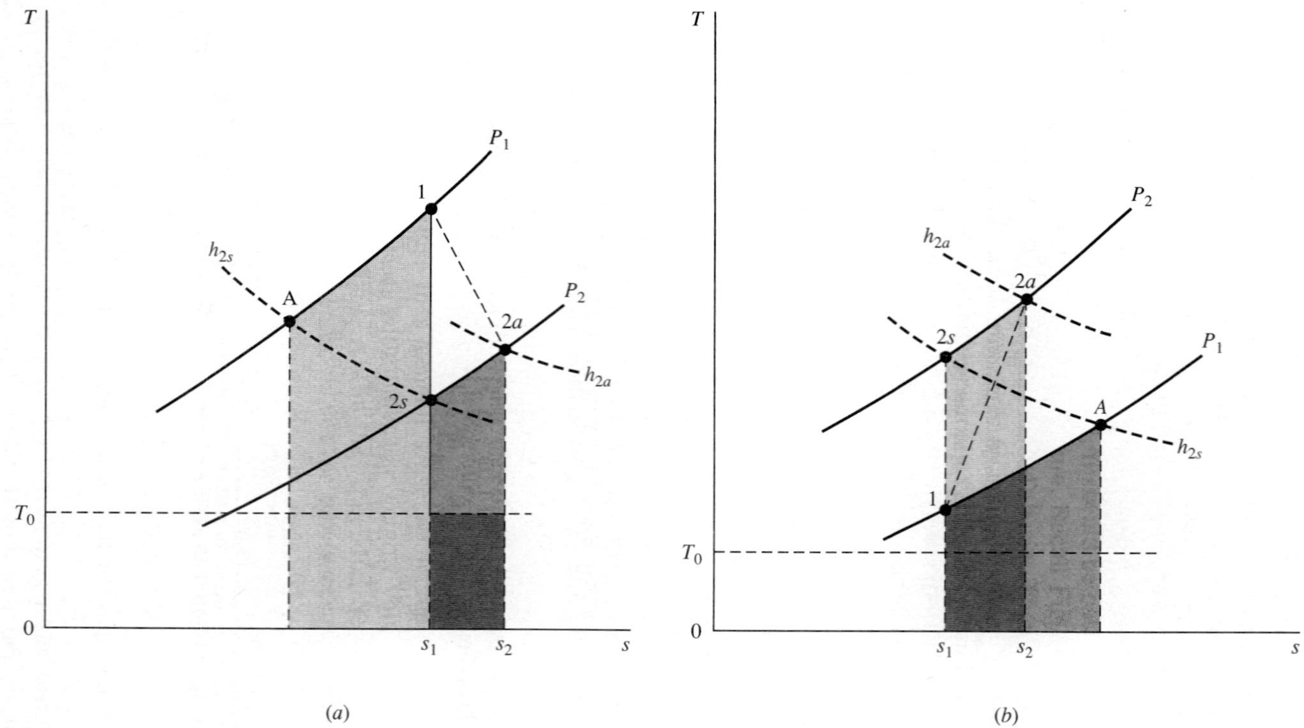

FIGURE 3-9
(*a*) Area representation of isentropic work and the loss in work due to irreversibilities in an adiabatic turbine; (*b*) area representation of isentropic work and the increase in work due to irreversibilities in an adiabatic compressor.

T_0 and between s_1 and s_2. Note that when the temperature of the fluid remains above T_0, then the exergy loss (irreversibility) is smaller than the work loss. This implies that further work output theoretically is possible for the fluid leaving at state $2a$, relative to a fluid leaving at state $2s$. This is true, since $T_{2a} > T_{2s}$ due to "frictional heating." The fluid leaving the turbine at T_{2a} theoretically could be cooled to T_{2s}, and the heat removed could be sent to a Carnot-type heat engine which rejects to the ambient at T_0 while producing some additional work. This additional theoretical quantity of work is represented by the area in Fig. 3-9(a) between lines s_1 and s_2, beneath constant-pressure line P_2 and above line T_0. Hence the minimum loss of work capability is still $T_0(s_2 - s_1)$ when the gas expands irreversibly.

The turbine steady-state availability balance on a unit-mass basis in terms of input and output is

$$\psi_e - \psi_i = w_{\text{act}} + \sum_i q_i \left(1 - \frac{T_0}{T_i}\right) - i \tag{3-57}$$

For adiabatic turbines the maximum output w_{rev} is diminished by the amount $i = T_0 \sigma_{\text{CV}}$. The only desired output on the right-hand side of Eq. (3-57) is the shaft work. Thus *turbine effectiveness* ϵ_t is defined as a measure of how well the stream availability of the fluid is converted into shaft-work output. That is,

$$\epsilon_t \equiv \frac{w_{\text{shaft}}}{\psi_e - \psi_i} = \frac{w}{w - i} \qquad \text{(turbine)} \tag{3-58}$$

where w is negative in value. Equation (3-57) also shows why heat losses from a turbine are highly undesirable, since $\Phi_{Q,i}$ losses detract from the work output.

Example 3-7. Steam enters a turbine at 40 bars, 500°C, and 140 m/s and leaves as a saturated vapor at 100°C. The measured work output is 746.0 kJ/kg. The average temperature T_b at the outer surface of the turbine where heat transfer occurs to the environment is taken as the average of the inlet and outlet steam temperatures. The potential-energy change is negligible. (a) Determine (1) the specific availability change and (2) the specific irreversibility, both in kJ/kg, for the process within the turbine, and (3) the effectiveness of the process. (b) Now, enlarge the control volume so that the local environment at 25°C is included. Find the specific availability change and the irreversibility in this new situation.

Solution

(a) (1) The specific stream availability change is given by Eq. (3-50). Using Δh, Δs, and Δke data from Prob. 2-3, we find that

$$\Delta\psi = \Delta h - T_0\,\Delta s + \Delta\text{ke} = -769.2 - 298(0.2648) - 6.6 = -854.7 \text{ kJ/kg}$$

This is the maximum possible work output for the given end states while exchanging heat solely with the environment at 25°C ($\approx$ 298 K).

(2) The specific irreversibility of the process is given by Eq. (3-53). Thus

$$i_{\mathrm{CV}} = -\Delta\psi + q\left(1 - \frac{T_0}{T_b}\right) + w_{\mathrm{act},sf}$$

$$= -(-854.7) + (-29.8)\left(1 - \frac{298}{573}\right) + (-746.0) = 94.4 \text{ kJ/kg}$$

This also is the value obtained from $i = T_0\sigma_m$. From Prob. 2-3, $\sigma_m = 0.3168$ kJ/kg · K. Hence $i_{\mathrm{CV}} = 298(0.3168) = 94.4$ kJ/kg.

(3) The change in steam availability is -854.7 kJ/kg, the availability ϕ_Q associated with the heat transfer is -14.3 kJ/kg, and the irreversibility i within the turbine is 94.4 kJ/kg. Therefore the turbine effectiveness ϵ_t is

$$\epsilon_t = \frac{w_{\mathrm{shaft}}}{\Delta\psi} = \frac{746.0}{854.7} = 0.873$$

This same result could be obtained from the relationship

$$\epsilon = 1 - \frac{\text{availability destruction and loss}}{\text{availability input}}$$

$$= 1 - \frac{i + \phi_Q}{\Delta\psi} = 1 - \frac{94.4 + 14.3}{854.7} = 0.873$$

Although the quantity ϕ_R associated with q has the potential to produce more work, it is a loss as far as turbine performance is concerned. This loss eventually represents an irreversibility of the surroundings.

(*b*) The specific stream availability change remains the same at -854.8 kJ/kg. The specific irreversibility for the enlarged control volume is found by rewriting Eq. (3-0) as

$$i_{\mathrm{tot}} = T_0\sigma_{m,\mathrm{tot}} = -\Delta\psi + q\left(1 - \frac{T_0}{T_R}\right) - w_{\mathrm{act},sf}$$

$$= -(-854.7) + 0 - (-746.0) = 108.7 \text{ kJ/kg}$$

Recall from Example 2-3 that $\sigma_{m,\mathrm{tot}} = 0.3648$ kJ/kg·K. Hence $i = T_0\sigma_{m,\mathrm{tot}} = 298(0.3648) = 108.7$ kJ/kg. Since the irreversibility within the control volume is 94.4 kJ/kg, then $i_Q = 108.7 - 94.4 = 14.3$ kJ/kg. This is the same value given by Eq. (3-21),

$$i_Q = T_0\sigma_Q = T_0 q\left(\frac{1}{T_0} - \frac{1}{T_b}\right) = 298(-29.8)\left(\frac{1}{298} - \frac{1}{573}\right) = 14.3 \text{ kJ/kg}$$

Thus the irreversibility due to heat transfer in this case is about 15 percent of that within the control volume due to frictional losses.

One method for illustrating energy and availability accounting graphically is the use of diagrams called *Sankey* or *band diagrams*. The width of the arrows entering or leaving a control volume is a quantitative measure of the designated parameter. In Fig. 3-10 a band diagram is shown for the turbine data of Example

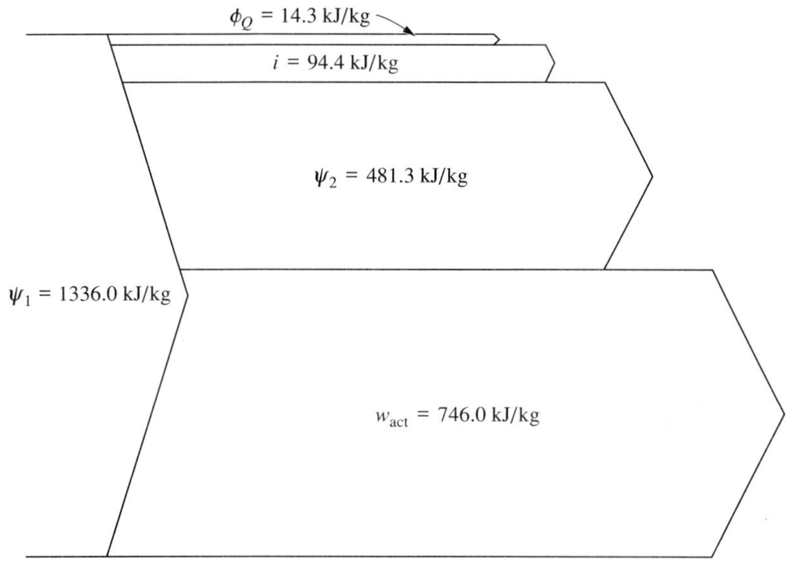

FIGURE 3-10
Band diagram showing the availability balance for an irreversible turbine. (Data refer to Example 3-5.)

3-5. Band or Sankey diagrams are especially useful in illustrating energy or exergy quantities for cyclic processes.

If the working fluid in an adiabatic turbine (or compressor) can be modeled as an ideal gas with constant c_p and k values, then it can be shown that the exergy loss i due to internal irreversibilties is only a function of the pressure ratio and the adiabatic efficiency for given values of c_p, k, and T_0. The irreversibility i equals $T_0\,\Delta s$ for an adiabatic process. Substitution of the isentropic relation $P_2/P_1 = (T_2/T_1)^{k/(k-1)}$ into the Δs relation for an ideal gas yields

$$\Delta s = c_p \ln \frac{T_{2a}}{T_1} - R \ln \frac{P_2}{P_1} = c_p \ln \frac{T_{2a}}{T_{2s}} \qquad \text{(ideal gas)}$$

Use of the turbine efficiency relation $\eta_t = (T_1 - T_{2a})/(T_1 - T_{2s})$ allows replacement of T_{2a} in the Δs expression above. The result is

$$\Delta s = c_p \ln \left[\frac{T_1}{T_{2s}} (1 - \eta_t) + \eta_t \right]$$

Making use again of the isentropic temperature-pressure relation, the irreversibility of a turbine expansion process for an ideal gas is given by

$$i_t = T_0\,\Delta s = T_0 c_p \ln \left[(1 - \eta_t) \left(\frac{P_1}{P_2} \right)^{(k-1)/k} + \eta_t \right] \qquad \text{(ideal gas)} \qquad (3\text{-}59)$$

Note that the internal availability loss does not depend in this case on the temperature level of the process.

3-7-2 Nozzles

The steady-state energy balance for nozzle flow with negligible heat transfer and potential-energy change is

$$\mathrm{ke}_2 - \mathrm{ke}_1 = \frac{V_2^2}{2} - \frac{V_1^2}{2} = h_1 - h_2$$

and the first-law nozzle efficiency η_n is

$$\eta_n = \frac{\mathrm{ke}_{2a} - \mathrm{ke}_1}{\mathrm{ke}_{2s} - \mathrm{ke}_1} = \frac{h_1 - h_{2a}}{h_1 - h_{2s}}$$

Because the equations which represent nozzle and turbine performances are analogous, the *Ts* diagram in Fig. 3-9(*a*) is applicable to both devices. Recall that the shaded area below the pressure line P_1 and between states 1 and A equals $h_{2s} - h_1$. Hence this area is also a measure of the isentropic change in kinetic energy ($\mathrm{ke}_{2s} - \mathrm{ke}_1$) for a nozzle. In addition, Eq. (3-59) is equally valid for ideal-gas flow through nozzles if η_t is replaced by η_n.

The decrease in the exit kinetic energy as a result of an irreversible expansion, in comparison to an isentropic expansion, is also depicted graphically in Fig. 3-9(*a*). This decrease, $\mathrm{ke}_{2a} - \mathrm{ke}_{2s}$, equals $h_{2s} - h_{2a}$. But as noted in the discussion on turbine performance, the area beneath the constant-pressure line P_2 for process 2*a*-2*s* is a measure of $(h_{2s} - h_{2a})$. Thus this area depicts $(\mathrm{ke}_{2a} - \mathrm{ke}_{2s})$. Also, the exergy loss i of the irreversible expansion 1-2*a* is measured by the area $T_0(s_2 - s_1)$ in Fig. 3-9(*a*). The availability balance for a nozzle is $\psi_1 = \psi_2 + i$. Consequently, the second-law nozzle efficiency ϵ_n on an input/output basis is

$$\epsilon_n = \frac{\psi_2}{\psi_1} = \frac{\psi_1 - i}{\psi_1} \tag{3-60}$$

The loss in a subsonic nozzle is usually small.

3-7-3 Compressors and Pumps

In Sec. 3-7-1 it was shown that both the isentropic turbine work between pressures P_1 and P_2, and the decrease in work output due to internal irreversibilities could be represented by areas on a *Ts* diagram. The same type of depiction is applicable to adiabatic compression processes. The paths of the isentropic compression 1-2*s* and the internally irreversible compression 1-2*a* are shown in Fig. 3-9(*b*). Again, the constant-entropy lines h_{2s} and h_{2a} are also shown. By reasoning similar to the turbine analysis the shaded area beneath the constant-pressure curve P_1 between states 1 and A represents the value of the isentropic work $w_{c,s} = h_{2s} - h_1$.

Since $w_{c,a} = h_{2a} - h_1$, the additional work input when an irreversibility is present is given by $h_{2a} - h_{2s}$. This quantity of energy is depicted by the total shaded area beneath the pressure line P_2 and between the specific entropy values of s_1 and s_{2a}. The exergy loss i is measured by the rectangular area below the T_0 isotherm and between s_1 and s_{2a}. Thus the additional work required, $h_{2a} - h_{2s}$, could be significantly larger than the irreversibility, $i = T_0(s_{2a} - s_1)$. However, not all of this additional work input is lost, because the actual exit gas temperature T_{2a} is higher than the isentropic outlet temperature T_{2s}. After leaving the compressor at (T_{2a}, P_2) the gas could be cooled reversibly to (T_{2s}, P_2), and the heat removed could be used in a Carnot-type heat engine which rejects heat at T_0. The work theoretically regained is depicted in Fig. 3-9(*b*) as the area between s_1 and s_{2a} that is beneath the P_2 line and above the T_0 line. The heat rejected by the heat engine would be measured by the area $T_0(s_{2a} - s_1)$. This additional work capability, of course, is not normally regained in practice.

In Sec. 3-7-1 an expression was developed for the irreversibility within a turbine for a fluid modeled as an ideal gas with constant c_p and k values. A similar relation can be developed for compressor flow by the same procedure, except that the compressor adiabatic efficiency η_c is used in place of η_t. The result is

$$i_c = T_0 c_p \ln\left[\frac{1}{\eta_c} - \left(\frac{1}{\eta_c} - 1\right)\left(\frac{P_1}{P_2}\right)^{(k-1)/k}\right] \qquad \text{(ideal gas)} \qquad (3\text{-}61)$$

Analogous to an ideal-gas expansion through turbines, the internal exergy loss is not dependent on the temperature level for a compression of an ideal gas.

The availability balance for a compressor or pump is written as

$$w = \psi_e - \psi_i + - \sum_i q_i \left(1 - \frac{T_0}{T_i}\right) + i$$

For a compressor or pump the work input is used to increase the stream availability. Thus the appropriate effectiveness of these two devices is

$$\epsilon_c = \epsilon_p \equiv \frac{\psi_e - \psi_i}{w_{\text{shaft}}} = \frac{w - i}{w} \qquad \text{(compressor or pump)} \qquad (3\text{-}62)$$

where w is positive. In both the above equations ψ may be replaced by $b = h - T_0 s$. For adiabatic pumps and compressors it is clear that the work added not only increases ψ, but also is used to overcome internal irreversibilities as measured by $T_0 \sigma_{\text{CV}}$.

The "frictional heating" q_F and the irreversibility i for an incompressible fluid passing adiabatically through a pump may be quantified in the following manner. Since $q_F = w_a - w_s$, $w_s = v\,\Delta P$, and $\eta_P = w_s/w_a$, we may write for the frictional heating that

$$q_F = \left(\frac{1}{\eta_P} - 1\right) v\,\Delta P \qquad \text{(pump)}$$

Recall that a pumping process is essentially one at constant temperature. Therefore the fraction of q_F which could be converted into work output is $q_F(1 - T_0/T)$. The inconvertible part is the irreversibility associated with the pump, and is given by

$$i = q_F\left(\frac{T_0}{T}\right) = \left(\frac{1}{\eta_p} - 1\right)\frac{T_0 v\,\Delta P}{T} \qquad \text{(pump)} \tag{3-63}$$

When $T = T_0$, $i = q_F$ since no work output is possible from a heat engine when the supply and ambient temperatures are equal. When fluids are pumped at temperatures higher than T_0, the exergy loss i decreases as the fluid temperature increases. The irreversibility of the fluid is also given by $T_0\,\Delta s$.

Example 3-8. Water at 0.1 MPa undergoes a pressure rise of 1.0 MPa in a pump with an adiabatic first-law efficiency of 70 percent. Determine the irreversibility of the process in kJ/kg by use of (*a*) Eq. (3-63) and (*b*) steam table entropy data, and (*c*) the second-law efficiency for an inlet temperature of 50°C if T_0 is 20°C.

Solution

(*a*) At 50°C, $v \approx v_f = 1.0121\ \text{cm}^3/\text{g}$. On the basis of Eq. (3-63)

$$i = \left(\frac{1}{\eta_p} - 1\right)\frac{T_0 v\,\Delta P}{T} = \left(\frac{1}{0.7} - 1\right)\frac{293(1.0121)(1.0)}{323}\,\frac{\text{cm}^3\cdot\text{MPa}}{\text{g}} = 0.393\ \text{kJ/kg}$$

(*b*) From a computer-based steam table we find that at $P_1 = 0.1$ MPa and $P_2 =$ 1.1 MPa

$$T_1 = 50°\text{C},\ h_1 = 209.396\ \text{kJ/kg},\ s_1 = 0.7038\ \text{kJ/kg·K},\ \psi_1 = 5.9994\ \text{kJ/kg}$$

$$T_{2s} = 50.04°\text{C},\ h_{2s} = 210.433\ \text{kJ/kg},\ s_2 = 0.7038\ \text{kJ/kg·K},\ v_2 = 1.0117\ \text{cm}^3/\text{g}$$

The actual outlet enthalpy for $\eta_p = 0.70$ is

$$h_{2a} = h_1 + \frac{h_{2s} - h_1}{\eta_p} = 209.396 + \frac{210.433 - 209.346}{0.70} = 210.877\ \text{kJ/kg}$$

At this enthalpy and a pressure of 1.1 MPa the steam table yields

$$T_{2a} = 50.15°\text{C},\ s_{2a} = 0.7052\ \text{kJ/kg·K},\ v = 1.0117\ \text{cm}^3/\text{g},\ \psi_{2a} = 7.0616\ \text{kJ/kg}$$

Therefore

$$i = T_0\,\Delta s = 293.15(0.7052 - 0.7038) = 0.410\ \text{kJ/kg}$$

There is roughly a 4 percent difference in the answers for i. Most of this is due to the lack of significant figures in the evaluation of Δs from the tables. Note that the tabular data predict an actual temperature rise of 0.15°C. Hence the assumption leading to Eq. (3-63) that T is essentially a constant is quite good.

(*c*) Using tabular data, we find the second-law efficiency by

$$\epsilon_p = \frac{\psi_{2a} - \psi_1}{w_{\text{act}}} = \frac{7.0616 - 5.9994}{210.877 - 209.396} = \frac{1.062}{1.481} = 0.717$$

or

$$\epsilon_p = \frac{w - i}{w} = \frac{1.481 - 0.410}{1.481} = 0.723$$

The small difference is due to round-off error.

3-7-4 Throttling

A throttling process is one for which $h_1 = h_2$ and the availability balance reduces to $i = -\Delta\psi$. Since the process is adiabatic, $\sigma_{\text{CV}} = s_2 - s_1$ and $\delta i = T_0\, ds$. The basic property identity, $dh = T\, ds + v\, dP$, reduces to $ds = -v\, dP/T$ for the process. Hence

$$\delta i = \frac{-T_0 v\, dP}{T} \tag{3-64}$$

and

$$\delta \dot{I} = \frac{-\dot{m} T_0 v\, dP}{T} = \frac{-\dot{V} T_0\, dP}{T}$$

where $\dot{V}$ is the volume flow rate. According to Eq. (3-64), the irreversibility during throttling for a given dP is proportional to (v/T). The effect of (v/T) on i depends on the nature of the substance. For example, when an incompressible fluid is throttled, the temperature change is extremely small and v is a constant. Under this condition the temperature T in Eq. (3-64) is essentially T_1. Hence the specific irreversibility becomes

$$i = T_0\, \Delta s \approx \frac{T_0 v(P_1 - P_2)}{T_1} \qquad \text{(incompressible)}$$

This equation shows a strong dependence of i on the temperature. When a liquid is throttled, the fluid temperature should be maintained as high as possible in order to minimize the availability loss.

When an ideal gas is throttled, the quantity v/T in Eq. (3-64) may be replaced by R/P. Hence

$$i = -T_0 \int \frac{R\, dP}{P} = RT_0 \ln \frac{P_1}{P_2} \tag{3-65}$$

Interestingly, i for the throttling of an ideal gas is *independent* of the temperature level. This is not true for a real gas. As a first approximation, for a real gas $Pv = ZRT$. If we replace v/T in Eq. (3-64) by ZR/P, then

$$i = -\int_1^2 \frac{ZRT_0}{P}\, dP \approx Z_{\text{avg}} RT_0 \ln \frac{P_1}{P_2}$$

Qualitatively the irreversibility of a real gas is less than that for an ideal gas for the same pressure values when Z_{avg} is less than unity. Also, Z_{avg} is a function of temperature. Thus the level of temperature affects i, unlike the situation for an ideal gas.

If the second-law efficiency of a throttling process is defined in terms of output/input, then

$$\epsilon = \frac{\psi_2}{\psi_1} = \frac{\psi_1 - i}{\psi_1} \tag{3-66}$$

The magnitude of ϵ depends on the temperature and pressure levels involved and on the nature of the fluid.

Example 3-9. Steam is throttled from 100 to 60 bars for initial temperatures of (1) 480°C and (2) 360°C. (*a*) Determine the irreversibility, in kJ/kg, for the two cases. (*b*) Discuss how the average temperature level of the fluid affects the magnitude of the irreversibility if $T_0 = 25$°C. (*c*) Determine the effectiveness of process (1).

Solution
(*a*) The steam table data for the four states are listed below.

(1) In: $P_1 = 100$ bars, $T_1 = 480$°C, $h_1 = 3321.4$ kJ/kg, $s_1 = 6.5282$ kJ/kg · K
Out: $P_2 = 60$ bars, $T_2 = 458.1$°C, $h_2 = 3321.4$ kJ/kg, $s_2 = 6.7464$ kJ/kg · K
(2) In: $P_1 = 100$ bars, $T_1 = 360$°C, $h_1 = 2962.1$ kJ/kg, $s_1 = 6.0060$ kJ/kg · K
Out: $P_2 = 60$ bars, $T_2 = 323.0$°C, $h_2 = 2962.1$ kJ/kg, $s_2 = 6.2007$ kJ/kg · K

The irreversibility $i = T_0 \Delta s$. Therefore, for the two cases,

$$(1)\ i = 298.15(6.7464 - 6.5282) = 65.06 \text{ kJ/kg}$$
$$(2)\ i = 298.15(6.2007 - 6.0060) = 58.05 \text{ kJ/kg}$$

If steam were an ideal gas, the irreversibility would be 70.27 kJ/kg in both cases.

(*b*) The average temperature for case (1) is 469°C and for case (2) it is 341.5°C. Thus for the conditions chosen the irreversibility increases with increasing average temperature. At the reduced temperatures and pressures for the two processes, the Z chart indicates that Z_{avg} for case (1) is greater than that for case (2).

(*c*) From the data for part (1)

$$\psi_1 = (3321.4 - 105.0) - 298.15(6.5282 - 0.3674) = 1379.6 \text{ kJ/kg}$$
$$\psi_2 = (3321.4 - 105.9) - 298.15(6.7464 - 0.3674) = 1314.5 \text{ kJ/kg}$$

Therefore $\epsilon = 1314.5/1379.6 = 0.953$. In this case the loss in availability is reasonably small.

3-7-5 Heat Exchange without Mixing

Figure 3-11 shows a steady-state heat exchanger with mass flow rates $\dot{m}_h$ and $\dot{m}_c$ for the hot and cold streams, respectively. Three types of availability losses occur in such devices. These are: (1) loss due to heat exchange between the fluids across a finite temperature difference, (2) loss due to fluid friction, and (3) loss due to

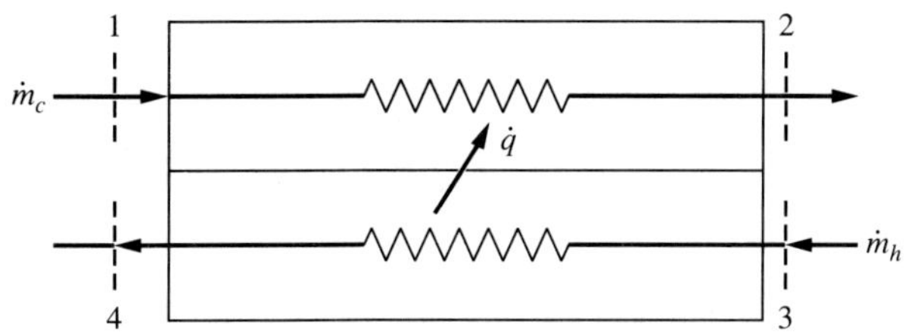

FIGURE 3-11
Schematic for a two-stream heat exchanger without mixing.

heat exchange with the environment. External heat loss (through an insulated outer surface) and changes in kinetic and potential energies are usually neglected. The energy and availability rate equations for this situation are

$$0 = \dot{m}_c(h_2 - h_1) + \dot{m}_h(h_4 - h_3)$$
$$0 = \dot{m}_c(\psi_2 - \psi_1) + \dot{m}_h(\psi_4 - \psi_3) + \dot{I}_{CV}$$

When case (1) above is the only loss, the irreversibility due solely to heat exchange can be shown on a $T\dot{S}$ diagram, similar to that for heat exchange between reservoirs shown in Fig. 3-4. The specific isobaric processes shown on the $T\dot{S}$ diagram of Fig. 3-12 represent the cooling of a hot gas for stream h from state 3

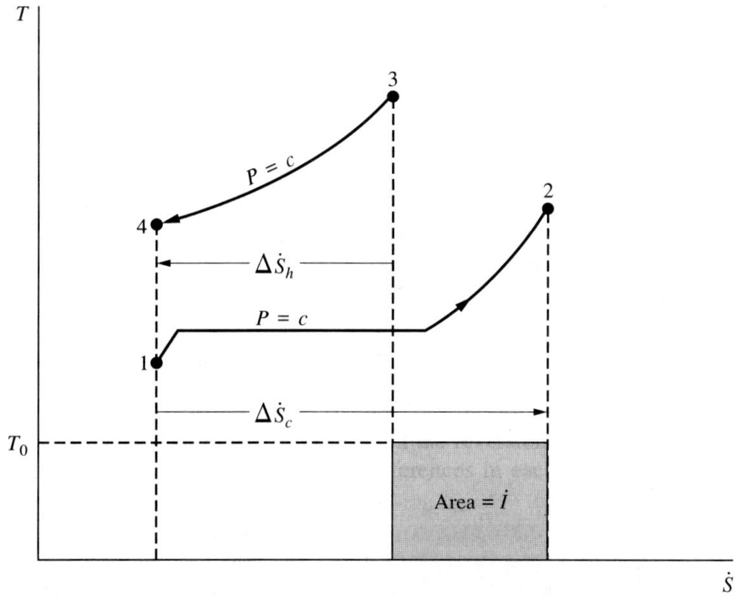

FIGURE 3-12
Graphical representation of the irreversibility $\dot{I}$ due to heat transfer across a finite temperature difference in a steady-state heat exchanger without mixing.

to state 4, and the heating of compressed liquid water to a superheated vapor for cold stream c from state 1 to state 2. If the heat exchanger is perfectly insulated, the areas under isobaric lines 1-2 and 3-4 are equal. Since $\dot{I} = T_0 \sum \Delta \dot{S}$, the gray shaded area on the figure represents the internal irreversibility $\dot{I}$. As the process profiles on the $T\dot{S}$ plot approach each other (as is possible under countercurrent flow), the area representation of $\dot{I}$ decreases. This occurs because the irreversibility associated with a finite temperature difference between the streams is reduced. The irreversibility can also be represented by a vertical line length on an $\dot{H}\dot{S}$ diagram [4].

The effectiveness of such a heat exchanger may be defined in two forms. The first of these, expressed as product/supply, is measured by the increase in the availability of the cold stream divided by the negative change in availability of the hot stream on a rate basis. That is,

$$\epsilon = \frac{\dot{m}_c(\psi_2 - \psi_1)}{-\dot{m}_h(\psi_4 - \psi_3)} \qquad \text{(heat exchanger)} \qquad (3\text{-}67)$$

Recall that for a fluid at $T < T_0$ the fluid gains availability when a heat loss occurs. In a heat exchanger in this situation the heating fluid gains availability while the stream being heated loses availability.

A second form of heat exchanger effectiveness is a measure of the availability outputs divided by the availability inputs. In this format the availability balance above can be rearranged into the form

$$\dot{m}_c\psi_1 + \dot{m}_h\psi_3 = \dot{m}_c\psi_2 + \dot{m}_h\psi_4 + \dot{I}_{\text{CV}}$$

In this format states 1 and 3 are inputs and states 2 and 4 are outputs. Hence the effectiveness may also be defined as

$$\epsilon = \frac{\dot{m}_c\psi_2 + \dot{m}_h\psi_4}{\dot{m}_c\psi_1 + \dot{m}_h\psi_3} \qquad \text{(heat exchanger)} \qquad (3\text{-}68)$$

The first form of ϵ for a heat exchanger is usually preferred.

3-7-6 Energy Exchange by Mixing Pure Fluids

Common examples include the direct mixing of two gas streams or two liquid streams and the open feedwater heater in steam power cycles. A schematic for the overall process is shown in Fig. 3-13. In an open feedwater heater the fluids are the same substance. Hence the mass, energy, and availability balances, when neglecting external heat transfer and changes in kinetic and potential energies, are

$$\begin{aligned} 0 &= \dot{m}_1 + \dot{m}_2 + \dot{m}_3 \\ 0 &= \dot{m}_3 h_3 - \dot{m}_1 h_1 - \dot{m}_2 h_2 \\ 0 &= \dot{m}_3\psi_3 - \dot{m}_1\psi_1 - \dot{m}_2\psi_2 + \dot{I}_{\text{CV}} \end{aligned}$$

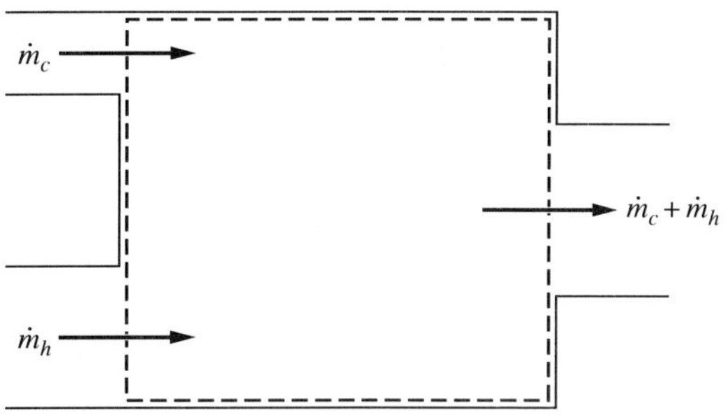

FIGURE 3-13
Schematic of a direct-mixing heat exchanger.

If we arbitrarily define state 1 in the figure as the colder entering fluid with mass rate $\dot{m}_c$, and make use of the mass balance, then the availability balance may be rewritten as

$$\dot{m}_h(\psi_3 - \psi_2) = \dot{m}_c(\psi_3 - \psi_1) + \dot{I}_{\mathrm{CV}}$$

The effectiveness may be defined as the gain in availability of the colder entering stream divided by the change in the availability of the hotter entering stream which loses energy to the colder stream. Thus

$$\epsilon = \frac{\dot{m}_1(\psi_3 - \psi_1)}{\dot{m}_2(\psi_2 - \psi_3)} \qquad \text{(mixing)} \tag{3-69}$$

Like the previous discussion for heat exchangers, an alternate definition for ϵ is possible. The availability balance can also be written as

$$\dot{m}_1\psi_1 + \dot{m}_2\psi_2 = (\dot{m}_1 + \dot{m}_2)\psi_3 + I_{\mathrm{CV}}$$

This is another input/output format. In this case the effectiveness for the mixing process becomes

$$\epsilon = \frac{\text{output}}{\text{input}} = \frac{(\dot{m}_1 + \dot{m}_2)\psi_3}{\dot{m}_1\psi_1 + \dot{m}_2\psi_2} \tag{3-70}$$

Either format is acceptable.

PROBLEMS

3-1. A transfer of 1000 kJ of heat occurs between thermal reservoirs at 600 and 360 K. (a) Determine the magnitude, in kilojoules, and the direction of the availability transfer for the two reservoirs. (b) Now, consider the transfer of the same energy between thermal reservoirs at 290 and 260 K. Determine the magnitude and direction of the availability transfers for these two reservoirs. (c) Compare the direction of the heat transfer versus the availability transfer for the two processes. T_0 is 300 K.

3-2. Consider the transfer of 10 kJ of heat between thermal reservoirs at 500 and 400 K. (a) Determine the magnitude, in kilojoules, and the direction of the availability transfer for the two reservoirs. (b) Now, consider the transfer of 10 kJ of heat between thermal reservoirs at 280 and 250 K. Determine the magnitude and direction of the availability transfers for these two reservoirs. (c) Compare the direction of heat transfer versus availability transfer for the two processes. The environmental temperature is 298 K.

3-3. In case A two thermal reservoirs at 500 and 360 K exchange a quantity of heat Q. In case B the high- and low-temperature reservoirs are at T_B and 450 K, and in case C these temperatures are T_C and 600 K, respectively. Determine the percent increase in ΔT permitted between the two reservoirs (a) in case B relative to case A and (b) case C relative to case A so that the exergy loss will be the same in all three cases.

3-4. A 0.50-m^3 partially evacuated rigid tank contains atmospheric air at 0.04 MPa and 298 K. The environmental state is 0.10 MPa and 298 K. Determine the availability of the air within the tank, in kilojoules.

3-5. A rigid tank contains 500 kg of liquid water initially at 87°C. Due to a heat loss to the environment, the temperature falls to 47°C. Determine the irreversibility of the overall process, in kilojoules, if $T_0 = 20$°C.

3-6. Steam is maintained in a closed system at 80 bars and 440°C.

a. Determine the availability, in kJ/kg, if T_0 is 25°C and P_0 is 1 bar.

b. The steam is now cooled in a rigid tank until the pressure drops to 40 bars by transferring heat to the environment. Determine the percent loss in availability associated with the change in the state of the fluid, in kJ/kg.

3-7. Air at 1.1 bars is cooled at constant pressure from 440 to 20°C, with all the heat given up by the closed system appearing in the environment, which has a temperature of 20°C and a pressure of 1.0 bar. Determine (a) the availability change associated with the heat-transfer process and (b) the availability change of the air, in kJ/kg.

3-8. Nitrogen with a mass of 0.040 kg initially at 2.0 bars and 27°C is contained in a well-insulated tank. A battery passes a current through a resistor within the tank until the pressure is 2.8 bars. Neglecting energy storage in the resistor and walls of the tank, determine (a) the actual work performed, (b) the minimum work required for the given change of state, and (c) the irreversibility of the process, all answers in kilojoules, and (d) the second-law efficiency. Finally, (e) determine the irreversibility and effectiveness if heat transfer from a reservoir at 600 K is used instead of electrical work. Let $T_0 = 27$°C and $P_0 = 1$ bar. Use tabular ideal-gas data for nitrogen.

3-9. A rigid tank contains 0.30 kg of carbon dioxide (CO_2) at 1.1 bars and 27°C. An electrical resistor-battery system is used to raise the temperature of the gas adiabatically to 267°C. The gas is then returned to its initial state by rejecting heat to the surroundings at 17°C and 1.0 bar. Determine (a) the electrical resistor work done on the gas initially, in kilojoules, (b) the reversible work for the CO_2-atmosphere combination during the work process, in kilojoules, and (c) the effectiveness of the work process. Then find (d) the reversible work associated with the cooling process, in kilojoules, and (e) the fraction of the original work input that could be regained by use of a Carnot heat engine during the heat-rejection process. Neglect energy storage in the resistor and walls, and use the CO_2 ideal-gas table for data.

3-10. Refrigerant-134a (R-134a) is compressed in a piston-cylinder device from a saturated vapor at −8°C to a final pressure of 9 bars. The process is adiabatic and the compressor efficiency is 78 percent. Determine (a) the actual work required, in kJ/kg, (b) the actual outlet temperature, in degrees Celsius, (c) the minimum work required, in kJ/kg, for the actual state found in part (b), (d) the irreversibility, in kJ/kg, and (e) the effectiveness. The environmental state is 20°C and 1 bar.

3-11. One kilogram of air initially at 1 bar, 27°C, is contained in a well-insulated tank. An impeller inside the tank is turned by an external mechanism until the pressure is 1.2 bars. T_0 is 27°C and P_0 = 1 bar. Determine (a) the actual work performed, in kilojoules, (b) the minimum required work input for the given change of state, in kilojoules, (c) the irreversibility by two methods, in kJ/kg, and (d) the second-law effectiveness. Now, remove the insulation and use heat transfer from a thermal reservoir at 500 K instead of work to change the state of the air. For this alternate process, (e) determine the irreversibility and the second-law effectiveness.

3-12. A storage battery is capable of delivering 1 kW·h of energy. What volume of air stored in a tank at 27°C and 20 bars is required theoretically to have the same work capability, in cubic meters? The state of the environment is 27°C and 1 bar.

3-13. A storage battery is capable of delivering 8 kW·h of energy. Find the volume of air required in a tank at 27°C and 40 bars theoretically to have the same work capability, in cubic meters. The state of the environment is 27°C and 1 bar.

3-14. A tank with a volume of 0.40 m^3 contains nitrogen gas at 10 bars and 300 K. The surrounding atmosphere is at 0.98 bar and 300 K.

a. Determine the availability (exergy) of the nitrogen, in kilojoules.
b. If the nitrogen now undergoes a free expansion until its volume is tripled, determine the change in the availability (exergy) of the gas and the final value of the exergy, in kilojoules.
c. Determine the effectiveness of the process.
d. If the original free expansion occurs until the volume is six times greater, determine the same quantities as in parts (b) and (c).

3-15. A tank with a volume of 0.30 m^3 contains air at 6 bars and 300 K. The surrounding atmosphere is at 0.95 bar and 300 K. Determine

a. the availability of the air, in kilojoules
b. the change in the availability of the air if it now undergoes a free expansion until its volume is doubled, in kilojoules
c. the irreversibility of the process in part (b), in kilojoules
d. the second-law effectiveness of the process
e. Now, if the original air undergoes a free expansion until the volume is tripled, determine the final availability of the air, in kilojoules, and the second-law effectiveness.

3-16. Steam is contained in a closed, insulated rigid tank at a state of 1.0 bar and 100°C. Paddle-wheel work is done on the steam until the pressure reaches 1.5 bars. If the values of T_0 and P_0 are 20°C and 1 bar, respectively, determine (a) the paddle-wheel work required, (b) the change in exergy of the fluid, (c) the irreversibility or lost work of the process, all in kJ/kg, and (d) the second-law effectiveness of the process. Finally, (e) determine the second law effectiveness if heat transfer from a reservoir at 600 K is used instead of paddle-wheel work.

3-17. Saturated-water vapor is contained in a rigid tank at 90°C. Energy is added until the pressure reaches 1.5 bars by means of (a) heat addition from a reservoir at 900 K and (b) electrical resistor work. The environmental state is 25°C and 1.0 bar. Determine the second-law effectiveness of the two processes.

3-18. A kilogram of copper is cooled (1) from 25 to 10°C and (2) from 50 to 35°C. In either case assume the specific heat of the metal is constant at 0.384 kJ/kg · °C, T_0 = 27°C and P_0 = 1 bar.

a. Determine the exergy change of the copper, in kilojoules, for the two cases.

b. One of your answers in part (a) should be positive, and the other will be negative. Explain why in one case the exergy increased due to the cooling process, whereas in the other case the exergy decreased.

3-19. Develop an equation for the dimensionless nonflow availability, ϕ/cT_0, of an incompressible substance as a function of the dimensionless temperature, T/T_0. Then plot ϕ/cT_0 versus $\ln(T/T_0)$ for T/T_0 values of 0.01, 0.04, 0.1, 0.4, 1, 3, and 5. Assume the specific heat c is a constant.

3-20. Water in a closed tank undergoes a temperature rise from T_i to $T_i + \Delta T$ by using an electrical resistance heater, as in a home water heater. Assume the water of mass m is incompressible and that the specific heat c is a constant.

a. Derive an equation for the second-law effectiveness of the process in terms of T_i, T_0, ΔT, and the energy efficient η. The energy efficient accounts for a heat loss Q from the tank due to imperfect insulation such that $\eta = \Delta U_{water}/W_{elec}$.

b. A water heater receives 4000 W·h of electrical work, 5 percent of which appears as a heat loss through the heater walls. As a result, the water temperature rises from 20 to 50°C. For environmental conditions of 20°C and 1.0 bar, determine (1) the availability change for the water, in kilojoules, and (2) the effectiveness of the process.

3-21. Reconsider Example 3-5 by changing 18 kg to 2 kg of lead, other conditions remaining the same.

a. Determine the total change in availability and the second-law efficiency.

b. Compare the second-law efficiency to that obtained in Example 3-5, and comment.

3-22. A tank which is thermally insulated on the outside is divided internally into two sections by a thermally insulating wall. One section of the tank initially contains an oil with a density of 761 kg/m^3 and a specific heat of 2.60 kJ/kg · K. The volume of the oil is 1.80 m^3 and its initial temperature is 230°C. On the other side of the tank there is water initially at 10°C and 1.0 bar with a volume of 1.40 m^3. The insulating wall separating the two substances is now replaced by a thermally conducting wall. The environmental state is 10°C and 1.0 bar. Find

a. the total entropy change for the process due to heat transfer between the substances, in kJ/K

b. the change in availability of the oil and of the water, in kilojoules

c. the net change in availability for the overall process, in kilojoules

d. the irreversibility of the overall process, in kilojoules

e. the second-law effectiveness of the process

3-23. A tank which is thermally insulated on the outside is divided internally into two compartments by a wall. One compartment contains oil with a specfic gravity of 0.87 and a specific heat of 2.260 kJ/kg · °C. The volume of the oil is 2.90 m^3 and its initial temperature is 180°C. The other compartment holds water initially at 20°C with a volume of 1.60 m^3. Heat is now exchanged between the two fluids until thermal equilibrium is reached. If the environmental conditions are 20°C and 1 atm, determine

a. the entropy generation for the process, in kJ/K
b. the changes in exergy (availability) of the oil and the water, in kilojoules
c. the irreversibility or lost work for the process, in kilojoules

3-24. Steam enters a turbine at 100 bars, 400°C, and expands to 10 bars, 200°C, in a steady-state adiabatic process. The ambient state is 1 bar and 27°C. Disregard changes in kinetic and potential energy. Determine, in kJ/kg,

a. the actual work
b. the maximum work output for the given end states
c. the work output if the process were isentropic from the same initial state to the same final pressure
d. the irreversibility of the actual process
e. Then determine the effectiveness of the process.

3-25. Air is expanded adiabatically in a steady-flow turbine from 6.0 bars, 560 K, to 1.0 bar. Changes in kinetic and potential energies are negligible, and the environment conditions are 1 bar and 27°C. If the adiabatic turbine efficiency is 88.0 percent, use the air table for data to determine

a. the actual work of expansion, in kJ/kg
b. the actual outlet temperature, in kelvins
c. the optimum (maximum) work output for the same end state as in part (b)
d. the irreversibility or lost work, in kJ/kg
e. the second-law efficiency

3-26. Steam enters a turbine at 30 bars, 400°C, and expands to 1 bar, 120°C, in a steady-state adiabatic process. The ambient pressure and temperature are 1 bar and 27°C. Disregard changes in kinetic and potential energies. Determine (a) the actual work delivered, (b) the maximum possible work output for the given states, (c) the theoretical work if the expansion were isentropic from the initial state to the final given pressure, and (d) using two methods, the irreversibility, all answers in kJ/kg. Then (e) find the effectiveness.

3-27. Nitrogen gas enters an insulated nozzle at 140 kPa with a negligible velocity. At the exit of the nozzle the state of the fluid is 120 kPa and 27°C, the flow area is 10.0 cm^2, and $\dot{m}$ is 0.20 kg/s. Find (a) the irreversibility of the process, in kJ/kg, and (b) the effectiveness of the process. T_0 is 25°C and P_0 is 1 bar.

3-28. Steam enters an adiabatic nozzle at 30 bars and 320°C and leaves at 15 bars and 535 m/s. Neglecting the inlet velocity, determine (a) the irreversibility, in kJ/kg, and (b) the second-law efficiency. T_0 is 25°C and P_0 is 1 bar.

3-29. Air is compressed adiabatically in steady flow from 1 bar, 17°C, to 4 bars. Neglect changes in kinetic and potential energies. The required work input is 180 kJ/kg. Determine (a) the irreversibility of the process, in kJ/kg, (b) the change in stream availability, in kJ/kg, and (c) the second-law efficiency. The atmospheric state is 1 bar and 17°C.

3-30. Air is compressed adiabatically in steady flow from 2 bars, 27°C, to 4 bars. Neglect changes in kinetic and potential energies. The measured work input is 80.6 kJ/kg, $P_0 = 1$ bar, and $T_0 = 27$°C. Determine

a. the irreversibility of the process, in kJ/kg
b. the change in stream availability, in kJ/kg
c. the effectiveness of the process

3-31. Derive equations for the frictional heating q_F in (a) a turbine and (b) a compressor where the fluid is modeled as a ideal gas with a constant specific-heat ratio k. Both equations should be expressed in terms of k, R, T_1, (P_2/P_1), and either η_t or η_c.

3-32. Water enters a pump with an adiabatic efficiency of 75 percent at 1 bar. The environmental temperature is 20°C. If the inlet temperature is (a) 20°C, and (b) 50°C, determine (1) the "frictional heating," (2) the irreversibility, based on $\delta w_{\text{rev}} = v\,dP$, (3) the irreversibility directly from accurate steam table data, and (4) the second-law effectiveness for an exit pressure of 7.0 bars.

3-33. Reconsider the water pump in Example 3-8. Determine the same quantities for inlet fluid temperatures of (a) 20°C and (b) 90°C, and (c) compare and comment on the variation of the irreversibility and effectiveness as a function of temperature.

3-34. Saturated liquid R-134a at 6 bars is throttled to 2 bars. The state of the environment is 1 bar and 27°C. Determine

a. the entropy generation, in kJ/kg · K
b. the irreversibility or lost work for the process, in kJ/kg
c. the second-law efficiency

3-35. Carbon dioxide enters an expansion valve at 40 bars and 400 K and is throttled to 20 bars. Determine (a) the irreversibility of the process, in kJ/kg, and (b) the second-law efficiency. The ambient state is 1 bar and 27°C.

3-36. Liquid water undergoes two throttling processes from 50 to 25 bars. The initial temperatures are (a) 180°C, and (b) 100°C. Estimate the irreversibility in each case, in kJ/kg, and determine whether the greater loss occurs at higher or lower temperatures. T_0 is 290 K and P_0 is 1 bar.

3-37. Refrigerant-134a is throttled from 10 to 2.8 bars for initial temperatures of (1) 50°C and (2) 100°C. Determine (a) the initial Z values and (b) the irreversibility, in kJ/kg, for both cases. (c) Discuss the variation of the irreversibility with temperature in terms of the Z data found. T_0 is 25°C.

3-38. Saturated R-134a with 50 percent quality enters an expansion valve at 8 bars and leaves at 2 bars. Determine (a) the entropy generation, in kJ/kg · K, (b) the irreversibility of the adiabatic throttling process, in kJ/kg, and (c) the second-law efficiency. The ambient state is 1 bar and 25°C.

3-39. Steam is condensed on the outside of some heat exchanger tubes by passing air through the inside of the tubes. The air enters at 1.30 bars, 20°C, and 15 m/s, and exits at 90°C. The steam enters at 3 bars and 200°C with a mass flow rate of 7.0 kg/min and leaves as a saturated liquid at 3 bars. Determine (a) the total entropy change for the heat exchanger process, in kJ/K · min, (b) the availability change for each fluid and for the overall process, in kJ/min, and (c) the second-law efficiency. Then (d) repeat parts (a) and (b) if the air leaves the exchanger at 1.20 bars, rather than at the initial pressure of 1.30 bars. (e) Determine the irreversibility, in kJ/min, for the processes described in parts (a) and (d). Assume T_0 is 20°C.

3-40. In the condenser of a steam power plant the steam from the turbine enters the condenser at 0.10 bar with a quality of 95 percent and leaves at the same pressure as a saturated liquid. The steam is condensed by transferring heat to a stream of cooling water which enters at 1.3 bars and 5°C and leaves the heat exchanger at 1.2 bars and 25°C. The environmental temperature is 5°C. Determine

a. the change in stream availability (exergy) of the condensing steam, in kJ/kg
b. the mass of cooling water circulated through the condenser per unit mass of steam condensed, kg/kg

c. the change in stream exergy of the cooling water, in kJ/kg *steam*
d. the ratio of the change in exergy, cooling water to steam flow
e. the irreversibility or lost work for the process, in kJ/kg *steam*

3-41. A steady-flow heat exchanger is employed to cool oil from 120 to 50°C. The oil flow rate is 800 kg/h and its specific heat is 2.30 kJ/kg · K. The other fluid is water which enters and leaves at 7 bars with a flow rate of 1500 kg/h. $T_0 = 10°C$.

a. If the water enters at 10°C and the velocities of both flow streams are small, determine the total entropy change for the heat exchange process in kJ/K · h.
b. If the heat exchanger in part (a) is operated in parallel flow, determine the availability change of each fluid and the total availability change in kJ/h.
c. Determine the effectiveness of the process.
d. Consider now that the inlet water temperature is raised to 40°C. Determine the total availability change for the process in kJ/h.

3-42. Water and steam enter an insulated mixing chamber in steady flow. The liquid water enters at 20°C and 3 bars with a rate of 100 kg/min. The steam enters at 320°C and 3 bars, and the mixture leaves the chamber at 130°C and 3 bars. Determine (a) the total availability change for the process, in kJ/min, (b) the irreversibility of the adiabatic process, in kJ/min, and (c) the second-law efficiency. $T_0 = 20°C$ and $P_0 = 1$ bar. (See Prob. 2-20.)

3-43. Steam at 7 bars and 320°C and liquid water at 7 bars and 40°C enter an adiabatic mixing tank in steady flow, and the mixture leaves as a saturated vapor at 7 bars. Determine (a) the change in stream availability (exergy) for the overall process, in kJ/min, and (b) the effectiveness of the process if $T_0 = 22°C$ and the steam enters at 200 kg/min.

3-44. Two streams of nitrogen enter an adiabatic mixing chamber at 20 bars. One stream is a saturated liquid and the second stream is a superheated vapor at 300 K. The mixture leaves as a saturated vapor at 20 bars. Determine

a. the availability ψ of each flow stream, in kJ/kg
b. the second-law effectiveness of the process by two methods

3-45. Show that the minimum work required to evacuate a tank of volume V filled with atmospheric air at P_0 and T_0 is P_0V.

3-46. Reconsider the transient flow Prob. 1-31. An insulated tank with a volume of 0.50 m^3 contains air at 1.0 bar and 25°C. The tank is connected through a valve to a line which carries air continuously at 7.0 bars and 177°C. The valve is opened and air is allowed to flow into the tank until the pressure reaches 5.0 bars. Use constant specific-heat data to determine (a) the amount of mass which has entered, in kg, (b) the final temperature of the air in the tank, in degrees Celsius, and (c) the irreversibility of the process, in kilojoules, if $T_0 = 25°C$.

3-47. When an ideal gas flows from a line into an evacuated insulated tank, $h_{\text{line}} = u_{\text{tank}}$. For constant specific heats this reduces to $T_{\text{tank}} = kT_{\text{line}}$. Now, consider atmospheric air at T_0 and P_0 flowing into an evacuated insulated tank of volume V. The process stops when the pressure in the tank reaches P_0.

a. Derive an expression solely in terms of P_0, V, and k for the reversible useful work $W_{\text{rev},u}$ associated with the process.
b. Estimate the reversible useful work per unit volume of tank, in kJ/m^3, for atmospheric air at 1 bar and 25°C.

REFERENCES

1. Gouy, G., "Sur l'Energie Utilisable," *J. Phys.*, **8**:501–518, 1989.
2. Stodola, A., *Steam Turbines*, translated by L. C. Lowenstein, Van Nostrand, New York, 1905.
3. Rant, Z., "Exergie ein Neues Wort für Technische Arbeitfähigkeit," *Ingenieurwes.*, **22**:36, 1956.
4. Szargut, J., D. R. Morris, and F. R. Steward, *Exergy Analysis of Thermal, Chemical, and Metallurgical Processes*, Hemisphere Publishing Corp., New York, 1988.

CHAPTER 4

AVAILABILITY ANALYSIS OF CYCLES

The availability analysis of processes involving closed and steady-state systems is introduced in Chap. 3. In this chapter availability analysis is applied to simple theoretical and practical cycles.

4-1 SIMPLE CYCLES

In Chap. 2 the concept of entropy generation was applied to the study of heat engines, refrigerators, and heat pumps operating between temperature (heat) reservoirs. Entropy production σ occurs in these devices due to either irreversible heat transfer or irreversibilities within the device itself. A comparison of the σ values for the sources of irreversibility can be used to determine which source is the major contributor to the degradation of performance of the device. The next major step is to analyze these uncomplicated cyclic closed systems in terms of the availability Φ and the irreversibility I.

Recall that the total availability Φ_Q associated with various heat transfers Q_i is given by

$$\Phi_Q \equiv \sum_{i=1}^{n} Q_i \left(1 - \frac{T_0}{T_i}\right) \tag{3-16}$$

and for the special application of heat transfer Q_R to or from a single reservoir at $T_i = T_R$,

$$\Phi_{Q,R} \equiv Q_R \left(1 - \frac{T_0}{T_R}\right) \qquad \text{(heat reservoir)} \tag{3-18}$$

Thus any heat transfer across a system boundary at T_i may affect the system's potential to produce work if $T_i \neq T_0$. In terms of Eq. (3-12) the irreversibility for a cyclic closed system becomes

$$I = W_{\text{act}} - W_{\text{rev}} = W_{\text{act}} + \sum \Phi_{Q,i} \tag{4-1}$$

where the "+" sign is required, because $W_{\text{rev}} = -\sum \Phi_{Q,i}$, and the summation normally involves only two terms.

4-1-1 Heat-Engine Cycle

Figure 4-1(*a*) is a schematic of a heat engine receiving heat Q_H from a reservoir at T_H and rejecting heat Q_L to a reservoir at T_L. The temperatures within the engine at which heat is received and rejected are noted as T_A and T_B, respectively. An energy balance on the engine itself requires that

$$Q_H + Q_L + W = \Delta E = 0$$

and the second law for the engine, since $\Delta S_{\text{eng}} = 0$, is

$$\sigma_{\text{eng}} = \Delta S_{\text{eng}} - \sum_i \frac{Q_i}{T_i} = -\frac{Q_H}{T_A} - \frac{Q_L}{T_B} \geq 0 \tag{4-2}$$

where Q_H, Q_L, and W are measured, signwise, relative to the engine.

In the case where the heat transfer is reversible, then $T_H = T_A$ and $T_L = T_B$. The engine itself may be reversible or irreversible. If reversible, then the device

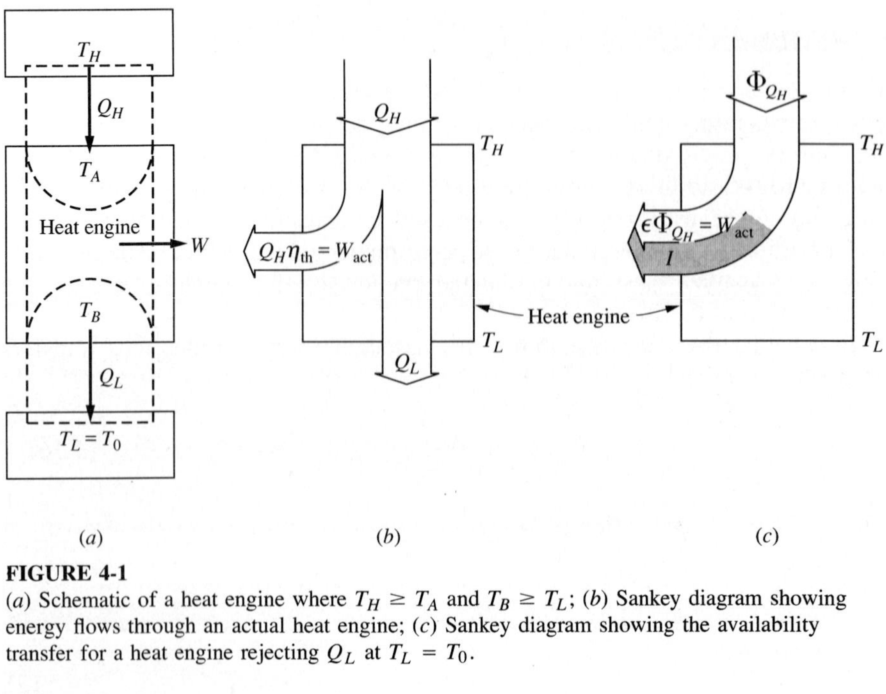

FIGURE 4-1
(*a*) Schematic of a heat engine where $T_H \geq T_A$ and $T_B \geq T_L$; (*b*) Sankey diagram showing energy flows through an actual heat engine; (*c*) Sankey diagram showing the availability transfer for a heat engine rejecting Q_L at $T_L = T_0$.

is the well-known Carnot heat engine, for which $\sigma_{eng} = 0$ and two familiar expressions result.

$$\frac{Q_L}{T_L} = \frac{-Q_H}{T_H} \qquad \text{and} \qquad \eta_{th} \equiv \frac{W_{out}}{Q_{in}} = \eta_{Carnot} = 1 - \frac{T_L}{T_H}$$

However, for reversible heat transfer and an irreversible engine, $\sigma_{eng} > 0$ and $I_{eng} > 0$. A combination of Eqs. (3-18) and (4-1) shown above yields

$$I_{eng} = W_{act} + Q_H\left(1 - \frac{T_0}{T_H}\right) + Q_L\left(1 - \frac{T_0}{T_0}\right)$$

$$= W_{act} + Q_H\left(1 - \frac{T_0}{T_H}\right) = W_{act} + \Phi_{Q,H} \tag{4-3}$$

where it is assumed that heat rejection is to the environment, so $T_L = T_0$, and W_{act} has a negative value.

Sankey or *band* diagrams which illustrate energy and availability accounting for the heat engine are shown in Figs. 4-1(*b*) and 4-1(*c*). In Fig. 4-1(*b*) the energy accounting shows Q_H entering and an equal quantity Q_L plus W leaving. In Fig. 4-1(*c*) an availability accounting relative to Eq. (4-3) is shown. Exergy $\Phi_{Q,H}$ enters and W_{act} exits. The gray shaded region represents the destruction of exergy or availability, I. Note that the availability transfer associated with Q_L is zero because $T_L = T_0$.

The second-law effectiveness of a cyclic heat engine ϵ_{HE} may be defined as the net work output of the cycle divided by the availability input $\Phi_{Q,H}$ associated with heat transfer Q_H from the reservoir at T_H; that is,

$$\epsilon_{HE} \equiv \frac{W_{act,out}}{\Phi_{Q,H}} \tag{4-4}$$

where both $W_{act,out}$ and $\Phi_{Q,H}$ are taken as positive. In the special case where both heat-transfer processes are reversible but the engine is irreversible, use of Eq. (4-3) in Eq. (4-4) leads to the relation

$$\epsilon_{HE} = \frac{\Phi_{Q,H} - I_{HE}}{\Phi_{Q,H}} \tag{4-5}$$

where $I_{HE} = T_0\sigma_{HE}$. Replacing W_{cycle} by $\eta_{th,act}Q_H$ and $\Phi_{Q,H}$ by Eq. (3-18), we find that

$$\epsilon_{HE} = \frac{\eta_{th,act}Q_H}{Q_H\left(1 - \dfrac{T_0}{T_H}\right)} = \frac{\eta_{th,act}}{1 - \dfrac{T_0}{T_H}} = \frac{\eta_{th,act}}{\eta_{Carnot}} \tag{4-6}$$

since $(1 - T_0/T_H)$ is the Carnot engine efficiency. This value of ϵ accounts solely for irreversibilities within the heat engine, because the heat-transfer processes were assumed to be reversible.

A second case is the situation where the heat-transfer processes are also irreversible; that is, $T_H > T_A$ and $T_B > T_L$. Thus we must account for three irreversibilities. The engine irreversibility expressed by Eq. (4-3) must be modified to the form

$$I_{\text{eng}} = W_{\text{act}} + Q_H\left(1 - \frac{T_0}{T_A}\right) + Q_L\left(1 - \frac{T_0}{T_B}\right) \tag{4-7}$$

where again $T_L = T_0$, and Q_H and Q_L are measured relative to the engine. The irreversibilities of the heat-transfer processes may be evaluated from Eq. (3-21b). For the irreversibility $I_{Q,H}$ associated with the heat transfer Q_H between T_H and T_A

$$I_{Q,H} = T_0 Q_H\left(\frac{1}{T_A} - \frac{1}{T_H}\right) \tag{4-8}$$

where Q_H is a positive value. Similarly, $I_{Q,L}$ for the heat transfer Q_L between T_B and T_L is

$$I_{Q,L} = T_0 Q_L\left(\frac{1}{T_L} - \frac{1}{T_B}\right) \tag{4-9}$$

where Q_L is a positive value. The total irreversibility for the overall process is the sum of the above three terms. The total irreversibility is also found from the relation $I_{\text{tot}} = T_0\sigma_{\text{tot}}$, or by the equivalent relation

$$I_{\text{tot}} = T_0(\Delta S_H + \Delta S_L + \Delta S_{\text{sys}}) = T_0(\Delta S_H + \Delta S_L) \tag{4-10}$$

where $\Delta S_{\text{sys}} = 0$, because the system is a cyclic device.

Example 4-1. A heat engine operates between two heat reservoirs, of which the temperatures are 1000 and 300 K. First, consider the situation where the cyclic system is also at 1000 K while receiving heat and 300 K while rejecting heat [recall Fig. 3-4(*a*)]. In addition, the measured efficiency is 50 percent. If the heat supplied during one cycle of operation is 100 kJ, (*a*) determine the irreversibility of the engine, in kilojoules, for one cycle. Now, consider the same situation, except that the engine temperature T_A during heat addition is 900 K and T_B during heat rejection is 350 K. The actual thermal efficiency is now 40 percent. Determine the irreversibility, in kilojoules, for (*b*) the heat engine, (*c*) the high-temperature heat supply process, and (*d*) the low-temperature heat-rejection process.

Solution

(*a*) On the basis of Eq. (4-3) the irreversibility of a heat engine with reversible heat supply and heat rejection is given by Eq. (4-3),

$$I_{\text{eng}} = W_{\text{act}} + Q_H\left(1 - \frac{T_0}{T_H}\right)$$

For the given data, and noting that $W_{act} = \eta_{th} Q_H = 0.5(100) = 50$ kJ,

$$I_{eng} = -50 + 100\left(1 - \frac{300}{1000}\right) = -50 + 70 = 20 \text{ kJ}$$

This answer can also be found from the relation $I_{eng} = T_0 \sigma_{eng}$. Using the σ data from Example 2-4, $I_{eng} = 300(0.0667) = 20$ kJ. In this particular case, $T_0 \sigma_{eng} = T_0 \, \Delta S_{tot}$.

(*b*) In the case of irreversible heat transfer the irreversibility of the engine is expressed in terms of Eq. (4-7), namely,

$$I_{eng} = W_{act} + Q_H\left(1 - \frac{T_0}{T_A}\right) + Q_L\left(1 - \frac{T_0}{T_B}\right)$$

$$= -0.4(100) + 100\left(1 - \frac{300}{900}\right) - 60\left(1 - \frac{300}{350}\right)$$

$$= -40 + 66.67 - 8.57 = 18.1 \text{ kJ}$$

(*c*) The irreversibility of the high-temperature heat-transfer process is given by Eq. (4-8), namely,

$$I_{Q,H} = T_0 Q_H\left(\frac{1}{T_A} - \frac{1}{T_H}\right) = 300(100)\left(\frac{1}{900} - \frac{1}{1000}\right) = 3.33 \text{ kJ}$$

(*d*) For the low-temperature process Eq. (4-9) yields

$$I_{Q,L} = T_0 Q_L\left(\frac{1}{T_L} - \frac{1}{T_B}\right) = 300(60)\left(\frac{1}{300} - \frac{1}{350}\right) = 8.57 \text{ kJ}$$

The total irreversibility for the cycle is $I_{tot} = 18.1 + 3.33 + 8.57 = 30.0$ kJ. The heat engine itself, in this case, contributes close to 60 percent of the total, with nearly 30 percent due to the low-temperature heat-transfer process.

Similar to part (*a*), the irreversibility for the three parts also can be found from the σ data in Examples 2-5 and 2-6. That is,

$$I = T_0 \sigma_{tot} = 300(0.0603 + 0.0111 + 0.0286)$$

$$= 18.1 + 3.33 + 8.57 = 30 \text{ kJ}$$

A third method is to apply the relation $I = T_0 \, \Delta S_{tot}$ to the engine-reservoir system.

$$I = T_0 (\Delta S_H + \Delta S_L + \Delta S_{sys}) = T_0\left(\frac{Q_H}{T_H} + \frac{Q_L}{T_L} + 0\right)$$

$$= 300\left(-\frac{100}{1000} + \frac{60}{300}\right) = 30 \text{ kJ}$$

Since $\Phi_{Q,H} = 70$ kJ, $\epsilon = 40/70 = 0.571$ (57.1 percent).

4-1-2 Refrigeration Cycle

Figure 4-2(a) is a schematic of a $2T$-refrigeration cycle receiving heat Q_L from a reservoir at T_L and rejecting heat Q_H to a reservoir at T_H. The temperatures of regions within the refrigerator where heat is received and rejected are noted as T_B and T_A, respectively, on the diagram. Like the heat engine discussed in Sec. 4-1-1, the first and second laws for the cyclic refrigerator are

$$Q_H + Q_L + W = \Delta E = 0$$

and

$$\sigma_{\text{refrig}} = -\frac{Q_H}{T_A} - \frac{Q_L}{T_B} \geq 0 \tag{4-2}$$

where Q_H, Q_L, and W are measured, signwise, relative to the refrigerator. For given values of T_B, T_A, and Q_L the values of Q_H and W depend on the entropy generation σ_{refrig} within the device.

When the heat transfers Q_H and Q_L are reversible, then $T_B = T_L$ and $T_A = T_H$. The performance of the refrigerator itself depends on σ_{refrig}. For reversible operation $\sigma_{\text{refrig}} = 0$ and the device is the well-known Carnot refrigerator for which

$$\frac{Q_L}{T_L} = \frac{-Q_H}{T_H} \quad \text{and} \quad \text{COP}_{\text{refrig}} \equiv \frac{Q_L}{W} = \text{COP}_{\text{Carnot}} = \frac{T_L}{T_H - T_L}$$

However, for reversible heat transfers but an irreversible refrigerator Eq. (4-1) becomes

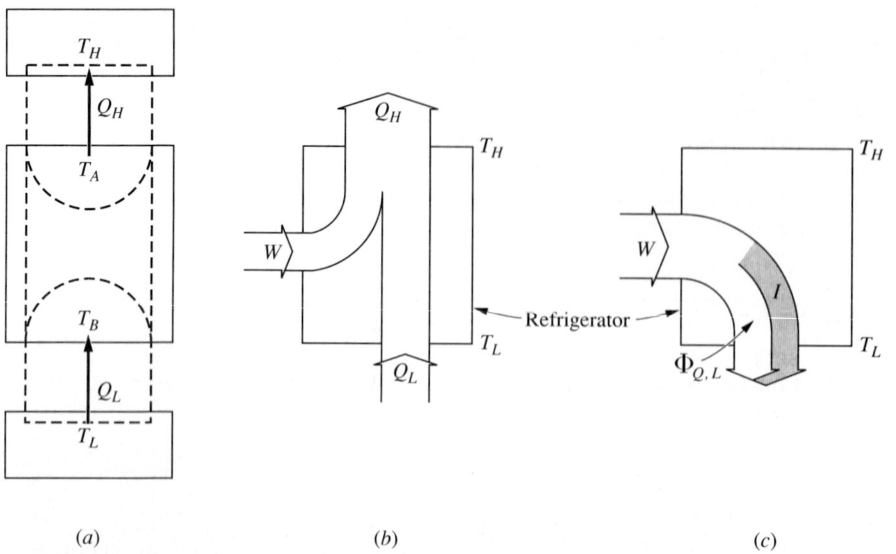

FIGURE 4-2
(a) Schematic of a refrigerator where $T_A \geq T_H$ and $T_L \geq T_B$; (b) Sankey diagram showing energy flows through the device; (c) availability transfer diagram for a refrigerator rejecting heat Q_H at $T_H = T_0$.

$$I_{\text{refrig}} = W_{\text{act}} + Q_L\left(1 - \frac{T_0}{T_L}\right) + Q_H\left(1 - \frac{T_0}{T_0}\right)$$

$$= W_{\text{act,refrig}} + Q_L\left(1 - \frac{T_0}{T_L}\right) \tag{4-11}$$

where W_{act} and Q_L have positive values and $T_H = T_0$. The energy and availability diagrams for this situation appear as Figs. 4-2(*b*) and 4-2(*c*). With reference to Fig. 4-2(*c*), exergy in the form of W_{act} enters while exergy or availability $\Phi_{Q,L}$ leaves. The quantitative difference between these two quantities is the irreversibility I, shown as the gray shaded region. (Recall that the direction of heat transfer and exergy transfer are opposite when $T < T_0$. Note also that the availability associated with the heat Q_H rejected to the environment at T_H is zero.)

The second-law effectiveness of a refrigerator is commonly defined as the ratio of exergy delivered to the cold region at T_L divided by the actual exergy (work) input. Symbolically,

$$\epsilon_{\text{refrig}} \equiv \frac{\Phi_{Q,L}}{W_{\text{act,in}}} \tag{4-12}$$

where both terms are taken as positive. In the special case of reversible heat transfer but an irreversible refrigeration device the effectiveness becomes

$$\epsilon_{\text{refrig}} = \frac{\Phi_{Q,L}}{W_{\text{act,in}}} = \frac{\Phi_{Q,L}}{\Phi_{Q,L} + I_{\text{refrig}}}$$

where $\Phi_{Q,L}$ is positive and Q_L is negative relative to the reservoir at T_L. Replacing $W_{\text{act,in}}$ by $Q_L/\text{COP}_{\text{act}}$ (by definition) and $\Phi_{Q,L}$ by Eq. (3-18), we find that

$$\epsilon_{\text{refrig}} = -Q_L\left(1 - \frac{T_0}{T_L}\right)\frac{\text{COP}_{\text{act}}}{Q_L} = \left(\frac{T_0}{T_L} - 1\right)\text{COP}_{\text{act}} = \frac{\text{COP}_{\text{act}}}{\text{COP}_{\text{Carnot}}} \tag{4-13}$$

where $\text{COP}_{\text{Carnot}} = T_L/(T_0 - T_L)$, since $T_0 = T_H$ in this case. This value of ϵ accounts solely for irreversibilities within the refrigerator, since the heat-transfer processes are assumed reversible.

As a second situation, consider the heat-transfer processes also to be irreversible so that $T_A > T_H$ and $T_L > T_B$. In terms of Eq. (4-1) the irreversibility of the refrigerator is expressed by

$$I_{\text{refrig}} = W_{\text{act}} + Q_H\left(1 - \frac{T_0}{T_A}\right) + Q_L\left(1 - \frac{T_0}{T_B}\right) \tag{4-14}$$

where again $T_H = T_0$, and Q_H and Q_L are measured relative to the refrigerator. For the two heat-transfer processes Q_H and Q_L the irreversibilities are given by

$$I_{Q,H} = T_0 Q_H\left(\frac{1}{T_H} - \frac{1}{T_A}\right) \quad \text{and} \quad I_{Q,L} = T_0 Q_L\left(\frac{1}{T_B} - \frac{1}{T_L}\right) \tag{4-15}$$

where Q_H and Q_L are positive values. The total irreversibility of the overall process is the sum of the above three terms. I_{tot} can also be found by Eq. (4-10). The effectiveness in this situation is still found by Eq. (4-12).

4-1-3 Heat Pump Cycle

A schematic of a heat pump removing heat Q_L from the environment at T_L and delivering heat Q_H at T_H to a building or process is shown in Fig. 4-3(*a*). When the heat-transfer processes are reversible, $T_B = T_L$ and $T_A = T_H$. For reversible operation the Carnot relations hold, namely,

$$\frac{Q_L}{T_L} = \frac{-Q_H}{T_H} \qquad \text{and} \qquad \text{COP}_{HP} \equiv \frac{Q_H}{W} = \frac{T_H}{T_H - T_L}$$

For the case of an irreversible heat pump with reversible heat transfer

$$I_{HP} = W_{\text{act},HP} + Q_H\left(1 - \frac{T_0}{T_H}\right) \tag{4-16}$$

where W_{act} has a positive value and $T_L = T_0$. The energy and exergy diagrams for the irreversible heat pump are shown as Figs. 4-3(*b*) and 4-3(*c*). With reference to this diagram, exergy in the form of W_{act} enters and exergy $\Phi_{Q,H}$ leaves at temperature T_H. The difference between these two quantities, shown by the gray shaded arrow, is the irreversibility I within the heat pump. Note that the availability associated with the heat transfer Q_L at T_L is zero. The value of Q_L

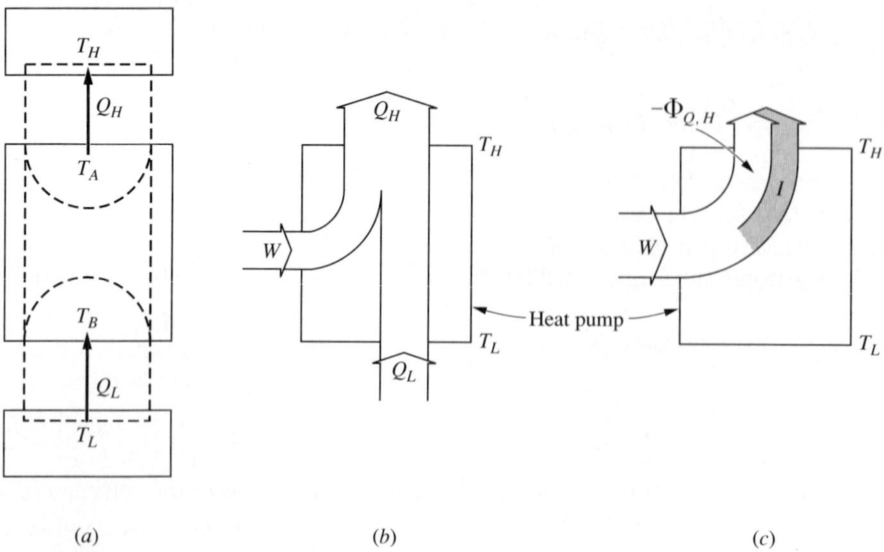

FIGURE 4-3
(*a*) Schematic of a heat pump where $T_A \geq T_H$ and $T_L \geq T_B$; (*b*) diagram showing energy flows; (*c*) availability-flow diagram for a heat pump receiving heat Q_L and $T_L = T_0$.

depends (for given T_L, T_H, and Q_H) on the entropy production or irreversibility within the heat pump.

The second-law effectiveness of a heat pump is defined as the ratio of the exergy entering the reservoir at T_H over the exergy (work) input. That is,

$$\epsilon_{HP} \equiv \frac{\Phi_{Q,H}}{W_{\text{act,in}}} \tag{4-17}$$

where both terms are taken as positive. In the special case where the heat transfer is reversible but the heat pump device itself is irreversible the above equation becomes

$$\epsilon_{HP} \equiv \frac{\Phi_{Q,H}}{W_{\text{act,in}}} = \frac{\Phi_{Q,H}}{\Phi_{Q,H} + I_{HP}} \tag{4-18}$$

where again $\Phi_{Q,H}$ is taken as positive. Note that $W_{\text{act,in}} = Q_H/\text{COP}_{\text{act}}$ and $\Phi_{Q,H} = Q_H(1 - T_0/T_H)$. Substitution of these two relations into the above equation yields

$$\epsilon_{HP} = Q_H\left(1 - \frac{T_0}{T_H}\right)\frac{\text{COP}_{\text{act}}}{Q_H} = \text{COP}_{\text{act}}\left(1 - \frac{T_0}{T_H}\right) = \frac{\text{COP}_{\text{act}}}{\text{COP}_{\text{Carnot}}} \tag{4-19}$$

The preceding equation applies solely to an irreversible heat pump with reversible heat transfer.

In actual heat pump applications a finite temperature difference exists for heat transfer such that $T_A > T_H$ and $T_L > T_B$. For this situation the irreversibility of the heat pump is

$$I_{HP} = W_{\text{act}} + Q_H\left(1 - \frac{T_0}{T_A}\right) + Q_L\left(1 - \frac{T_0}{T_B}\right) \tag{4-20}$$

and irreversibilities for the two heat-transfer processes become

$$I_{Q,H} = T_0 Q_H\left(\frac{1}{T_H} - \frac{1}{T_A}\right) \quad \text{and} \quad I_{Q,L} = T_0 Q_L\left(\frac{1}{T_B} - \frac{1}{T_L}\right) \tag{4-21}$$

where Q_H and Q_L are positive values. The effectiveness in this latter situation is given by Eq. (4-17).

4-2 CYCLE APPLICATIONS OF AVAILABILITY

The energy and availability analyses of several common power and refrigeration cycles are developed in this section.

4-2-1 Gas-Turbine Cycle

The equipment schematic and Ts diagram for an open gas-turbine cycle are shown in Fig. 4-4. Irreversible compressor and turbine performances are shown as

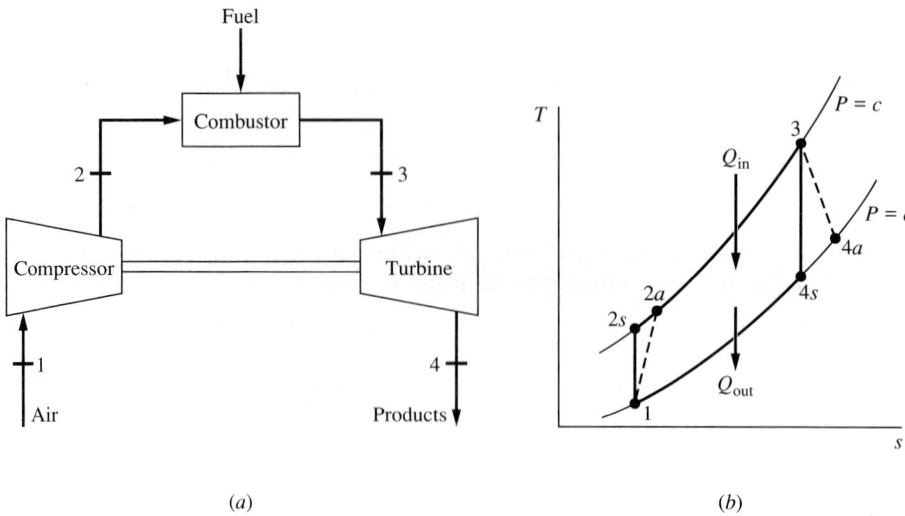

FIGURE 4-4
(a) Schematic of equipment; (b) Ts diagram for an open gas-turbine cycle with irreversible compressor and turbine.

processes 1-2a and 3-4a on the Ts plot. The basic energy relations for the steady-flow compressor, combustor, and turbine are

$$w_c = h_{2a} - h_1 \qquad q_{\text{comb}} = h_3 - h_{2a} \qquad w_t = h_{4a} - h_3$$

where kinetic and potential energies are neglected. The compressor and turbine first-law adiabatic efficiencies are

$$\eta_c = \frac{h_{2s} - h_1}{h_{2a} - h_1} \qquad \text{and} \qquad \eta_t = \frac{h_3 - h_{4a}}{h_3 - h_{4s}}$$

where subscripts a and s represent actual and isentropic states. The availability balances for the compressor, combustor, turbine, and the second-law efficiency are

$$w_c = (b_{2a} - b_1) + i_c \qquad \text{(compressor)} \tag{4-22a}$$

$$\sum_i q_i \left(1 - \frac{T_0}{T_i}\right) = b_3 - b_{2a} + i_{\text{comb}} \qquad \text{(combustor)} \tag{4-22b}$$

$$b_{4a} - b_3 = w_t - i_t \qquad \text{(turbine)} \tag{4-22c}$$

$$\epsilon_{\text{cycle}} = \frac{w_{\text{net}}}{\Delta\Psi_{\text{comb}}} \tag{4-22d}$$

where Δb could be replaced by $\Delta\psi$. The combustion of fuel and air is modeled as heat transfer through the combustor walls, and the irreversibility within the combustor is frequently neglected as a first approximation. This model will be greatly improved by considering the availability of the fuel itself in Chap. 12. The use of these relations is illustrated in the example below.

TABLE 4-1
Property data for an open gas-turbine power cycle

State	T_a, K	P, bars	h, kJ/kg	s, kJ/kg · K	b, kJ/kg
1	295	1.0	295.2	0	295.2
2	543	6.4	547.2	0.0866	521.4
3	1080	6.4	1138.0	0.8387	887.8
4	732	1.0	784.0	0.9360	468.9

Example 4-2. An irreversible gas-turbine power plant operates between pressures of 1.0 and 6.4 bars with compressor and turbine inlet temperatures of 22 and 807°C, respectively. Determine (*a*) the compressor and turbine work, (*b*) the heat supplied to the combustor, (*c*) the change in availability across the three devices, and (*d*) the irreversibility in all three devices, all answers in kJ/kg air.

Solution. Table 4-1 summarizes the major property values at the four states of interest for $T_0 = 25°C$. The entropy and availability values are calculated from the ideal-gas relations

$$s_e - s_i = s_e^o - s_i^o - R \ln(P_e/P_i)$$
$$b_e - b_i = h_e - h_i - T_0(s_e - s_i)$$

where h and s^o data come directly from the air table in the Appendix. The value of s_1 arbitrarily has been set equal to zero. Values for ψ would require that $(h_0 - T_0 s_0)$ be subtracted from each δ value. An energy accounting of inputs and outputs is shown in Table 4-2. The irreversibility within the compressor and turbine are found

TABLE 4-2
Energy and availability accounting for an open gas-turbine power cycle

A. Energy accounting (in kJ/kg)

Energy in:		Energy out:	
Air in at state 1	295.2	Air out at state 4	748.0
Compressor work	252.0	Turbine work	390.0
Heat in	590.8		
Total	1138.0	Total	1138.0

B. Availability accounting (in kJ/kg)

Process	q	w	Δb or b	i
Compressor	—	252.0	226.2	25.8
Combustor	590.8	—	366.4	—
Turbine	—	−390.0	−418.9	29.0
Inlet fluid	—	—	295.2	—
Outlet fluid	—	—	−468.9	—
	590.8	138.0	0.0	54.8

from $i = T_0 \Delta s$ for the adiabatic devices. The value of i for the combustor is zero, since the flow is assumed to be frictionless.

The energy analysis in part A of Table 4-2 accounts for inputs and outputs for the cycle. It does indicate that significant energy leaves with the turbine exhaust, and that the first-law efficiency is 23.4 percent. However, it gives no information about the effect of irreversibilities or availability losses on performance. An availability accounting for the cycle is shown in part B of Table 4-2. The data are more meaningful if they are applied to the steady-state availability balance. In an input/output format

$$\sum_i q_i \left(1 - \frac{T_0}{T_i}\right) + b_i = b_e - (w_t + w_c) + i_c + i_t$$

Substitution of values yields

$$366.4 + 295.2 = 468.9 + 138.0 + 25.8 + 29.0$$

The availability analysis is much more revealing than the energy balance. The input of 661.6 kJ/kg of availability appears primarily as 468.9 kJ/kg in the turbine exhaust stream, compared to 138.0 kJ/kg of net work output. The destruction of availability in the compressor and turbine is fairly insignificant, totaling 54.8 kJ/kg. The net availability of the inlet and exhaust streams is 47 percent and the irreversibilities are 15 percent of the availability added in the combustor. Thus far too much work capability is lost in the turbine exhaust stream. A graphic illustration of the availability accounting for the cycle is shown in Fig. 4-5. The second-law efficiency for the cycle, on a desired output/required input basis, is

$$\epsilon_{\text{cycle}} = \frac{w_{\text{net}}}{\Delta\Psi_{\text{comb}}} = \frac{138.0}{366.4} = 0.377$$

Only about one-third of the availability input with the heat transfer is converted into useful work output.

It is well known that the thermal efficiency of a gas-turbine cycle can be enhanced by use of a regenerator within the cycle. A schematic of this process and the Ts plot are shown in Fig. 4-6. The "waste" energy in the turbine exhaust is used to increase the energy and temperature of the compressor exhaust stream before it enters the combustor. The subsequent decrease in heat input in the combustor, coupled with no change in net work output could lead to a significant increase in thermal efficiency and second-law efficiency. The only additional energy relations required are the effectiveness and first-law analysis for the regenerator. In terms of the states on the Ts plot, these are

$$\eta_{\text{reg}} = \frac{h_x - h_2}{h_4 - h_2} \qquad \text{and} \qquad h_x - h_2 = h_5 - h_6$$

The availability balance for the regenerator is

$$0 = (b_3 - b_2) + (b_6 - b_5) + i_{\text{reg}} \tag{4-23}$$

These relations are now applied to the preceding example.

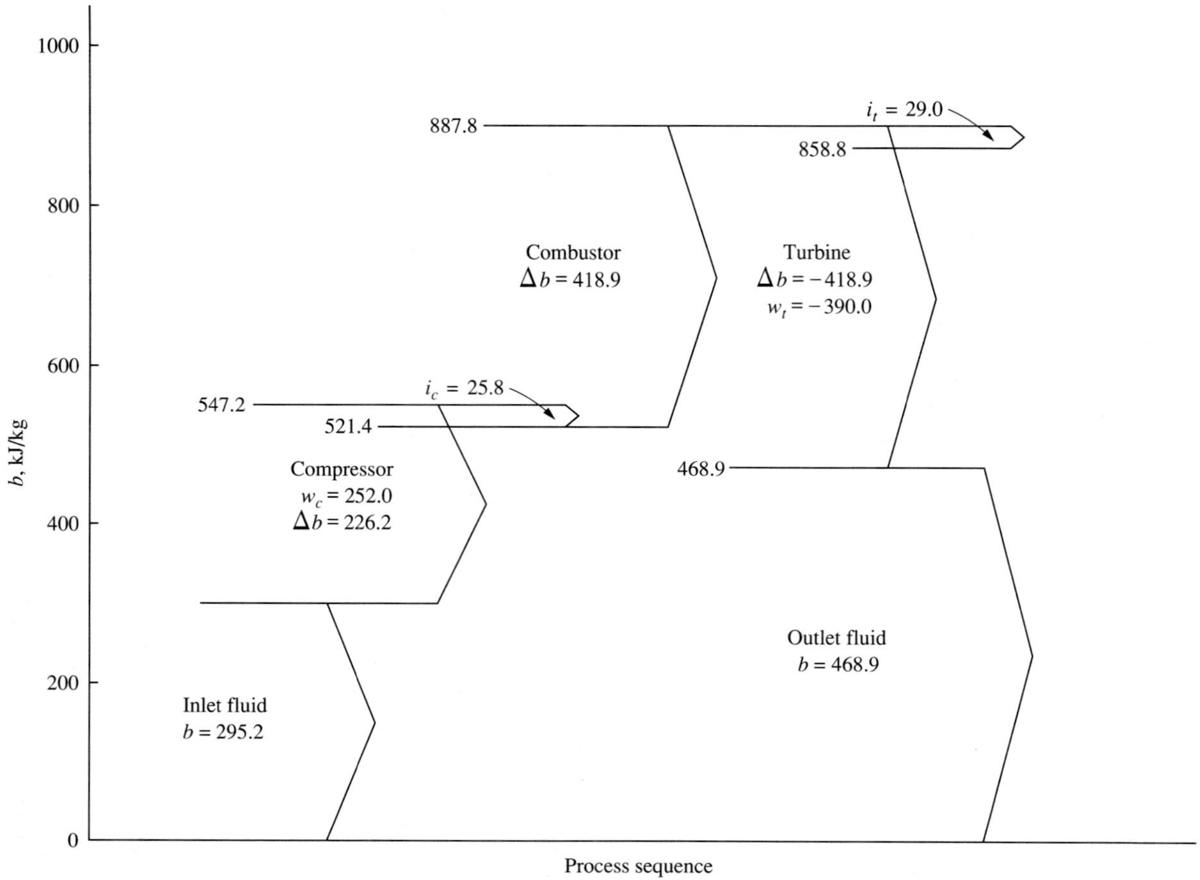

FIGURE 4-5
Availability accounting for an open gas-turbine power cycle with irreversible compressor and turbine. (Data refer to Example 4-2.)

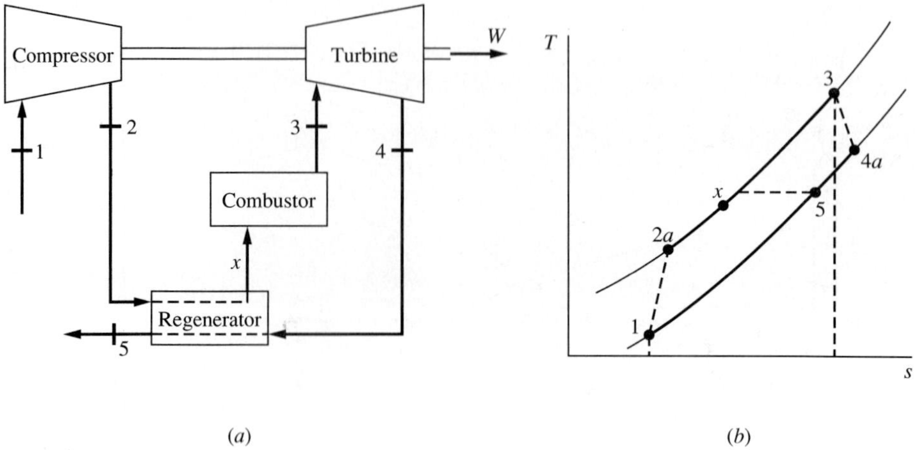

FIGURE 4-6
(*a*) Schematic of equipment; (*b*) *Ts* diagram for a regenerative gas-turbine cycle with irreversible compressor and turbine.

Example 4-3. A regenerator with 75 percent effectiveness is added to the gas-turbine cycle examined in Example 4-2. Complete the energy and availability analyses for this cycle.

Solution. Table 4-3 summarizes property values for the cycle. The energy and availability accountings are shown in Table 4-4. The thermal efficiency is now 0.313 versus 0.234 without a regenerator. Also, the exhaust enthalpy has been reduced from 748.0 to 597.4 kJ/kg as the exhaust temperature is lowered from 732 to 591 K. The availability analysis is more revealing. Again, written as input = output,

$$\sum q_i \left(1 - \frac{T_0}{T_i}\right) + b_i = b_e - (w_t + w_c) + i_c + i_t + i_{\text{reg}}$$

Substitution of data leads to

$$289.2 + 295.2 = 386.3 + 138.0 + 25.8 + 29.0 + 5.3$$

TABLE 4-3
Property data for a regenerative gas-turbine power cycle

State	T_a, K	P, bars	h, kJ/kg	s, kJ/kg·K	b, kJ/kg
1	295	1.0	295.2	0	295.2
2	543	6.4	547.2	0.0866	521.4
x	685	6.4	697.8	0.3323	598.7
3	1080	6.4	1138.0	0.8387	887.9
4	732	1.0	748.0	0.9360	468.9
5	591	1.0	597.4	0.7082	386.3

TABLE 4-4
Energy and availability accounting for a regenerative gas-turbine power cycle

A. Energy accounting (in kJ/kg)

Energy in:		**Energy out:**	
Air in at state 1	295.2	Air out at state 5	597.4
Compressor work	252.0	Turbine work	390.0
Heat in	440.2		
Total	987.4	Total	987.4

B. Availability accounting (in kJ/kg)

Process	q	w	Δb **or** b	i
Compressor	—	252.0	226.2	25.8
Combustor	440.2	—	289.2	—
Turbine	—	−390.0	−418.9	29.0
Inlet fluid	—	—	295.2	—
Outlet fluid	—	—	−386.3	—
Regenerator	—	—	−5.3	5.3
	440.2	138.0	0.0	60.1

In this case the net availability of the inlet and exhaust streams is only 32 percent of the availability added in the combustor. This is down from 47 percent without a regenerator. The irreversibilities of the cycle are 20 percent and the net work is 48 percent of the availability input. This latter value also is the effectiveness ϵ for the cycle. Thus a sizable increase in ϵ has occurred with the installation of the regenerator, since the effectiveness without a regenerator is 38 percent. A graphical availability accounting is shown in Fig. 4-7.

4-2-2 Simple Steam Power Cycle

The equipment schematic and Ts diagram for a simple Rankine steam power cycle with irreversible turbine and pump are shown in Fig. 4-8. On the basis of the numbered states on the Ts diagram the basic energy relations for the steady-state boiler, turbine, condenser, and pump are

$$q_{in} = h_3 - h_2 \qquad w_t = h_4 - h_3 \qquad q_{cond} = h_1 - h_4 \qquad w_p = h_2 - h_1$$

where kinetic and potential energies are neglected. The turbine and pump adiabatic efficiencies are

$$\eta_t = \frac{h_3 - h_{4a}}{h_3 - h_{4s}} \qquad \text{and} \qquad \eta_p = \frac{h_{2s} - h_1}{h_{2a} - h_1}$$

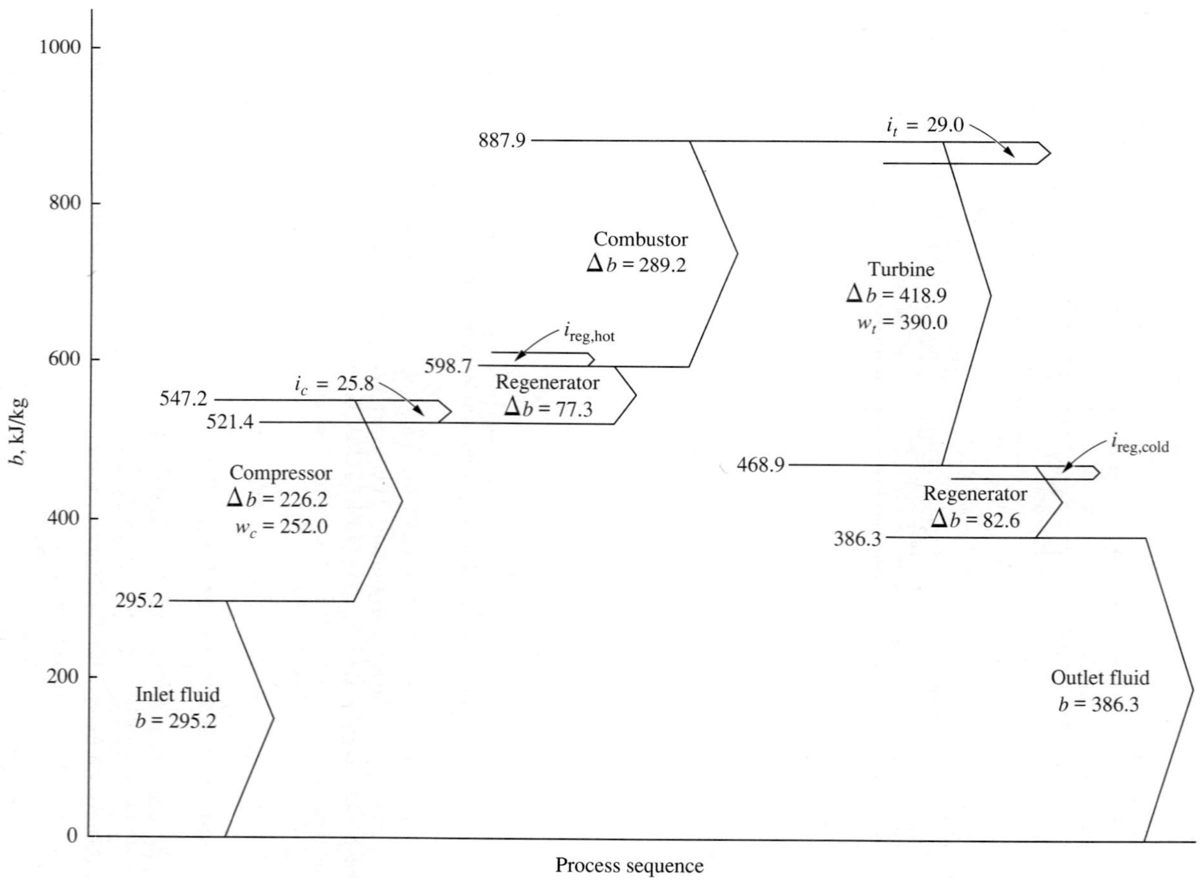

FIGURE 4-7
Availability accounting for a regenerative open gas-turbine power cycle with irreversible compressor and turbine. (Data refer to Example 4-3.)

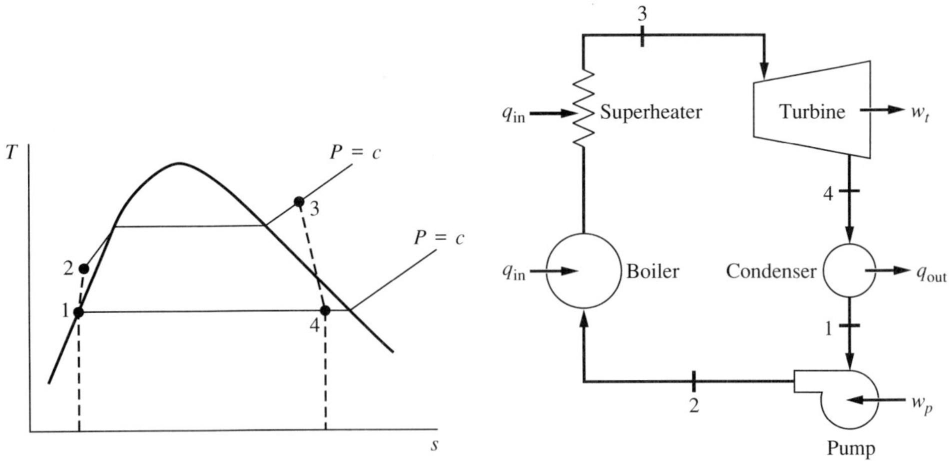

FIGURE 4-8
The Ts diagram and equipment schematic for a Rankine cycle with superheating.

where subscripts a and s represent actual and isentropic states. The availability balances for the four devices on an input/product format are

$$\sum_i q_i \left(1 - \frac{T_0}{T_i}\right) + b_2 = b_3 + i_{\text{boiler}} \qquad \text{(boiler-superheater)} \quad (4\text{-}24a)$$

$$b_3 = b_{4a} - w_t + i_t \qquad \text{(turbine)} \quad (4\text{-}24b)$$

$$b_{4a} = b_1 - \sum_i q_i \left(1 - \frac{T_0}{T_i}\right) + i_{\text{cond}} \qquad \text{(condenser)} \quad (4\text{-}24c)$$

$$w_p + b_1 = b_{2a} + i_p \qquad \text{(pump)} \quad (4\text{-}24d)$$

$$\epsilon = \frac{w_{\text{net}}}{\Delta\Psi_{\text{comb}}} \qquad (4\text{-}24e)$$

In all of these relations Δb may be replaced by $\Delta\psi$. The irreversibility for the boiler flow and the condenser flow will be neglected. The expression for the effectiveness will be improved by considering the exergy of the fuel itself in Chap. 12. The use of these relations is illustrated in the example below.

Example 4-4. A simple steam power cycle generates steam in the boiler superheater at 140 bars and 560°C and condenses steam at 0.06 bar. The cooling water required in the condenser experiences a temperature rise from 18 to 28°C. The adiabatic efficiency of the turbine is 85 percent and that of the pump is 70 percent. Make suitable energy and availability analyses for the cycle.

Solution. Table 4-5 summarizes the major property values at the four states of interest for the dead state at T_0 = 25°C. The states are numbered according to the

TABLE 4-5
Property data for a steam power cycle

State	T, °C	P, bars	h, kJ/kg	s, kJ/kg · K	b, kJ/kg	ψ, kJ/kg
1	36.2	0.06	151.5	0.5210	−3.8	0.8
2	36.5	140.	171.6	0.5404	10.5	15.1
3	560.0	140.	3486.0	6.5941	1520.0	1524.6
4	36.2	0.06	2249.0	7.3012	72.1	76.7
0	25.0	1.00	104.9	0.3674	−4.6	0.0

diagram for the cycle. An energy accounting is shown in part A of Table 4-6. The major items based on these data are the back-work ratio, $w_p/w_t = 0.016$, and the thermal efficiency, $\eta_{\text{th}} = w_{\text{net}}/q_{\text{in}} = 1216.9/3314.4 = 0.367$. The first item indicates that a very small amount of turbine work is used to drive the pump. The second item reveals that 63 percent of the heat added is removed in the condenser cooling stream.

An availability analysis reveals more information on the condenser. With reference to part B of Table 4-6, the effectiveness of the cycle in terms of useful product/input is 81 percent [$\epsilon = (w_t - w_p)/\phi_{Q,\text{in}} = 1216.9/1509.5 = 0.806$]. The loss of availability in the condenser is minimal, amounting to 5 percent of the availability input in the boiler. A tremendous energy loss occurs, but only a small availability loss occurs in the condenser. This is due to the low temperature of the water (36°C) as it condenses, relative to the environment at 25°C.

The irreversibilities of the boiler and condenser are taken as zero in Table 4-6. The *internal* losses are zero, because frictional losses are neglected. External irreversibilities exist in both cases, however, due to heat transfer across a finite temperature gradient. The irreversibility for the condenser can be evaluated, since the inlet and outlet temperatures for the cooling water have been specified as 18 and 28°C. An energy balance around the condenser shows that

TABLE 4-6
An energy and availability accounting for a steam power cycle

A. Energy accounting (in kJ/kg)

Energy in:		Energy out:	
Heat added in boiler	3314.4	Heat out in condenser	2097.5
Pump work	20.1	Turbine work	1237.0
Total	3334.5	Total	3334.5

B. Availability accounting (in kJ/kg)

Process	q	w_a	Δb	i
Boiler	3314.4	—	1509.5	—
Turbine	—	−1237.0	−1447.9	210.9
Condenser	−2097.5	—	−75.9	—
Pump	—	20.1	14.3	5.8
Total	1216.9	−1216.9	0.0	216.7

$$\frac{\dot{m}_{\text{coolant}}}{\dot{m}_{\text{steam}}} = \frac{-\Delta h_{\text{steam}}}{\Delta h_{\text{coolant}}} = \frac{-(151.5 - 2249)}{(117.43 - 75.58)} = 50.1$$

where the water coolant enthalpies are the saturated-liquid values h_f at the given temperatures. The irreversibility of the adiabatic condenser on a kilogram of steam basis is

$$i = \sum T_0 \,\Delta s = 289.15[(0.5210 - 7.3012) + 50.1(0.4093 - 0.2679)]$$
$$= 91.4 \text{ kJ/kg steam}$$

where the coolant entropy values are s_f data from the saturation table. On the basis of part B of Table 4-6 the condenser irreversibility (loss in work capability) is only about 43 percent of the loss in the turbine. Nevertheless, this loss is worth reducing through the use of feedwater heaters. The boiler loss of availability is more difficult to ascertain, since the end states of the combustion gases during the heat-transfer process may not be well known. In addition, another irreversibility is associated with the combustion process itself. This latter topic is discussed in Chap. 11. An availability flow diagram for the simple steam power cycle without condenser and boiler irreversibilities is shown in Fig. 4-9, in accordance with the availability equations developed above for the separate equipment. The widths of the arrow are only qualitative. The pump i contribution is too small to be shown, and the width of the pump work arrow is greatly overemphasized. The effectiveness of the cycle is modeled by

$$\eta_{\text{II}} = \epsilon = \frac{w_{\text{net}}}{\Delta\Psi_{\text{comb}}} = \frac{w_t - w_p}{\Delta\Psi_{\text{comb}}}$$

The outlet arrow at the bottom, right of the figure, shows $w_{\text{net}} = \epsilon\phi_{\text{Q,in}}$. For this example $\epsilon = 1216.9/1509.5 = 0.806$, or 80.6 percent.

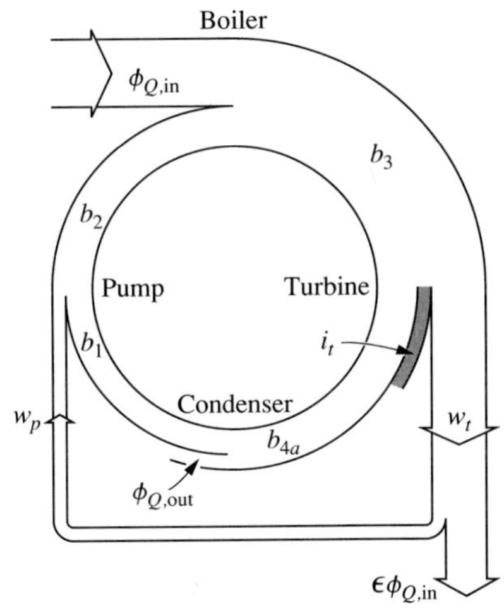

FIGURE 4-9
Availability flow diagram for a simple steam power (Rankine) cycle showing transfers and losses.

4-2-3 Vapor-Compression-Refrigeration Cycle

An equipment schematic and Ts diagram for a vapor-compression-refrigeration cycle with an irreversible compressor is shown in Fig. 4-10. Using the state notation found on the diagrams, we can write the following steady-state energy relations for the compressor, condenser, throttling valve, and evaporator:

$$q_c + w_c = h_{2a} - h_1 \qquad q_{\text{cond}} = h_3 - h_{2a} \qquad h_3 = h_4 \qquad q_{\text{evap}} = h_1 - h_4$$

where kinetic- and potential-energy terms are neglected. The availability balances for the four processes on an input/product basis are

$$\psi_1 + w_c = \psi_2 - \sum q_i \left(1 - \frac{T_0}{T_i}\right) + i_c \qquad \text{(compressor)} \qquad \text{(4-25a)}$$

$$\psi_2 = \psi_3 - \sum q_i \left(1 - \frac{T_0}{T_i}\right) + i_{\text{cond}} \qquad \text{(condenser)} \qquad \text{(4-25b)}$$

$$\psi_3 = \psi_4 + i_{\text{valve}} \qquad \text{(throttle)} \qquad \text{(4-25c)}$$

$$\psi_4 = \psi_1 - \sum q_i \left(1 - \frac{T_0}{T_i}\right) + i_{\text{evap}} \qquad \text{(evaporator)} \qquad \text{(4-25d)}$$

$$\epsilon_{\text{refrig}} = -\frac{\phi_{Q,\text{evap}}}{w_c} \qquad \text{(4-26)}$$

A $q_i(1 - T_0/T_i)$ term has been included in the compressor balance to account for unavoidable heat transfer. The flow through the lines connecting the equipment is assumed to be adiabatic and frictionless. Also, ψ may be replaced by b in the above relations. The example below illustrates energy and availability analysis for a refrigerator.

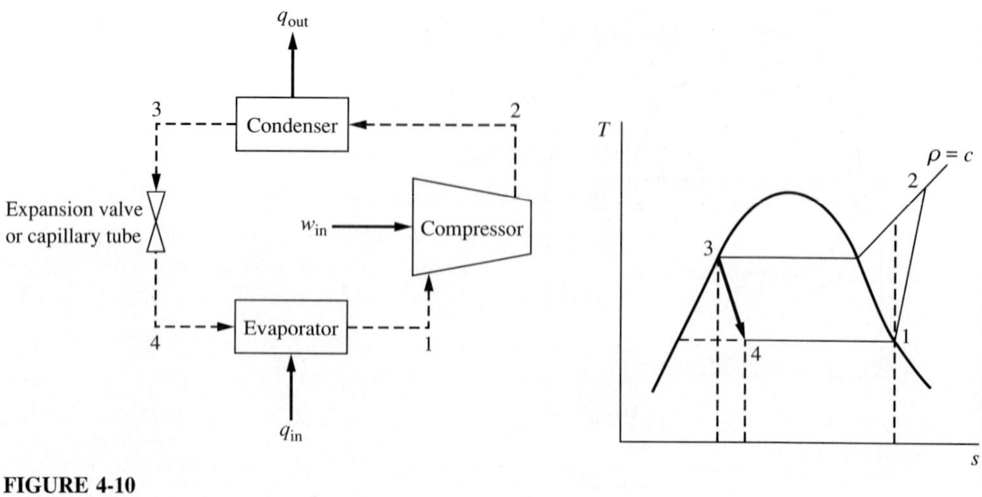

FIGURE 4-10
A schematic of the equipment and a Ts diagram for a vapor-compression-refrigeration cycle.

TABLE 4-7
Property data for a vapor-compression-refrigeration cycle

State	T,°C	P, MPa	h, kJ/kg	s, kJ/kg·K	ψ, kJ/kg
1	−10.09	0.20	241.30	0.9253	18.28
2	43.18	0.80	277.08	0.9481	47.26
3	31.33	0.80	93.42	0.3459	43.15
4	−10.09	0.20	93.42	0.3632	37.99
0	25.00	0.10	274.39	1.0976	0.00

Example 4-5. Saturated refrigerant-134a (R-134a) vapor enters the adiabatic compressor of a vapor-compression-refrigeration cycle at 0.20 MPa and leaves the condenser as a saturated liquid at 0.80 MPa. The temperature of the vapor leaving the irreversible compressor is 43.2°C and the ambient temperature is 25°C. Determine the energy and availability changes and the irreversibilities around the cycle.

Solution. Table 4-7 summarizes the major property data of the four states and the dead state. An energy accounting for the cycle is shown in part A of Table 4-8. As noted, the compressor operation is assumed to be adiabatic. The only important information to be gained is that the COP value is 147.88/35.78 = 4.13. The availability accounting leads to several interesting points. First, if the flow in the condenser is assumed frictionless, then from the availability balance shown earlier

$$\sum q_i\left(1-\frac{T_0}{T_i}\right)=\phi_Q=\psi_3-\psi_2=-4.11 \text{ kJ/kg}$$

TABLE 4-8
An energy and availability accounting for a vapor-compression-refrigeration cycle

A. Energy accounting (in kJ/kg)

Energy in:		Energy out:	
Compressor work in	35.78	Heat out in condenser	188.66
Heat added in evaporator	147.88		
Total	183.66		

B. Availability accounting (in kJ/kg)

Process	q	w	$\Delta\psi$	i
Compressor	—	35.78	28.98	6.80
Condenser	−183.66	—	−4.11	—
Throttle valve	—	—	−5.16	5.16
Evaporator	147.88	—	−19.71	—
Total	−35.78	35.78	0.00	11.96

Thus the direction of both q and ϕ_Q is out of the condenser. In the case of a frictionless evaporator

$$\sum q_i\left(1 - \frac{T_0}{T_i}\right) = \phi_Q = \psi_1 - \psi_4 = -19.71 \text{ kJ/kg}$$

while q is positive. Hence q and ϕ_Q pass in the opposite direction at the boundary of the evaporator. This is not unusual if one recalls the discussion in Sec. 4-3 on availability transfer associated with heat transfer. When the system temperature T is less than T_0, the system *loses* availability when heat is added.

Part B of Table 4-8 indicates that the irreversibilities in the compressor and in the throttle device are of the same magnitude. If the condenser flow is irreversible, then $\Delta\psi$ would be a measure of both i and ϕ_Q in the condenser. Likewise for irreversible evaporator flow, $\Delta\psi$ measures the sum of availability destruction and outflow. An exergy flow diagram which accounts for conversion and partial destruction of availability around a vapor-compression-refrigeration cycle is shown in Fig. 4-11. In the condenser section of the diagram the quantity $\phi_{Q,\text{cond}}$ is taken to be zero, since the heat removed ends up in the environment. Finally, the second-law efficiency in this example is $\epsilon = -\phi_{Q,\text{evap}}/w_c = 19.71/35.78 = 0.55$.

The preceding discussion and numerical example for a vapor-compression-refrigeration cycle consider irreversibilities solely within the equipment itself. To simplify the analysis, we have neglected flow losses in the evaporator and the condenser. Even if we accept this model, irreversibilities external to these two

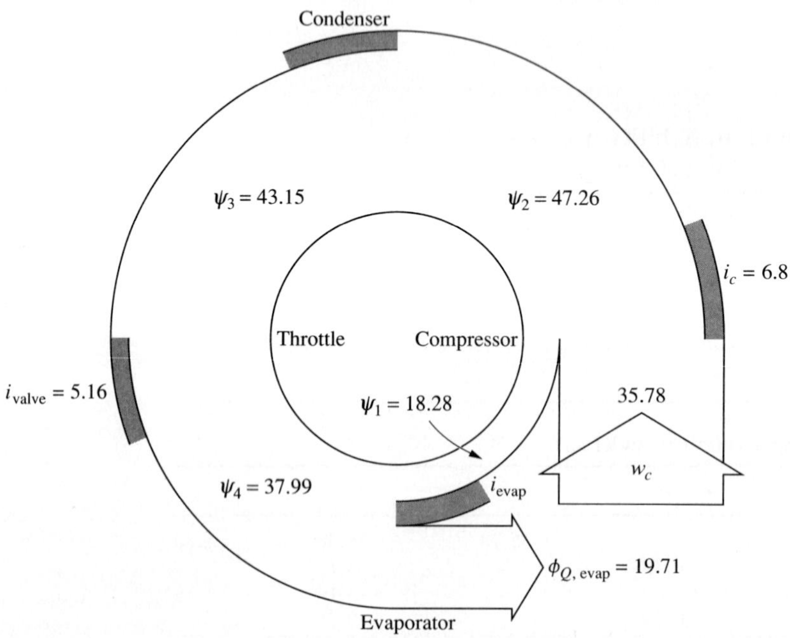

FIGURE 4-11
Availability flow diagram for a simple vapor-compression-refrigeration cycle showing transfers and losses for Example 4-5.

devices due to irreversible heat transfer should be considered. To this end, we shall assume the source of heat transfer to the evaporator fluid, the cold refrigerated region, is at a constant temperature T_{cold}. Then application of Eq. (3-45) to the evaporator heat-transfer process reveals that

$$\dot{I}_{\text{tot}} = \dot{W}_{\text{act}} + \dot{m}(b_{\text{in}} - b_{\text{out}}) + \dot{Q}_R\left(1 - \frac{T_0}{T_R}\right)$$

or for a unit mass through the evaporator

$$i_{Q,\text{evap}} = b_{\text{in}} - b_{\text{out}} + q_{\text{evap}}\left(1 - \frac{T_0}{T_{\text{cold}}}\right) \tag{4-27}$$

where $\Delta\psi$ could replace Δb. This equation is equivalent to that for I_Q for the heat-transfer process itself between the refrigerant and the cold region.

When Eq. (3-45) cited above is applied to the heat-transfer process in the condenser, it reduces to

$$i_{Q,\text{cond}} = b_{\text{in}} - b_{\text{out}} = h_{\text{in}} - h_{\text{out}} - T_0(s_{\text{in}} - s_{\text{out}}) \tag{4-28}$$

where Δb can be replaced by $\Delta\psi$. There is no reservoir term, as for the evaporator, because the heat is rejected to the atmosphere at T_0.

Example 4-6. Reconsider the vapor-compression-refrigeration cycle analyzed in Example 4-5. If the cold region is maintained at 0°C, determine the irreversibilities for the heat-transfer processes in the evaporator and condenser. Compare these values to those for the compressor and throttle device found in Example 4-5.

Solution. Based on the data in Tables 4-7 and 4-8, the irreversibility of the evaporator heat-transfer process is

$$i_{Q,\text{evap}} = 37.99 - 18.28 + 147.88\left(1 - \frac{298.15}{273.15}\right) = 19.71 - 13.53 = 6.18 \text{ kJ/kg}$$

For the heat transfer from the condenser

$$i_{Q,\text{cond}} = 47.26 - 43.15 = 4.11 \text{ kJ/kg}$$

These irreversibility values compare to 6.80 and 5.16 kJ/kg for the compressor and throttle valve, respectively. Thus the irreversibility of all four processes are of similar magnitude. An increase in the reversibility of the two heat-transfer processes requires smaller temperature differences in each device.

PROBLEMS

4-1. A 2T-heat engine, which is internally reversible, operates between a heat source at 900 K and an environment at T_0 = 300 K. Each of the two temperature differences associated with the heat supply to and heat rejection from the heat engine is 20 K. Calculate (a) the rates of exergy loss for the two heat-transfer processes, in kJ/s, and (b) the second-law effectiveness, if the power output is 5.0 MW.

4-2. A refrigerator operates between a cold space at 255 K and the atmosphere at 300 K. The actual heat removal at 255 K is 1000 kJ/min and the actual COP is 4.00 for the refrigeration device. (a) If the heat-transfer processes are reversible, determine the irreversibility, in kJ/min, and the effectiveness of the refrigeration device. Now, allow a 5°C difference for each heat-transfer process and an increase in the work input of 20 percent. (b) Determine the irreversibilities of the refrigeration device, the high-temperature heat exchange, and the low-temperature heat exchange, in kJ/min, and the system effectiveness.

4-3. An industrial heat pump operates with a COP of 3.60 between supply and rejection temperatures of $T_0 = 280$ K and $T_u = 360$ K. The heat exchange processes are assumed to be reversible. Determine (a) the second-law efficiency and (b) the irreversibility of the heat pump for a heat rejection at 360 K of 1080 kJ/min, in kJ/min.

4-4. An actual heat engine receives Q_H of 1000 kJ from $T_H = 1000$ K and rejects a Q_L of 400 kJ to a reservoir $T_L = 300$ K. $T_0 = 300$ K. The heat-transfer processes are reversible. Determine

a. the entropy production σ for the engine, in kJ/K
b. the irreversibility of the engine, in kilojoules
c. the second-law efficiency for the device
d. the first-law efficiency by means of Eq. (4-6)

4-5. An actual refrigerator absorbs 500 kJ of heat from a sink at 250 K and rejects 660 kJ of heat to the environment at 300 K. $T_0 = 300$ K. The heat-transfer processes are reversible. Determine

a. the entropy generation σ for the refrigerator, in kJ/K
b. the irreversibility of the device, in kilojoules
c. the exergy deposited in the T_L cold space, in kilojoules
d. the second-law efficiency for the device
e. the actual COP of the device

4-6. An actual heat pump supplies 800 kJ of heat to a reservoir at 450 K and receives 450 kJ of heat from the atmosphere at 300 K. The heat transfers are reversible. Determine

a. the entropy production σ for the heat pump, in kJ/K
b. the irreversibility of the heat pump, in kilojoules
c. the availability deposited in the high-temperature region, in kilojoules
d. the second-law efficiency of the device
e. the actual COP of the device

4-7. A heat engine operates between two heat reservoirs with temperatures of 560 and 280 K. Let $T_0 = 280$ K.

a. The heat engine itself has a temperature of 560 K during heat addition from the 560 K reservoir and a temperature of 280 K during heat rejection to the 280 K reservoir. The thermal efficiency of the actual unit is 40 percent and the heat-transfer rate from the 560 K reservoir is 1000 kJ/min. Determine the irreversibility of the engine, in kJ/min.

b. Now, for the same reservoir temperatures, heat-transfer rate, and engine thermal efficiency the engine itself receives heat at 540 K and rejects heat at 300 K. Determine the irreversibility, in kJ/min, for (1) the heat engine, (2) the high-temperature heat supply process, and (3) the low-temperature heat-rejection process, and (4) find the effectiveness. (5) Which process has the largest irreversibility?

c. Finally, for the same reservoir temperatures, heat-transfer rate, and engine temperatures cited in part (b) the thermal efficiency is increased to 0.42. For these conditions, determine the irreversibility in kJ/min, for (1) the heat engine, (2) the high-temperature heat supply process, and (3) the low-temperature heat-rejection process. (4) Which process now has the largest irreversibility?

4-8. A refrigeration unit operates between two heat reservoirs with temperatures of 240 and 320 K. The refrigeration unit itself has a temperature of 230 K during heat addition from the 240 K reservoir and a temperature of 340 K during heat rejection to the 320 K reservoir. The COP of the actual unit is 2.00 and the heat-transfer rate from the 240 K reservoir is 1000 kJ/min. T_0 is 300 K.

a. Determine the irreversibility, in kJ/min, for (1) the refrigeration unit, (2) the low-temperature heat-transfer process, and (3) the high-temperature heat-transfer process, and (4) find the effectiveness. (5) Which process has the largest irreversibility?

b. Now, for the same temperatures and heat-transfer rate, reduce the actual COP to 1.50. Again, determine the irreversibility for (1) the refrigeration unit, (2) the low-temperature heat-transfer process, and (3) the high-temperature heat-transfer process. (4) Which process has the largest irreversibility?

4-9. A heat pump unit operates between two heat reservoirs with temperatures of 270 and 330 K. The heat pump itself has a temperature of 255 K during heat addition from the 270 K reservoir and a temperature of 350 K during heat rejection to the 330 K reservoir. The COP of the actual unit is 3.40 and the heat transfer from the 270 K reservoir is 2000 kJ/min.

a. Determine the irreversibility, in kJ/min, for (1) the heat pump unit, (2) the low-temperature heat-transfer process, and (3) the high-temperature heat-transfer process, and (4) find the effectiveness. (5) Which process has the largest irreversibility?

b. Now, for the same temperatures and a heat-transfer rate of 2000 kJ/min, reduce the actual COP to 3.10. Again, determine the irreversibility for (1) the heat pump unit, (2) the low-temperature heat-transfer process, and (3) the high-temperature heat-transfer process. (4) Which process has the largest irreversibility?

4-10. A heat pump operates between two heat reservoirs with temperatures of 260 and 340 K. The heat pump unit itself has a temperature of 250 K during heat addition from the 260 K reservoir and a temperature of 350 K during heat rejection to the 340 K reservoir. The COP of the heat pump itself is 2.40, and the heat-transfer rate to the 340 K reservoir is 10,000 kJ/min. If $T_0 = 260$ K, determine the irreversibility, in kJ/min, for

a. the heat pump unit itself
b. the low-temperature heat-transfer process
c. the high-temperature heat-transfer process
d. Then find the second-law efficiency for the device.
e. Finally, find what COP of the heat pump is necessary for the irreversibility of the low-temperature process to be one-half that of the heat pump.

4-11. An irreversible gas-turbine power plant has a compressor with an inlet state of 1 bar and 22°C and an outlet state of 4 bars and 470 K. The turbine has an inlet state of 4 bars and 1200 K and an exhaust state of 1 bar and 900 K. The environment is at 1 bar and 22°C. Neglect kinetic- and potential-energy changes. Determine

a. the actual compressor and turbine work, in kJ/kg
b. the change in stream availability for the compressor and the turbine, in kJ/kg

c. the irreversibility for the compressor and the turbine, in kJ/kg
d. the ratio of the stream availability of the turbine exhaust to the net work output of the cycle
e. the cycle effectiveness

4-12. With reference to Prob. 4-11, consider that the turbine exhaust stream goes to a regenerative heat exchanger and is cooled to 580 K and 1 bar. Neglect pressure losses in the heat exchanger. Determine

a. the stream availability of the turbine exhaust stream as it leaves the regenerative heat exchanger, in kJ/kg
b. the fraction of the stream availability which remains, compared to the value at the turbine exhaust
c. the ratio of the stream availability at the regenerator exit to the net work output of the cycle
d. the irreversibility of the regenerative heat exchanger, in kJ/kg

4-13. With reference to Prob. 4-12, assume that frictional resistance to flow causes a finite pressure drop in the heat exchanger. The final state of the turbine stream leaving the heat exchanger is now 0.97 bar, and the final pressure of the air from the compressor which leaves the heat exchanger is 3.95 bars. Neglecting the effect of these pressure changes on compressor and turbine performances, determine

a. the stream availability of the turbine exhaust leaving the regenerator, in kJ/kg
b. the irreversibility or lost work of the regenerative heat exchanger under the new conditions, in kJ/kg

4-14. The pressure ratio of an air-standard gas-turbine cycle is 6.0/1, and the inlet conditions to the compressor are 105 kPa and 27°C. The turbine is limited to a temperature of 1100 K, and the compressor and turbine adiabatic efficiencies are 80 and 85 percent, respectively. The environmental state is 100 kPa and 27°C. Determine

a. the actual compressor and turbine work, in kJ/kg
b. the change in stream availability for the compressor and the turbine, in kJ/kg
c. the irreversibility for the compressor and the turbine, in kJ/kg
d. the ratio of the stream availability of the turbine exhaust to the net work output of the cycle
e. the cycle effectiveness

4-15. With reference to Prob. 4-14, suppose that the turbine exhaust is now passed through a heat exchanger (regenerator) so that it preheats the stream leaving the compressor to a state of 730 K and 630 kPa before it enters the combustor. Determine

a. the stream availability of the turbine exhaust after it passes through the regenerator and enters the environment, in kJ/kg
b. the ratio of the exhaust stream availability in part (a) to the net work output of the cycle
c. the irreversibility, in kJ/kg, for the overall heat exchange process in the regenerator

4-16. With reference to Prob. 4-15, the pressure ratio across the compressor and turbine remain at 6.0/1. However, assume that due to friction in the regenerator the final state of the gas leaving the heat exchanger is 100 kPa and the final pressure of the air entering the combustor is 624 kPa. Neglect any pressure change across the combustor. Determine (a) the availability of the exhaust gas leaving the regenerator and (b) the irreversibility of the overall regeneration process, in kJ/kg, if the temperatures used in Prob. 4-15 remain the same.

4-17. A gas-turbine cycle operating on air as the working fluid has a compressor with an inlet environmental state of 295 K and 1 bar and an exit state of 540 K and 6.4 bars. The turbine inlet state is 1080 K and the exit state is 730 K. Neglect kinetic- and potential-energy changes. Determine

a. the actual compressor and turbine work, in kJ/kg
b. the change in stream availability for the compressor and the turbine, in kJ/kg
c. the irreversibility for the compressor and the turbine, in kJ/kg
d. the ratio of the stream availability of the turbine exhaust to the net work output of the cycle
e. the cycle effectiveness

4-18. With reference to Prob. 4-17, suppose that the exhaust stream from the turbine goes to a regenerative heat exchanger and is cooled to 600 K and 1 bar while heating air from the compressor at 6.4 bars.

a. Find the stream availability of the exhaust gas leaving the heat exchanger, in kJ/kg.
b. Find the irreversibility of the regenerative process, in kJ/kg.

4-19. With reference to Problem 4-18, suppose frictional resistance to flow causes a finite pressure drop. The final state of the exhaust gas leaving the heat exchanger is 0.96 bar, and the final pressure of the air from the compressor and leaving the heat exchanger is 6.34 atm. Neglecting the effect of these pressure changes on compressor and turbine performance, find

a. the new stream availability of the regenerative exhaust gas, in kJ/kg
b. the irreversibility for the new regenerative process, in kJ/kg

4-20. A simple steam power system has a turbine inlet state of 6.0 MPa and 540°C and an exhaust pressure of 0.008 MPa. The cooling water in the condenser experiences a temperature rise from 20 to 27°C and T_0 is 20°C. The adiabatic efficiencies of the turbine and pump are 85 and 70 percent, respectively. Determine, in kJ/kg of steam,

a. the availability changes across the boiler, turbine, condenser, and pump
b. the irreversibilities in the turbine, pump, and condenser.
c. Then find the effectiveness of the cycle.

4-21. The simple steam power cycle in Prob. 4-20 is modified in the following manner. Steam entering the turbine at 6.0 MPa and 540°C is expanded to 0.50 MPa, reheated reversibly to 500°C, and expanded to 0.008 MPa. Determine, in kJ/kg,

a. the availability changes across the two turbine stages, condenser, reheat section, boiler, and pump
b. the irreversibilities in the two turbine stages, condenser, and pump.
c. Then find the effectiveness of the cycle.

4-22. The simple steam power cycle in Prob. 4-20 is modified in the following way. Steam entering the turbine is expanded to 0.50 MPa, where a portion is bled to a single open heater and the remainder is expanded to 0.008 MPa. Determine, in kJ/kg of steam through the boiler,

a. the availability changes across the two turbine stages, condenser, feedwater heater, boiler, and pumps
b. the irreversibilities in the two turbine stages, condenser, feedwater heater, and pumps.
c. Then find the effectiveness of the cycle.

4-23. A simple steam power system has a turbine inlet state of 120 bars and 600°C and an exhaust pressure of 0.08 bar. The cooling water in the condenser experiences a temperature rise from 22 to 28°C and T_0 is 22°C. The adiabatic efficiencies of the turbine and pump are 85 and 70 percent, respectively. Determine, in kJ/kg,

a. the availability changes across the boiler, turbine, condenser, and pump
b. the irreversibilities in the turbine, pump, and condenser.
c. Then find the effectiveness of the cycle.

4-24. The simple steam power cycle in Prob. 4-23 is modified in the following manner. Steam entering the turbine at 120 bars and 600°C is expanded to 10 bars, reheated reversibly to 540°C, and expanded to 0.08 bar. Determine, in kJ/kg,

a. the availability changes across the two turbine stages, condenser, reheat section, boiler, and pump
b. the irreversibilities in the two turbine stages, condenser, and pump.
c. Then find the effectiveness of the cycle.

4-25. The simple steam power cycle in Prob. 4-23 is modified in the following way. Steam entering the turbine is expanded to 10 bars, where a portion is bled to a single open heater and the remainder is expanded to 0.08 MPa. Determine, in kJ/kg of steam through the boiler,

a. the availability changes across the two turbine stages, condenser, feedwater heater, boiler, and pumps
b. the irreversibilities in the two turbine stages, condenser, feedwater heater, and pumps.
c. Then find the effectiveness of the cycle.

4-26. Saturated-R-134a vapor enters the compressor of a vapor-compression-refrigeration cycle at 1.4 bars; saturated liquid enters the expansion valve at 9 bars. T_0 is 25°C. If the compressor outlet temperature is 50°C and the cold region is at -10°C, determine, in kJ/kg,

a. the availability changes across the compressor, condenser, expansion valve, and evaporator
b. the irreversibilities associated with the compressor, expansion valve, evaporator, and condenser.
c. Then find the effectiveness of the cycle.

4-27. Saturated-R-134a vapor enters the compressor of a vapor-compression-heat pump cycle at 2.8 bars; saturated liquid enters the expansion valve at 10 bars. T_0 is 25°C. If the compressor outlet temperature is 50°C and the cold region is at 5°C, determine, in kJ/kg,

a. the availability changes across the compressor, condenser, expansion valve, and evaporator
b. the irreversibilities associated with the compressor, expansion valve, evaporator, and condenser.
c. Then find the effectiveness of the cycle.

CHAPTER 5

EQUATIONS OF STATE

Of broad interest in the analysis of energy systems are the properties—internal energy, enthalpy, and entropy, among others. Fundamentally, the evaluation of these properties requires knowledge of the PvT behavior of pure substances and mixtures. In this chapter PvT relations are examined for both the superheat and saturation states. The use of these relations to evaluate properties such as u, h, and s is demonstrated in the next chapter.

5-1 INTRODUCTION

On the basis of the state postulate for simple compressible substances any intensive property is solely a function of two other independent, intensive properties. Of great importance in thermodynamics are equilibrium relationships involving the pressure P, the temperature T, and the specific volume v. Such relations are known as *equations of state* and have the functional form

$$f(P, v, T) = 0 \tag{5-1}$$

In some cases it may be preferable to replace the specific volume by the density ρ in the above expression. [In general, a functional relationship among *any* three properties could be called an *equation of state*. However, by common usage the expression "equation of state" usually refers only to relations represented by Eq. (5-1).] Equations of state are particularly important for the gas and liquid phases. In a broad sense the actual functional relationship among the three properties in Eq. (5-1) may appear as an algebraic equation, a graph, or a set of tabular data. All are important formats for the presentation of PvT data. The initial discussion is directed toward algebraic equations of state.

An equation of state serves two useful purposes. Its most obvious use is in the prediction of the PvT behavior of a substance over the desired range of values. The equilibrium states of a simple compressible substance can be represented by a surface on PvT cartesian coordinates. The projection of a typical surface onto the Pv plane is shown in Fig. 5-1. The figure is presented using the reduced pressure P_r, the reduced temperature T_r, and the reduced volume v_r as parameters. These quantities are defined as

$$P_r \equiv \frac{P}{P_c} \qquad v_r \equiv \frac{v}{v_c} \qquad T_r \equiv \frac{T}{T_c} \tag{5-2}$$

where subscript "c" represents the critical state. Note that logarithmic scales are used in the coordinates to emphasize the wide range of pressures and specific volumes (or densities) of interest. Although the range of interest in T_r in practical processes is fairly small, the range of P_r and v_r might typically be several orders of magnitude, or greater. At relatively low densities (v_r much greater than 1), the figure shows no rapid variation in pressure with specific volume along any isotherm. As a result, a fairly simple mathematical expression can be used to represent an isotherm with reasonable accuracy. However, as the specific volume decreases to values equal to or less than the critical volume of a substance ($v_r \leq 1$), the pressure changes rapidly for small changes in density for a given temperature. Hence a fairly complex analytical equation of state will be required

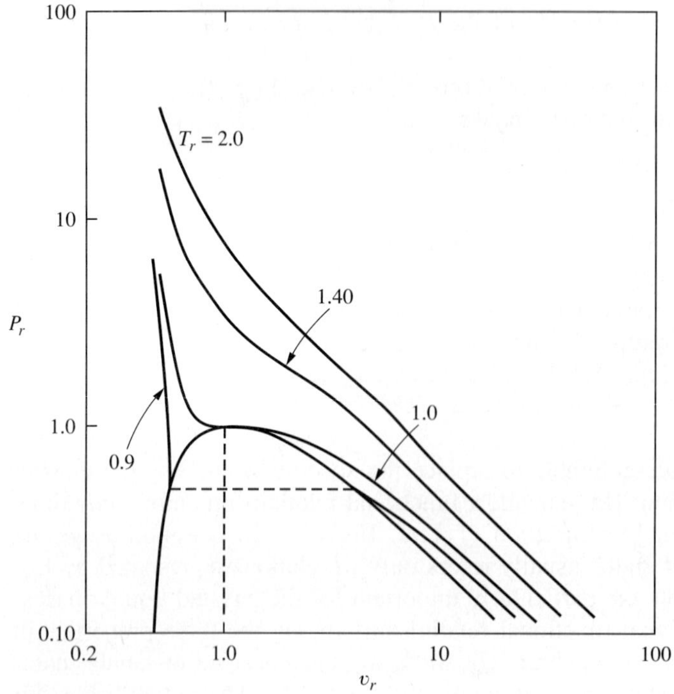

FIGURE 5-1
Experimental $P_r - v_r$ relationships for argon along isotherms T_r.

to represent with reasonable accuracy the pressure variation at constant temperature for a range of reduced volumes, for example, from 100 to 0.40 (a 250-fold variation).

This picture is complicated even further when the reduced temperature is less than unity. As is well known, an isotherm becomes horizontal in the two-phase region separating the saturated-liquid and saturated-vapor lines. For the study of liquid-vapor equilibrium it would be important for an equation of state to represent the saturation states as well as the gas and liquid single-phase regions. This places an additional requirement on an equation of state.

The other major use of PvT data is in the evaluation of thermodynamic property data which are not directly measurable. Properties of interest include u, h, s, c_p, c_v, the Gibbs function g, the Helmholtz function a, and functions related to the fugacity f. The theoretical relationships for these properties contain first and second derivatives involving P, v, and T. Hence it is important that the mathematical format of an equation of state lend itself to the required differentiation and subsequent integration. In addition, careful examination of Pv and PT diagrams shows that *all* substances have certain PvT characteristics that are unique. An equation of state should, as much as possible, exhibit these same unique characteristics if it is to predict successfully fundamental property values. This makes the choice of the format of an algebraic equation of state even more restrictive.

In the development of an equation of state there are three primary decisions to be made:

1. the range of values of interest, especially with respect to density or specific volume
2. the precision or accuracy desired for the PvT data as output
3. the amount and kind of data that will be required as input for evaluation of parameters within the equation

With respect to these items, a fairly simple equation will often suffice for low densities, whereas a more complicated equation may be required if both low and high densities are to be represented with reasonable accuracy. However, a more complex equation of state containing more adjustable constants is not always better than a simpler model. In some instances only a very limited amount of experimental data may be available. In other situations the equation needs to represent a large amount of experimental data within the precision of the measurements. The equations of state representing these two extremes would be quite different.

5-2 CHARACTERISTICS OF *PvT* BEHAVIOR

Before we examine specific equations of state, it is important to establish some criteria which these models should fulfill. As noted previously, all substances exhibit certain PvT behavior which is unique and identifiable. (Recall Fig. 5-1.)

At the critical state the critical isotherm passes through a point of horizontal slope and inflection. This requires that

$$\left(\frac{\partial P}{\partial v}\right)_{T_c} = 0 \tag{5-3}$$

and

$$\left(\frac{\partial^2 P}{\partial v^2}\right)_{T_c} = 0 \tag{5-4}$$

Hence an equation of state which models reality should fulfill the two conditions established by Eqs. (5-3) and (5-4).

A second well-known characteristic of gases is the shape of isometric (constant-volume) lines on a *PT* diagram, as shown in Fig. 5-2. All the isometric lines are nearly straight (linear). In particular, the critical isometric is essentially straight for all substances. In addition, the slope of the critical isometric is equal to the slope of the vapor-pressure curve right at the critical state. Matching the slope of the vapor-pressure curve at the critical state is another possible criterion for an equation of state. If the vapor-pressure curve of Fig. 5-2. is replotted for various substances as $\log_{10} P_r$ versus $1/T_r$, Fig. 5-3 results. Curves drawn through experimental data indicate that the lines are nearly straight. Also, the negative of the slope of these lines is directly related to $(dP/dT)_{\text{crit}}$, which is the slope of the vapor-pressure curve at the critical state. Recall that, mathematically,

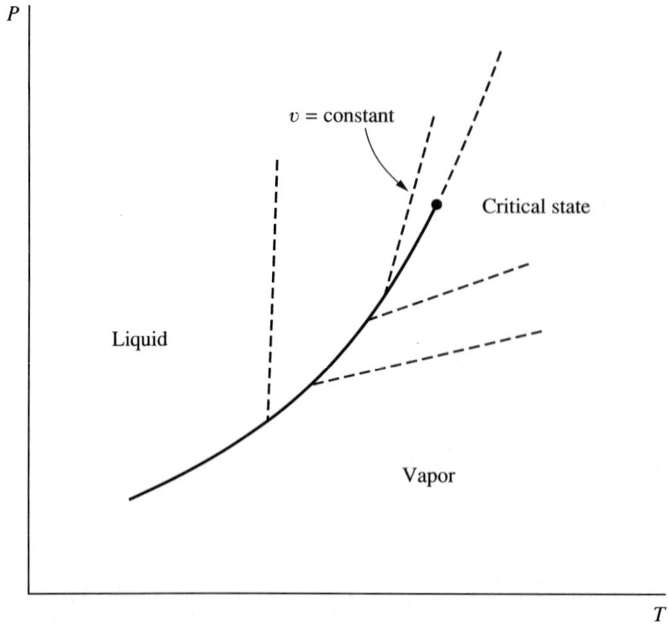

FIGURE 5-2
PT diagram showing the vapor-pressure curve for a pure substance and constant-volume lines in the single-phase regions.

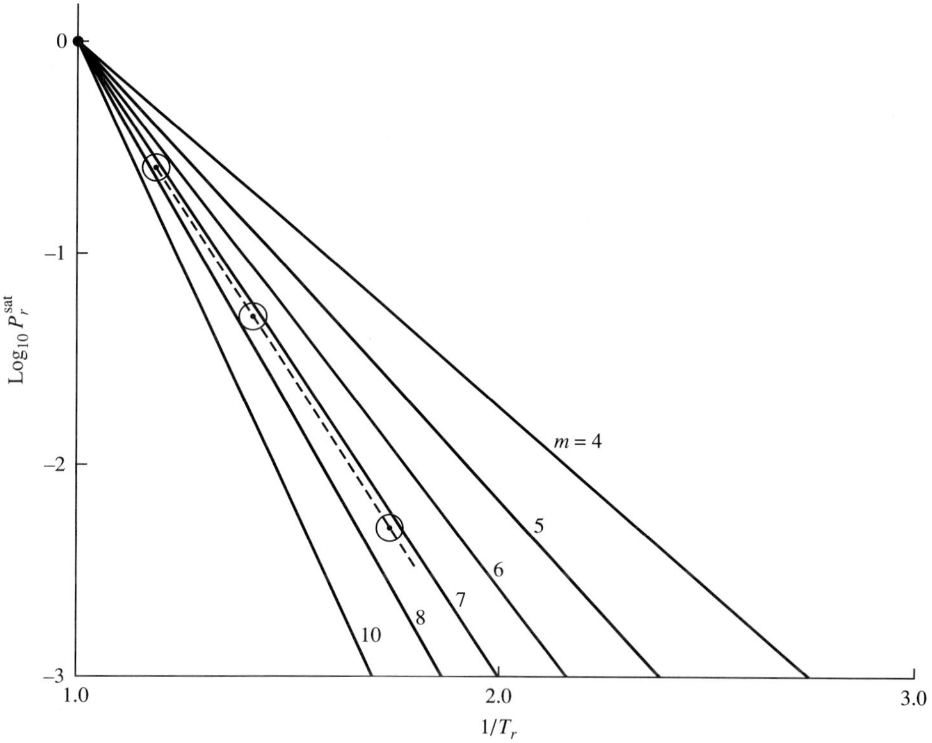

FIGURE 5-3
Generalized vapor-pressure plot.

$$\frac{d \ln P_r^{\text{sat}}}{d(1/T_r)} = -\frac{T_r^2}{P_r}\frac{dP_r}{dT_r}$$

In terms of the slope at the critical state

$$\frac{d \log_{10} P_r^{\text{sat}}}{d(1/T_r)} = \frac{-1}{2.303}\left(\frac{dP_r}{dT_r}\right)_{\text{crit}} = -0.434\frac{T_c}{P_c}\left(\frac{dP}{dT}\right)_{\text{crit}} = -0.434m \quad (5\text{-}5)$$

where
$$m \equiv \left(\frac{dP_r}{dT_r}\right)_{\text{crit}} = \frac{T_c}{P_c}\left(\frac{dP}{dT}\right)_{\text{crit}} \quad (5\text{-}6)$$

The lines in Fig. 5-3 represent various values of m. The value of m for a given substance is uniquely fixed by its vapor-pressure data. However, when dP_r/dT_r is found from an equation of state (written on a reduced basis) for $v_r = 1.0$, this derivative should have the same value as that of m determined from vapor-pressure data. If it does not, then the equation of state probably may not be a good representation of actual PvT data. The slope of the critical isometric on a PT diagram is given by mP_c/T_c.

Another useful point arises from Fig. 5-3. If linearity is assumed throughout the whole range of liquid-vapor data, the complete vapor-pressure curve of any substance can be predicted solely from data at the critical state and one other vapor-pressure point.

Example 5-1. (*a*) Estimate the constants A and B in the linear vapor-pressure equation $\log_{10} P_r = A + (B/T_r)$ for steam, using data at the critical state and another saturation state near the critical state. (*b*) Estimate the value of m for steam.

Solution

(*a*) The pressure and temperature at the critical state are 221.2 bars and 647.30 K. At 50 bars the saturation temperature is 264.0°C or 537 K. In terms of reduced pressure and temperature these values are (1, 1) and (0.226, 0.830). Consequently, for the two sets of data

$$\log_{10} 1 = A + \frac{B}{1}$$

and
$$\log_{10} 0.226 = A + \frac{B}{0.830}$$

From the first relation $A = -B$. Hence the second equation becomes

$$-0.646 = A - 1.205A$$

$$A = 3.15 = -B$$

Therefore
$$\log_{10} P_r = 3.15 - \frac{3.15}{T_r}$$

(*b*) The slope of $\log_{10} P_r$ versus $1/T_r$ is given by the value of B. Thus Eq. (5-5) leads to the relation

$$-3.15 = -0.434m$$

or
$$m = 7.2$$

The value of 7.2 for water vapor can be checked by plotting saturation data for water at 1, 10, and 50 bars. The saturation line for water through these three data points is shown as a dashed line in Fig. 5-3. The line, essentially linear for this wide range of pressures, lies between lines representing m values of 7 and 8 on the figure.

Experimental evidence indicates that there are other general PvT characteristics which gases exhibit. Among these are the following:

1. As the pressure approaches zero, gases behave as ideal gases and the compressibility factor Z ($\equiv Pv/RT$) approaches one. That is,

$$\lim_{P \to 0} \left(\frac{Pv}{RT} \right) = \lim_{P \to 0} Z = 1$$

This trend is demonstrated by the fact that all T_r lines on a Z versus P_r plot (a generalized compressibility chart) approach a Z value of 1 as $P_r \to 0$.

2. Also, as the pressure approaches zero, the difference between the specific volume as predicted by the ideal-gas equation and the actual specific volume approaches a constant, nonzero value known as the residual volume α. The value of α is solely a function of the temperature. Hence

$$\lim_{P\to 0}\left(\frac{RT}{P} - v\right) = \alpha = f(T)$$

Figure 5-4 shows a plot of $(RT/P - v)$ versus P for CO and CO_2. The extrapolation of the data at a given temperature to zero pressure provides the value of the residual volume α at that temperature. An equation of state should be capable of predicting α values with reasonable accuracy.

3. With reference to Fig. 5-1, it is noted that the critical isotherm is essentially vertical as the pressure becomes very large. That is, the specific volume becomes a constant at the critical temperature and high pressure. This constant generally is in the vicinity of 0.26 of the critical specific volume; consequently,

$$\lim_{P\to\infty} (v)_{T_c} \simeq 0.26 v_c$$

The Redlich-Kwong equation, to be discussed later, was devised empirically to satisfy this criterion.

4. The critical compressibility factor, Z_c, of any substance generally fits into the small range from 0.22 to 0.30. Hydrocarbons have factors normally in the 0.26 to 0.28 range, whereas most polar compounds have Z factors in the 0.24 to 0.26 range. Therefore an equation of state should predict that

$$Z_c \equiv \frac{P_c v_c}{RT_c} \simeq 0.22 - 0.30$$

for all gases.

5. As mentioned earlier, the critical isometric line on a PT diagram is straight. Mathematically this requires that

$$\left(\frac{\partial^2 P}{\partial T^2}\right)_{v_c} = 0$$

in the superheat region.

There are a number of other characteristics of PvT data which generally are exhibited by all substances. One would not expect a simple PvT equation to satisfy all these criteria. However, one would hope that any equation of state would reproduce a number of these characteristics with reasonable accuracy. The necessity of fulfilling these criteria limits the mathematical nature of any empirical equation of state. The following section reviews some of the more important equations of state, from either a historical or modern computational viewpoint.

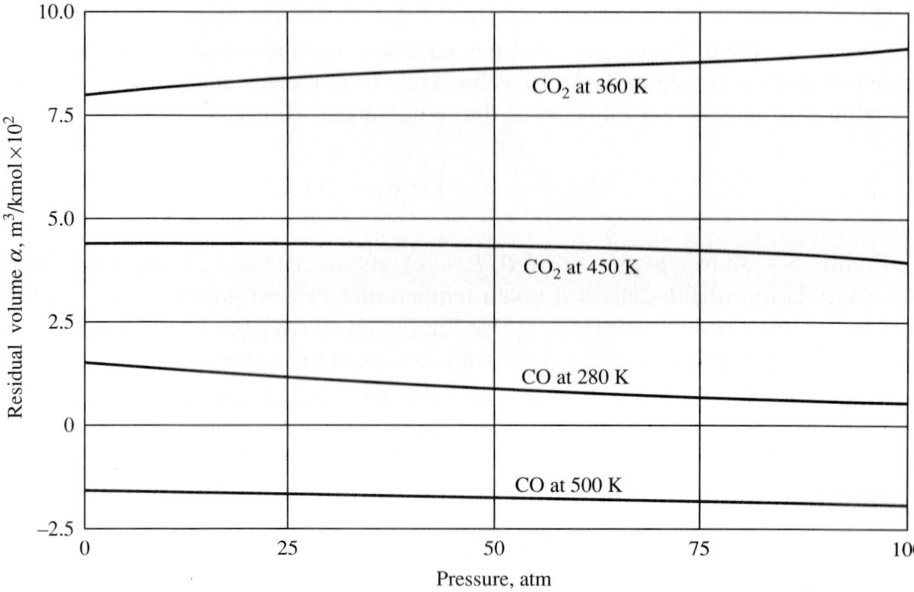

FIGURE 5-4
Isotherms for the residual volume α versus pressure for CO and CO_2.

5-3 THE COMPRESSIBILITY FACTOR

As stated in the preceding section, as the pressure approaches zero in value all gases behave as ideal gases, for which

$$Pv = RT \tag{5-7}$$

The simplest method mathematically to correct the ideal-gas equation so that it predicts real-gas behavior is to write it in the format

$$Pv = ZRT \tag{5-8}$$

where Z is known as the compressibility factor. Since $v = RT/P$ for an ideal gas, Eq. (5-8) can be written as

$$Z = \frac{v_{\text{act}}}{v_{\text{ideal}}} \tag{5-9}$$

Thus Z is a measure of the actual specific volume, compared to the ideal-gas specific volume at the same temperature and pressure.

The compressibility factor is a property which varies with the nature of the gas, as well as with two independent intensive properties such as temperature and pressure. Figure 5-5 shows the variation of Z with pressure for nitrogen for a few chosen temperatures. As noted from the figure, an actual gas might have a specific volume which is anywhere from at least one-half to twice that of an ideal gas at the same temperature and pressure. For gases in general the range is somewhat greater than that.

For most pure gases not enough PvT data are available to construct a plot such as the one in Fig. 5-5. To overcome this difficulty, we shall present some

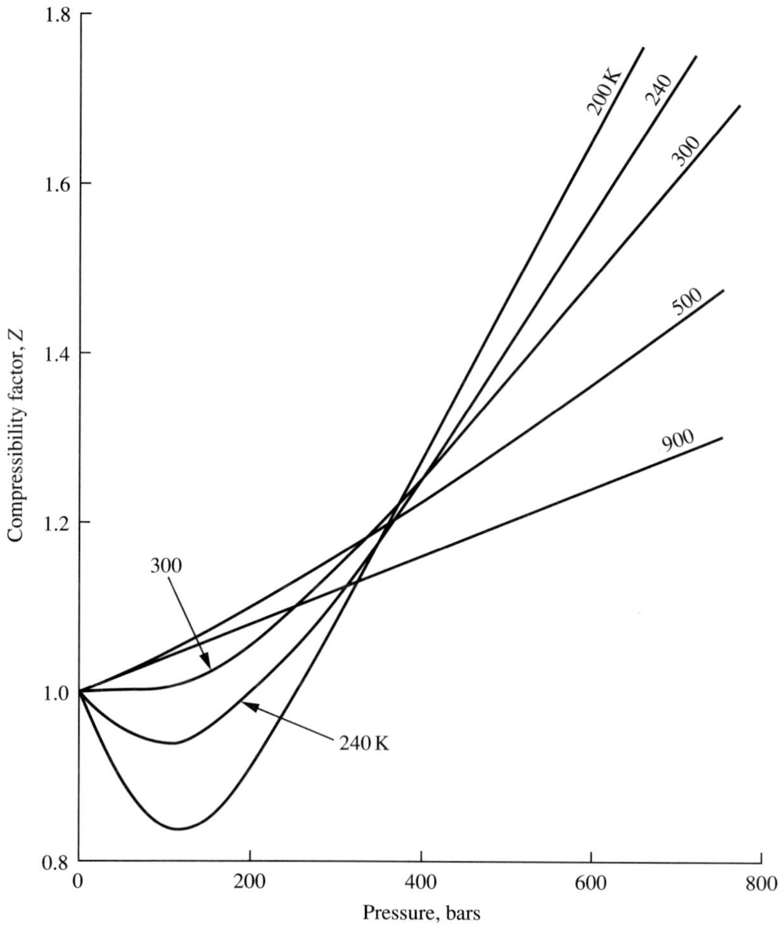

FIGURE 5-5
Compressibility plot for nitrogen gas.(*Source:* Based on data from *Thermodynamic Properties of Nitrogen,* V. V. Sychev et al., National Standard Reference Data Service of the USSR, Hemisphere Publishing Corp., New York, 1987.)

alternate methods for estimating values of Z later in this chapter. The concept of the compressibility factor is an important one, because it introduces a single dimensionless parameter to predict real-gas behavior, and its limiting value at zero pressure is unity. In addition, many other more complex equations of state can be rewritten into the format of the parameter Z.

5-4 TWO-PARAMETER CUBIC EQUATIONS OF STATE

When very little is known about the PvT data for a substance, polynomial equations which are cubic in the specific volume are quite useful. They are the simplest expressions capable of predicting both vapor and liquid behavior. In addition, such equations are a reasonable compromise between simplicity and generality. When P or T is the dependent variable in equations of state in this and later sections, a calculator is usually sufficient for the numerical evaluation. However, when the specific volume is the dependent variable, computer software lends itself to fast and accurate numerical solutions.

5-4-1 van der Waals Equation

The first cubic equation of state was proposed by J. D. van der Waals in 1873. For an ideal gas the volume occupied by the particles is assumed to be infinitesimal compared to the total volume, and attractive and repulsive forces are negligible. In actuality, the attractive forces among particles, frequently termed the *van der Waal forces,* reduces the pressure below the value predicted by the ideal-gas equation. This pressure correction, sometimes called the internal pressure, can be shown by kinetic theory to be inversely proportional to the square of the specific volume. Thus

$$P_{\text{act}} = P_{\text{ideal}} - \frac{a}{v^2}$$

or

$$P_{\text{ideal}} = P_{\text{act}} + \frac{a}{v^2}$$

where a is a proportionality constant.

In addition, the particle size is finite and repulsive forces are present when particles find themselves close together. Thus the volume available to the molecules is somewhat less than the total volume. On a specific-volume basis the quantity v must be replaced by $(v - b)$ in the ideal-gas equation. Hence the theoretical van der Waals equation of state is expressed by

$$P = \frac{RT}{v-b} - \frac{a}{v^2} \tag{5-10}$$

The quantities a and b usually are assumed to be constant for a given substance, but they are different for various substances for a given set of units.

The constants in any equation of state may be evaluated by a fit to available PvT data. On the absence of sufficient data, however, the constants in simple equations of state may be evaluated by application of one or more of the criteria discussed in the preceding section. The procedure used by van der Waals involved the use of Eqs. (5-3) and (5-4), namely,

$$\left(\frac{\partial P}{\partial v}\right)_{T_c} = 0 \quad \text{and} \quad \left(\frac{\partial^2 P}{\partial v^2}\right)_{T_c} = 0$$

Application of these two requirements to Eq. (5-10), and the use of Eq. (5-10) itself at the critical state, yields the following three equations:

$$\left(\frac{\partial P}{\partial v}\right)_{T_c} = \frac{-RT_c}{(v_c - b)^2} + \frac{2a}{v_c^3} = 0 \tag{5-11}$$

$$\left(\frac{\partial^2 P}{\partial v^2}\right)_{T_c} = \frac{2RT_c}{(v_c - b)^3} - \frac{6a}{v_c^4} = 0 \tag{5-12}$$

$$P_c = \frac{RT_c}{v_c - b} - \frac{a}{v_c^2} \tag{5-13}$$

For a given substance the values of T_c, P_c, and v_c are known. Thus the use of the critical state criteria leads to three equations in terms of two unknowns, a and b.

Any two of the three preceding equations could be selected for the solution. However, it is found that only the use of Eqs. (5-11) and (5-12) leads to a critical isotherm which has zero values of $(\partial P/\partial v)_{T_c}$ and $(\partial^2 P/\partial v^2)_{T_c}$ at the critical state. In this case

$$a = \frac{9RT_c v_c}{8} \tag{5-14}$$

and

$$b = \frac{v_c}{3} \tag{5-15}$$

When these two equations are placed into Eq. (5-13), which still is a valid equation, the result is

$$P_c = \frac{3RT_c}{8v_c} \tag{5-16}$$

Although T_c and v_c are the independent variables in Eqs. (5-14) and (5-15), the use of Eq. (5-16) permits a and b to be expressed also in terms of v_c and P_c, or T_c and P_c. An examination of experimental data [1] indicates that the best choice of independent variables is T_c and P_c. The values of a and b then become

$$a = \frac{27R^2 T_c^{\ 2}}{64P_c} \tag{5-17}$$

and

$$b = \frac{RT_c}{8P_c} \tag{5-18}$$

Values of a and b for a few common substances are given in Table A-23 in the Appendix. Equations (5-17) and (5-18) give values of a and b that are different from those obtained from Eqs. (5-14) and (5-15) for a given substance. This occurs because Eq. (5-16) is not valid when actual data are substituted. Rearrangement of Eq. (5-16) shows that

$$Z_c \equiv \frac{P_c v_c}{RT_c} = 0.375$$

The factor Z_c, known as the critical compressibility factor, is always 3/8 for a van der Waals gas, whereas for most real gases it ranges between 0.22 and 0.30.

This discrepancy between the theoretical and actual values of Z_c leads to another problem. The use of Eqs. (5-17) and (5-18) in van der Waals equation will lead to a critical isotherm which does not pass through the critical state on a Pv diagram. In Fig. 5-6 the critical isotherm for an actual gas is compared to that predicted from the van der Waals equation using actual values of P_c and T_c

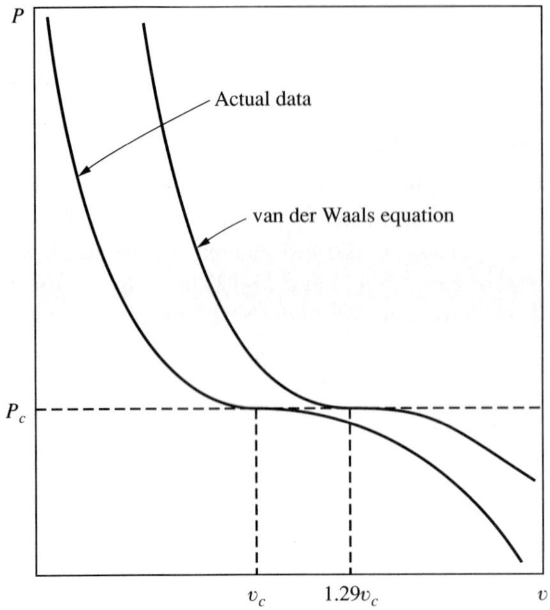

FIGURE 5-6
Critical isotherms for argon.

for the gas. The general shape is correct, but the van der Waals isotherm exhibits a critical volume greater than the true value. The reason is that Z_c for a van der Waals (vdW) gas is too large. With reference to Fig. 5-6, it may be shown that

$$\frac{v_{c,\text{vdW}}}{v_{c,\text{act}}} = \frac{Z_{c,\text{vdW}}}{Z_{c,\text{act}}}$$

For argon, for example, with a Z_c value of 0.291, the critical volume ratio in the above equation is 1.29. Thus the critical state in Fig. 5-6 for the van der Waals isotherm is moved roughly 30 percent to the right from the true position if the gas were argon.

Although the van der Waals equation is not very good in predicting real-gas behavior except at low pressures, the preceding discussion demonstrates one important point: The two constants in any cubic equation of state may be determined more accurately by using experimental PvT data to fit the constants. However, in the absence of such data the criteria of PvT behavior at the critical state as expressed by Eqs. (5-3) and (5-4) is commonly used to evaluate a and b.

For given values of P and T the van der Waals equation will yield three volume roots, two of which may be complex. A root is physically significant only if it is real, positive, and greater than b. Three specific situations exist:

1. When $T > T_c$, there are two imaginary roots and one real positive root for v. The pressure has a monotonically decreasing value as v increases, as shown by the curve marked $T > T_c$ on Fig. 5-7.

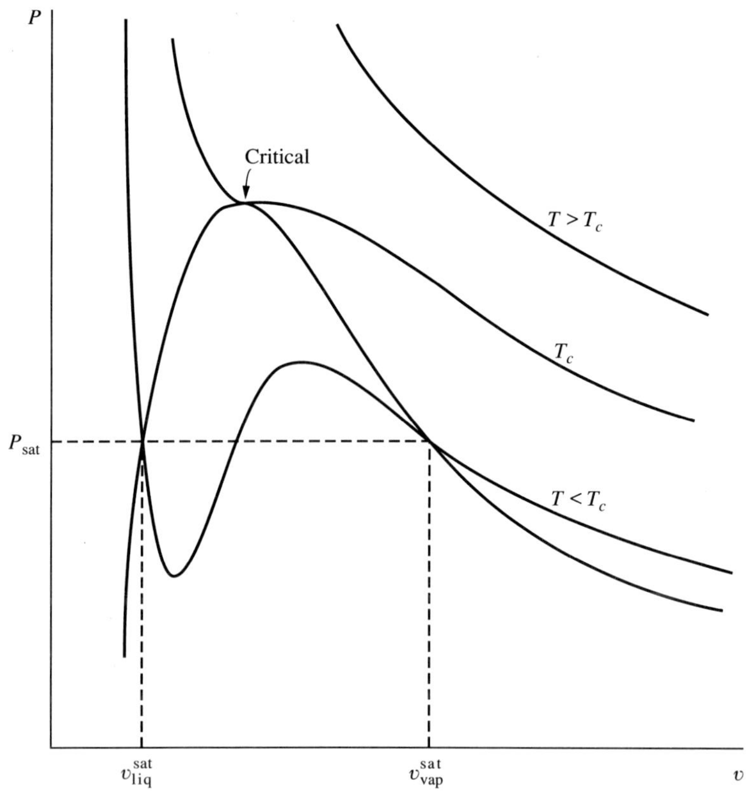

FIGURE 5-7
Isotherms as given by a cubic equation of state.

2. For the critical isotherm, where $T = T_c$, again only one real positive root exists, except at the critical pressure. At P_c the equation has three real and equal roots, all equal to v_c. The isotherm also has a horizontal inflection at P_c, as shown in Fig. 5-7.
3. When $T < T_c$, one real positive root occurs at high pressures. Over a range of lower pressures, however, the three roots are real and unequal. The isotherm exhibits both a minimum and a maximum, as shown by the curve marked $T < T_c$ on Fig. 5-7.

We recognize that this latter curve is not representative of experimental isotherms when the temperature is less than the critical value. In actuality, liquid and vapor phases coexist at some pressure and the isotherm becomes horizontal within the two-phase region. The horizontal dashed line in Fig. 5-7 represents one possible position of the actual isotherm. The outer intersections of the horizontal line with the van der Waals isotherm gives the approximate positions of the saturated-liquid and saturated-vapor states. The midintersection point has no physical significance. Thus the van der Waals equation (and other cubic equations of state) are capable of predicting both liquid and vapor behavior. (The method used to set the position of the dashed line is discussed in Sec. 6-10.) The van der Waals equation (and any other cubic equation of state) can be transformed into the form $Z = f(T_r, P_r)$. In this case

$$Z^3 - \left(\frac{P_r}{8T_r} + 1\right)Z^2 + \left(\frac{27P_r}{64T_r^2}\right)Z - \frac{27P_r^2}{512T_r^3} = 0 \tag{5-19}$$

where again $P_r = P/P_c$ and $T_r = T/T_c$. Either Eq. (5-10) or Eq. (5-19) may be used to find the relationship among P, v, and T for a substance modeled by the van der Waals equation.

5-4-2 Redlich-Kwong Equation

Although the van der Waals equation was developed in 1873, no major improvement in cubic equations of state occurred until the publication of the Redlich-Kwong (RK) equation in 1949 [2]. It has proved to be one of the more useful and widely accepted cubic equations of state and is considerably more accurate than the van der Waals equation. The two-parameter RK equation of state is given by

$$P = \frac{RT}{v - b} - \frac{a}{T^{1/2}v(v + b)} \tag{5-20}$$

Unlike the van der Waals equation, both terms on the right contain a temperature function. The RK equation does quite well at small to moderate densities and is much better when $T > T_c$. When experimental PvT data are not available for evaluating a and b, the criteria of the critical state are used. Setting $(\partial P/\partial v)_{T_c} = (\partial^2 P/\partial v^2)_{T_c} = 0$, and also employing Eq. (5-20), one finds that

$$a = \Omega_a \frac{R^2 T_c^{2.5}}{P_c} = 0.42748 \frac{R^2 T_c^{2.5}}{P_c} \tag{5-21}$$

$$b = \Omega_b \frac{RT_c}{P_c} = 0.08664 \frac{RT_c}{P_c} \tag{5-22}$$

$$Z_{c,\mathrm{RK}} = \frac{P_c v_c}{RT_c} = \frac{1}{3} \tag{5-23}$$

$$Z^3 - Z^2 + \left[\Omega_a \frac{P_r}{T_r^{2.5}} - \Omega_b \frac{P_r}{T_r}\left(1 + \Omega_b \frac{P_r}{T_r}\right)\right]Z - \Omega_a \Omega_b \frac{P_r^2}{T_r^{3.5}} = 0 \tag{5-24}$$

Values of a and b are listed in Table A-23 in the Appendix for a few substances. Note that the critical compressibility factor based on the critical state criteria is 1/3. This value is better than the van der Waals value of 3/8, but it is still somewhat higher than the typical range of experimental values.

Similar to the discussion of the van der Waals equation, the RK equation predicts liquid and vapor specific volumes when the temperature is below the critical value. The van der Waals and RK equations are easily solved when the independent variables are either v and T, or v and P. However, when P and

T are known, a cubic equation in v must be solved. As noted earlier, either programmable electronic calculators or computer software should be used to find the roots of the cubic equation.

5-5 THE CORRESPONDING STATES PRINCIPLE

With referral to Eqs. (5-19) and (5-24), both the van der Waals and RK equations of state have been written in the form

$$Z = f(T_r, P_r) \tag{5-25}$$

This functional relationship was first recognized by van der Waals in 1873. Thus the prediction of a unique relationship among Z, T_r, and P_r is known as the van der Waals two-parameter *theorem of corresponding states*. It states that any pure gas at the same reduced temperature and pressure should have the same compressibility factor. Expressions such as Eq. (5-25) are known as *generalized* equations, because of their general applicability to all substances. Since the van der Waals and RK equations apply to both the liquid and gas phases, the theorem expressed by Eq. (5-25) should also apply to both single phases, up to and including the saturated-vapor and saturated-liquid states.

The theorem of corresponding states has been tested by evaluating Z versus T_r and P_r from experimental PvT data for a sizable number of substances. Figure 5-8 shows Z plotted as a function of T_r and P_r for 10 common substances [3]. The overall average deviation from the mean isotherm lines is on the order of 1 percent. However, for many other substances the accuracy sometimes proves to be less than desirable for engineering calculations. The well-known Nelson-Obert [4] charts for Z as a function of (T_r, P_r), based on the average PvT behavior of 30 fluids, predict Z within 5–6 percent except near the critical state (Z less than 0.5), or near the saturation line. Figure A-36 in the Appendix shows a modified form of a Nelson-Obert chart for the low-pressure range ($P_r \leq 1$), where the saturation line is shown for both the vapor and liquid regions. Reasonable estimates of Z for either the saturated-vapor or saturated-liquid states can be made from this chart. Figures A-37 and A-38 in the Appendix are Nelson-Obert charts for the midpressure range, where P_r varies from 0 to 10, and for the high-pressure range, where P_r varies from 10 to 40. The two-parameter corresponding states correlation may be considerably in error for strongly polar substances. (A polar substance is one for which the centers of positive and negative charge do not coincide.) In such cases the estimate for Z may be off by as much as 15–20 percent. When used for hydrogen, helium, and neon (quantum gases), one suggestion for improving the correlation is to redefine the reduced pressure and temperature for these three gases in the following manner:

$$P_r = \frac{P}{P_c + C} \quad \text{and} \quad T_r = \frac{T}{T_c + C} \tag{5-26}$$

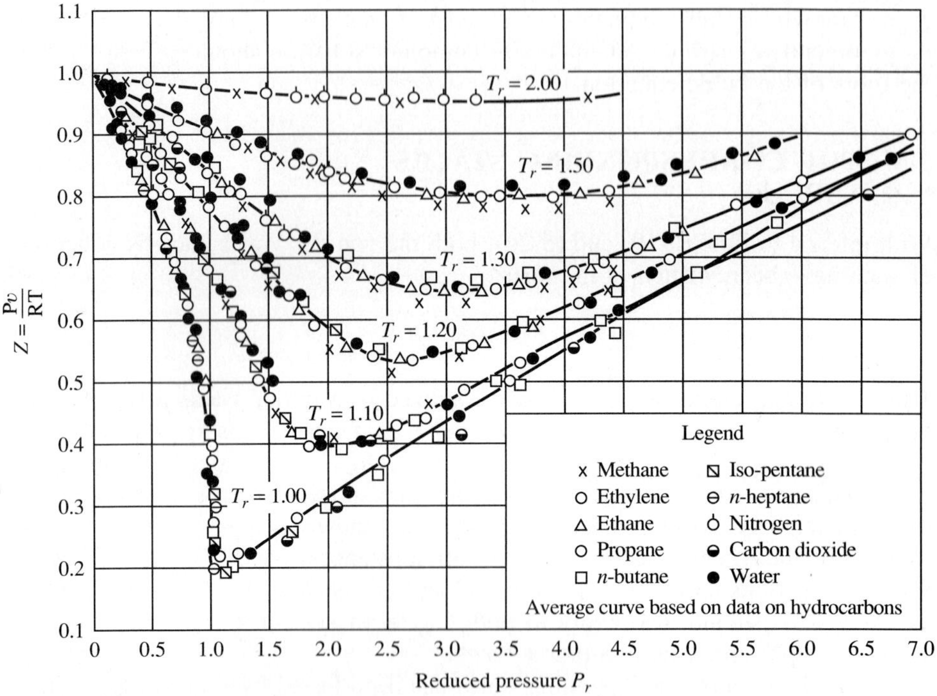

FIGURE 5-8
Experimental data correlation for a generalized Z chart. [K. Wark., *Thermodynamics,* 5th ed., McGraw-Hill, Inc., New York, 1988. *Source:* Gour-Jen Su: "Modified Law of Corresponding State," *Ind. Eng. Chem.* (Intern. ed.), **38**:803, 1946.] Reprinted with permission by the American Chemical Society.

If P is expressed in atmospheres and T in kelvins, then the value of C in both equations is 8.

A generalized Z chart is a graphical representation of the relation $Z = f(T_r, P_r)$. The variable v is introduced into generalized Z charts in the following manner. Substitution of the definitions of P_r, T_r, and v_r given by Eq. (5-2) into Eq. (5-8) which defines Z yields

$$Z = \frac{P_r v_r}{T_r}\frac{P_c v_c}{RT_c} = \frac{P_r}{T_r}(v_r Z_c) \tag{5-27}$$

where the critical compressibility factor $Z_c = P_c v_c / RT_c$ appears. Recall that Z_c ranges from 0.22 to 0.30 for most substances. Hence the above equation predicts that v_r will not correlate very well with T_r and P_r; however, the parameter $v_r Z_c$ should. Consequently, the variable v is introduced on the Nelson-Obert chart in the format of the pseudoreduced volume, v'_r, where

$$v'_r \equiv v_r Z_c = \frac{vP_c}{RT_c} = Z\frac{T_r}{P_r} \tag{5-28}$$

Equation (5-28) shows that although isometric lines of the form $v_r = v/v_c$ cannot be added to a Nelson-Obert-type chart, a redefined reduced volume line in the form of v_r' can be added. Lines of constant v_r' are also shown in Figs. A-36, A-37, and A-38 in the Appendix.

5-5-1 Three-Parameter Corresponding States Principle

Statistical theory of particle behavior shows that the two-parameter principle of corresponding states is fulfilled only if the intermolecular potentials of various types of particles are identical, except for the specific energy scale ϵ and the distance σ factors which are characteristic of each substance. Thus any two macroscopic parameters, such as T_c and P_c, could be used to correlate PvT data. However, the failure of the two-parameter corresponding states principle to predict Z accurately except for a small number of substances indicates one or more additional parameters are required if better accuracy is desired for a larger number of substances. When actual experimental Z data for a number of fluids are compared, they tend to fall into three rough classes:

1. Molecules which are spherical with respect to geometry and force field agree quite closely.
2. Molecules with nonsphericity and/or a weak to moderate dipole moment show similar behavior, but variations are more pronounced within the group and in comparison to those in class (1) above.
3. Molecules with extreme nonsphericity and/or strong dipole moments vary most widely from the norm.

This qualitative classification indicates that the degree of "sphericity" of a molecule might be used to account for the wide range of PvT behavior observed. Thus many more substances might conform more closely to the corresponding states principle if a third parameter, characteristic of molecular structure, were included in the correlation.

5-5-2 The Critical Compressibility as a Third Parameter

One approach is to group fluids into several classes, where the critical compressibility is the third parameter. This method, first proposed by Lydersen, Greenkorn, and Hougen [5] presents separate Z-T_r-P_r charts or tables for particular values of Z_c. That is, the charts represent the relation $Z = f(T_r, P_r, Z_c)$. This approach does provide a significant improvement in accuracy.

5-5-3 The Acentric Factor as the Third Parameter

In the class structure described above molecules with a spherical geometry and force field and negligible quantum effects were called *simple fluids* by Pitzer [6].

The heavier permanent gases argon, krypton, and xenon fulfill these characteristics. (In addition, the structure of gases such as methane, nitrogen, carbon monoxide, and hydrogen sulfide is close to the permanent gases.) Fluids which fit class (2) described in Sec. 5-5-1 are termed *normal* fluids. Because simple fluid behavior is easily measured, a third parameter might be developed which measures the deviation from simple fluid behavior. That is, it measures nonsphericity or acentricity. Whatever characteristic is used, it should be easily measured. A property which fits this nicely is the vapor pressure. Recall Fig. 5-3, which presented vapor-pressure data on a reduced (generalized) basis. These curves are approximately linear, and thus can be represented by a single slope. If the two-parameter corresponding states principle were generally valid, then the slope of the lines in the figure would be the same for all substances. This obviously is not true, except for the simple fluids (Ar, Kr, Xe). The vapor-pressure data for these fluids do lie on the same curve on this plot. Thus the position of the reduced vapor-pressure curve of a fluid relative to the position for a simple fluid is a sensitive property upon which a third parameter may be based.

Pitzer et al. [6] attributed this difference in vapor-pressure variation to the degree of the sphericity of the molecular force field of a substance. They defined a third parameter ω, the *acentric* factor for a fluid, as the difference between the value of $\log_{10} P_r^{\text{sat}}$ for the fluid and that for a simple fluid at a fixed value of T_r^{sat}. It was noted that $\log_{10} P_r^{\text{sat}} = -1$ at $T_r^{\text{sat}} = 0.7$ for simple fluids. (In terms of Fig. 5-3, $m = 5.4$ for simple fluids.) With this data point as a convenient reference, the definition of ω becomes

$$\omega \equiv -1.0 - \log_{10} (P_r^{\text{sat}})_{T_r=0.7} \tag{5-29}$$

The evaluation of ω only requires T_c, P_c, and the vapor pressure at $T_r = 0.7$. This choice for a definition of ω makes $\omega = 0$ for a simple fluid and positive for nearly all other fluids. In a general sense, ω increases with polarity and with hydrocarbons it increases with molecular weight. Values of ω for a number of substances are listed in Table A-2 in the Appendix. Several references [7, 8] provide an extensive list of acentric values for substances.

Example 5-2. Determine the value of ω for water based on Eq. (5-29).

Solution. The critical data for water are 647.3 K and 220.9 bars. At a reduced temperature of 0.7 the actual temperature is

$$T = T_r T_c = 0.7(647.3) = 453.1 \text{ K} = 180°\text{C}$$

From the saturation table for water the vapor pressure at 180°C is 10.02 bars. Hence on the basis of Eq. (5-29)

$$\omega = -1.00 - \log_{10} \frac{10.02}{220.9} = 0.343$$

This value is in substantial agreement with that found in Table A-2.

The three-parameter corresponding states principle of Pitzer states that

$$Z = f(T_r, P_r, \omega)$$

Unlike the method used by Lydersen et al. when Z_c is the third parameter, Pitzer and coworkers developed a linear correlation for Z, which has the form

$$Z = Z^{(0)} + \omega Z^{(1)} \tag{5-30}$$

where $Z^{(0)}$ and $Z^{(1)}$ are both functions of T_r and P_r. In the case of simple fluids where $\omega = 0$, Eq. (5-30) reduces to $Z = Z^{(0)}$. Thus the correlation for Z^0 as a function of T_r and P_r is based solely on data for argon, krypton, and xenon. Analytical expressions for $Z^{(0)}$ and $Z^{(1)}$ are not practical. Hence the data are presented in graphical and tabular format. Data for $Z^{(0)}$ and $Z^{(1)}$ appear in Tables A-24 and A-25 in the Appendix as generated by Lee and Kesler [9], who revised and extended the data of Pitzer.

The Lee and Kesler correlation is not particularly successful for polar compounds. The method has been extended [10] to polar compounds by the addition of a third term $YZ^{(2)}$ to Eq. (5-30), where Y is a parameter unique to each polar compound. In addition, water was used as a reference fluid to establish values of $Z^{(2)}$ as a function of T_r and P_r.

Example 5-3. Estimate the compressibility factor for nitrogen at 50 bars and 200 K by means of (*a*) Lee-Kessler data and (*b*) the superheat table.

Solution

(*a*) The critical temperature and pressure of nitrogen are 126.2 K and 33.9 bars, respectively. Therefore,

$$T_r = \frac{200}{126.2} = 1.59 \qquad \text{and} \qquad P_r = \frac{50}{33.9} = 1.48$$

From Tables A-24 and A-25 the values of $Z^{(0)}$ and $Z^{(1)}$ are close to 0.895 and 0.130, respectively. On the basis of $\omega = 0.039$ from Table A-2, Eq. (5-30) yields

$$Z = 0.895 + 0.039(0.130) = 0.900$$

(*b*) From the superheat table for nitrogen at 200 K and 50 bars, $v = 10.7\ \text{cm}^3/\text{g} = 0.0107\ \text{m}^3/\text{kg}$. Therefore

$$Z = \frac{Pv}{RT} = \frac{50\ \text{bars} \times 0.0107\ \text{m}^3/\text{kg}}{200\ \text{K}} \times \frac{28\ \text{kg}\cdot\text{K}}{0.08314\ \text{bar}\cdot\text{m}^3} = 0.901$$

The value from the three-parameter principle of corresponding states, which depends on the accuracy of reading $Z^{(0)}$ and $Z^{(1)}$ from the figures, is very close to the tabulated value. Note that the third parameter correction is very small in this case. Nitrogen behaves nearly as a simple fluid, as indicated by its small value of ω relative to other substances.

5-6 THREE-PARAMETER CUBIC EQUATIONS OF STATE

As discussed in the preceding section, the addition of a third parameter greatly improves the accuracy of compressibility factor data. The use of the acentric factor ω has also been extended to cubic equations of state. Two widely used equations in this format are the Redlich-Kwong-Soave (RKS) equation and the

Peng-Robinson (PR) equation. The primary use of these two equations is in the prediction of vapor-pressure and vapor-liquid equilibria data. Although similar in format, the RKS equation has an advantage in that it is somewhat easier to solve.

5-6-1 Redlich-Kwong-Soave Equation

The modification of the Redlich-Kwong equation made by Soave [11] in 1972 involves the cohesive energy term, or last term, in Eq. (5-20). The temperature quantity $T^{1/2}$ was altered to include both the temperature and the acentric factor ω. The RKS equation is

$$P = \frac{RT}{v-b} - \frac{a\alpha}{v(v+b)} \tag{5-31}$$

where
$$a = \frac{0.42748 R^2 T_c^2}{P_c} \quad \text{and} \quad b = \frac{0.08664 R T_c}{P_c} \tag{5-32}$$

$$\alpha = [1 + S(1 - T_r^{1/2})]^2 \tag{5-33}$$

$$S = 0.480 + 1.574\omega - 0.176\omega^2 \tag{5-34}$$

where S is a correlation against the Pitzer acentric factor. The value of b is the same as for the RK equation, while the factor a requires T_c^2 rather than $T_c^{2.5}$. The factor α is an empirical function determined from the vapor-pressure data of pure hydrocarbons. The value of α in the RK equation is simply $T^{-1/2}$. A refit of the original correlation for S has been done by Graboski and Daubert [12], and their equation is

$$S = 0.48508 + 1.5517\omega - 0.15613\omega^2 \tag{5-35}$$

The RKS equation was developed to predict reasonable vapor pressures for pure fluids. Thus, there is no reason to suspect it is more accurate than the RK equation in predicting PvT behavior in the superheat region. One use of the RKS equation or its modification has been the prediction of the thermodynamic properties of refrigerants [13].

5-6-2 Peng-Robinson Equation

This equation of state was developed primarily for vapor-liquid equilibrium predictions and uses the same format for α as appears in the RKS equation. The Peng-Robinson (PR) equation [14] is

$$P = \frac{RT}{v-b} - \frac{a\alpha}{v(v+b) + b(v-b)} \tag{5-36}$$

where the equation for S in Eq. (5-33) is given by

$$S = 0.37464 + 1.54226\omega - 0.26992\omega^2 \tag{5-37}$$

The equation for α was again developed so as to reproduce hydrocarbon vapor-pressures data. Since the denominator of the last term in Eq. (5-36) differs from the

RKS equation, the relations for a and b determined from the critical state criteria are altered. These become

$$a = \frac{0.45724R^2T_c^2}{P_c} \quad \text{and} \quad b = \frac{0.07780RT_c}{P_c} \tag{5-38}$$

The PR equation has about the same accuracy as the RKS equation in predicting compressibility factors or specific volumes.

5-6-3 General Comments on Cubic Equations of State

The van der Waals, Redlich-Kwong, Soave, and Peng-Robinson cubic equations of state discussed earlier are very similar in their analytical format. They are called cubic equations of state, because, if expanded, each would contain volume terms raised to the first, second, and third powers. These equations, and many other cubic equations of state, can be expressed by a general relation in the format [15]

$$P = \frac{RT}{v - b} - \frac{\gamma}{(v + c)(v + d)} \tag{5-39}$$

where b, c, and d are constants, and γ is a function of temperature and has dimensions of Pv^2. The quantities v and R may be expressed on either a mass or molar basis. As we have seen, the constants in cubic PvT equations are invariably evaluated in terms of critical data by employing Eqs. (5-3) and (5-4). Table 5-1 summarizes the values of γ, b, c, and d in Eq. (5-39) for the four cubic equations previously discussed.

As a result of the data in Table 5-1, these four equations can be converted to a dimensionless format in terms of the reduced parameters P_r and T_r, and the pseudoreduced volume $v_r' = P_c v/RT_c$. Substitution of these reduced parameters into Eq. (5-39) gives the general reduced equation,

$$P_r = \frac{T_r}{v_r' - B^*} - \frac{A^*}{(v_r' + C^*)(v_r' + D^*)} \tag{5-40}$$

where we have introduced the dimensionless parameters

$$A^* \equiv \frac{P_c\gamma}{R^2T_c^2} \qquad B^* \equiv \frac{P_c b}{RT_c} \qquad C^* \equiv \frac{P_c c}{RT_c} \qquad D^* \equiv \frac{P_c d}{RT_c} \tag{5-41}$$

Finally, as we have seen already for the van der Waals and RK relations, cubic PvT equations can also be expressed as cubic equations for Z as a function of P_r and T_r. On the basis of the definition

$$Z \equiv \frac{Pv}{RT} = \frac{P_r v_r'}{T_r} \tag{5-42}$$

TABLE 5-1
Constants for four cubic equations of state in terms of critical data

Equation	c	d	b	γ
van der Waals	0	0	$\dfrac{RT_c}{8P_c}$	$\dfrac{27}{64}\dfrac{R^2T_c^2}{P_c}$
Redlich-Kwong	b	0	$\dfrac{0.08664RT_c}{P_c}$	$\dfrac{0.42748RT_c^{2.5}}{P_cT^{0.5}}$
Soave	b	0	$\dfrac{0.08664RT_c}{P_c}$	$\dfrac{0.42748\alpha R^2T_c^2}{P_c}$
Peng-Robinson	$\left(1+\sqrt{2}\right)b$	$\left(1-\sqrt{2}\right)b$	$\dfrac{0.07780RT_c}{P_c}$	$\dfrac{0.45724\alpha R^2T_c^2}{P_c}$

where $\alpha \equiv \left[1 + S\left(1 - T_r^{0.5}\right)\right]^2$ and $S = f(\omega)$.

Eq. (5-40) can be rewritten as

$$Z^3 + Z^2\left[\left(C^* + D^* - B^*\right)\frac{P_r}{T_r} - 1\right] + Z\left[\frac{A^*P_r}{T_r^2} - \left(1 + \frac{B^*P_r}{T_r}\right)\left(C^* + D^*\right)\frac{P_r}{T_r} + \frac{C^*D^*P_r^2}{T_r^2}\right] - \left[\frac{A^*B^*P_r^2}{T_r^3} + \left(1 + \frac{P_rB^*}{T_r}\right)\left(\frac{C^*D^*P_r^2}{T_r^2}\right)\right] = 0 \quad (5\text{-}43)$$

When P_r and T_r are given, Eq. (5-43) can be solved for Z by either the direct solution for a cubic equation or some iterative technique.

5-7 OTHER EQUATIONS OF STATE

Cubic equations of state are extremely useful for predicting vapor and liquid PvT data, vapor-pressure data, and vapor-liquid equilibria data. However, their accuracy is limited by the use of only two or three adjustable parameters. To obtain greater overall accuracy, it is necessary to use equations of greater complexity. A few of the dozens of equations of state available in the literature, some quite complex, are discussed below.

5-7-1 Virial Equations

Examination of molecular behavior by the methods of statistical mechanics reveals that a suitable theoretical equation relating Z to PvT data should take the form of a series expansion in $1/v$. The result is written in the form

$$Z = \frac{Pv}{RT} = 1 + \frac{B}{v} + \frac{C}{v^2} + \frac{D}{v^3} + \cdots \quad (5\text{-}44)$$

This format is called a *virial* equation of state. The constants B, C, . . . , are called *virial coefficients* and are only a function of temperature for a given gas. B is known as the second virial coefficient; C, the third virial coefficient, and so on. This format of the virial equation expresses $Z = f(v, T)$, and the equation is explicit in pressure. The Benedict-Webb-Rubin (BWR) equation discussed in the following subsection is in the format of a virial expansion with an added term to increase its accuracy.

An alternate form of the above equation is quite useful when we need an equation that is explicit in volume. This format for Z is written as a series expansion in P, namely,

$$Z = \frac{Pv}{RT} = 1 + B_P P + C_P P^2 + D_P P^3 + \cdots \tag{5-45}$$

The virial coefficients in the two equations are directly related. As a consequence, B_P, C_P, and so on, are also only a function of temperature. It may be shown that

$$B_P = \frac{B}{RT} \qquad C_P = \frac{C - B^2}{R^2T^2} \qquad D_P = \frac{D - 3BC + 2B^3}{R^3T^3} \tag{5-46}$$

It should be noted that B_P is the slope of an isotherm on a Z-P diagram as the pressure approaches zero. On the basis of a Z chart it is found that B_P is a relatively large negative number at a low temperature. It increases, becomes zero, and then becomes positive as the temperature increases further. The general trend of B, plotted in a nondimensional manner versus T_r, is shown in Fig. 5-9. The temperature for which B is zero, known as the Boyle temperature, is around 2.6 on a reduced basis.

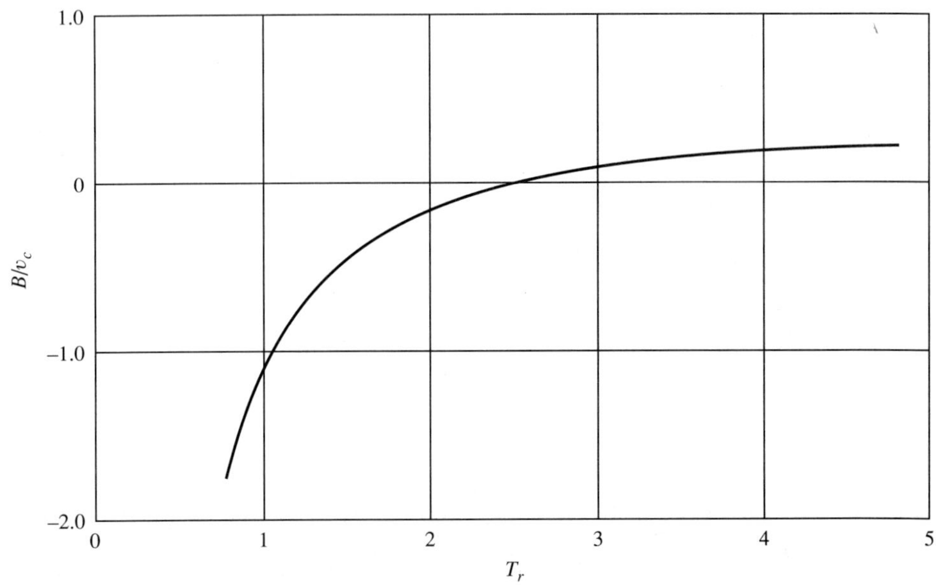

FIGURE 5-9
Dimensionless plot of the second virial coefficient B versus temperature for the relation $Z = 1 + \frac{B}{v} + \frac{C}{v^2} + \cdots$, based on the data for ethane, methane, nitrogen, and argon.

The virial coefficients are directly related to the intermolecular forces between molecules according to statistical mechanics. The coefficient B arises due to interactions between pairs of molecules, C is related to interactions among three molecules, and so on. Unfortunately, the computations required to obtain virial coefficients from intermolecular force potentials have been successful only when simplified potentials are employed. Even then, results are available only for the second and third virial coefficients. Consequently, if a virial-type equation is desired, the coefficients typically are fitted from experimental data.

For gases and vapors at low to moderate pressures only two or three terms of a virial expansion are required. As a guideline, if the pressure is below 50 bars, the virial equation truncated to three terms gives good results. In this situation

$$Z = \frac{Pv}{RT} = 1 + \frac{B}{v} + \frac{C}{v^2} \tag{5-47}$$

or

$$Z = \frac{Pv}{RT} = 1 + B_P P + C_P P^2 \tag{5-48}$$

When the virial equation is truncated to the same number of terms, the format given by Eq. (5-47) will invariably be more accurate than Eq. (5-48). Data for values of B and C (or B_P and C_P) are available in the literature for a number of gases. However, there is much more data on B than on C.

If only modest pressures are involved, the virial expressions can be truncated to two terms and reasonable accuracy can be retained. In this case Eq. (5-45) is preferred, and it becomes

$$Z = \frac{Pv}{RT} = 1 + B_P P = 1 + \frac{BP}{RT} \tag{5-49}$$

In this case Z is a linear function of pressure. On the basis of Fig. A-36 in the Appendix note that each isotherm is essentially linear in the low-pressure portion of the plot. The section of approximate linearity depends on the temperature for a given substance. The value of the second virial coefficient is negative when the reduced temperature T_r is less than roughly 2.5. Thus Eq. (5-49), except at high temperatures, predicts that the compressibility factor is less than unity.

In the absence of experimental data for B_P its value can be estimated from the three-parameter corresponding states principle. First, Eq. (5-49) may be written in the reduced format

$$Z = 1 + \left(\frac{BP_c}{RT_c}\right)\frac{P_r}{T_r} \tag{5-50}$$

Pitzer and coworkers proposed that the quantity in the parentheses in Eq. (5-50) could be expressed as a linear function of ω in a fashion similar to Eq. (5-30) for Z. That is,

$$\frac{BP_c}{RT_c} = B^0 + \omega B^1 \tag{5-51}$$

whereas

$$Z = Z^{(0)} + \omega Z^{(1)} \tag{5-30}$$

Substitution of Eq. (5-51) into Eq. (5-50) yields

$$Z = 1 + B^0 \frac{P_r}{T_r} + \omega B^1 \frac{P_r}{T_r}$$

A comparison of this result with Eq. (5-30) above shows that

$$Z^{(0)} = 1 + B^0 \frac{P_r}{T_r}$$

and

$$Z^{(1)} = B^1 \frac{P_r}{T_r}$$

Thus B^0 and B^1 can be evaluated from $Z^{(0)}$ and $Z^{(1)}$ data in regions where $Z^{(0)}$ and $Z^{(1)}$ are approximately linear functions of P_r. Examination of plots of $Z^{(0)}$ and $Z^{(1)}$ versus P_r and T_r do show that $Z^{(0)}$ and $Z^{(1)}$ are approximately linear in that region of pressure for which the truncated virial equation is expected to be valid. A fit of $Z^{(0)}$ and $Z^{(1)}$ data shows that the following equations [16] for nonpolar molecules

$$B^0 = 0.083 - \frac{0.422}{T_r^{1.6}} \tag{5-52}$$

$$B^1 = 0.139 - \frac{0.172}{T_r^{4.2}} \tag{5-53}$$

are good representations. Similar to the situation for Eq. (5-30), the above generalized second virial coefficient correlations are most accurate when applied to nonpolar or slightly polar compounds. For polar compounds a modification has been proposed by Tsonopoulos [17] which involves the addition of a third term B^2 to Eq. (5-51).

Example 5-4. Estimate the compressibility factor for nitrogen at 50 bars and 200 K by means of the generalized second virial coefficient correlation. Compare to the result of the generalized compressibility factor correlation in Example 5-2.

Solution. In Example 5-2 it was found that $T_r = 1.59$, $P_r = 1.48$, and $\omega = 0.039$. Values of B_P^0 and B_P^1 are found from Eqs. (5-52) and (5-53).

$$B_P^0 = 0.083 - \frac{0.422}{(1.59)^{1.6}} = -0.118$$

$$B_P^1 = 0.139 - \frac{0.172}{(1.59)^{4.2}} = 0.114$$

The compressibility factor then becomes

$$Z = 1 + (B^0 + \omega B^1)\frac{P_r}{T_r}$$

$$Z = 1 + [-0.118 + 0.039(0.114)]\frac{1.48}{1.59} = 0.894$$

This compares with a value of 0.901 determined for values of $Z^{(0)}$ and $Z^{(1)}$ from the Lee-Kesler correlations.

5-7-2 Benedict-Webb-Rubin Equation

One of the most widely used complex equations is that of Benedict, Webb, and Rubin. This empirical equation contains eight constants and is written as

$$P = \frac{RT}{v} + \frac{B_0RT - A_0 - C_0/T^2}{v^2} + \frac{bRT - a}{v^3} + \frac{a\alpha}{v^6} + \frac{c}{v^3T^2}\left(1 + \frac{\gamma}{v^2}\right)\exp\left(\frac{-\gamma}{v^2}\right) \tag{5-54}$$

where A_0, B_0, C_0, a, b, c, α, and γ are all constants for a given gas. Constants derived from experimental PvT data are available for a number of light hydrocarbons and a few other common gases [18, 19]. Table A-23 in the Appendix lists values of the constants for a few substances. The Benedict-Webb-Rubin (BWR) equation has proven to be quite accurate over wide ranges of pressure and temperature. However, predictions of PvT behavior for polar compounds based on the BWR equation and modifications of it are frequently not as accurate as desired. A generalized format of the BWR equation used within the context of Pitzer's three-parameter Z correlation was used by Lee and Kesler [9] to develop the $Z^{(0)}$ and $Z^{(1)}$ tables discussed earlier. Another important modification of the BWR equation is noted below.

5-7-3 Modified Benedict-Webb-Rubin Equation

In 1973 Jacobsen and Stewart [20] proposed a modification of the BWR equation of state. The modified Benedict-Webb-Rubin (MBWR) equation of state has the format

$$P = \sum_{n=1}^{9} \frac{a_n}{v^n} + \exp\left(-\frac{v_c}{v}\right)^2 \sum_{n=10}^{15} \frac{a_n}{v^{2n-17}} \tag{5-55}$$

where the a_n coefficients are temperature dependent. The number of adjustable constants increases to 32 in the MBWR equation of state, compared to eight in the original BWR equation. This increased complexity leads to a more accurate representation of experimental data over wide ranges of temperature and pressure. High-speed computers are used to resolve the more involved fitting procedure for complex equations of state, which may include 50 or more constants. The MBWR equation is widely accepted for the PvT representation of hydrocarbons and cryogenic fluids, and has also been applied to halocarbon refrigerants [21,22].

5-7-4 Martin-Hou Equation

Many equations of state are developed to represent a fairly restricted class of substances. For example, the BWR equation was developed primarily for hydrocarbons. In 1955 Martin and Hou [23] presented an equation designed to represent all gases through the use of 11 specific constants. These constants are a function of the critical properties P_c, T_c, and Z_c, as well as the slope of the critical isometric. The Martin-Hou equation is

$$P = \frac{RT}{v - b} + \frac{A_2 + B_2T + C_2e^{-cT}}{(v - b)^2} + \frac{A_3 + B_3T + C_3e^{-cT}}{(v - b)^3} + \frac{A_4}{(v - b)^4} + \frac{A_5 + B_5T + C_5e^{-cT}}{(v - b)^5} \tag{5-56}$$

The equation is expected to reproduce experimental data within 1 percent for densities up to $1.5\rho_c$ and temperatures up to $1.5T_c$. It has been applied to both nonpolar and polar gases, including the properties of refrigerant-134a [24].

5–7–5 Carnahan-Starling-DeSantis Equation

An important model for the PvT behavior of pure refrigerants and their mixtures is the Carnahan-Starling-DeSantis (CSD) equation of state. Based on the original Carnahan-Starling equation [25],

$$\frac{Pv}{RT} = \frac{1 + y + y^2 - y^3}{(1 - y)^3}$$

where $y = b/4v$, a modification by DeSantis et al. [26], leads to

$$\frac{Pv}{RT} = \frac{1 + y + y^2 - y^3}{(1 - y)^3} - \frac{a}{RT(v + b)} \tag{5-57}$$

where the second term on the right accounts for long-range attractive forces between molecules. The first term represents the behavior of rigid, noninteracting spherical molecules. Both parameters a and b are temperature dependent in the following manner:

$$a = a_0 \exp\left(a_1T + a_2T^2\right) \qquad \text{and} \qquad b = b_0 + b_1T + b_2T^2$$

The relative simplicity of the CSD equation relative, for example, to the MBWR equation, permits fitting the equation with a minimum of experimental data. The CSD equation may lack good accuracy in predicting other property data for some highly polar fluids [22]. One reported use of the CSD equation of state is by Morrison and McLinden [27].

PROBLEMS

5-1. Steam at 120 bars and 400°C is a real gas. Determine the specific volume, in m^3/kg, by means of (a) the ideal-gas equation, (b) van der Waals equation, (c) the Redlich-Kwong (RK) equation, (d) the two-parameter Z chart, and (e) the superheat table.

5-2. Determine the specific volume, in cm^3/g, of steam at 100 bars and 360°C by using (a) the ideal-gas equation, (b) the van der Waals equation, (c) the Redlich-Kwong equation, (d) the two-parameter compressibility chart, and (e) the tabulated superheat value.

5-3. Determine the specific volume, in m^3/kg, of CO_2 at 30 bars and 300 K by means of (a) the ideal-gas equation, (b) the van der Waals equation, (c) the Redlich-Kwong equation, (d) the two-parameter compressibility chart, and (e) the tabulated superheat value.

5-4. The specific volume of R-134a at 60°C is 23.01 cm^3/g. Determine the pressure in bars by means of (a) the ideal-gas equation, (b) the van der Waals equation, (c) the Redlich-Kwong equation, (d) the two-parameter corresponding states principle, and (e) the superheat tables.

5-5. Ethane gas (C_2H_6) is maintained at 280 K with a density of 33.0 kg/m^3. Determine the pressure, in megapascals (MPa), of the gas by means of (a) the ideal-gas equation, (b) the Redlich-Kwong equation, and (c) the two-parameter Z chart.

5-6. The specific volume of R-134a at 80°C is 0.01435 m^3/kg. Determine the pressure of the fluid by means of (a) the ideal-gas equation, (b) the van der Waals equation, (c) the Redlich-Kwong equation, (d) the two-parameter compressibility factor, and (e) the tabulated superheat value, in bars.

5-7. Estimate the value of the acentric factor, ω, for CO_2 using tabular vapor pressure and critical data. The vapor pressures of CO_2 at 216.54 and 220.0 K, respectively, are 5.173 and 6.000 bars. (*Note:* Do not linearly interpolate the vapor-pressure data, but assume that the vapor pressure p_v is related to the absolute temperature by the relation $\ln p_v = (A/T) + B$.)

5-8. Estimate the value of the acentric factor, ω, for nitrogen, using tabular vapor-pressure and critical data. The vapor pressures of nitrogen at 85 and 90 K, respectively, are 2.291 and 3.608 bars. (*Note:* Do not linearly interpolate the vapor-pressure data, but assume that the vapor pressure P_v is related to the absolute temperature by the relation, $\ln P_v = A/T + B$.)

5-9. Use the vapor-pressure data for R-134a and critical data to determine the value of its acentric factor. [*Note:* Do not linearly interpolate vapor-pressure data, but assume the vapor pressure p_v is related to the absolute temperature by the relation $\ln p_v = (A/T) + B$.] Compare the answer to the value listed in Table A-2 in the Appendix.

5-10. Determine the Z factor for CO_2 at 50 bars and 300 K by means of (a) the two-parameter generalized chart, (b) the three-parameter corresponding states principle based on $Z^{(0)}$ and $Z^{(1)}$ data, and (c) the superheat table.

5-11. Determine the Z factor for nitrogen at 200 K and 100 bars by means of (a) the generalized two-parameter Z chart, (b) the acentric factor, $Z^{(0)}$, and $Z^{(1)}$ data, if $\omega = 0.040$, (c) the Peng-Robinson equation of state, and (d) the superheat table.

5-12. Butane gas (C_4H_{10}) is maintained at 187°C with a specific volume of 0.02784 m^3/kg. Determine the pressure of the gas by means of (a) the ideal-gas equation, (b) the Redlich-Kwong-Soave equation, and (c) the Benedict-Webb-Rubin equation, in bars.

5-13. Carbon dioxide is maintained at 320 K with a specific volume of 0.02771 m^3/kg. Determine the pressure of the gas in bars by means of (a) the ideal-gas equation, (b) the Redlich-Kwong-Soave equation, and (c) the Benedict-Webb-Rubin equation. If the literature value is 20.0 bars, determine the percent error for the three equations of state.

5-14. Nitrogen gas is maintained at 100 bars and 200 K. Determine the specific volume, in m^3/kg, by using

a. the truncated virial equation $Pv = RT + bP$, where b is given as

$$b = 39.5 - \frac{1 \times 10^4}{T} - \frac{1.084 \times 10^6}{T^2},$$

where b is in cm^3/gmol and T is in kelvins

b. the tabular superheat data

c. the ideal-gas equation of state

d. What percent errors are made when using the virial-type equation and the ideal-gas equation, compared to the tabular value?

5-15. The specific volume of nitrogen at 200 K is 0.00502 m^3/kg. Determine the pressure of the gas, in bars, by means of (a) the ideal-gas equation, (b) the Redlich-Kwong-Soave equation, (c) the Benedict-Webb-Rubin equation, (d) the two-parameter Z-factor method, and (e) the superheat table.

5-16. Calculate the specific volume, in m^3/kg, of nitrogen at 102 bars and -45°C by (a) the ideal-gas equation; (b) the two-parameter compressibility factor; (c) the virial equation represented by $Pv = a + bP + cP^2 + dP^3$, where the virial coefficients at this temperature are $b = -2.34 \times 10^{-2}$, $c = 3.61 \times 10^{-5}$, and $d = 5.18 \times 10^{-7}$, where P is in bars, T is in kelvins, and v is in m^3/kmol; (d) the virial expansion represented by $Z = 1 + (B/v) + (C/v^2) + (D/v^3)$, where $B = -25.0$, $C = 1.72 \times 10^3$, $D = 18.5 \times 10^3$, and v is in cm^3/gmol; and (e) interpolating in the superheat table.

5-17. The constants for the virial equation in the form, $Pv = RT[1 + (B/v) + (C/v^2) + \cdots]$, have been found experimentally for nitrogen at -100°C. The values are: $B = -0.05185$ m^3/kmol and $C = 0.002125$ $m^6/(kmol)^2$. Determine the compressibility factor at 50 bars and -100°C by using

a. the virial expansion given above

b. the three-parameter compressibility correlation

5-18. Find the compressibility factor for R-134a at 16.0 bars and 100°C by means of (a) the generalized two-parameter Z chart, (b) the three-parameter Z correlation using $Z^{(0)}$, $Z^{(1)}$ and acentric factor data, (c) a truncated virial expansion in v, using the equation for B versus T_r found in Sec. 5-7-1, (d) the Peng-Robinson equation of state, and (e) the PvT data from a superheat table.

5-19. The constants for the virial equation of state in the form $Pv = RT[1 + (B/v) + (C/v^2) + (D/v^3)]$ have been determined experimentally for nitrogen at 200 K. The values are $B = -37.6$ cm^3/gmol, $C = 1.92 \times 10^3$ $(cm^3/gmol)^2$, and $D = 5.8 \times 10^3$ $(cm^3/gmol)^3$. Find the values of v, in cm^3/gmol, and Z by using (a) the virial equation; (b) the three-parameter Z chart, and (c) the superheat tables, for $P = 100$ bars.

5-20. The constants in the Carnahan-Starling-DeSantis equation of state for refrigerant-12 (CCl_2F_2) are

$$a_0 = 3524.12 \qquad a_1 = -2.77230 \times 10^{-3} \qquad a_2 = -0.67318 \times 10^{-6}$$
$$b_0 = 0.15376 \qquad b_1 = 1.84195 \times 10^{-4} \qquad b_2 = -5.03644 \times 10^{-8}$$

where a is in $J \cdot L/gmol^2$ and b is in L/gmol. Find the specific volume, in m^3/kg, at 0.80 MPa and 40°C.

REFERENCES

1. Martin, J. J., "Equations of State," *Ind. Eng. Chem,* **59** (12):34, 1967.
2. Redlich, O., and J. N. S. Kwong, "On the Thermodynamics of Solutions. V. An Equation of State. Fugacities of Gaseous Solutions," *Chem. Rev.,* **44**:233, 1949.
3. Su, Gour-Jen, "Modified Law of Corresponding States," *Ind. Eng. Chem.* (Intern. ed.), **38**:803, 1946.
4. Nelson, L. C., and E. F. Obert, "Generalized *PvT* Properties of Gases," *Trans. ASME,* **76**:1057, 1954.
5. Lydersen, A. L., R. A. Greenkorn, and O. A. Hougen, "Generalized Thermodynamic Properties of Pure Fluids," *Univ. Wisconsin Coll. Eng., Eng. Exp. Stn. Report 4,* Madison, WI, October 1955.
6. Pitzer, K. S., D. Z. Lippmann, R. F. Curl, Jr., C. M. Huggins, and D. E. Petersen, "The Volumetric and Thermodynamic Properties of Fluids. II. Compressibility Factor, Vapor Pressure and Entropy of Vaporization," *J. Am. Chem. Soc.,* **77**:3433, 1955.
7. Reid, R. C., J. M. Prausnitz, and B. E. Poling, *The Properties of Gases and Liquids,* 4th ed., McGraw-Hill, Inc., New York, 1986.
8. Ambrose, D., *Vapor-Liquid Critical Properties,* NPL Report Chem. 107, 1980.
9. Lee, B. I., and M. G. Kesler, "A Generalized Thermodynamic Correlation Based on Three-Parameter Corresponding States," *AIChE J.,* **21**:510, 1975.
10. Wu., G. Z., and L. I. Stiel, *AIChE J.,* **31**:1632, 1985.
11. Soave, G., "Equilibrium Constants from a Modified Redlich-Kwong Equation of State," *Chem. Eng. Sci.,* **27**:1197, 1972.
12. Graboski, M. S., and T. E. Daubert, *Ind. Eng. Chem. Process Des. Develop.,* **17**:443, 1978.
13. Camporese, R., G. Bigolaro, and L. Rebellato, "Calculation of Thermodynamic Properties of Refrigerants by the Redlich-Kwong-Soave Equation of State," *Int. J. Refrig.,* **8**:147, 1985.
14. Peng, D. Y., and D. B. Robinson, "A New Two-Constant Equation of State," *Ind. Eng. Chem. Fundam.,* **15**:59, 1976.
15. Brown, C. L., Professor Emeritus, Purdue University, private communication, 1989.
16. Van Ness, H. C., and M. M. Abbott, *Classical Thermodynamics of Nonelectrolyte Solutions,* McGraw-Hill, Inc., New York, pp. 126–133, 1982.
17. Tsonopoulos, C., "An Empirical Correlation of Second Virial Coefficients," *AIChE J.,* **20**:263, 1974.
18. Cooper, H. W., and J. C. Goldfrank, *Hydrocarbon Process. Petrol. Refiner.,* **46**(12):141, 1967.
19. Perry, R. H., D. W. Green, and J. O. Mahoney, *Perry's Chemical Engineer's Handbook,* 6th ed., McGraw-Hill, Inc., New York, 1984.
20. Jacobsen, R. T., and R. B. Stewart, "Thermodynamic Properties of Nitrogen Including Liquid and Vapor Phases from 63 K to 2000 K with Pressures to 10,000 Bar," *J. Phys. Chem. Ref. Data,* **2**:757–922, 1973.
21. ASHRAE, *Thermodynamic Properties of Refrigerants,* American Society of Heating, Refrigerating, and Air-Conditioning Engineers, Inc., Atlanta, GA, 1986.
22. McLinden, M. O., et al., "Measurement and Formulation of the Thermodynamic Properties of Refrigerants 134a and 123," *ASHRAE Trans.,* **95**(Pt 2):263, 1989.
23. Martin, J. J., and Y. C. Hou, "Development of an Equation of State for Gases," *AIChE J.,* **1**:142, 1955.
24. Wilson, D. P., and R. S. Basu, "Thermodynamic Properties of a New Stratospherically Safe Working Fluid—Refrigerant-134a," *ASHRAE Trans.,* **94**(Pt. 2):2095, 1988.
25. Carnahan, N. F., and K. E. Starling, "Equation of State for Nonattracting Rigid Spheres," *J. Chem. Phys.,* **51**(2):635–636, 1969.
26. DeSantis, R., F. Gironi, and L. Marrelli, "Vapor-Liquid Equilibrium for a Hard-Sphere Equation of State," *Ind. Eng. Chem. Fundam.,* **15**(2):183–189, 1976.
27. Morrison, G., and M. D. McLinden, *NBS Technical Note 1226,* National Bureau of Standards, Washington, DC, 1985.

CHAPTER 6

THERMODYNAMIC PROPERTY RELATIONS

The importance of the PvT relations presented in the preceding chapter for the evaluation of other property data will now be demonstrated.

6-1 INTRODUCTION

In the solution of engineering and scientific problems it is essential to be able to determine the values of thermodynamic properties. A great deal of empirical data have been obtained over the past decades. Of these, the correlation among pressure, specific volume, and temperature and the relationship between the specific heats and temperature at low pressures are most easily measured. The evaluation of such properties as the internal energy, the enthalpy, and the entropy is made on the basis of calculations involving the directly measurable data mentioned earlier. Consequently, one of the major tasks of thermodynamics is to provide basic equations to evaluate properties such as u, h, and s, and others, from measurable property data.

A second point to recognize is that in many cases insufficient data are available to calculate properties, even if the basic mathematical relations have been developed. Hence approximation techniques are needed, since in numerous situations insufficient time or money may require that alternative methods of evaluating properties be available. A case in point is the use of the generalized compressibility factor Z, which provides a means of correlating PvT data in the absence of sufficient direct experimental information. This corresponding states method is extended to other properties in this chapter. Before we develop some general thermodynamic relations for the qualitative and quantitative insight they provide, a brief review of some rules of partial differential calculus is in order.

6-2 FUNDAMENTALS OF PARTIAL DERIVATIVES

Many of the expressions developed in this chapter involve a dependent variable expressed as a function of two independent variables, since we are primarily interested in simple compressible systems. With this in mind, consider three thermodynamic variables represented by x, y, and z. Their functional relationship may be expressed in either the form $f(x, y, z) = 0$ or in the form $x = x(y, z)$. The total differential of the dependent variable x is given by the equation

$$dx = \left(\frac{\partial x}{\partial y}\right)_z dy + \left(\frac{\partial x}{\partial z}\right)_y dz \tag{a}$$

Similar expressions for dy and dz may also be written, depending on which variable is selected as the dependent variable.

The reason for not considering a partial derivative as a fraction is easily seen by recalling that the equilibrium states of a simple system may be represented by a three-dimensional surface. Such a surface is shown in Fig. 6-1 for a single-phase region, where the thermodynamic properties are symbolized by x, y, and z. As an example, this surface could represent the superheat region of a PvT diagram. Sections of the surface have been cut away around the equilibrium state D so that the curvature of the surface is more clearly seen. Consider a plane of constant z which intersects the surface. The curve of intersection is labeled with the points C, D, and E on the diagram. The quantity $(\partial x/\partial y)_z$ is the slope of the surface at any stage along this curve. The value of the slope of the tangent AB is the

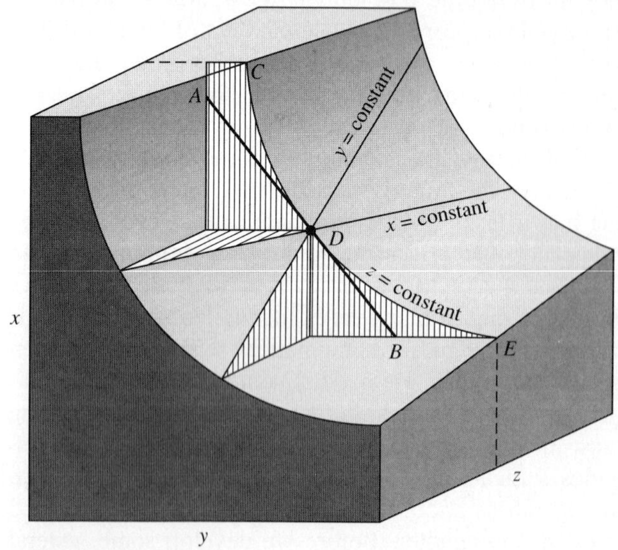

FIGURE 6-1
A representation of a partial derivative for a thermodynamic surface.

value of $(\partial x/\partial y)_z$ at state D. This partial derivative exists, of course, when any other plane of constant z intersects the equilibrium surface as long as the surface is continuous in the z direction at the state of interest. It is apparent that similar interpretations of quantities $(\partial x/\partial z)_y$ and $(\partial y/\partial z)_x$ can be made when planes of constant y and x, respectively, intersect the surface of equilibrium states.

A mathematical test is available from partial differential calculus to determine whether the total differential of a function is an exact differential. This test is a necessary and sufficient condition for the total differential to be exact. Since differential equations relating thermodynamic properties are exact, the test for exactness provides a means of developing additional relations among intrinsic properties. If in Eq. (a) we denote the coefficient of dy by M and the coefficient of dz by N, the above equation becomes

$$dx = M\,dy + N\,dz \tag{b}$$

Partial differention of M and N with respect to z and y, respectively, leads to

$$\left.\frac{\partial M}{\partial z}\right)_y = \frac{\partial^2 x}{\partial y \partial z} \qquad \text{and} \qquad \left.\frac{\partial N}{\partial y}\right)_z = \frac{\partial^2 x}{\partial z \partial y}$$

If these partial derivatives exist, it is known from the calculus that the order of differentiation is immaterial so that

$$\left.\frac{\partial M}{\partial z}\right)_y = \left.\frac{\partial N}{\partial y}\right)_z \tag{6-1a}$$

When Eq. (6-1a) is satisfied for any function x, then dx is an exact differential. Equation (6-1a) is known as the *test for exactness*. Its usefulness in terms of intrinsic property relations is demonstrated in the next section.

Another important expression from partial differential calculus is the *cyclic relation*, which is

$$\left(\frac{\partial w}{\partial x}\right)_y \left(\frac{\partial x}{\partial y}\right)_w \left(\frac{\partial y}{\partial w}\right)_x = -1 \tag{6-1b}$$

Equation (6-1b) is important in interrelating thermodynamic variables. It also illustrates the fallacy of handling partial derivatives as fractions. When the variables held constant on the partial derivatives are different, the quantities within the partials cannot be canceled.

6-3 SOME FUNDAMENTAL RELATIONS FOR SIMPLE COMPRESSIBLE SYSTEMS

In Chap. 2 several important equations were developed for the entropy change of simple compressible systems of fixed chemical composition by combining the first and second laws of thermodynamics. One result is the first $T\,ds$ equation,

$$T\,ds - P\,dv = du \tag{6-2}$$

Since $h = u + Pv$, then $du = dh - P\,dv - v\,dP$. Replacement of du in Eq. (6-2) by this relationship leads to the second $T\,ds$ equation, namely,

$$T\,ds + v\,dP = dh \tag{6-3}$$

Two additional equations of interest may be formed by defining two other properties of matter. The *Helmholtz function a* is defined by the equation

$$a \equiv u - Ts \tag{6-4}$$

Hence

$$da = du - T\,ds - s\,dT \tag{6-5}$$

If Eq. (6-2) is substituted into Eq. (6-5), then

$$da = -P\,dv - s\,dT \tag{6-6}$$

The *Gibbs function g* is defined by the equation

$$g \equiv h - Ts \tag{6-7}$$

Therefore

$$dg = dh - T\,ds - s\,dT \tag{6-8}$$

Substitution of dh from Eq. (6-3) into Eq. (6-8) yields

$$dg = v\,dP - s\,dT \tag{6-9}$$

To summarize these four important relationships among properties of simple compressible systems, they are collected together as a set and presented here:

$$du = T\,ds - P\,dv \tag{6-2}$$

$$dh = T\,ds + v\,dP \tag{6-3}$$

$$da = -P\,dv - s\,dT \tag{6-6}$$

$$dg = v\,dP - s\,dT \tag{6-9}$$

Note that the variables on the right-hand sides of these equations include only P, v, T, and s. These equations are sometimes referred to as the *gibbsian equations*.

One of the most important sets of thermodynamic relations which arise from the gibbsian, or $T\,ds$, equations is obtained by applying the test for exactness to the preceding expressions for du, dh, da, and dg. Application of Eq. (6-1a) to the four equations yields the following set of relations among partial derivatives:

$$\left(\frac{\partial T}{\partial v}\right)_s = -\left(\frac{\partial P}{\partial s}\right)_v \tag{6-10}$$

$$\left(\frac{\partial T}{\partial P}\right)_s = \left(\frac{\partial v}{\partial s}\right)_P \tag{6-11}$$

$$\left(\frac{\partial P}{\partial T}\right)_v = \left(\frac{\partial s}{\partial v}\right)_T \tag{6-12}$$

$$\left(\frac{\partial v}{\partial T}\right)_P = -\left(\frac{\partial s}{\partial P}\right)_T \tag{6-13}$$

This set of equations is referred to as the *Maxwell relations*.

6-4 GENERALIZED RELATIONS FOR CHANGES IN ENTROPY, INTERNAL ENERGY, AND ENTHALPY

One of the most important functions of thermodynamics is to provide fundamental equations for the evaluation of properties or the change in properties under the most general considerations. These equations, for example, should be independent of the type or the phase of the substance under consideration. Again, we restrict ourselves to simple compressible thermodynamic systems. Of primary interest in this section are equations for the changes in entropy, internal energy, and enthalpy. Relationships for the specific heats appear in Sec. 6-5.

It is convenient to begin with a derivation of an equation for the change in entropy of a simple compressible system. Once this is accomplished, equations for du and dh may be obtained directly by employing the $T\,ds$ equations already developed. We select the independent properties to be any pair of the group (P, v, T). For practical reasons we are usually concerned only with the two pairs of variables (T, v) and (T, P). If the entropy is chosen to be a function of T and v, we may write that the total differential of s is given by

$$ds = \left(\frac{\partial s}{\partial T}\right)_v dT + \left(\frac{\partial s}{\partial v}\right)_T dv \tag{6-14}$$

It is now desirable to express ds solely in terms of measurable quantities. This requires replacing the partial derivatives in Eq. (6-14) with other terms which include only the variables P, v, T, and the specific heats c_v and c_p.

The first partial derivative on the right-hand side of Eq. (6-14) is related to c_v and T. If $u = u(T, v)$, then

$$du = \left(\frac{\partial u}{\partial T}\right)_v dT + \left(\frac{\partial u}{\partial v}\right)_T dv = c_v\,dT + \left(\frac{\partial u}{\partial v}\right)_T dv$$

When du in the above expression is replaced by Eq. (6-2), then, upon rearrangement

$$ds = \frac{c_v\,dT}{T} + \frac{1}{T}\left[\left(\frac{\partial u}{\partial v}\right)_T + P\right] dv \tag{6-15}$$

Equations (6-14) and (6-15) are both valid expressions for ds. Hence the coefficients in front of the dT terms on the right must be equal. Therefore

$$\left(\frac{\partial s}{\partial T}\right)_v = \frac{c_v}{T} \tag{6-16}$$

The second partial derivative is contained in a Maxwell relation represented by Eq. (6-12). By making the proper substitutions into Eq. (6-14), we find that

$$ds = \frac{c_v\,dT}{T} + \left(\frac{\partial P}{\partial T}\right)_v dv \tag{6-17}$$

This is the result we sought, since the right-hand side of the equation is now expressed solely in terms of measurable quantities.

An equivalent equation for ds in terms of the variables T and P may be obtained by starting with the relation

$$ds = \left(\frac{\partial s}{\partial T}\right)_P dT + \left(\frac{\partial s}{\partial P}\right)_T dP \tag{6-18}$$

Again, the second partial derivative may be replaced by a Maxwell relation. In addition, it can be shown in a manner similar to the development of Eq. (6-16) that

$$\left(\frac{\partial s}{\partial T}\right)_P = \frac{c_p}{T} \tag{6-19}$$

As a consequence, the substitution of Eqs. (6-13) and (6-19) into Eq. (6-18) yields the desired result:

$$ds = \frac{c_p\,dT}{T} - \left(\frac{\partial v}{\partial T}\right)_P dP \tag{6-20}$$

These equations are *generalized* relations, since they are not restricted to any particular substance or to any particular phase of a substance. They are restricted, however, to simple compressible systems.

A generalized equation for the internal energy change is now possible by recalling the first $T\,ds$ equation developed previously, $du = T\,ds - P\,dv$ [see Eq. (6-2)]. The quantity ds is eliminated from this equation by substituting Eq. (6-17) for it. After separation of variables

$$du = c_v\,dT + \left[T\left(\frac{\partial P}{\partial T}\right)_v - P\right] dv \tag{6-21}$$

The generalized equation for the change in enthalpy is found similarly by employing the second $T\,ds$ equation, namely, $dh = T\,ds + v\,dP$ [see Eq. (6-3)]. Substitution of Eq. (6-20) for ds and subsequent rearrangement lead to

$$dh = c_p\,dT + \left[v - T\left(\frac{\partial v}{\partial T}\right)_P\right] dP \tag{6-22}$$

Integration of Eqs. (6-17), (6-20), (6-21), and (6-22) for actual systems requires experimental knowledge of the PvT behavior of the substance in the region of interest plus experimental information on the relationship between the specific heats and temperature.

6-5 GENERALIZED RELATIONS FOR c_p AND c_v

Two expressions for the specific heats have already been developed in this chapter:

$$c_v = T\left(\frac{\partial s}{\partial T}\right)_v \tag{6-16}$$

$$c_p = T\left(\frac{\partial s}{\partial T}\right)_P \tag{6-19}$$

These expressions are generalized relations for c_v and c_p, and they may be used in any single-phase region where PvT data are available. An alternative method of evaluating specific-heat data is based on the experimental fact that it is relatively easier to measure specific-heat data at quite low pressures than at elevated pressures. Such data are called *zero-pressure* specific heats, $c_{p,0}$. Consequently, we are primarily concerned with determining in what manner specific-heat values vary with increasing pressure (or decreasing specific volume) at constant temperature. Such an evaluation must again be based solely on the use of measured PvT data. In a mathematical sense we seek expressions for the two terms $(\partial c_v/\partial v)_T$ and $(\partial c_p/\partial P)_T$. Generalized equations for these two expressions are obtained by starting with Eqs. (6-17) and (6-20),

$$ds = \frac{c_v\,dT}{T} + \left(\frac{\partial P}{\partial T}\right)_v dv \tag{6-17}$$

$$ds = \frac{c_p\,dT}{T} - \left(\frac{\partial v}{\partial T}\right)_P dP \tag{6-20}$$

Since these equations are expressions for exact differentials, the test for exactness may be employed. When it is applied to Eq. (6-17), the following relation results:

$$\left(\frac{\partial c_v}{\partial v}\right)_T = T\left(\frac{\partial^2 P}{\partial T^2}\right)_v \tag{6-23}$$

If we start with Eq. (6-20), it can be shown by an analogous procedure that

$$\left(\frac{\partial c_p}{\partial P}\right)_T = -T\left(\frac{\partial^2 v}{\partial T^2}\right)_P \tag{6-24}$$

To obtain the value of c_p, for example, at an elevated pressure, Eq. (6-24) is integrated from zero pressure to the desired value. Hence

$$c_p - c_{p,0} = -T\int_0^P \left(\frac{\partial^2 v}{\partial T^2}\right)_P dP \tag{6-25}$$

where $c_{p,0}$ again is the zero-pressure, or ideal-gas, specific heat. The integration of the right-hand side requires a knowledge of the PvT behavior of the substance in either tabular or analytical form.

Another thermodynamic relation of interest is the difference between the constant-pressure and constant-volume heats, that is, $c_p - c_v$. One reason for this interest is that c_p values usually are easier to measure than c_v values. Hence c_v values in theory could be evaluated solely from c_p and PvT data. Since the change in any property value is not dependent on the method of evaluation, we may equate the two equations for ds previously presented. Equating Eqs. (6-17) and (6-20), we find

$$\frac{c_v\,dT}{T} + \left(\frac{\partial P}{\partial T}\right)_v dv = \frac{c_p\,dT}{T} - \left(\frac{\partial v}{\partial T}\right)_P dP$$

or

$$\frac{c_p - c_v}{T}\,dT = \left(\frac{\partial P}{\partial T}\right)_v dv + \left(\frac{\partial v}{\partial T}\right)_P dP$$

Differentiation with respect to pressure at constant volume yields

$$\frac{c_p - c_v}{T}\left(\frac{\partial T}{\partial P}\right)_v = \left(\frac{\partial v}{\partial T}\right)_P$$

or upon rearrangement

$$c_p - c_v = T\left(\frac{\partial v}{\partial T}\right)_P \left(\frac{\partial P}{\partial T}\right)_v \tag{6-26}$$

An equivalent form for $c_p - c_v$ can be obtained by replacing $(\partial P/\partial T)_v$ in terms of the cyclic rule $(\partial P/\partial T)_v = -(\partial v/\partial T)_P(\partial P/\partial v)_T$. Use of this expression in Eq. (6-26) leads to the relation

$$c_p - c_v = -T\left(\frac{\partial v}{\partial T}\right)_P^2 \left(\frac{\partial P}{\partial v}\right)_T \tag{6-27}$$

A number of important qualitative results stem from Eq. (6-27).

First, on the basis of experimental data it is known that $(\partial P/\partial v)_T$ is always negative for all substances in all phases. Since the first partial in Eq. (6-27) is a squared term, $c_p - c_v$ must always be positive, or zero. This quantity becomes zero on two occasions. The first occurs when T is absolute zero on the thermodynamic scale if the remaining terms remain finite at this state. Consequently, the specific heats at constant pressure and constant volume are identical at the absolute zero of temperature. The specific heats will also be equal if the value of $(\partial v/\partial T)_p$ is ever zero. This occurs, for example, in the case of liquid water at roughly 4°C, where the fluid is at its state of maximum density. Also, note that even at temperatures above zero the difference in specific heats will generally be small for liquids and solids. This is true, since the value of $(\partial v/\partial T)_P$ is very small for most condensed phases. Hence one frequently speaks of the specific heat of a liquid or a solid without specifying the type, since the c_v and c_p values are not significantly different in many cases. The data quoted are usually c_p values. The fact that $c_p \geq c_v$ also leads to the generalization that constant-volume lines always have steeper slopes than constant-pressure lines at the same point on a *Ts* diagram.

The property variation of solids and liquids is often expressed in terms of the volumetric expansion coefficient β and the isothermal coefficient of compressibility K_T. These two quantities are defined as

$$\beta = \frac{1}{v}\left(\frac{\partial v}{\partial T}\right)_P \tag{6-28a}$$

and

$$K_T = -\frac{1}{v}\left(\frac{\partial v}{\partial P}\right)_T \tag{6-28b}$$

Substitution of these two equations in Eq. (6-27) leads to

$$c_p - c_v = \frac{vT\beta^2}{K_T} \tag{6-29}$$

The use of β and K_T is quite helpful in many calculations, since their values may often be assumed to be constant during a given process. The slow variation of these two properties with temperature is illustrated by the data of Table 6-1, which shows data for solid copper and water.

TABLE 6-1
β, K_T, and ρ at 1 bar versus temperature for (a) copper and (b) water

(a) Copper				(b) Water			
T, K	$\beta \times 10^6$, K^{-1}	$K_T \times 10^6$, bar^{-1}	ρ, kg/m^3	T, °C	$\beta \times 10^6$, K^{-1}	$K_T \times 10^6$, bar^{-1}	ρ, kg/m^3
100	31.5	0.721	9009	0	−68.14	50.89	999.8
150	41.0	0.733	8998	10	87.90	47.81	999.7
200	45.6	0.748	8973	20	20.66	45.90	998.2
250	48.0	0.762	8951	30	303.1	44.77	995.7
300	49.2	0.776	8930	40	385.4	44.24	992.2
500	54.2	0.837	8837	50	457.8	44.18	988.0

6-6 RESIDUAL PROPERTY FUNCTIONS

The use of the first and second laws in the design or analysis of thermodynamic processes requires information on Δu, Δh, Δs, and other thermodynamic properties. In some cases the ideal-gas and incompressible substance models may be used effectively for property predictions. In these situations the generalized equations introduced earlier reduce to simple and well-known equations when the appropriate PvT relations are employed. If these two models are not appropriate, then more complex PvT relationships must be used. As noted in Chap. 5, cubic PvT equations are noted for their reasonable simplicity and fair accuracy. They also require little thermodynamic data to establish the adjustable constants. When desirable, more complex equations can be constructed to represent the PvT behavior of a substance with a high degree of accuracy. However, they require a considerable amount of experimental PvT data in order to evaluate the increased number of adjustable constants.

Even after the development of a PvT equation of state based on experimental data, several other items must be considered. The specific format employed to evaluate u, h, s, and other thermodynamic property data depends on whether the PvT equation is explicit in specific volume or pressure. In practice, equations of state are developed in either the format of $v = f(P, T)$ or $P = f(T, v)$ [or $P = f(T, \rho)$]. Modern equations of state are usually explicit in pressure. However, another modern tool used to estimate property data when experimental data are lacking relies on an equation of state explicit in specific volume. Thus it is important to present the general theory for both methods.

In the preceding section methods were developed for direct numerical calculation of the effect of pressure on property changes between two given states. However, it has been found more convenient to evaluate changes in properties such as u, h, and s in states other than the ideal-gas state by means of a *residual function*. For any specific property y a residual function y^R is defined as

$$y^R \equiv y^* - y \qquad \text{or} \qquad y^R \equiv y - y^* \tag{6-30}$$

where y is the desired value of y at (T, P), and y^* is the property value the fluid would have at the same pressure and temperature if the fluid were an ideal gas. The state represented by $(^*)$ is a hypothetical one, but the introduction of it proves to be quite useful. Thus the ideal-gas state serves as a reference for the real state.

Rearrangement of Eq. (6-30) shows that $y = y^* - y^R$. Thus the change in any property between two actual states 1 and 2 is given by

$$\begin{aligned} y_2 - y_1 &= (y_2^* - y_2^R) - (y_1^* - y_1^R) \\ &= y_1^R - y_2^R + (y_2^* - y_1^*) \end{aligned} \tag{6-31}$$

The last term on the right is simply the change in y if the substance were an ideal gas. The first two terms require knowledge of the residual property values at the two actual states. To illustrate the use of residual functions, the evaluation of enthalpy and entropy changes between two real-gas states is discussed below.

6-6-1 Evaluation of Enthalpy Changes

Consider a change of state for a real gas as represented by states 1 and 2 on the *P-h* and *h-s* diagrams in Fig. 6-2. In terms of these two states Eq. (6-31) may be written as

$$h_2 - h_1 = \left(h_1^* - h_1\right)_{T_1,P_1} - \left(h_2^* - h_2\right)_{T_2,P_2} + \left(h^*_{P_2,T_2} - h^*_{P_1,T_1}\right) \tag{6-32}$$

The first and second terms on the right are the *residual* enthalpy terms h_1^R and h_2^R at states (T_1, P_1) and (T_2, P_2), respectively. The third term is simply the enthalpy change of an ideal gas between T_1 and T_2. This is given by the integral of $c_{p,0}\, dT$. Hence the above equation reduces to

$$h_2 - h_1 = h_1^R - h_2^R + \int_{T_1}^{T_2} c_{p,0}\, dT \tag{6-33}$$

Methods for the evaluation of the residual property terms on the right will be discussed in later sections. The third term is found by substituting an equation for the zero-pressure specific heat and integrating. Alternatively, standardized ideal-gas tables containing h data already represent the integration process for selected temperatures.

6-6-2 Evaluation of Entropy Changes

A general equation for the entropy change between two real-gas states has the same format as Eq. (6-32), namely,

$$s_2 - s_1 = \left(s_1^* - s_1\right)_{T_1,P_1} - \left(s_2^* - s_2\right)_{T_2,P_2} + \left(s^*_{P_2,T_2} - s^*_{P_1,T_1}\right) \tag{6-34}$$

The first and second terms are the *residual* entropy terms s_1^R and s_2^R for states (T_1, P_1) and (T_2, P_2), respectively. The third term is the entropy change for an ideal gas between the given end states. This is given by the well-known ideal-gas equation,

$$s^*_{P_2,T_2} - s^*_{P_1,T_1} = \int_{T_1}^{T_2} \frac{c_{p,0}}{T}\, dT - R \ln \frac{P_2}{P_1}$$

This expression also requires an equation for $c_{p,0}$ as a function of temperature. Standardized ideal-gas tables (such as the JANAF Tables) list this integral in terms of $s_2^o - s_1^o$. Consequently, Eq. (6-34) may be rewritten in the format

$$s_2 - s_1 = s_1^R - s_2^R + \int_{T_1}^{T_2} \frac{c_{p,0}}{T}\, dT - R \ln \frac{P_2}{P_1} \tag{6-35}$$

The last two terms represent the entropy change of an ideal gas between states 1 and 2, and the first two terms correct for the nonideality of the real-gas states. Methods for evaluating s^R are discussed in later sections.

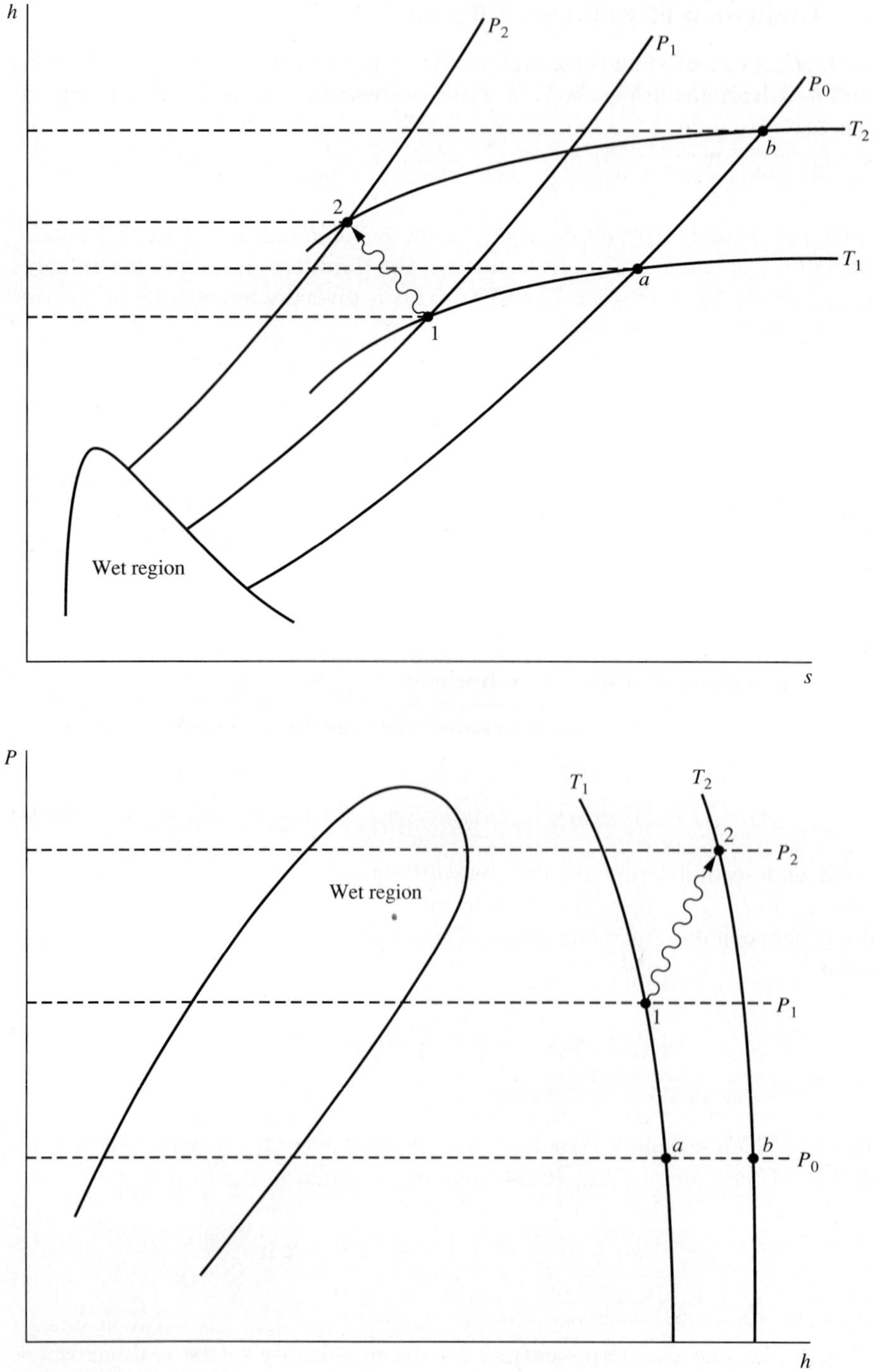

FIGURE 6-2
P-h and *h-s* diagrams illustrating the use of the residual function in evaluating property changes between two real-gas states.

6-6-3 Property Evaluation of Tabular Data

In constructing tables of real-gas data, we modify Eqs. (6-33) and (6-35) in the following manner. One may select state 1 in these equations to be an ideal-gas reference state with an enthalpy h_0^* and an entropy s_0^* at T_0 and P_0. In this reference state h_1^R and s_1^R are zero. Therefore Eq. (6-33) becomes, in general,

$$h = h_0^* + \int_{T_0}^{T} c_{p,0}\, dT - h^R \qquad \text{(6-36)}$$

and the general form of Eq. (6-35) is

$$s = s_0^* + \int_{T_0}^{T} \frac{c_{p,0}}{T}\, dT - R \ln \frac{P_2}{P_0} - s^R \qquad \text{(6-37)}$$

The only information required for the use of these two equations are PvT data and ideal-gas specific-heat data. Equations derived from the first and second laws do not allow the calculation of absolute values for enthalpy and entropy. Hence the reference state (T_0, P_0) may be selected for convenience and values of h_0^* and s_0^* must be arbitrarily set.

6-6-4 Departure Functions

An alternative to the use of the residual function in determining property values or property changes is found in the departure function, y^D. By definition, for any property y,

$$y^D \equiv y_0^* - y \qquad \text{or} \qquad y^D \equiv y - y_0^* \qquad \text{(6-38)}$$

where y is the desired value of y at (T, P), and y_0^* is the property value of the fluid at the same temperature T and a low enough pressure P_0 so that the gas behaves as an ideal gas. With reference to Fig. 6-2, this change of state is given by processes 1-a and 2-b. The enthalpy change for process 1-2 can be found from the alternate process path 1-a-b-2. Symbolically,

$$h_2 - h_1 = \left(h_1^* - h_1\right)_{T_1,P_1} + \left(h_{a,P_0}^* - h_{1,P_1}^*\right)_{T_1} + \left(h_{b,T_2}^* - h_{a,T_1}^*\right)_{P_0}$$
$$- \left(h_{b,P_0}^* - h_{2,P_2}^*\right)_{T_2} - \left(h_2^* - h_2\right)_{T_2,P_2}$$

where the first and last terms on the right are h_1^R and h_2^R, respectively. The sum of the first two terms on the right is the departure enthalpy function h^D for state 1, and the sum of the last two terms is $-h^D$ for state 2. That is,

$$h_1^D = h_1^R + \left(h_{a,P_0}^* - h_{1,P_1}^*\right)_{T_1}$$

and

$$h_2^D = h_2^R + \left(h_{b,P_0}^* - h_{2,P_2}^*\right)_{T_2}$$

But the value of the last term in both the above equations is zero, because the enthalpy of an ideal gas is not a function of pressure. Hence $h^D = h^R$, and Eq. (6-33) is also valid when supplied with h^D values found in tables and graphs.

The above conclusion is not valid when entropy departure values are available. Replacing h with s in the above equations, we find that

$$s_1^D = s_1^R + \left(s_{a,P_0}^* - s_{1,P_1}^*\right)_{T_1}$$

and

$$s_2^D = s_2^R + \left(s_{b,P_0}^* - s_{2,P_2}^*\right)_{T_2}$$

where the last term in the two above equations is the entropy change of an ideal gas due to a change in pressure at constant temperature. Substitution of these last two equations into Eq. (6-35) for s_1^R and s_2^R yields

$$s_2 - s_1 = s_1^D - s_2^D + \int_{T_1}^{T_2} \frac{c_{p,0}}{T}\, dT \tag{6-39}$$

Tables and plots of s^D data require that a value of P_0 be arbitrarily chosen.

6-7 RESIDUAL PROPERTIES AND THE GIBBS FUNCTION

The starting point for the development of basic equations involving the residual functions for properties such as u, h, s, a, and g depends on the nature of the PvT equation of state one wishes to apply. Pressure and temperature are directly measurable and controllable; consequently, they are common choices for the independent variables of a thermodynamic system. From first- and second-law considerations the property most closely related to temperature and pressure is the Gibbs function (recall that $dg = v\,dP - s\,dT$). Thus in the case where $v = f(P, T)$ is known, the Gibbs function g is the logical choice to begin the development of residual properties. An important example of this latter choice is in the development of correlations involving the reduced properties $P_r = P/P_c$ and $T_r = T/T_c$, and the equation of state $Pv = ZRT$.

Rather than develop an expression for dg^R as a function of dP and dT, it is better to work with its dimensionless format, $d(g^R/RT)$. Recall the mathematical identity

$$d\left(\frac{y}{x}\right) = \frac{dy}{x} - \frac{y\,dx}{x^2}$$

Consequently,

$$d\left(\frac{g}{RT}\right) = \frac{dg}{RT} - \frac{g\,dT}{RT^2}$$

But $dg = v\,dP - s\,dT$ at constant composition, and $g \equiv h - Ts$. Substitution yields

$$d\left(\frac{g}{RT}\right) = \frac{v}{RT}\,dP - \frac{h}{RT^2}\,dT \tag{6-40}$$

This equation is a general result. For the case of an ideal-gas state (ig) it becomes

$$d\left(\frac{g^{ig}}{RT}\right) = \frac{v^{ig}}{RT}\,dP - \frac{h^{ig}}{RT^2}\,dT \tag{6-41}$$

Recall from Eq. (6-30) that $y^R \equiv y^* - y = y^{ig} - y$. Thus subtraction of Eq. (6-40) from Eq. (6-41) yields

$$d\left(\frac{g^R}{RT}\right) = \frac{v^R}{RT}\,dP - \frac{h^R}{RT^2}\,dT \tag{6-42}$$

This result is a fundamental *residual property relation* for systems of constant composition.

The final step is to express g^R/RT as a function of P, v, and T. This is achieved by integration of Eq. (6-42) at constant temperature. The lower-pressure limit is taken to be zero pressure, and the upper limit is the actual system pressure P. Because all gases behave as ideal gases at zero pressure, the value of the lower limit for g^R/RT is zero. Thus

$$\frac{g^R}{RT} = \int_0^P \frac{v^R}{RT}\,dP = \int_0^P \left(\frac{1}{P} - \frac{v}{RT}\right) dP \qquad \text{(constant } T\text{)} \tag{6-43}$$

where use has been made of the identity $v^R \equiv v^* - v = RT/P - v$. This equation expresses g^R/RT solely as a function of P, v, and T. Equally important, v appears explicitly on the right-hand side of the equation. Thus use of an appropriate equation in the form $v = f(P, T)$ allows integration of Eq. (6-43).

The importance of Eqs. (6-42) and (6-43) cannot be underestimated. Knowledge of (g/RT) as a function of T and P enables one to evaluate all other thermodynamic properties by straightforward mathematical operations. For example, at constant pressure Eq. (6-42) becomes

$$\frac{h^R}{RT} = -T\left[\frac{\partial(g^R/RT)}{\partial T}\right]_P \qquad \text{(constant } P\text{)} \tag{6-44}$$

where the residual enthalpy term has been made dimensionless. From the definition of g we may write $g^R = h^R - Ts^R$. Therefore in dimensionless format the residual entropy becomes

$$\frac{s^R}{R} = \frac{h^R}{RT} - \frac{g^R}{RT} = -T\left[\frac{\partial(g^R/RT)}{\partial T}\right]_P - \frac{g^R}{RT} \qquad \text{(constant } T\text{)} \tag{6-45}$$

It is apparent that suitable equations can also be developed for the residual internal energy u^R and the residual Helmholtz function a^R, since $u \equiv h - Pv$ and $a \equiv u - Ts$. For example, $u^R = h^R - (Pv)^R$, or

$$\frac{u^R}{RT} = \frac{h^R}{RT} - \frac{P^*v^* - Pv}{RT} = \frac{h^R}{RT} - 1 + Z \tag{6-46}$$

where $Z = Pv/RT$. Thus the residual Gibbs function g^R serves as a generating function for other residual properties. It is important to keep in mind that any residual property relates the real-gas state to its ideal-gas state at the same pressure and temperature.

6-7-1 General Relations for h^R, s^R, and u^R Based on Z

An important example of evaluating residual functions by means of the above relations is the use of the corresponding states principle. The compressibility factor $Z = Pv/RT$ is introduced into the integrand of Eq. (6-43) by noting that

$$v^R \equiv v^{ig} - v = \frac{RT}{P} - \frac{ZRT}{P} = \frac{RT}{P}(1 - Z)$$

Therefore Eq. (6-43) for g^R/RT becomes

$$\frac{g^R}{RT} = \int_0^P (1 - Z)\frac{dP}{P} \qquad \text{(constant } T) \tag{6-47}$$

The use of Eq. (6-47) in Eq. (6-44) yields

$$\frac{h^R}{RT} = T\int_0^P \left(\frac{\partial Z}{\partial T}\right)_P \frac{dP}{P} \qquad \text{(constant } T) \tag{6-48}$$

and on the basis of Eq. (6-47), Eq. (6-45) becomes

$$\frac{s^R}{R} = T\int_0^P \left(\frac{\partial Z}{\partial T}\right)_P \frac{dP}{P} - \int_0^P (1 - Z)\frac{dP}{P} \qquad \text{(constant } T) \tag{6-49}$$

Equations for u^R/RT and a^R/RT are obtainable directly from the three preceding equations.

Solutions to the integrals in Eqs. (6-48) and (6-49) are presented in the literature in two formats. These are a two-parameter format found in terms of the reduced temperature and pressure T_r and P_r, and a three-parameter format in terms of T_r, P_r, and ω (the acentric factor). For the two-parameter format Eqs. (6-48) and (6-49) are rewritten in the reduced form

$$\frac{h^R}{RT_c} = T_r^2\int_0^{P_r} \left(\frac{\partial Z}{\partial T_r}\right)_{P_r} d\ln P_r \qquad \text{(constant } T) \tag{6-50}$$

and

$$\frac{s^R}{R} = \frac{h^R}{RT_cT_r} - \int_0^{P_r} (1 - Z)\frac{dP_r}{P_r} \qquad \text{(constant } T) \tag{6-51}$$

Values of Z and $(\partial Z/\partial T_r)_{P_r}$ are evaluated directly from a generalized chart, or a set of generalized data. The results for h^R/RT_c and s^R/R have been published and appear in several graphical formats. Like the Z chart, these are called *two-parameter generalized* charts and examples appear as Figs. A-39 and A-40 in the Appendix.

6-7-2 Enthalpy and Entropy Changes Based on the Acentric Factor

The acentric factor has been used previously to correlate Z data and saturation data. The general three-parameter method has been extended to enthalpy and entropy data. The correlations have the same format as for the compressibility factor, namely,

$$Z = Z^{(0)} + \omega Z^{(1)} \tag{5-30}$$

Hence the enthalpy and entropy residual functions take on the format

$$\frac{h^R}{R_u T_c} = \left(\frac{h^R}{R_u T_c}\right)^{(0)} + \omega \left(\frac{h^R}{R_u T_c}\right)^{(1)} \tag{6-52}$$

and

$$\frac{s^R}{R_u} = \left(\frac{s^R}{R_u}\right)^{(0)} + \omega \left(\frac{s^R}{R_u}\right)^{(1)} \tag{6-53}$$

Equations (6-52) and (6-53) evolve directly from Eqs. (5-30), (6-50), and (6-51). First, differentiation of Eq. (5-30) with respect to reduced temperature yields

$$\left(\frac{\partial Z}{\partial T_r}\right)_{P_r} = \left(\frac{\partial Z^{(0)}}{\partial T_r}\right)_{P_r} + \omega \left(\frac{\partial Z^{(1)}}{\partial T_r}\right)_{P_r}$$

Use of the three-parameter expressions for Z and $(\partial Z/\partial T_r)_{P_r}$ in Eqs. (6-50) and (6-51) leads to

$$\frac{h^R}{R_u T_c} = T_r^2 \int_0^{P_r} \left(\frac{\partial Z^{(0)}}{\partial T_r}\right)_{P_r} \frac{dP_r}{P_r} + \omega T_r^2 \int_0^{P_r} \left(\frac{\partial Z^{(1)}}{\partial T_r}\right) \frac{dP_r}{P_r}$$

and
$$\frac{s^R}{R_u} = \int_0^{P_r} \left[T_r \left(\frac{\partial Z^{(0)}}{\partial T_r}\right)_{P_r} + Z^{(0)} - 1\right] \frac{dP_r}{P_r} + \omega \int_0^{P_r} \left[T_r \left(\frac{\partial Z^{(1)}}{\partial T_r}\right)_{P_r} + Z^{(1)}\right] \frac{dP_r}{P_r}$$

A comparison of these last two equations with Eqs. (6-52) and (6-53) shows that the first terms on the right represent $(h^R/R_u T_c)^{(0)}$ and $(s^R/R_u)^{(0)}$, respectively. Similarly, the second terms which follow ω are represented in the literature by $(h^R/R_u T_c)^{(1)}$ and $(s^R/R_u)^{(1)}$, respectively. These terms containing integrals are evaluated numerically or graphically from the $Z^{(0)}$ and $Z^{(1)}$ data in Figs. A-24 and A-25 in the Appendix for selected values of T_R and P_R. The Lee-Kesler tables of data for the three-parameter enthalpy and entropy residual properties appear in Tables A-26 through A-29 in the Appendix.

Data evaluated from Eqs. (6-52) and (6-53) are employed in Eqs. (6-33) and (6-35) in a manner similar to the use of the two-parameter residual data. The three-parameter approach should increase the accuracy of estimating property data over the two-parameter approach, especially for substances with large values of the acentric factor. However, even the three-parameter approach may be fairly inaccurate for highly polar compounds.

Example 6-1. Estimate the change in enthalpy, in kJ/kmol, and entropy, in kJ/kmol·K, for CO_2 when it undergoes an isothermal change at 300 K from 20 to 50 bars.
(*a*) Employ the three-parameter residual functions of Lee and Kesler. (*b*) Check against the data in Table A-21.

Solution
(*a*) The critical data for CO_2 are 304 K and 73.9 bars, and $\omega = 0.239$. Hence

$$P_{r,1} = \frac{P}{P_c} = \frac{20}{73.9} = 0.27, \qquad P_{r,2} = \frac{50}{73.9} = 0.68,$$

$$T_{r,1} = T_{r,2} = \frac{T}{T_c} = \frac{300}{304} = 0.99$$

From Tables A-26 through A-29 one finds the following approximate values:

$$\left(\frac{h_1^R}{R_u T_c}\right)^{(0)} = 0.30, \quad \left(\frac{h_1^R}{R_u T_c}\right)^{(1)} = 0.30, \quad \left(\frac{h_2^R}{R_u T}\right)^{(0)} = 0.90, \quad \left(\frac{h_2^R}{R_u T}\right)^{(1)} = 0.86$$

and

$$\left(\frac{s_1^R}{R_u}\right)^{(0)} = 0.22, \quad \left(\frac{s_1^R}{R_u}\right)^{(1)} = 0.25, \quad \left(\frac{s_2^R}{R_u}\right)^{(0)} = 0.63, \quad \left(\frac{s_2^R}{R_u}\right)^{(1)} = 0.75$$

Equation (6-33) then shows that

$$\begin{aligned} h_2 - h_1 &= h_1^R - h_2^R + \int_{T_1}^{T_2} c_{p,0}\, dT \\ &= \{[0.30 + 0.239(0.30)] - [0.90 + 0.239(0.86)]\}(8.314)(300) \\ &= -1830 \text{ kJ/kmol} \end{aligned}$$

Likewise, Eq. (6-35) leads to

$$\begin{aligned} s_2 - s_1 &= s_1^R - s_2^R + \int_{T_1}^{T_2} \frac{c_{p,0}}{T}\, dT - R \ln \frac{P_2}{P_1} \\ &= \left\{[0.22 + 0.239(0.25)] - [0.63 + 0.239(.75)] - \ln \frac{50}{20}\right\} 8.314 \\ &= -12.0 \text{ kJ/kmol} \cdot \text{K} \end{aligned}$$

(*b*) Employing the data from Table A-21, we obtain

$$\Delta h = (362.64 - 404.77)(44.01) = -1854 \text{ kJ/kmol}$$

$$\Delta s = (1.3145 - 1.5955)(44.01) = -12.37 \text{ kJ/kmol·K}$$

Thus the three-parameter residual functions permit a reasonably accurate evaluation of Δh and Δs in this case when compared to data based on an accurate equation of state.

6-8 RESIDUAL PROPERTIES AND THE HELMHOLTZ FUNCTION

As noted earlier, the starting point for the development of basic equations involving properties such as u, h, and s depends on the nature of the PvT equation of state one wishes to apply. Most equations of state are explicit in pressure, that is, they appear in the format $P = f(v, T)$. Recall that the Helmholtz function a has temperature and specific volume as a set of independent variables. Hence in this case the Helmholtz function is the basic property from which other residual property relations can be easily generated. A residual property relates a real and a hypothetical state at the same temperature. Consequently, for the Helmholtz function a the general relation $da = -P\,dv - s\,dT$ reduces to $da_T = -P\,dv$. Integration between the real and hypothetical states defined by Eq. (6-30) yields

$$a^* - a = -\int_v^{v^*} P\,dv = -\int_\infty^{v^*} P\,dv - \int_v^\infty P\,dv \qquad \text{(constant } T\text{)} \qquad (6\text{-}54)$$

where the basic integral between the limits of v to v^* has been rewritten as the sum of two integrals with limits of ∞ to v^* and v to ∞. The state where v is infinite at a given temperature represents the situation of an ideal gas at zero pressure. Hence both the limits of the first integral in Eq. (6-54) represent ideal-gas states, and P in this integral can be replaced by RT/v. One difficulty remains in the above equation. When the limit of either integral on the right-hand side is infinite, the value of the integral is without bound.

To avoid this difficulty, we add the integral of RT/v between the limits of v and ∞ and subtract it from the right-hand side. That is,

$$a^* - a = -\int_v^\infty \left(P - \frac{RT}{v}\right) dv - \int_\infty^{v^*} \frac{RT}{v} dv - \int_v^\infty \frac{RT}{v} dv$$

$$= -\int_v^\infty \left(P - \frac{RT}{v}\right) dv - \int_v^{v^*} \frac{RT}{v} dv$$

or upon exchanging the limits of integration, we have

$$a^* - a = \int_\infty^v \left(P - \frac{RT}{v}\right) dv + RT \ln \frac{v}{v^*} \qquad \text{(constant } T\text{)} \qquad (6\text{-}55)$$

Unlike the integrals in Eq. (6-54), the integral in Eq. (6-55) remains finite in value as P approaches 0. Hence the first term remains finite when the limit of infinity is applied. Finally, recall that the compressibility factor $Z \equiv v/v^{ig}$, where v and v^{ig} are measured at the same temperature and pressure. Thus the preceding equation becomes

$$a^* - a = \int_\infty^v \left(P - \frac{RT}{v}\right) dv + RT \ln Z \qquad \text{(constant } T\text{)} \tag{6-56}$$

The first term on the right can be evaluated for any equation of state in the format $P = f(v, T)$. Keep in mind that T is a constant during the integration. It is very convenient to express residual properties in a dimensionless format. Since a has the units of specific energy, the dimensionless molar format usually used is

$$\frac{a^R}{R_u T} = \frac{a^* - a}{R_u T} = \frac{1}{R_u T} \int_\infty^v \left(P - \frac{R_u T}{v}\right) dv + \ln Z \tag{6-57}$$

The use of R_u requires that a and v be expressed in molar units. The two equations above are now used to generate other thermodynamic properties.

Based on the relation, $da = -P\,dv - s\,dT$, it is noted that $s = -(\partial a/\partial T)_v$. Hence in terms of a residual function format

$$s^* - s = -\frac{\partial}{\partial T}(a^* - a)_v \tag{6-58}$$

Substitution of Eq. (6-56) into Eq. (6-58) leads to

$$s^* - s = -\frac{\partial}{\partial T}\left[\int_\infty^v \left(P - \frac{RT}{v}\right) dv + RT \ln Z\right]_v$$

or

$$s^* - s = -\int_\infty^v \left[\left(\frac{\partial P}{\partial T}\right)_v - \frac{R}{v}\right] dv - R \ln Z \tag{6-59}$$

and in a dimensionless molar format

$$\frac{s^* - s}{R_u} = -\frac{1}{R_u} \int_\infty^v \left[\left(\frac{\partial P}{\partial T}\right)_v - \frac{R_u}{v}\right] dv - \ln Z \tag{6-60}$$

The term $(\partial P/\partial T)_v$ is obtained from a selected equation in the form $P = f(T, v)$.

The residual enthalpy function is generated by noting that

$$h \equiv u + Pv = a + Ts + Pv$$

and therefore $\qquad h^* - h = a^* - a + T(s^* - s) + P^*v^* - Pv$

where $P^*v^* = RT$. Substitution of Eqs. (6-56) and (6-59) into the above expression yields, after cancellation of common terms,

$$h^* - h = -\int_\infty^v \left[T\left(\frac{\partial P}{\partial T}\right)_v - P\right] dv + P^*v^* - Pv \tag{6-61}$$

and in a dimensionless molar form

$$\frac{h^* - h}{R_u T} = -\frac{1}{R_u T}\int_{\infty}^{v}\left[T\left(\frac{\partial P}{\partial T}\right)_v - P\right]dv - Z + 1 \tag{6-62}$$

Finally, to eliminate T from the left-hand side, we may rewrite Eq. (6-62) as

$$\frac{h^* - h}{RT_c} = -\frac{T_r}{R_u T}\int_{\infty}^{v}\left[T\left(\frac{\partial P}{\partial T}\right)_v - P\right]dv + T_r(1 - Z) \tag{6-63}$$

Two other important properties are u and g. Since $g = a + Pv$ and $u = h - Pv$, we find that

$$\frac{g^* - g}{R_u T} = \frac{a^* - a}{R_u T} + 1 - Z \tag{6-64}$$

and on the basis of Eq. (6-46)

$$\frac{u^* - u}{R_u T} = \frac{h^* - h}{R_u T} + Z - 1 \tag{6-65}$$

Thus with any equation of state in the format $P = f(v, T)$ we can use the above equations to generate values of important properties relative to the hypothetical ideal-gas state at the same pressure and temperature.

For the purpose of illustrating the calculation of residual properties based on a PvT equation explicit in pressure we shall derive relations for h^R/R_uT and s^R/R_u for a fluid modeled as a Redlich-Kwong (RK) gas. The three equations we need are

$$P = \frac{RT}{v - b} - \frac{a}{T^{1/2}v(v + b)} \qquad \text{(Redlich-Kwong)} \tag{5-20}$$

$$\frac{h^R}{R_u T} = -\frac{1}{R_u T}\int_{\infty}^{v}\left[T\left(\frac{\partial P}{\partial T}\right)_v - P\right]dv - Z + 1 \tag{6-62}$$

$$\frac{s^R}{R_u} = -\frac{1}{R_u}\int_{\infty}^{v}\left[\left(\frac{\partial P}{\partial T}\right)_v - \frac{R_u}{v}\right]dv - \ln Z \tag{6-60}$$

Differentiation of the RK equation will show that

$$T\left(\frac{\partial P}{\partial T}\right)_v - P = \frac{3a}{2T^{1/2}v(v + b)}$$

Substitution of this relation into Eq. (6-62) for the residual enthalpy leads to

$$\frac{h^R}{R_u T} = -\frac{3a}{2R_u T^{3/2}}\int_{\infty}^{v}\frac{dv}{v(v + b)} - Z + 1$$

$$= 1 - Z + \frac{3a}{2bR_u T^{3/2}}\ln\left(1 + \frac{b}{v}\right) \qquad \text{(Redlich-Kwong)} \tag{6-66}$$

To determine s^R/R, one first finds that for a RK fluid

$$\left(\frac{\partial P}{\partial T}\right)_v - \frac{R_u}{v} = \frac{R_u b}{v(v-b)} + \frac{a}{2T^{3/2}v(v+b)}$$

Substitution of this relation into Eq. (6-60) for the residual entropy yields

$$\begin{aligned}
\frac{s^R}{R_u} &= -\frac{1}{R_u}\int_{\infty}^{v}\left[\frac{R_u b}{v(v-b)} + \frac{a}{2T^{3/2}v(v+b)}\right]dv - \ln Z \\
&= -\ln\left(1-\frac{b}{v}\right) + \frac{a}{2bR_uT^{3/2}}\ln\left(1+\frac{b}{v}\right) - \ln Z \\
&= -\ln\left[Z\left(1-\frac{b}{v}\right)\right] + \frac{a}{2bR_uT^{3/2}}\ln\left(1+\frac{b}{v}\right) \qquad \text{(Redlich-Kwong)} \qquad (6\text{-}67)
\end{aligned}$$

The relationship for the internal energy is found by applying Eqs. (6-65) and (6-66). The result is

$$\begin{aligned}
\frac{u^* - u}{R_uT} &= \frac{h^* - h}{R_uT} + (Z-1) \\
&= \frac{3a}{2bR_uT^{3/2}}\ln\left(1+\frac{b}{v}\right) \qquad \text{(Redlich-Kwong)} \qquad (6\text{-}68)
\end{aligned}$$

Finally, the residual Gibbs function requires the use of Eq. (6-64), or starts from the definition, $g \equiv h - Ts$. In terms of g^R this latter relation may be written as

$$\frac{g^* - g}{R_uT} = \frac{h^* - h}{R_uT} - \frac{s^* - s}{R_u}$$

Substitution of Eqs. (6-66) and (6-67) on the right-hand side of this expression leads to the desired result for a RK gas, namely,

$$\frac{g^* - g}{R_uT} = (1-Z) + \ln\left[Z\left(1-\frac{b}{v}\right)\right] + \frac{a}{bR_uT^{3/2}}\ln\left(1+\frac{b}{v}\right) \qquad (6\text{-}69)$$

To evaluate the residual properties derived above for a given state, we use the RK PvT equation first to relate P, v, T, and Z. These data are then used in Eqs. (6-66), (6-67), (6-68), and (6-69) to evaluate the residual values of the specific enthalpy, entropy, internal energy, and Gibbs function for that state.

Example 6-2. Estimate the change in enthalpy, in kJ/kmol, and entropy, in kJ/kmol·K, for CO_2 when it undergoes an isothermal change at 300 K from 20 to 50 bars. Employ the residual property functions for a Redlich-Kwong gas.

Solution. The residual enthalpy of a RK gas is given by Eq. (6-66), namely,

$$\frac{h^R}{R_uT} = \frac{3a}{2bR_uT^{3/2}} \ln\left(1 + \frac{b}{v}\right) - Z + 1$$

The values of a and b for CO_2 are 64.64 bar $(m^3/kmol)^2\ K^{1/2}$ and 0.02969 $m^3/kmol$, respectively. The specific volumes in the two states are found from

$$P = \frac{RT}{v-b} - \frac{a}{T^{1/2}v(v+b)} \tag{5-20}$$

At states 1 and 2 the specific volumes are found to be 1.119 and 0.3475 $m^3/kmol$, and the Z values are 0.897 and 0.697, respectively. Substitution of these data into Eq. (6-66) shows that

$$\frac{h_1^R}{R_uT} = \frac{3(64.64)}{2(0.02969)(0.08314)(300)^{3/2}} \ln\left(1 + \frac{0.02969}{1.119}\right) - 0.897 + 1$$
$$= 8.325(0.0262) + 0.103 = 0.321$$

By a similar calculation $h_2^R/R_uT = 0.986$. Equation (6-33) then yields

$$h_2 - h_1 = h_1^R - h_2^R + \int_{T_1}^{T_2} c_{p,0}\, dT$$
$$= (0.321 - 0.986)(8.314)(300) = -1660 \text{ kJ/kmol}$$

Tabular data in the Appendix based on a more accurate PvT equation of state indicate a Δh value of around -1870 kJ/kmol. Thus the use of the RK equation leads to roughly a 10 percent error.

The residual entropy of a RK gas is given by Eq. (6-67), namely,

$$\frac{s^R}{R_u} = -\ln\left[Z\left(1 - \frac{b}{v}\right)\right] + \frac{a}{2bR_uT^{3/2}} \ln\left(1 + \frac{b}{v}\right)$$

For state 1

$$\frac{s^R}{R_u} = -\ln(0.897)\left(1 - \frac{0.02969}{1.119}\right)$$
$$+ \frac{64.64}{2(0.02969)(0.08314)(300)^{3/2}} \ln\left(1 + \frac{0.02969}{1.119}\right)$$
$$= 0.1356 + 2.775(0.0262) = 0.208$$

By a similar evaluation, $s_2^R/R_u = 0.678$. Equation (6-35) then yields

$$s_2 - s_1 = s_1^R - s_2^R + \int_{T_1}^{T_2} \frac{c_{p,0}}{T}\, dT - R_u \ln\frac{P_2}{P_1}$$
$$= (0.208 - 0.678)(8.314) - 8.314 \ln(50/20)$$
$$= -3.91 - 7.62 = -11.53 \text{ kJ/kmol·K}$$

From the superheat table in the Appendix the value is around -12.4 kJ/kmol·K. This corresponds to a 7 percent lower value by using the RK equation.

6-9 FLOW AVAILABILITY FROM RESIDUAL FUNCTIONS

An important application of residual function data is to the evaluation of stream availability or exergy ψ at a specified T and P. It is defined as $\psi \equiv (h - h_0) - T_0(s - s_0)$. The residual exergy ψ^R at a state (T, P) is defined as $\psi^R \equiv \psi^* - \psi$, where ψ^* is the exergy value at (T, P) if the gas were an ideal gas at that temperature and pressure. From these two definitions we find that

$$\psi^R = (\psi^* - \psi)_{T,P} = [(h^* - h) - (h_0^* - h_0)] - T_0[(s^* - s) - (s_0^* - s_0)]$$

Division by RT_c and subsequent rearrangement yield

$$\left(\frac{\psi^* - \psi}{RT_c}\right)_{T,P} = \left[\frac{h^* - h}{RT_c} - \frac{T_0}{T_c}\left(\frac{s^* - s}{R}\right)\right] - \left[\frac{h_0^* - h_0}{RT_c} - \frac{T_0}{T_c}\left(\frac{s_0^* - s_0}{R}\right)\right]$$

The first and third terms on the right are in the format of a tabulated residual enthalpy function h^R/RT_c, whereas the second and fourth terms are formatted as a tabulated residual entropy function s^R/R. Thus the equation for a dimensionless exergy function ψ^R/RT_c may be written as

$$\frac{\psi^R}{RT_c} = \frac{\psi^* - \psi}{RT_c} = \left[\frac{h^R}{RT_c} - \frac{T_0}{T_c}\frac{s^R}{R}\right]_{T,P} - \left[\frac{h^R}{RT_c} - \frac{T_0}{T_c}\frac{s^R}{R}\right]_{T_0,P_0} \tag{6-70}$$

For a steady-state process from state 1 to state 2

$$\psi_2 - \psi_1 = -RT_c\left(\frac{h_2^R}{RT_c} - \frac{h_1^R}{RT_c}\right) + RT_0\left(\frac{s_2^R}{R} - \frac{s_1^R}{R}\right) \tag{6-71}$$

In the last two equations the values of h^R/RT_c and s^R/R are found from two-parameter or three-parameter generalized charts as a function of reduced temperature T_r and pressure P_r.

Example 6-3. Nitrogen gas is maintained at 200 K and 50 bars, while the environmental state is 300 K and 1 bar. Calculate the exergy of the gas, in kJ/kg, by means of (*a*) superheat data, (*b*) ideal-gas data, and (*c*) the corresponding states principle.

Solution

(*a*) Based on data in the nitrogen superheat Table A-18,

$$\begin{aligned}\psi &= (h - h_0) - T_0(s - s_0)\\ &= (282.0 - 411.2) - 300(5.169 - 6.844) = 383.3 \text{ kJ/kg}\end{aligned}$$

This value should be highly accurate.

(*b*) Based on data in the ideal-gas nitrogen Table A-6 in the Appendix,

$$\psi = (h - h_0) - T_0\left[s^0 - s_0^0 - R\ \ln(P/P_0)\right]$$

$$= (5812.8 - 8724.1) - 300(179.842 - 191.646) + 300(8.3145)\ \ln(50/1)$$

$$= 10{,}390 \text{ kJ/kmol} = 370.9 \text{ kJ/kg}$$

The ideal-gas model predicts a value which is only 3 percent too low at this state, in spite of the fact that $Z = 0.90$.

(*c*) The generalized data, based on two-parameter correlations, are shown below for the actual state and the dead state

$$P_r = 1.48 \qquad T_r = 1.59 \qquad \frac{h^R}{RT_c} = 0.62 \qquad \frac{s^R}{R} = 0.30$$

$$P_{r,0} = 0.03 \qquad T_{r,0} = 2.38 \qquad \frac{h_0^R}{RT_c} \approx 0 \qquad \frac{s_0^R}{R} \approx 0$$

On the basis of Eq. (6-70)

$$\psi^* - \psi = 8.3145(126.2)\left[0.62 - \frac{300}{126.2}(0.30)\right] \approx -98 \text{ kJ/kmol} = -3.5 \text{ kJ/kg}$$

The value of ψ^* is simply the value of ψ in part (*b*) for the ideal-gas model. Therefore

$$\psi = \psi^* - (\psi^* - \psi) = 370.9 + 3.5 = 384.4 \text{ kJ/kg}$$

Thus the generalized method prediction is in good agreement with superheat data.

6-10 PROPERTIES OF THE SATURATION STATE

Phase changes are of primary interest in many engineering areas. In this text we shall restrict ourselves primarily to the details of liquid-vapor equilibrium, although many of the general results apply to solid-liquid or solid-gas equilibrium. The discussion in this chapter is restricted to a pure component, whereas Chap. 8 is devoted to multicomponent liquid-vapor phase equilibrium. We seek fundamental relationships among and approximate evaluation techniques for the basic properties—P, T, u, h, s, a, and g—of a two-phase system. The basic starting point is to determine a criterion for phase equilibrium of a pure component at a given pressure or temperature.

The familiar phase transitions—vaporization, melting, and sublimation—have two well-known characteristics. First, the temperature and pressure remain constant during the phase change. Second, an energy interaction such as heat is required. Historically this quantity of energy has been called the *latent heat*, for example, the latent heat of vaporization. When heat is used, the energy equation for a closed, constant-pressure, simple compressible system reduces to $q_P = \Delta u + P\Delta v = \Delta h$. The entropy change for such a system is $\Delta s = \int \delta q/T = q_p/T$. The combination of these two equations shows that $\Delta s = \Delta h/T$, or

$$\Delta h - T\Delta s = 0 \qquad \text{(phase change)}$$

In addition, recall that $g \equiv h - Ts$, or $\Delta g_T = \Delta h - T\Delta s$. From the preceding result we conclude that $\Delta g_T = 0$ for a phase change, or that

$$g^\alpha = g^\beta \qquad \text{(phase change)} \tag{6-72}$$

where α and β are any two *saturation* phases of a pure component. Thus the equality of the specific Gibbs function for each phase is the criterion for phase equilibrium. (This criterion is a special case of a more general one developed in Chap. 8.) Equation (6-72) leads to an important relationship for the slope of the vapor-pressure curve, as shown below. The equation is also the starting point for establishing numerical values for Δh and Δs based on a reliable PvT equation of state.

Note from the preceding development that both s and v change during familiar phase transitions. In addition, from the relationship $dg = v\,dP - s\,dT$, we see that

$$v = \left(\frac{\partial g}{\partial p}\right)_T \qquad \text{and} \qquad s = -\left(\frac{\partial g}{\partial T}\right)_P$$

Because v and s change discontinuously, the first-order derivatives of the Gibbs function change discontinuously. Any phase change which satisfies the last statement is known as a *phase change of the first order*. (A phase change for which the second-order derivatives of the Gibbs function change discontinuously is known as a phase change of the second order.) Figure 6-3 shows the temperature variations of g, v, and s for a first-order phase change. Because $c_p = T(\partial s/\partial T)_p$, the c_p value of the two-phase mixture is infinite when both phases are present. Note, however, that c_p of either phase remains finite when only the single phase is present at the transition (saturation) temperature.

As illustrated in Fig. 6-4 on PT and Pv diagrams, we now consider the situation where the temperature (and pressure) of a two-phase equilibrium system is raised by a differential amount dT (and dP). For the initial (i) equilibrium state, on the basis of Eq. (6-72), $g_i^L = g_i^V$. For the final equilibrium state, $g_i^L + dg^L = g_i^V + dg^V$, where L and V in this instance represent the saturated liquid and vapor phases, respectively. One must conclude for the change dT that

$$dg^L = dg^V \tag{6-73}$$

As noted earlier, Eq. (6-9) dictates that for a simple compressible fluid $dg = v\,dP - s\,dT$. Use of this relation in the above equation yields

$$v^L\,dP - s^L\,dT = v^V\,dP - s^V\,dT$$

or upon rearrangement

$$\left.\frac{dP}{dT}\right)_{\text{sat}} = \frac{s^V - s^L}{v^V - v^L} \tag{6-74}$$

where "sat" is a reminder that this is an equation for the rate of change of vapor pressure with saturation temperature. Thus the slope of the vaporization line on a PT diagram is related to the discontinuities in the values of the specific entropy

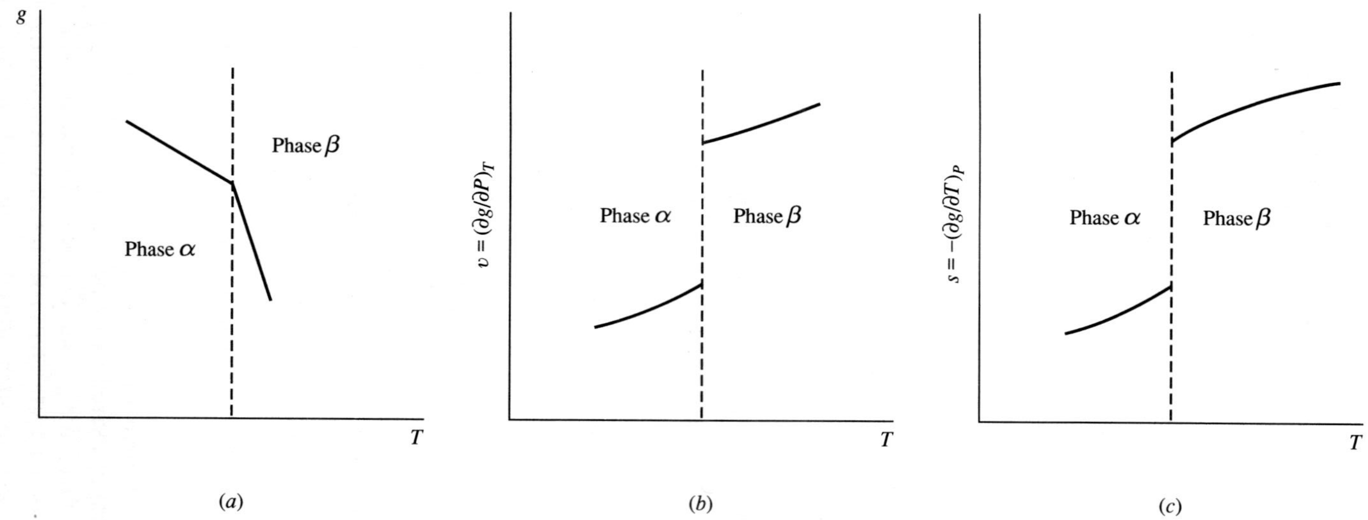

FIGURE 6-3
Variations of g, v, and s with temperature during first-order phase changes.

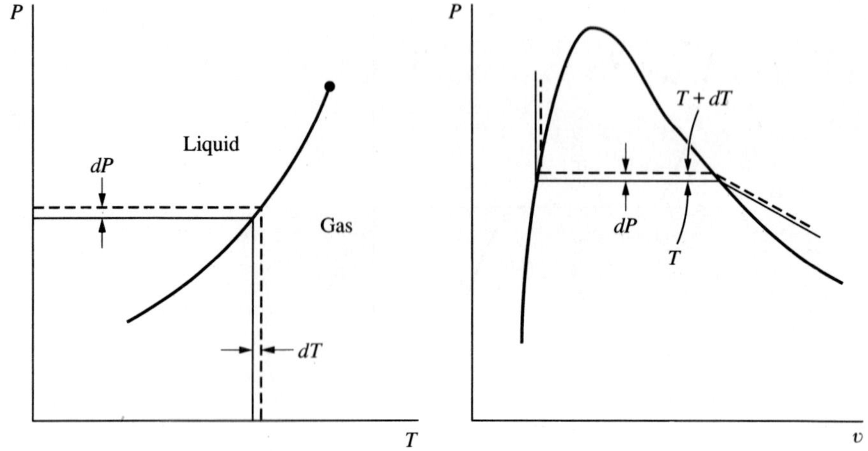

FIGURE 6-4
Process involving a differential change in saturation temperature and pressure.

and specific volume which occur during the isothermal change of state. Finally, recall from above that $\Delta h = T\,\Delta s$ for a phase change. As a result,

$$\left.\frac{dP}{dT}\right)_{\text{sat}} = \frac{h^V - h^L}{T(v^V - v^L)} \tag{6-75}$$

or, more generally, for any set of two phases

$$\left.\frac{dP}{dT}\right)_{\text{sat}} = \frac{\Delta h}{T\Delta v} \tag{6-76}$$

Equation (6-76) is the general format of the *Clapeyron* equation, which is an exact thermodynamic relation. Equation (6-75) is a specific application to liquid-vapor equilibrium.

The enthalpy of vaporization can be measured calorimetrically and experimental values are available in the literature for many substances. However, the data frequently are for a single temperature which is not the temperature of interest. The Clapeyron equation in the form of Eq. (6-76) relates the enthalpy (latent heat) of vaporization to vapor-pressure and volumetric data. If these data are available at the temperature of interest, the Clapeyron equation can be used to predict the enthalpy of vaporization, as well as that for sublimation or melting. A major computational problem when using Eq. (6-76) is that dP/dT must be evaluated accurately from measured vapor-pressure data.

Approximation methods for Δh must be used in the absence of volumetric data at the temperature of interest. The method below is restricted to liquid-gas and solid-gas transformations. In such cases the volume of the gas phase is usually much greater than that of the liquid or solid phase (except as the critical state is approached). In addition, it is assumed that the vapor behaves as an ideal gas.

Both of these conditions are better met at low pressures. Thus Δv in Eq. (6-76) is replaced by RT/P, and the equation becomes

$$\left(\frac{dP}{P}\right)_{\text{sat}} = \frac{\Delta h \, dT}{RT^2}$$

or

$$d \ln P^{\text{sat}} = -\frac{\Delta h}{R} d\left(\frac{1}{T}\right) \tag{6-77}$$

In terms of the enthalpy change for the phase change

$$\Delta h = -R \frac{d \ln P^{\text{sat}}}{d(1/T)} \tag{6-78}$$

Any one of the three forms shown above is the approximation equation known as the *Clausius-Clapeyron* equation. In this case the enthalpy of vaporization (or sublimation) is solely related to vapor-pressure data, and volumetric data are not required.

Over a not too large temperature interval the enthalpy of vaporization may be assumed to be constant. On this basis integration of Eq. (6-77) leads to

$$\ln P^{\text{sat}} = -\frac{\Delta h}{RT} + \text{const} \tag{6-79}$$

or

$$\ln \frac{P_2^{\text{sat}}}{P_1^{\text{sat}}} = -\frac{\Delta h}{R}\left(\frac{1}{T_2} - \frac{1}{T_1}\right) \tag{6-80}$$

Equation (6-80) indicates that an average enthalpy of vaporization between T_1 and T_2 could be obtained from a knowledge of the vapor pressures at these two temperatures. Equation (6-79) suggests that a plot of $\ln P^{\text{sat}}$ versus $1/T$ would produce a straight line over the selected small temperature range. Surprisingly, experimental data for many substances plot nearly as a straight line over the entire range of temperatures from the triple state to the critical state. This occurs in spite of the fact that h_{fg} decreases rapidly as the critical state is approached.

The success of this type of plot lies in a reexamination of the development of the Clausius-Clapeyron equation. Rather than replace Δv in the Clapeyron equation by the ideal-gas specific volume, RT/P, a more accurate equation would be $\Delta v = (Z_g - Z_f)RT/P = \Delta ZRT/P$, where Z_g and Z_f are the saturated vapor and liquid compressibility factors, respectively. This leads to a *modified* Clausius-Clapeyron equation in the form

$$d \ln P^{\text{sat}} = -\frac{\Delta h}{R\Delta Z} d\left(\frac{1}{T}\right) \tag{6-81}$$

In this format the variation of $\ln P^{\text{sat}}$ with $(1/T)$ depends on the ratio $\Delta h/\Delta Z$. This ratio tends to remain roughly constant, with a minimum value at $T_r = 0.85$ for many substances. A widely used correlation for nonassociating liquids by Watson [1] indicates that $h_{fg} = \alpha(1 - T_r)^{0.38}$. In addition, Liley [2] has found that Z_{fg} may be correlated by the relation $Z_{fg} = \beta(1 - T_r)^{0.38}$ with an average

1.4 percent error for $\beta = 1.371$ and T_r values between 0.70 and 0.95. When these two experimental correlations are substituted into Eq. (6-81), we find that

$$d \ln P^{\text{sat}} = \frac{\alpha}{\beta R}\frac{dT}{T^2} = \gamma\frac{dT}{T^2} \tag{6-82}$$

Thus the variations of h_{fg} and Z_{fg} with temperature are such that the linear variation of $\ln P^{\text{sat}}$ with $1/T$ (or $\ln P_r^{\text{sat}}$ with $1/T_r$, as noted in Fig. 5-3) is a reasonably good approximation for many fluids. When accurate vapor-pressure data are plotted versus $1/T$, the data have the shape of a shallow S-curve. The preceding discussion considered theoretical and approximate equations relating the enthalpy of vaporization to vapor-pressure data. We shall now examine separate correlations for vapor-pressure data and the enthalpy of vaporization.

6-10-1 Vapor-Pressure Correlations

The Clausius-Clapeyron equation suggests that vapor-pressure data might be fit reasonably well with an equation of the type

$$\ln P^{\text{sat}} = A - \frac{B}{T} \tag{6-83}$$

where A is a constant and $B = \Delta h/R$. Both A and B are different for various species. The data for a large number of substances have been fitted [3] to Eq. (6-83). The temperature range over which the equation is applicable is also listed. This model of linearity is quite useful when there is a need to interpolate from a limited amount of measured vapor-pressure data. Also, recall that the definition of the acentric factor ω is based on the format of Eq. (6-83) as applied to simple fluids.

In order to maintain relatively high accuracy, the temperature range over which Eq. (6-83) is applicable is usually made relatively small. Frequently for engineering purposes less accuracy is permissible and a larger temperature range is desirable. A modification of Eq. (6-83) can be made by noting that the critical temperature and pressure, as well as the normal boiling point T_b of a fluid, are easily measured. [The normal boiling point is the saturation temperature at 1 atm (or 1 bar).] The constants A and B can be ascertained solely from these data. At the normal boiling point $\ln P = 0$, and hence $B = -AT_b$. Substitution of this relation back into Eq. (6-83) yields

$$\ln P = A\left(1 - \frac{T_b}{T}\right)$$

Application of critical state data to the above equation shows that $A = T_c \ln P_c/(T_c - T_b)$. Substitution of this result into the preceding equation, and subsequent rearrangement, yields

$$\ln P^{\text{sat}} = \frac{T_c \ln P_c}{T_c - T_b}\left(1 - \frac{T_b}{T}\right) \tag{6-84}$$

where P^{sat} and P_c are in bars when T_b, in kelvins, is measured at 1 bar. This equation is reasonably accurate over a fairly wide range of temperatures and requires a minimum of input data.

Another expression used for presenting vapor-pressure data is the *Antoine* [4] equation, given by

$$\ln P^{\text{sat}} = A - \frac{B}{T + C} \tag{6-85}$$

This equation is quite satisfactory, and the constants are readily found in references such as Boublik et al. [5] and Reid et al. [6]. However, the applicable temperature range is small and usually is chosen so that the pressure range represented is from about 0.01 to 2 bars. Note that the Antoine equation is not dimensionless. When using tables of the Antoine constants A, B, and C, one must ascertain whether $\ln P$ or $\log P$ is used and what pressure unit is implied. In addition, tables for C in some handbooks require that T be expressed in degrees Celsius. Antoine constants in this latter temperature format have been reported for nearly 700 organic compounds by Yaws and Yang [7]. The Antoine equation does not reproduce the correct shape of the vapor-pressure curve beyond the temperature range indicated in tables. Hence it should never be used for extrapolating data. Nevertheless, the Antoine equation does tend to account for the small curvature found in accurate $\ln P$ versus $1/T$ plots.

To improve the accuracy of Eq. (6-83) over a wide range of temperature requires an increase in complexity of the equation. For example, the Clausius-Clapeyron equation is derived by assuming Δh is a constant. Over a wider temperature range one might model Δh as a linear function of temperature. This leads to a vapor-pressure-temperature equation of the format

$$\ln P^{\text{sat}} = A - \frac{B}{T} + C \ln T$$

In spite of the additional term, there are few occasions [8] where the above equation might be preferred over Eq. (6-83). To improve the accuracy of curve fitting of vapor-pressure data even further, more terms are required. Three empirical vapor-pressure correlations in frequent use are listed below.

$$\ln P^{\text{sat}} = A + \frac{B}{T} + C \ln T + DT^n \tag{6-86}$$

$$\ln P^{\text{sat}} = A + \frac{B}{T} + CT + DT^2 + FT^3 \tag{6-87}$$

$$\ln P^{\text{sat}} = A + \frac{B}{T} + CT + DT^2 + E \ln T \tag{6-88}$$

Other formats are discussed in the literature [8]. Whatever format is chosen for fitting experimental data, the number of terms on the right is usually restricted to four or five.

The curve fitting of experimental vapor-pressure data allows one to interpolate data with excellent accuracy. Another possible use of vapor-pressure equations is to evaluate dP/dT at a given temperature. This information, then, could be used in conjunction with Eq. (6-76), the Clapeyron equation, to evaluate Δh at that temperature if volumetric data were also available. This approach has one major mathematical difficulty if high accuracy is desired. Although the vapor-pressure equation itself may be highly accurate, there is no reason that dP/dT based on an empirical equation truly represents the slope of the actual vapor-pressure curve at the given temperature.

The law of corresponding states was first applied in Chap. 5 to relate P, v, and T in the gas phase in terms of reduced properties. This was extended to properties such as u, h, s, and c_p earlier in this chapter. In both cases a two-parameter approach and a three-parameter approach were shown. The corresponding states principle is now extended to the liquid-vapor region. On the basis of the state postulate the number of independent, intensive variables is reduced to one for the two-phase system. A two-parameter format of the corresponding states principle can be deduced from the Clausius-Clapeyron equation in the format of Eq. (6-83), namely,

$$\ln P^{\text{sat}} = A - \frac{B}{T} \tag{6-83}$$

If the term $\ln P_c$ is added and subtracted from the left-hand side, and the term B/T is multiplied and divided by T_c, then the above equation becomes

$$\ln P_r^{\text{sat}} = A' - \frac{B'}{T_r^{\text{sat}}} = A'\left(1 - \frac{1}{T_r^{\text{sat}}}\right)$$

where $P_r^{\text{sat}} = P^{\text{sat}}/P_c$ and $T_r^{\text{sat}} = T^{\text{sat}}/T_c$. The above equation predicts that on reduced coordinates the vapor-pressure data for all liquids should approximate a single straight line. Data presented by Guggenheim [9] are shown in Fig. 6-5 for seven *simple* fluids. (The acentric factors for the seven are all less than 0.05.) The line in the diagram is for argon, krypton, and xenon. All data are represented quite well by this single line for the reduced temperature range from roughly 0.55 to 1.0. For this line $A' = B' = 5.3$. Thus

$$\ln \frac{P^{\text{sat}}}{P_c} = 5.3\left(1 - \frac{T_c}{T^{\text{sat}}}\right) \qquad \left(\text{for } 0.55 < \frac{T}{T_c} < 1\right) \tag{6-89}$$

This equation is of limited accuracy for nonsimple fluids, as might be expected.

To increase the accuracy of a reduced-state vapor-pressure correlation to include normal fluids requires that a second adjustable parameter be included. One method of correlating saturation data is the three-parameter format, $P^{\text{sat}} = f(T^{\text{sat}}, \omega)$. A correlation for the reduced vapor pressure as a function of the reduced saturation temperature and the acentric factor was first proposed by Pitzer et al. [10] in the format

$$\log P_r^{\text{sat}} = (\log P_r)^{(0)} + \omega\left(\frac{\partial \log P_r}{\partial \omega}\right)_T \tag{6-90}$$

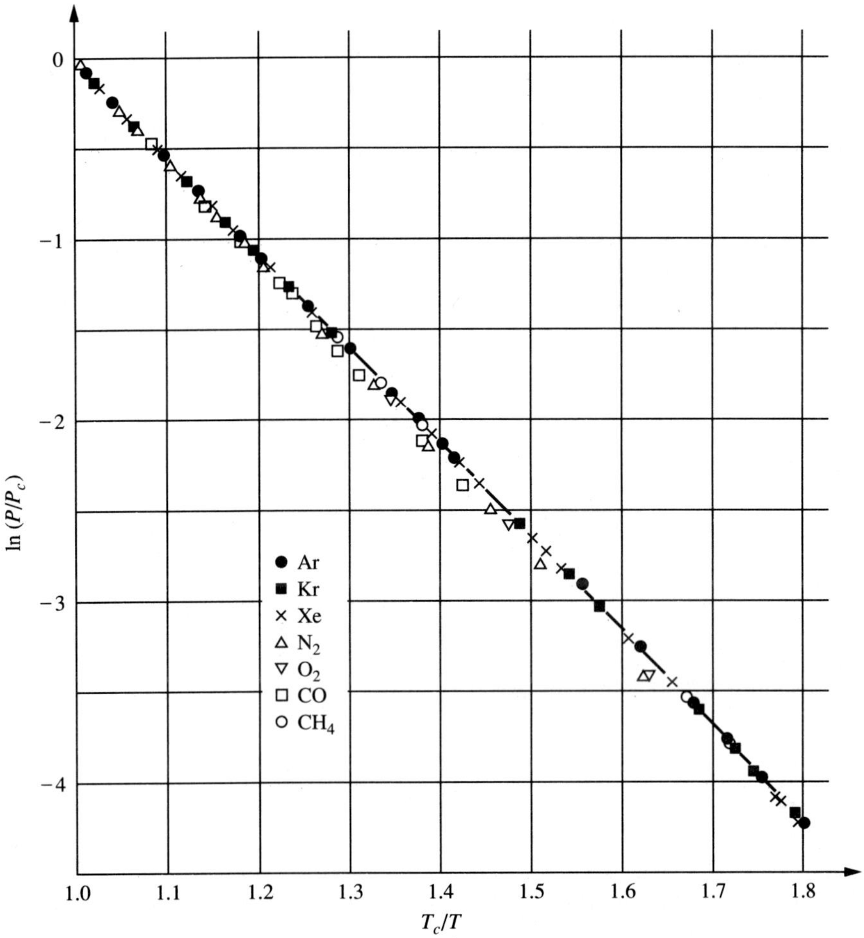

FIGURE 6-5
Application of the law of corresponding state to liquid-vapor equilibrium for simple fluids. (*Source*: E. Q. Guggenheim, *Thermodynamics*, North-Holland and Interscience Publishers, New York, 1957.)

where the logarithm is to the base 10. The values of the two functions on the right are found in Table A-30 in the Appendix as a function of T_r. From the definition of the acentric factor at a reduced temperature of 0.70

$$(\log p_r)^{(0)} = -1.000 \qquad \text{and} \qquad \left(\frac{\partial \log p_r}{\partial \omega}\right)_T = -1.000$$

Note that these values appear in the table when $T_r = 0.70$. The Pitzer table of saturation data also includes values of the compressibility factors $Z^{(0)}$ and $Z^{(1)}$ for the liquid and vapor phases. This permits the estimation of the specific volume of the saturated liquid and vapor phases from the relation $v = ZRT/P$, where again $Z = Z^{(0)} + \omega Z^{(1)}$. (The values for the liquid phase are not too accurate at the lower values of T_r, where $Z^{(0)}$ and $Z^{(1)}$ are quite small.)

Several authors have expressed the Pitzer vapor-pressure correlation in the format

$$\ln P_r^{\text{sat}} = f^{(0)} + \omega f^{(1)} \tag{6-91}$$

The functions $f^{(0)}$ and $f^{(1)}$ appear in tabular form in several articles [9–11]. Lee and Kesler [12] have expressed the functions analytically such that Eq. (6-91) becomes

$$\ln P_r^{\text{sat}} = 5.92714 - \frac{6.09648}{T_r} - 1.28862 \ln T_r + 0.169347\, T_r^6$$

$$+ \omega \left(15.2518 - \frac{15.6875}{T_r} - 13.4721 \ln T_r + 0.43577\, T_r^6\right) \tag{6-92}$$

Rather than use the values of ω listed in the Appendix, it is recommended when using Eq. (6-92) that the acentric factor be calculated from the relation

$$\omega = \frac{\alpha}{\beta} \tag{6-93}$$

where
$$\alpha = -\ln P_c - 5.92714 + 6.09648\Theta^{-1} + 1.28862 \ln \Theta - 0.169347\Theta^6$$
$$\beta = 15.2518 - 15.6875\Theta^{-1} - 13.4721 \ln \Theta + 0.43577\Theta^6$$

$\Theta = T_b/T_c$, T_b is the normal boiling point, and P_c is in atmospheres. The Lee-Kesler equation satisfies three important criteria. First, it fulfills the definition of the Pitzer acentric factor, Eq. (4-23). Second, the curve passes through the critical state; that is, $P_r = 1$ when $T_r = 1$. Finally, the curvature of the saturation line is zero right at the critical state. Tables for the Lee-Kesler quantities $f^{(0)}$ and $f^{(1)}$ as expressed by Eq. (6-92) appear in the literature [8] at intervals of 0.001 in T_r. Another set of expressions for $f^{(0)}$ and $f^{(1)}$ have been developed by Dong and Lienhard, [13], which yields Eq. (6-91) in the form

$$\ln P_r^{\text{sat}} = 5.37270\left(1 - \frac{1}{T_r}\right)$$

$$+ \omega(7.49408 - 11.18177\, T_r^3 + 3.68769\, T_r^6 + 17.92998 \ln T_r) \tag{6-94}$$

Its major advantage over the Lee-Kesler equation is that it predicts data better over a wider range of ω. Dong and Lienhard reported a 0.45-rms (root mean square) error for this equation for the range $-0.3 < \omega < 0.9$. This equation satisfies the definition of ω, as well as the critical state condition.

6-10-2 Estimation of Enthalpy of Vaporization Data

Correlations are available which are fairly accurate in predicting the enthalpy of vaporization, h_{fg}. The normal boiling point T_b is defined as the saturation temperature for a pressure of 1 atm or 1.01325 bars. One method of estimating $h_{fg,b}$ at the normal boiling point is due to Fishtine [14],

$$\frac{h_{fg,b}}{T_b} = k(36.6 + 8.31 \ln T_b) \tag{6-95}$$

where $k = 1$ for nonpolar compounds and $1 < k < 1.38$ for polar and hydrogen-bonded compounds. The units for $h_{fg,b}$ are J/gmol and T_b is in kelvins. Reidel [15] has proposed the following dimensionless relation for the enthalpy of vaporization $h_{fg,b}$ at the reduced normal boiling point, $T_{b,r}$

$$\frac{h_{fg,b}}{RT_c} = \frac{1.093 T_{b,r}(\ln P_c - 1.013)}{0.930 - T_{b,r}} \tag{6-96}$$

where T_c is in kelvins and P_c is in bars. The error in predicting $h_{fg,b}$ by the above equation usually does not exceed 2 percent. If the enthalpy of vaporization is known at one temperature, the value at another temperature for a pure liquid may be estimated from the widely used Watson correlation for nonassociating liquids [16],

$$\frac{h_{fg,2}}{h_{fg,1}} = \left(\frac{1 - T_{r,2}}{1 - T_{r,1}}\right)^{0.38} \tag{6-97}$$

A method for evaluating h_{fg} from any empirical PvT equation of state will be given later.

A three-parameter corresponding states approach to the enthalpy of vaporization usually begins with the evaluation of the entropy of vaporization, s_{fg}. The entropy change for the vaporization process is theoretically related to vapor-pressure data by the Clapeyron equation,

$$\frac{dP}{dT} = \frac{\Delta s}{\Delta v}$$

Since v of each phase may be represented by $v = ZRT/P$, then

$$\Delta s = \frac{dP}{dT}\frac{\Delta ZRT}{P} = -\frac{R\Delta Z}{T_r}\frac{d \ln P_r}{d(1/T_r)} \tag{6-98}$$

where $\Delta Z = Z_g - Z_f$. Knowledge of compressibility data and the temperature derivatives of the vapor pressure, then, leads to entropy of vaporization data. Similar to compressibility factor data, a suitable three-parameter correlation for Δs of vaporization has the form

$$\Delta s = s_{fg} = \Delta s_v^{(0)} + \omega \, \Delta s_v^{(1)} \tag{6-99}$$

Data by Pitzer et al. for the two parameters on the right-hand side of the above equation are also listed in Table A-30 in the Appendix. The parameters are shown in dimensionless format in the table, which is a modification of the original data.

The enthalpy of vaporization h_{fg} equals Ts_{fg}. Consequently, use of Eq. (6-99) for s_{fg} leads to the following h_{fg} relation by Pitzer et al.:

$$h_{fg} = RT\left(\frac{\Delta s_v^{(0)}}{R} + \omega\,\frac{\Delta s_v^{(1)}}{R}\right) \tag{6-100}$$

This relation leads to reasonable values of the enthalpy of vaporization for normal fluids.

An alternate method for estimating h_{fg} is to apply any accurate vapor-pressure equation directly to Eq. (6-98), which is rewritten here as

$$h_{fg} = \frac{RT^2(Z_g - Z_f)}{P}\frac{dP}{dT} = \frac{RTT_r\Delta Z}{P_r}\frac{dP_r^{\text{sat}}}{dT_r} \tag{6-101}$$

Either Eq. (6-92) or (6-94) could be used to find $1/P_r(dP_r/dT_r)$ in the above relation at a given temperature. The difference $(Z_g - Z_f)$ is available from a two-parameter generalized compressibility chart such as Fig. A-36 in the Appendix, or the $Z^{(0)}$ and $Z^{(1)}$ data of Table A-30. An analytical correlation for ΔZ published in 1946 by Haggenmacher [17] is also useful.

$$Z_g - Z_f = \left(1 - \frac{P_r}{T_r^3}\right)^{1/2} \tag{6-102}$$

This relation is fairly accurate from the freezing point to the critical point.

Equation (6-100) may be made dimensionless by multiplying through by $1/T_c$. As a result,

$$\frac{h_{fg}}{RT_c} = T_r\left(\frac{\Delta s_v^{(0)}}{R} + \omega\,\frac{\Delta s_v^{(1)}}{R}\right) \tag{6-103}$$

Hence h_{fg}/RT_c is solely a function of ω and T_r, and the units for h_{fg} depend solely on the units chosen for RT_c. Figure 6-6 has been constructed on the basis of Eq. (6-103) and the tabular $\Delta s_v/R$ data of Pitzer et al. These data show that the dimensionless parameter h_{fg}/RT_c is essentially a linear function of ω for fixed values of T_r. That is,

$$\frac{h_{fg}}{RT_c} = a + b\omega \qquad (\text{fixed } T_r)$$

where a and b are functions of T_r. The curves by Reid shown in Fig. 6-6 for $0.6 < T_r \leq 1.0$ are reasonably well represented by the expression [6]

$$\frac{h_{fg}}{RT_c} = 7.08(1 - T_r)^{0.354} + 10.95\omega(1 - T_r)^{0.456} \tag{6-104}$$

Equation (6-104) permits an estimation of h_{fg} for normal fluids which does not require interpolation of Pitzer's tabular data.

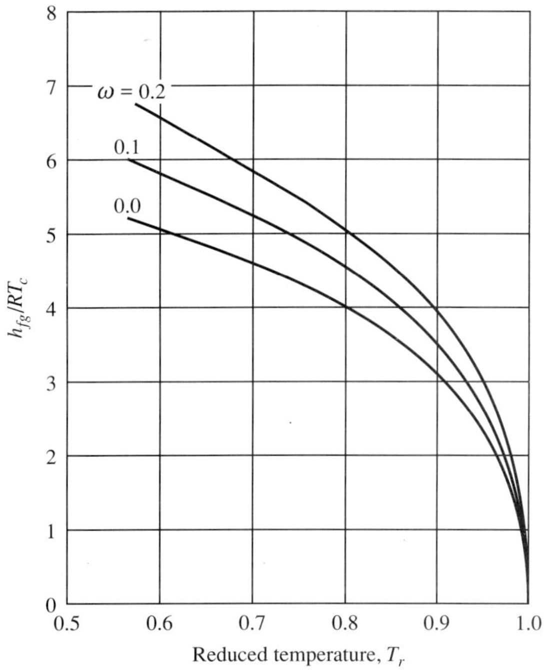

FIGURE 6-6
Plot of data from Pitzer et al. correlation for enthalpies of vaporization. (*Source*: Reid, R. C, J. M. Prausnitz, and B. E. Poling, *The Properties of Gases and Liquids*, 4th ed., p. 221, McGraw-Hill, Inc., New York, 1987.) By permission of McGraw-Hill, Inc.

A dimensionless correlation for the entire temperature range from the critical state (T_c) to the triple state (T_t) has been proposed for the enthalpy of vaporization by Torquato and Stell [18] and Torquato and Smith [19]. The correlation involves two dimensionless parameters

$$\tau = \frac{T_c - T}{T_c - T_t} \qquad \text{and} \qquad \lambda = \frac{h_{fg}(T)}{h_{fg}(T_t)} \tag{6-105}$$

and the correlation itself is

$$\lambda = 0.60176\tau^{1/3} + 3.45913\tau^{0.79} + 4.62671\tau^{1.208} - 6.89614\tau - 1.10643\tau^2 + 0.31522\tau^3 \tag{6-106}$$

This equation represents the solid line shown in Fig. 6-7 which fits the data for water. Data for five other substances are in reasonable agreement. A major disadvantage of Eq. (6-106) is that knowledge of h_{fg} at the triple state is required.

Example 6-4. Use (*a*) Eq. (6-100), (*b*) Eq. (6-101), and (*c*) Eq. (6-104) to estimate the enthalpy of vaporization of R-134a at 20°C, in kJ/kg. Compare it to the tabulated value.

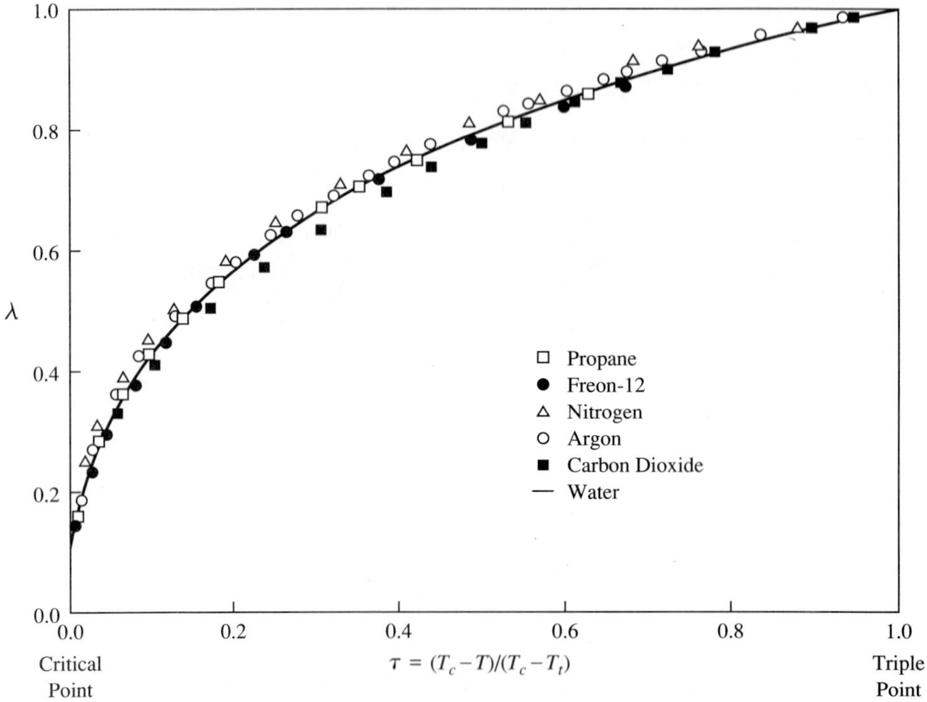

FIGURE 6-7
Correlation of the enthalpy of vaporization of six fluids. (*Source:* After Torquato and Smith, "The Latent Heat of Vaporization of a Widely Diverse Class of Fluids," Journal of Heat Transfer, ASME, Vol. 106, February 1984, pp. 252-254.)

Solution. From Table A-2 in the Appendix, $T_c = 374$ K, $M = 102.03$, and $\omega = 0.326$. Thus $T_r = 293/374 = 0.783$.

(*a*) From Table A-30 for the Pitzer data, $\Delta s_v^{(0)}/R = 5.28$ and $\Delta s_v^{(1)}/R = 6.93$. Use of these data in Eq. (6-100) yields

$$h_{fg} = 8.314(293)[5.28 + 0.326(6.93)] = 18{,}370 \text{ kJ/kmol} = 180.0 \text{ kJ/kg}$$

(*b*) Based on the Lee-Kesler vapor-pressure equation at 293.15 K, from Eq. (6-92) the value of $1/P_r(dP_r/dT_r) = 11.580$, and the two-parameter Z chart shows that $Z_g \approx 0.84$ and $Z_f \approx 0.035$. Substitution of these values into Eq. (6-101) yields

$$h_{fg} = \frac{8.314(293.15)(0.783)(0.805)(11.580)}{102.03} = 174 \text{ kJ/kg}$$

(*c*) From Eq. (6-104)

$$\frac{h_{fg}}{RT_c} = 7.08(1 - 0.783)^{0.354} + 10.95(0.326)(1 - 0.783)^{0.456} = 5.90$$

$$h_{fg} = 5.90(8.314)(374)/102.03 = 179.8 \text{ kJ/kg}$$

The literature value is 181.1 kJ/kg, and the estimated values are in error by roughly −0.6, 3.9, and −0.7 percent, respectively. The larger error of part (*b*) is not unexpected, since it is based on several approximations.

6-10-3 Phase Equilibrium Properties from Equations of State

It has been shown for a phase change that the values of the specific Gibbs function for each phase of a pure component are equal at equilibrium. For liquid-vapor equilibrium, for example,

$$g^L = g^V \tag{6-72}$$

In terms of a residual function the above equation may be written as

$$\left(\frac{g - g^*}{R_u T}\right)_L = \left(\frac{g - g^*}{R_u T}\right)_V \tag{6-107}$$

When an equation of state in the form $P = f(V, T)$ is employed, the method of evaluating $(g - g^*)/R_u T$ requires the use of Eq. (6-64), namely,

$$\frac{g - g^*}{R_u T} = \frac{a - a^*}{R_u T} + (Z - 1) \tag{6-64}$$

Substitution of Eq. (6-57) for $(a - a^*)/R_u T$ yields

$$\frac{g - g^*}{R_u T} = -\frac{1}{R_u T}\int_\infty^v \left(P - \frac{R_u T}{v}\right) dv - \ln Z + (Z - 1) \tag{6-108}$$

This is the general expression required when one chooses an equation of state explicit in pressure. Equations (6-107) and (6-108), along with a suitable PvT equation of state, lead to predictions of the equilibrium properties of a single component, two-phase mixture.

As an illustration, consider the use of the Redlich-Kwong (RK) equation to evaluate liquid-vapor saturation data. The equation for $(g - g^*)/R_u T$ in this case is Eq. (6-69) in Sec. 6-8. That is, for a Redlich-Kwong fluid

$$\frac{g - g^*}{R_u T} = (Z - 1) - \ln\left[Z\left(1 - \frac{b}{v}\right)\right] - \frac{a}{bR_u T^{1.5}} \ln\left(1 + \frac{b}{v}\right) \tag{6-69}$$

This equation is valid for the liquid and vapor phases at equilibrium.

A procedure for setting up a computer program to find the equilibrium properties of a two-phase state employing the Redlich-Kwong equation is as follows:

1. For a given saturation temperature an initial guess for the saturation pressure is chosen. (Correlations by Pitzer, Lee-Kesler, or Dong-Lienhard are useful in selecting an initial value for P^{sat}.)
2. Next, evaluate the specific volume and compressibility factor for the two phases based on the RK equation of state. (This requires a computer technique for the solution of cubic equations.)
3. Item (2) leads to values of v_L, Z_L, v_V, Z_V, b/v_L, and b/v_V. These values allow the evaluation of Eq. (6-69) for $(g^* - g)/R_u T$ for the two phases.

4. Another P^{sat} is chosen close to the original value and the calculation is repeated until satisfactory agreement occurs between the residual Gibbs function values.
5. Once a satisfactory solution for the saturation pressure is found, then h^R/R_uT for both phases can be calculated by means of Eq. (6-66), namely,

$$\frac{h^R}{R_uT} = (1 - Z) + \frac{3a}{2bR_uT^{1.5}} \ln\left(1 + \frac{b}{v}\right) \tag{6-66}$$

From these latter calculations one determines the enthalpy of vaporization by

$$h_{fg} = R_uT\left(\frac{h_L^R}{R_uT} - \frac{h_V^R}{R_uT}\right) \tag{6-109}$$

6. Finally, $s_{fg} = h_{fg}/T$.

The procedure outlined above may be used with any appropriate equation of state which is explicit in pressure. A computer program greatly expedites the accurate determination of P^{sat} and other saturation properties. As noted in Sec. 6-6-3, the tabulation of values for h_f, h_g, s_f, and s_g in a saturation table requires the arbitrary selection of datum or reference states for the two properties.

6-11 THE JOULE-THOMSON COEFFICIENT

The reduction in fluid temperature required in practical refrigeration and liquefaction systems is achieved primarily by two devices. These are an expansion valve and an expansion engine (or a combination of these two systems). The reduction in temperature when a gas undergoes an adiabatic expansion in an engine is well known. In this section we shall describe the effect of an adiabatic expansion on temperature in the absence of work. Consider the flow of a fluid down a duct which contains a restriction to the flow. This restriction might be some type of porous plug, such as steel wool or cotton, or a partially opened valve. The effect of this restriction is a significant pressure drop across the restriction. A schematic of such a device is shown in Fig. 6-8(*a*). The shaft work within the control volume is zero, and the flow passage is heavily insulated. In addition, the changes in the kinetic and potential energies of the flow stream across the porous plug may be made negligible. Hence the conservation of energy principle for a control volume around the restriction reduces to

$$h_1 = h_2$$

A process for which the inlet and outlet enthalpies are the same is called *throttling*. The effect of throttling has a number of important scientific and engineering applications.

Based on some original experiments by Joule and Thomson [20], the outcome of allowing a fluid to flow through a restriction from a higher to a lower pressure is frequently called the *Joule-Thomson effect*. The importance of the Joule-Thomson effect is several-fold. First, we shall find that under certain con-

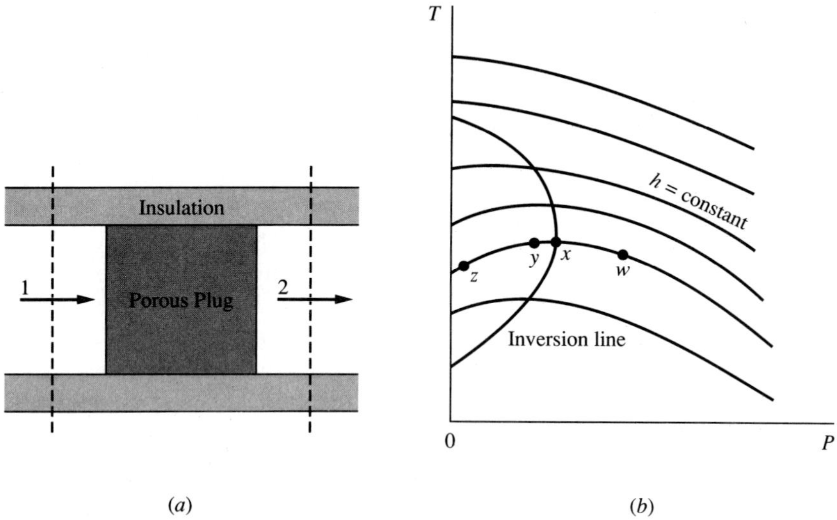

FIGURE 6-8
(*a*) Equipment for Joule-Thomson experiment; (*b*) a plot of data resulting from typical throttling measurements.

ditions the result of throttling is a reduction in the temperature of the fluid. Thus low temperatures may be achieved with a device which has no moving parts. In fact, under proper conditions it is possible that one or more of the components of a gas stream might pass into the liquid (or solid) phase during the throttling process. Such liquefaction provides a means of separating components of a gas mixture. Second, other properties may be related or evaluated from measurements of the Joule-Thomson effect. These include, for example, specific volumes, specific heats, and enthalpies.

It is an experimental fact that throttling of a fluid leads to a final temperature T_2, which may be higher, the same, or lower than the initial value, depending on the values of P_1, T_1, and P_2. However, each exit state has the same enthalpy as the initial upstream state at (T_1, P_1). The effect of P_2 on T_2 is clearly seen by plotting a sufficient number of exit measurements and by drawing a line of constant enthalpy through the data points. The points w, x, y, and z shown on Fig. 6-8(*b*) lie on a constant h, or *isenthalpic*, line. Several other isenthalpic lines are also shown. The slope of a constant-enthalpy line at any state on the TP plot is defined as the Joule-Thomson coefficient, μ_{JT}, at that state. That is,

$$\mu_{JT} \equiv \left(\frac{\partial T}{\partial P}\right)_h \tag{6-110}$$

It is a mathematical measure of the effect of a pressure drop on the temperature during throttling.

Figure 6-8(*b*) shows that a number of the constant-enthalpy lines have a state of maximum temperature. The line shown in the figure which passes through these states of maximum temperature is called the *inversion line*, and the value of the temperature for that state is the *inversion temperature*. Due to the parabolic shape of the inversion line on the TP plot, a pressure line will cut the inversion curve at two different states; hence one speaks of the upper and lower inversion temperatures for a given pressure. This line has an important physical significance. The Joule-Thomson coefficient is negative for states to the right of the inversion line on a TP plot. That is, in this particular region the temperature will increase as the pressure decreases through the throttling device. However, to the left of the inversion curve the Joule-Thomson coefficient is positive, which means that a drop in temperature will occur for expansions in this region. Hence on the throttling of a fluid the final temperature may be greater than, equal to, or less than the initial temperature, depending on the final pressure for any given set of initial conditions. For example, in Fig. 6-8(*b*) a typical initial state might be point w. Expansion to the inversion curve (point x) results in a rise in temperature of the fluid. If further expansion to point y is permitted, some internal cooling will occur, but this is not sufficient to lower the temperature to that of the initial state. However, if expansion to point z is possible, the final temperature will be brought to a lower value than that for the initial state.

The upper part of the inversion curve passes through zero pressure at some finite temperature for all substances. Consequently, many enthalpy lines at high temperatures never pass through the inversion line, as seen in Fig. 6-8(*b*). For these enthalpy lines the Joule-Thomson coefficient is always negative throughout the range of pressures and a decrease in temperature is impossible. Examples are hydrogen and helium, which have negative coefficients at ordinary temperatures and low pressures. Hence for these two gases the temperature must be artificially lowered considerably before throttling can be employed for an additional lowering of the temperature. For most substances, however, at ordinary temperatures the Joule-Thomson coefficient is negative at high pressures and becomes positive at lower pressures. For a given pressure drop the maximum cooling effect is attained only if the initial state lies on or to the left of the inversion line on a TP plot. If the initial state lies to the right of the inversion curve, part of the expansion results in a temperature rise, which is undesirable.

Data for the Joule-Thomson coefficient of a substance is frequently plotted against temperature for various pressures. Such a plot (or table of data) shows clearly the regions of pressure and temperature where a "cooling" effect is possible. Data for the Joule-Thomson coefficient of nitrogen and argon determined from experimental work are shown in Fig. 6-9. Data for air are shown graphically in Fig. 6-10 and are listed in Table 6-2.

The Joule-Thomson coefficient may be expressed in terms of its relation to the variables P, v, T, and c_p. If we apply the cyclic relation Eq (6-1b) to the variables h, T, and P, then

$$\left(\frac{\partial h}{\partial T}\right)_p \left(\frac{\partial T}{\partial P}\right)_h \left(\frac{\partial P}{\partial h}\right)_T = -1$$

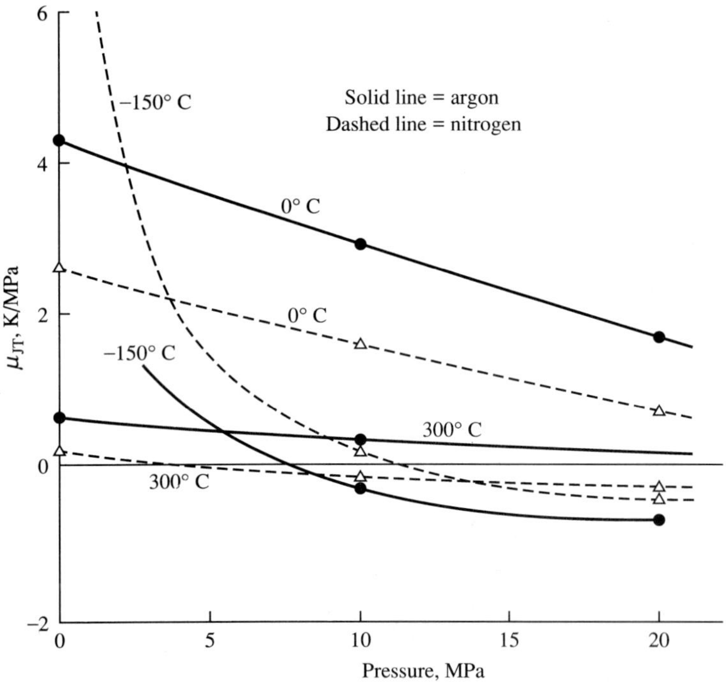

FIGURE 6-9
Joule-Thomson coefficients for argon and nitrogen. (*Source:* The Smithsonian Physical Tables, p. 279, 1954.)

or on the basis of the definitions of μ_{JT} and c_p

$$\mu_{JT} = -\frac{1}{c_p}\left(\frac{\partial h}{\partial P}\right)_T \tag{6-111}$$

Recall the generalized equation for the enthalpy, namely,

$$dh = c_p\, dT + \left[v - T\left(\frac{\partial v}{\partial T}\right)_P\right] dP \tag{6-22}$$

The term in brackets is $(\partial h/\partial P)_T$. Substitution of this term in Eq. (6-111) yields

$$\mu_{JT} = \left(\frac{\partial T}{\partial P}\right)_h = \frac{1}{c_p}\left[T\left(\frac{\partial v}{\partial T}\right)_P - v\right] \tag{6-112}$$

This equation, in conjunction with an appropriate equation of state, permits evaluation of the Joule-Thomson coefficient over an extended range of pressures and temperatures. This includes the evaluation of the inversion curve itself. By

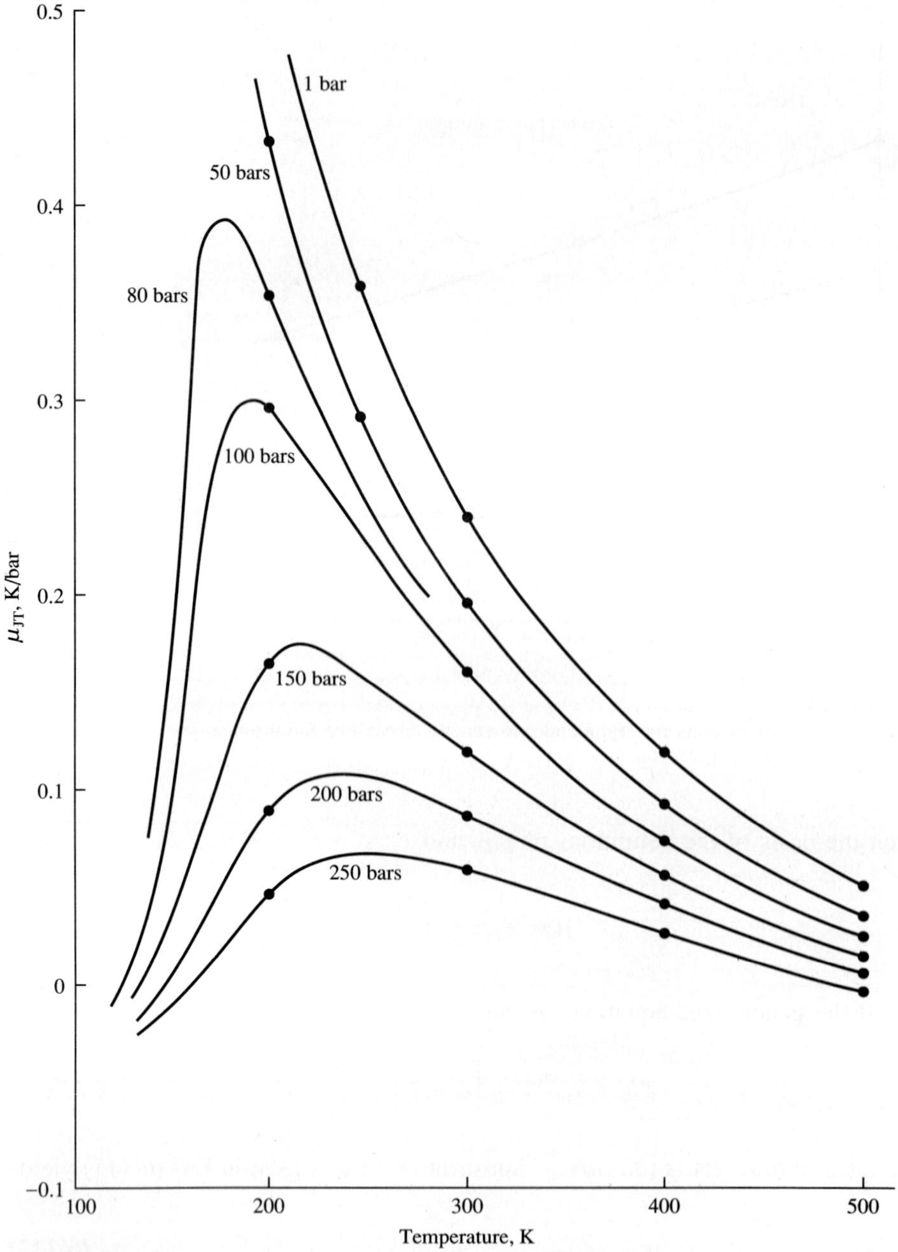

FIGURE 6-10
Joule-Thomson coefficient for air.

TABLE 6-2
Joule-Thomson coefficient for air, in K/bar

	Pressure, bar					
Temperature, K	**1**	**50**	**100**	**150**	**200**	**500**
150	0.804	0.644	0.132	0.042	0.010	−0.037
200	0.500	0.412	0.290	0.168	0.092	−0.026
250	0.336	0.273	0.215	0.155	0.105	−0.016
275	0.280	0.227	0.159	0.136	0.097	−0.014
300	0.234	0.190	0.152	0.117	0.085	−0.013
350	0.165	0.133	0.107	0.083	0.061	−0.015
400	0.116	0.091	0.072	0.055	0.040	−0.020
500	0.051	0.037	0.025	0.016	0.007	−0.031
800	−0.030	−0.034	−0.037	−0.040	−0.043	−0.055

setting Eq. (6-112) equal to zero, we find that the *general criterion* for the inversion curve in terms of PvT data is

$$\left(\frac{\partial v}{\partial T}\right)_p = \frac{v}{T} \qquad \text{or} \qquad \left(\frac{\partial v_r}{\partial T_r}\right)_{P_r} = \frac{v_r}{T_r}$$

where the second equation is simply the reduced format. The prediction of the Joule-Thomson inversion curve is a very sensitive test of an equation of state.

A semiquantitative explanation of the trends shown in Fig. 6-8(b) is found by employing the equation of state $Pv = ZRT$ in Eq. (6-112). The result is

$$\mu_{\text{JT}} = \frac{RT^2}{Pc_p}\left(\frac{\partial Z}{\partial T}\right)_P$$

Thus cooling occurs when the sign on $(\partial Z/\partial T)_p$ is positive. On the basis of the Z chart:

1. In the low-pressure region ($P_r < 10$), Z increases with an increase in T_r until T_r reaches 5. Beyond this value of T_r, Z decreases as T_r increases and a rise in temperature occurs upon throttling.
2. In the high-pressure region ($P_r > 10$), Z always decreases with increasing T_r and a temperature rise always occurs.

This general analysis is in agreement with the generalized $P_r - T_r$ diagram shown in Fig. 6-11, based on experimental data. The Joule-Thomson inversion phenomenon is correlated reasonably well by the corresponding states principle. Again, "cooling" is only possible when P_r is less than roughly 11-12 and T_r is less than 5. The pressure maximum occurs when $T_r \approx 2.25$. This figure also

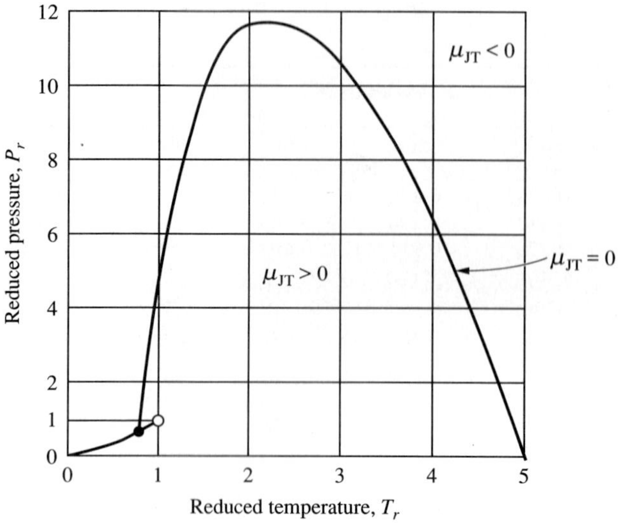

FIGURE 6-11
The inversion curve on reduced coordinates.

shows that the low-temperature end of the inversion curve eventually will reach the saturation line for the fluid, and then stop. The lower inversion temperature, if extrapolated to zero pressure, is about $0.8T_c$. An empirical curve fit of the inversion curve in reduced format reported by Miller [21] is

$$P_r = 24.21 - \frac{18.54}{T_r} - 0.825T_r^2 \tag{6-113}$$

This equation is based on a least square fit of experimental data.

As noted at the beginning of this section, a second importance of the Joule-Thomson coefficient is its relationship to other thermodynamic properties. For example, Eq. (6-112) relates μ_{JT} to the specific heat at constant pressure. There also is a direct relationship between μ_{JT} and the residual enthalpy h^R (first described in Sec. 6-6). Recall that $h^R \equiv h^* - h$, or $h = h^* - h^R$. Because h^* (the ideal-gas enthalpy) is not a function of pressure, we may write

$$\left(\frac{\partial h}{\partial P}\right)_T = \left(\frac{\partial h^*}{\partial P}\right)_T - \left(\frac{\partial h^R}{\partial P}\right)_T = -\left(\frac{\partial h^R}{\partial P}\right)_T$$

If this identity is substituted into Eq. (6-111), we find that

$$\mu_{JT} = \frac{1}{c_p}\left(\frac{\partial h^R}{\partial P}\right)_T \tag{6-114}$$

Thus a direct relationship exists among μ_{JT}, c_p, and h^R at a given state. Data on any two of these functions lead to the evaluation of the third quantity.

PROBLEMS

6-1. Prove that the constant-pressure lines in the wet region of an hs diagram are straight and not parallel and that the slope of a constant-pressure line in the superheat region increases with temperature.

6-2. From any of the four Maxwell relations, derive the other three, making use of Eq. (6-1b).

6-3. What can be concluded qualitatively about the change in enthalpy of a fluid during an isentropic compression?

6-4. Show that the slope of a constant-volume line is greater than that for a constant-pressure line through the same state point in the gas region of a Ts diagram. Also, show that the slope of a constant-pressure line in the gas region of a Ts diagram normally increases with increasing temperature.

6-5. Derive expressions for (a) $(\partial u/\partial P)_T$ and (b) $(\partial u/\partial v)_T$ that involve only P, T, and v.

6-6. **a.** Derive the relation $c_p = T(\partial P/\partial T)_s(\partial v/\partial T)_P$ by starting from the basic definition, $c_p = (\partial h/\partial T)_p = T(\partial s/\partial T)_p$, and using a cyclic relation and a Maxwell relation where needed.

b. Determine the value of c_p for steam at 1.0 MPa and 360°C by employing the equation derived in part (a), and by curve-fitting tabular data, in kJ/kg·K.

c. Repeat the calculation of part (b), but now use the defining equation for c_p; that is, $c_p = (\partial h/\partial T)_P$. Again, curve-fit tabular data to evaluate partial derivatives.

6-7. Determine the value of c_p for steam at (a) 60 bars and 400°C and (b) 120 bars and 480°C by employing the equation derived in Prob. 6-6. Compare with the value obtained from the derivative $c_p = (\partial h/\partial T)_P$. For both methods curve-fit tabular superheat data to evaluate numerically the partial derivatives.

6-8. Derive the relation $(\partial^2 g/\partial T^2)_P = -c_p/T$.

6-9. Derive the relation $(\partial^2 a/\partial T^2)_v = -c_v/T$.

6-10. On the basis of Eqs. (6-16) and (6-19), show that the slope of a constant-volume line is greater than that for a constant-pressure line through the same state point in the gas region of a Ts diagram.

6-11. Prove that for any homogeneous system $c_p/c_v = (\partial P/\partial v)_s(\partial P/\partial v)_T$.

6-12. Check the validity of the relation, $(\partial s/\partial P)_T = -(\partial v/\partial T)_P$, for steam at 30 bars and 400°C, in units of (kJ/kg·K)/bar. Use a polynomial least square fit of five superheat data to find $v = f(T)$ and $s = f(P)$, and then find the derivatives on both sides of the equation.

6-13. Check the validity of the Maxwell relation, $(\partial v/\partial T)_P = -(\partial s/\partial P)_T$, by finding the approximate values of the derivatives for refrigerant-134a at 6 bars and 50°C. Use tabular data, and report the values in m^3/kg · K. Use a polynomial least square fit of superheat data to determine the values of the partial derivatives.

6-14. Check the validity of the Maxwell relation, $(\partial v/\partial T)_p = -(\partial s/\partial P)_T$, by finding the approximate values of the derivatives for steam at 80 bars and 400°C. Use tabular data, and report values in cm^3/g·K. Use a polynomial least square fit of superheat data to determine the values of the partial derivatives.

6-15. **a.** Derive an expression for the isothermal change in internal energy for a substance which follows the Redlich-Kwong equation of state. The result should be given in terms of T, v, a, b, and a constant.

b. On the basis of the equation derived in part (a), evaluate the internal energy change of carbon dioxide, in kJ/kg, as it is expanded isothermally at 350 K from 14.47 to 20.02 cm^3/g.

c. Check the answer against tabular superheat data for CO_2, in kJ/kg.

6-16. Refrigerant-134a is compressed isothermally at 60°C from 8.0 to 16.0 bars.

a. Employing the generalized equation for du as a function of T and v, derive an expression for the isothermal change in internal energy for a substance which follows the Redlich-Kwong equation. The result should be expressed in terms of T, v, a, b, and a constant.

b. On the basis of the equation derived in part (a), evaluate the internal energy change for R-134a, in kJ/kg.

c. Check the answer against tabular superheat data.

6-17. a. Develop an expression for the isothermal change in internal energy for a substance which is represented by the Redlich-Kwong equation of state. The result should be given in terms of T, v, a, b, and a constant.

b. Estimate the change in the internal energy of nitrogen during an isothermal compression at 250 K and from 73.6 to 14.4 cm^3/g.

c. Check the answer against tabular superheat data, in kJ/kg.

6-18. Estimate the value of $(c_p - c_p^*)$ for air at 300 K and 40 atm, in kJ/kgmol·K, using

a. the generalized expression, assuming the validity of the truncated virial equation, $Pv = RT + bP$. The second virial coefficient B is given by

$$b = 37.28 - 8.43 \times 10^3/T - 1.75 \times 10^6/T^2 + 7.35 \times 10^7/T^3,$$

where b is in cm^3/gmol and T is in kelvins.

b. the data from NBS Circular 564, which lists (c_p/R) at 0.01 and 40 atm as 3.501 and 3.722, respectively

c. a generalized chart for data

6-19. Consider the application of the virial equation, $Pv = RT + bP$ for determining c_p data. The equation for b for nitrogen gas is given by $b = 39.5 - (1.00 \times 10^4)/T - (1.084 \times 10^6)/T^2$, where b is in cm^3/gmol and T is in kelvins.

a. Use these equations to estimate the value of $(c_p - c_p^*)$ at 300 K and 20 bars, in kJ/kg·K.

b. Compare part (a) with an estimate based on the definition of c_p and tabular superheat data.

6-20. Assume that the PvT relationship for diatomic nitrogen can be represented by the truncated virial equation, $Pv = RT + bP$. The coefficient b is given by

$$b\,(\text{cm}^3/\text{gmol}) = 39.5 - 1.00 \times 10^4/T - 1.084 \times 10^6/T^2$$

where T is expressed in kelvins.

a. Use these equations to estimate the value of $(c_p - c_p^*)$ at 250 K and 50 bars, in kJ/kg·K.

b. Compare with an estimate based on tabular h superheat data and the definition of c_p.

6-21. Nitrogen gas is compressed isothermally at 250 K from 1.0 bar to 150 bars. Determine the enthalpy change for the process, in kJ/kmol, by

a. employing the truncated virial equation, $Pv = RT + bP$, where $b = 39.5 - (1 \times 10^4)/T - (1.084 \times 10^6)/T^2$, with b in cm^3/gmol and T in kelvins

b. employing the two-parameter generalized enthalpy chart
c. employing the three-parameter data of Lee-Kesler
d. using tabular superheat data

6-22. At reasonably low pressures the PvT relationship for air can be represented by $Pv = RT + bP$. The coefficient b, based on experimental data, is given by

$$b\,(\text{cm}^3/\text{gmol}) = 37.28 - 8.43 \times 10^3/T - 1.75 \times 10^6/T^2 + 7.35 \times 10^7/T^3$$

where T is in K. Consider a change of state from 7.0 to 40.0 bars at a constant temperature of 200 K. Determine the change in enthalpy, in kJ/kmol, based on (a) the two-parameter generalized chart, (b) the generalized equation for dh and the equation of state shown above, and (c) the ideal-gas equation of state. (*Note:* The tabular value for Δh in NBS Circular 564 is roughly -490 kJ/kmol.)

6-23. Assume that the PvT relationship for diatomic nitrogen can be represented by the truncated virial equation, $Pv = RT + bP$. The coefficient b, based on experimental data, is given by

$$b\,(\text{cm}^3/\text{gmol}) = 39.5 - 1.00 \times 10^4/T - 1.084 \times 10^6/T^2$$

where T is expressed in kelvins. Nitrogen undergoes a change of state from 3.4 bars to 40.7 bars at a constant temperature of 250 K. Determine the change in enthalpy, in kJ/kmol, based on

a. the generalized three-parameter chart
b. the generalized equation for dh and the virial equation of state
c. tabular superheat data

6-24. Carbon dioxide gas is compressed isothermally at 350 K from 15 to 150 bars. Estimate the change in entropy in kJ/kg·K by means of

a. the generalized two-parameter chart
b. the generalized three-parameter data
c. the assumption that the gas is an ideal gas

6-25. Acetylene (C_2H_2) is expanded from 98°C and 125 bars to 20°C and 25 bars in a steady-flow process. The molar specific heat at zero pressure is given in the Appendix. Determine the enthalpy change of the gas, in kJ/kmol, using (a) ideal-gas relations and (b) the three-parameter generalized data by Lee-Kesler.

6-26. Ethylene (C_2H_4) at 67°C and 1 bar is compressed isothermally in a reversible, nonflow process to 255 bars. Determine (1) the change in entropy, in kJ/kmol·K, and (2) the change in internal energy, (3) the heat transferred, and (4) the work required, in kJ/kmol, on the basis of (a) the two-parameter generalized chart and (b) the residual property equations for a Redlich-Kwong gas.

6-27. Ethane (C_2H_6) is compressed isothermally and reversibly at 93°C from 10 bars to 98 bars in a steady-flow process.

a. Calculate the heat transfer by means of (1) the two-parameter generalized chart, (2) the three-parameter data of Lee-Kesler, and (3) the ideal-gas equation, in kJ/kg.
b. Calculate the work required, in kJ/kg, by means of the three same techniques.

6-28. Ethane is compressed from 32°C and 19.5 bars to 63°C and 146 bars in a steady-flow process. The molar specific heat at zero pressure $c_{p,0}$, in kJ/kmol·K, is given by $c_{p,0} = 16.75 + 0.123T$, where T is in kelvins. Determine the entropy change of the gas, in kJ/kmol·K, using (a) ideal-gas relations, (b) a two-parameter generalized chart, and (c) the three-parameter generalized data by Lee-Kesler.

6-29. Ethane (C_2H_6) is compressed from 32°C and 19.5 bars to 185°C and 98 bars in a steady-flow process. The molar specific heat in kJ/kmol·K is given by $c_{p,0} = 16.8 + 0.123T$, where T is in kelvins. Determine the enthalpy change, in kJ/kmol, if

a. the gas is an ideal gas
b. generalized two-parameter data are used
c. generalized three-parameter data are used

6-30. Steam at 280 bars and 520°C is expanded adiabatically and reversibly in a steady-flow process to 100 bars. Determine

a. the final temperature, in degrees Celsius
b. the change in enthalpy, in kJ/kg, using (1) the two-parameter generalized chart and (2) tabular superheat data.

6-31. Ethylene (C_2H_4) is expanded adiabatically in a gas turbine from 180°C and 256 bars to 67°C and 41 bars. The inlet flow rate is 1.0 m^3/min, and a mean $c_{p,0}$ value of 52.7 kJ/kmol·K may be assumed. Estimate the shaft output power, in kilowatts, by means of (a) the ideal-gas model, (b) the two-parameter, and (c) the three-parameter corresponding states principle.

6-32. Carbon dioxide gas is expanded isothermally at 320 K from 40 to 20 bars.

a. Determine the values of Z and v (in m^3/kg) for the two states using the two-parameter Z chart. Then, using the Redlich-Kwong-Soave equation, determine
b. the values of Z and v, in m^3/kmol, for the two states
c. the values of h^R/RT for the two states and $(h_2 - h_1)$, in kJ/kmol
d. the values of s^R/R for the two states and $(s_2 - s_1)$, in kJ/kmol·K, for the process

6-33. Refrigerant-134a (1,1,1,2-tetrafluoroethane) is compressed isothermally at 80°C from 1.0 to 16.0 bars. Determine

a. the Z factors at the initial and final states, using the two-parameter compressibility chart. Then, using the RK equation, determine
b. the Z factors at the initial and final states
c. the values of h^R/RT and s^R/R for the initial and final states
d. the change in enthalpy, in kJ/kg, and entropy, in kJ/kg·K, for the process

6-34. The Horvath-Lin equation (*Can. J. Chem. Engr.*, **55**, August 1977) is designed to predict both the liquid and the vapor region for which $P < P_c$ and $Z_c \leq 0.275$. The equation is defined by:

$$P = \frac{RT}{v - b} - \frac{a}{Tv(v + c)}$$

$$a = \alpha \frac{R^2 T_c^3}{P_c} \qquad b = \beta \frac{RT_c}{P_c} \qquad c = \gamma b = \gamma\beta \frac{RT_c}{P_c}$$

where

$$\gamma = \frac{Z_c^{-4.72}}{360}$$

$$f = [1 + (\gamma + 1)^{1/3} + (\gamma + 1)^{2/3}]^{-1}$$

$$\alpha = \frac{(1 + \gamma f)^2 (1 - 2f - \gamma f^2)}{(1 - f)^4 (2 + \gamma f)^2}$$

$$\beta = \frac{1 - 2f - \gamma f^2}{(2 + \gamma) + (2 + 4\gamma) f + (\gamma + 2\gamma^2) f^2}$$

Refrigerant-134a is compressed isothermally from 4 to 10 bars at 50°C.

a. Use the Horvath-Lin equation to derive expressions for $(h^* - h)$ and $(s^* - s)$, which may include the quantities a, b, c, Z, R, T, v_1, and v_2.

b. Evaluate the constants γ, f, α, and β.

c. Evaluate v, in m^3/kg, and Z for the two given states.

d. Evaluate the residual enthalpy, $h^* - h$, for the two states and $(h_2 - h_1)$, in kJ/kg.

e. Evaluate the residual entropy, $s^* - s$, for the two states and $(s_2 - s_1)$, in kJ/kg·K.

6-35. Estimate the enthalpy of vaporization of R-134a at (a) −12°C and (b) 36°C by (1) first employing the Clapeyron equation. Determine dP^{sat}/dT from a least square fit of saturation data in the form $\ln P^{\text{sat}} = A - B/T$. Then employ (2) the modified Clausius-Clapeyron equation involving Z data. Finally, (3) compare with tabular data, in kJ/kg.

6-36. Estimate the enthalpy of vaporization of nitrogen at (a) 90 K and (b) 25.0 bars by (1) first employing PvT data from the saturation tables in the Clapeyron equation. Determine dP^{sat}/dT from a least square fit of saturation data in the form $\ln P^{\text{sat}} = A - B/T$. Then (2) use the modified Clausius-Clapeyron equation, which includes $(Z_g - Z_f)$. Finally, (3) compare the answers with tabular data, in kJ/kg.

6-37. The specific volumes of liquid water and solid water (ice) at 0°C are 1.0002 and 1.0908 cm^3/g, respectively. Estimate the value of the slope of the melting-freezing curve on a PT diagram for water, in bar/K, if the enthalpy of melting at this temperature is 332.5 kJ/kg.

6-38. Estimate the enthalpy of vaporization of water at (a) 180°C and (b) 250°C by employing (1) the Clapeyron equation, fitting by least squares $\ln P = A + B/T$ and (2) the Z-modified Clausius-Clapeyron equation. (3) Compare with tabulated saturation data.

6-39. The vapor pressure of water can be represented with reasonable accuracy between 5 and 50 bars by the equation, $\ln P = (-4692/T) + 0.0124 \ln T + 12.58$, with P in bars and T in kelvins. Compute the enthalpy of vaporization at (a) 160°C and (b) 200°C using values for the specific volume from a saturation table. Then compare answers against the table.

6-40. The vapor pressure of ammonia (NH_3) can be represented by the equation $\log_{10} P = 22.779 - (1.831 \times 10^3/T) - 6.401 \log_{10} T - 7.466 \times 10^{-4} T + 4.782 \times 10^{-6} T^2$, where P is in bars and T is in kelvins. Calculate the enthalpy of vaporization, h_{fg}, of ammonia at 5°C, and compare to the value of 1244 kJ/kg listed in the ammonia tables of the National Bureau of Standards. The density of the liquid and vapor phases at 5°C are 0.633 and 0.00403 g/cm^3, respectively.

6-41. The liquid-vapor saturation line for diatomic nitrogen over the range from the triple state to the normal (1 atm) boiling point may be represented by $\ln P = A - BT - C/T$, where P is in bars, T is in kelvins, $A = 11.302$, $B = 0.01443$, and $C = 786.7$.

a. Derive an expression for the enthalpy of vaporization, h_{fg}, in terms of the constants A, B, and C, the temperature, and the change in specific volume, v_{fg}.

b. Evaluate h_{fg} at 80 K, in kJ/kg, using the tabular value of v_{fg}.

c. Compare the answer in part (b) with the value found in the saturation table.

6-42. Helium 4 boils at 4.22 K at 1 bar, and the enthalpy of vaporization is 82.3 kJ/kmol. By producing a vacuum over the liquid phase, the fluid boils at a lower temperature.

Estimate what pressure, in millibars, is necessary to produce a temperature of (a) 2 K and (c) 0.5 K.

6-43. Determine the vapor pressure of carbon dioxide (CO_2) at 260 K, in bars, based on the reduced-vapor-pressure correlations of (a) Dong-Lienhard and (b) Lee-Kesler.

6-44. Determine the vapor pressure of nitrogen at 90°C on the basis of the reduced-vapor-pressure correlations of (a) Lee and Kesler and (b) Dong and Lienhard, in bars. Then compare to data in a saturation table.

6-45. Determine the vapor pressure of butane (C_4H_{10}) at 300 K, in bars, based on the reduced-vapor-pressure correlations of (a) Lee-Kesler and (b) Dong-Lienhard. A typical literature value is 2.582 bars.

6-46. On the basis of reduced properties and the Pitzer acentric factor, determine from tabular correlations the following properties of carbon dioxide at 260 K:

a. the vapor pressure, in bars
b. the specific volume of the saturated liquid and saturated vapor, in m^3/kg
c. the enthalpy of vaporization, in kJ/kg
d. Compute the percent error when compared with the tabulated saturation values.

6-47. On the basis of reduced properties and the Pitzer acentric factor, determine from tabular correlations the following properties of R-134a at 24°C:

a. the vapor pressure, in bars
b. the specific volume of the saturated liquid and vapor, in m^3/kg
c. the enthalpy of vaporization, in kJ/kg
d. Compare these results with tabular saturation data. Then, using the correlation of Reid [Eq. (6-104)], calculate the enthalpy of vaporization.

6-48. On the basis of reduced properties and the Pitzer acentric factor, determine from tabular correlations the following properties of nitrogen at 90 K:

a. the vapor pressure, in bars
b. the specific volume of the saturated liquid and vapor, in m^4/kg
c. the enthalpy of vaporization, in kJ/kg. Compare to tabular saturation data.
d. Then, using the correlation given by Torquato and Smith, calculate the enthalpy of vaporization.
e. Finally, using the correlation of Reid [Eq. (6-104)], calculate the enthalpy of vaporization.

6-49. On the basis of (a) Eqs. (6-96) and (6-97), (b) Eq. (6-100), (c) Eq. (6-104), and (d) Eq. (6-105), estimate the enthalpy of vaporization for nitrogen at 120 K, in kJ/kg. (e) Compare the results to the value listed in a saturation table.

6-50. On the basis of (a) Eqs. (6-96) and (6-97), (b) Eq. (6-100), and (c) Eq. (6-104), estimate the enthalpy of vaporization for butane (C_4H_{10}) at 300 K, in kJ/kg. The normal boiling point temperature is 272.6 K at 1 atm. (A typical literature value for h_{fg} is 359–360 kJ/kg.)

6-51. On the basis of (a) Eqs. (6-96) and (6-97), (b) Eq. (6-100), and (c) Eq. (6-104), estimate the enthalpy of vaporization for propane at 300 K, in kJ/kg. The normal boiling point temperature is 231.3 K. (The values in the literature are around 330–332 kJ/kg.)

6-52. Refrigerant-134a is in liquid-vapor equilibrium at 25°C. At the given temperature, and employing the Redlich-Kwong-Soave (RKS) equation,

a. Find a saturation pressure, in bars, and the values of v_g and v_f, in $m^3/kmol$, and Z_g and Z_f which reasonably satisfy the Gibbs function criterion for phase equilibrium.

b. Evaluate the residual enthalpies of the vapor and liquid phases, in kJ/kmol, and the enthalpy of vaporization, in kJ/kg, on the basis of data from part (a).
c. Estimate the percent error in P^{sat}, v_g, v_f, and h_{fg}, using a saturation temperature table for comparison.

6-53. Refrigerant-134a is in liquid-vapor equilibrium at 0°C. On the basis of the Redlich-Kwong-Soave equation of state, determine

a. the saturation pressure, in bars, which reasonably satisfies the Gibbs function criterion for phase equilibrium, and the values of Z_f and Z_g and the saturated specific volumes of the liquid and gas phases, in m^3/kg
b. the residual enthalpies of the vapor and liquid phases and the enthalpy of vaporization, in kJ/kg
c. the percent error in P^{sat}, v_g, v_f, and h_{fg} found above, compared to published saturation data

6-54. Carbon dioxide is in vapor-liquid equilibrium at 260 K. Employing the Redlich-Kwong-Soave equation,

a. Find the saturation pressure, in bars, v_g and v_f, in m^3/kmol, and Z_g and Z_f which reasonably satisfies the Gibbs function criterion for phase equilibrium.
b. Evaluate the residual enthalpies of the vapor and liquid phases, in kJ/kmol, and the enthalpy of vaporization, in kJ/kmol, on the basis of data from part (a).
c. Estimate the percent error in v_g, v_f, and h_{fg} found above by using data from the saturation table in the Appendix.

6-55. The PvT relationship for ammonia at low pressures is reasonably represented by the equation $v = RT/P - 0.0030 - 0.34/(0.01T)^3 - 60/(0.01T)^{11}$, where P is in atmospheres, T is in kelvins, and v is in m^3/kg. If the value of c_p at 5 bars and 37°C is 2.45 kJ/kg·K, estimate the value of the Joule-Thomson coefficient at this state, in K/bar.

6-56. On the basis of the generalized equation, estimate from superheat specific volume and enthalpy data the Joule-Thomson coefficient for steam at 80 bars and 400°C, in °C/bar.

6-57. At relatively low pressures the equation of state for a gas may be represented by $Pv = RT + bP$, and b is modeled by $b = C_1 + C_2T^{-1} + C_3T^{-2}$.

a. Derive an equation for the inversion temperature in terms of the constants C_1, C_2, and C_3.
b. For nitrogen the values of the constants are $C_1 = 39.5$ cm^3/g, $C_2 = -1 \times 10^4$ (cm^3/g)(K), and $C_3 = -1.084 \times 10^6$ (cm^3/g)(K^2). Estimate the maximum inversion temperature of nitrogen, in kelvins.

6-58. Using tabular data, estimate the Joule-Thomson coefficient for R-134a, in K/bar, at 80°C and 14.0 bars on the basis of the relations (a) $(1/c_p)\ [T(\partial v/\partial T) - v]$ and (b) $(\partial T/\partial P)_h$.

6-59. The equation of state for dry air may be expressed by $Pv = RT + bP$, and $b = 37.28 - 8.43 \times 10^3/T - 1.75 \times 10^6/T^2 + 7.35 \times 10^7/T^3$, where b is in cm^3/gmol and T is in kelvins.

a. If the measured Joule-Thomson coefficient at 1 atm and 0°C is 0.275°C/atm, calculate the value of c_p at this state, in J/g·K.
b. Compare to c_p data at low pressures.

6-60. On the basis of the generalized equation for $(\partial T/\partial P)_h$, estimate the Joule-Thomson coefficient in °C/bar for steam at 30 bars and 320°C, using tabular superheat data.

6-61. The PvT relationship for ammonia (NH_3) at low pressures may be represented by the relation $v = RT/P - 0.003 - 0.34 \times 10^6/T^3 - 60 \times 10^{22}/T^{11}$, where P is in atmospheres, v is in m^3/kg, and T is in kelvins. If the Joule-Thomson coefficient at 10 atm and 400 K is 1.05 K/atm, estimate the value of c_p at this state, in kJ/kg·K.

REFERENCES

1. Watson, K. M., "Thermodynamics of the Liquid State," *Ind. Eng. Chem.*, **35**:398, 1943.
2. Liley, P. E., "The Vapor Pressure Equation $\ln P = a + b/T$ Revisited," *Intl. J. Mech. Engr. Ed.*,**20**(1):75, 1992.
3. *Handbook of Chemistry and Physics,* 70th ed., CRC Press, Cleveland, OH, 1989.
4. Antoine, C., *Compt. Rend.*, **107**:681, 836, 1143; 1888.
5. Boublik T., V. Fried, and E. Ha'la, *The Vapor Pressures of Pure Substances*, Elsevier, Amsterdam, 1984.
6. Reid, R. C., J. M. Prausnitz, and B. E. Poling, *The Properties of Gases and Liquids*, 4th ed., McGraw-Hill, Inc., New York, 1987.
7. Yaws, C. L., and H. C. Yank, "To Estimate Vapor Pressure Easily," *Hydrocarbon Process.*, **68**:65, 1989.
8. Ambrose, D., *Equations for the Correlation and Estimation of Vapor Pressures*, National Physical Laboratory Report CHEM **114**, August 1980.
9. Guggenheim, E. A., *Thermodynamics*, North-Holland Publishing Co., Amsterdam, 1957.
10. Pitzer, K. S., D. Z. Lippmann, R. F. Curl, C. M. Huggins, and D. E. Petersen, "The Volumetric and Thermodynamic Properties of Fluids. II. Compressibility Factor, Vapor Pressure, and Enthalpy of Vaporization," *J. Am. Chem. Soc.*, **77**:3433, 1955.
11. Carruth, G. F., and R. Kobayaski, "Extension of Low Reduced Temperatures of Three-Parameter Corresponding States: Vapor Pressures, Enthalpies, and Entropies of Vaporization, and Liquid Fugacity Coefficients," *Ind. Engr. Chem. Fundam.*, **11**:509,1972.
12. Lee, B. I., and M. G. Kessler, "A Generalized Thermodynamic Correlation Based on Three-Parameter Corresponding States," *AIChE J.*, **21**:510, 1975.
13. Dong, W. G., and J. H. Lienhard, "Corresponding States Correlation of Saturated and Metastable Properties," *Can. J. Chem. Engr.*, **64**:158, 1986.
14. Fishtine, S. H., "Reliable Latent Heats of Vaporization," *Ind. Eng. Chem.*, **55**:(4) 20, (5) 49, (6) 47, 1963.
15. Reidel L., *Chem. Ing. Tech.*, **26**:679, 1954.
16. Watson, K. M., "Thermodynamics of the Liquid State," *Ind. Eng. Chem.*, **35**:398, 1943.
17. Haggenmacher, J. E., "The Heat of Vaporization as a Function of Pressure and Temperature," *J. Am. Chem. Soc.*, **68**:1633, 1946.
18. Torquato, S., and G. R. Stell, "An Equation for the Latent Heat of Vaporization," *Ind. Eng. Chem. Fundam.*, **21**:202, 1982.
19. Torquato, S., and P. Smith, "The Latent Heat of Vaporization of a Widely Diverse Class of Fluids," *J. Heat Transfer*, **106**:252, 1984.
20. Thomson, R., and J. B. Joule, *Philos. Trans. R. Soc. London*, 1853.
21. Miller, Donald G., "Joule-Thomson Inversion Curve, Corresponding States, and Simpler Equations of State," *Ind. Eng. Chem. Fundam.*, **9**(4):585, 1970.

CHAPTER 7

THE THIRD LAW OF THERMODYNAMICS

The second-law analysis of chemically reactive systems requires the assignment of an "absolute" entropy value to each reacting species. The use of conventional expressions for the change in entropy of a substance due to pressure and temperature changes alone is insufficient because of the formation and destruction of the reacting species. The third law provides the basis for the establishment of "absolute" entropy values. This chapter also investigates the behavior of the specific heats of solids at very low temperatures.

7-1 INTRODUCTION

A significant corollary of the second law of thermodynamics is the expression for the entropy change of a closed system. For the system as a whole, by definition,

$$dS = \left(\frac{\delta Q}{T}\right)_{\text{int rev}}$$

or for a finite change of state for a unit mass

$$\Delta s = \int_1^2 \left(\frac{\delta q}{T}\right)_{\text{int rev}}$$

For simple compressible substances we may replace δq in the latter expression by $dh - v\,dP$, so that

$$\Delta s = \int \frac{dh}{T} - \int \frac{v\,dP}{T} \tag{6-3}$$

Then for special classes of substances such as an ideal gas or an incompressible substance specific relationships among h, P, v, and T permit direct integration of the above equation. On the other hand, for other substances such as a real gas it is necessary first to measure P, v, T, and c_p at a large number of states. Then somewhat complicated mathematical techniques are carried out on a computer to evaluate s data for the substance. These values of s are completely arbitrary, because they are measured relative to some reference state where s has been set equal to zero. Superheat tables for various substances are examples of this technique. When entropy changes are calculated from such tables, the arbitrarily chosen reference state cancels out. There are many engineering applications for which the evaluation of Δs by equation or from tabular data is meaningful to the analysis.

There is, however, an important situation for which such Δs evaluations are meaningless. This is the case of chemically reactive systems. We cannot talk about the entropy change of a substance if it is formed or consumed during a reaction. For the formation of a set of products in a given state from an initial set of reactants in another state the overall entropy change must be written as

$$\Delta S = \sum (N_i s_i)_{\text{prod}} - \sum (N_i s_i)_{\text{reac}}$$

where the subscripts "prod" and "reac" refer to the products and reactants of the chemical reaction. This equation forces one to address a new problem. Rather than evaluation of Δs for a specific substance, we must now identify what we mean by s_i for each species in the reaction.

A method of determining s_i for a species is found by considering the entropy change for the heating of the substance at constant pressure from its solid state to its gaseous state. At constant pressure the entropy change of any single phase (solid, liquid, or gas) due to a temperature variation is given by

$$\Delta s = \int_{T_1}^{T_2} \frac{dh}{T} = \int_{T_1}^{T_2} c_p \frac{dT}{T} \tag{7-1}$$

In addition, a phase change is isothermal in nature. Hence the entropy change during a phase change is simply

$$\Delta s = \int \frac{dh}{T} = \frac{\Delta h_{ab}}{T} \qquad \text{(phase change)} \tag{7-2}$$

where Δh_{ab} is the energy required to carry out the phase change from phase a to phase b at that temperature. The general variation of s during the heating of a solid to its gaseous state is represented on the Ts plot of Fig. 7-1. To include some special cases that occur in nature, the plot shows two solid phases, α and β. In this case the solid α phase at absolute zero of temperature is heated along

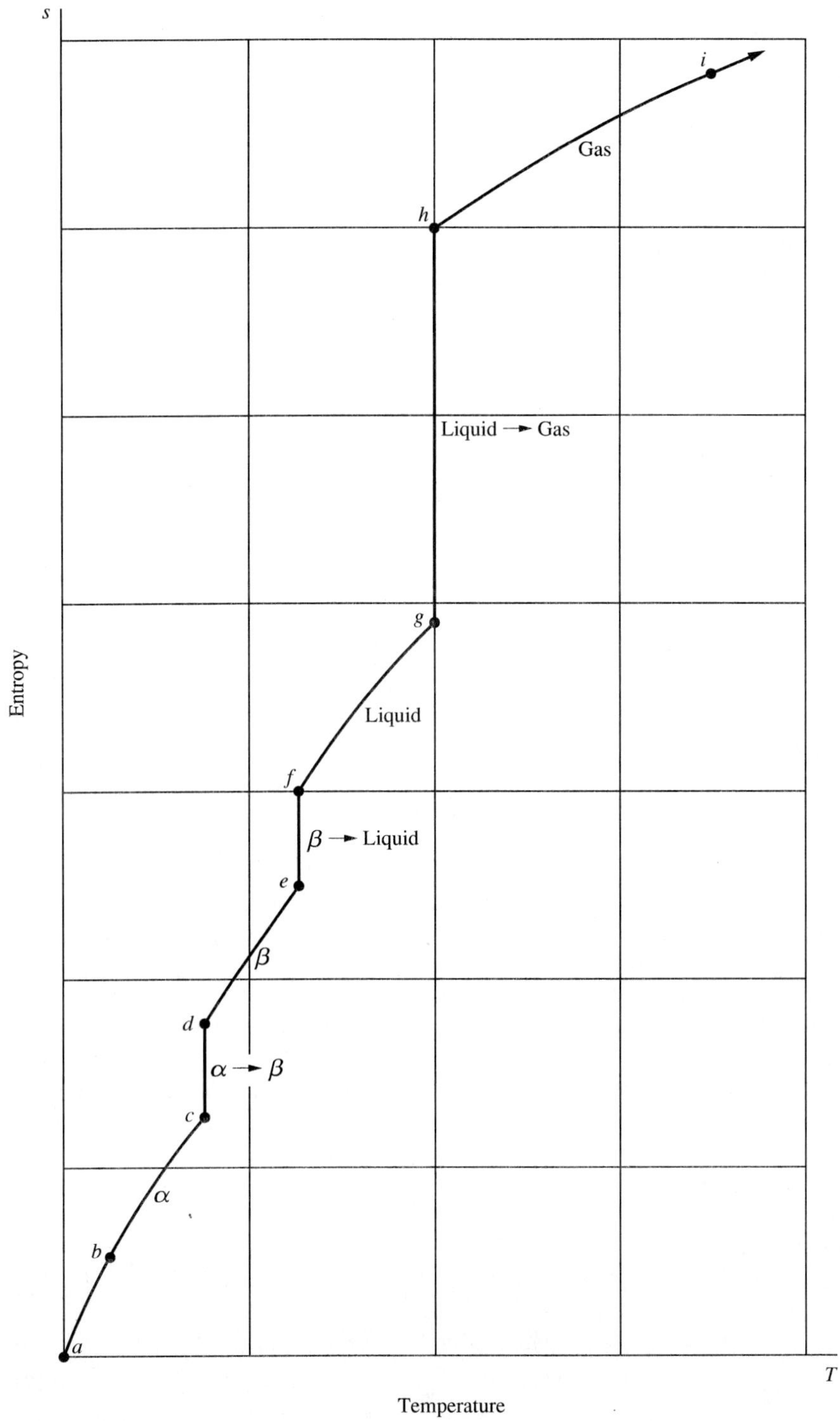

FIGURE 7-1
General variation of entropy with temperature for various phases and phase changes.

path a-b-c until a solid phase transformation α-β occurs along the isothermal path c-d. Then solid β is heated along path d-e, until the solid β phase melts from state e to state f at constant temperature. The liquid then is heated along path f-g, and then vaporized along path g-h. Finally, the gas phase is heated from state h to some arbitrary state i. The value of the entropy at any state along this path would include its zero-temperature entropy s_0 at state a, plus the sum of all the j temperature-related paths, plus the sum of all the k isothermal changes. In general, on the basis of Eqs. (7-1) and (7-2),

$$s = s_0 + \sum_j \int c_p \frac{dT}{T} + \sum_k \frac{\Delta h_{ab}}{T} \tag{7-3}$$

Thus the entropy of a pure substance depends on specific-heat data for various phases, enthalpy (latent heat) data for various isothermal phase changes, and the selection of a reference entropy s_0 at the absolute zero temperature.

The numerical evaluation of Eq. (7-3) presents two difficulties. First, the proper method of selection of a reference entropy value s_0 at absolute zero temperature is not self-evident. Second, specific-heat data of solids are difficult to measure at temperatures approaching absolute zero. For example, the data for path a-b in Fig. 7-1 may not be available. Hence some reasonably accurate method for extrapolating specific-heat data in this low-temperature region is required. The next four sections of this chapter address these two points.

7-2 THIRD LAW OF THERMODYNAMICS

The second law of thermodynamics introduces a new property, the entropy s, which lacks a datum value s_0. In the absence of chemical reactions or phase changes a zero reference value can be chosen at an arbitrary state when compiling tabular data for a given substance. However, in the analysis of chemical reactions it is important to establish an entropy datum s_0 based on physical considerations, if possible. The third law of thermodynamics is concerned primarily with physical events at very low temperatures which suggest a method of setting the entropy value at absolute zero temperature. This requires an investigation of other property data at very low temperatures.

In the early 1900s a great deal of effort was directed toward examining the thermodynamic behavior of substances at low temperatures. Of special interest for establishing an entropy datum was the variation of ΔG and ΔH with temperature for isothermal chemical reactions. These two parameters may be related to each other (and to the entropy function) in the following way. For a system of fixed composition

$$G = H - TS \tag{a}$$

$$dG = V\,dP - S\,dT \tag{b}$$

and

$$(\partial G/\partial T)_P = -S \tag{c}$$

The combination of Eqs. (a) and (c) leads to

$$G = H + T\left(\frac{\partial G}{\partial T}\right)_P \tag{7-4}$$

Equation (7-4) is one form of the Gibbs-Helmholtz equation. For an isothermal change of state, such as a chemical reaction, the above relationship should hold for both the initial and final states. Designating these states as 1 and 2, respectively, we find that

$$G_2 - G_1 = H_2 - H_1 + T\left(\frac{\partial(G_2 - G_1)}{\partial T}\right)_P$$

or,

$$\Delta G_T = \Delta H_T + T\left(\frac{\partial \Delta G}{\partial T}\right)_P \tag{7-5}$$

Equation (7-5) is another form of the Gibbs-Helmholtz equation. It shows that ΔG and ΔH for an isothermal change of state are uniquely related to $(\partial\,\Delta G/\partial T)_P$.

Another important relation is found by applying Eq. (a) to an isothermal change of state. The result is

$$\Delta G_T = \Delta H_T - T\,\Delta S_T \tag{d}$$

A comparison of Eq. (d) with Eq. (7-5) shows that

$$\left(\frac{\partial\,\Delta G}{\partial T}\right)_P = -\Delta S_T \tag{7-6}$$

Consequently, not only is $(\partial\,\Delta G/\partial T)_P$ a measure of the difference between ΔG_T and ΔH_T, by Eq. (7-5), but this same parameter also equals the negative change in entropy for an isothermal reaction. Hence the variation of ΔG with temperature is directly related to the entropy change of an isothermal process, including processes at absolute zero temperature.

An examination of Eq. (7-5) reveals that unless $(\partial\,\Delta G/\partial T)_P$ is infinite at 0 K, then ΔG_T would equal ΔH_T at absolute zero. In 1902 Richards [1] presented data from galvanic cell experiments which indicated that ΔG_T and ΔH_T approached each other as the temperature decreased. This information of itself is not significant. What is more important is the manner in which ΔG varies as the temperature approaches zero. For example, consider the two possibilities shown in Fig. 7-2. Both curves satisfy the condition that $\Delta G = \Delta H$ at absolute zero (the behavior of ΔH with temperature is not shown). However, curve (2) in conjunction with Eq. (7-6) would lead to the conclusion that $\Delta S = 0$ as $T \to 0$. Richard's data suggested that the slope of ΔG versus T approached zero as $T \to 0\,\text{K}$. This and other information led Nernst [2] to make the original formulation of the third law. Because it appeared that

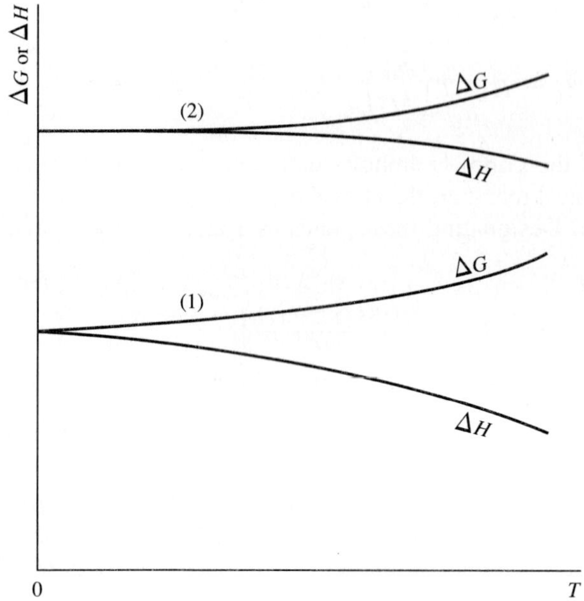

FIGURE 7-2
Typical plots of the isothermal enthalpy change and Gibbs function change at low temperatures.

$$\lim_{T\to 0}(\Delta G - \Delta H)_T = \lim_{T\to 0}\left(\frac{\partial\, \Delta G}{\partial T}\right)_P = 0$$

then it was possible also to state on the basis of Eq. (7-6) that

$$\lim_{T\to 0} \Delta S_T = 0 \tag{7-7}$$

This mathematical formulation is frequently referred to as the *Nernst-Simon statement of the third law*. It states that the entropy change associated with any isothermal reversible process of a condensed system approaches zero as the temperature approaches zero.

In addition to the initial studies of chemical reactions, other evidence supports the Nernst-Simon statement of the third law. The behavior of the melting curve of helium at very low temperatures is one example. On the basis of the Clapeyron equation (Sec. 6-10)

$$\frac{dP}{dT} = \frac{s^f - s^i}{v^f - v^i}$$

where the superscripts "i" and "f" represent the initial and final states, respectively. But $(v^f - v^i)$ is never zero for a first-order phase change. Consequently, if the third law requires that

$$\lim_{T \to 0} (s^f - s^i) = 0$$

then the Clapeyron equation requires that

$$\lim_{T \to 0} \frac{dP}{dT} = 0$$

Thus the slope of the saturation line on a P-T diagram should approach zero as the temperature approaches zero. This relation is supported by the data of Simon and Swenson for the melting curve of He^4. The slope of the curve, dP/dT, does approach zero very rapidly as the temperature is decreased below 1.4 K.

Another check on the Nernst-Simon statement is found in the study of solid-solid (allotropic) phase transformations. A pure substance may exist in more than one crystalline solid phase. Phosphine (PH_3), for example, exists in three solid phases (α, β, and γ) when the temperature is below 49.43 K. On rapid cooling below this temperature the α phase is transformed into the γ phase at 30.29 K. The substance then remains in the γ phase as it is cooled further to absolute zero. If cooled slowly, the α phase is transformed into the β phase at 49.43 K. On further cooling the β phase remains until absolute zero is reached. Experimentally the changes in entropy from 0 to 49.43 K have been determined separately for the $\alpha - \gamma$ rapid cooling process and the β slow cooling process. In addition, the change in entropy for the allotropic transformation, $\alpha \to \beta$, at 49.43 K must also be determined. Because entropy is a property, the following condition must be fulfilled:

$$\Delta s(\gamma \text{ at } 0 \text{ K to } \alpha \text{ at } 49.43 \text{ K}) + \Delta s(\alpha \to \beta \text{ at } 49.43 \text{ K}) \\ + \Delta s(\beta, 49.43 \text{ to } 0 \text{ K}) + \Delta s(\beta \to \gamma \text{ at } 0 \text{ K}) = 0$$

That is, the summation of entropy changes around a cycle must equal zero. The first three terms have been evaluated from experimental data by Stephenson and Giauque. Their values are 34.02, −15.73, and −18.33 J/gmol · K, respectively, and the sum of these values is −0.04 J/gmol · K. Therefore from the above equation

$$\Delta s(\beta \to \gamma \text{ at } 0 \text{ K}) \approx 0$$

Thus the data support, within experimental accuracy, the statement that the change in entropy for an isothermal reversible process at 0 K is zero. Other measurements for allotropic phase changes give added support to the theory.

Another interesting sidelight regarding the entropy function at 0 K should be examined. One of the fundamental Maxwell property relations is $(\partial s/\partial P)_T = -(\partial v/\partial T)_P$. Also, recall that the coefficient of volumetric expansion β is defined as

$$\beta = \frac{1}{v}\left(\frac{\partial v}{\partial T}\right)_P$$

Substitution of this latter expression into the Maxwell relation yields

$$\left(\frac{\partial s}{\partial P}\right)_T = -v\beta$$

As the temperature approaches zero, v remains finite for a solid. However, all data for solids indicate that β approaches zero as T approaches zero. Therefore

$$\lim_{T\to 0}\left(\frac{\partial s}{\partial P}\right)_T = 0$$

This result states that s is not a function of pressure at 0 K.

All the preceding third-law discussions have focused primarily on the entropy change for isothermal reversible processes at absolute zero. Of equal importance in the study of chemically reactive systems, for example, is the value of the entropy itself at absolute zero. As early as 1912 Planck [3] suggested that the value of s at 0 K should be taken as zero. Also, it was felt that the specification of a crystalline structure at absolute zero was essential to the establishment of any base value for the entropy function. Hence glasses and solutions were excluded from consideration in the creation of an "absolute entropy" scale. The statement regarding crystalline structure is too restrictive in light of modern quantum mechanics. In addition, the effects of isotropic variation and nuclear spin effects are usually ignored. Under these constraints one can establish a reference state for entropy which leads to meaningful results. In effect, the entropy of a system which is in internal thermodynamic equilibrium disappears at absolute zero temperature. This choice has no real theoretical significance on a macroscopic basis but, rather, is simply a convenient convention. It establishes a "conventional scale" for "absolute" entropies and is important in the second-law analyses of systems involving chemical reactions and phase changes. However, there is a physical significance to the "zero entropy" at absolute zero temperature on a microscopic basis. This point is discussed in Chap. 13.

Since the early 1900s scientists have sought means to lower the temperature of a system to absolute zero. Liquid hydrogen was prepared by Dewar in 1898 at a temperature of 20.4 K. Liquid helium was produced by Kamerlingh-Onnes in 1908 at 4.2 K. By rapid adiabatic vaporization of liquid helium a temperature around 1 K is possible. Much lower temperatures than this are possible by means of adiabatic magnetization of paramagnetic compounds. A natural question is whether the temperature of a substance can be reduced to absolute zero. The answer is provided by an alternative form of the third law known as the "unattainability" statement of the third law, which states that it is impossible for any process involving a finite number of steps to reduce the temperature of any system to absolute zero. Thus just as in the case of the second law, the third law has several equivalent statements.

7-3 THE SPECIFIC HEAT OF A SOLID

As early as 1819, on the basis of Dulong and Petit, it was known that the specific heat c_v of some solid elements around room temperature was close to $3R_u$. In 1864 Kopp noted that some elements with a low atomic weight (such as carbon, boron, silicon, and beryllium) have a specific heat considerably less than $3R_u$ at room temperature. In 1871 Boltzmann derived the value of $3R_u$ on the basis of classical kinetic theory. Whatever the model, the first step is to evaluate the specific internal energy. Then c_v is determined from its definition, namely, $c_v = (\partial u/\partial T)_v$. In the Boltzmann model atoms were considered to occupy a definite site in the crystalline lattice structure and to vibrate about the site in simple harmonic motion. The mean of each of the potential and kinetic energies is equal to $kT/2$ for each vibrational mode of such an atom, where k is Boltzmann's constant. Each atom has a mode of vibration in each of the three orthogonal directions. Thus the internal energy of a gram mole of substance is $3N_A(2)(kT/2) = 3N_AkT$, where N_A is Avogadro's number. But, by definition, $N_Ak = R_u$. Hence the Boltzmann model of N_A independent harmonic oscillators yields a value of $3R_uT$ for the internal energy and a value of $3R_u$ for the specific heat c_v of a solid. This is known as the *classical* value; that is,

$$c_{v,\text{classical}} = 3R_u \qquad \text{(solid)} \tag{7-8}$$

The classical value is the limiting value which most solids approach as the temperature is increased. Unfortunately, this model is unable to predict the variation of c_v with temperature.

On the lower end of the temperature scale it is found that the specific heats of all solids decrease as the temperature is lowered. In fact, the specific heat falls rapidly toward zero as the temperature approaches 0 K. The general variation of c_v versus T is shown in Fig. 7-3. Figure 7-3(*a*) shows data for solid elements, whereas the data for Fig. 7-3(*b*) are for nonmetallic compounds. Both sets of curves exhibit the same trends. Following the advent of quantum theory in the early 1900s, a number of theories have been developed to predict the noted variation of c_v with temperature for solids. Two of the earliest and best-known theories were developed by Einstein and Debye. A discussion of their work follows.

7-4 EINSTEIN THEORY OF SPECIFIC HEAT

In the classical equipartition of energy theorem it is assumed that each independent mode of vibration contains energy equal to kT. As noted in the preceding section, for three directions of vibration and N_A atoms the total energy is simply $3R_uT$. Einstein [4] sensed that the decrease of c_v with temperature could be explained by applying the quantization of energy concept to the calculation of the energy of atomic oscillations. In his model all modes of vibration are taken to have a single frequency ν, as shown in Fig. 7-4. However, the energy of a harmonic oscillator is quantized and given by

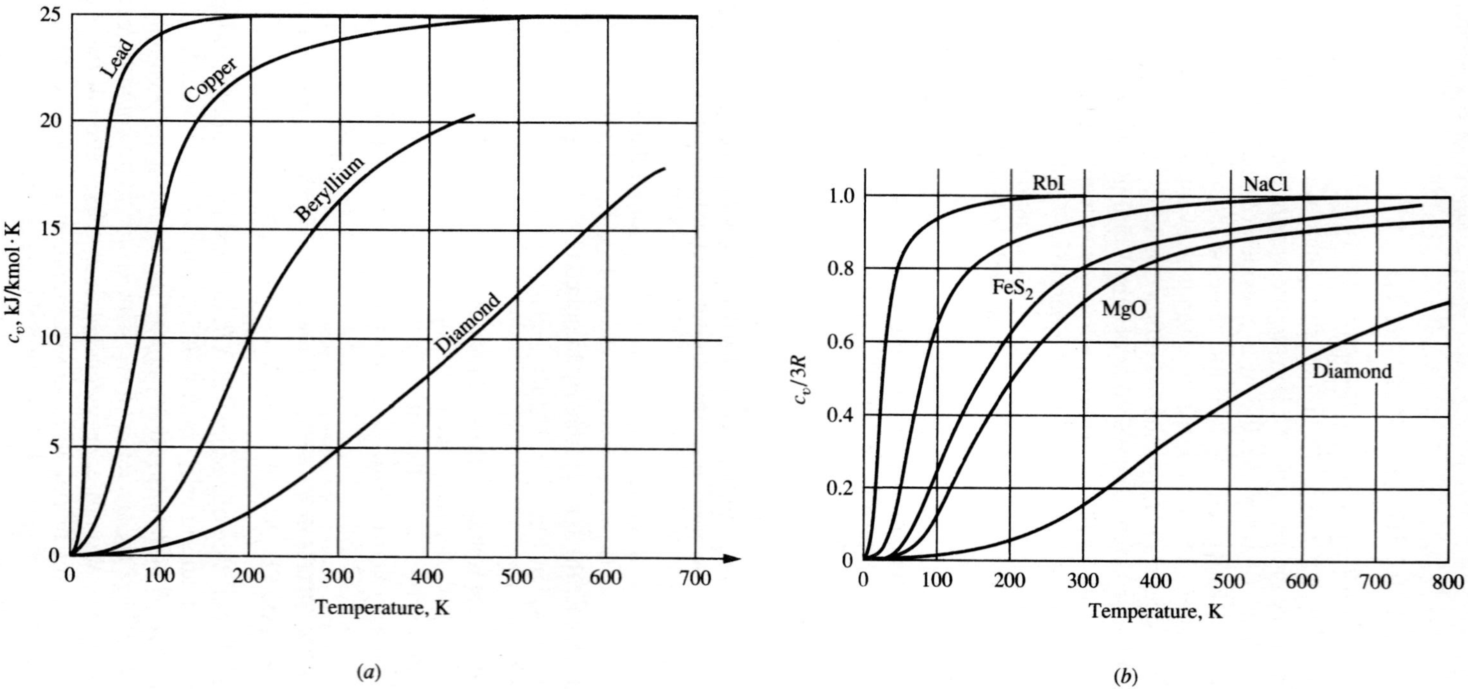

FIGURE 7-3
Typical variation of the specific heat of solids with temperature; (*a*) Elements, and (*b*) compounds.

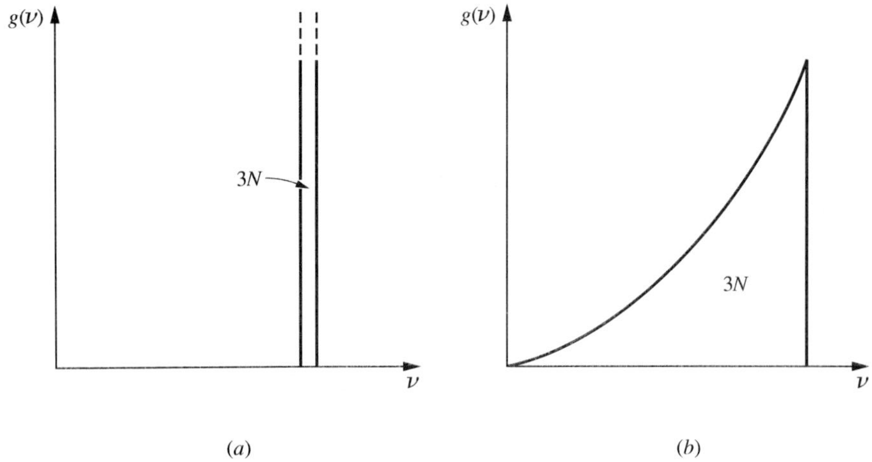

FIGURE 7-4
Frequency distribution of lattice vibrations for: (*a*) Einstein approximation, and (*b*) Debye approximation.

$$\epsilon_{\text{vib}} = \left(n + \frac{1}{2}\right)h\nu \tag{7-9}$$

where $n = 0, 1, 2, \ldots,$. The constant h is Planck's constant. In addition, quantum theory dictates that the average energy of an oscillator with frequency ν is given by

$$\bar{\epsilon}_{\text{vib},\nu} = \sum \frac{\epsilon_n e^{-\epsilon_n/kT}}{\sum e^{-\epsilon_n/kT}} \tag{7-10}$$

where each summation is taken over n from 0 to ∞, and k is Boltzmann's constant.

After substituting Eq. (7-9) into Eq. (7-10) and carrying out the summations, one finds that

$$\bar{\epsilon}_\nu = \epsilon_0 + \frac{kTx}{e^x - 1} \tag{7-11}$$

where $x = h\nu/kT$, and ϵ_0 is the ground-state energy. Equation (7-11) describes the energy of a planckian oscillator of frequency ν. For N_A atoms oscillating in three orthogonal directions the molar internal energy becomes

$$u_\nu = u_0 + \frac{3R_u T x}{e^x - 1} \tag{7-12}$$

Because $c_v \equiv (\partial u/\partial T)_v$, the molar specific heat at constant volume for the Einstein model is

$$c_{v,\text{vib}} = \frac{3R_u x^2 e^x}{(e^x - 1)^2} \tag{7-13}$$

If we define an Einstein characteristic temperature Θ_E as

$$\Theta_E \equiv \frac{h\nu_E}{k} \tag{7-14}$$

where ν_E is the single frequency associated with a particular solid substance, then by the Einstein model

$$\frac{c_v}{3R_u} = \frac{y^2 e^y}{(e^y - 1)^2} \tag{7-15}$$

where $y = \Theta_E/T$. The function $y^2 e^y (e^y - 1)^{-2}$ is known as the Einstein function and its value as a function of y is found in standard mathematical tables.

It may be shown that $c_{v,\text{vib}}$ for the Einstein model approaches $3R_u$ as the temperature increases, in agreement with classical theory. Also, c_v falls to zero for this model at 0 K. However, when ν is chosen to give a good fit with experimental c_v for a substance at moderate temperatures, the agreement at very low temperatures is poor. For the Einstein model c_v approaches zero temperature exponentially, which is faster than experimental data indicate.

Einstein recognized that the assumption of a single frequency ν for all oscillators in a solid is unsatisfactory. Nevertheless, the model was the first attempt to use the quantum theory to attempt to explain the variation of c_v with temperature for solids. An improved model involves the same quantum principles, but the single frequency is replaced by a spectrum of frequencies.

7-5 DEBYE'S THEORY OF SPECIFIC HEAT

In 1912 Debye [5] proposed that a solid be treated as a continuous elastic medium with a spectrum of vibrational frequencies for the longitudinal and transverse waves. In addition, he assumed that this spectrum extended only from zero to a maximum value ν_m so that the total number of vibrations is consistent with the $3N_A$ modes of vibration in a mole of a solid. A study of longitudinal and transverse waves in an isotropic solid leads to a simple expression for $dg(\nu)$, the number of vibrations in the frequency range from ν to $\nu + d\nu$. This expression is

$$dg(\nu) = \frac{9N_A}{\nu_m^3}\nu^2 d\nu \tag{7-16}$$

Thus the distribution of frequencies in the Debye model is a squared function of ν. The Debye approximation is shown in Fig. 7-4(*b*). To determine the molar internal energy, we must integrate the product of the energy of a Planckian oscillator [Eq. (7-11)] and the total number of modes with frequency ν [Eq. (7-16)] over the frequency range from zero to ν_m. Thus

$$u = \int_0^{\nu_m} \bar{\epsilon}_\nu \, dg = \int_0^{\nu_m} \left(\frac{h\nu}{2} + \frac{h\nu}{\left(e^{\frac{h\nu}{kT}} - 1\right)} \right) \left(\frac{9N_A \nu^2 d\nu}{\nu_m^3} \right)$$

This equation can be rewritten in the form

$$u = u_0 + \frac{9R_u T}{x_m^3} \int_0^{\nu_m} \frac{x^3 \, dx}{e^x - 1} \tag{7-17}$$

The quantity u_0 is the ground-level energy, $x = h\nu/kT$, and $x_m = h\nu_m/kT$. The *Debye characteristic temperature* Θ_D is defined as

$$\Theta_D \equiv \frac{h\nu_m}{k} \tag{7-18}$$

so that $x_m = \Theta_D/T$.

An equation for c_v based on the Debye model is developed from $c_v = (\partial u/\partial T)_v$. Thus

$$c_v = \frac{-9R_u x_m}{e^{x_m} - 1} + \frac{36R_u}{x_m^3} \int_0^{x_m} \frac{x^3 \, dx}{e^x - 1} \tag{7-19}$$

The integral shown in Eqs. (7-17) and (7-19) is frequently referred to as Debye's integral. Another form of Eq. (7-19) often cited is

$$c_v = \frac{9R_u}{x_m^3} \int_0^{x_m} \frac{x^4 e^x \, dx}{(e^x - 1)^2} \tag{7-20}$$

The integrals in Eqs. (7-17), (7-19), and (7-20) cannot be evaluated directly but must be solved by numerical methods. Values of c_v/R_u are then tabulated versus a dimensionless temperature, T/Θ_D, as shown in Table 7-1.

A plot of the Debye specific-heat function, based on data from Table 7-1, is shown in Fig. 7-5. Experimental data for aluminum, copper, and silver are also shown as data points in the figure. The data are based on the Θ_D values for the three substances listed in the figure. The fit is quite good over the temperature range shown, and the same general agreement would be found for T/Θ_D values greater than 0.5. Table 7-2(a) lists the Debye temperatures of some metals, and Table 7-2(b) lists Θ_D values for some nonmetals and common compounds.

Although the Debye integral cannot be evaluated directly, its value can be approximated closely in two special cases. As shown by the data in Table 7-1, c_v/R_u reaches 95 percent of its maximum value when T/Θ_D is unity. Hence c_v/R_u increases very slowly in the high-temperature range. For this high-temperature region the integrand of Eq. (7-17) is rewritten as a series of terms by expanding $(e^x - 1)$ in a power series. As a result, for the first few terms in the series

TABLE 7-1

Debye specific-heat function

Values of c_v/R_u as a function of Θ_D/T

Θ_D/T	0.0	0.1	0.2	0.3	0.4	0.5	0.6	0.7	0.8	0.9
0	3.000	2.999	2.994	2.987	2.976	2.963	2.947	2.928	2.906	2.882
1	2.855	2.826	2.795	2.761	2.726	2.688	2.648	2.608	2.565	2.521
2	2.476	2.430	2.383	2.335	2.287	2.238	2.188	2.138	2.088	2.038
3	1.988	1.938	1.889	1.840	1.790	1.742	1.694	1.647	1.600	1.554
4	1.509	1.465	1.421	1.379	1.337	1.296	1.256	1.217	1.179	1.142
5	1.106	1.071	1.037	1.004	0.971	0.940	0.909	0.880	0.851	0.824
6	0.797	0.771	0.746	0.722	0.698	0.675	0.653	0.632	0.611	0.592
7	0.573	0.554	0.536	0.519	0.503	0.487	0.471	0.456	0.442	0.428
8	0.415	0.402	0.389	0.377	0.366	0.355	0.344	0.333	0.323	0.314
9	0.305	0.296	0.287	0.278	0.270	0.262	0.255	0.248	0.241	0.234
10	0.228	0.221	0.215	0.209	0.203	0.198	0.193	0.188	0.183	0.178
11	0.173	0.169	0.164	0.160	0.156	0.152	0.148	0.144	0.141	0.138
12	0.134	0.131	0.128	0.125	0.122	0.119	0.116	0.114	0.111	0.109
13	0.106	0.104	0.101	.0991	.0969	.0948	.0927	.0907	.0888	.0869
14	.0851	.0833	.0815	.0798	.0782	.0766	.0750	.0735	.0721	.0706
15	.0692	.0679	.0665	.0652	.0640	.0628	.0616	.0604	.0593	.0581

When $\Theta_D/T \geq$ then 16, $c_v/R_u = 233.8(T/\Theta_D)^3$.

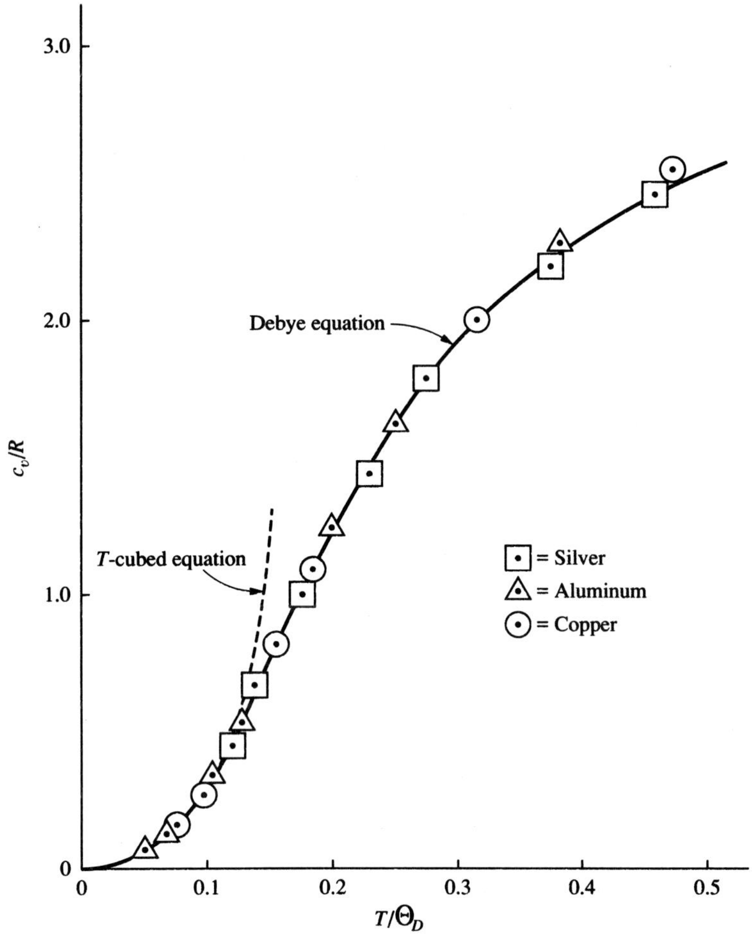

FIGURE 7-5
Comparison of experimental c_v values for aluminum, copper, and silver at low temperatures with the curve from the Debye theory.

$$\frac{x^3}{e^x - 1} = x^2 - \frac{x^3}{2} + \frac{x^4}{12} - \frac{x^6}{720} + \frac{x^8}{30,240} - \cdots$$

and Debye's integral becomes

$$\int_0^{x_m} \frac{x^3\,dx}{e^x - 1} = \frac{x_m^3}{3} - \frac{x_m^4}{8} + \frac{x_m^5}{60} - \frac{x_m^7}{5040} + \frac{x_m^9}{272,160} - \cdots$$

Substitution of this result into Eq. (7-17) yields

$$u = 3R_uT\left(1 + \frac{x_m^2}{20} - \frac{x_m^4}{1680} + \frac{x_m^6}{90,720} - \cdots\right)$$

TABLE 7-2
Table of Debye characteristic temperatures and electronic constants: Θ_D in kelvins, γ in J/kmol·K^2

(a) Metals			(b) Nonmetals and compounds	
Element	**Θ_D**	**γ**	**Substance**	**Θ_D**
Be	1440	0.17	Argon	93
Mg	400	1.3	Silicon	636
Al	428	1.35	C(diamond)	2230
Ca	230	2.9	NaCl	321
Cr	630	1.40	KCl	233
Mn	410	14	CaF_2	510
Fe	467	5.0	SiO_2	470
Ni	450	7.1	Fe_2O_3	660
Cu	315	0.688	MgO	946
Zn	310	0.65	FeS_2	637
Ag	212	0.650		
Pt	240	6.8		
Au	164	0.75		
Pb	105	3.0		

where $x_m = \Theta_D/T$. Finally, differentiation of the above equation with respect to temperature leads to

$$\frac{c_v}{3R_u} = 1 - \frac{1}{20}\left(\frac{\Theta_D}{T}\right)^2 + \frac{1}{560}\left(\frac{\Theta_D}{T}\right)^4 - \frac{1}{18{,}144}\left(\frac{\Theta_D}{T}\right)^6 + \cdots \tag{7-21}$$

Equation (7-21) represents the Debye function quite well for T/Θ_D values of 0.5 or greater. This is the range not plotted in Fig. 7-5. As the temperature increases to large values, Eq. (7-21) fulfills the condition that c_v approaches $3R_u$.

The other special case where the Debye function can be closely approximated by an equation is the situation where T approaches absolute zero. This temperature region is of particular interest. Accurate measurements of c_v are seldom available for many substances in the temperature region below 20–40 K. Hence it is important to have a method for extrapolating c_v data down to 0 K. When T is small, the parameter x_m is very large. As a result, the upper limit of the integral in Eq. (7-19) may be set equal to infinity without altering the value of the integral. The resulting integral has a value of $\pi^4/15$. Therefore at extremely low temperatures the Debye specific-heat function may be represented by $c_v = 36\pi^4 R_u/15x_m^3$, or

$$c_v = 233.8R_u\left(\frac{T}{\Theta_D}\right)^3 \tag{7-22}$$

This famous equation developed in 1912 is known as Debye's *T-cubed equation.* Its range of validity is restricted to Θ_D/T values greater than around 15 (note the comment at the bottom of Table 7-1). Observed values of c_v satisfy Eq. (7-22) fairly well for a number of elements and crystalline compounds, within the limit of experimental error. A line representing the T-cubed equation is shown on Fig. 7-5. When T/Θ_D is greater than roughly 0.1, the figure shows that the T-cubed equation overpredicts the value of c_v/R_u.

The Debye specific-heat function for crystalline solids has been relatively successful in spite of the relative simplicity of the model. One inadequacy of the model is that Θ_D is not really a constant, but varies slowly with temperature. For example, Θ_D might be calculated from a c_v value at a low temperature using Eq. (7-22), and from another c_v value at a moderately high temperature using Eq. (7-21). The two Θ_D values probably will not be the same. In addition, because the model is based on an isotropic elastic medium, Θ_D can be evaluated from elastic properties. Commonly, Θ_D based on elastic data will not be in too good an agreement with values based on thermal (c_v) measurements. The theory also appears to work better for compounds when the lattice structure is cubic. Compounds with this structure include sodium and potassium chloride, calcium fluoride, and magnesium chloride, among others. Poor values of c_v result when the Debye function is applied to compounds forming rhombic structures, such as sulfur and graphite. In this latter case better results are predicted if T/Θ_D is replaced in the correlations by $(T/\Theta_D)^n$. Nevertheless, the Debye theory has been quite successful in predicting the variation of the specific heat of solids with temperature for a large number of substances.

7-6 THE ELECTRONIC SPECIFIC HEAT OF A METAL

The free, or conducting, electrons in a metal behave essentially as a gas. Consequently, it is common to speak of them collectively as an electron gas. At very low and at high temperatures these electrons contribute significantly to the specific heat of a metal. In 1928 Sommerfeld applied Fermi-Dirac statistics to this physical situation. As an approximation, he found that the average energy of an electron could be expressed by

$$\epsilon_{\text{avg}} = \frac{3\epsilon_F}{5}\left(1 + \frac{5\pi^2}{12}\left(\frac{kT}{\epsilon_F}\right)^2 + \cdots\right)$$

where k is Boltzmann's constant and the Fermi energy ϵ_F is defined as

$$\epsilon_F = \frac{h^2}{8m}\left(\frac{3n}{\pi V}\right)^{2/3} \tag{7-23}$$

where m is the mass of an electron. The total internal energy is $n\epsilon_{\text{avg}}$. Consequently, the specific heat at constant volume contributed by the electron gas would be

$$c_{v,\text{elec}} = \left(\frac{\partial u}{\partial T}\right)_v = \frac{n\pi^2 k^2 T}{2\epsilon_F}$$

On a molar basis the electron-gas specific heat becomes

$$\frac{c_{v,\text{elec}}}{R_u} = \frac{\pi^2 kT}{2\epsilon_F} \tag{7-24}$$

It is convenient to define a *Fermi characteristic temperature* Θ_F by the quantity ϵ_F/k. As a result, the molar specific heat of an electron gas based on the Sommerfeld approximation may be written in the form

$$\frac{c_{v,\text{elec}}}{R_u} = \frac{\pi^2}{2}\left(\frac{T}{\Theta_F}\right) \tag{7-25}$$

or

$$c_{v,\text{elec}} = \gamma T \tag{7-26}$$

where $\gamma = \pi^2 R_u/2\Theta_F$. Thus the electronic specific heat of an electron gas in a solid is a linear function of the temperature. Due to Sommerfield's approximation, linearity holds only when $T < 0.2\Theta_F$. However, values of Θ_F are very large. Hence the linear approximation is a valid one for solids in most applications.

The Fermi characteristic temperature of copper is about 60,000 K. Thus Eq. (7-25), when applied to copper, becomes

$$c_{v,\text{elec}}/R_u = 8.2 \times 10^{-5} T \qquad \text{(copper)}$$

and Eq. (7-26) becomes

$$c_{v,\text{elec}} = 6.8 \times 10^{-4} T \qquad \text{(copper)}$$

where γ has a value of 6.8×10^{-4} $\text{J/gmol}\cdot\text{K}^2$ and T is in degrees kelvin. At 300 K the electron-gas contribution to the specific heat would be, roughly, $0.025R_u$. This compares with a vibrational specific heat of around $3R_u$. Consequently, the contribution by the electron gas is negligible in this temperature range. However, at extremely low temperatures, of the order of 10 K or less, the effect indicated by Eq. (7-26) must be added to the Debye T-cubed term found in Sec. 7-5, since both are of the same order of magnitude. In addition, at extremely high temperatures the contribution by the electron gas will increase c_v beyond its vibrational value of $3R_u$. Table 7-2(a) lists some values of γ for some common elements.

7-7 ABSOLUTE ENTROPY EVALUATION

Concepts presented in Secs. 7-1, 7-2, and 7-5 are required for the evaluation of the absolute entropy of a pure substance from thermal data. First, Eq. (7-3) is the basic expression for determining the entropy of a substance. From Sec. 7-1

$$s = s_0 + \sum_j \int c_p \frac{dT}{T} + \sum_k \frac{\Delta h_{ab}}{T} \tag{7-3}$$

The second term summation on the right accounts for the variation of s with temperature for various phases. The third term summation accounts for the entropy changes associated with isothermal phase changes. Thus specific-heat data for various phases and enthalpy (latent heat) data for isothermal phase changes are essential.

Second, in Sec. 7-2 it is pointed out that a conventional entropy scale is established by setting the entropy at 0 K equal to zero. In Eq. (7-3) the term s_0 becomes zero. Third, for many substances the c_p (or c_v) values of a solid phase are determined for temperatures down only to around 20–40 K, due to experimental difficulties. Hence the entropy contribution at temperatures below this range must be found by extrapolation, using the Debye T-cubed equation. Fourth, the $c_p\,dT/T$ summation term in Eq. (7-3) may need to include two integrals for the solid phase, because some solids exist in more than one crystalline form. (Recall the example of phosphine, PH_3, discussed in Sec. 7-2.) Finally, it is common to report the entropy of substances in tabulations at 1 atm and 25°C. This is a standard reference state.

Consider the situation where a pure substance exists as a gas at 1 atm and 25°C. Also, assume that the substance exists in two solid phases, α and β. The entropy variation with temperature in this situation has been shown in Fig. 7-1. The series of terms in Eq. (7-3) is to be evaluated along the path *a-b-c-d-e-f-g-h-i* shown in the figure. For the substance under consideration Eq. (7-3) becomes

$$s = \int_a^b \left(\frac{c_p\,dT}{T}\right)_\alpha + \int_b^c \left(\frac{c_p\,dT}{T}\right)_\alpha + \frac{\Delta h_{\alpha\to\beta}}{T_c} + \int_d^e \left(\frac{c_p\,dT}{T}\right)_\beta$$

$$+ \frac{\Delta h_{\beta\to\text{liq}}}{T_e} + \int_f^g \left(\frac{c_p\,dT}{T}\right)_{\text{liq}} + \frac{\Delta h_{\text{liq}\to\text{gas}}}{T_g} + \int_h^i \left(\frac{c_p\,dT}{T}\right)_{\text{gas}} \tag{7-27}$$

State b represents the lowest temperature at which a c_p value is available experimentally. Consequently, Debye's T-cubed equation must be used between states a and b. When the T-cubed equation is substituted into the first term on the right of Eq. (7-27), the term reduces to the quantity $c_p^+/3$, where c_p^+ is the specific-heat value measured at the lowest temperature.

An example of the use of Eq. (7-27) is the determination of the entropy of methane gas (CH_4) at 25°C and 1 atm. Solid methane undergoes an α-β phase transition at 20.4 K. Melting of the β phase occurs at 90.6 K, and vaporization occurs at 111.8 K and 1 atm. Specific-heat data in the original investigation were available down to 10.33 K, with extrapolation below this temperature down to absolute zero by means of the Debye T-cubed equation. A summary of the contributions of the various terms in Eq. (7-27) for methane, in J/gmol·K, and based on 1931 data, are found in Table 7-3. The sum of these values, the conventional or absolute entropy of methane based on calorimetric data, thus was found to be 181.54 J/gmol·K at 25°C and 1 atm. More recent data predict a value of 186.16 J/gmol·K, due to improved measurements. Nevertheless, the method

TABLE 7-3
Absolute entropy evaluation for methane

Region of contribution	Value
Extrapolation below 10.33 K	1.51
α-β phase change at 20.4 K	3.72
Melting at 90.6 K	10.33
Vaporization at 111.8 K	73.64
Heating of solid and liquid phases from 10.33 to 111.8 K	60.88
Heating of gas phase from 111.8 to 298.15 K	31.46

outlined by Eq. (7-27) remains a valid approach, especially if the substance is still a liquid at 25°C.

PROBLEMS

7-1. On the basis of the Debye table of specific heats

a. Estimate the temperature, in kelvins, at which the c_v value of (1) beryllium and (2) manganese is 15, 30, 50, 70, and 90 percent of the classical (maximum) value. (See Table 7-2 for Θ_D values of the two elements.)

b. Use the above data to plot c_v/R_u versus T for the two elements.

7-2. On the basis of the Debye table of specific heats,

a. Estimate the temperature, in kelvins, at which the c_v value of (1) aluminum and (2) diamond is 15, 30, 50, 70, and 90 percent of its classical (maximum) value.

b. Use the above data to plot c_v/R versus T for the two elements.

7-3. On the basis of data from the Debye table of specific heats

a. Estimate the temperature, in kelvins, at which the c_v value of (1) berylium and (2) iron is 10, 30, 50, 70, and 90 percent of the classical (maximum) value.

b. Plot c_v/R_u versus T for the two elements.

7-4. Listed below are values of c_p for silver, in J/gmol·K, versus temperature, in kelvins.

T	15	20	40	60	100
c_p	0.670	1.647	8.419	14.27	20.10

a. Use the table for the Debye specific-heat function (c_v/R_u versus Θ_D/T) to ascertain the value of Θ_D, in kelvins, at the given temperatures.

b. Compare these values to the single value shown in Table 7-2.

7-5. Listed below are values of c_p for copper, in J/gmol·K, versus temperature, in kelvins.

T	20	40	60	80	100
c_p	0.462	3.740	8.595	12.85	16.01

a. Use the table for the Debye specific-heat function (c_v/R_u versus Θ_D/T) to ascertain the value of Θ_D, in kelvins, at the given temperatures.

b. Compare these values to the single value shown in Table 7-2.

7-6. Listed below are values of c_p for gold, in J/gmol·K, versus temperature, in kelvins.

T	10	15	20	40
c_p	0.431	1.474	3.187	11.22

a. Use the table for the Debye specific-heat function (c_v/R_u versus Θ_D/T) to ascertain the value of Θ_D, in kelvins, at the given temperatures.

b. Compare these values to the single value shown in Table 7-2.

7-7. The specific heat of iridium ($M = 193.1$) is listed as 0.0903 and 0.113 kJ/kg·K at 100 and 150 K, respectively.

a. From the table of the Debye specific-heat function, determine the value of Θ_D for iridium at these two temperatures, in kelvins.

b. From the average of the two values found in part (a), estimate the specific heat at 25 K, in kJ/kg·K.

c. Using the data for 100 K, evaluate Θ_E for the Einstein function, in kelvins. Then estimate c_v at 25 K using the Einstein function, in kJ/kg·K.

7-8. The following table gives some enthalpy values for solid diatomic oxygen at atmospheric pressure:

T, K	0.0	0.5	1.0	1.5	2.0	2.5	3.0
h, kJ/kg $\times 10^6$	0.000	0.9547	15.27	77.32	244.40	596.6	1237.1

a. Calculate the enthalpies at (1) 4.0 and (2) 5.0 K, in kJ/kg, on the basis of the Debye T-cubed law.

b. Estimate the specific heat at constant pressure at (1) 1.5 K and (2) 2.5 K, in kJ/kg·K.

c. Estimate the specific heat at constant pressure at (1) 1.5 K and (2) 2.5 K, based on the definition of $c_p = (\partial h/\partial T)_P$, in kJ/kg·K.

d. Estimate the entropy at (1) 1.5 K and (2) 2.5 K, in kJ/kg·K, on the basis of third-law considerations.

7-9. The following table lists some enthalpy values for germanium:

T, K	0	1.0	2.0	3.0	4.0
h, kJ/kg $\times 10^6$	0.00	0.132	2.11	10.7	34.3

a. If the specific heat $c_v = KT^3$ in this temperature range, determine the value of the constant K for the four values of h.

b. On the basis of part (a), estimate the value of h, in kJ/kg, for a temperature of 6 K. (The experimental value is about 179×10^{-6}.)

c. Estimate the specific heat at 2.0 K, in kJ/kg·K, based on third-law considerations. (The listed value is 4.23×10^{-6}.)

d. Estimate the specific heat at 2.0 K, in kJ/kg·K, based on the definition of c_p as a function of h. Why does this answer differ from that found in part (c)?

e. Estimate the entropy at 2.0 K, in kJ/kg·K, on the basis of third-law considerations.

7-10. The c_v data below, in kJ/kg·K, are for nickel as a function of temperature. The total specific heat may be expressed by $c_v = \gamma T + KT^3$. (a) Obtain an equation which expresses c_v/T versus T^2 by means of a linear least square fit of the data. From the equation obtained in part (a), determine (b) a mean Debye temperature, in kelvins, and (c) the value of the constant for the electronic specific heat, in kJ/kmol·K^2. (d) Plot the seven data points on a c_v/T versus T^2 diagram, and show the position of the line obtained in part (a):

T, K	1	2	3	4	6	8	10
$c_v \times 10^6$	120	242	369	505	820	1190	1620

7-11. The data in Table P7-11 are the specific-heat values for germanium. (a) Plot c_v/T versus T^2. By a least square fit of the data in the plot to an equation of the type $c_v = aT^3 + \gamma T$, find (b) the values of a, in J/gmol·K^4, and γ, in J/gmol·K^2, and (c) the Debye characteristic temperature, in kelvins. For the data shown T is in kelvins and c_v is in J/gmol·K.

TABLE P7-11

T	c_v	T	c_v	T	c_v
4.555	0.003831	2.999	0.001086	1.837	0.000252
4.237	0.003153	2.608	0.000701	1.661	0.000186
3.491	0.001826	2.297	0.000474	1.488	0.000132
3.288	0.001429	2.002	0.000336		

7-12. The specific heat of gold ($M = 197.2$) is listed as 19.63 and 23.59 J/gmol·K at 80 and 150 K, respectively.

a. From the table of the Debye specific-heat function, determine the value of Θ_D for gold at these two temperatures, in kelvins.

b. From the average of the two values found in part (a), estimate the specific heat at 25 K, in J/gmol·K.

c. Using the data at 80 K, evaluate the Einstein characteristic temperature Θ_E for gold, in kelvins. Then estimate c_v at 25 K using the Einstein function, in J/gmol·K.

7-13. The value of c_p for triammonium hydrogen phosphate at 13.8 K has been found to be 3.626 kJ/kmol·K. Find

a. the Debye characteristic temperature, in kelvins, on the basis of the Debye function table

b. the approximate upper temperature limit for validity of the T^3 law and
c. the absolute entropy of the substance at this temperature, in kJ/kmol·K, on the basis of the third-law convention

7-14. The value of c_p for triammonium hydrogen phosphate at 8.75 K has been found to be 0.926 kJ/kmol·K.

a. Estimate the absolute entropy of the substance at this temperature, in kJ/kmol·K on the basis of the Debye model at low temperatures.
b. Determine Θ_D, in kelvins.
c. Is the low-temperature T^3 model appropriate at this state?

7-15. Listed below are some c_v values for potassium chloride (KCl) as a function of temperature at low temperatures:

T, K	1.466	1.656	2.324	2.925	3.316	3.567	3.93	4.23	4.293
c_v	0.468	0.690	2.006	3.856	5.709	7.11	9.27	11.52	12.02

a. Make a plot of c_v/T versus T^2 based on these data, noting that the units of c_v are in mJ/gmol·K. Then determine the value of Θ_D for the substance, in kelvins, on the basis of a least square fit of the data.
b. On the basis of your answer to part (a), discuss whether any of the data points should be excluded from the analysis.

7-16. At low temperatures one must consider both the vibrational and electronic contributions to the specific heat and entropy of certain solids. Based on the Debye characteristic temperature and the electronic constant for chromium (Cr), determine

a. the temperature, in kelvins, at which the vibrational and electronic specific heats are equal
b. the temperature, in kelvins, at which the vibrational and electronic entropy contributions are equal.

7-17. At low temperatures one must consider both the vibrational and electronic contributions to the specific heat and the entropy of metals. Using the data for magnesium found in Table 7-2, determine

a. the temperature at which the vibrational and electronic specific heats are equal
b. the temperature at which the vibrational and electronic entropy contributions are equal, in kelvins.

7-18. At low temperatures one must consider both the vibrational and electronic contributions to the specific heat and entropy of certain solids. For aluminum the Debye temperature Θ_D may be taken to be 390 K, while the constant γ for the linearly varying electronic specific heat is 1.45×10^{-3} J/gmol·K^2. Determine

a. the temperature, in kelvins, at which the vibrational and electronic specific heats are equal
b. the temperature at which the vibrational and electronic entropy contributions are equal, in kelvins.

7-19. Methylammonium chloride exists in two different crystalline solid phases, β and γ, below 220.4 K. The phase transition between them occurs at 220.4 K. On the basis of the following 1-atm data, determine the entropy of solid-phase transition from a β to a γ solid at 220.4 K, in J/gmol:

$$\Delta s \text{ } (\beta \text{ solid, from 0 to 12.04 K}) = 0.280 \text{ J/gmol·K}$$
$$\Delta s \text{ } (\beta \text{ solid, from 12.04 to 220.4 K}) = 93.412 \text{ J/gmol·K}$$
$$\Delta s \text{ } (\gamma \text{ solid, from 0 to 19.55 K}) = 1.849 \text{ J/gmol·K}$$
$$\Delta s \text{ } (\gamma \text{ solid, from 19.55 to 264.5 K}) = 115.520 \text{ J/gmol·K}$$
$$\Delta s \text{ } (\gamma \text{ solid, from 220.4 to 264.5 K}) = 15.439 \text{ J/gmol·K}$$

7-20. Giauque and Stephenson measured the properties of sulfur dioxide shown in Table P7-20 (T is in kelvins and c_p is in J/gmol·K). The enthalpy of fusion is 7401 J/gmol at the normal melting point of 197.64 K and the enthalpy of vaporization is 24,937 J/gmol at 263.08 K, which is the boiling point at 1 atm.

a. Calculate the entropy of gaseous sulfur dioxide at its boiling point and 1 atm, in J/gmol·K and J/g·K.

b. If the specific heat of the gas phase is represented by the expression

$$c_p/R = 3.108 + 6.969 \times 10^{-3}T - 4.582 \times 10^{-6}T^2 + 1.035 \times 10^{-9}T^3$$

where T is in kelvins and c_p is in J/gmol·K, calculate the entropy at 298 K and 1 atm, in J/gmol·K and J/g·K.

TABLE P7-20

T	c_p	T	c_p	T	c_p
15	3.47	60	36.07	150	57.82
20	6.95	70	40.04	160	59.96
25	11.46	80	43.18	170	62.13
30	15.86	90	45.73	180	64.52
35	20.29	100	48.07	190	67.03
40	24.18	110	50.08	200 (liq)	87.74
45	27.66	120	51.88	220	87.28
50	30.79	130	53.68	240	86.86
55	33.56	140	55.69	260	86.44

Source: Giauque, W. F., and C. C. Stephenson, "Sulfur dioxide: The heat capacity of solid and liquid. Vapor pressure. Heat of vaporization. The entropy values from thermal and molecular data," *J. Am. Chem. Soc.*, **60:** 1389-94, 1938.

7-21. Giauque and Clayton measured calometrically the specific-heat data for diatomic nitrogen, and their results are listed in Table P7-21. Note that there is a solid-solid phase transition at 35.61 K. The left set of data is for the solid and liquid phases, up to the normal boiling point, whereas the right set of data is for the gas phase. Below the left set of data are the enthalpy values for the various phase changes. The original specific-heat data have been converted to units of J/gmol·K. Determine

a. the Debye characteristic temperature for nitrogen, in kelvins

b. the absolute entropy of the gas phase at 77.32 K, in J/gmol·K

c. the absolute entropy of the gas phase at 298.15 K, in J/gmol·K

TABLE P7-21

T, K	c_p	T, K	c_p	T, K	c_p
15.82	13.07	34.42	43.01	61.40	46.40
17.66	16.06	35.05	43.89	63.14	Melting point
19.51	19.15	35.33	44.64	65.02	55.77
21.94	22.79	35.61	Transition	68.41	56.27
24.49	26.49	39.13	37.44	70.28	56.27
24.85	26.69	43.27	39.02	72.69	56.74
27.14	30.00	48.07	40.80	74.57	56.86
28.32	31.55	51.88	42.22	76.58	57.24
29.89	34.05	53.55	42.93	77.32	Boiling point
31.29	36.16	55.88	43.68		
32.84	39.32	57.99	44.56		

Enthalpy of transition at 35.61 K = 228.9 J/gmol

Enthalpy of fusion at 63.14 K = 720.9 J/gmol

Enthalpy of vaporization at 77.32 K = 5,576.9 J/gmol

T, K	$c_{p/R}$	T, K	$c_{p/R}$
100	3.613	180	3.519
110	3.570	190	3.516
120	3.553	200	3.515
130	3.542	220	3.512
140	3.534	240	3.510
150	3.529	260	3.509
160	3.525	280	3.508
170	3.521	300	3.508

Source: Giauque, W. F., and J. O. Clayton, "The heat capacity and entropy of nitrogen. Heat of vaporization. Vapor pressure of solid and liquid," *J. Am. Chem. Soc.*, **55**: 4875-89, 1933.

7-22. The experimental data of Giauque and Wiebe (with converted units) are presented in Table P7-22 for the specific heat of hydrogen chloride (HCl) up to 188.07 K and 1 atm. This temperature is the normal boiling point of the substance. The specific heat of the gas phase may be represented by the equation

$$c_p = 28.167 + 1.810 \times 10^{-3}T + 1.547 \times 10^{-6}T^2$$

where T is expressed in kelvins and c_p in kJ/kmol·K. From these data for a pressure of 1 atm, and by appropriate data fitting and numerical integration, estimate the entropy change (a) from 0 to 17.3 K, (b) from 17.3 to the gas phase at 188.07 K, (c) from 188.07 to 298.15 K in the gas phase, and (d) the absolute entropy at 298.15 K.

TABLE P7-22

T, K	c_p	T, K	c_p	T, K	c_p
17.29	4.314	63.31	23.22	117.30	42.43
21.34	6.849	67.85	24.47	123.92	43.76
24.71	8.644	72.63	25.77	131.18	44.64
26.85	9.749	75.46	26.48	138.79	45.81
28.10	10.42	77.65	27.30	148.90	47.47
31.89	12.31	80.25	28.08	155.06	48.74
35.82	14.20	82.63	28.84	158.91	Fusion
39.95	15.87	84.69	29.51	163.72	58.12
44.20	17.29	87.70	30.66	171.45	58.37
48.62	18.71	88.79	31.00	171.74	58.37
51.23	19.57	92.83	32.58	178.64	58.62
56.01	21.21	98.36	Transition	185.20	58.87
58.94	21.89	103.01	40.33	188.07	Vaporization

Δh (transition) = 1190 kJ/kmol

Δh (fusion) = 1992 kJ/kmol

Δh (vaporization) = 16,150 kJ/kmol

Source: Giauque, W. F., and R. Wiebe, "The entropy of hydrogen chloride. Heat capacity from 16 K to boiling point. Heat of vaporization. Vapor pressure of solid and liquid," *J. Am. Chem. Soc.*, **50:** 101-22, 1927.

7-23. Giauque and Powell measured the properties shown in Table P7-23 for diatomic chlorine, where T is in kelvins and c_p is in J/gmol · K. The enthalpy of fusion is 6406 J/gmol at the normal melting point of 172.12 K and the enthalpy of vaporization is 20410 J/gmol at 239.05 K, which is the boiling point at 1 atm.

a. Calculate the entropy of gaseous chlorine at its boiling point and 1 atm, in J/gmol · K and J/g · K, using data fitting and numerical integration.

b. If the specific heat of the ideal-gas phase is represented by the expression

$$c_p/R = 2.881 + 6.055 \times 10^{-3}T - 7.361 \times 10^{-6}T^2 + 1.876 \times 10^{-9}T^3$$

where T is in K and c_p is in J/gmol · K, calculate the entropy at 298 K and 1 atm, in J/gmol · K and J/g · K.

TABLE P7-23

T	c_p	T	c_p	T	c_p	T	c_p
15	3.72	60	33.47	130	47.24	150	67.03
20	7.74	70	33.81	140	49.08	190	66.90
25	12.09	80	38.20	150	51.04	200	66.73
30	16.69	90	40.63	160	53.05	210	66.48
35	20.79	100	42.26	170	55.10	220	66.27
40	23.97	110	43.81	172.12	55.52	230	65.98
45	26.74	120	45.48	172.12	67.07	239	65.69
50	29.25						

Source: Giauque, W. F., and T. M. Powell, "Chlorine. The heat capacity, vapor pressure, heats of fusion and vaporization, and entropy," *J. Am. Chem. Soc.*, **61:** 1970-4, 1939.

REFERENCES

1. Richards, Th. W., *Zeitsch. Physik. Chem.*, **42**:129, 1902.
2. Nernst, H. W., *The New Heat Theorem*, E. P. Dutton & Co., New York, 1926.
3. Planck, M., *Treatise on Thermodynamics*, 3rd ed., Dover Publications, New York, 1945.
4. Einstein, A., *Ann. Phys.*, **22**:180, 1907.
5. Debye, P., *Ann. Phys.*, **39**:789, 1912.

CHAPTER 8

THERMODYNAMIC PROPERTIES OF HOMOGENEOUS MIXTURES

In this chapter we shall develop general methods for evaluating both intensive and extensive properties of mixtures. Of major interest are the definitions and uses of two intensive properties called the partial molal property and the chemical potential.

8-1 INTRODUCTION

There are many practical applications of thermodynamics to processes which involve phases containing mixtures, either gas or liquid, of two or more components. Any thermodynamic analysis of such systems, frequently called "solutions," must take into account the influence of composition on property values. As might be expected, the addition of composition as a variable greatly increases the complexity of property evaluation. Nevertheless, the development of the general theory is reasonably straightforward. The practical application of the theory is often difficult unless certain restrictions are applied. The principal special case in the thermodynamics of solutions is the "ideal solution." Although not an adequate representation for many engineering problems, the ideal-solution concept is important for two reasons. First, for some systems it is a fair approximation to reality. Second, the model acts as a reference to which real solutions may be compared. The following section develops general equations for the important extensive properties of any homogeneous phase in terms of two other extensive properties of the phase and its composition.

8-2 FUNDAMENTAL PROPERTY RELATIONSHIPS FOR SYSTEMS OF VARIABLE COMPOSITION

In Chap. 6 the four gibbsian equations for du, dh, da, and dg were of fundamental importance. On an extensive basis these equations are

$$dU = T\,dS - P\,dV \tag{6-2}$$
$$dH = T\,dS + V\,dP \tag{6-3}$$
$$dA = -P\,dV - S\,dT \tag{6-6}$$
$$dG = V\,dP - S\,dT \tag{6-9}$$

These equations are applicable to changes in closed simple systems between equilibrium states in the absence of chemical reactions. They are generally applied only to homogeneous systems; they may be applied to heterogeneous systems, however, if the composition of each phase remains fixed and the pressure and temperature are the same for each phase. For many multiphase systems the criterion of fixed composition in each phase cannot be met. For example, a change in either pressure or temperature of the total system may alter the composition of the phases present. In other instances it may be desirable to study systems which change composition because of mass transfer of one or more components between a given phase and the environment. Finally, one may also wish to account for the presence of chemical reactions within the system, which directly alter the composition of the system.

Consider a homogeneous phase in which there are a number of different substances (components). If the composition of this phase remains fixed, the internal energy as given by Eq. (6-2) above is only a function of S and V. For a system of variable composition, however, the internal energy is also a function of the moles of each of the substances present at a given time. This is expressed mathematically by

$$U = U(S, V, N_1, N_2, \ldots, N_i) \tag{8-1}$$

where the quantities N_i represent the moles of each chemical species within the system at the equilibrium state. A change of state may be brought about by altering the entropy or the volume or by changing the composition. The total differential of U is given by

$$dU = \left(\frac{\partial U}{\partial S}\right)_{V,N} dS + \left(\frac{\partial U}{\partial V}\right)_{S,N} dV + \sum_i \left(\frac{\partial U}{\partial N_i}\right)_{S,V,N_j} dN_i \tag{8-2}$$

In Eq. (8-2) the subscript "N" on the first two partial derivatives indicates that the moles of every species are held fixed during the change in either S or V. In the summation term the subscript "N_j" on the partial derivative indicates the constancy of all mole numbers except N_i, along with fixed values of S and V. When the composition is held fixed, the summation term in Eq. (8-2) is zero. Under this condition Eq. (8-2) should be identical with Eq. (6-2) above. Consequently, we find that

$$\left(\frac{\partial U}{\partial S}\right)_{V,N} = T \qquad \text{and} \qquad \left(\frac{\partial U}{\partial V}\right)_{S,N} = -P \tag{8-3}$$

Also, for conciseness, the partial derivatives which appear in the summation in Eq. (8-2) will be defined as the chemical potential μ_i. Hence

$$\mu_i = \left(\frac{\partial U}{\partial N_i}\right)_{S,V,N_j} \tag{8-4}$$

This quantity was first introduced by J. Willard Gibbs. On this basis Eq. (8-2) can now be written as

$$dU = T\,dS - P\,dV + \sum_i \mu_i\,dN_i \tag{8-5}$$

The chemical potential as defined by Eq. (8-4) may be given the following interpretation: If an infinitesimal quantity of the ith chemical species is added to a homogeneous system and the entropy and volume are held fixed, the chemical potential is a measure of the rate of increase in the internal energy of the mass of the system per unit addition of the ith substance. The importance of the chemical potential will become apparent in further discussions.

The chemical potential may also be introduced in terms of the Gibbs function G, the Helmholtz function A, and the enthalpy function H. Recall that the Gibbs function change for a system of fixed composition is given by $dG = V\,dP - S\,dT$. In order to account for variable composition, an equation analogous to Eq. (8-1) is

$$G = G(P, T, N_1, N_2, \ldots, N_i) \tag{8-6}$$

Therefore the total differential becomes

$$dG = \left(\frac{\partial G}{\partial P}\right)_{T,N} dP + \left(\frac{\partial G}{\partial T}\right)_{P,N} dT + \sum_i \left(\frac{\partial G}{\partial N_i}\right)_{P,T,N_j} dN_i$$

$$= V\,dP - S\,dT + \sum_i \left(\frac{\partial G}{\partial N_i}\right)_{P,T,N_j} dN_i \tag{8-7}$$

For a comparative purpose Eq. (8-5) may be transformed into an equation for dG by adding the quantity $d(PV - TS)$ to both sides of Eq. (8-5). As a consequence,

$$dU + d(PV - TS) = T\,dS - P\,dV + \sum_i \mu_i\,dN_i + d(PV - TS)$$

or

$$dG = V\,dP - S\,dT + \sum_i \mu_i\,dN_i \tag{8-8}$$

If the right-hand side of Eq. (8-8) is compared now term by term with the right-hand side of Eq. (8-7), it is apparent that the chemical potential is also given by

$$\mu_i = \left(\frac{\partial G}{\partial N_i}\right)_{P,T,N_j} \tag{8-9}$$

This form of the equation for the chemical potential will be particularly useful, because the pressure and temperature are experimental variables which are relatively easy to control.

From the basic definitions of H and A it can be shown that the chemical potential is defined by the following set of identities:

$$\mu_i = \left(\frac{\partial U}{\partial N_i}\right)_{S,V,N_j} = \left(\frac{\partial G}{\partial N_i}\right)_{P,T,N_j} = \left(\frac{\partial H}{\partial N_i}\right)_{S,P,N_j} = \left(\frac{\partial A}{\partial N_i}\right)_{T,V,N_j} \tag{8-10}$$

In addition, the total differentials of H and A are now represented by

$$dH = T\,dS + V\,dP + \sum_i \mu_i\,dN_i \tag{8-11}$$

$$dA = -S\,dT - P\,dV + \sum_i \mu_i\,dN_i \tag{8-12}$$

It might be noted at this point that the chemical potential is an intensive property of a phase. (This will be proved shortly by applying Euler's theorem on mathematically homogeneous functions to thermodynamic relations.) It is a function of the composition of the phase, but not of the overall quantity. Therefore it may be expressed as a function of other intensive properties. This basic relationship will be developed shortly. Also, bear in mind that the term $T\,dS$ which appears in Eq. (8-5) no longer represents a quantity of heat if dN_i is due to transfer of mass across the boundary of the phase. The next section will show that the chemical potential, when expressed in the form of Eq. (8-9), is simply one example of a whole class of intensive properties of great usefulness in the analysis of mixtures.

8-3 PARTIAL MOLAR PROPERTIES

In general, the change in any extensive property Y can be expressed as a function of the change in temperature, pressure, and the moles of each independent component. That is,

$$dY = \left(\frac{\partial Y}{\partial T}\right)_{P,N} dT + \left(\frac{\partial Y}{\partial P}\right)_{T,N} dP + \sum_i \left(\frac{\partial Y}{\partial N_i}\right)_{T,P,N_j} dN_i \tag{8-13}$$

The quantity $(\partial Y/\partial N_i)_{T,P,N_j}$ is defined as the partial property, or the partial molar property $\bar{y}_i$ of the ith component in a mixture. Thus

$$\bar{y}_i \equiv \left(\frac{\partial Y}{\partial N_i}\right)_{T,P,N_j} \tag{8-14}$$

For a system of C components a partial molar property is a function of $C + 1$ intensive variables. Also, the partial molar property $\bar{y}_i$ of a given component may be interpreted as the rate of change of the extensive property Y with respect to the addition of an infinitesimal amount dN_i of that component to the phase as the temperature, pressure, and quantities of all other components remain fixed. Therefore Eq. (8-13) may be written as

$$dY = \left(\frac{\partial Y}{\partial T}\right)_{P,N} dT + \left(\frac{\partial Y}{\partial P}\right)_{T,N} dP + \sum_i \bar{y}_i \, dN_i \tag{8-15}$$

This equation illustrates the general relationship between partial molar properties and the change in the values of extensive properties in multicomponent systems. It is pertinent, therefore, to inquire further into the characteristics of partial molar properties.

Extensive properties are properties of a phase which are directly proportional to the amount of mass, or number of moles, in the phase. Consider a phase of volume V which contains N_i moles of the ith chemical species. In general, we may write for a given phase

$$V = V(T, P, N_1, N_2, \ldots, N_i) \tag{8-16}$$

Because the components of a phase are equally distributed at equilibrium, a volume kV (where k is any arbitrary positive constant) contains kN_i moles of each component. For the kth portion of the phase Eq. (8-16) becomes

$$kV = V(T, P, kN_1, kN_2, \ldots, kN_i) \tag{8-17}$$

By comparing these last two equations, we obtain the relation

$$V(T, P, kN_1, kN_2, \ldots, kN_i) = kV(T, P, N_1, N_2, \ldots, N_i) \tag{8-18}$$

Note that the factor k multiplies only the extensive variables, that is, N_1, N_2, and so on, on the left-hand side of Eq. (8-18).

A function $f(x, y, \ldots)$ is called, mathematically, homogeneous of the mth degree in $x, y, \ldots$, if the following identity exists:

$$f(kx, ky, \ldots) = kf(x, y, \ldots) \tag{8-19}$$

For example, it may be shown that the functions $e^{y/x}$, $(x^2 + y^2)^{1/2}$ and $(x^2 + y^2)$ are homogeneous to the zeroth, first, and second degrees, respectively. When m is set equal to unity, it is seen that Eq. (8-18) for the volume of the kth portion of a phase is in the same form as Eq. (8-19) for a function of first degree. Hence all extensive properties are homogeneous functions of the first degree with respect to the number of moles of each component.

Further information on extensive properties and partial molar properties is obtained by deriving a simple form of Euler's theorem on homogeneous functions. Equation (8-19) is totally differentiated with respect to k. Since $f(x, y, \ldots)$ is not a function of k, we obtain

$$\frac{df}{dk} = mk^{m-1}f(x, y, \ldots) \tag{8-20}$$

In addition, the left-hand side may be expressed as

$$\frac{df}{dk} = \frac{\partial f}{\partial(kx)}\frac{\partial(kx)}{\partial k} + \frac{\partial f}{\partial(ky)}\frac{\partial(ky)}{\partial k} + \cdots$$

$$= x\frac{\partial f}{\partial(kx)} + y\frac{\partial f}{\partial(ky)} + \cdots$$

Therefore, in general,

$$mk^{m-1}f(x, y, \ldots) = x\frac{\partial f}{\partial(kx)} + y\frac{\partial f}{\partial(ky)} + \cdots \tag{8-21}$$

When k is unity, expression (8-21) reduces to

$$mf(x, y, \ldots) = x\frac{\partial f}{\partial x} + y\frac{\partial f}{\partial y} + \cdots \tag{8-22}$$

Equation (8-22) is a simplified form of *Euler's theorem* on homogeneous functions. Equations (8-19) and (8-22) are of particular interest from a thermodynamic point of view.

As noted above, all extensive properties are homogeneous functions of the first degree, where $m = 1$. Equation (8-1) expresses the internal energy U as a function of S, V, and all the N_i's. Thus application of Eq. (8-22) to the internal energy function leads to

$$U = S\left(\frac{\partial U}{\partial S}\right)_{V,N_i} + V\left(\frac{\partial U}{\partial V}\right)_{S,N_i} + \sum_i N_i\left(\frac{\partial U}{\partial N_i}\right)_{S,V,N_j}$$

When Eqs. (8-3) and (8-10) are substituted into this relation, we find that

$$U = TS - PV + \sum_i N_i\mu_i \tag{8-23}$$

This relation is known as *Euler's equation* in thermodynamics because it is a direct application of Euler's theorem on homogeneous functions.

In thermodynamics the function $f(x, y, \ldots)$ represents any extensive property of a phase, and $x, y, \ldots$ are the moles $N_1, N_2, \ldots, N_i$ of the components present in the phase. Consequently, the partial derivatives in Eq. (8-22) are the partial molar properties of the phase. For example, if $f(x, y, \ldots)$ is the volume and $x, y, \ldots$ are the number of moles in the phase, then

$$\frac{\partial f}{\partial x} = \left(\frac{\partial V}{\partial N_i}\right)_{T,P,N_j} = \bar{v}_i$$

As noted above, the quantity m is unity for any extensive property relation. Therefore Eq. (8-22) for the specific case of the volume becomes

$$V = N_1\bar{v}_1 + N_2\bar{v}_2 + \cdots = \sum_i N_i\bar{v}_i \tag{8-24}$$

In general terms, this means that any extensive property of a phase may be found by summing the product of the moles of each component times its partial molar value for the equilibrium state. Thus

$$Y = \sum_i N_i\bar{y}_i \tag{8-25}$$

where Y is any extensive property of a phase, and $\bar{y}_i$ is the partial molar value for the ith component. The partial molar properties must be used in general to evaluate the properties of a multicomponent phase, instead of the specific properties of the pure components. As a specific example of Eq. (8-25), it is a well-known experimental fact that the volume obtained by mixing two miscible liquids is frequently not the sum of the volumes of the individual liquids before mixing. For example, the volume of an ethyl alcohol-water mixture is less than that of the total volume of the unmixed liquids by an easily observable amount. The deviation, of course, varies with the composition of the mixture at a given temperature and pressure. In general, the specific properties of the pure components y_i should not be used in place of $\bar{y}_i$. For any phase of a multicomponent system we may now write

$$U = \sum_i N_i\bar{u}_i \qquad H = \sum_i N_i\bar{h}_i \tag{8-26}$$

$$V = \sum_i N_i\bar{v}_i \qquad S = \sum_i N_i\bar{s}_i \tag{8-27}$$

Similar equations would be valid for other extensive properties. From Eq. (8-25) it is seen that a mean (specific) molar property y_m for a mixture can be defined as

$$y_m = \sum_i x_i\bar{y}_i \tag{8-28}$$

where x_i is the mole fraction of the ith component. For example, equations for u_m, s_m, h_m, and v_m are readily obtained by dividing Eqs. (8-26) and (8-27) by the total number of moles of the mixture. On the basis of either Eq. (8-25) or Eq. (8-28) we note that a partial molar property $\bar{y}_i$ behaves as an intensive property. A proof of this will be presented shortly.

In the thermodynamic analysis of solutions, it is important to keep straight the various symbols used. In this text any extensive property of a phase is represented by a capital letter, such as Y. Partial properties are denoted by an overbar on a lowercase letter, and a subscript identifies the component, such as $\bar{y}_i$. In addition, it is necessary to have a symbol for a specific property of any component of the solution as it exists in the pure state at the temperature and pressure of the phase. These molar (or unit-mass) properties are represented by a lowercase letter with a subscript, such as y_i. In summary,

Solution extensive properties	Y;	for example, U, H, S, G
Partial properties	$\bar{y}_i$;	for example, $\bar{u}_i$, $\bar{h}_i$, $\bar{s}_i$, $\bar{g}_i$
Pure component properties	y_i;	for example, u_i, h_i, s_i, g_i

For a single-component system $Y = N_i y_i$ in terms of the above notation. For the same system Eq. (8-25) reduces to $Y = N_i \bar{y}_i$. Hence for a phase containing a single component the partial molar property equals the pure specific property. That is,

$$\bar{y}_i = y_i \qquad \text{(single-component phase)}$$

An important example of this is the Gibbs function. Equation (8-25) applied to the Gibbs function gives

$$G = \sum_i N_i \bar{g}_i \tag{8-29}$$

For a single-component phase this equation reduces to $G = N\mu$. Thus $\mu = G/N = g$ for a pure component.

Further information on partial molar properties is found by differentiation of Eq. (8-25) with respect to N_j at constant temperature, pressure, and moles of all other components. The result is

$$\sum_i N_i \frac{\partial^2 Y}{\partial N_i \, \partial N_j} = 0 \tag{8-30}$$

The proof of this statement is left to the reader. This relation can be written in two forms in terms of partial molar properties. One of these is

$$\sum_i N_i \frac{\partial \bar{y}_i}{\partial N_j} = 0 \tag{8-31}$$

or in an expanded format

$$N_1 \left(\frac{\partial \bar{y}_1}{\partial N_1} \right)_{T,P,N_j} + N_2 \left(\frac{\partial \bar{y}_2}{\partial N_1} \right)_{T,P,N_j} + \cdots = 0 \tag{8-32}$$

Recognizing that the order of differentiation is immaterial, one can also write Eq. (8-30) as

$$\sum_i N_i \frac{\partial \bar{y}_j}{\partial N_i} = 0 \tag{8-33}$$

or in an expanded format

$$N_1 \left(\frac{\partial \bar{y}_1}{\partial N_1} \right)_{T,P,N_j} + N_2 \left(\frac{\partial \bar{y}_1}{\partial N_2} \right)_{T,P,N_j} + \cdots = 0 \tag{8-34}$$

It is understood that the partial derivatives in Eqs. (8-30) to (8-33) have been taken at constant temperature and pressure, if not expressed explicitly. Of major importance is the fact that relations (8-31) and (8-32) indicate that the partial molar properties of different components in a phase at equilibrium are not independent.

When the degree of a mathematically homogeneous equation is zero ($m = 0$), Eqs. (8-22) and (8-19) become

$$x\frac{\partial f}{\partial x} + y\frac{\partial f}{\partial y} + \cdots = 0 \tag{8-35}$$

and

$$f(kx, ky, \ldots) = f(x, y, \ldots) \tag{8-36}$$

If f in this case represents $\bar{y}_j$ and $x, y, \ldots$ again are the moles of the components, then Eq. (8-34) satisfies Eq. (8-35). That is, the partial molar property $\bar{y}_j$ is a homogeneous function of zeroth degree in N_i. Consequently, any partial molar property must satisfy Eq. (8-36). But this equation simply states that a homogeneous function of zeroth degree is not a function of the size of the independent variables if they are all changed in the same proportion. Hence a partial molar property is an *intensive* property of a phase. A partial molar property is a function of the composition of a mixture, but not of the overall quantity.

Two important relations for pure substances are

$$h \equiv u + Pv \qquad \text{and} \qquad g \equiv h - Ts$$

Analogous equations are valid for mixtures in terms of partial molar properties. One starts with a relationship among extensive properties for the phase as a whole. This relationship is then differentiated with respect to the moles N_i of a given component. For example,

$$H = U + PV$$

Differentiation with respect to N_i at constant T, P, and N_j yields

$$\left(\frac{\partial H}{\partial N_i}\right)_{T,P,N_j} = \left(\frac{\partial U}{\partial N_i}\right)_{T,P,N_j} + P\left(\frac{\partial V}{\partial N_i}\right)_{T,P,N_j}$$

This result can be written in the notation

$$\bar{h}_i = \bar{u}_i + P\bar{v}_i \tag{8-37}$$

By the same method, the definition $G = H - TS$ becomes

$$\bar{g}_i = \mu_i = \bar{h}_i - T\bar{s}_i \tag{8-38}$$

A number of methods for evaluating partial molar properties are discussed in the literature. One of these methods is discussed in a following section for a binary mixture.

An illustration of the difference between $\bar{y}_i$ and y_i is given by evaluating the change in an extensive property when two or more pure components are mixed at constant temperature and pressure. Any extensive property of unmixed components is given by

$$Y^* = \sum_i N_i y_i \tag{8-39}$$

where the "*" symbol in this case denotes a system of pure, unmixed components. Equation (8-25) represents the extensive property after mixing. The change in the extensive property Y for a mixing process at constant temperature and pressure becomes

$$\Delta Y = Y - Y^* = \sum_i N_i(\bar{y}_i - y_i) \tag{8-40}$$

Particularly meaningful examples of this equation are those involving the volume and enthalpy functions. In terms of the volume

$$\Delta V = \sum_i N_i(\bar{v}_i - v_i) \tag{8-41}$$

As we have already noted in the case of ethyl alcohol-water mixtures, it is an experimental fact that ΔV has a negative value. As a result, we must conclude that $\bar{v}_i$ does not equal v_i for each component. Similarly for the enthalpy function,

$$\Delta H = \sum_i N_i(\bar{h}_i - h_i) \tag{8-42}$$

The quantity ΔH in the equation is called the *enthalpy,* or *heat, of mixing*. Again, it is experimentally observed that heat must often be added or removed from a solution upon mixing the pure components in order to maintain the temperature constant. For an exothermic mixing process (such as the addition of sulfuric acid to water) ΔH is negative. Thus $\bar{h}_i$ does not equal h_i, in general. We shall have an occasion in a later section to examine a special type of solution for which both ΔV and ΔH of mixing are zero.

For second-law considerations of adiabatic mixing processes the entropy of mixing is important. On the basis of Eq. (8-40)

$$\Delta S = \sum_i N_i(\bar{s}_i - s_i) \tag{8-43}$$

Regardless of the type of system under investigation, the entropy of mixing should always be positive for this irreversible adiabatic process, in compliance with the second law. The example below illustrates the concept of the volume change on mixing for two common substances.

Example 8-1. The partial molar volumes of ethyl alcohol and water in a mixture which is 59 percent ethyl alcohol by weight are found experimentally to be 57.3 and 17.15 $cm^3/gmol$, respectively. Determine the volume of 100 g of mixture by employing (*a*) the partial molar volume values and (*b*) the specific volumes of the pure components, all at 25°C.

Solution

(*a*) For 100 g of solution the moles of each component are

$$\text{Moles ethyl alcohol} = \frac{59}{46.07} = 1.282$$

$$\text{Moles water} = \frac{41}{18.02} = 2.275$$

On the basis of Eqs. (8-26)

$$V = \sum_i N_i \bar{v}_i = 1.282(57.3) + 2.275(17.15) = 112.4 \text{ cm}^3$$

(*b*) At 25°C the specific gravity of pure ethyl alcohol and of water is 0.78507 and 0.99708, respectively. Therefore the total volume of 100 g of solution based on pure components is

$$V^* = \sum_i V_i = \sum_i \frac{m_i}{\rho_i} = \frac{59}{0.78507} + \frac{41}{0.99708} = 116.3 \text{ cm}^3$$

Therefore a 3.5 percent error would occur if the final volume upon mixing were based on the pure specific volumes rather than on the partial molar volumes.

Finally, there are several additional equations involving the partial molar Gibbs function that are quite useful. Equation (8-29) is a completely general expression for G for any homogeneous phase at a given temperature and pressure. Differentiation at constant temperature and pressure, but varying composition, yields

$$dG_{T,P} = \sum_i N_i \, d\bar{g}_i + \sum_i \bar{g}_i \, dN_i \tag{8-44}$$

where $\bar{g}_i = \mu_i$. In addition, Eq. (8-8) dictates that at constant temperature and pressure

$$dG_{T,P} = \sum_i \mu_i \, dN_i \tag{8-8}$$

A comparison of the two preceding equations shows that both of them can be generally valid only if

$$\sum_i N_i \, d\mu_i = 0 \tag{8-45}$$

This is known as the *Gibbs-Duhem* equation. On a mole-fraction basis

$$\sum_i x_i \, d\mu_i = 0 \tag{8-46}$$

This is a widely used form of the Gibbs-Duhem equation. It is applied particularly to problems involving phase equilibrium.

Having established the concept of partial molar properties, we are now in a position to develop two equations of considerable usefulness. These equations relate the change in the chemical potential with changes in temperature and pressure to partial molar properties. The quantity dG is an exact differential. Therefore the test for exactness may be used on the right-hand side of Eq. (8-8). The two important results are

$$\left(\frac{\partial \mu_i}{\partial T}\right)_{P,N} = -\left(\frac{\partial S}{\partial N_i}\right)_{T,P,N_j} = -\bar{s}_i \tag{8-47}$$

and

$$\left(\frac{\partial \mu_i}{\partial P}\right)_{T,N} = \left(\frac{\partial V}{\partial N_i}\right)_{T,P,N_j} = \bar{v}_i \tag{8-48}$$

where the subscript "N" again denotes the constancy of all mole numbers. The chemical potential can be considered a function of the temperature, pressure, and $C - 1$ components if expressed in terms of intensive variables. Consequently, the total differential of μ_i is

$$d\mu_i = \left(\frac{\partial \mu_i}{\partial T}\right)_{P,x_j} dT + \left(\frac{\partial \mu_i}{\partial P}\right)_{T,x_j} dP + \sum_i \left(\frac{\partial \mu_i}{\partial x_i}\right)_{T,P,x_j} dx_i \tag{8-49}$$

where the summation on the right is only over $(C-1)$ components. By substituting Eqs. (8-47) and (8-48) into Eq. (8-49), we find that

$$d\mu_i = -\bar{s}_i\, dT + \bar{v}_i\, dP + \sum_i \left(\frac{\partial \mu_i}{\partial x_i}\right)_{T,P,x_j} dx_i \tag{8-50}$$

For a system undergoing composition changes at constant temperature and pressure Eq. (8-50) reduces to

$$d\mu_i = \sum_i \left(\frac{\partial \mu_i}{\partial x_i}\right)_{T,P,x_j} dx_i \tag{8-51}$$

In Eqs. (8-50) and (8-51) the summation again is over $C - 1$ components.

8-4 PARTIAL MOLAR PROPERTY EVALUATION FOR BINARY PHASES

Partial molar properties can be ascertained from experiments measuring changes in extensive properties which occur when a pure substance is added to a solution. The development which follows is in terms of the volume function, but the general approach applies equally to the enthalpy, entropy, Gibbs function, and even the compressibility factor Z.

On the basis of the definition of the partial molar volume, $(\partial V/\partial N_i)_{T,P,N_j}$, a direct method of evaluation is to plot V versus N_i. The slope of the resulting curve at a given value of N_i is a measure of $\bar{v}_i$. The slope can be ascertained from the procedures of either numerical or graphical differentiation. The difficulty with this method is that the experimental data must be taken with great precision if the slope is to be found with reasonable accuracy.

An alternative technique known as the *tangent method* is quite useful. In this particular case we seek a relationship for the unknown partial molar volume, $\bar{v}_i$, in terms of the measured specific volume of the mixture, v_m, at a measured mole fraction, x_i. The relationship between the measured data and the desired value of the partial molar property is found to be particularly simple for binary (two-component) systems.

The mean or specific molar property y_m of a mixture was defined previously as

$$y_m = \sum_i x_i \bar{y}_i \tag{8-28}$$

If this equation is applied to the volume of a mixture with components 1 and 2, and it is noted that $x_1 = 1 - x_2$, then

$$v_m = (1 - x_2)\bar{v}_1 + x_2\bar{v}_2 \tag{8-52}$$

Next, differentiation of this equation with respect to x_2 yields

$$\frac{\partial v_m}{\partial x_2} = (1 - x_2)\left(\frac{\partial \bar{v}_1}{\partial x_2}\right)_{T,P} - \bar{v}_1 + \bar{v}_2 + x_2\left(\frac{\partial \bar{v}_2}{\partial x_2}\right)_{T,P} \tag{8-53}$$

Another equation relating the same variables is obtained from a modified form of Eq. (8-35). On a mole fraction basis and for a binary system Eq. (8-35) may be written in the form

$$(1 - x_2)\left(\frac{\partial \bar{v}_1}{\partial x_2}\right)_{T,P} + x_2\left(\frac{\partial \bar{v}_2}{\partial x_2}\right)_{T,P} = 0 \tag{8-54}$$

When Eq. (8-54) is subtracted from Eq. (8-53) we find that

$$\left(\frac{\partial v_m}{\partial x_2}\right)_{T,P} = \bar{v}_2 - \bar{v}_1 \tag{8-55}$$

Finally, by solving Eqs. (8-52) and (8-55) simultaneously in order to eliminate $\bar{v}_2$, we arrive at the desired expression

$$\bar{v}_1 = v_m - x_2\left(\frac{\partial v_m}{\partial x_2}\right)_{T,P} \tag{8-56}$$

If $\bar{v}_1$ were eliminated in a similar fashion, the final result would be

$$\bar{v}_2 = v_m + (1 - x_2)\left(\frac{\partial v_m}{\partial x_2}\right)_{T,P} \tag{8-57}$$

Consequently, $\bar{v}_1$ and $\bar{v}_2$ are expressed as functions of the mixture specific volume and the mole fraction of one of the components, as desired.

Both Eqs. (8-56) and (8-57) contain a partial derivative which must be evaluated. Fortunately, there is a simple graphical or geometrical interpretation of

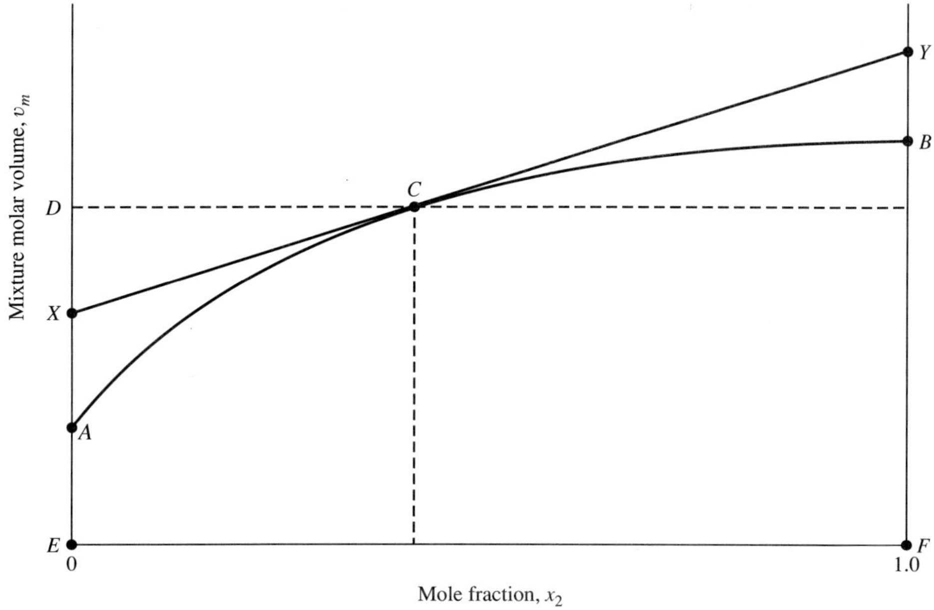

FIGURE 8-1
Mixture molar volume-composition diagram for a binary system.

these two equations. Figure 8-1 is a plot of v_m versus the mole fraction x_2 obtained from typical experimental measurements. The curve ACB has been drawn through the experimental data. Then a tangent XCY is drawn to the experimental curve at the desired mole fraction. It is easily seen that the first term on the right-hand side of Eq. (8-56) is the length DE, and the second term on the right is represented by the length DX. Therefore, according to Eq. (8-56), $\bar{v}_1$ is represented by the length XE. But this is simply the value of v_m at the intercept of the tangent line where $x_1 = 1.0$. In a similar fashion, it may be shown by means of Eq. (8-57) that the value of v_m at the intercept Y where $x_2 = 1.0$ is a measure of $\bar{v}_2$. Thus partial molar volumes of a binary mixture are given by the intercepts of the tangent drawn to the experimental curve of v_m versus x_2 at the given value of x_2.

The value of $\bar{v}_i$ obtained by this or other methods is an absolute value. This would also be true of a partial compressibility factor, $\bar{Z}_i$. However, this method would lead only to changes in enthalpy, entropy, or Gibbs function, because only changes in the mixture property can be measured and plotted. For example, in the case of enthalpy one would measure the "enthalpy of mixing" versus the mole fraction. As a result, the intercept of the tangent at $x_2 = 0$ on a Δh_m versus x_2 plot would lead to a value of $(\bar{h}_1 - h_1)$, and not $\bar{h}_1$ itself, where h_1 is the pure specific enthalpy at the given pressure and temperature. Nevertheless, Eqs. (8-56) and (8-57) are of general validity. They can be generalized for any extensive property Y. Hence

$$\bar{y}_1 = y_m - x_2\left(\frac{\partial y_m}{\partial x_2}\right)_{T,P} \tag{8-58}$$

and

$$\bar{y}_2 = y_m + (1 - x_2)\left(\frac{\partial y_m}{\partial x_2}\right)_{T,P} \tag{8-59}$$

where y_m is the measurable mixture specific property. It might also be noted that the above analysis could have been done entirely on a mass basis rather than on a molar basis.

The method of tangents discussed above requires an accurate placement of the tangent itself. Another approach is to use a least square method to fit an equation to the data in the vicinity of the state of interest. Differentiation of the equation leads to a reasonable value of the required partial derivative, $(\partial y_m/\partial x_2)$, shown above.

8-5 FUGACITY AND FUGACITY COEFFICIENT

It has been demonstrated in preceding sections that the chemical potential plays an extremely important role in defining equilibrium conditions in certain situations. These include the studies of phase equilibrium, membrane equilibrium, and chemical equilibrium. In order to predict property values of interest in such studies, one must be able to evaluate the chemical potential or functions directly related to it.

Equations relating the chemical potential to other measurable properties are developed in the following manner. Recall that the Gibbs function is of particular interest in chemical thermodynamics due to its direct functional relationship to temperature and pressure. For a pure, simple compressible substance this is expressed by

$$dg = v\,dP - s\,dT \tag{6-9}$$

Also, recall that for a pure substance the chemical potential equals the pure specific Gibbs function. That is, μ equals g. Therefore for a pure substance

$$d\mu = v\,dP - s\,dT \tag{8-60}$$

For any pure substance undergoing an isothermal change of state this equation reduces to

$$d\mu_T = v\,dP \tag{8-61}$$

For an ideal gas, $v = RT/P$. Therefore Eq. (8-57) takes the form

$$d\mu_{T,\text{id}} = \frac{RT\,dP}{P} = RT\,d\ln P \qquad \text{(ideal gas)} \tag{8-62}$$

This equation enables us to evaluate the chemical potential of a pure gas which approximates ideal-gas behavior, relative to some arbitrary reference state. As the pressure increases at a given temperature, the equation $Pv = RT$ may be in considerable error. Thus Eq. (8-62) also may not be acceptable for predicting real-gas behavior if reasonable accuracy is desired.

One method of overcoming this difficulty is to use a more accurate equation of state relating v to P and T for the real-gas state. Such an equation could be substituted directly into the $v\,dP$ term in Eq. (8-61), as we did above for the ideal-gas state. In order to achieve the desired accuracy, however, one might find it necessary to use different equations of state for various substances. Thus the equation for the chemical potential would have different formats, depending on the equation of state used. Experience has shown that there is a more fruitful approach to the problem.

In 1901 G. N. Lewis [1] suggested that it would be more convenient if the mathematical format of Eq. (8-62) were retained for real-gas behavior. To accomplish this, the pressure P in Eq. (8-62) is replaced by a new thermodynamic function called the *fugacity*, f. Thus, by definition, for a real gas

$$d\mu_{T,\text{real}} = RT\,d\ln f \tag{8-63}$$

This equation relates only the change in the chemical potential to a change in the fugacity function. The absolute value of the fugacity can be established if the following important condition or restriction on the fugacity is considered. If Eq. (8-63) is to be of general validity, then it should apply to the ideal-gas state as well as to the real-gas state. For an ideal gas we can equate Eqs. (8-62) and (8-63). As a result,

$$RT\,d\ln f = RT\,d\ln P \qquad \text{(ideal gas)}$$

Indefinite integration of this expression leads to

$$\ln f = \ln P + \ln c$$

or

$$f = cP \qquad \text{(ideal gas)}$$

where c is a constant. The only state where all real gases are truly ideal gases is at zero pressure. As we approach this state, f and P should take on the same value and c must be unity. This condition is expressed formally by the statement

$$\lim_{P\to 0}\frac{f}{P} = 1 \tag{8-64}$$

This relationship completes the definition of the fugacity. Note that the fugacity function is expressed in pressure units. In practice, the fugacity may have essentially the same value as the pressure at system pressures well above zero. As long as the gas behaves ideally (within the accuracy desired), the fugacity equals the pressure. On the other hand, many gases are not ideal gases even at atmospheric pressure and a given temperature. So even at pressures close to atmospheric, one may have to use the fugacity concept to evaluate accurately the chemical potential.

One method of evaluating the fugacity of a substance is suggested by combining Eqs. (8-61) and (8-63). The isothermal variation of fugacity with pressure is then given by

$$d \ln f = \frac{v\,dP}{RT} \tag{8-65}$$

or upon integration between arbitrary pressure limits of P^* and P

$$RT \ln \frac{f}{f^*} = \int_{P^*}^{P} v\,dP \tag{8-66}$$

This equation, in conjunction with an appropriate equation of state, may be used to evaluate the fugacity of a real gas. The accuracy of the result depends on the accuracy of the chosen equation of state. Note that Eq. (8-66) requires that the equation of state be explicit in the variable v; that is, $v = f(P, T)$. However, many PvT equations of state presented in the literature are explicit in pressure. That is, they are of the form $P = f(v, T)$. In order to use an equation of state in this latter format, the first step is to transform the independent variable in Eq. (8-66) from P to v. This change in the independent variable is accomplished by integration by parts. This technique allows the integral in Eq. (8-66) to be written in the form

$$\int_{P^*}^{P} v\,dP = Pv - P^*v^* - \int_{v^*}^{v} P\,dv \tag{8-67}$$

where v^* is the specific volume at P^* and T. The last term on the right of the above equation requires an equation of state in the format $P = f(v, T)$.

Some two-constant equations of state are reasonably accurate for predicting the PvT behavior of a real gas. An example of this is the Redlich-Kwong (RK) equation, which is in the format $P = f(v, T)$. The equation itself is

$$P = \frac{RT}{v - b} - \frac{a}{T^{0.5}v(v + b)} \tag{5-20}$$

When this equation of state is used in conjunction with the integration by parts shown above and Eq. (8-66), then integration yields

$$RT \ln \frac{f}{f^*} = Pv - P^*v^* - RT \ln \frac{(v - b)}{(v^* - b)} - \frac{a}{bT^{0.5}} \ln \frac{(v + b)v^*}{(v^* + b)v}$$

Now, as $P^* \to 0$, then $v^*/(v^* + b) \to 1$, $P^*v^* \to RT$, $v^* \to RT/P^*$, and $f^* \to P^*$. As a result of these simplifications, the equation for the fugacity reduces to

$$\ln f = \frac{Pv}{RT} - 1 + \ln \frac{RT}{v - b} - \frac{a}{bRT^{1.5}} \ln \frac{v + b}{v}$$

Finally, when P in the above equation is replaced by the equation of state, then the first two terms on the right reduce to

$$\frac{Pv}{RT} - 1 = \frac{b}{v - b} - \frac{a}{RT^{1.5}(v + b)}$$

Therefore for a real gas modeled by the RK equation

$$\ln f_{\mathrm{RK}} = \frac{b}{v-b} + \ln\frac{RT}{v-b} - \frac{a}{RT^{1.5}}\left(\frac{1}{v+b} + \frac{1}{b}\ln\frac{v+b}{v}\right) \tag{8-68}$$

Note that if the state of the system is specified by P and T, then the equation of state must be solved first for v before the value of f can be determined from Eq. (8-68). In comparison, if the van der Waals equation of state, namely,

$$P = \frac{RT}{v-b} - \frac{a}{v^2} \tag{5-10}$$

were used, the result by an analysis similar to that for an RK gas would be

$$\ln f_{\mathrm{vdW}} = \frac{b}{v-b} + \ln\frac{RT}{v-b} - \frac{2a}{RTv} \tag{8-69}$$

The following example illustrates the evaluation of the fugacity for a gas which is modeled by the Redlich-Kwong and van der Waals equations of state.

Example 8-2. Determine the fugacity, in bars, for Refrigerant-134a (R-134a) at 90°C and 10 bars by means of (a) the Redlich-Kwong and (b) the van der Waals equations of state.

Solution

(a) The use of Eq. (8-68) requires values of a and b for R-134a for a Redlich-Kwong gas. From standard tables (or by direct evaluation) these are found to be

$$a = 197.10 \text{ bars } (\text{m}^3/\text{kmol})^2(\text{K})^{0.5} \qquad \text{and} \qquad b = 0.06634 \text{ m}^3/\text{kmol}$$

In addition, the specific volume at 10 bars and 90°C must be determined from the equation of state. Substituting values of a and b given above into the Redlich-Kwong equation, one finds that $v = 2.724$ m^3/kmol. As a result, Eq. (8-68) becomes

$$\ln f_{\mathrm{RK}} = \frac{0.06634}{2.724 - 0.06634} + \ln\frac{0.08314(363)}{2.724 - 0.06634} - \frac{197.10}{0.08314(363)^{1.5}}$$

$$\times\left(\frac{1}{2.724 + 0.06634} + \frac{1}{0.06634}\ln\frac{2.724 + 0.06634}{2.724}\right)$$

or $\quad \ln f_{\mathrm{RK}} = 0.0250 + 2.4298 - 0.2472 = 2.208$

Then

$$f = 9.09 \text{ bars}$$

For comparison to results from a later example, the value of f/P at this state is $9.09/10 = 0.909$, when R-134a is modeled as a Redlich-Kwong gas.

(b) Equation (8-69) is applicable for a van der Waals gas. The values of a and b for this type of gas are

$$a = 10.05 \text{ bars } (\text{m}^3/\text{kmol})^2 \qquad \text{and} \qquad b = 0.0957 \text{ m}^3/\text{kmol}$$

In addition, the specific volume calculated from the van der Waals equation of state at the given pressure and temperature is 2.764 $m^3/kmol$. Substituting these three values into Eq. (8-69) yields

$$\ln f_{\text{v dW}} = \frac{0.0957}{2.764 - 0.0957} + \ln \frac{0.08314(363)}{2.764 - 0.0957} - \frac{2(10.05)}{0.08314(363)(2.764)}$$

or $$\ln f_{\text{v dW}} = 0.0359 + 2.4258 - 0.2409 = 2.221$$

As a result,

$$f_{\text{v dW}} = 9.21 \text{ bars} \qquad \text{and} \qquad (f/P)_{\text{v dW}} = 0.921$$

In a later example we shall compare these two methods against three others which are reasonably accurate over a wide range of conditions. It should be noted that a more accurate value of v (from the R-134a superheat table) is close to 2.70 $m^3/kmol$. Therefore both the Redlich-Kwong and the van der Waals equations are predicting reasonably accurate values of the specific volume at this particular state. Consequently, the f and f/P values determined above are probably fairly accurate.

Another approach to the evaluation of the fugacity of a real gas is through a function known as the fugacity coefficient. The quantity f/P [as first noted in Eq. (8-64)] is defined as the *fugacity coefficient* ϕ for a pure substance. That is,

$$\phi \equiv \frac{f}{P} \tag{8-70}$$

Rather than evaluate the fugacity directly, it is found more convenient in some situations first to determine f/P. Values of f/P as a function of the system pressure and temperature may be determined from PvT data for the substance. If enthalpy and entropy data have already been evaluated from experimental PvT data, then the tabulated h and s data themselves can be used to estimate the fugacity coefficient for a given state. In order to relate ϕ to P and T, logarithmic differentiation of Eq. (8-70) is carried out. The result is

$$d \ln \phi = d \ln f - d \ln P \tag{8-71}$$

The first term on the right is given by Eq. (8-65), namely,

$$d \ln f = \frac{v\, dP}{RT} \tag{8-65}$$

When this result is substituted into Eq. (8-71), we find that

$$d \ln \phi = \left(\frac{v}{RT} - \frac{1}{P} \right) dP \tag{8-72}$$

This equation is now integrated from zero pressure, where Eq. (8-64) indicates that $\phi = 1$, to a state of pressure P. The resulting equation is

$$\ln \frac{f}{P} = \int_0^P \left(\frac{v}{RT} - \frac{1}{P} \right) dP \tag{8-73}$$

Equation (8-73) is of fundamental importance, because it relates PvT data to the logarithm of the fugacity coefficient.

The actual integration of Eq. (8-73) for real gases may be carried out in a number of ways. Graphical integration of experimental PvT data is always a possibility. In addition to the direct use of PvT data in this equation, an alternative method is to recast Eq. (8-73) in terms of the compressibility factor Z. By definition, $Z = Pv/RT$. Therefore

$$\frac{v}{RT} - \frac{1}{P} = \frac{Z-1}{P}$$

Thus Eq. (8-73) becomes

$$\ln \frac{f}{P} = \int_0^P \frac{Z-1}{P}\, dP \tag{8-74}$$

Use of Eq. (8-74) requires that Z data be calculated from experimental PvT data. Then $(Z - 1)/P$ is plotted versus P at a given temperature. The area under the curve from zero pressure to any specified pressure P is a measure of $\ln f/P$ at the specified temperature and pressure.

An alternative form of Eq. (8-74) in terms of the reduced pressure P_r $(= P/P_C)$ is

$$\ln \frac{f}{P} = \int_0^{P_r} \frac{Z-1}{P_r}\, dP_r \tag{8-75}$$

where the integration is carried out at a constant value of $T_r = T/T_c$. A general correlation of Z versus P_r and T_r can now be used to evaluate the integral. Fortunately, this integral must also be evaluated when values of the residual entropy, $s^* - s$, are determined [see Eq. (6-51)]. Hence creation of a generalized entropy chart also provides data for the creation of a generalized fugacity coefficient chart. The result of such an evaluation is found in Reference [3].

Under certain constraints Eq. (8-75) can be approximated by the following technique. Examination of a generalized Z chart shows that lines of constant T_r are fairly linear when P_r is less than roughly 0.4 in value. Thus the slope of an isotherm on a Z-P_r plot at reasonably low pressures is given by $(Z - 1)/P_r$. Because this quantity is a constant, the right-hand side of Eq. (8-75) is easily integrated. The result is

$$\ln \frac{f}{P} = Z - 1 \qquad \text{or} \qquad \frac{f}{P} = \exp(Z - 1)$$

In addition, the exponential quantity can be expanded as a series of terms such that

$$\frac{f}{P} = 1 + (Z-1) + \frac{(Z-1)^2}{2!} + \frac{(Z-1)^3}{3!} + \cdots$$

Finally, when $Z \geq 0.9$, only the first two terms on the right-hand side are significant, within an error of less than 1 percent. Consequently, when $P_r \leq 0.4$ and $Z \geq 0.9$, the fugacity coefficient is closely approximated by the relation

$$\frac{f}{P} = Z \tag{8-76}$$

This relation predicts reasonably accurate values of f/P at low pressures. In addition, only the critical pressure and temperature of the gas are required.

Tables for the properties of gases in the superheat region have been prepared for a limited number of substances. The method of generating such tables from experimental PvT and specific-heat data has been discussed in Sec. 6-8. These tables usually present the properties v, h, and s as a function of pressure and temperature. Tabular data of this type can be used to estimate the fugacity of a pure substance with reasonable accuracy. The method requires integration of Eq. (8-63) between a suitable reference state and the actual state of interest. Thus Eq. (8-63) becomes

$$\mu_{T,P} - \mu_{\text{ref},T} = RT \ln \frac{f}{f_{\text{ref}}} \tag{8-77}$$

Recall that the chemical potential equals the specific Gibbs function, g, for a pure substance. Therefore Eq. (8-77) may be written in the form

$$\ln \frac{f}{f^{**}} = \frac{g - g^{**}}{RT}$$

where the reference state is now symbolized by a superscript "**." The reference state is chosen to be at a pressure low enough so that the gas behaves as an ideal gas at the given temperature. This choice of the reference state is convenient, because f equals P for an ideal gas. Therefore $f^{**} = P^{**}$. Also, g may be replaced by $(h - Ts)$, by definition. Hence the above equation may be written as

$$\ln \frac{f}{P^{**}} = \frac{1}{R}\left[\frac{h - h^{**}}{T} - (s - s^{**})\right] \tag{8-78}$$

In practice, the values of h and s are read from the superheat table at the pressure and temperature for which f is to be evaluated. The values of h^{**} and s^{**} are read at the lowest possible pressure (P^{**}) available in the table and the specified temperature. If the gas is not an ideal gas at the lowest available pressure data, then the value of f calculated will be somewhat in error. To reduce this error, one may find sufficient data in some superheat tables to extrapolate h and s values to a lower pressure. In such cases the extrapolated values of h and s are the desired values of h^{**} and s^{**}. This allows a more accurate evaluation of f/P.

Example 8-3. (*a*) Determine the fugacity and the fugacity coefficient for Refrigerant 134a (R-134a) at 90°C and 10 bars from tabular h and s data. (*b*) Estimate the fugacity coefficient for the same state from a generalized chart.

Solution

(*a*) At 10 bars and 90°C the values of h and s are 324.01 kJ/kg and 1.0707 kJ/kmol·K, respectively. Next, we assume that R-134a is an ideal gas at 90°C and the lowest tabulated pressure for this temperature. In this case the appropriate pressure is 0.60 bars in the superheat table. [At this pressure and temperature the value of $Z = Pv/RT$ is close to 0.995, based on PvT data from the superheat table. Hence R-134a behaves nearly as an ideal gas at this low-pressure state.] Then appropriate values of h^{**} and s^{**} are found to be 335.00 kJ/kg and 1.3224 kJ/kg·K, respectively. Hence on the basis of Eq. (8-78)

$$\ln \frac{f}{P^{**}} = \frac{102.03}{8.3145}\left[\frac{324.01 - 335.00}{363.15} - (1.0707 - 1.3224)\right] = 2.717$$

and

$$\frac{f}{P^{**}} = 15.14$$

Because P^{**} is 0.6 bar and P is 10.0 bars, then

$$f = 15.14(0.6) = 9.08 \text{ bars}$$

and

$$\frac{f}{P} = \frac{9.08}{10.0} = 0.908$$

(*b*) The critical pressure and temperature of R-134a are 40.64 bars and 374.3 K, respectively. The reduced pressure and temperature at 10.0 bars and 90°C (363 K) are 0.246 and 0.970, respectively. The generalized fugacity coefficient chart indicates a value for f/P of slightly above 0.92 for the specified state. Therefore at 10 bars the value of f is roughly 9.2 bars. This value is slightly higher than the range of values between 9.08 and 9.2 calculated by the three preceding methods. Nevertheless, all four methods are in substantial agreement.

In an earlier discussion [Sec. 5-5] the Pitzer acentric factor ω was introduced as a third parameter for correlating PvT data. The acentric factor is supposed to account for the deviation in behavior of normal fluids from that of a simple fluid. Recall that the compressibility factor Z is correlated with the acentric factor according to the expression

$$Z = Z^{(0)} + \omega Z^{(1)} \tag{5-30}$$

where $Z^{(0)}$ is the value of Pv/RT for a simple fluid. Correlations for the enthalpy residual function, $(h - h^*)$, and the entropy residual function, $(s - s^*)$, have also been presented by Pitzer, and others, in terms of the acentric factor.

Because of the relationship between Z and f/P shown by Eq. (8-72), it is reasonable to expect that the fugacity coefficient can be correlated in terms of the acentric factor and the reduced pressure and temperature. The format used is

$$\log \frac{f}{P} = \left(\log \frac{f}{P}\right)^{(0)} + \omega \left(\log \frac{f}{P}\right)^{(1)} \tag{8-79}$$

where the common (base 10) logarithm is used in preparing tables. The tabular presentation of $\log(f/P)^{(0)}$ and $\log(f/P)^{(1)}$ data by Lee and Kesler [2] appears in the Appendix as Tables A-31 and A-32. The example below illustrates the method.

Example 8-4. Determine the fugacity coefficient and the fugacity of R-134a at 90°C and 10 bars by applying data from the Lee-Kesler correlation.

Solution. From Example (8-3) the reduced pressure and temperature at the given state were found to be 0.246 and 0.970, respectively. Linear interpolation in the Lee-Kesler tables shows that $\log(f/P)^{(0)} = -0.0408$ and $\log(f/P)^{(1)} = -0.0111$. Substitution of these values and the value of 0.326 for ω into Eq. (8-79) yields

$$\log \frac{f}{P} = -0.0408 + 0.326(-0.0111) = -0.0444$$

Hence

$$\frac{f}{P} = 0.903 \qquad \text{and} \qquad f = 9.03$$

This value is in substantial agreement with preceding calculations for this same quantity.

The chemical potential of a pure real gas relative to its value in an arbitrary reference state is found through integration of Eq. (8-63). The result is

$$\mu_{T,P} - \mu_{\text{ref},T} = RT \ln \frac{f}{f_{\text{ref}}} \tag{8-80}$$

In a similar fashion, integration of Eq. (8-62) for the chemical potential of a pure ideal gas yields

$$\mu_{\text{id},T,P} - \mu_{\text{id,ref},T} = RT \ln \frac{P}{P_{\text{ref}}} \qquad \text{(ideal gas)} \tag{8-81}$$

The standard reference state for a pure real gas and for a pure ideal gas is usually chosen to be the state of an ideal gas at unit pressure. The pressure unit historically has been a standard atmosphere. (Other pressure references, such as 1 bar, could be chosen.) As a result, the reference value of μ in the two preceding equations is equal to g_T^o, where the superscript "o" signifies an ideal gas at unit pressure, usually 1 atm. (Note, however, that this reference state might be a hypothetical state. Some substances might not behave as ideal gases at 1 atm, and other substances might not even exist as a gas at the given temperature and 1-atm pressure.) As a result of this choice of a standard reference state, the chemical potential of a pure ideal gas $\mu_{T,P,\text{id}}$ is given by

$$\mu_{T,P,\text{id}} = g_{T,P,\text{id}} = g_T^o + RT \ln P \qquad \text{(ideal gas)} \tag{8-82}$$

where the value of P must be measured in the same unit that is chosen for the standard reference state (usually atmospheres or bars).

In addition, recall that in any ideal-gas state the fugacity equals the pressure. Thus the reference state is also one of unit fugacity; that is, $f_{\text{ref}} = 1$. Consequently, Eq. (8-80) for a pure real gas may be written in the form

$$\mu_{T,P} = g_T^o + RT \ln f \qquad \text{(real gas)} \tag{8-83}$$

where, again, the pressure unit on f is the same as that used to define the standard reference state. Finally, if Eq. (8-82) for an ideal gas is subtracted from Eq. (8-83) for a real gas, then we find that

$$\mu_{T,P} - \mu_{T,P,\text{id}} = RT \ln \frac{f}{P} = RT \ln \phi \tag{8-84}$$

Thus the difference in value of the chemical potential of a real gas and a hypothetical ideal gas at the given temperature and pressure is directly related to the fugacity coefficient at that state. In the limit as $P \to 0$, $\phi \to 1$ and $\mu_{\text{real}} \to g_{\text{id}}$.

8-6 FUGACITY OF LIQUIDS AND SOLIDS

Equations for the fugacity of a compressed liquid or a solid are developed from one of the basic equations presented in Sec. 8-5, namely,

$$d \ln f = \frac{v\, dP}{RT} \tag{8-65}$$

For example, the fugacity of a compressed liquid relative to that of a saturated liquid at the given temperature is found by integrating the above equation from the saturation vapor pressure to the actual pressure at the given temperature. The result is

$$\ln \left(\frac{f}{f_{\text{sat}}} \right)_{\text{liq}} = \frac{1}{RT} \int_{P_{\text{sat}}}^{P} v\, dP \tag{8-85}$$

This equation would apply equally well for the solid phase, where v would become the solid specific volume. For most liquids (and solids) the specific volume is fairly insensitive to pressure at relatively low pressures. Therefore, as a good approximation,

$$\ln \left(\frac{f}{f_{\text{sat}}} \right)_{\text{liq}} \approx \frac{v(P - P_{\text{sat}})}{RT} \tag{8-86}$$

Now, consider the case where the actual pressure P is not too much greater than the saturation pressure at the given temperature. Note also that v is a small number for liquids and solids, compared to the gas phase. Under these conditions the right-hand side of Eq. (8-86) is close to zero in value and the fugacity of the substance in the compressed liquid state is essentially the same as that for the saturated liquid. That is, as a further approximation,

$$f^{\text{liq}} \approx f_{\text{sat}}^{\text{liq}} \tag{8-87}$$

This result is not too surprising, because other properties of the compressed liquid state, such as v, h, and s, may be approximated by values of the saturated liquid.

Next, consider a pure liquid in equilibrium with its vapor. Recall from Sec. 8-5 that

$$d\mu = v\,dP - s\,dT \tag{8-60}$$

and

$$d\mu = RT\,d\ln f \tag{8-63}$$

Since vaporization occurs at constant temperature and pressure, Eq. (8-60) dictates that $d\mu = 0$ during such a phase change. Thus the chemical potentials (and also the specific Gibbs functions) of each phase are equal when two pure phases are in equilibrium. In this situation Eq. (8-63) requires that $(d \ln f) = 0$, or

$$f_{\text{eq}}^{\text{vap}} = f_{\text{eq}}^{\text{liq}} \tag{8-88}$$

Hence the fugacities of the two phases are also equal. This result would also apply to vapor-solid and liquid-solid transformations. (In Sec. 9-2 it will be shown that the equality of the fugacities given by Eq. (8-88) is a special case of a more general principle.) On the basis of Eq. (8-88) we find that Eq. (8-87) reduces to

$$f^{\text{liq}} \approx f_{\text{sat}}^{\text{vap}} \tag{8-89}$$

The approximations used to develop Eq. (8-89) are minor ones so that the equation yields answers that are reasonably good. Thus the fugacity of a compressed liquid or a solid frequently may be estimated from a knowledge of a saturated-vapor fugacity. The fugacity of the saturated vapor can be determined by methods discussed previously in this chapter. Finally, if the actual pressure is not too high so that the vapor behaves essentially as an ideal gas, then

$$f^{\text{liq}} \approx P^{\text{sat}} \tag{8-90}$$

That is, the fugacity of a liquid (or a solid) is measured reasonably well by its vapor pressure at the given temperature if the pressure is fairly low. It should also be noted that liquid fugacities can be estimated from correlations involving the corresponding states principle.

8-7 FUGACITY RELATIONS FOR REAL-GAS MIXTURES

For a pure component the chemical potential is expressed either in terms of P for an ideal gas or f for a real gas. Now, consider a mixture of gases. The chemical potential of constituent i in a mixture is based on the pressure or fugacity of the individual component. For an ideal gas the development is straightforward. Equation (8-48) states that

$$\left(\frac{\partial \mu_i}{\partial P}\right)_{T,N} = \bar{v}_i \tag{8-48}$$

For an ideal-gas mixture

$$V = (N_1 + N_2 + \cdots + N_i + \cdots)\frac{RT}{P}$$

The partial molar volume of an ideal gas is found by differentiating the above equation with respect to N_i, holding P, T, and all other N values constant. The result is

$$\bar{v}_i = \left(\frac{\partial V}{\partial N_i}\right)_{T,P,N_j} = \frac{RT}{P} \tag{8-91}$$

When this relation is substituted into Eq. (8-48), we find that for an ideal gas

$$\left(\frac{\partial \mu_i}{\partial P}\right)_{T,N} = \frac{RT}{P}$$

or

$$d\mu_i = RT\, d \ln P$$

However, for an ideal-gas mixture, $p_i = x_i P$. Logarithmic differentiation at constant x_i demonstrates that $d \ln p_i = d \ln P$. Therefore the above equation can be written as

$$d\,\mu_{i,\text{id},T} = RT\, d \ln p_i \qquad \text{(ideal gas)} \tag{8-92}$$

This equation is analogous to Eq. (8-62) for a pure ideal gas. Integration at constant temperature leads to

$$\mu_{i,\text{id},T} = \mu_{\text{ref}} + RT \ln \frac{p_i}{p_{i,\text{ref}}} \qquad \text{(ideal gas)} \tag{8-93}$$

where μ_{ref} depends on the nature of the gas and the temperature. This equation is analogous to Eq. (8-81) for a pure ideal gas. As discussed in Sec. 8-5, the standard reference state for a gas usually is selected as one of a pure ideal gas at unit pressure (normally 1 atm). As a result, the value of μ_{ref} in Eq. (8-93) becomes g_T^o. On the basis of this standard reference state Eq. (8-93) becomes

$$\mu_{i,T} = g_T^o + RT \ln p_i \qquad \text{(ideal gas)} \tag{8-94}$$

This equation will be of importance in several following sections.

In a mixture of nonideal gases it is convenient to define the chemical potential of component i in a fashion similar to that used for a pure real gas. [Recall Eq. (8-80).] Hence, by definition, for a real gas in a mixture

$$\mu_i = \bar{g}_i = \mu_{i,\text{ref}} + RT \ln \frac{\bar{f}_i}{\bar{f}_{i,\text{ref}}} \tag{8-95}$$

where $\bar{f}_i$ is the fugacity of component i in the mixture. We must also specify, similar to Eq. (8-64), that

$$\lim_{P\to 0} \frac{\bar{f}_i}{y_i P} = \lim_{P\to 0} \frac{\bar{f}_i}{p_i} = 1 \tag{8-96}$$

where $p_i = y_i P$, and y_i is the mole fraction. Equation (8-95) may also be written as

$$d\mu_i = d\,\bar{g}_i = RT\, d \ln \bar{f}_i \tag{8-97}$$

where μ_i, $\bar{g}_i$, and $\bar{f}_i$ are the chemical potential, partial molar Gibbs function, and fugacity of component i in a mixture at the given temperature and pressure.

Thus Eq. (8-97) for a real gas in a mixture is analogous to Eq. (8-63) for a pure real gas. Take note also of the similarities between Eqs. (8-93) and (8-95).

In order to evaluate $\bar{f}_i$, recall that $(\partial \mu_i/\partial P)_T = \bar{v}_i$, or

$$d\mu_{i,T} = \bar{v}_i \, dP$$

Combination of this relation with Eq. (8-97) yields

$$RT \, d \ln \bar{f}_i = \bar{v}_i \, dP$$

or upon integration

$$RT \ln \frac{\bar{f}_i}{\bar{f}_i^*} = \int_{P^*}^{P} \bar{v}_i \, dP \tag{8-98}$$

where the superscript "*" is used to designate a state of low pressure. In addition, by combining Eqs. (8-61) and (8-62), we find that for an ideal gas

$$RT \, d \ln P = v \, dP$$

or

$$RT \ln \frac{P}{P^*} = \int_{P^*}^{P} \frac{RT}{P} \, dP \tag{8-99}$$

Subtraction of Eq. (8-99) from Eq. (8-98) and rearrangement yield

$$RT \ln \frac{\bar{f}_i}{P} = RT \ln \frac{\bar{f}_i^*}{P^*} + \int_{P^*}^{P} \left(\bar{v}_i - \frac{RT}{P} \right) dP$$

Note that on the basis of Eq. (8-96), as $P^* \rightarrow 0$,

$$\lim_{P \to 0} \frac{\bar{f}_i^*}{P^*} = \lim_{P \to 0} \frac{y_i \bar{f}_i^*}{y_i P^*} = y_i$$

Therefore when $P^* = 0$, the preceding equation reduces to

$$\ln \bar{f}_i = \ln y_i P + \int_0^P \left(\frac{\bar{v}_i}{RT} - \frac{1}{P} \right) dP \tag{8-100}$$

In comparison, Eq. (8-73) showed earlier that for a pure real gas

$$\ln f = \ln P + \int_0^P \left(\frac{v}{RT} - \frac{1}{P} \right) dP \tag{8-73}$$

Note the term-by-term similarity between Eq. (8-100) for component i in a real-gas mixture and Eq. (8-73) for a pure real gas.

Recall that the fugacity coefficient ϕ of a pure substance is

$$\phi \equiv \frac{f}{P} \tag{8-70}$$

In a similar fashion, the fugacity coefficient $\bar{\phi}_i$ for a component in a solution is defined as

$$\bar{\phi}_i = \frac{\bar{f}_i}{y_i P} \tag{8-101}$$

where y_i again is the mole fraction. On the basis of this definition Eq. (8-100) may be written as

$$\ln \frac{\bar{f}_i}{y_i P} = \ln \bar{\phi}_i = \int_0^P \left(\frac{\bar{v}_i}{RT} - \frac{1}{P} \right) dP \tag{8-102}$$

Thus Eq. (8-102) may be used to calculate component fugacities or fugacity coefficients in a real-gas mixture if adequate data on the partial molar volumes as a function of pressure and composition are available. If such data are not available, certain approximations must be made.

The fugacity $\bar{f}_i$ of component i in a mixture is a function of pressure, temperature, and composition. The variation of $\bar{f}_i$ with composition at a fixed pressure and temperature is given by an important relationship known as the Duhem-Margules relation. Consider a liquid-vapor system in equilibrium which contains two components (1 and 2). When an infinitesimal change in composition occurs in a phase at constant pressure and temperature, the Gibbs-Duhem equation requires that

$$x_1\, d\mu_1 + x_2\, d\mu_2 = 0 \tag{8-46}$$

Use of Eq. (8-97) in the Gibbs-Duhem equation yields

$$x_1\, d \ln \bar{f}_1 + x_2\, d \ln \bar{f}_2 = 0$$

At constant pressure and temperature the fugacity $\bar{f}_i$ is solely a function of composition. For a two-component system it is only a function of x_i. Hence, mathematically, $d \ln \bar{f}_i = (\partial \ln f_i / \partial x_i)\, dx_i$. Using this relation in the above equation, we find that

$$x_1 \frac{\partial \ln \bar{f}_1}{\partial x_1} dx_1 + x_2 \frac{\partial \ln \bar{f}_2}{\partial x_2} dx_2 = 0$$

Also, for a two-component system $dx_2 = -dx_1$. Hence

$$x_1 \left(\frac{\partial \ln \bar{f}_1}{\partial x_1} \right)_{T,P} = x_2 \left(\frac{\partial \ln \bar{f}_2}{\partial x_2} \right)_{T,P} \tag{8-103}$$

or noting that $dx/x = d\ \ln x$, we have

$$\left(\frac{\partial \ln \bar{f}_1}{\partial \ln x_1} \right)_{T,P} = \left(\frac{\partial \ln \bar{f}_2}{\partial \ln x_2} \right)_{T,P} \tag{8-104}$$

These equations are forms of the *Duhem-Margules* relations. They are of general applicability, because their derivation is based solely on thermodynamic considerations.

At relatively low pressures the behavior of the vapor phase may be approximated as that of an ideal gas. On this basis the fugacity $\bar{f}_i$ of each component may be replaced by its partial vapor pressure, p_i. As a result, Eq. (8-104) becomes

$$\left(\frac{\partial \ln p_1}{\partial \ln x_1}\right)_{T,P} = \left(\frac{\partial \ln p_2}{\partial \ln x_2}\right)_{T,P} \tag{8-105}$$

Another useful format of the above equation is

$$\left(x_1 \frac{\partial \ln p_1}{\partial x_1}\right)_{T,P} = -x_2 \left(\frac{\partial \ln p_2}{\partial x_1}\right)_{T,P} \tag{8-106}$$

Equation (8-103) could also be written in terms of p_i, when applicable. The Duhem-Margules relations indicate that the variations of f_i (or p_i) with composition for each component in a binary mixture are not independent. This information is especially useful when plots of f_i (or p_i) versus x_i are prepared. The Duhem-Margules equation, particularly in the form of Eq. (8-106), implies a definite relationship between the gradients of the fugacity or vapor-pressure curves. For example, a plot of p_1 and p_2 versus x_1 would exhibit a trend similar to that shown in Fig. 8-2. The Duhem-Margules relation is quite useful for checking the consistency of experimental data.

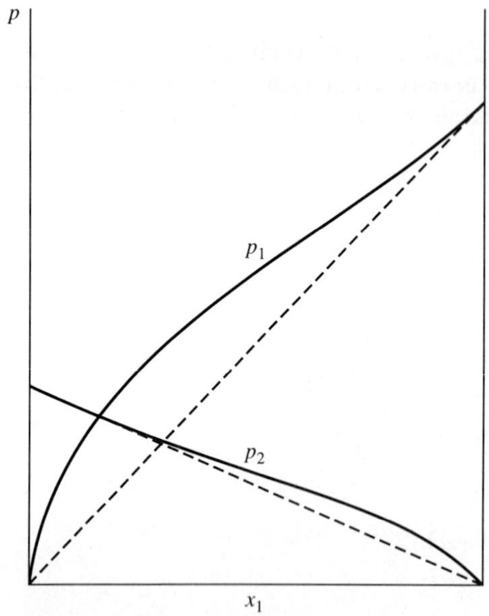

FIGURE 8-2
A plot of vapor pressure versus liquid composition for a two-component, two-phase system, which illustrates the Duhem-Margules relation.

8-8 IDEAL SOLUTIONS

Recall from Sec. 8-7 that the chemical potential of a gas in an ideal-gas mixture is given by

$$\mu_{i,\text{id}} = g_i^o + RT \ln p_i \tag{8-94}$$

Due to the choice of the standard state, the quantity p_i usually is measured in atmospheres or bars. From the definition of the partial pressure, that is, $p_i = x_i P$, the above equation can be transformed into

$$\begin{aligned} \mu_i &= g_i^o + RT \ln P + RT \ln x_i \\ &= \mu_i^* + RT \ln x_i \end{aligned} \tag{8-107}$$

where μ_i^* in this special case is defined as $g_i^o + RT \ln P$. This quantity is simply the Gibbs function of a pure substance at the given temperature and total pressure. More generally, we shall define a single-phase mixture to be an *ideal solution* if the chemical potential of each of the components in the mixture is given by

$$\mu_i = \mu_i^* + RT \ln x_i \tag{8-108}$$

where μ_i^* is the chemical potential of the pure substance at the temperature and pressure of the mixture. That is, by definition, a solution is called an *ideal solution* if the chemical potential of *every* component is a linear function of the logarithm of its mole fraction. As we have already seen, an ideal-gas mixture fulfills this relationship for all possible compositions of the mixture. A number of liquid solutions also closely approximate Eq. (8-108). Note, however, that it is not necessary that Eq. (8-108) hold over the entire range of possible compositions. A mixture in a phase may approximate an ideal solution in one region of compositions, and not in others. In addition, properties of nonideal solutions are frequently evaluated relative to what the property values would be if the solution hypothetically were an ideal solution. Consequently, it is important to derive a number of relations which are valid for ideal solutions. (It might be pointed out that there are a number of equivalent ways for defining ideal solutions. The actual definition used is not important although the one given above is of great convenience.)

In Sec. 8-3 the enthalpy, or heat, of mixing was defined as

$$\Delta H = \sum_i N_i(\bar{h}_i - h_i) \tag{8-42}$$

This quantity exists because $\bar{h}_i$ and h_i generally do not have the same value for a given component. The relationship between these two quantities for an ideal solution is developed in the following manner. From the differential calculus

$$\left(\frac{\partial(\mu_i/T)}{\partial T}\right)_{P,N} = \frac{1}{T}\left(\frac{\partial \mu_i}{\partial T}\right) - \frac{\mu_i}{T^2}$$

If Eqs. (8-38) and (8-47) are substituted into this mathematical equality, we obtain

$$\left(\frac{\partial(\mu_i/T)}{\partial T}\right)_{P,N} = -\frac{\bar{s}_i}{T} - \frac{\mu_i}{T^2} = -\frac{\bar{h}_i}{T^2} \tag{8-109}$$

In a similar fashion, it may be shown that for a pure component at the same temperature and pressure

$$\frac{\partial(g_i/T)}{\partial T} = -\frac{h_i}{T^2} \tag{8-110}$$

Now, we introduce the restriction of an ideal solution by dividing Eq. (8-108) by the temperature and by differentiating the resulting expression with respect to temperature at constant pressure and composition. In this way we obtain the following relation for an ideal solution (noting that $\mu_i^* = g_i$):

$$\frac{\partial(\mu_i/T)}{\partial T} - \frac{\partial(\mu_i^*/T)}{\partial T} = \frac{\partial(\mu_i/T)}{\partial T} - \frac{\partial(g_i/T)}{\partial T} = 0$$

When Eqs. (8-109) and (8-110) are substituted into the above expression, we find that for ideal solutions

$$\bar{h}_i = h_i \qquad \text{(ideal solution)} \tag{8-111}$$

Hence for an ideal solution the partial molar enthalpy for a given component equals its specific enthalpy in the pure state at the same pressure and temperature. Thus Eq. (8-44) for the enthalpy of mixing becomes

$$\Delta H_{\text{mix}} = 0 \qquad \text{(ideal solution)} \tag{8-112}$$

That is, the enthalpy, or heat, of mixing is zero for components forming an ideal solution.

In a similar fashion, if Eq. (8-108) is differentiated with respect to the pressure while holding the temperature and composition constant, then

$$\frac{\partial \mu_i}{\partial P} = \frac{\partial \mu_i^*}{\partial P}$$

However, from Eq. (8-30) recall that $\partial \mu_i/\partial P = \bar{v}_i$. Thus

$$\bar{v}_i = v_i \qquad \text{(ideal solution)} \tag{8-113}$$

for an ideal solution. The partial molar volume equals the pure specific volume, and in terms of the volume on mixing

$$\Delta V_{\text{mix}} = 0 \qquad \text{(ideal solution)} \tag{8-114}$$

Thus there is neither a change in volume nor enthalpy on mixing components which form an ideal solution.

The specific-enthalpy and specific-volume relationship for a mixture are given generally by

$$h_{\text{mixt}} = \sum_i x_i \bar{h}_i \qquad \text{and} \qquad v_{\text{mixt}} = \sum_i x_i \bar{v}_i$$

When Eqs. (8-111) and (8-113) for an ideal solution are substituted into these relations, we find that for a binary mixture of components A and B

$$h_{\text{mixt}} = \sum_i x_i h_{i,\text{pure}} = x_A h_A + (1 - x_A) h_B \qquad \text{(ideal solution)}$$

and

$$v_{\text{mixt}} = \sum_i x_i v_{i,\text{pure}} = x_A v_A + (1 - x_A) v_B \qquad \text{(ideal solution)}$$

These two relationships have simple interpretations on the $h - x_B$ and $v - x_B$ diagrams shown in Fig. 8-3. The two terms on the right of either equation above are represented by vertical distances at a given value of x_B. The sum of these distances in Fig. 8-3(*a*) is the value of h_{mixt} on the solid line connecting h_A and h_B. The same interpretation holds for v data in Fig. 8-3(*b*).

The entropy of mixing and the change in the Gibbs function on mixing also may be determined. Differentiation of Eq. (8-108) with respect to the temperature, at constant pressure and composition, and subsequent use of Eq. (8-29) lead to

$$\bar{s}_i = s_i - R \ln x_i \qquad \text{(ideal solution)} \tag{8-115}$$

The partial molar entropy of any component in an ideal solution thus is related to its specific entropy in the pure state for that phase at the given temperature and pressure and to its mole fraction x_i. The entropy of mixing is found by $\sum_i N_i(\bar{s}_i - s_i)$. The use of Eq. (8-115) in this relation yields

$$\Delta S_{\text{mix}} = -R \sum_i N_i \ln x_i \qquad \text{(ideal solution)} \tag{8-116}$$

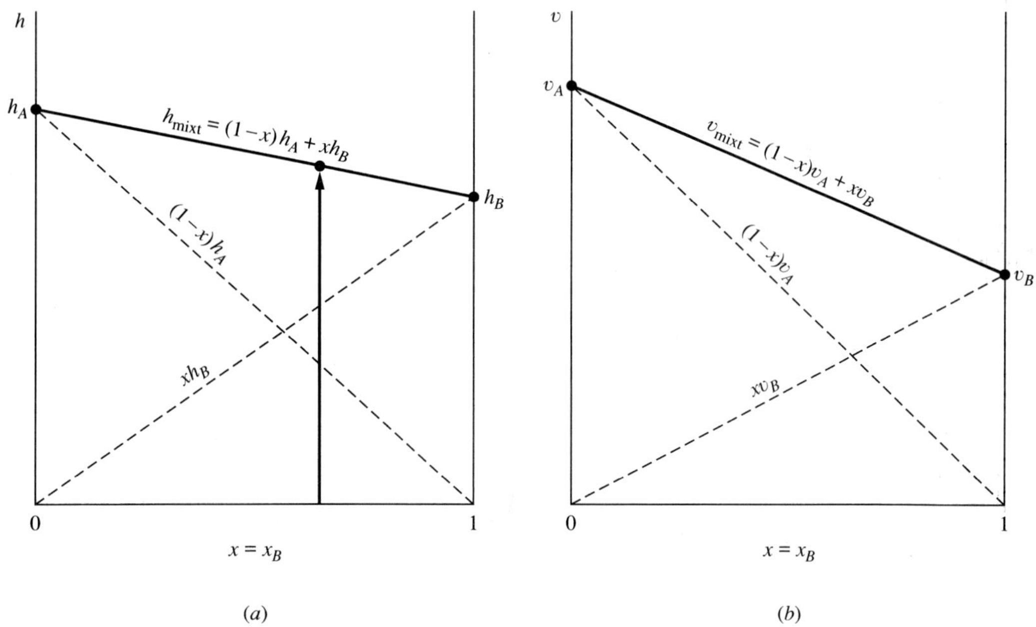

FIGURE 8-3
(*a*) The molal enthalpy-composition; (*b*) the molal volume-composition diagram for a binary ideal solution.

This is the same result as the equation for the mixing of ideal gases. In order to determine the change of the Gibbs function on mixing, recall the defining equation

$$\Delta G_{\text{mix}} = \Delta H_{\text{mix}} - T\Delta S_{\text{mix}}$$

Since ΔH_{mix} is zero for an ideal solution, then

$$\Delta G_{\text{mix}} = -T\Delta S_{\text{mix}} = RT\sum_i N_i \ln x_i \qquad \text{(ideal solution)} \qquad (8\text{-}117)$$

In a multicomponent solution the value of x_i is always less than unity. Therefore the summation term in either Eq. (8-116) or Eq. (8-117) is always negative in value. Consequently, the entropy of mixing is always positive, which is in agreement with the requirement of the second law. On the other hand, the change in the Gibbs function on mixing is always negative. This is in agreement with the minimization requirement for G at constant temperature and pressure discussed in Sec. 9-1.

Finally, another important property of an ideal solution is the fugacity of component i in the mixture. The mixture fugacity $\bar{f}_i$ can be related to the fugacity of the component in its pure state at the given mixture temperature and pressure $f_{i,P}$ in the following manner. If Eq. (8-108) is differentiated at constant temperature, we find that

$$d\mu_i = RT\, d\ln x_i \qquad \text{(ideal solution)} \qquad (8\text{-}118)$$

This equation is restricted to an ideal solution. In addition, the change in the chemical potential is always given by Eq. (8-97), namely,

$$d\mu_i = RT\, d\ln \bar{f}_i \qquad (8\text{-}97)$$

because this is the defining equation relating μ_i and $\bar{f}_i$ for component i in any phase. When Eqs. (8-97) and (8-118) are equated, we find that for an ideal solution

$$d\mu_i = RT\, d\ln \bar{f}_i = RT\, d\ln x_i$$

Integration of this equation between two states of different composition yields

$$\frac{\bar{f}_{i,2}}{\bar{f}_{i,1}} = \frac{x_{i,2}}{x_{i,1}}$$

Finally, let one of the states be that of pure component i, where $x_i = 1$ and $\bar{f}_i = f_i$. Hence the above equation generally may be written for an ideal liquid (L) solution in the form

$$\bar{f}_i^L = x_i f_{i,P}^L \qquad \text{(ideal-liquid solution)} \qquad (8\text{-}119)$$

and for the vapor (V) phase in the form

$$\bar{f}_i^V = y_i f_{i,P}^V \qquad \text{(ideal-gaseous solution)} \qquad (8\text{-}120)$$

where x_i is the liquid mole fraction and y_i is the vapor mole fraction in this and later sections of the text. These two equivalent relations, valid when a mixture is modeled as an ideal solution, are statements of the *Lewis-Randall rule*. Note that $f_{i,P}$ is the fugacity of the pure component (liquid or vapor) at the *total* pressure and the temperature of the mixture. The value of $f_{i,P}$ may be determined by methods previously discussed in Secs. 8-5 and 8-6.

[Another method of deriving the Lewis-Randall rule is to subtract Eq. (8-73) from (8-102). The resulting integral is found to be zero in value when use of Eq. (8-113), $\bar{v}_i = v_i$, is made. The remaining terms then reduce to the format of Eq. (8-119).]

All the previously derived relations for the properties of components of an ideal solution apply to any solid, liquid, or gaseous solution as long as that solution conforms to the basic definition provided by Eq. (8-108). An ideal solution could be defined by any of the basic property relationships developed in this section, for example, Eqs. (8-111), (8-113), or (8-119) and (8-120). An acceptance of any one of these expressions leads to the others by proper manipulation of equations.

8-9 HEAT AND WORK INTERACTIONS FOR GASEOUS IDEAL SOLUTIONS

The application of the first and second laws to closed and open systems requires property data. Values of property data depend on what model is used regarding the behavior of the substance of interest. As an illustration, it is informative to compare a gaseous ideal-solution model with an ideal-gas mixture model. Consider a steady-flow compressor (or expander) which operates in an isothermal manner. The minimum work input (or maximum work output) will occur when the device is internally reversible. The value of the heat and work interactions under these conditions can be determined from first- and second-law considerations. The heat transfer to or from a nonreacting gas mixture is given, in general, by

$$Q_{\text{rev},T=c} = T\Delta S = T\sum_i N_i(\bar{s}_{i,2} - \bar{s}_{i,1}) \tag{8-121}$$

The work then is evaluated from the steady-flow energy equation, namely,

$$w = \Delta h + \Delta\text{ke} + \Delta\text{pe} - q$$

Thus a knowledge of Δs and Δh values is required for the gas mixture. However, property data for gas mixtures, in general, are not readily available. Hence some approximation techniques must be used. One choice might be to assume ideal-gas behavior. Only limited data are required under this approximation, but if the pressure is sufficiently high, the answers could be considerably in error. A less restrictive approach is to assume that the nonreacting gas mixture behaves as an ideal solution.

For an ideal solution Eq. (8-115) requires that

$$\bar{s}_i = s_i - R\ln x_i \tag{8-115}$$

Therefore

$$\Delta\bar{s}_i = (s_{i,2} - R \ln x_i) - (s_{i,1} - R \ln x_i) = s_{i,2} - s_{i,1}$$

and

$$Q_{\text{rev},T=c} = T \sum_i N_i(s_{i,2} - s_{i,1}) \qquad \text{(ideal solution)} \tag{8-122}$$

The heat transfer per mole of mixture is simply

$$q_{\text{rev},T=c} = T \sum_i x_i(s_{i,2} - s_{i,1}) \qquad \text{(ideal solution)} \tag{8-123}$$

where the s_i values must be on a molar basis and are the values for the pure state at the given mixture temperature and pressure. [If s_i is on a mass basis, then mass fractions must be used in Eq. (8-123) rather than mole fractions.] The rate of heat transfer $\dot{Q}$ can be found by multiplying both sides of Eq. (8-123) by the molar flow rate, $\dot{N}$. As a result,

$$\dot{Q}_{\text{rev},T=c} = \dot{N}T \sum_i x_i(s_{i,2} - s_{i,1}) \qquad \text{(ideal solution)} \tag{8-124}$$

An equation similar to Eq. (8-124) could also be written in terms of the total mass flow rate and mass fractions. In comparison, for the more restrictive assumption of an ideal-gas mixture the value of Δs_i in either Eq. (8-123) or (8-124) is given by

$$\Delta s_{i,T=c} = -R \ln \frac{p_{i,2}}{p_{i,1}} = -R \ln \frac{P_2}{P_1}$$

for an isothermal process. Note that much different data are employed for the two approximation techniques discussed above.

Once the heat transfer is determined, the value of the work interaction is found from the steady-flow energy equation. This requires evaluation of Δh_m. In the case of an ideal solution

$$\Delta H_m = \sum_i N_i(\bar{h}_{i,2} - \bar{h}_{i,1}) = \sum_i N_i(h_{i,2} - h_{i,1}) \tag{8-125}$$

or on an intensive basis

$$\Delta h_m = \sum_i x_i(h_{i,2} - h_{i,1}) \qquad \text{(ideal solution)} \tag{8-126}$$

where molar data must be used in these two equations. (Similar equations exist for the use of data on a mass basis.) The enthalpy change is not equal to zero for an ideal solution, because $h_{i,2}$ does not necessarily equal $h_{i,1}$. The two quantities are measured as pure at the same temperature, but at different pressures. However, for an ideal-gas mixture the pressure has no influence on enthalpy values. Therefore $h_{i,2} = h_{i,1}$ in this case and the overall enthalpy change is zero. A significant difference in evaluating Δh_m exists, depending on whether an ideal solution or an ideal-gas mixture is assumed for the model. As a result, the values of both heat transfer and work could be quite different when evaluated by these two approaches.

The equations for heat and work interactions involving ideal solutions require h and s data for the pure substances at the mixture temperature and pressure. In some cases such data are available in superheat tables. In the absence of tabular data the two properties could be estimated from generalized charts and tables discussed in an earlier chapter.

Example 8-5. A gas mixture consisting of 70 mol percent methane and 30 mol percent nitrogen is compressed reversibly and isothermally in a steady-flow process from 10 to 100 bars at 250 K. The volume flow rate at the inlet is 0.20 m^3/min. Calculate (1) the heat transfer, in kJ/min, and (2) the power requirement, in kilowatts, if one assumes the mixture to behave as (*a*) an ideal solution and (*b*) an ideal-gas mixture.

Solution

(*a*) Data for the pure substances at the mixture state are used when the assumption of an ideal solution is made. In this case tabular superheat data for the two species are available. Appropriate data from such tables are listed in Table 8-1.

(1) The heat transfer can be determined through the use of Eq. (8-124). Before we can proceed, however, the molar flow rate $\dot{N}_m$ must be found from the basic relation, $\dot{N}_m = \dot{V}_m / v_m$. The molar specific volume v_m of the ideal solution at the inlet is found by

$$v_m = \sum_i x_i v_i$$

$$= [0.7(125.4)(16.04) + 0.3(73.6)(28.01)] \frac{1}{10^3} = 2.026 \text{ m}^3/\text{kmol}$$

Note that the specific-volume data have been converted to a molar basis. Therefore

$$\dot{N}_m = \frac{0.20 \text{ m}^3}{\text{min}} \frac{\text{kmol}}{2.026 \text{ m}^3} = 0.0987 \text{ kmol/min}$$

Equation (8-124) is now used to evaluate $\dot{Q}$. The molar masses again are used to convert the entropy data to a molar basis.

$$\dot{Q} = 0.0987(250)[0.7(8.364 - 10.002)(16.04) + 0.3(5.190 - 5.962)(28.01)]$$
$$= -614 \text{ kJ/min}$$

TABLE 8-1
Data for Example 8-5

	Pure gas at 10 bars			Pure gas at 100 bars		
	v	h	s	v	h	s
Methane	125.4	1077.9	10.002	8.88	928.5	8.364
Nitrogen	73.6	356.3	5.962	7.12	330.4	5.190

v is in cm^3/g, h is in kJ/kg, and s is in kJ/kg·K.

(2) The power input is found from the first law. First, the change in enthalpy for the mixture is

$$\Delta h_m = \sum_i x_i \Delta h_i = 0.7(928.5 - 1077.9)(16.04) + 0.3(330.3 - 356.3)(28.01)$$
$$= -1895 \text{ kJ/kmol}$$

Placing this result into the first law, we find that

$$\dot{W} = \dot{N}\Delta h - \dot{Q} = 0.0987(-1895) - (-614) = 427 \text{ kJ/min} = 7.12 \text{ kW}$$

If tabular superheat data were not available, v, h, and s data could be approximated, for example, by use of generalized charts.

(*b*) (1) The assumption of an ideal-gas mixture is a more restrictive assumption than an ideal solution. The molar flow rate is still found from the relation $\dot{N}_m = \dot{V}_m/v_m$. However, the molar specific volume for an ideal-gas mixture is

$$v_m = \frac{RT}{P} = \frac{0.08314(250)}{10} = 2.079 \text{ m}^3/\text{kmol}$$

Hence the molar flow rate in this case is

$$\dot{N}_m = \frac{0.20 \text{ m}^3}{\text{min}} \frac{\text{kmol}}{2.079 \text{ m}^3} = 0.0962 \text{ kmol/min}$$

The entropy change for a mole of mixture is simply

$$\Delta s = -R \ln \frac{P_2}{P_1} = -8.314 \ln \frac{100}{10} = -19.14 \text{ kJ/kmol·K}$$

The use of Eq. (8-124) in this case yields for the heat-transfer rate

$$\dot{Q} = 0.0962(250)(-19.14) = -460 \text{ kJ/min}$$

Thus the ideal-gas assumption leads to a $\dot{Q}$ value which is 25 percent lower than that based on an ideal solution.

(2) The isothermal change in enthalpy for an ideal-gas mixture is zero. Therefore the first law for the steady-flow process reduces to

$$\dot{W} = -\dot{Q} = 460 \text{ kJ/min} = 7.67 \text{ kW}$$

The ideal-gas approach leads to a power input which is 8 percent greater than that predicted by assuming a gas-phase ideal solution.

8-10 THE ENTHALPY-COMPOSITION DIAGRAM

In process design calculations it is frequently useful to have data on the enthalpy of mixtures as a function of composition, temperature, and pressure. To simplify the situation, we shall restrict our discussion to binary (two-component) systems. The general approach we follow would be applicable to more complex systems.

For binary systems the desired data may easily be presented on an *enthalpy-composition* diagram. Temperature and pressure then are represented as families of lines on the diagram. Initially, we shall hold the pressure constant so that it will not be necessary to plot P as an independent variable. In Sec. 8-3 the enthalpy-composition diagram was shown for an ideal solution. In this case the enthalpy of mixing is zero. To draw a more general diagram for nonideal solutions, we must include the enthalpy of mixing in the data to be plotted. Hence it is appropriate to examine the enthalpy of mixing concept in greater detail.

As developed in Sec. 8-3, the enthalpy of mixing is given by

$$\Delta H_{\text{mix}} = \sum_i N_i(\bar{h}_i - h_i) \qquad \text{or} \qquad \Delta h_{\text{mix}} = \sum_i x_i(\bar{h}_i - h_i)$$

where x_i is the mole fraction. These equations apply when two or more streams are mixed isothermally in a steady-state open system. When ideal gases at the same temperature are mixed, the enthalpy of mixing is zero, since an ideal-gas mixture is an ideal solution. For the mixing of real gases at the same temperature (in the absence of chemical reactions) the enthalpy of mixing is usually quite small. However, in the case of the mixing of liquids, this effect can be significant. In many instances a hand placed on the outside of the mixing vessel can easily detect the temperature rise when liquid mixing occurs. With some mixtures the energy release is great enough to cause boiling. In other cases the temperature actually falls during the mixing process.

The enthalpy of mixing is equal to the quantity of heat which must be removed, or added, in order to keep the temperature constant during mixing. For tabulation purposes it is expressed either per unit mass or per unit mole of mixture. As would be anticipated, the enthalpy of mixing depends on the relative fraction of each component added. Hence it is useful (and sometimes necessary) to have information on Δh_{mix} for a wide range of compositions. Figure 8-4 illustrates the enthalpy of mixing for a mixture of water and ethanol at a given temperature as a function of composition. Note that at 0°C the enthalpy of mixing is negative at all compositions. Heat would have to be removed to keep the temperature constant, or under adiabatic conditions the temperature would rise when preparing a mixture of these two components. Not only is the enthalpy of mixing a function of composition, but it is frequently a strong function of temperature as well. This is illustrated also by the data of Fig. 8-4. Note that at the two higher temperatures the sign on Δh_{mix} changes as the composition changes. This brief information on Δh_{mix} is sufficient to proceed in the general development of an enthalpy-composition diagram.

To simplify notation in the following development, the subscript "1" will denote the initial state before mixing and "2" will signify the state after mixing. The composition of a system containing substances A and B may be represented by either mole fractions or mass fractions, whichever is convenient for a given problem. The mole fraction of B in the liquid phase will be taken as x, and that of A as $(1 - x)$. The symbol y will be used if it is necessary to designate the mole fraction of B in the vapor phase. Capital letters will denote mass fractions. For

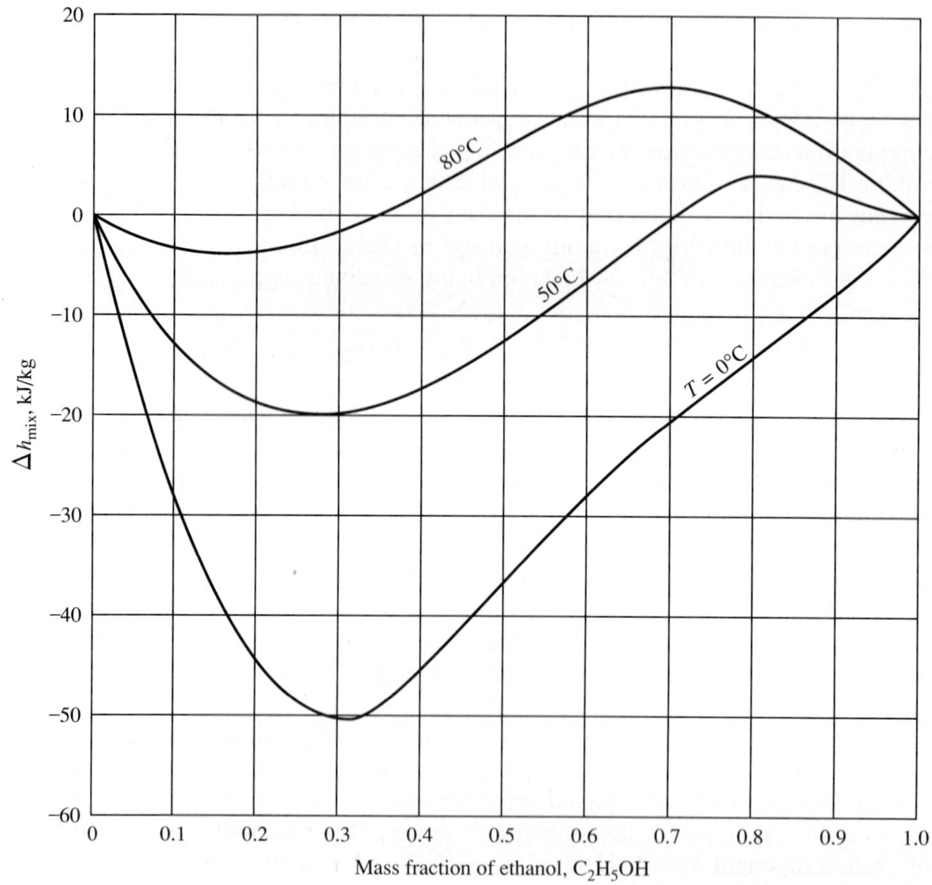

FIGURE 8-4
Enthalpy of mixing for liquid ethanol-water mixtures. (*Source:* Based on Fig. II.8, F. Bosjakovic, *Technical Thermodynamics*, Holt, Rinehart and Winston, New York, 1965. Reproduced by permission of the publisher.)

example, the mass fraction of B in the liquid phase is X, and in the vapor phase, Y. In any general development of theory it is unimportant whether mass fractions or mole fractions are used, since only the symbols are altered.

On the basis of the conservation of energy principle for a steady-state control volume the enthalpy of mixing is given by

$$\Delta h_{\text{mix}} = q_{\text{mix}} = h_2 - h_1$$

In addition, on the basis of the notation introduced above

$$h_1 = (1 - x)h_A + xh_B$$
$$h_2 = h_{\text{mixt}}$$

where we have chosen arbitrarily to write the equations on a mole basis. Combining these three equations, we find that

$$\Delta h_{\text{mix}} = q_{\text{mix}} = h_{\text{mixt}} - [(1-x)h_A + xh_B]$$

or

$$h_{\text{mixt}} = [(1-x)h_A + xh_B] + \Delta h_{\text{mix}} \tag{8-127}$$

Hence the enthalpy of any binary mixture can be determined from the enthalpies of the pure components and the enthalpy of mixing at the given temperature and pressure. For liquids the enthalpy function is rather insensitive to pressure variations.

The first term on the right of Eq. (8-127) is the contribution of an ideal solution, as discussed in Sec. 8-8. The graphical interpretation of this term is shown in Fig. 8-5(*a*), which is a repeat of Fig. 8-3(*a*). Keep in mind that the positions of point h_A (at $x = 0$) and point h_B (at $x = 1$) on the vertical axis are completely arbitrary, since the enthalpy of a pure component is based on some arbitrary reference state. Thus h_A could lie, numerically, above or below h_B. Figure 8-5(*b*) shows the result when Δh_{mix} in Eq. (8-127) is added, algebraically, to the term $[(1-x)h_A + xh_B]$ at a given composition x and temperature. In the figure we have assumed Δh_{mix} to be positive at all compositions. On the basis of Fig. 8-4 the curve representing h_{mixt} could have a much different shape from that shown in Fig. 8-5(*b*). The curve shown for h_{mixt} represents a single temperature. Isotherms for other temperatures are ascertained in a similar manner. Data for such diagrams can be presented also on a mass basis, as noted below.

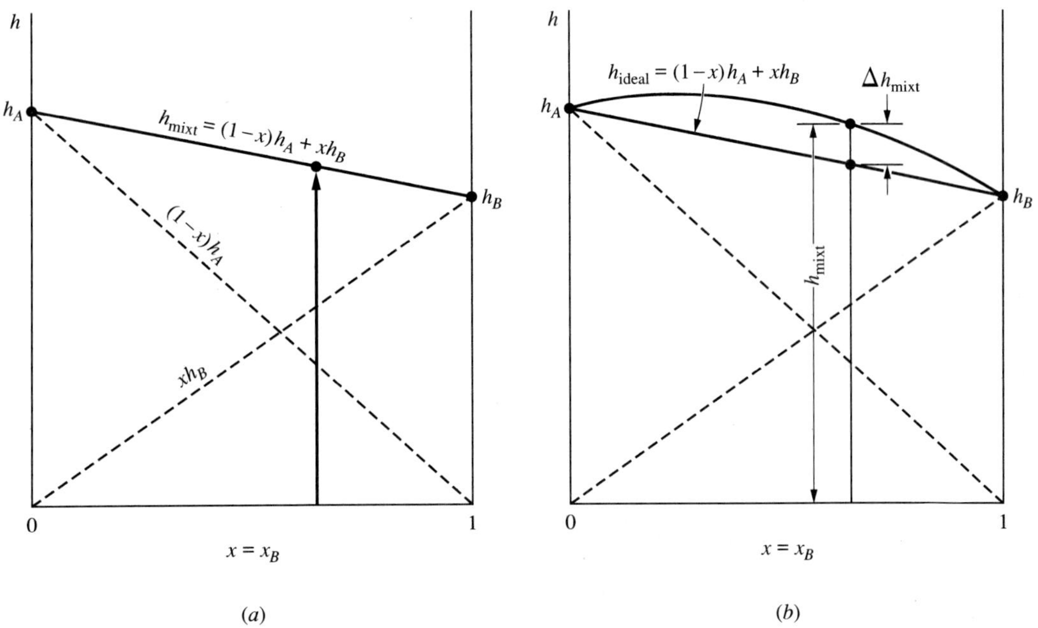

FIGURE 8-5
The molal enthalpy-composition diagram for: (*a*) an ideal solution; (*b*) a real solution with a finite enthalpy of mixing.

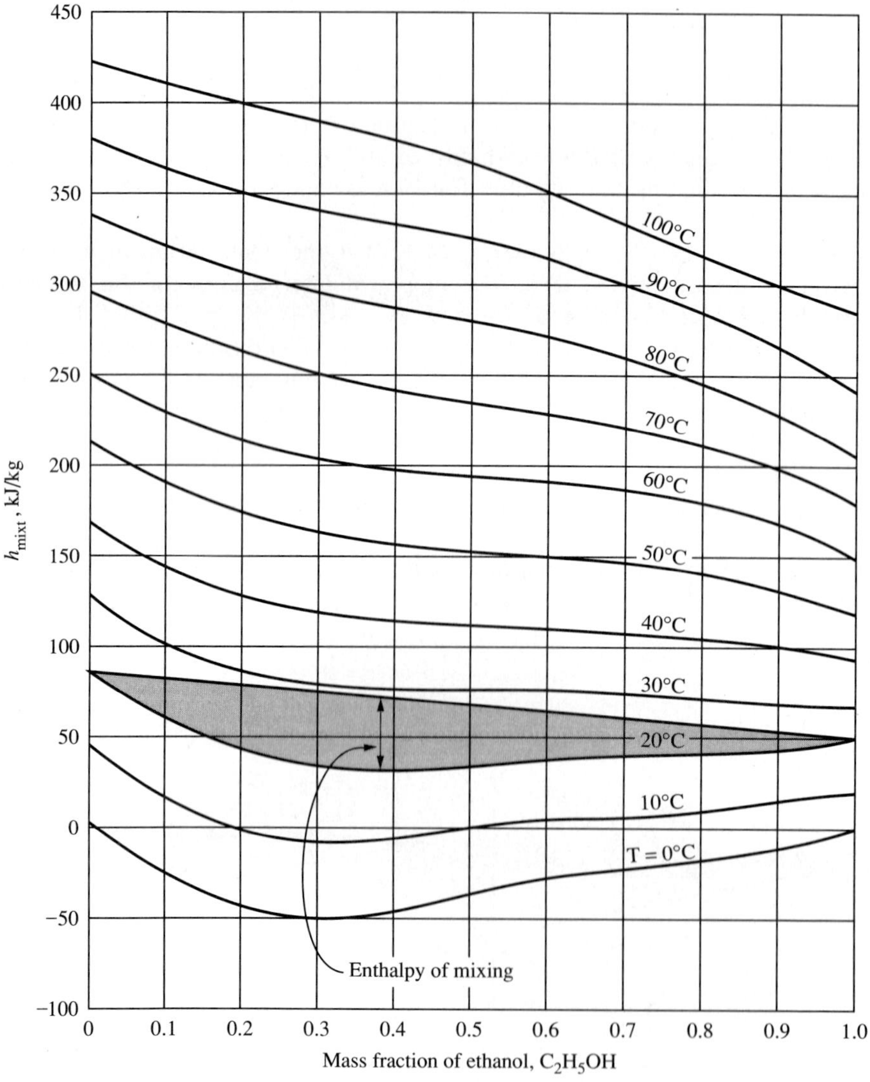

FIGURE 8-6
Enthalpy-composition diagram for liquid ethanol-water mixtures. (*Source:* Based on Fig. II.11, F. Bosjakovic, *Technical Thermodynamics*, Holt, Rinehart and Winston, New York, 1965. Reproduced by permission of the publisher.)

Figure 8-6 is an enthalpy-mass fraction (h-X) diagram for ethanol-water mixtures at temperatures between 0 and 100°C. In this particular case both pure component enthalpies have been chosen to be zero at 0°C. The curves for 0, 50, and 80°C in Fig. 8-6 are in agreement with data for Δh_{mix} given in Fig. 8-4. If the ends of any one of these isotherms in Fig. 8-6 are connected by a straight line, the vertical distance between this line and the actual isotherm is a measure of Δh_{mix} at the chosen composition and temperature. This point is also illustrated on Fig. 8-6 by the data for 20°C.

8-11 MIXING OF TWO FLUID STREAMS

One use of the enthalpy-composition diagram for binary mixtures is demonstrated by the following practical situation. Two fluid streams of low velocity enter a steady-state, steady-flow mixing chamber as shown in Fig. 8-7. The fluid properties such as v, T, and h may be quite different for the inlet flow streams, indexed as 1 and 2. Their molal compositions x_1 and x_2 in terms of species B also may be different. The molal flow rates $\dot{N}_1$ and $\dot{N}_2$ are chosen such that the exit molal flow rate $\dot{N}_3$ is unity.

In order to ascertain the exit enthalpy and temperature of the mixture, we shall set up mass and energy balances on the control volume. The conservation of total mass for the given time interval requires that

$$\dot{N}_1 + \dot{N}_2 = \dot{N}_3 = 1 \tag{a}$$

With respect to substance B in the flow streams we require that

$$x_1\dot{N}_1 + x_2\dot{N}_2 = x_3\dot{N}_3 = x_3(1) \tag{b}$$

Finally, the conservation of energy principle must be applied. The general steady-state form is

$$\dot{Q} + \dot{W}_{\text{shaft}} = \sum_i \dot{N}_i(h + \text{ke} + \text{pe})_i$$

If we assume initially that the mixing chamber is adiabatic, and neglect kinetic- and potential-energy terms, then

$$\dot{N}_1 h_1 + \dot{N}_2 h_2 = \dot{N}_3 h_3 = 1(h_3) \tag{c}$$

The combination of Eqs. (a) and (b) yields

$$\dot{N}_1 = \frac{x_2 - x_3}{x_2 - x_1} \qquad \text{and} \qquad \dot{N}_2 = \frac{x_3 - x_1}{x_2 - x_1} \tag{d}$$

The use of Eq. (d) in Eq. (c) leads to the expression for the exit enthalpy h_3, namely,

$$h_3 = \frac{x_2 - x_3}{x_2 - x_1}(h_1) + \frac{x_3 - x_1}{x_2 - x_1}(h_2)$$

$$= h_1 + \frac{x_3 - x_1}{x_2 - x_1}(h_2 - h_1) \tag{8-128}$$

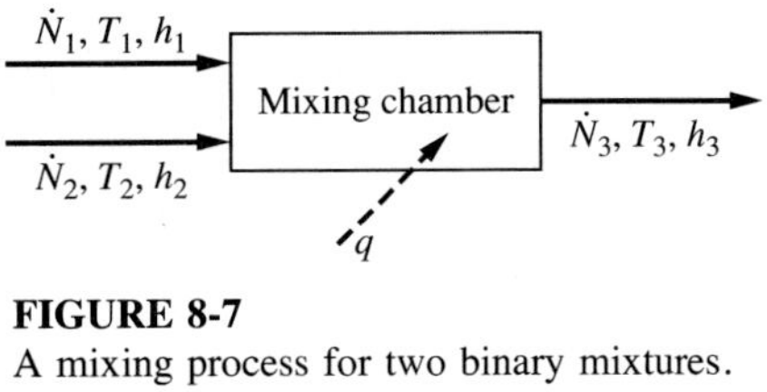

FIGURE 8-7
A mixing process for two binary mixtures.

Thus h_3 varies linearly with the mole fraction x_3 of species B. This relationship for the final enthalpy after mixing has a simple graphical interpretation in terms of an h-x diagram.

In Fig. 8-8(a) the initial states 1 and 2 are shown on h-x coordinates for the entering flow streams. A straight line is drawn to connect the initial states. State point M is placed on this line at the composition x_3 of the final mixture. For similar triangles 1-M-N and 1-2-P we find that

$$\frac{\overline{MN}}{h_2 - h_1} = \frac{x_3 - x_1}{x_2 - x_1}$$

or for the line segment $\overline{MN}$

$$\overline{MN} = \frac{x_3 - x_1}{x_2 - x_1}(h_2 - h_1)$$

Substitution of this relation into Eq. (8-128) shows that in terms of an h-x diagram

$$h_3 = h_1 + \overline{MN}$$

Thus the final mixture enthalpy, h_3, is given by the intersection of the line representing x_3 of the final mixture (mole fraction of B in the final mixture) with the line connecting the inlet states 1 and 2. That is, h_3 is represented by state M on the diagram. The value of x_3 may be found directly from Eq. (b). The mixing temperature T_M (or T_3) can be found by interpolating between known isotherms on the h-x diagram.

Another interesting relationship is valid, which also determines the position of state M on line 12. From similar triangles, again,

$$\frac{\overline{1M}}{\overline{M2}} = \frac{x_3 - x_1}{x_2 - x} = \frac{(x - x_1)/(x_2 - x_1)}{(x_2 - x_3)/(x_2 - x_1)}$$

Comparing this result with Eq. (d) leads us to conclude that

$$\frac{\overline{1M}}{\overline{M2}} = \frac{\dot{N}_2}{\dot{N}_1} \tag{e}$$

Hence the ratio of lengths $\overline{1M}$ to $\overline{M2}$ must be equal to the ratio of the inlet molal flow rates $\dot{N}_2$ to $\dot{N}_1$ mixed in the chamber. This fact is also noted in Fig. 8-8(a). If line 12 is divided into the ratio $\dot{N}_2/\dot{N}_1$, state M for the final mixture is uniquely determined. The line $\overline{1M2}$ is commonly called the *mixing line*. Equation (8-128) and its graphical interpretation in terms of Fig. 8-8(a) is called the *mixing rule* for the adiabatic mixing of two flow streams of different binary composition, temperature, and pressure. The mixing rule is of general applicability, since no statement on the nature of the two components or the nature of the phase involved is required.

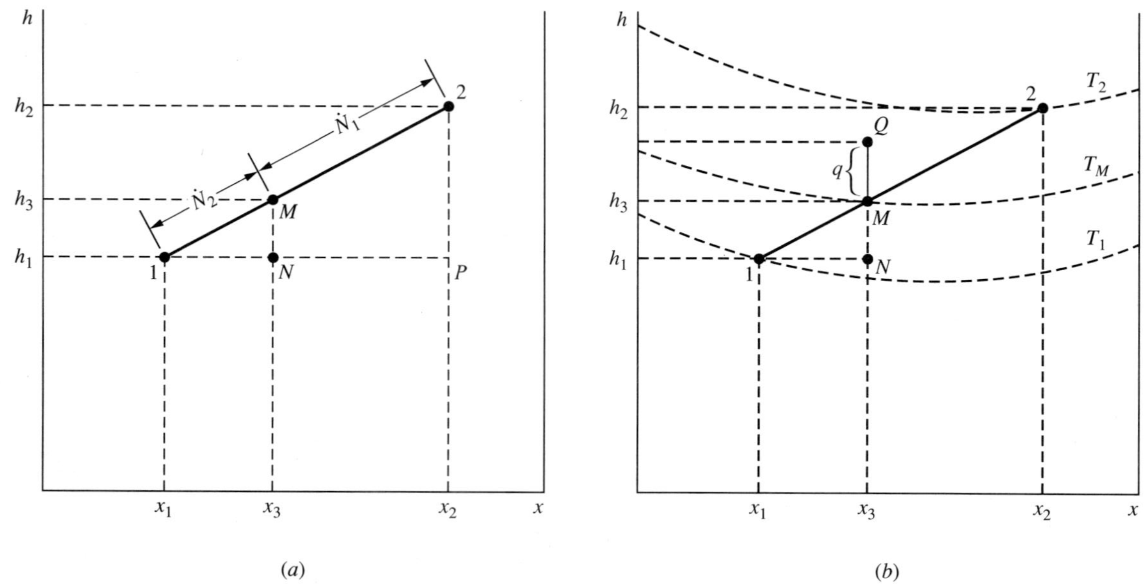

FIGURE 8-8
Graphical interpretation of the mixing of two binary mixtures on an enthalpy-composition diagram (*a*) without and (*b*) with heat transfer.

The problem of nonadiabatic mixing is solved in a similar manner. All the equations previously used for the adiabatic case remain the same except for the energy balance. This becomes

$$\dot{q} + \dot{N}_1 h_1 + \dot{N}_2 h_2 = \dot{N}_3 h_3 = (1) h_3$$

where q is the heat added or removed per unit mole leaving. Consequently, analogous to Eq. (8-128),

$$h_3 = h_1 + \frac{x_3 - x_1}{x_2 - x_1}(h_2 - h_1) + q \tag{8-129}$$

for mixing with heat transfer. Figure 8-8(*b*) illustrates the use of Eq. (8-129). If the q term were missing, the final enthalpy would lie at point M in Fig. 8-8(*b*), as noted earlier in Fig. 8-8(*a*). With heat transfer into the control volume, the final mixture enthalpy h_3, noted by state point Q, lies a distance q above the state point M. When heat transfer is out of the system, state Q will lie vertically below state M. The relevant isotherms are also shown. The temperature T of the exit flow stream may be found, whether point M or Q, by interpolation between isotherms. It must be kept in mind that the results of the last two sections apply equally well to enthalpy-mass fraction (h-X) and enthalpy-mole fraction (h-x) diagrams.

PROBLEMS

8-1. The following data represent the mixture molar specific volume v_m versus mole fraction for a binary liquid solution composed of ethane (C_2H_6) and n-pentane (C_5H_{12}) at 105°C and 140 bars:

Mole fraction of ethane	0.2	0.3	0.4	0.5	0.6	0.7	0.8
Molal volume, $cm^3/gmol$	54.18	52.42	50.92	49.67	48.63	48.37	49.76

a. Estimate graphically the partial molal volumes of ethane and n-pentane for a solution of 40 mol percent ethane, in $cm^3/gmol$.
b. Fit the data for mole fractions of 0.2 through 0.6 by the least square method to the equation: $v_m = a + bx + cx^2$. Then calculate the partial molar volumes of the two components at 40 mol percent ethane from v_m and $\partial v_m/\partial x$ data.
c. Show that the calculated $\bar{v}_i$ values lead to a line which is tangent to the plot at $x = 0.40$ drawn in part (a).

8-2. Consider the molal volume data shown in Prob. 8-1 for a binary liquid solution composed of ethane (C_2H_6) and n-pentane (C_5H_{12}) at 105°C and 140 bars.

a. Estimate graphically the partial molal volumes of ethane and n-pentane for a 50 mol percent ethane solution, in $cm^3/gmol$.
b. Fit the data for mole fractions 0.2 to 0.6 by the least square method to the equation: $v_m = a + bx + cx^2$. Then calculate the partial molar volumes of the two components at 50 mol percent from v_m and $\partial v_m/\partial x$ data.
c. Show that the calculated $\bar{v}_i$ values lead to a line which is tangent to the plot at $x = 0.50$ drawn in part (a).

8-3. The following data represent the mixture molar volume v_m versus mole fraction y for a binary gaseous solution composed of carbon dioxide (CO_2) and methane (CH_4) at 37.8°C and 8.618 MPa:

Mole fraction of CO_2	0.0000	0.1528	0.3944	0.5939	0.7961	1.0000
Molal volume, $m^3/kmol$	0.2668	0.2590	0.2425	0.2205	0.1879	0.0935

a. Estimate graphically by the method of intercepts the partial molal volumes of the two components for a solution of 64 mol percent CO_2, in $m^3/kmol$.
b. Fit the middle four data points by the least square method to the equation: $v_m = a + by + cy^2$. Then calculate the partial molal volumes of the two components at 64 mol percent CO_2 dioxide from v_m and $\partial v_m/\partial y$ data, and compare to part (a).
c. Determine the partial molar volume of each component if it behaves as an ideal gas, in $m^3/kmol$.
d. Determine the volume change when 1 kmol of mixture is formed from the pure states of the components, in cubic meters.

8-4. The following data represent the mixture molar volume v_m, in $m^3/kmol$, versus mole fraction x of CO_2 for a binary liquid solution composed of CO_2 and n-decane ($C_{10}H_{22}$) at 71.3°C and 34.5 MPa:

Mole fraction	0.4	0.5	0.6	0.7	0.8
Molar volume	0.1370	0.1224	0.1081	0.0943	0.0808

a. Fit the data points to the equation $v_m = a + bx + cx^2$. Then calculate the partial molar volumes of the two components at 75 mol percent CO_2 from v_m and $\partial v_m/\partial x$ data.
b. Do the values found in part (a) reasonably agree with the graphical method of intercepts?

8-5. An equimolar gaseous mixture of carbon dioxide and methane is maintained at 37.8°C and 8.618 MPa. The partial molar volumes of the two components are 0.176 and 0.288 $m^3/kmol$ and the pure specific volumes are 0.0935 and 0.2668 $m^3/kmol$, respectively. Determine

a. the molar volume v_m of the actual mixture, in $m^3/kmol$
b. the percent error in v_m if it were assumed that the gas mixture behaved as an ideal gas
c. the value of the compressibility factor Z_m for the mixture
d. the volume change when the two pure gases are mixed, in $m^3/kmol$ of mixture

8-6. The partial molar volumes, in $m^3/kmol$, versus mole fraction of CO_2 are given for CO_2 and n-decane ($C_{10}H_{22}$) binary liquid mixture at 71.3°C and 345 bars (see Table P8-6). Find (a) the total volume, in cubic meters, and (b) the molar specific volume, in $m^3/kmol$ of mixture, when 11.0 kg of CO_2 and 40.0 kg of $C_{10}H_{22}$ are mixed at the given conditions. The $\bar{v}$ data at the required mole fraction should be found by curve-fitting the tabular data in the region of interest to the form $\bar{v} = a + bx + cx^2$.

TABLE P8-6

$x(CO_2)$	0.2	0.3	0.4	0.5	0.6	0.7	0.8
$\bar{v}(CO_2)$	0.04771	0.04827	0.04902	0.05002	0.05227	0.05364	0.05396
$\bar{v}(C_{10}H_{22})$	0.19616	0.19597	0.19566	0.19478	0.19210	0.18910	0.18835

8-7. Partial molar volumes may be found from data on the specific volume of a binary mixture as a function of mole fraction. Show that the same graphical method may be used to determine partial molar compressibility factors if the overall mixture molar-compressibility factor Z_m is defined as

$$Z_m = y_1\bar{Z}_1 + y_2\bar{Z}_2$$

where y_1 and y_2 are the gaseous mole fractions of components 1 and 2, and $\bar{Z}_i$ is the partial molal Z factor for the ith component. Also, note that $PV = NZ_mRT$, and $\bar{v}_i = \bar{Z}_iRT/P$

8-8. Consider the data given below for Z_m versus gaseous mole fraction y of methane (CH_4) in a binary solution of CH_4 and n-pentane (C_5H_{12}) at 38°C and 275 bars:

y	0.2	0.3	0.4	0.5	0.6	0.7	0.8	0.9
Z_m	1.0775	1.0182	0.9623	0.9097	0.8651	0.8358	0.8291	0.8491

a. Fit the data for mole fractions of 0.4 through 0.7 by the least square method to the equation $Z_m = a + by + cy^2 + dy^3$. Then find analytically the values of $\bar{Z}_i$ for the two components at a y value of 0.6.

b. Plot a Z_m versus y diagram and check whether your answers to part (a) reasonably agree with the graphical solution for the partial molar-compressibility factors.

c. Find the $\bar{v}_i$ values of the two components for the 60 mol percent solution, in m^3/kmol, based on the $\bar{Z}_i$ data found in part (a).

d. Calculate the partial molar volumes for 60 mol percent CH_4 if it is assumed that the mixture behaves as an ideal-gas mixture, in m^3/kmol.

e. Determine the total volume occupied by 1 kmol of mixture of the system described above using first the results of part (c), and then of part (d), in cubic meters.

8-9. Consider the mixture-compressibility factor data for Z_m versus gaseous mole fraction y of CH_4 shown in Prob. 8-8 for a binary solution of CH_4 and n-pentane (C_5H_{12}) at 38°C and 275 bars.

a. Fit the data for mole fractions of 0.3 through 0.7 by the least square method to the equation $Z_m = a + by + cy^2 + dy^3$. Then find analytically the values of $\bar{Z}_i$ for the two components at a y value of 0.50.

b. Plot a Z_m versus y diagram and check whether your answers to part (a) reasonably agree with the graphical solution for the partial molar-compressibility factors.

c. Find the $\bar{v}_i$ values of the two components for the 50 mol percent solution, in $m^3/kmol$, based on the $\bar{Z}_i$ data found in part (a).
d. Calculate the partial molar volumes for 50 mol percent CH_4 if it is assumed that the mixture behaves as an ideal-gas mixture, in $m^3/kmol$.
e. Determine the total volume occupied by 1 kmol of mixture of the system described above using the results first of part (c), and then of part (d), in cubic meters.

8-10. The data given below are for Z_m versus gaseous mole fraction y of CO_2 in a binary solution of methane (CH_4) and carbon dioxide (CO_2) at 37.8°C and 8.618 MPa:

y	0.0000	0.1528	0.3944	0.5939	0.7961	1.0000
Z_m	0.8892	0.8634	0.8084	0.7350	0.6262	0.3117

a. Fit the middle four data points by the least square method to the equation $Z_m = a + by + cy^2 + dy^3$.
b. Find analytically the values of $\bar{Z}_i$ for the two components at a mole fraction of 0.40.
c. Plot a Z_m versus y diagram and check to see if your answers to part (b) reasonably agree with the graphical solution for the partial molar-compressibility factors.
d. Find the $\bar{v}_i$ values of the two components for a mole fraction of 0.40, in $m^3/kmol$, based on $\bar{Z}_i$ data found in part (b).
e. Find the $\bar{v}_i$ values if the mixture is an ideal gas, in $m^3/kmol$.

8-11. The PvT relation for CH_4 at 20°C is given by

$$Pv/RT = 1 - 2.02 \times 10^{-3}P + 3.72 \times 10^{-6}P^2 + 43.6 \times 10^{-12}P^4$$

where P is expressed in atmospheres. (a) Find analytically the fugacity coefficient for CH_4 at 20°C and 230 atm. Check the answer in part (a) by using (b) a two-parameter generalized method and (c) a three-parameter generalized correlation.

8-12. Determine the value of the fugacity coefficient f/P for ethylene (C_2H_4), at 10°C and 30.7 bars by the following methods:

a. the gas is an ideal gas
b. the gas follows the virial equation of state, $Z = 1 + B_P P + C_P P^2 + \cdots$, where $B_P = -6.629 \times 10^{-3}$ (bar^{-1}) and $C_P = -3.03 \times 10^{-5}$ (bar^{-2})
c. the two-parameter generalized fugacity coefficient chart
d. the Lee-Kesler three-parameter data

8-13. Determine the value of f/P for nitrogen at 250 K and (1) 200 bars and (2) 100 bars by the following methods:

a. the gas is an ideal gas
b. the gas follows the equation of state, $Pv = RT + bP$, where $b = 39.5 - 1 \times 10^4/T - 1.084 \times 10^6/T^2$, T is in kelvins and b is in $cm^3/gmol$
c. the two-parameter generalized fugacity coefficient chart
d. the gas follows van der Waals equation of state

8-14. Determine the value of f/P for nitrogen at 200 K and 100 bars by the following methods:

a. the gas is an ideal gas
b. the gas follows the equation of state, $Pv = RT + bP$, where $b = 39.5 - 1 \times 10^4/T - 1.084 \times 10^6/T^2$, with T in kelvins and b in $cm^3/gmol$
c. the two-parameter generalized fugacity coefficient chart

8-15. The virial PvT relation for CH_4 at 20°C is given by

$$Pv/RT = Z = 1 - 2.02 \times 10^{-3}P + 3.72 \times 10^{-6}P^2 + 43.6 \times 10^{-12}P^4$$

where P is expressed in atmospheres.

a. Determine analytically the fugacity coefficient for CH_4 at 20°C and 183 atm, employing the virial equation of state.
b. Determine the fugacity coefficient from the Lee-Kesler three-parameter data.

8-16. The virial expansion for CH_4 at 20°C is given by

$$Pv/RT = 1 - 2.02 \times 10^{-3}P + 3.72 \times 10^{-6}P^2 + 43.6 \times 10^{-12}P^4$$

where P is expressed in atmospheres. (a) Use the virial expansion to determine analytically the fugacity coefficient for CH_4 at 20°C and 320 atm. Check the answer in part (a) by using (b) the two-parameter generalized chart and (c) the Lee-Kesler data.

8-17. Determine the value of f/P for nitrogen at 200 K and 100 bars by

a. employing enthalpy and entropy data from the nitrogen tables for a real gas
b. employing Lee-Kesler three-parameter data

8-18. Estimate the value of the fugacity coefficient for refrigerant-134a (R-134a) at 16 bars and 70°C through the use of

a. tabular data from the superheat tables
b. the Lee-Kesler data

8-19. Estimate the value of f/P for R-134a at 14.0 bars and 80°C through the use of

a. tabular superheat data
b. three-parameter Lee-Kesler data

8-20. Estimate the fugacity of liquid water at 40°C and 150 bars, in bars, by using

a. the method described in Sec. 8-6
b. the integration of $v\,dP$ based on compressed liquid data
c. the tables of Lee-Kesler data

8-21. Estimate the fugacity of water at 80°C and 200 bars, in bars, by using

a. the method described in Sec. 8-6
b. the integration of $v\,dP$ based on tabular compressed liquid data
c. the Lee-Kesler tables

8-22. Water exists at 20°C and 100 bars at the outlet of a pump. Estimate the fugacity of the liquid, in bars, by using

a. the method described in Sec. 8-6
b. the integration of $v\,dP$ based on compressed liquid data
c. the Lee-Kesler tables

8-23. Estimate the fugacity of water, in bars, at 115°C and 88.4 bars by means of

a. the analytical method described in Sec. 8-6

b. the three-parameter Lee-Kesler data

8-24. The component pressures of ethyl alcohol-water mixtures are given in Table P8-24 for various compositions at 25°C. Determine by a least square fit of the data the value of each term in the Duhem-Margules relationship for an ethyl alcohol mole fraction of 0.4. Is the relationship valid at this composition and temperature, within the accuracy of your evaluation? Data are in mm Hg.

TABLE P8-24

x(ethyl alcohol)	p(water)	p(ethyl alcohol)
0	23.75	0
0.1	21.7	17.8
0.2	20.4	26.8
0.3	19.4	31.2
0.4	18.35	34.2
0.5	17.3	36.9
0.6	15.8	40.1

8-25. The component pressures of ethyl alcohol-water mixtures are given in Prob. 8-24 for various compositions at 25°C. Determine by a least square fit of the data the value of each term in the Duhem-Margules relationship in the form of Eq. (8-100) for an ethyl alcohol mole fraction of 0.3. Is the relationship valid at this composition and temperature, within the accuracy of your evaluation?

8-26. The component pressures of ethyl alcohol-water mixtures are given in Prob. 8-24 for various compositions at 25°C.

a. Fit the data in the region of interest by the least square method to obtain equations for p(alcohol) and $p(H_2O)$ as a function of x(alcohol). Determine by a least square fit of appropriate data the value of each term in the Duhem-Margules relationship (Eq. 8-100) for an ethyl alcohol mole fraction of 0.50.

b. Is the relationship valid at this composition and temperature, within the accuracy of your evaluation?

8-27. Assuming an ideal solution, find the ratio between the fugacity of pure benzene and the fugacity of benzene in a solution containing 30 g of diethyl ether [$(C_2H_5)_2O$], 25 g of carbon disulfide (CS_2), and 20 g of benzene (C_6H_6).

8-28. Assuming an ideal solution, find the ratio between the fugacity of pure $(C_2H_5)_2O$ and the fugacity of $(C_2H_5)_2O$ in a solution containing 20 g of diethyl ether [$(C_2H_5)_2O$], 20 g of carbon disulfide (CS_2), and 20 g of benzene (C_6H_6).

8-29. Assuming an ideal solution, find the ratio between the fugacity of pure CS_2 and the fugacity of CS_2 in a solution containing 40 g of diethyl ether [$(C_2H_5)_2O$], 30 g of carbon disulfide (CS_2), and 20 g of benzene (C_6H_6).

8-30. A gas mixture contains 30 mol percent CO_2 and 70 mol percent refrigerant-14 (R-14) ($CF_4, M = 88.01$) at 300 K and 5.0 MPa. The experimental values of the pure specific volumes at this state are 0.00779 and 0.00469 m^3/kg, respectively. Find the specific volume of the mixture, in m^3/kg, if the mixture behaves as (a) an ideal-gas mixture, (b) an ideal solution, and (c) fulfilling the equation $Pv_m = Z_m RT$, where $T_{c,m} = \sum y_i T_{c,i}$ and $P_{c,m} = \sum y_i P_{c,i}$. For R-14, $T_c = 227.4$ K and $P_c = 37.4$ bars.

8-31. A gas mixture contains 40 mol percent water vapor and 60 mol percent propane (C_3H_8, $M = 44.09$) at 500 K and 20 bars. The values of the specific volume in the pure state at the given temperature and pressure are 0.1043 and 0.0355 m^3/kg, respectively. Find the specific volume of the mixture, in m^3/kg, if the mixture is assumed to behave as (a) an ideal-gas mixture, (b) an ideal solution, and (c) fulfilling the equation $Pv_m = Z_m RT$, where $T_{c,m} = \sum y_i T_{c,i}$ and $P_{c,m} = \sum y_i P_{c,i}$.

8-32. A gas mixture contains 30 mol percent ethylene (C_2H_4) and 70 mol percent propane (C_3H_8) at 350 K and 20.0 bars. The values of the specific volume in the pure state at the given temperature and pressure are 0.0482 and 0.0253 m^3/kg, respectively. Estimate (1) the Z factor for the mixture and (2) the specific volume of the mixture, in m^3/kg, if the mixture behaves as (a) an ideal-gas mixture and (b) an ideal solution.

8-33. A mixture of 60 mol percent methane (CH_4) and 40 mol percent ethane (C_2H_6) is compressed reversibly and isothermally in a steady-flow compressor from 2.0 to 7.0 MPa at 320 K. The inlet volume flow rate is 10 m^3/min. Determine (a) the power required in kilowatts and (b) the rate of heat transfer, in kJ/min, if the mixture behaves (1) as an ideal gas and (2) as an ideal solution. Data on the pure states are given in Table P8-33. (The enthalpy and entropy data have arbitrary base values, v is in m^3/kg, h is in kJ/kg, and s is in kJ/kg·K.)

TABLE P8-33

	Pure gas at 2.0 MPa			Pure gas at 7.0 MPa		
	v	h	s	v	h	s
Methane	0.08078	1229.0	10.184	0.02174	1185.4	9.430
Ethane	0.03848	1366.1	6.843	0.00391	1138.1	5.896

8-34. A mixture of 40 mol percent C_2H_6 and 60 mol percent C_3H_8 is compressed reversibly and isothermally in steady flow from 2.0 to 20.0 bars at 500 K. The inlet volume flow rate is 50 m^3/min. Determine (1) the power required, in kilowatts, and (2) the rate of heat transfer, in kJ/min, if the mixture behaves as (a) an ideal gas and (b) an ideal solution. Data on the pure states are given in Table P8-34. (The enthalpy and entropy data have arbitrary base values, and v is in m^3/kg, h is in kJ/kg, and s is in kJ/kg·K).

TABLE P8-34

	Pure gas at 2.0 bars			Pure gas at 20.0 bars		
	v	h	s	v	h	s
Ethane	0.6901	1198.56	4.8706	0.00609	1071.47	3.3989
Propane	0.4691	1029.78	3.5105	0.00355	849.71	2.3702

8-35. A mixture of 40 mol percent methane (CH_4) and 60 mol percent ethane (C_2H_6) is compressed reversibly and isothermally in a steady-flow compressor from 3.0 to 8.0 MPa at 350 K. The inlet volume flow rate is 10 m^3/min. Determine (1) the power required, in kilowatts, and (2) the rate of heat transfer, in kJ/min, if the mixture behaves as (a) an ideal gas and (b) an ideal solution. Data on the pure states at 350 K are given in Table P8-35. (The enthalpy and entropy data have arbitrary base values, v is in m^3/kg, h is in kJ/kg, and s is in kJ/kg·K.)

TABLE P8-35

	Pure gas at 3.0 MPa			Pure gas at 8.0 MPa		
	v	h	s	v	h	s
Methane	0.05889	1294.2	10.174	0.02131	1258.7	9.585
Ethane	0.02755	1413.2	6.887	0.00667	1292.2	6.343

8-36. Consider the enthalpy-composition data or liquid ethanol-water mixtures at various temperatures shown in Fig. 8-6. Employ data from the diagram to answer the following questions.

a. At a temperature of 0°C estimate the enthalpy of the mixture for the actual solution at an ethanol mass fraction of 0.30, in kJ/kg.

b. At a temperature of 0°C estimate the enthalpy of the mixture at an ethanol mass fraction of 0.30 if one assumes the solution is an ideal solution, in kJ/kg.

c. Estimate from the diagram the enthalpy of mixing at a mass fraction ethanol of 0.30 and a temperature of 30°C.

d. At what mass fraction of ethanol is the enthalpy of mixing equal to zero at 50°C (other than mass fractions of 0 and 1.0)?

e. Estimate at what initial temperature pure water should be mixed adiabatically with pure ethanol at 0°C to give a final mixture of 40 mass percent ethanol at 30°C.

f. Pure water at 60°C is mixed adiabatically with pure ethanol at 20°C. Algebraically find what mass fraction of ethanol is used so that the final mixture enthalpy is 150 kJ/kg.

g. Calculate from a slope on the diagram and a basic thermodynamic relation between h_{mixt} and $\bar{h}$ data the partial molal enthalpy of ethanol at a mass fraction of 0.30 and a temperature of 100°C. Compare to the slope-intercept method.

8-37. Consider the enthalpy-composition data or liquid ethanol-water mixtures at various temperatures shown in Fig. 8-6. Employ data from the diagram to answer the following questions.

a. At a temperature of 50°C estimate the enthalpy of the mixture for the actual solution at an ethanol mass fraction of 0.70, in kJ/kg.

b. At a temperature of 0°C estimate the enthalpy of the mixture at an ethanol mass fraction of 0.60, if one assumes that the solution is an ideal solution, in kJ/kg.

c. Estimate from the diagram the enthalpy of mixing at a mass fraction ethanol of 0.50 and a temperature of 10°C.

d. At what mass fraction of ethanol is the enthalpy of mixing equal to zero at 60°C (other than mass fractions of 0 and 1.0)?

e. Estimate algebraically at what initial temperature pure water should be mixed adiabatically with pure ethanol at 20°C to give a final mixture of 45 mass percent ethanol at 40°C. Check graphically by means of the mixing rule.

f. Pure water at 90°C is mixed adiabatically with pure ethanol at 60°C. Algebraically find what mass fraction of ethanol is used so that the final mixture enthalpy is 290 kJ/kg. Check graphically by means of the mixing rule.

g. Calculate from a slope on the diagram and a basic thermodynamic relation between h_{mixt} and $\bar{h}$ data the partial molal enthalpy of ethanol at a mass fraction of 0.60 and a temperature of 70°C. Compare to the slope-intercept method.

REFERENCES

1. Lewis, G. N., "The Law of Physico-Chemical Change," *Proc. Am. Acad., Arts Sci.* **37**:49, 1901.
2. Lee, B. I., and M. G. Kesler, "A Generalized Thermodynamic Correlation Based on Three-Parameter Corresponding States," *AIChE J.* **21**:510, 1975.
3. Moran, M. J., and H. N. Shapiro, *Fundamentals of Engineering Thermodynamics,* 2nd ed., John Wiley & Sons, New York, 1992.

CHAPTER 9

MULTIPHASE-MULTICOMPONENT SYSTEMS

In Chap. 8 property relations were developed for mixtures in a given phase. These concepts will now be extended to mixtures for which various species exist in two or more phases in equilibrium.

9-1 EQUILIBRIUM CRITERIA

In the preceding chapters a number of thermodynamic relationships among properties have been developed which apply to systems in equilibrium. It is now important to determine the criteria which must be satisfied in order for equilibrium states to exist. One criterion already developed involves the entropy function. We shall use this criterion as a starting point to develop other criteria which are extremely useful in the analysis of processes involving mass transfer among phases and chemical reactions.

Consider a closed system which contains an arbitrary number of components and phases. The temperature and pressure are uniform throughout the system, but the system initially is in nonequilibrium. The effect of mass transfer among the phases and chemical reactions among the components will tend to bring the system ever closer to equilibrium. For the present situation the surroundings will be considered to remain in thermal and mechanical equilibrium with the system as the system proceeds toward equilibrium. Thus heat and work effects between the system and its surroundings are reversible. The system and its surroundings, taken together, constitute an isolated system. For an isolated system the second law of thermodynamics requires that $dS_{\text{isol}} \geq 0$. In this case we may write

$$dS^t + dS_{\text{surr}} \geq 0 \tag{a}$$

where S^t represents the total entropy of all the phases within the system. In addition, the entropy change of the surroundings is given simply by

$$dS_{\text{surr}} = \frac{\delta Q_{\text{surr}}}{T_{\text{surr}}} = \frac{-\delta Q}{T} \tag{b}$$

where δQ is with respect to the system. Because the heat transfer is reversible, $T_{\text{surr}} = T$, and the quantities δQ_{surr} and δQ are equal in magnitude but opposite in sign. Substitution of Eq. (b) into Eq. (a) yields, after rearrangement,

$$\delta Q \leq T\,dS^t \tag{c}$$

In addition, the conservation of energy principle for a simple compressible system with reversible boundary work dictates that

$$dU^t = \delta Q + \delta W = \delta Q - P\,dV^t$$

or

$$\delta Q = dU^t + P\,dV^t \tag{d}$$

Substitution of Eq. (d) into Eq. (c) leads to the desired result, namely,

$$dU^t + P\,dV^t - T\,dS^t \leq 0 \tag{9-1}$$

The variables in Eq. (9-1) are all properties solely of the system. Hence it applies to *any* changes in state of a closed system of uniform pressure and temperature. The conditions of mechanical and thermal reversibility assumed in the derivation no longer apply. The inequality applies to an incremental change of the closed system between nonequilibrium states. Of major importance is the fact that the direction of the inequality dictates the direction of change that leads toward equilibrium. The equality applies to changes between equilibrium states, that is, to reversible processes. Consequently, Eq. (6-2) derived earlier is just a special case of Eq. (9-1).

Equation (9-1), when expressed as an inequality, is not particularly useful in this general format. However, restricted forms of the equation are of great utility. For example, consider any closed system held at constant total volume and total entropy. Under this condition Eq. (9-1) dictates that

$$(dU^t)_{S,V} \leq 0 \tag{9-2}$$

Thus any closed system held at constant entropy and volume must have a decrease in internal energy as it proceeds toward equilibrium, and at equilibrium the internal energy is a minimum. In a similar fashion, for processes at constant total volume and internal energy Eq. (9-1) becomes

$$(dS^t)_{U,V} \geq 0 \tag{9-3}$$

This result is simply another mathematical format for the increase in the entropy principle, namely, $dS_{\text{isol}} \geq 0$. Any isolated system necessarily is constrained to having a constant volume and internal energy. Hence Eq. (9-3) provides no new information. Several modifications of Eq. (9-1), however, do lead to practical equilibrium criteria.

Reconsider Eq. (9-1) under the conditions that the temperature and pressure remain constant during a process. Mathematically, Eq. (9-1) now may be written in the format

$$dU^t_{T,P} + d(PV^t)_{T,P} - d(TS^t)_{T,P} \le 0$$

or

$$d(U^t + PV^t - TS^t)_{T,P} \le 0$$

But the Gibbs function in this case is $G^t = U^t + PV^t - TS^t$. Thus

$$dG^t_{T,P} \le 0 \tag{9-4}$$

This equation states that the Gibbs function always decreases for a spontaneous (irreversible), isothermal, isobaric change of a closed system. As the process approaches equilibrium, the Gibbs function attains a minimum value, and in the limiting case of equilibrium, dG is zero. Equation (9-4) is of great practical value, because there are many situations involving either mass transfer or chemical reactions for which the temperature and pressure are maintained at certain values. Another set of experimental conditions of practical value is one of constant volume and temperature. Under these conditions Eq. (9-1) can be written in the form

$$dU^t_{T,V} + 0 - d(TS^t)_{T,V} \le 0$$

or

$$d(U^t - TS^t)_{T,V} \le 0$$

But the Helmholtz function is defined as $A = U - TS$. Therefore

$$dA^t_{T,V} \le 0 \tag{9-5}$$

Thus the Helmholtz function must decrease when a spontaneous (irreversible) process occurs in a closed system at constant temperature and volume. Ultimately it reaches a minimum value and the system reaches a state of thermodynamic equilibrium. Equation (9-4), the equilibrium criterion in terms of the Gibbs function, will be of significant value in future sections.

9-2 PHASE EQUILIBRIUM AND MASS TRANSFER

One of the many areas in which the chemical potential plays an important role is in the study of mass transfer of one or more components between two or more phases. In general, a given multicomponent, multiphase closed system would have a constant overall composition and mass. However, unless equilibrium exists for each phase within the total system, the phases are open to mass transfer among themselves, until total equilibrium is attained. The direction of mass transfer between any two phases and the mathematically sufficient condition for equilibrium can be ascertained in the following manner.

To simplify the analysis initially, consider two phases (α and β), each of which contains C components. These two phases will be considered initially to be in thermal and mechanical equilibrium; that is, they have the same temperature and pressure. Nevertheless, they are not necessarily in phase equilibrium. For systems at constant temperature and pressure we normally apply the general criterion

$$dG_{T,P} \leq 0 \tag{9-4}$$

The change in the Gibbs function for each phase is obtained from Eq. (8-7)

$$dG = V\,dP - S\,dT + \sum_i \mu_i\,dN_i \tag{8-7}$$

where $i = 1, 2, \ldots, C$. Substitution of Eq. (8-7) into the general criterion for the given situation yields

$$\sum_i \mu_i^\alpha\,dN_i^\alpha + \sum_i \mu_i^\beta\,dN_i^\beta \leq 0 \tag{9-6}$$

In addition, in the absence of chemical reactions, the moles of each of the i species must be conserved. Thus for the mass transfer process

$$dN_i^\alpha + dN_i^\beta = 0 \tag{9-7}$$

or

$$dN_i^\beta = -dN_i^\alpha \tag{9-8}$$

The substitution of Eq. (9-8) into Eq. (9-6) yields

$$\sum_i (\mu_i^\alpha - \mu_i^\beta)\,dN_i^\alpha \leq 0 \tag{9-9}$$

The application of Eq. (9-9) to equilibrium and nonequilibrium situations is described below.

When the system is in total equilibrium, the equality sign in Eq. (9-9) applies. However, the quantities dN_i^α in the summation are nonzero and independent of each other. Therefore the only way for the left-hand side of this equation to equal zero is for each term in parentheses to equal zero. That is,

$$\mu_i^\alpha = \mu_i^\beta$$

at equilibrium for each of the C components. This result, valid for two phases in equilibrium, may be extended to the general case of P phases by successively considering the phases in pairs. The general result is

$$\begin{aligned}
\mu_1^1 &= \mu_1^2 = \mu_1^3 = \cdots = \mu_i^P \\
\mu_i^1 &= \mu_i^2 = \mu_i^3 = \cdots = \mu_i^P \\
\mu_C^1 &= \mu_C^2 = \mu_C^3 = \cdots = \mu_C^P
\end{aligned} \tag{9-10}$$

Thus the chemical potential of any given component must be the same in every phase at phase equilibrium. This is both a necessary and mathematically sufficient condition for equilibrium. However, it must not be interpreted that the chemical potential of every component is the same. If all C components exist in all P phases, Eq. (9-10) dictates that there are $C(P-1)$ equations relating the chemical potentials in the overall system.

The criterion for phase equilibrium can also be expressed in terms of the fugacities of the components. Recall that the chemical potential of a component in a solution is defined in terms of the fugacity $\bar{f}_i$ of that component by

$$d\mu_i = RT\,d\ln\bar{f}_i \tag{8-97}$$

Indefinite integration gives

$$\mu_i = RT\ln\bar{f}_i + C \tag{9-11}$$

where C is a constant which depends on some arbitrary reference state. If the same reference state is chosen for all phases, then the set of equations shown in Eq. (9-10) becomes

$$\begin{aligned} \bar{f}_1^1 &= \bar{f}_1^2 = \bar{f}_1^3 = \cdots = \bar{f}_1^P \\ \bar{f}_i^1 &= \bar{f}_i^2 = \bar{f}_i^3 = \cdots = \bar{f}_i^P \\ \bar{f}_C^1 &= \bar{f}_C^2 = \bar{f}_C^3 = \cdots = \bar{f}_C^P \end{aligned} \tag{9-12}$$

Thus the fugacity of a given component must be the same in each phase. This criterion frequently is more useful than that of the chemical potential.

The inequality sign in Eq. (9-9) applies to irreversible or spontaneous changes of state. In this circumstance for the transfer of component i between the α and β phases

$$(\mu_i^\alpha - \mu_i^\beta)\,dN_i^\alpha < 0 \tag{9-13}$$

Now, consider that a differential quantity of mass dN_i^α passes from phase β to phase α. That is, dN_i^α is a positive quantity with respect to phase α. In this situation Eq. (9-13) dictates that

$$\mu_i^\beta - \mu_i^\alpha > 0 \tag{9-14}$$

Therefore, in general, mass transfer tends to occur from regions of higher to regions of lower chemical potential. Thus the chemical potential is the potential, or driving force, for mass transfer from which the term logically follows. Diffusion of chemical species through a system is due to differences in the chemical potential of a given species. In some instances, however, a membrane may separate two regions, or phases. Only those substances which can freely pass through the membrane will do so, and in the direction of decreasing chemical potential.

9-3 THE PHASE RULE FOR NONREACTIVE COMPONENTS

When dealing with closed systems containing a single component, we know that any two intensive properties fix the state if one phase exists and only one intensive property fixes the state if two phases exist in equilibrium. There is a general rule, known as Gibbs' phase rule, which allows one to determine the number of independent, intensive properties required to fix or establish the intensive state of any system. This rule was first deduced by J. Willard Gibbs in 1875. Consider a simple, closed system which contains P homogeneous phases in contact. In equilibrium the system contains C nonreacting chemical species which are assumed to be distributed throughout all phases. Each phase is at the same pressure and temperature. The number of independent, intensive properties F required to fix the state of the system will now be determined.

Any intensive variable of a given phase depends on the pressure, temperature, and relative masses of the C components. The relative masses, on an intensive basis, are specified by the mole fractions, x_i. Only $(C - 1)$ mole fractions must be specified per phase for a C component system, because $\sum_i x_i = 1$. Hence for each phase we need to specify $[(C-1)+2]$, or $(C+1)$, intensive variables. If there are P phases present which are spatially separate and independent, the total number of intensive variables would be $P(C + 1)$. However, since these phases are in contact and in equilibrium, there are certain other relationships which reduce the number of intensive variables determined above. The actual number of independent, intensive properties equals the total number determined above minus the number of independent equations relating certain intensive variables of the system.

There are three sets of conditions which must be considered when evaluating the number of independent equations.

1. The phases are all in thermal equilibrium, and hence have a common temperature. This requires that
$$T^\alpha = T^\beta = \cdots = T^P$$
Thus there are $(P - 1)$ interrelating equations involving the temperature.
2. The phases are also in mechanical equilibrium (no internal membranes) and have a common pressure. This requires that
$$P^\alpha = P^\beta = \cdots = P^P$$
Thus there are $(P - 1)$ more interrelating equations involving the pressure.
3. Finally, phase equilibrium requires that the chemical potential of a given component must be the same in every phase. On the basis of Eq. (9-10)
$$\mu_1^\alpha = \mu_1^\beta = \cdots = \mu_1^P$$
$$\mu_i^\alpha = \mu_i^\beta = \cdots = \mu_i^P$$
$$\mu_C^\alpha = \mu_C^\beta = \cdots = \mu_C^P$$

These relationships supply $C(P-1)$ more equations. Therefore the total number of independent equations relating intensive variables is $[2(P-1)+C(P-1)]$. Subtracting the number of independent equations from the total number of intensive properties required, we find that the required number of independent, intensive properties required to fix the intensive state is

$$F = P(C+1) - [2(P-1) + C(P-1)]$$

or

$$F = C + 2 - P \tag{9-15}$$

This is known as *Gibbs' phase rule*, and F is frequently termed the *degrees of freedom* of the system. When F number of intensive variables have been selected for a system of C components and P phases, the state of the system is determinate.

For a single-component system the phase rule reduces to $F = 3 - P$. In this situation for a single-phase system the degrees of freedom are 2 in number. When two phases coexist, the number of independent, intensive variables reduces to 1. These two cases agree with the discussion at the beginning of this section. In the latter case, if pressure is selected, the temperature and other intensive properties of each phase are then fixed. When three phases coexist, there is no degree of freedom for a single-component system. This is true at any triple state of a pure substance. In a two-phase, two-component system the value of F is 2. In this case one might select the temperature and the mole fraction of one of the components as the independent variables, or pressure and one mole fraction.

The derivation of the phase rule above required that each component be present in every phase. However, this is not a necessary condition. The absence of one or more components in various phases does not alter the validity of the phase rule. This is true since every time a component is removed from a phase, the three conditions for equilibrium involving the temperature, pressure, and chemical potential must also be removed from the list of independent equations. It must be kept in mind, however, that the phase rule as expressed by Eq. (9-15) is restricted to the analysis of simple systems.

9-4 VAPOR-LIQUID EQUILIBRIUM AND RAOULT'S LAW

In Chap. 8 we examined some of the basic property relationships for solutions. These relationships are very useful for evaluating either a property or a property change of a given phase containing a number of components. Such an evaluation depends on the temperature, pressure, and composition of the phase. For single-phase systems the values of these variables are usually known. However, the analysis of multicomponent, multiphase systems is not quite as straightforward. The reason is that the composition of two or more phases in equilibrium is not the same when two or more components are present. Thus one of the important tasks of thermodynamics is to predict the equilibrium composition of each phase of such systems. In this section we shall restrict ourselves to liquid-vapor systems for which both phases fulfill the condition of an ideal solution.

First, it is useful to determine the number of independent and dependent variables associated with a multicomponent two-phase system. This is accomplished through the use of Gibbs' phase rule. In general, for any nonreacting system the number of components is C and the number of phases is P. Because the phase rule requires intensive properties, we shall describe each phase in terms of its temperature and pressure, and the mole fractions of each component. The sum of the mole fractions of C components in a given phase is unity. Thus the number of mole fractions required for a given phase is $(C - 1)$. For a two-phase system this number is simply $2(C - 1)$. Because the temperature and pressure are the same in each phase, the total number of intensive variables for a two-phase system is $2(C - 1) + 2$, or simply $2C$. On the other hand, application of the phase rule shows that the number of independent intensive properties $F = C + 2 - P = C + 2 - 2 = C$. Hence once C independent variables are fixed, there remain C dependent variables which in theory can be determined. Because the composition of each phase will be different, it is important to distinguish symbolically the difference between the mole fraction of a given component in the two phases. For clarity, the mole fraction of the ith component will be x_i in the liquid phase and y_i in the gas phase. Hence in a liquid-vapor mixture the list of $2C$ variables of interest includes temperature, pressure, x_i values, and y_i values.

A formal procedure for determining the relations between the C dependent and C independent variables is as follows. When two phases coexist in equilibrium, the two phases are of equal temperature and pressure. In addition, the chemical potential of a given component is the same in each phase. That is, $\mu_i^V = \mu_i^L$, where V and L represent the vapor and liquid phases, respectively. Moreover, recall Eq. (9-12) which requires that

$$\bar{f}_i^V = \bar{f}_i^L \tag{9-16}$$

when the two phases are in equilibrium. In addition, Eq. (8-95) states that the fugacity coefficient $\bar{\phi}$ of a component in a solution is given for any phase α by $\bar{\phi}_i^\alpha = \bar{f}_i^\alpha / x_i P$. For the vapor and liquid phases individually we write

$$\bar{\phi}_i^V = \frac{\bar{f}_i^V}{y_i P} \qquad \text{(vapor)}$$

and

$$\bar{\phi}_i^L = \frac{\bar{f}_i^L}{x_i P} \qquad \text{(liquid)}$$

where y and x now specifically refer to the vapor and liquid phases, respectively. By substituting these two equations into the preceding equation, we find that after canceling common quantities

$$\bar{\phi}_i^V y_i = \bar{\phi}_i^L x_i \tag{9-17}$$

where i represents any of the C components. This equation is of great importance, because it relates the composition of species i in the two phases. That is, it relates y_i to x_i.

Equation (9-17) represents a general solution to the analysis of liquid-vapor equilibrium for a multicomponent system. However, there are several difficulties associated with the equation. First, this equation is not explicit in y_i and x_i, since the $\bar{\phi}_i$ terms are also a function of composition, as well as of temperature and pressure. Secondly, it is difficult to evaluate the $\bar{\phi}_i$ terms, especially for the liquid phase. Techniques with various degrees of approximation, however, have been developed which are reasonably successful in overcoming the above difficulties. Some of these methods are beyond the scope of this text. Instead, we shall begin by assuming that both the liquid and vapor phases behave as ideal solutions. In this special case the values of $\bar{f}_i$ can be approximated by the Lewis-Randall rule discussed in Sec. 8-8. Substitution of Eq. (8-119) for $\bar{f}_i^L$ in Eq. (9-16) yields

$$\bar{f}_i^V = x_i f_{i,P}^L \tag{9-18a}$$

and if the Lewis-Randall rule is used in the form of Eq. (8-120) for $\bar{f}_i^V$, then the preceding equation becomes

$$y_i f_{i,P}^V = x_i f_{i,P}^L \tag{9-18b}$$

where the subscript "P" is a reminder that both f_i terms are measured at the system total pressure. Equation (9-18b) is now explicit in y_i and x_i, since the values of f_i for the liquid and vapor phases are both solely a function of temperature and pressure, and not of composition. These fugacities can be ascertained by methods described earlier. Thus a solution to the equilibrium composition calculation is at hand. However, a further simplification is possible which has little effect on the accuracy of the method. Recall that for a pure compressed liquid the fugacity can be approximated by

$$f_{i,P}^L \approx f_i^{V,\text{sat}} \qquad \text{(compressed liquid)} \tag{8-89}$$

Thus Eq. (9-18a) becomes

$$\bar{f}_i^V = x_i f_i^{V,\text{sat}} \tag{9-19}$$

and Eq. (9-18b) is written in the form

$$y_i f_{i,P}^V = x_i f_i^{V,\text{sat}} \tag{9-20}$$

This equation requires a knowledge of the fugacity $f_{i,P}^V$ in the gas phase at the system pressure and temperature, and of the fugacity $f_i^{V,\text{sat}}$ of the saturated vapor at the system temperature. Because both fugacity values can be evaluated by methods described in Sec. 8-5, Eq. (9-20) permits the evaluation of the liquid and vapor compositions of a binary mixture in equilibrium.

A further simplification of Eq. (9-20) is possible if the components in the vapor phase are assumed to behave as ideal gases. Recall that the fugacity of an ideal gas equals its pressure. Hence

$$f_{i,P}^V = p_i \text{ (at } P) = P$$

and $$f_i^{V,\text{sat}} = p_i^{\text{sat}}$$

Therefore Eq. (9-20) becomes

$$y_i P = x_i p_i^{\text{sat}} \tag{9-21}$$

Since $p_i = y_i P$ for an ideal-gas mixture,

$$p_i = x_i p_i^{\text{sat}} \tag{9-22}$$

These last two equations are statements of *Raoult's law*. They are based on the major assumptions that both phases behave as ideal solutions and the vapor phase behaves as an ideal-gas mixture. The equations require vapor-pressure data for the pure components as a function of temperature and knowledge of the total pressure. Equation (9-21) relates explicitly the mole fractions y_i and x_i at equilibrium. The second format of the law indicates that the partial (component) pressure of any component in the vapor phase is equal to the product of its mole fraction in the liquid phase and its pure vapor pressure at the given temperature. Thus Eq. (9-22) predicts that the vapor pressure of a pure species is depressed (lowered) when a second species is introduced into the liquid phase. Experimentally, Raoult's law is found to be quite accurate when the substances are physically and chemically similar, association phenomena are not present, and the pressure is relatively low. Equations (9-19) and (9-20) are used, rather than Eqs. (9-21) and (9-22), when the gases in the vapor phase are nonideal gases. For many liquid-vapor mixtures at low pressure there would be little difference between the fugacity of a component and its partial pressure. In such cases the use of Eq. (9-21) is a considerable simplification. Discussion of a more rigorous method for predicting liquid-vapor equilibrium by means of Eq. (9-16) is usually found in textbooks on chemical engineering thermodynamics [1].

Example 9-1. Five grams of sucrose ($C_{12}H_{22}O_{11}$) is added to exactly 100 g of water at 50°C. Estimate the vapor pressure of the solution based on Raoult's law, in bars.

Solution. The vapor pressure of pure water at 50°C is given as 0.1235 bar. From Raoult's law $p_i = x_i p_i^{\text{sat}}$. The mole fraction of water is

$$x_{H_2O} = \frac{100/18.02}{100/18.02 + 5.000/342.30} = \frac{5.5494}{5.5494 + 0.01461} = 0.99738$$

Therefore the partial vapor pressure is

$$p_{H_2O} = 0.99738(0.1235) = 0.1232 \text{ bar}$$

A method which sometimes requires less calculational accuracy is based on the relation $p_i^{\text{sat}} - p_i = x_N p_i^{\text{sat}}$, where x_N is the mole fraction of all nonvolatile solutes. The equation leads to the change in the vapor pressure, but it frequently requires less significant figures in the calculation.

9-5 PHASE EQUILIBRIUM OF IDEAL BINARY SOLUTIONS

As an application of Raoult's law to liquid-vapor equilibrium, consider a binary (two-component) system. The components will be designated by subscripts "1" and "2." Hence the list of variables of interest includes T, P, x_1, x_2, y_1, and y_2. Raoult's law [Eq. (9-21)] immediately provides two relationships involving all these variables. The two relationships are

$$y_1 P = x_1 p_1^{\text{sat}} \tag{9-23}$$

and

$$y_2 P = x_2 p_2^{\text{sat}} \tag{9-24}$$

The saturation pressure of a pure component is solely a function of temperature. Hence the variable T is contained implicitly within the quantities p_1^{sat} and p_2^{sat}. Saturation data are generally available in equation format, tabular format, or can be estimated by empirical techniques.

In addition to the above two formats of Raoult's law, two other equations are available which relate mole fractions. These simply are

$$x_1 + x_2 = 1.0$$

and

$$y_1 + y_2 = 1.0$$

Thus four equations are available containing six unknowns. By selecting values for two of these variables, the values of the remaining quantities can be calculated. This is in agreement with the general discussion earlier, which stated that C independent variables may be arbitrarily selected. In this particular case $C = 2$, because we have chosen a binary system. In practice, the C number of independent variables are chosen in the following fashion. Either temperature or pressure is selected, plus either one liquid or one vapor mole fraction.

The major purpose of Raoult's law is to provide a method for determining the composition of the two-phase system at various pressures and temperatures. In addition to determining values for the liquid and vapor phase mole fractions, it is also helpful to plot the results. Three major types of diagrams are possible. Of immediate interest are *P-x-y* and *T-x-y* diagrams. On a *P-x-y* diagram one plots the quantities P, p_1, and p_2 on the ordinate of the plot, and x and y values simultaneously on the abscissa. Such a plot is valid solely at the given temperature. A change in temperature will change the position of lines on this plot. Similarly, on a *T-x-y* diagram one plots the temperature on the ordinate, and x and y simultaneously on the abscissa. This latter diagram must be plotted at a given total pressure. A change in total pressure will alter the position of lines on this plot. Finally, a *P-T* (phase) diagram is frequently informative. In this case the quantities x and y appear as parameters within the plot.

First, consider a *P-x-y* diagram at a given temperature. On the basis of Eq. (9-22) we find that $p_i = x_i p_i^{\text{sat}}$, where i represents either component 1 or 2. Because the p_i^{sat} values are fixed by the selected temperature, then it is apparent that p_i varies linearly with x_i. In addition, the total pressure can be found from

Dalton's rule for ideal-gas mixtures. That is, the total pressure is the sum of the component (partial) pressures, or

$$P = x_1 p_1^{\text{sat}} + (1 - x_1)p_2^{\text{sat}} = p_2^{\text{sat}} + x_1(p_1^{\text{sat}} - p_2^{\text{sat}}) \tag{9-25}$$

Thus the total pressure also must vary linearly with the liquid mole fraction, x_i. These results are shown in Fig. 9-1(a), where p_1, p_2, and the total pressure P are plotted against x_1 at a selected temperature. (Although not shown, note that x_2 could also appear on the abscissa, and would change from 1 to 0 as x_1 changes from 0 to 1.)

It is also important to show the variation of pressure with the vapor mole fraction y_1. First, rewrite Eq. (9-23) for Raoult's law in the format

$$\frac{y_1}{x_1} = \frac{p_1^{\text{sat}}}{P}$$

Raoult's law in this format clearly shows that y_1 never equals x_1, except when x_1 equals 0 or 1. Note that p_1^{sat} is always greater than P for the example shown in Fig. 9-1(a), except when x_1 is unity. Therefore y_1 is greater than x_1 for a given system pressure, except when x_1 is zero or unity. This makes physical sense, because component 1 is more volatile (has a greater vapor pressure) than component 2 at the given temperature for this example. A specific relationship between y_1 and

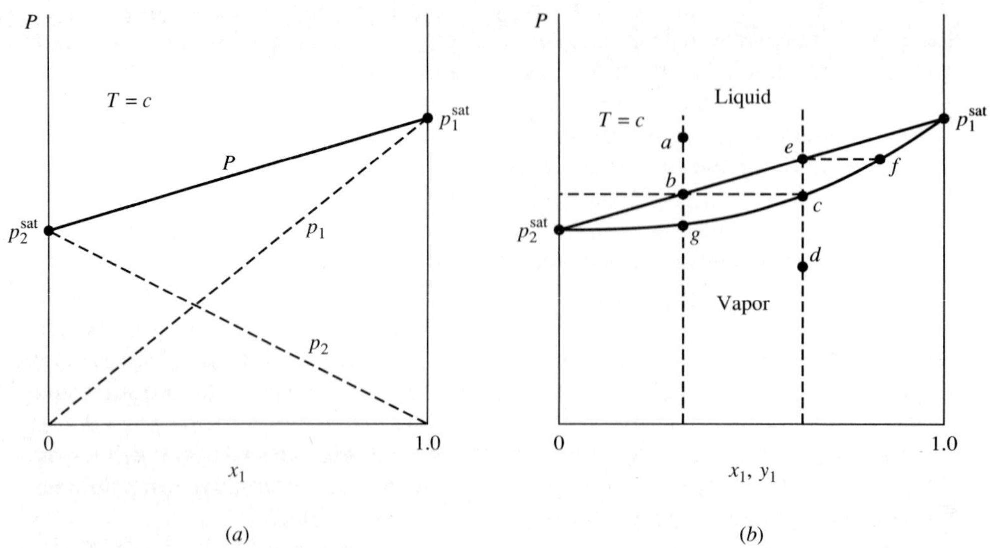

FIGURE 9-1
Variation of vapor pressures and total pressure with liquid mole fraction for a binary two-phase solution fulfilling Raoult's law at given temperature.

x_1 and the two saturation pressures can be developed by combining Eqs. (9-23), (9-24), and (9-25). If we define a new variable β as

$$\beta = \frac{p_1^{\text{sat}}}{p_2^{\text{sat}}}$$

where β is solely a function of temperature, then

$$x_1 = \frac{y_1}{\beta - y_1(\beta - 1)} \tag{9-26}$$

or

$$y_1 = \frac{x_1\beta}{1 + x_1(\beta - 1)} \tag{9-27}$$

A combination of the latter result with Raoult's law in the form of Eq. (9-23) leads to an equation for the total pressure as a function of the vapor mole fraction y_1. The result is

$$P = \frac{p_1^{\text{sat}}}{\beta - \beta y_1 + y_1} \tag{9-28}$$

A major consequence of Eq. (9-28) is that the total pressure is not a linear function of the vapor mole fraction y_1. This nonlinear relationship is shown in Fig. 9-1(*b*), where the P-y_1 curve is plotted in addition to the P-x_1 curve. The P-x_1 line is the *saturated liquid* line, and the P-y_1 line is the *saturated vapor* line for the binary mixture. Any state which lies above the P-x_1 line at the given temperature is in a compressed liquid phase. Likewise, any state which lies below the P-y_1 line is in a superheated vapor state at the given temperature. Thus the region above the P-x_1 line is marked as a liquid region and the region below the P-y_1 line is marked as a vapor region.

Consider a liquid of overall composition x_1, such as state *a* in Fig. 9-1(*b*), which has its pressure lowered at a constant temperature. A pressure will be reached where the first bubble of vapor appears. This is process *a*-*b* shown in the figure. The vapor formed has the composition of state *c*, and this saturated vapor state lies on the horizontal line passing through the saturated-liquid state *b*. The P-x line is frequently termed the *bubble point* line, since it represents states where boiling begins as the pressure is lowered for a given overall composition and temperature. If the pressure is lowered below that of state *b*, the system eventually will become a saturated vapor at state *g* with the same overall composition as the initial liquid state.

A similar analysis may be made for the situation where the pressure is increased on a superheated vapor at a given temperature. Consider state *d* in Fig. 9-1(*b*). When the system pressure reaches that of state *c*, the fluid is a saturated vapor. A small droplet of liquid condenses with the composition of state *b*. The saturated-liquid line is also called the *dew point* line, because it represents states when condensation begins at a given temperature as the pressure is increased.

The method of evaluating data for a P-x_1-y_1 plot is straightforward. One either specifies T-x_1 data and evaluates P-y_1 data, or specifies T-y_1 data and evaluates P-x_1 data. In either case the specification of the temperature permits us to determine values of p_1^{sat} and p_2^{sat} from tabular saturation data or equations relating p^{sat} to T^{sat}. As a result, either Eq. (9-26) or Eq. (9-27) is appropriate to determine the compositions of the liquid and vapor phases. To determine the system pressure,

1. when x_1 and x_2 are known, Raoult's law is used to calculate p_1 and p_2. Then the total pressure is the sum of p_1 and p_2.
2. when y_1 and y_2 are known, Raoult's law shows that $x_i = y_i P / p_i^{\text{sat}}$. These x_i values are substituted into the relation, $x_1 + x_2 = 1$. This leads to a value of the system pressure P.

The following example illustrates the necessary calculations for preparing a *P-x-y* diagram.

Example 9-2. A liquid-vapor mixture contains n-hexane and n-heptane at 27°C. Estimate the composition of the liquid and vapor phases as a function of total pressure if the phases behave as ideal solutions. Plot the results on a *P-x-y* diagram, where hexane is component 1 and heptane is component 2.

Solution. The vapor pressure of pure hexane and heptane can be calculated from the following equations:

$$\ln P_1 = 18.057 - \frac{3837.4}{T}$$

and

$$\ln P_2 = 18.217 - \frac{4231.0}{T}$$

In these equations the pressure is expressed in millibars and the temperature T must be in degrees kelvin. At 27°C or 300 K the pure vapor pressures of hexane and heptane are thus 193.6 and 61.2 millibars, respectively. The variation of P

TABLE 9-1
Calculated *P-x-y* data for mixture of *n*-hexane and *n*-heptane

x_1	y_1	P
0	0	61.2
0.1	0.260	74.5
0.2	0.442	87.6
0.3	0.575	101.
0.4	0.678	114.
0.5	0.760	127.
0.6	0.826	141.
0.7	0.881	154.
0.8	0.927	167.
0.9	0.966	180.
1.0	1.0	193.6

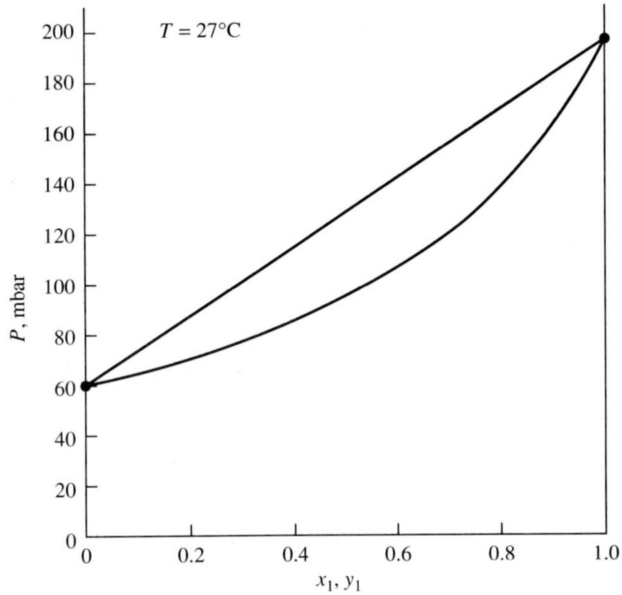

FIGURE 9-2
P-x-y plot for Example 9-2.

versus x_1 will be linear between these two values of pressure. [See Eq. (9-25).] The variation of the total pressure with vapor mole fraction is found directly from Raoult's law for a given value of the liquid mole fraction x_1 and the total pressure. [An alternative method of calculation is to use Eq. (9-26) to find y_1 as a function of x_1. Then use the calculated values of y_1 in Eq. (9-28) to find values of the total pressure.] The value of β in this problem is 3.163. On this basis the data of Table 9-1 has been calculated. These data are plotted in Fig. 9-2.

A *T-x-y* diagram is plotted at a fixed value of the total pressure. Rearrangement of Eq. (9-25) will lead to the following relationship among T, P, and x_1, namely,

$$x_1 = \frac{P - p_2^{\text{sat}}}{p_1^{\text{sat}} - p_2^{\text{sat}}} \tag{9-29}$$

In this equation T appears implicitly in the saturation pressure terms. When the temperature is selected so that p_1^{sat} equals P, then Eq. (9-29) indicates that $x_1 = 1$ (and $y_1 = 1$). If T is such that $p_2^{\text{sat}} = P$, then both x_1 and y_1 equal zero. Thus on a *T-x-y* diagram the range of temperatures of interest lies between the saturation temperatures of the two components, T_1^{sat} and T_2^{sat}. Selection of a set of temperatures between T_1^{sat} and T_2^{sat} leads to a set of x_1 values. Once the values of x_1 are calculated as a function of temperature by means of Eq. (9-29), the y_1 values can be found from either Eq. (9-23) or Eq. (9-27). Thus the calculation of x_1 and y_1 values needed for a *T-x-y* diagram is relatively straightforward. Since the saturation pressures in Eq. (9-29) are a nonlinear function of temperature, a line representing x_1 versus temperature at a given pressure on a T-x-y diagram is nonlinear. (Recall that x_1 plotted versus P on a P-x-y diagram is linear.) Also, the line representing T versus y_1 is nonlinear.

Figure 9-3 is an example of a T-x-y diagram for a binary liquid-vapor system which fulfills Raoult's rule and component 1 is more volatile than component 2. All states which lie above the T-y_1 line are vapor states, and those which lie below the T-x_1 line are liquid states. The T-x_1 line is the saturated-liquid or bubble-point line, and the T-y_1 line represents the saturated-vapor or dew-point line. If the total pressure is altered, the position of the lines will necessarily shift, because T_1^{sat} and T_2^{sat} are determined by the pressure.

Consider a binary liquid mixture at state a in Fig. 9-3. A rise in temperature at constant pressure will lead to state b, where incipient boiling occurs. State c represents the composition of the vapor first formed. As the temperature increases further, the liquid composition slowly changes from state b to state g along the saturated-liquid line, while the composition of the vapor in equilibrium varies from state c to state d. When the last drop of liquid disappears, the vapor at state d has the same overall composition as the liquid initially at state a. A further increase in temperature will simply superheat the vapor at constant overall composition. A similar analysis would hold for the cooling of a superheated vapor initially at state e. A liquid at state h appears, and on further cooling the liquid and vapor in equilibrium finally reach states g and d, respectively. Any further decrease in temperature leads to a subcooled liquid of the same composition as the original vapor. The following example illustrates the necessary calculations for preparing a T-x-y diagram.

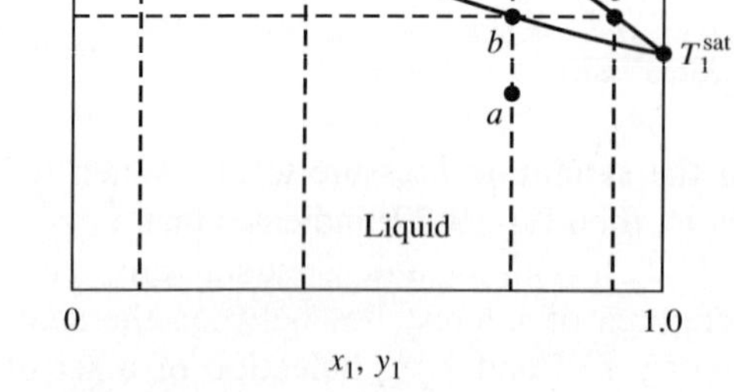

FIGURE 9-3
Variation of liquid and vapor composition with temperature for a binary two-phase solution fulfilling Raoult's law at constant pressure.

Example 9-3. A liquid-vapor mixture contains n-hexane and n-heptane at a total pressure of 2.0 bars. Estimate the composition of the liquid and vapor phases as a function of temperature if the phases behave as ideal solutions. Plot the results on a T-x-y diagram, where hexane is component 1 and heptane is component 2.

Solution. Equations for the vapor pressure of each component as a function of temperature are given in Example 9-2. For a pressure of 2000 mbar the saturation temperatures of hexane and heptane are found to be 367 and 399 K, respectively. This is the range of temperature of interest, then, for liquid-vapor equilibrium at 2.0 bars. Equation (9-29), in conjunction with the vapor-pressure equations in Example 9-2, is used to determine values of x_1 versus temperature. The vapor mole fraction is then calculated from Raoult's law, Eq. (9-23), where P is 2000 mbar. The resulting data are found in Table 9-2. These data are plotted in Fig. 9-4 in the form of a temperature-composition (T-x-y) diagram. Similar plots could be made at other pressures. However, at higher pressures both Dalton's rule and Raoult's law may become unreliable. In such cases models for real gases should be used.

As discussed earlier, the evaluation of the dew point of a superheated vapor or the bubble point of a compressed liquid is easily determined when the temperature and composition are known. However, when pressure and initial composition are known, the dew point or bubble point is found by computer or hand iteration, or by direct computer solution. For example, when pressure and x_1 are known, then Eq. (9-25) applies, namely,

$$P = x_1 p_1^{\text{sat}} + (1 - x_1) p_2^{\text{sat}} \tag{9-25}$$

This is an equation implicit in temperature, in terms of p_1^{sat} and p_2^{sat}. Rather than randomly guessing values of temperature, one can expedite the iteration process by restating Eq. (9-25) in the format

$$p_1^{\text{sat}} = \frac{\beta P}{\beta x_1 + x_2} \tag{9-30}$$

TABLE 9-2
Calculated T-x-y data for a mixture of n-hexane and n-heptane

T, °C	p_1^{sat}	p_2^{sat}	$P - p_2^{\text{sat}}$	$p_1^{\text{sat}} - p_2^{\text{sat}}$	x_1	y_1
367	2000	803	1197	1197	1.0	1.0
370	2177	882	1118	1295	0.863	0.939
375	2500	1027	973	1473	0.661	0.826
380	2860	1191	809	1669	0.485	0.694
385	3261	1377	623	1884	0.331	0.540
390	3705	1585	415	2120	0.196	0.363
395	4197	1818	182	2379	0.077	0.161
399	4626	2000	0	2626	0.0	0.0

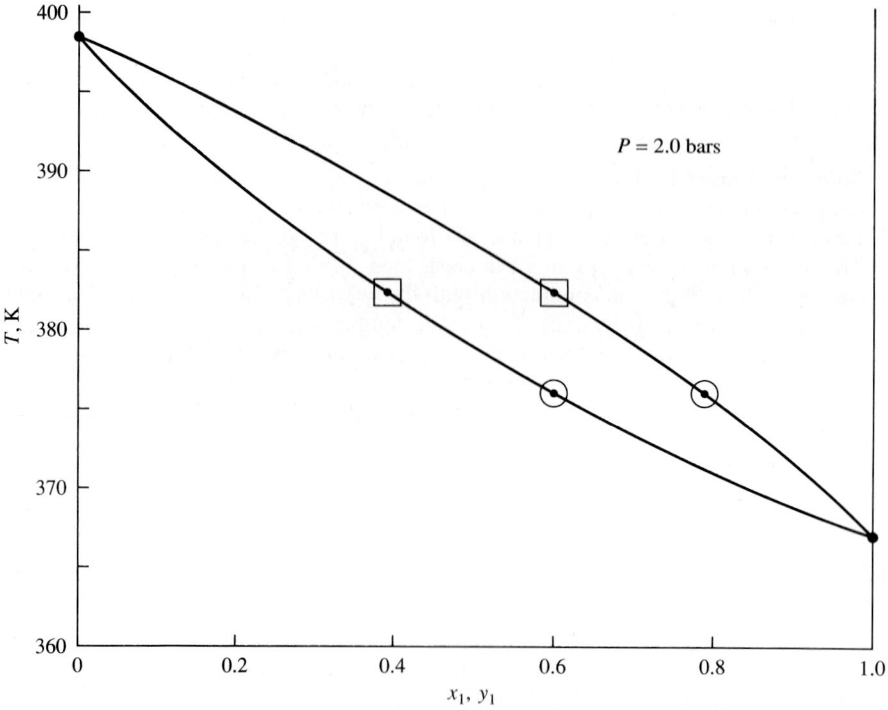

FIGURE 9-4
T-x-y diagram for Example 9-3.

When pressure and y_1 are known, the iteration process is improved by use of the relation

$$p_1^{\text{sat}} = P(\beta y_2 + y_1) \tag{9-31}$$

The use of the above two equations for determining the bubble point and dew point of a binary mixture by iteration when pressure and composition are known is illustrated in the example below.

When a software package for solving equations simultaneously is available, the bubble point is found directly by employing Eq. (9-25) in conjunction with the two nonlinear vapor-pressure equations. In the case of the dew-point temperature, Eq. (9-31) is solved simultaneously with the two vapor-pressure equations.

Example 9-4. A mixture of *n*-hexane and *n*-heptane is at a total pressure of 2.0 bars. (*a*) The mixture initially is a liquid at 350 K and the composition is 60 percent *n*-hexane (component 1). Determine the temperature, in degrees kelvin, and the vapor composition at its bubble point. (*b*) The mixture initially is a vapor at 420 K and the composition is 60 percent *n*-hexane. Determine the temperature, in degrees kelvin, and the liquid composition at its dew point. (*c*) Plot the results on Fig. 9-4 used for Example 9-3.

Solution. Equations for the vapor pressure of each component as a function of temperature are given in Example 9-2. For a pressure of 2000 mbar the saturation temperatures of n-hexane and n-heptane were found in Example 9-3 to be 367 and 399 K, respectively. Therefore at 350 K the mixture is a subcooled liquid and at 420 K the mixture is a superheated vapor. The bubble-point and dew-point temperatures sought must lie between 367 and 399 K.

(a) An effective iteration procedure for determining the bubble-point temperature is based on Equation (9-25). The temperature appears implicitly in the terms p_1^{sat} and β, since β is a function of p_1^{sat}. Although p_1^{sat} and p_2^{sat} change rapidly with small changes in temperature, the value of β changes slowly with temperature. This permits a rapid convergence on the desired solution.

The procedure is to guess an initial bubble-point temperature. Then the vapor-pressure equations are used to evaluate p_1^{sat}, p_2^{sat}, and β. In addition, p_1^{sat} is also calculated by means of Eq. (9-30). If the initial guess is correct, this latter p_1^{sat} value should match that based on the vapor-pressure equation. If not, the p_1^{sat} value calculated from Eq. (9-30) is used in the vapor-pressure equation to estimate a new temperature value. This new temperature is used to evaluate p_2^{sat} from its vapor-pressure equation, and hence a new value of β. Then Eq. (9-30) is used again to calculate a new value of p_1^{sat}. This procedure is repeated until the p_1^{sat} value based on its vapor-pressure equation agrees with that calculated from Eq. (9-30) to a predetermined accuracy (or the guessed and calculated temperature values converge).

In this problem we shall assume an initial temperature of 370 K. From the vapor-pressure equations $p_1^{sat} = 2177$ mbar, $p_2^{sat} = 882$ mbar, and $\beta = 2.469$. From Eq. (9-30) $p_1^{sat} = 2625$ mbar, which is a substantial change from the initial value. The vapor-pressure equation for p_1^{sat} indicates that a temperature of 376.8 K is a better guess. For this temperature $p_2^{sat} = 1083.8$ mbar and $\beta = 2.422$. Note that although p_1^{sat} and p_2^{sat} change considerably with a 6.8 degree temperature change, the value of β changes by 2 percent. When $\beta = 2.422$, Eq. (9-30) indicates that $p_1^{sat} = 2614$ mbar and the corresponding temperature is 376.6 K. Thus values for p_1^{sat} and the temperature are converging rapidly. A third guess of $T = 376.6$ K leads to $p_2^{sat} = 1078.7$, $\beta = 2.423$, and $p_1^{sat} = 2613.9$ mbar. Table 9-3 summarizes the results. Thus the bubble-point temperature is 376.6 K. The vapor composition at this temperature is found from Eq. (9-23). Hence

$$y_1 = \frac{x_1 p_1^{sat}}{P} = \frac{0.6(2614)}{2000} = 0.784$$

TABLE 9-3
Iterative calculations for the bubble-point temperature for a n-hexane and n-heptane mixture

T_{guess}	p_1^{sat}	p_2^{sat}	β	$p_{1,calc}^{sat}$	T_{calc}
370	2177	882	2.469	2625	376.8
376.8	2625	1083.8	2.422	2614	376.6
376.6	2614	1078.7	2.423	2613.9	376.6

TABLE 9-4
Iterative calculations for the dew-point temperature of a *n*-hexane and *n*-heptane mixture

T_{guess}	p_1^{sat}	p_2^{sat}	β	$p_{1,calc}^{sat}$	T_{calc}
375	2500	1027	2.434	3147	383.6
383.6	3147	1324	2.377	3102	383.1
383.3	3102	1302	2.381	3104.9	383.12
383.12	3104.9	1304	2.3806	3104.5	383.11

The values of x_1 and y_1 at 376.6 K are plotted as circles in Fig. 9-4. Note that they lie precisely on the saturated-liquid and saturated-vapor lines previously calculated. (*b*) The dew-point temperature and liquid composition for an initial overall vapor composition of $y_1 = 0.6$ are easily determined in a similar fashion by means of Eq. (9-31). We assume initially that the dew-point temperature is 375 K. From the vapor-pressure equations we find that $p_1^{sat} = 2500$ mbar, $p_2^{sat} = 1027$ mbar, and $\beta = 2.434$. On the basis of Eq. (9-31) $p_1^{sat} = 3147$ mbar, which is considerably different from the initial value. On the basis of $p_1^{sat} = 3147$ mbar we find that $T = 383.6$ K, $p_2^{sat} = 1324$ mbar, and $\beta = 2.377$. In this case Eq. (9-31) yields a value of 3102 mbar for p_1^{sat}. The results of two more iterations are summarized in Table 9-4. The last iteration step probably is not needed, since the accuracy is not warranted. This calculation also illustrates the small change in β during the iteration process and the rapid convergence. On the basis of a dew point of 383.1 K the droplet of liquid formed has a composition given by

$$x_1 = \frac{y_1 P}{p_1^{sat}} = \frac{0.6(2000)}{3102} = 0.387$$

The values of x_1 and y_1 for the dew point at 383.1 K are plotted as squares in Fig. 9-4.

9-6 EFFECT OF TOTAL PRESSURE ON VAPOR PRESSURE

The vapor-pressure data found in conventional tables represent the saturation vapor pressure as a function of temperature for pure substances. However, there are processes of engineering interest where noncondensable gases exist with the vapor above the liquid phase. If these gases increase the total system pressure above the saturation pressure for the given temperature, the vapor pressure of the liquid itself is altered. The effect of pressure on the vapor pressure is frequently called the *Poynting effect*. For the purpose of the following derivation it is convenient to represent the pure saturation vapor pressure by $p_{v,o}$ and the actual vapor pressure by p_v.

Starting with a pure component two-phase system, we shall consider the addition of a small amount of noncondensable gases to the system which raises the total pressure by the amount dP, as shown in Fig. 9-5. In the initial pure equilibrium state the chemical potential of the substance is the same in each phase (recall Sec. 9-2). After the addition of the noncondensable gases, the chemical potential of the condensable substance will be altered, but its liquid and vapor values must still be equal in the new equilibrium state. Consequently, the change in the chemical potential of the liquid and its vapor phase must be the same due to the increase in total pressure. This is expressed mathematically by

$$d\mu_{\text{liq}} = d\mu_{\text{vap}} \tag{a}$$

The changes in the chemical potential of the two phases may be evaluated in the following manner. If the noncondensable gases are essentially insoluble in the liquid phase, then the liquid phase may be treated as pure, for which

$$d\mu_{\text{liq}} = dg_{\text{liq}} = v\,dP - s\,dT$$

Since the change in pressure occurs isothermally,

$$d\mu_{\text{liq}} = v\,dP \qquad (T = \text{constant}) \tag{b}$$

The change in the chemical potential of the vapor in the gas mixture above the liquid is given by Eq. (8-87), namely,

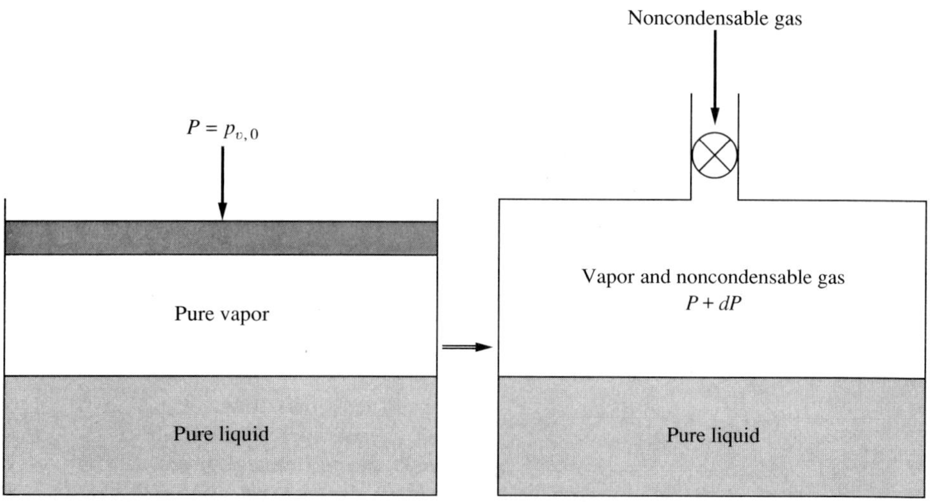

FIGURE 9-5
Effect on the vapor pressure by the introduction of a noncondensable gas which raises the total pressure at constant temperature.

$$d\mu_i = RT\, d\ln \bar{f}_i \tag{8-97}$$

where species i in this case represents the vapor. Due to the difficulty in evaluating $\bar{f}_i$, it is assumed that the gas mixture behaves as an ideal solution. On the basis of the Lewis-Randall rule [see Eq. (8-108)] $\bar{f}_i = y_i f_i$, where f_i is measured at the temperature and pressure of the system. Thus the above equation becomes

$$d\mu_i = RT\, d\ln(y_i f_i)$$

The pure fugacity f_i of the vapor could be determined by one of the methods developed in Sec. 8-5. Rather than pursue that path, it is convenient at this point to assume further that the gas phase is an ideal-gas mixture. If the vapor behaves as an ideal gas, then $f_i = P$. In addition, recall that $y_i P = p_i$. As a result, the preceding equation may be written as

$$d\mu_{\text{vap}} = RT\, d\ln p_v \qquad (T = \text{constant}) \tag{c}$$

where p_v is the vapor pressure of the liquid phase.

Equations (b) and (c) may now be substituted into Eq. (a). Under the restrictions that the gas phase behaves as an ideal solution and an ideal-gas mixture we find that

$$v_{\text{liq}}\, dP = RT\, d\ln p_v \tag{d}$$

This equation is now integrated from an initial state of a pure liquid and vapor in equilibrium to a final state where noncondensable gases have raised the system pressure to a value P. The integral of Eq. (d) can be written in the form

$$\int_{p_{v,o}}^{p_v} d\ln p_v = \frac{v_{\text{liq}}}{RT}\int_{p_{v,o}}^{P} dP$$

where $p_{v,o}$ represents the pure gas phase and v_{liq} is assumed to be constant. Integration yields

$$\ln\frac{p_v}{p_{v,o}} = \frac{v_{\text{liq}}}{RT}(P - p_{v,o}) \tag{9-32}$$

Thus the effect of increasing the pressure over a liquid by introducing noncondensable gases is to increase the vapor pressure of the liquid at the given temperature. As mentioned above, this increase is called the *Poynting effect*. For a given value of P the increase in the vapor pressure becomes larger for larger values of the liquid specific volume and molar mass. Hence some substances are more sensitive to this effect than others.

Example 9-5. At 20°C the vapor pressure of water is 0.02339 bar. Determine (*a*) the vapor pressure when the total pressure due to the presence of inert gases is 10 bars and (*b*) the total pressure necessary to increase the vapor pressure by 3.0 percent.

Solution
(*a*) Substitution of values into Eq. (9-32) yields

$$\ln \frac{p_v}{0.02339} = \frac{1.0018(10 - 0.02339)(18.02)}{0.08314(293)(10^3)} = 0.00739$$

$$p_v = 1.0074(0.02339) = 0.02356 \text{ bar}$$

The change amounts to slightly less than 1 percent. The effect of total pressure on the vapor pressure is quite small in this case.

(*b*) The use of Eq. (9-32) in this new case leads to

$$\ln 1.03 = \frac{1.0018(18.02)}{0.08314(293)(10^3)}(P - 0.02339)$$

$$P = 0.02339 + 39.887 = 39.91 \text{ bars}$$

Thus an increase of 3 percent in the vapor pressure requires an increase in the total pressure of over 1700 times the initial value of the vapor pressure.

9-7 ELEVATION OF THE BOILING POINT

When a solute is added to a pure solvent, Raoult's law indicates that the partial vapor pressure of the solvent is lowered, since the mole fraction of the solvent is now less than unity. The effect of the addition of a nonvolatile solute on the vapor pressure of a solution with one volatile component is shown in Fig. 9-6. The upper curve represents the vapor pressure of the pure solvent as a function of temperature, and the lower (dashed) curve is the variation of the partial vapor pressure with temperature when a fixed amount of nonvolatile solute is present. The saturation vapor pressure and corresponding temperature for the pure volatile solvent are indicated by p^{sat} and $T_{b,o}$, respectively. One arbitrary saturation state is W in the figure. On the addition of solute, the vapor pressure at the same temperature becomes p. Subscripts have been omitted with respect to the volatile component i, since we are interested solely in the behavior of the single volatile component, or solvent.

The addition of a nonvolatile solute to a pure substance generally would lead to a nonideal solution. However, for extremely dilute solutions it is quite likely that the solution would behave as an ideal solution. In this case we may apply Raoult's law, which states that $p = xp^{\text{sat}}$, where x is the mole fraction of the volatile component. If the quantity p^{sat} is subtracted from each side of the equation, we obtain

$$p^{\text{sat}} - p = (1 - x)p^{\text{sat}} = x_N p^{\text{sat}} \tag{9-33}$$

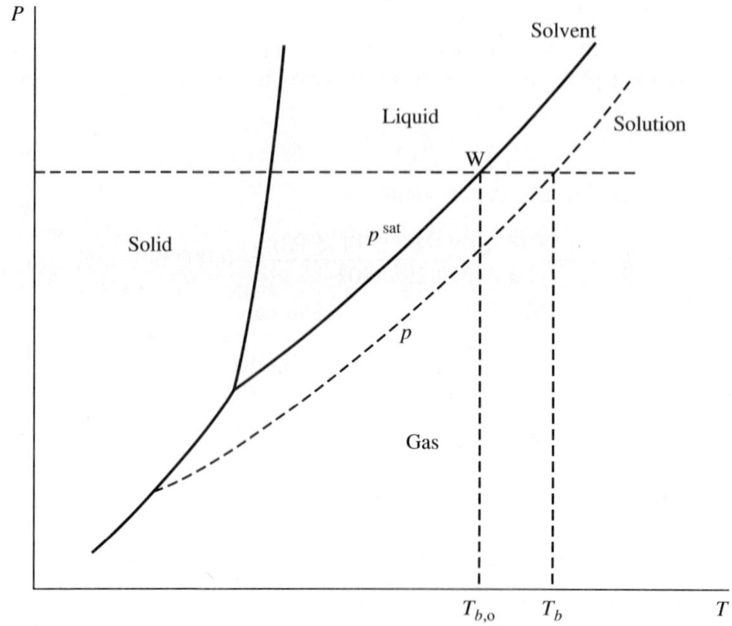

FIGURE 9-6
The effect of a nonvolatile solute on the boiling point and vapor pressure of a solvent.

where x_N is the mole fraction of the nonvolatile component, or solute. The depression of the vapor pressure ($p^{sat} - p$) then is given by the mole fraction of the solute added times the vapor pressure of the pure solvent at that temperature. Since the vapor pressure is lowered by the addition of the nonvolatile solute to a solvent, the temperature of the solution will necessarily have to be raised in order for its vapor pressure to match the original vapor pressure of the pure solvent. This difference in boiling-point temperatures at the same vapor pressure is known as the elevation of the boiling point. This is shown as the quantity $T_b - T_{b,o}$ in Fig. 9-6, where T_b is the temperature required to raise the vapor pressure of the volatile component back to p^{sat} after the solute is added. The value of the elevation of the boiling point is found in the following manner.

The requirement for equilibrium of volatile component i between the liquid and vapor phases is the equality of the chemical potential of i in the two phases. (Recall Sec. 9-2.) Because the temperature is the same in both phases, this equality may be written in the form

$$\frac{\mu_i^{liq}}{T} = \frac{\mu_i^{vap}}{T}$$

A small addition of nonvolatile solute to the liquid phase would change both μ_i and T in the two phases. However, the general format of the above equation would still be valid. Consequently, for this small variation in x_N it must be true that

$$d\left(\frac{\mu_i^{\text{liq}}}{T}\right) = d\left(\frac{\mu_i^{\text{vap}}}{T}\right) \tag{9-34}$$

In the liquid phase, in general, μ_i/T depends on temperature, pressure, and mole fraction. In the vapor phase this quantity depends only on temperature and pressure, because only component i is present. In this particular analysis the pressure is held constant. Therefore the total differentials in Eq. (9-34) then can be expressed as

$$\frac{\partial(\mu_i^{\text{liq}}/T)}{\partial T}\,dT + \frac{1}{T}\frac{\partial \mu_i^{\text{liq}}}{\partial x_i}\,dx_i = \frac{\partial(\mu_i^{\text{vap}}/T)}{\partial T}\,dT \tag{9-35}$$

The first and third quantities in this equation are easily evaluated by recalling Eq. (8-109); that is,

$$\left(\frac{\partial(\mu_i/T)}{\partial T}\right)_{P,N} = -\frac{\bar{h}_i}{T^2} \tag{8-109}$$

The second term in Eq. (9-35) is evaluated by noting that if the solution is ideal, by definition,

$$\mu_i^{\text{liq}} = \mu_i^* + RT\ln x_i \tag{8-108}$$

where the reference term μ_i^* depends only on the temperature and pressure. Differentiation of Eq. (8-108) with respect to x_i yields

$$\left(\frac{\partial(\mu_i^{\text{liq}}/T)}{\partial x_i}\right)_{T,P} = \frac{R}{x_i} \tag{9-36}$$

Substitution of Eqs. (8-109) and (9-36) into Eq. (9-35) yields

$$-\frac{\bar{h}_i^{\text{liq}}}{T^2}\,dT + \frac{R}{x_i}dx_i = -\frac{\bar{h}_i^{\text{vap}}}{T^2}\,dT$$

Since both phases behave as an ideal solution, $\bar{h}_i = h_i$. Making this substitution and rearranging the equation, we find that

$$\frac{dx_i}{x_i} = -\frac{(h_i^{\text{vap}} - h_i^{\text{liq}})}{RT^2}\,dT \tag{9-37}$$

Equation (9-37) relates the change in boiling-point temperature which occurs when a small amount of nonvolatile liquid is added to the liquid solution. The quantity $(h_i^{\text{vap}} - h_i^{\text{liq}})$ is the molar enthalpy (heat) of vaporization of volatile component i as it passes from the solution to the pure vapor phase. This quantity is not quite equal to the enthalpy of vaporization of the pure solvent, h_{fg}. However, h_{fg} is a close approximation in most cases. In addition, since the boiling-point elevation is usually not more than several degrees, it is appropriate to assume that the enthalpy of vaporization is constant. As a result, the integration of Eq. (9-37) takes on the form

$$\int_{1.0}^{x_i} d \ln x_i = \int_{T_{b,o}}^{T_b} -\frac{h_{fg}}{RT^2}\, dT$$

or

$$\ln x_i = -\frac{h_{fg}(T_b - T_{b,o})}{RT_b T_{b,o}} \tag{9-38}$$

One further minor approximation is to note that T_b and $T_{b,o}$ are essentially the same value on an absolute temperature basis. Consequently, Eq. (9-38) may be written as

$$T_b - T_{b,o} = -\frac{RT_{b,o}^2 \ln x_i}{h_{fg}} \tag{9-39}$$

This equation gives quite good evaluations of the boiling-point depression of nonelectrolytes. However, substances which are electrolytes produce more than one solute particle per molecular weight when dissolved. The boiling-point elevation of such substances is observed to be greater than that predicted by Eq. (9-39).

One further approximation can be made. When the liquid solution is fairly dilute (small amount of nonvolatiles added), then x_i approaches unity. Mathematically, then, $\ln x_i \approx (x_i - 1) = -x_{\text{solute}}$. As a result,

$$T_b - T_{b,o} \approx \frac{RT_{b,o}^2 x_{\text{solute}}}{h_{fg}} \tag{9-40}$$

The boiling-point elevation, then, is primarily a function of the mole fraction of the nonvolatile solute, x_{solute}, added to the pure volatile solvent at a given temperature. This is a remarkable result, because the elevation of the boiling point is not a function of the actual solute used. It is solely a function of the quantity used, and the identity of the solute is unimportant. A property which varies according to the quantity of a substance in a system, without regard to its identity, is known as a *colligative* property. Other examples follow in this chapter.

Example 9-6. Calculate the elevation of the boiling point of water initially pure at 100°C when the mole fraction of the solute is 0.010.

Solution. The enthalpy of vaporization of pure water at 100°C is 2257.0 kJ/kg. The value of R in Eq. (9-39) refers to the gas phase, which is composed solely of water vapor. Hence

$$T_b - T_{b,o} = \frac{8.314(373.15)^2}{18.02(2257.0)} \ln 0.990 = 0.286°\text{C}$$

The approximate answer given by Eq. (9-40) is 0.285°C. Thus a solution with 1 mole percent solute in this case has a boiling-point elevation of less than 0.3°C.

Example 9-7. Estimate the boiling-point elevation of water when 5.00 g of sucrose ($C_{12}H_{22}O_{11}$) is added per 100 g of pure water at 50°C and the pressure is maintained at the saturation pressure of 0.1235 bar.

Solution. From Example 9-1 it was found for the given masses that the mole fraction of water and sucrose in the solution are 0.99738 and 0.00262, respectively. Because the mole fraction of solute is quite small, we may employ Eq. (9-40) with good accuracy. Hence

$$T_b - T_{b,o} = \frac{8.314(323.15)^2(0.00262)}{2257.0(18.02)} = 0.0559°\text{C}$$

The boiling point is raised 0.056°C by the addition of sucrose when the pressure is maintained constant. Recall from Example 9-1 that for the same addition of sucrose at the same temperature the vapor pressure is lowered 0.0003 bar for a fixed temperature.

9-8 DEPRESSION OF THE FREEZING POINT

A pure liquid and solid in equilibrium have a well-defined melting or freezing point at a given pressure. (Handbooks usually quote data at 1 atm.) The variation of the freezing point with pressure is shown in Fig. 9-7 by the line marked

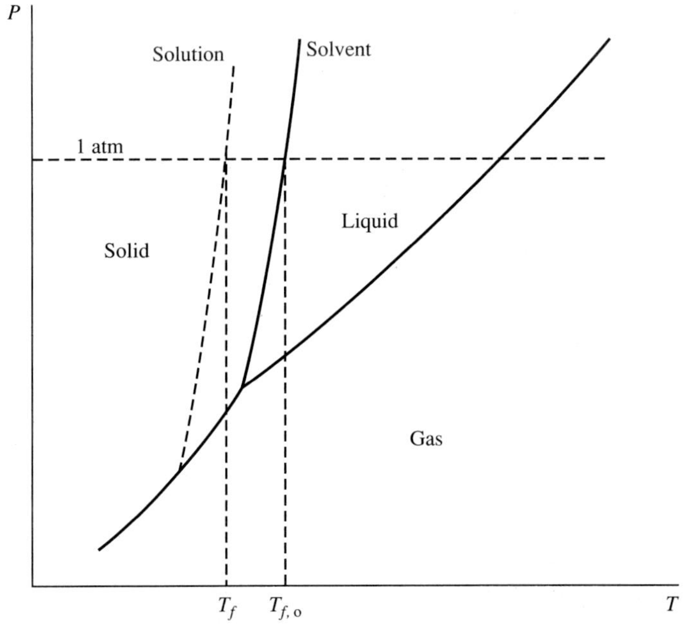

FIGURE 9-7
Phase diagram illustrating the depression of the freezing point by addition of a solute to a pure solvent.

"solvent" which separates the pure solid and liquid regions. This line generally has a very steep slope, and the slope is positive for those substances which contract in volume upon freezing. (Recall that water is a substance which expands upon freezing, and its slope on a phase diagram is negative.) It is well known that the freezing point of a pure substance may be lowered significantly by the addition of a nonvolatile substance, or solute. Ethylene glycol and other liquids are added to water in the radiator of automobiles to depress the freezing point to a value less than the anticipated lowest atmospheric temperature during the winter. Thus we are concerned in this case with the equilibrium between a liquid solution and one of its components as a pure solid phase. The effect of the addition of a soluble liquid to another pure liquid on the freezing-point temperature of the pure liquid is shown by the line marked "solution" on the left-hand side of Fig. 9-7. For example, at the 1-atm condition shown in the figure the freezing point of the pure liquid (solvent) is noted as $T_{f,o}$. For the solution at 1 atm the freezing point for the given amount of solute added is now T_f. The evaluation of the freezing-point depression $T_f - T_{f,o}$ (a negative quantity) is shown below.

The situation described above is identical in theory to the case of the boiling-point elevation. In this latter case, studied in the preceding section, we were concerned with equilibrium between a liquid solution and one of its components as a pure gas phase. Therefore a similar development of the theory will be valid. Equations (9-34) and (9-35) are still valid as the starting point in the development, except that the superscript "vap" must be replaced by superscript "solid." As a result, Eq. (9-35) is written in the form

$$\frac{dx_i}{x_i} = -\frac{(h_i^{\text{solid}} - h_i^{\text{liq}})}{RT^2}\,dT \tag{9-41}$$

Equation (9-41) relates the change in the freezing- (or melting-) point temperature when a small amount of nonvolatile liquid is added to the liquid solution. The quantity $(h_i^{\text{liq}} - h_i^{\text{solid}})$ is the molar enthalpy (heat) of melting of component i as it passes from the pure solid phase to the liquid solution phase. This quantity is not quite equal to the enthalpy of melting of the pure substance, h_{if}. However, h_{if} is usually a close approximation. In addition, it is appropriate to assume that h_{if} is independent of temperature. As a result, the integral form of Eq. (9-41) is

$$\int_{1.0}^{x_i} d\ln x_i = \int_{T_{f,o}}^{T_f} \frac{h_{if}}{RT^2}\,dT$$

or

$$\ln x_i = \frac{h_{if}(T_f - T_{f,o})}{RT_{f,o}T_f} \tag{9-42}$$

Note that h_{if} is the enthalpy of melting in the above expression, and has a positive value. Solving for the freezing-point depression $T_f - T_{f,o}$ (a negative value), we find that

$$T_f - T_{f,o} = \frac{RT_fT_{f,o}\ln x_i}{h_{if}} \tag{9-43}$$

Unfortunately, the final freezing point after addition of solute to the pure liquid appears on both sides of the equation. This obstacle can be overcome by rewriting the equation in the form

$$T_f - T_{f,o} = \frac{KT_{f,o}}{1-K} \tag{9-44}$$

where

$$K = \frac{RT_{f,o} \ln x_i}{h_{if}} \tag{9-45}$$

This form of the equation is fairly accurate if the liquid behaves as an ideal solution.

In the development of the boiling-point elevation discussed in Sec. 9-7 further approximations were made. These included assuming that the change in temperature due to the addition of the solute was small, and that the quantity of solute added was small. In the case of the freezing-point depression these two approximations frequently are not good. For example, when antifreeze is added to water, the solution typically contains up to 50 percent by volume of the antifreeze component. In addition, the desired change in temperature is around 20–25°C (40–50°F). However, in other cases where the added mass of solute and the change in temperature are reasonably small an equation equivalent to Eq. (9-40) can be derived. For the freezing-point depression this becomes

$$T_f - T_{f,o} \approx \frac{-RT_{f,o}^2 x_{\text{solute}}}{h_{if}} \tag{9-46}$$

Both Eqs. (9-44) and (9-46) indicate that the freezing-point depression is a colligative property. The change in freezing point due to the addition of a solute to a pure liquid is a function of the quantity of solute added but is not a function of the identity of the solute.

Example 9-8. Estimate the mole fraction of methyl alcohol required to lower the freezing point of water at 1 bar down to −12°C.

Solution. The freezing point is to be changed by −12°C. The enthalpy of melting of water at 0°C is 333.4 kJ/kg, and it will be assumed to be constant. From Eq. (9-43) we find that

$$\ln x_i = \frac{333.4(18.02)(-12)}{8.314(273.15)(261.15)} = -0.1216$$

$$x_i = 0.885 \qquad \text{and} \qquad x_{\text{solute}} = 0.115$$

The approximation formula given by Eq. (9-46) will give an answer of 0.115 for the mole fraction of solute required. Thus a mole percent of roughly 12 will lower the freezing point to −12°C. Note that it was not necessary to know what solute was added. If, however, it were desired to find the volume of solute added per volume of solvent, knowledge of the solute would be necessary.

9-9 OSMOTIC PRESSURE OF AN IDEAL SOLUTION

The phenomenon of osmosis is the process of a solvent diffusing through a membrane which is permeable to it but which is impermeable to the chemical species which make up the solute. It is only when a membrane exists which has this property that diffusion can occur. For example, certain semipermeable membranes permit the passage of water from a pure solution into a solution containing water and a dissolved solute, as shown in Fig. 9-8. The phenomenon exists because of the inequality of the chemical potential of water in the pure liquid state and in the mixture. Recall that the chemical potential of a substance i in an ideal solution is given by $\mu_i = \mu_i^* + RT \ln x_i$. It is easily seen that μ_i in a solution is lower than μ_i in the pure state, because of its smaller mole fraction. Chemical species diffuse from regions of higher to lower chemical potential. Thus the pure solvent will diffuse into the mixture while at the same time raising the chemical potential of the solvent in the liquid mixture. Only at infinite dilution will the chemical potentials of the solvent on both sides of the membrane approach each other, and diffusion stops. The chemical potential of the solute can have any value, however, since it is not free to pass through the semipermeable membrane.

9-9-1 Evaluation of the Osmotic Pressure

The passage of solvent i through the membrane can be prevented if its chemical potential in the mixture can be increased to the same value as that of the pure liquid

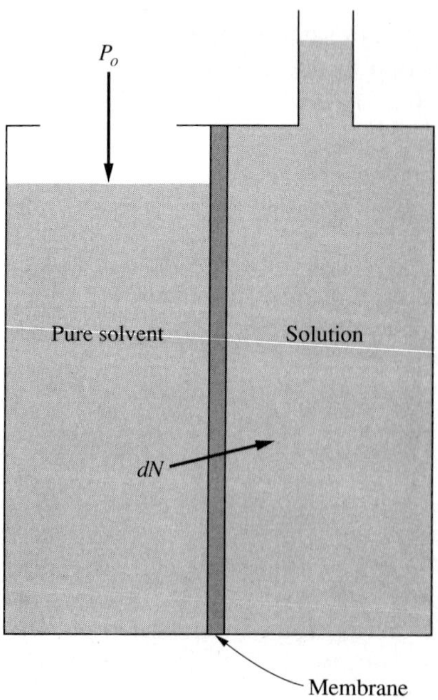

FIGURE 9-8
Diffusion of solvent in direction of lower chemical potential.

at temperature T and pressure P_o. Recall Eq. (8-48), namely, $(\partial \mu_i/\partial P)_{T,N} = \bar{v}_i$. Since $\bar{v}_i$ is always positive, then μ_i always increases with increasing pressure. Therefore an increase in μ_i in the liquid solution can be achieved by increasing the total pressure of the mixture solution. Of course, the semipermeable membrane must be strong enough to support this increase in pressure. The *osmotic pressure* is defined as the pressure difference which must be maintained across the membrane between the mixture and pure solution in order to maintain equilibrium at the same temperature. The osmotic pressure is a function of the solution and is not a function of the nature of the membrane. The osmotic pressure Π, then, is defined as

$$\Pi \equiv P - P_o \tag{9-47}$$

where P is the pressure maintained on the mixture solution at equilibrium.

The value of Π is ascertained in the following manner for an ideal solution. The chemical potential of the pure solvent at equilibrium is given by

$$\mu_i = \mu_{i,o}^* + RT \ln x_i \tag{9-48}$$

where x_i is unity and $\mu_{i,o}^*$ is a reference value for the pure liquid measured at pressure P_o. For the mixture in equilibrium with the pure solvent

$$\mu_i = \mu_i^* + RT \ln x_i \tag{9-49}$$

where μ_i^* is a reference value measured for the pure liquid at the pressure P and the same temperature. Upon equating these two equations, one finds that

$$\mu_{i,o}^* = \mu_i^* + RT \ln x_i$$

or

$$RT \ln x_i = \mu_{i,o}^* - \mu_i^* \tag{9-50}$$

The quantity $(\mu_{i,o}^* - \mu_i^*)$ represents the change in g for the pure substance at constant temperature as the pressure is altered from P to P_o. For a pure substance, $dg = v\,dP - s\,dT$; therefore for a pure liquid $d\mu_T = dg_T = v\,dP$. Thus

$$(\mu_{i,o}^* - \mu_i^*) = \int_P^{P_o} v\,dP$$

where v is the molar volume of the pure solvent. The volume of the liquid is assumed to be independent of pressure. Subsequent integration of the above expression and substitution into Eq. (9-50) yield

$$RT \ln x_i = v(P_o - P)$$

or on the basis of the definition of the osmotic pressure

$$\Pi = P - P_o = -\frac{RT \ln x_i}{v} \tag{9-51}$$

If x_N represents the mole fraction of the total solutes, then

$$\Pi = -\frac{RT}{v} \ln(1 - x_N) \tag{9-52}$$

Equations (9-51) and (9-52) indicate that the osmotic pressure is another colligative property. It depends on the quantity of solute, but not on its identity.

When the solution is quite dilute, $\ln(1 - x_N) \approx -x_N$; therefore an approximate relation for the osmotic pressure is

$$\Pi \approx \frac{RTx_N}{v} \tag{9-53}$$

where v again is the molar volume of the pure solvent. Further, since $v = V/N_i$ and for very dilute solution N_i/N is roughly unity, Eq. (9-53) can be written as

$$\Pi \approx \frac{N_N RT}{V} \tag{9-54}$$

This latter equation is of historical interest, since it is the form originally presented by van't Hoff. It is of reasonable accuracy only when the molality of a solution is less than unity.

The osmotic pressures exerted by even dilute solutions are extremely large. For example, dilute solutions of sucrose in water at normal temperatures have osmotic pressures in the range of 25–100 atm. The phenomenon of osmosis is extremely important, of course, in the study of biology and physiology. The mechanism of reverse osmosis has important applications in engineering. For example, by supplying a sufficiently high pressure on a saline water solution, it is commercially feasible to convert saltwater to freshwater by reversing the process of osmosis. It is necessary only to bring the chemical potential of the water in the solution to a high enough value so that it will diffuse through the membrane into the side containing pure water.

Example 9-9. A solution contains 5.0 g of sucrose in 100.0 g of pure water at 25°C. Predict the osmotic pressure of the solution in bars.

Solution. In Example 9-1 the mole fraction of water was found to be 0.99738. The molar specific volume of water at 25°C is 1.0029 cm^3/g. The use of Eq. (9-51) yields

$$\Pi = \frac{RT}{v}(-\ln x_i) = \frac{0.08314(298.15)(10^3)}{18.02(1.0029)}(-\ln 0.99738) = 3.60 \text{ bars}$$

The approximation equation (9-53) leads to the same numerical answer, because the mole fraction of the solute is extremely small.

9-9-2 Desalination

There is a tremendous need by populations throughout the world, as well as by industry, to enhance their fresh (potable) water supply. The two major sources of nonpotable water are brackish (slightly salty) groundwater and seawater. These two sources have *total dissolved solids* (TDS) content in the 2000–3000 and 35,000–45,000 ppm (parts per million) range, respectively. The World Health Organization has set a maximum TDS standard for the human population of 500 ppm. There are two major methods of desalting either source to reach the standard. One of these involves evaporation, where the saltwater is boiled and the water vapor produced is removed and condensed. The technique is known as *multistage flash distillation* (MSF). This is an energy-intensive process. The second method is *reverse osmosis* (RO), where membrane technology is employed. The osmotic pressures for 2000 ppm TDS brackish water and for seawater are roughly 1.3 and 25 bars [20 and 375 psi (pounds per square inch)], respectively. To produce freshwater from seawater at an economical rate, a typical pressure difference of 70 bars (1000 psi) is applied across the membrane.

Reverse osmosis has a number of advantages over MSF distillation. Ease of operation and energy efficiency are two major considerations. An RO plant typically uses one-third less energy than MSF distillation. A single pass through a RO unit is sufficient to reduce the TDS count from 40,000 ppm to less than 500 ppm. This is accomplished by introducing the feedwater under high pressure to a cylindrical module which contains a bundle of hollow fibers, each about the size of a human hair. A bundle may contain a million fibers. The feedwater flows around the bundle, and pure water passes through the fibers into the hollow cores and is removed. Since the concentrated seawater leaving a reverse osmosis unit is under high pressure, it is used in an energy recovery turbine incorporated in the feedwater pumping system which provides fresh seawater to the plant. This turbine/pump system may reduce the required pumping energy by as much as 40 percent. The brine concentrated by the process is returned to the sea.

By mid-1991 there were around 7500 desalination plants in the world that produced 3.5 billion gallons of water daily. The cost of desalted water could be three to 10 times the cost of conventional treatment systems. The cost is highly dependent on the cost of power to run desalination plants, as well as on the type of process used. About 60 percent of the world's desalination plants are on the Arabian Peninsula, with 11 percent in the United States in mid-1991. The amount of water produced by desalination in Saudi Arabia is around 500 million gallons daily. In this same time frame around 180 plants were operating in California. Many of these plants treat brackish water, but the increased treatment of seawater is a high priority. The use of RO will continue to grow, especially in areas worldwide where the scarcity of water is a way of life.

9-10 VAPOR PRESSURE OF SPHERICAL DROPLETS

The vapor-pressure data of pure substances tabulated in the literature are measured for a flat surface between the liquid and vapor phases. There are numerous engineering applications where liquids in droplet form are present. For example, the use of spray nozzles in many diverse industries is well documented. When a liquid appears in a droplet form, the phenomenon of surface tension plays an important role. Surface tension γ is defined as the work required per unit increase in area A_s of a liquid surface. For a given liquid it is a function not only of the temperature, but also of the radius of curvature. Surface tension data are typically reported for a surface of infinite curvature, that is, for a flat surface. Based on the definition of surface tension, the work required to increase the area of a surface is given simply by $\gamma\, dA_s$.

Consider a droplet of a pure liquid at temperature T and radius of curvature r. A differential quantity of mass dN will be transferred to the droplet, thereby increasing its surface area A_s and its volume V. The work done on the system is the sum of the boundary work due to a volume change and surface tension work due to a change in surface area. The first law for a reversible change takes on the form

$$dU = T\,dS - P\,dV + \gamma\,dA_s + \sum_i \mu_i\,dN_i \tag{a}$$

Thus Eq. (8-4) has been modified to account for the additional work done against the force associated with surface tension. The change in the Gibbs function for the liquid phase from its definition is

$$dG = dU + P\,dV + V\,dP - T\,dS - S\,dT \tag{b}$$

When Eq. (a) is substituted into Eq. (b), we find that

$$dG = V\,dP - S\,dT + \gamma\,dA_s + \sum_i \mu_i\,dN_i$$

or for a pure liquid phase

$$dG = V\,dP - S\,dT + \gamma\,dA_s + \mu\,dN$$

If the droplet is now assumed to be spherical, the quantity dA_s can easily be expressed in terms of dN. For a sphere $A_s = 4\pi r^2$, or $dA_s = 8\pi r\,dr$. The volume is given by $(\frac{4}{3})\pi r^3$ and the change in volume $dV = 4\pi r^2\,dr$. If the quantity dr is eliminated between the equations for dA_s and dV, we find that

$$dA_s = \frac{2\,dV}{r}$$

But $dV = v_{\text{liq}}\,dN$, where v_{liq} is the molar specific volume, and therefore $dA_s = 2v_{\text{liq}}\,dN/r$. Substitution of this result into the above equation for dG yields

$$dG = V\,dP - S\,dT + \frac{2\gamma v_{\text{liq}}\,dN}{r} + \mu\,dN$$

$$= V\,dP - S\,dT + \left(\frac{2\gamma v_{\text{liq}}}{r} + \mu^F\right) dN \tag{9-55}$$

where μ^F is the chemical potential for a flat surface. The chemical potential of a liquid with a curved surface, μ^C, is defined by the coefficient of dN in the above equation; that is,

$$\mu_{\text{liq}}^{C} = \left(\frac{\partial G}{\partial N_i}\right)_{T,P} \equiv \mu_{\text{liq}}^{F} + \frac{2\gamma v_{\text{liq}}}{r} \tag{9-56}$$

At equilibrium, $\mu^C = \mu^{\text{vap}}$. In the general case the chemical potential of the vapor in equilibrium with the droplet is given in terms of the fugacity by $d\mu = RT\, d \ln f$. For many engineering systems containing droplets the pressure will be relatively low so that the vapor can be assumed to behave as an ideal gas. For an ideal gas $d\mu_{ig} = RT d \ln p$. If this equation is integrated from a state where the vapor is in equilibrium with a flat surface to one where a curved surface exists, we find that

$$(\mu^C - \mu^F)_{ig} = RT \ln \frac{p^C}{p^F}$$

or

$$\mu_{ig}^{C} = \mu_{ig}^{F} + RT \ln \frac{p^C}{p^F} \tag{9-57}$$

where p^F and p^C are the vapor pressures of a flat surface and a curved surface, respectively. Equations (9-56) and (9-57) are now equated to each other. Note also that $\mu_{\text{liq}}^{F} = \mu_{ig}^{F}$. As a result,

$$\ln \frac{p^C}{p^F} = \frac{2\gamma v_{\text{liq}}}{rRT} \tag{9-58}$$

Note that all the terms on the right-hand side of the equation are positive for a liquid droplet (or convex meniscus in a capillary tube). Therefore the vapor pressure of a liquid droplet is always greater than that of a flat liquid surface of the same substance. Equation (9-58) indicates that the difference between the curved surface and flat surface vapor pressures increases with increasing surface tension and specific volume and with decreasing radius of curvature and temperature. For a given substance and within a small temperature range this difference primarily is a function of the radius of curvature. The smaller the droplet is, the larger the difference becomes. However, there is a practical limit to the droplet size formed by spray nozzles and other devices. This frequently is in the range of 10–50 μm.

For many droplets the change in vapor pressure is small. In such a case $\ln x \approx x - 1$. With this approximation, Eq. (9-58) can be written in the form

$$p^C \approx p^F \left(1 + \frac{2\gamma v_{\text{liq}}}{rRT}\right) \tag{9-59}$$

This approximation is reasonably accurate when the change in the vapor pressure is less than about 5 percent.

9-11 IDEAL SOLUBILITY OF A GAS IN A LIQUID

Consider a liquid i and a gas B which are brought into contact. From a thermodynamic viewpoint, if substance B is soluble in the liquid, the resulting solution may be treated as a liquid mixture with mole fractions x_i and x_B. If the liquid solution is an ideal solution over the whole range of composition, then Raoult's law is valid for component B; that is,

$$p_B = x_B p_B^{\text{sat}} \tag{9-22}$$

or

$$x_B = \frac{p_B}{p_B^{\text{sat}}} \tag{9-60}$$

The value of x_B in the above equation is frequently termed the *ideal solubility.* Thus it is possible to estimate the solubility of an ideal gas in an ideal solution at any temperature below the critical temperature of the gas, provided the vapor pressure of the liquid form of substance B is known. [For a real gas fugacity data replace pressure data in Eq. (9-60).] For example, at 20°C, the vapor pressure of pure carbon dioxide (CO_2) is 56.3 atm. The ideal solubility, then, is $1/56.3 = 0.0178$ at a total CO_2 pressure of 1 atm. On the basis of Raoult's law the solubility is independent of the nature of the solvent. Experimentally, at 20°C and 1 atm the actual solubility of CO_2 in chloroform, toluene, and ethyl alcohol is 0.0123, 0.0107, and 0.0070, respectively. Hence the ideal solubility is at least the right order of magnitude, and its value frequently is a good starting point if actual experimental information is not available. Note that the solubility of any gas is also independent of the presence of other dissolved gases. This fact has an important engineering application in the scrubbing of gases with water to remove certain components of the gas phase. (If the presence of other gases raises the total system pressure considerably, the effect of the total pressure on the vapor pressure of substance B may have to be considered. However, this effect in most practical situations is negligible.) The following two generalizations may be made: The higher the critical temperature and normal boiling point (boiling temperature at 1 atm) of a substance, the greater the solubility. This occurs because substances with high normal boiling points have relatively low vapor pressures. From another viewpoint, a substance which is more easily liquefied when the temperature is lowered will be more soluble in a given solvent at a particular temperature. Also, solubility decreases with increasing temperature, because this causes an increase in the saturation pressure. For example, nitrogen gas is soluble in water to the extent of $0.0232\ \text{cm}^3/\text{cm}^3$ at 0°C and 1 atm, whereas at 25°C the solubility has decreased to $0.0147\ \text{cm}^3/\text{cm}^3$ at the same pressure. These values are frequently called the absorption coefficients of the liquid for that particular gas, where the volume of the gas is reduced to 0°C and 1 atm and the partial pressure of the gas is maintained at 1 atm above the liquid.

To extend the method of calculating ideal solubilities of a gas to temperatures which lie above its critical temperature, it is necessary to extrapolate the vapor pressure of the liquid. This may be done with the integrated form of the Clausius-Clapeyron equation, where it is assumed that the enthalpy of vaporization remains constant. Knowledge of the vapor pressure at two temperatures allows one to estimate an average enthalpy of vaporization in that range of temperatures. Use of this average value and the vapor pressure at one temperature, then, leads to an estimate of a hypothetical vapor pressure at some temperature above the critical temperature. For example the solubility of methane, with a critical temperature of 191 K, may be desired at room temperature. Extrapolation by means of vapor-pressure data for methane yields a hypothetical vapor pressure around 300 atm at 25°C (298 K). Hence if the actual pressure of methane above a liquid is 1 atm, the ideal solubility is $1/300 = 0.0033$, approximately, in any solvent. The observed values in ethyl ether, hexane, and acetone are 0.0045, 0.0031, and 0.0022, respectively. (The value for water, however, is only 0.00002.) Even though most solutions of gases in liquids are not ideal, the general conclusions discussed in this section are in agreement with experimental data.

9-12 ABSORPTION REFRIGERATION

In any refrigeration process the energy removed from the cold region eventually must be rejected to another region which is at a considerably higher temperature. This second region is usually the surrounding environment. To carry out the heat-rejection process, the temperature of the fluid within the refrigeration cycle must be raised to a value above that of the environment. In a vapor-compression-refrigeration cycle the temperature of the vapor leaving the evaporator is raised by a compression process. The work input required in an ideal steady-flow-compression process is given by the integral of $v\,dP$. Once the pressure range is set for a given refrigerant, the main variable which controls the amount of work input is the specific volume of the fluid. In a vapor-compression-refrigeration cycle the value of v is relatively large, since the fluid is in the superheated region throughout the compression process. Therefore the work input is also relatively large. One method of overcoming this disadvantage is to design a refrigeration cycle in which the fluid is a liquid during the compression process. Then the work input will be significantly smaller.

The technique of *absorption refrigeration* is based on this approach. To accomplish this, however, the overall cycle becomes physically more complex. In addition, a two-component mixture, such as ammonia and water or lithium bromide and water, must be used as the circulating fluid in part of the cycle, rather than the single component typically used in a vapor-compression cycle. The lithium bromide-water system is limited to applications where the minimum fluid temperature is above 0°C, the freezing point of water. Absorption is the process of a vapor of one substance going into solution with another substance. In the ammonia-water system, for example, ammonia acts as the refrigerant and water as the absorbent and transport fluid. When one is selecting pairs of substances, it is important that the transport medium hold large quantities of the vapor (refrigerant) at a relatively low temperature. In addition, a sizable fraction of the refrigerant vapor should leave the liquid solution when energy is added and the temperature increased.

Two-component fluids have an important characteristic which must be recognized. When two phases are present at equilibrium, the composition of a given component is not the same in the two phases. The vapor phase will contain more of that component which is more volatile at the given temperature (see Sec. 9-5). For example, consider an ammonia-water mixture. At 43°C (110°F) the saturation pressure of ammonia is 17 bars (247 psia), whereas that of water is 0.09 bar (1.3 psia). Therefore ammonia has a much greater tendency to vaporize at a given temperature than water. Hence for an ammonia-water solution, the vapor phase contains much more ammonia (is richer in ammonia) than the liquid phase in equilibrium with it. This fact is extremely important in making mass and energy balances on equipment used in absorption refrigeration.

9-12-1 General Design Principles

A schematic diagram of a simple absorption-refrigeration cycle is shown in Fig. 9-9. A condenser, throttle valve, and evaporator are shown on the left-hand side of the diagram. These three pieces of equipment also are used in a conventional vapor-compression cycle. The compressor in that cycle, however, is now replaced by four pieces of equipment: an absorber, a pump, a generator, and a valve. For the purpose of discussion we consider ammonia and water as the two components in the cycle. Essentially pure ammonia vapor leaves the evaporator at state 1 and enters the absorber. The absorbing medium is a weak solution (low ammonia concentration) of ammonia and water which continually enters the absorber at state 5. There are two different energy effects during the absorption process. The two different substances attract each other, and thus a heat release occurs

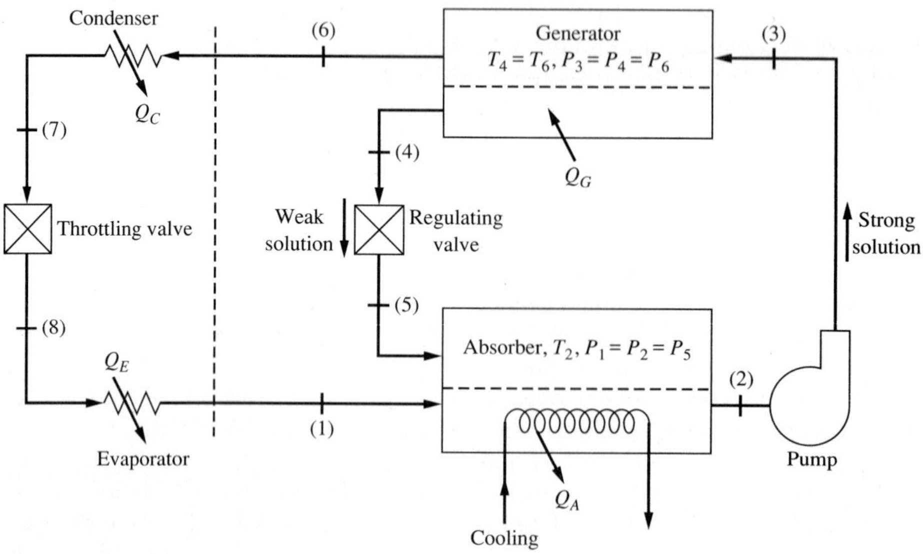

FIGURE 9-9
Schematic diagram of a simple absorption-refrigeration cycle.

in the formation of the solution. This energy release is called the *enthalpy* (heat) *of solution*. The enthalpy of condensation of the vapor must also be removed if the solution is to remain at a constant temperature. Thus the enthalpy change during absorption is greater than that for condensation of the pure vapor. Hence cooling water must be circulated through the absorber to keep the solution at a constant temperature. The temperature of the absorbing fluid must be kept at as low a temperature as possible, since the amount of pure refrigerant (ammonia) which can be absorbed decreases as the temperature increases. However, the absorber must operate at 10 to 20 degrees above the cooling water temperature to allow for adequate heat-transfer rates.

The liquid which leaves the absorber at state 2 is a rich or strong solution (high-ammonia concentration). This binary liquid mixture is now compressed by a pump to state 3, which is at the desired condenser pressure. The temperature rise of the binary mixture due to the pump work usually is quite small. Thus the strong solution is subcooled liquid as it enters the generator shown in Fig. 9-9. Heat Q_G must now be added to the solution in the generator to warm the incoming liquid to the saturation temperature and to drive out some of the ammonia from the solution. This energy requirement includes the enthalpy (heat) of vaporization of the ammonia, as well as plus or minus the enthalpy (heat) of solution. The nearly pure ammonia passes to the condenser at state 6 and eventually returns to the absorber at state 1. The weak solution left in the generator (state 4) now flows through a regulating valve, which drops the pressure of the solution to that in the absorber. It mixes with the solution remaining in the absorber, and cold vapor coming from the evaporator is added to the overall liquid solution. The strong solution is cooled, as before, and the cycle is repeated. Hence the ammonia-water solution which cycles through the absorber, pump, generator, and valve merely serves as a transport medium for the ammonia refrigerant.

Note that any absorption-refrigerator device requires an external heat source for the generation of refrigerant vapor. Thus absorption refrigeration is especially attractive if a low-temperature (100 to 200°C, or 200 to 400°F) source of thermal energy is readily available. Examples of thermal energy sources in this temperature range include waste energy from chemical or process steam plants, and solar and geothermal energies.

In practice, absorption-refrigeration units have at least two modifications that are not shown in Fig. 9-9. First, the cold, strong solution at state 3 needs warming before it enters the generator, and the warm, weak solution at state 4 needs cooling before it enters the absorber. Consequently, a heat exchanger is placed between the absorber and the generator, which permits heat transfer from the weak solution to the strong solution. Second, a major requirement is that the ammonia from the generator should be essentially free of water as it passes through the condenser-throttling-valve-evaporator loop. Any traces of water will freeze in the expansion valve and evaporator. Hence the vapor leaving the generator passes through a device called a rectifier before it enters the condenser. The rectifier separates any remaining water vapor from the vapor stream leaving the generator and returns the water to the generator.

9-12-2 Theoretical COP of an Absorption System

The coefficient of performance (COP) for any refrigeration cycle is the ratio of desired output to required input. For a vapor-compression cycle this is the ratio of the heat transfer to the evaporator from the cold region divided by the compressor work input. For the absorption cycle, however, the required energy input is both the heat transfer to the generator as well as the pump work. But the pump work may be assumed to be negligible. Hence on the basis of Fig. 9-9

$$\text{COP}_{\text{absorp}} = \frac{Q_E}{Q_G + W_p} \approx \frac{Q_E}{Q_G} \tag{9-61}$$

Because the required input is heat transfer to the generator, absorption-refrigeration cycles are called "heat-driven" systems, whereas vapor-compression devices are "work-driven" systems.

The maximum COP is attained when the system is totally reversible. This requires that the four heat-transfer processes shown in Fig. 9-9 be reversible and isothermal. The refrigeration effect Q_E occurs at $T_E = T_L$ (low temperature), and the heat supply Q_E in the generator occurs at $T_G = T_H$ (high temperature). In addition, the heat-rejection processes Q_C in the condenser and Q_A in the absorber occur at the ambient temperature $T_A = T_0$ (ambient temperature). If the pump work is neglected, then an energy balance on the entire cycle is simply

$$Q_G + Q_E = Q_C + Q_A \tag{a}$$

where all quantities have positive values.

The second law for the cycle requires that the sum of the entropy changes be zero in the totally reversible case. The fluid in the evaporator, condenser, generator, and absorber acts in a cyclic manner, and thus has a zero entropy change. However, the sources and sinks for heat transfer are noncyclic. Therefore

$$\Delta S_C + \Delta S_A + \Delta S_G + \Delta S_E \geq 0 \tag{9-62}$$

where the inequality is included as a reminder of irreversible operation. The entropy change of any reservoir is Q/T. Thus the above equation becomes

$$-\frac{Q_C + Q_A}{T_A} + \frac{Q_G}{T_H} + \frac{Q_E}{T_L} \geq 0 \tag{b}$$

where again the Q terms are taken as positive. Substitution of Eq. (a) into Eq. (b) yields, upon rearrangement,

$$\frac{Q_E}{Q_G} \leq \left(\frac{T_L}{T_A - T_L}\right)\left(\frac{T_H - T_A}{T_H}\right) \tag{c}$$

Finally, the use of the equality format of Eq. (c) in Eq. (9-61) leads to the overall COP of an absorption-refrigeration cycle under reversible conditions, namely,

$$\text{COP}_{\text{absorp,rev}} = \left(\frac{T_L}{T_0 - T_L}\right)\left(1 - \frac{T_0}{T_H}\right) \tag{9-63}$$

where the ambient temperature T_A has been replaced by T_0.

The result expressed by Eq. (9-63) is not unexpected. The first term on the right is the Carnot COP of a refrigerator operating between T_L and T_0. The second term is the Carnot efficiency of a heat engine operating between T_H and T_0. Thus all the equipment to the right of the dashed line in Fig. 9-9 effectively operates as a heat engine. It replaces the compressor of a simple vapor-compression-refrigeration cycle, where the efficiency of the prime mover is not considered in the definition of the COP. In practice, the COP of an absorption unit is usually less than 1.

Example 9-10. An ideal absorption-refrigeration system operates with $T_H = 130°\text{C}$, $T_L = -10°\text{C}$, and $T_0 = 22°\text{C}$. (*a*) Find the ideal COP. (*b*) Now, consider a vapor-compression system operating at the same T_L and T_0 values. However, the ideal compressor is driven by an ideal motor which receives its energy from an ideal Carnot power plant with a high-temperature source at 540°C (1000°F). Find the system and the overall COP in this case.

Solution

(*a*) Equation (9-63) shows that

$$\text{COP}_{\text{absorb,rev}} = \left(\frac{263}{295 - 263}\right)\left(1 - \frac{295}{403}\right) = 2.20$$

Since actual units have a COP around 1, irreversibilities greatly reduce the ideal COP.

(*b*) For a vapor-compression unit acting as an ideal Carnot refrigerator the COP is

$$\text{COP}_{\text{Carnot}} = \frac{T_L}{T_0 - T_L} = \frac{263}{295 - 265} = 8.22$$

This result looks very good in comparison to part (*a*). However, the delivery of energy as work input to the device has not been taken into account. For a Carnot heat engine operating between $T_H = 540°\text{C}$ and $T_0 = 22°\text{C}$

$$\eta_{\text{th,Carnot}} = 1 - \frac{T_0}{T_H} = 1 - \frac{295}{813} = 0.637$$

If the engine is directly connected to the refrigeration unit, then the overall COP becomes

$$\text{COP}_{\text{overall}} = 8.22(0.637) = 5.24$$

This value is more than twice as large as that of the absorption unit. However, a comparison of actual equipment tells a different story. In practice, the COP of a vapor-compression unit may be around 3, and the thermal efficiency of a power plant is around 35–40 percent. Thus the overall COP in reality is around 1 and is no different from the absorption unit in rough figures. The major point is that it is unfair to compare the overall COP of an absorption unit with the "local" COP of a vapor-compression unit. The latter system has not been charged for the energy cost of the work input.

Although the pump work for an absorption-refrigeration cycle is small, it must be driven by an external work device, such as an electric motor. This restricts its general use. An alternative to this is to use a three-fluid system, where hydrogen is added to ammonia and water. This allows the removal of the pump. This three-fluid system is popular in recreational vehicles. Propane is the sole energy source for Q_G and is already available in such vehicles for cooking purposes.

9-13 THE ENTHALPY-COMPOSITION DIAGRAM FOR MULTICOMPONENT PHASE CHANGES

The thermodynamic analysis of a two-component, two-phase system is addressed in Sec. 9-5 in terms of variations in temperature, pressure, and composition. Although the model used involved a liquid ideal solution, with a vapor phase modeled as an ideal gas, the trends in behavior are general for miscible fluids. The difference in compositions of the two phases in equilibrium is especially noted through the use of *P-x* and *T-x* diagrams in that section. An energy analysis of liquid-vapor phase changes for binary mixtures requires knowledge of the enthalpy for both phases as a function of temperature, pressure, and composition. One method of presentation of such data is by means of an *h-x* diagram. In Sec. 8-10 the enthalpy-composition diagram at a given pressure was constructed for binary liquid mixtures with complete miscibility. Figure 8-5 is an example of such a diagram. It is fairly easy conceptually to extend this type of diagram into the liquid-vapor and superheated-vapor regions.

Figure 9-10 shows details of the required diagram at a fixed pressure P. At the left and right ordinates of the *h-x* diagram the composition is pure A and pure B, respectively. Line G-H represents a subcooled-liquid temperature for the pressure P. As the temperature is raised, the saturated-liquid state $h_{f,B}$ is reached on the right coordinate. A vertical distance equal to the enthalpy of vaporization $h_{fg,B}$ is added to $h_{f,B}$, thus defining the position of the enthalpy of pure saturated vapor, $h_{g,B}$. The saturation temperature line $T_{\text{sat},B}$ is also shown running from $h_{g,B}$ down to $h_{f,B}$, and then to the left into the subcooled-mixture region. In a similar fashion, as the temperature is increased further above $T_{\text{sat},B}$, the state of pure saturated liquid for A, $h_{f,A}$ is reached on the left axis. (The diagram arbitrarily assumes A boils at a higher temperature than B for the given pressure.) The saturated-vapor state $h_{g,A}$ lies a distance $h_{fg,A}$ above $h_{f,A}$. The saturation temperature line $T_{\text{sat},A}$ runs from $h_{f,A}$ vertically to $h_{g,A}$, and then to the right into the superheated-vapor region. Several other temperature lines in the superheated-vapor and subcooled-liquid regions are shown. Although the temperature lines in the liquid region are drawn with considerable curvature, similar lines in the superheat region are essentially straight for the following reason. Recall that the

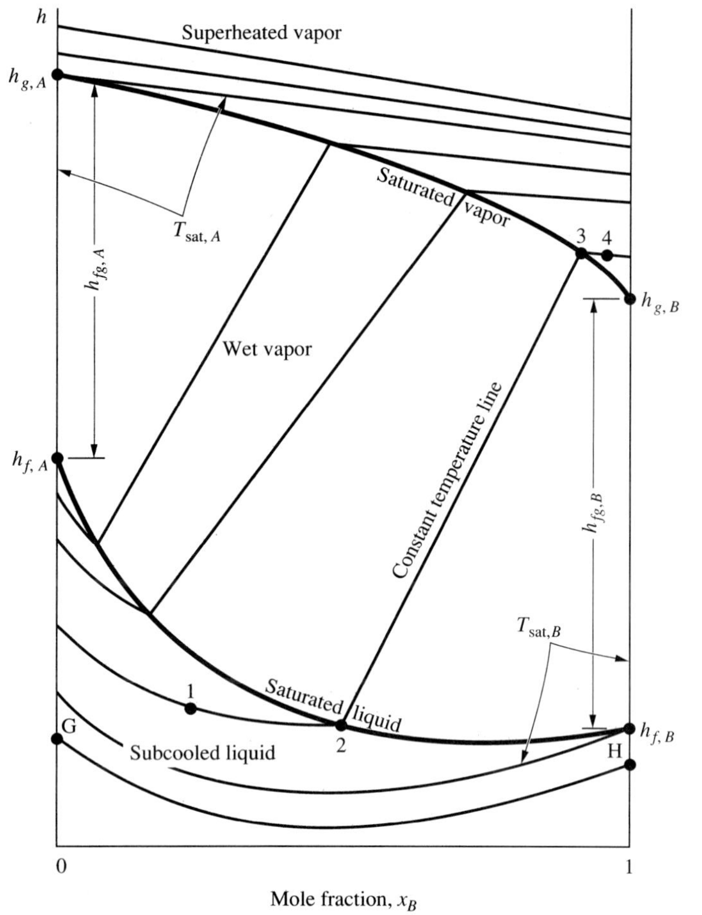

FIGURE 9-10
Enthalpy-composition diagram in the vaporization region.

enthalpy of a binary mixture in any single-phase region is given by Eq. (8-127), namely, $h_{\text{mixt}} = [(1 - x)h_A + xh_B] + \Delta h_{\text{mix}}$. In the gas phase Δh_{mix} is usually negligible. Hence the mixture enthalpy for the gas phase is a linear function of the mole fraction, x.

Finally, the vapor and liquid saturation lines for the given pressure are added to the h-x diagram, as shown in Fig. 9-10. The two saturation lines connect states $h_{g,A}$ to $h_{g,B}$ and $h_{f,A}$ to $h_{f,B}$. The saturated-vapor line represents h_g as a function of the vapor mole fraction y of species B, whereas the liquid line represents h_f as a function of the liquid mole fraction x of species B. (However, it is common practice to label the abscissa solely as x on enthalpy-composition diagrams.) The saturated-vapor line lies below the isotherm $T_{\text{sat},A}$, and the saturated-liquid line lies above the isotherm, $T_{\text{sat},B}$. Isotherms with values which lie between $T_{\text{sat},B}$ and $T_{\text{sat},A}$ can also be drawn between the saturated-liquid and -vapor lines. For

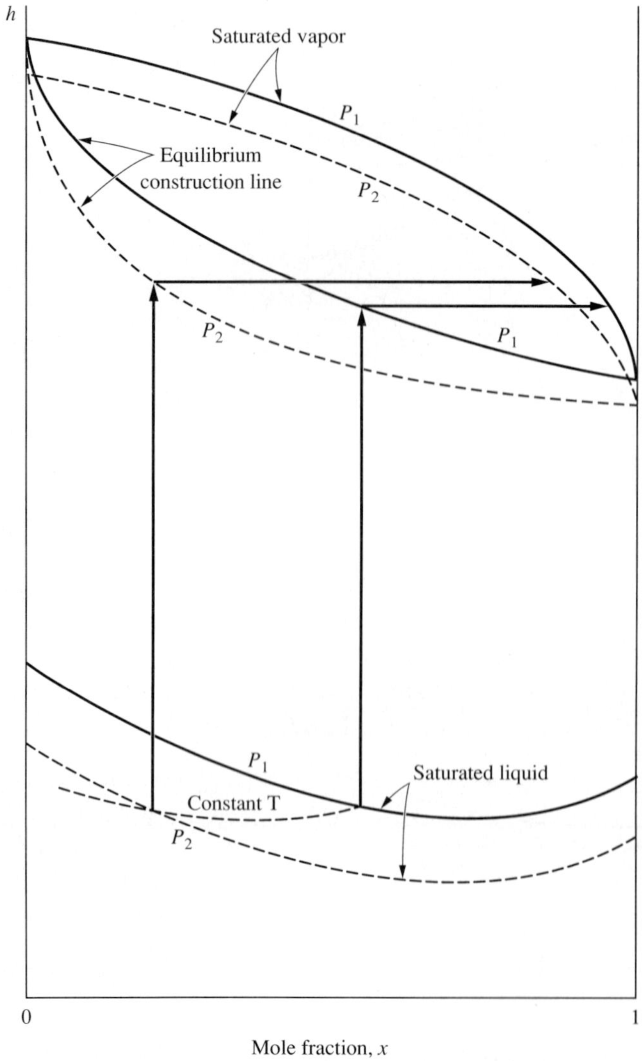

FIGURE 9-11
An enthalpy-composition diagram for a binary mixture showing several pressure and equilibrium construction lines.

example, consider the isotherm which follows the path 1-2-3-4 in Figure 9-10. Points 2 and 3 represent the saturated-liquid and -vapor states, whereas points 1 and 4 represent states in the subcooled-liquid and superheated-vapor regions, respectively. Each isotherm intersects the liquid and vapor lines at two different compositions, such as x_2 and x_3. The wet vapor isotherms become steeper, the closer they lie to the two ordinates. We could plot T and x at these intersections on a T-x-y diagram. If we did, a diagram similar to Fig. 9-3, discussed earlier for phase transformations, would result. Thus the h-x (or h-x-y) and the T-x-y

diagrams for a binary mixture at a given pressure are intimately connected. Note that on a T-x-y diagram the liquid and vapor lines meet at $x = 0, 1$. This is obviously not true for an h-x diagram. It should be emphasized that the general characteristics of an enthalpy-mole-fraction (h-x) diagram are the same as for an enthalpy-mass-fraction (h-X) diagram.

Figure 9-10 illustrates the position of temperature and saturation lines for vaporization at a single pressure. The location of the saturation lines would be different for different pressures. Thus a number of pairs of saturation lines are usually shown on an h-x (or h-X) diagram. The position of the isothermal lines in the subcooled-liquid region remain the same, independent of pressure. This occurs because the specific heat of the liquid phase is insensitive to pressure. However, the isotherms in the wet vapor and superheated-vapor regions are affected by the pressure. When data for several pressures are required, it becomes difficult to show all the necessary isotherms clearly in these latter two regions. (If the pressure range is low enough, the superheated-vapor phase will behave like an ideal gas. In this limited case the temperature lines in the superheat region will be independent of pressure, and may be easily shown, since only one set of lines is required.) In practice, temperature lines are often omitted and are replaced by *equilibrium construction lines* for each required pressure. Two such lines are shown in Fig. 9-11 for pressures P_1 and P_2, along with the two pairs of saturation lines. To find the saturated-vapor enthalpy and composition, one proceeds as follows. For a given liquid composition and saturation pressure a vertical (constant x) line is drawn to the equilibrium construction line for that pressure. Then a horizontal (constant h) line is drawn to the right until it intersects with the actual saturated-vapor line at the given pressure. This point of intersection is the state of the saturated vapor in equilibrium with the saturated liquid at the given temperature and pressure. The type of diagram outlined above is characteristic of the water-ammonia system used in absorption refrigeration.

PROBLEMS

9-1. At 20°C the vapor pressure of pure water is 2.339 kPa. Exactly 200 g of water at 20°C has m grams of sucrose ($C_{12}H_{22}O_{11}$) added to it. If sucrose has a negligible vapor pressure and the vapor pressure of the final solution is 2.327 kPa, estimate the grams of sucrose added. Assume the validity of Raoult's law.

9-2. At 25°C the vapor pressure of pure water is 3.169 kPa. Exactly 100 g of water at this temperature has m grams of urea [$CO(NH_2)_2$, $M = 60.06$] added to it. If urea is a nonvolatile compound (has a negligible vapor pressure) and the vapor pressure of the solution is found to be 3.154 kPa, estimate the grams of urea added, assuming an ideal solution.

9-3. At 25°C the vapor pressure of pure water is 0.03169 bar. Exactly 235 g of water has m grams of urea (CON_2H_4) added to it. If urea has a negligible vapor pressure and the vapor pressure of the solution is 0.03131 bar, estimate the grams of urea added. Assume the validity of Raoult's law.

9-4. A liquid-vapor system contains only n-pentane and n-hexane. The vapor pressures of pure pentane (n-C_5H_{12}) and pure hexane (n-C_6H_{12}) at 30°C are 818 and 250 mbar, respectively.

a. Determine the composition of the liquid phase (x) and the total pressure (P) for mole fractions of pentane in the vapor phase (y) of 0.2, 0.4, 0.5, 0.6, 0.7, 0.8, and 0.9 at the given temperature.

b. Plot a P-x and P-y diagram (simultaneously on the same diagram) for the data determined above, where the abscissa is both x and y. Thus both the liquid and vapor lines will appear on the diagram for x values from 0 to 1.0. Assume the validity of Raoult's law.

9-5. A liquid-vapor system contains only n-hexane and n-heptane. The vapor pressures of pure n-hexane (C_6H_{14}) and pure n-heptane (C_7H_{16}) at 20°C are 143.6 and 44.0 mbar, respectively.

a. Determine the composition of the liquid phase (mole fraction, x) and the total pressure P for mole fractions of hexane in the vapor phase, y, of 0, 0.2, 0.4, 0.6, 0.7, 0.8, 0.9, and 1.0 at the given temperature.

b. Plot simultaneously a P-x and a P-y diagram for the data, where the abscissa is both x and y. Thus both the liquid and vapor lines will appear on the diagram. Assume the validity of Raoult's law.

9-6. A liquid-vapor system contains only n-pentane and n-hexane. The vapor pressures of pure pentane (n-C_5H_{12}) and pure hexane (n-C_6H_{14}) at 20°C are 554 and 158 mbar, respectively.

a. Determine the composition of the liquid phase, x, and the total pressure, P, for mole fractions of pentane in the vapor phase, y, of 0.2, 0.4, 0.6, 0.7, 0.8, and 0.9 at the given temperature. Assume the validity of Raoult's law.

b. Plot a P-x-y diagram for the data, where the abscissa is both x and y of pentane. Thus both the liquid and vapor lines will appear on the diagram.

9-7. A mixture of 44.0 mol percent n-pentane and 56.0 mol percent n-hexane is maintained at 320 K. Assume the mixture behaves as an ideal solution. The vapor pressures of pure pentane (n-C_5H_{12}) and pure hexane (n-C_6H_{14}) are given by:

$$\ln p(n\text{-}C_5H_{12}) = 17.533 - \frac{3318.1}{T} \qquad \text{and} \qquad \ln p(n\text{-}C_6H_{14}) = 18.057 - \frac{3837.4}{T}$$

where p is in millibars and T in kelvins.

a. Estimate the initial boiling pressure, in millibars, if the mixture is initially a liquid at 1000 mbar.

b. Determine the composition, in percent, of the initial vapor formed.

c. Estimate the dew-point pressure, in millibars, as the pressure is raised at constant temperature from 500 mbar for the same initial composition.

d. Determine the composition, in percent, of the initial liquid formed.

e. Plot a P-x-y diagram (for pentane) which shows the position of the two initial states, the four saturation states determined above, and the states for $x = 0$ and $x = 1$.

9-8. A mixture of 54.0 mol percent n-pentane and 46.0 mol percent n-hexane is maintained at 330 K. Assume the mixture behaves as an ideal solution. Vapor-pressure equations for the two components are given in Prob. 9-7.

a. Estimate the initial boiling pressure, in millibars, if the mixture is initially a liquid at 1600 mbar.

b. Determine the composition, in percent, of the initial vapor formed.

c. Estimate the dew-point pressure, in millibars, as the pressure is raised at constant temperature from 700 mbar for the same initial composition.

d. Determine the composition, in percent, of the initial liquid formed.

e. Plot a P-x-y diagram (for hexane) which shows the position of the two initial states, the four saturation states determined above, and the states for $x = 0$ and $x = 1$.

9-9. A mixture of 64.0 mol percent n-pentane and 36.0 mol percent n-hexane is maintained at 300 K. Assume the mixture behaves as an ideal solution. Vapor-pressure equations for the two components are given in Prob. 9-7.

a. Estimate the initial boiling pressure, in millibars, if the mixture is initially a liquid at 550 mbar.

b. Determine the composition, in percent, of the initial vapor formed.

c. Estimate the dew-point pressure, in millibars, as the pressure is raised at constant temperature from 300 mbar for the same initial composition.

d. Determine the composition, in percent, of the initial liquid formed.

e. Plot a P-x-y diagram (for pentane) which shows the position of the two initial states, the four saturation states determined above, and the states for $x = 0$ and $x = 1$.

9-10. A liquid-vapor system contains only n-hexane and n-heptane. The vapor pressures of pure hexane and pure heptane are given by

$$\ln p(C_6H_{14}) = 18.057 - \frac{3837.4}{T} \quad \text{and} \quad \ln p(C_7H_{16}) = 18.217 - \frac{4231.0}{T}$$

where p is in millibars and T is in kelvins.

a. Determine the mole fraction of n-hexane in the liquid, x, and vapor, y, phases for a total pressure of 100 mbar and temperatures of 15, 20, 25, 30, and 35°C.

b. Plot a T-x and a T-y diagram simultaneously, where the abscissa is both x and y for hexane. Show the liquid and vapor lines for x values from 0 to 1.0. Assume the validity of Raoult's law.

9-11. Consider a liquid-vapor mixture of n-pentane and n-hexane. The vapor-pressure equations for the pure components are given in Prob. 9-7.

a. Determine the mole fractions of n-pentane in the liquid, x, and vapor, y, phases for a total pressure of 250 mbar and temperatures of 278, 283, 288, 293, 298, and 303 K. Assume the validity of Raoult's law.

b. Plot a T-x and a T-y diagram simultaneously, where the abscissa is both x and y for pentane. Show the position of the liquid and vapor lines for x values from 0 to 1.0.

9-12. A liquid-vapor system contains only n-hexane and n-heptane. The vapor pressure of pure hexane and pure heptane are given by

$$\ln p(C_6H_{14}) = 18.057 - \frac{3837.4}{T} \quad \text{and} \quad \ln p(C_7H_{16}) = 18.217 - \frac{4231.0}{T}$$

where p is in millibars and T is in kelvins.

a. Determine the mole fractions of n-hexane in the liquid, x, and vapor, y, phases for a total pressure of 300 mbar and temperatures of 40, 45, 50, 55, and 60°C.

b. Plot a T-x and a T-y diagram simultaneously, where the abscissa is both x and y for hexane. Show the liquid and vapor lines for x values from 0 to 1.0. Assume the validity of Raoult's law.

9-13. A liquid mixture of 68.0 mol percent *n*-hexane (C_6H_{14}) and 32.0 mol percent *n*-heptane (C_7H_{16}) is initially at 60°C and 800 mbar. Assume the mixture is an ideal solution. Vapor-pressure data for the two pure components are given in Prob. 9-12.

a. Estimate the initial boiling point, in kelvins, as the temperature is raised at constant pressure.

b. Determine the composition, in mole percent, of the initial vapor formed when the boiling point is reached.

Now, assume that the same initial mixture exists as a vapor at 360 K and 800 mbar.

c. Estimate the dew-point temperature, in kelvins, as the temperature is lowered at constant pressure.

d. Determine the composition, in mole percent, of the initial liquid formed at the dew point.

e. Plot a *T-x-y* diagram (for hexane) which shows the position of the two initial states, the four saturation states determined above, and the states for $x = 0$ and $x = 1$.

9-14. A mixture of *n*-pentane, C_5H_{12}, and *n*-hexane, C_6H_{14}, in the amount of 76.0 and 24.0 mol present, respectively, initially exists in the liquid phase at 17°C and 600 mbar. Assume the mixture forms an ideal solution. Vapor-pressure equations for the two pure components are found in Prob. 9-7.

a. Estimate the initial boiling-point temperature, in kelvins, as the temperature of the liquid is raised at constant pressure.

b. Determine the composition, in mole percent, of the initial vapor formed.

Now, assume that the same initial mixture exists as a vapor at 320 K and 600 mbar.

c. Estimate the initial dew-point temperature, in kelvins, as the temperature of the vapor mixture is lowered at constant pressure.

d. Determine the composition, in mole percent, of the initial liquid formed at the dew point.

e. Plot a *T-x-y* diagram (for pentane) which shows the position of the two initial states, the four saturation states determined above, and the states for $x = 0$ and $x = 1$.

9-15. A mixture of *n*-hexane and *n*-heptane in the amount of 72.0 and 28.0 mole percent, respectively, exists at 100 mbar. Assume the mixture is an ideal solution. The vapor-pressure equations for *n*-hexane and *n*-heptane are given in Prob. 9-12.

a. Estimate the initial boiling-point temperature, in kelvins, if the mixture is initially a liquid at 280 K.

b. Determine the composition, in mole percent, of the initial vapor.

Now, assume that the same initial mixture exists as a vapor at 320 K and 100 mbar.

c. Estimate the dew-point temperature, in kelvins, as the temperature is lowered from 310 K at constant pressure.

d. Determine the composition, in mole percent, of the initial liquid formed at the dew point.

e. Plot a *T-x-y* diagram (for hexane) which shows the position of the two initial states, the four saturation states determined above, and the states for $x = 0$ and $x = 1$.

9-16. At 10°C the vapor pressure of ammonia, NH_3, is 2.908 bars and the saturated-liquid specific volume is 0.001504 m^3/kg. Consider the introduction of an inert gas until the total pressure is 30.0 bars. Estimate

a. the vapor pressure of ammonia under the new total pressure, in bars
b. the percent change in the vapor pressure due to the inert gas
c. Then predict the total pressure required to increase the vapor pressure of ammonia at 10°C by 1.00 percent, in bars.

9-17. At 260 K the vapor pressure of pure carbon dioxide (CO_2) is 24.21 bars and the saturated-liquid specific volume is 0.001001 m^3/kg. An inert gas is introduced into the gas phase in order to increase the total pressure on the liquid phase to twice its original value. Determine

a. the vapor pressure of CO_2, in bars
b. the percent change in the vapor pressure after the inert gas is introduced.

9-18. Pure refrigerant-134a (R-134a) exists as a two-phase mixture at 20°C and 5.673 bars. Nitrogen gas is introduced into the vessel until the total pressure reaches 50.0 bars. Estimate

a. the vapor pressure of the refrigerant under the new total pressure, in bars
b. the percent change in the vapor pressure due to the presence of the inert gas.

9-19. Estimate (a) the total pressure required to increase the vapor pressure of potassium at 900 K by 0.10 percent, in bars, and (b) the percent increase in the system pressure to achieve the 0.10 percent increase in vapor pressure. The vapor pressure for potassium at 900 K is 0.254 bar.

9-20. Determine (a) the total pressure required to increase the vapor pressure of R-134a at 20°C by 1.00 percent, in bars, and (b) the percent increase in the system pressure required.

9-21. Estimate (a) the total pressure required to increase the vapor pressure of nitrogen (N_2) at 100 K by 1.00 percent, in bars, and (b) the percent increase in the system pressure required.

9-22. Calculate the boiling-point temperature, in degrees Celsius, for a solution composed of 100.0 g of water and 1.584 g of urea [$CO(NH_2)_2$, $M = 60.06$] for a total pressure of 1.0133 bar (1.0 atm).

9-23. Ethyl alcohol is added to water boiling at 100°C in order to elevate its boiling point by 0.60°C. Determine

a. the mole fraction of ethyl alcohol in the final solution
b. the mass ratio required, in kilograms, of alcohol per kilogram of water.

9-24. Methyl alcohol is added to water boiling at 95°C in order to elevate the boiling point of water by 1.00°C. Determine

a. the mole fraction of methyl alcohol in the final solution
b. the mass ratio required, in kilograms, of alcohol per kilogram of water.

9-25. The enthalpy of fusion of pure cyclohexane (C_6H_{12}) at 280 K is 2665 kJ/kmol. Estimate the freezing-point lowering, in degrees Celsius, when 1.00 gmol of a solute is added to exactly 1000 g of cyclohexane.

9-26. Determine the liters (L) of pure ethylene glycol (molar mass of 62.07) which must be added to a 14-L radiator to prevent freezing of the solution of water and glycol at −10°C. The density of ethylene glycol may be taken to be 1.1155 g/cm^3.

9-27. The enthalpy of fusion of pure dioxane ($C_4H_8O_2$) at 11.0°C is 145.81 kJ/kg. Calculate the freezing-point lowering, in degrees Celsius, when 1.00 gmol of a solute is added to 1000 g of dioxane.

9-28. With the aid of Raoult's law,

a. predict the osmotic pressure at 25°C of a 1-molal solution of sucrose in water, in bars. (*Note:* A 1.00-molal solution is one in which 1.00 gmol of solute is added to 1000 g of solvent.)

b. Then determine how much sucrose, $C_{12}H_{22}O_{11}$, in grams, must be dissolved in 1.00 L of water to give a solution with an osmotic pressure of 7.0 bars at 25°C.

9-29. Estimate the grams of sucrose ($C_{12}H_{22}O_{11}$) which must be dissolved in 1.0 L of water in order to prepare a solution with an osmotic pressure of 6.0 bars at 25°C.

9-30. Estimate how much dioxane ($C_4H_8O_2$), in grams, must be dissolved in 1.0 L of water at 25°C in order to prepare a solution with an osmotic pressure of 5.0 bars.

9-31. Ethyl alcohol (C_2H_5OH) is sprayed from a nozzle to form spherical droplets with a radius of 0.5×10^{-4} cm at 20°C. The surface tension is 0.02275 N/m, and the specific gravity is 0.7893 for the fluid. Estimate the percent change in vapor pressure of a curved surface compared to a flat surface.

9-32. Estimate the percent change in the vapor pressure of *n*-hexane (C_6H_{14}) when in the form of spherical droplets with a radius of 2.0×10^{-5} cm, relative to a flat surface condition, at 20°C. The surface tension of hexane in air is 0.01843 N/m and the specific gravity of hexane is 0.660.

9-33. Benzene (C_6H_6) is sprayed from a nozzle to form spherical droplets with a radius of 0.4×10^{-4} cm at 20°C. The surface tension is 0.0302 N/m and the density is 0.879 g/cm^3. Estimate the percent change in its vapor pressure, a curved surface compared to a flat surface.

REFERENCES

1. Smith, J. M., and H. C. Van Ness, *Introduction to Chemical Engineering Thermodynamics*, 4th ed., McGraw-Hill, Inc., New York, 1987.

CHAPTER 10

CHEMICAL REACTIONS

This chapter considers the thermodynamics of chemical reactions, including the application of thermomechanical availability to isothermal and adiabatic systems.

10-1 COMBUSTION STOICHIOMETRY

The focus of the next several sections will be toward combustion reactions. A *combustion reaction* involves a *fuel* such as a hydrocarbon with a generic formula C_xH_y and an oxidizer such as air or oxygen (although other fuel-oxidant combinations may be considered). Combustion reactions release energy, and such reactions are called *exothermic*. The most common oxidizer is atmospheric air. The *standard atmosphere,* based on worldwide measurements, has the molar and mass analysis on a dry basis found in Table 10-1. (Other gases in trace amounts would include krypton, hydrogen, and xenon. The value for carbon dioxide (CO_2) is increasing each decade.) Rather than work with this analysis, engineers more frequently use a simpler two-component model set at 21.0 percent oxygen (O_2) and 79.0 percent nitrogen (N_2) on a molar or volume basis. (Note from Table 10-1 that the major components of air are approximately 21 percent oxygen, 78 percent nitrogen, and 1 percent argon.) Since $0.79/0.21 = 3.76$, we may write for this simpler model of air

$$0.21 \text{ mol } O_2 + 0.79 \text{ mol } N_2 = 1.00 \text{ mol dry air}$$

$$1 \text{ mol } O_2 + 3.76 \text{ mol } N_2 = 4.76 \text{ mol dry air}$$

or

$$1 \text{ kg } O_2 + 3.31 \text{ kg } N_2 = 4.31 \text{ kg dry air}$$

The average molar mass of air is 28.97, which is frequently rounded off to 29.0 in calculations.

TABLE 10-1
Mole and mass fractions of gases found in the *standard atmosphere*

Species	Mole fraction	Mass fraction
Nitrogen (N_2)	0.780840	0.75520
Oxygen (O_2)	0.209476	0.23143
Argon (A)	0.009340	0.01288
Carbon dioxide (CO_2)	0.000314	0.00048
Neon (Ne)	0.000018	0.00001
Helium (He)	0.000005	0.00000
Total	0.999993	1.00000

Source: "U.S. Standard Atmosphere, 1962," Superintendent of Documents, U.S. Government Printing Office, Washington, DC, December, 1962.

The major products of hydrocarbon combustion are CO_2 and H_2O. The formation of NO and NO_2 from N_2 and O_2 is usually neglected as a first approximation in energy analysis. Some of the carbon in the fuel may appear as carbon monoxide (CO) if (1) the oxygen present is insufficient, (2) the mixing of fuel and oxidant is incomplete, and (3) the time for complete oxidation to CO_2 is insufficient on the basis of kinetic considerations. The combustion of C_xH_y is considered *complete* when all the carbon is converted to CO_2 and all the hydrogen is converted to H_2O. The *minimum* amount of air (or oxygen) required theoretically to achieve complete combustion of hydrocarbons is called the *theoretical* or *stoichiometric* air requirement. No excess oxygen appears in the products when the theoretical or stoichiometric amount of air is used and combustion is complete. A supply of 100 percent theoretical air is the minimum amount for complete combustion, whereas 150 percent theoretical air is equivalent to 50 percent *excess* air, or 1.5 times the stoichiometric requirement.

The amount of air (or oxygen) required in a stoichiometrically complete reaction is a standard in combustion studies. To obtain the moles of O_2 (or air) required for the 100 percent theoretical requirement, we write the stoichiometric equation for 1 mol C_xH_y as

$$C_xH_y + w(O_2 + 3.76\,N_2) \rightarrow a\ CO_2 + b\ H_2O + c\ N_2 \tag{10-1}$$

where w, a, b, and c represent unknown numbers of moles of air, CO_2, H_2O, and N_2, respectively. The conservation of atomic species principle applied to the four species present leads to

$$\begin{aligned} &\text{C balance:} & x &= a \\ &\text{H balance:} & y &= 2b \\ &\text{O balance:} & 2w &= 2a + b \\ &N_2 \text{ balance:} & 3.76w &= c \end{aligned}$$

As a result,

$$a = x \qquad b = \frac{y}{2} \qquad w = x + \frac{y}{4} \qquad c = 3.76\left(x + \frac{y}{4}\right)$$

Consequently, the *stoichiometric equation* for the complete combustion of C_xH_y with air is

$$C_xH_y + \left(x + \frac{y}{4}\right)(O_2 + 3.76\,N_2) \rightarrow x\ CO_2 + \frac{y}{2}\ H_2O + 3.76\left(x + \frac{y}{4}\right) N_2 \tag{10-2}$$

If the actual air used is α times the 100 percent theoretical amount so that excess O_2 appears in the products, then for complete combustion of C_xH_y in this case Eq. (10-2) becomes

$$C_xH_y + \alpha\left(x + \frac{y}{4}\right)(O_2 + 3.76\,N_2) \rightarrow x\ CO_2 + \frac{y}{2}\ H_2O + (\alpha - 1)\left(x + \frac{y}{4}\right) O_2 + 3.76\alpha\left(x + \frac{y}{4}\right) N_2 \tag{10-3}$$

As an example, the *stoichiometric* equation for the complete combustion of methane (CH_4) is, on the bases of Eq. (10-2),

$$CH_4 + 2\,O_2 + 7.52\,N_2 \rightarrow CO_2 + 2\,H_2O + 7.52\,N_2$$

The numbers which appear before each species in the stoichiometric equation are known as the *stoichiometric coefficients* of the various species, and usually they are represented by the symbol ν_i for the ith species. A general chemical reaction is then represented by

$$\nu_A\,A + \nu_B\,B \rightarrow \nu_E\,E + \nu_F\,F \tag{10-4}$$

where reactants A and B form products E and F according to the stoichiometric coefficients ν_A, ν_B, ν_E, and ν_F. (In general, any number of reactants and products might need to be considered.)

10-2 THERMOCHEMISTRY

This section involves the application of the conservation of energy principle to chemical reactions. Primary interest is directed to combustion processes.

10-2A General Energy Analysis

The study of the application of thermodynamics to chemical reactions (thermochemistry) falls into two broad categories. A large majority of engineering studies in thermochemistry is directed toward steady-state, steady-flow applications. A smaller but important group of applications involves constant-volume closed systems. The general principles for both categories will be presented in the next several sections.

10-2-1 Steady-State, Steady-Flow Processes

The conservation of energy principle for a steady-state combustion chamber with negligible kinetic and potential-energy effects has the form

$$Q + W = \Delta H = H_{\text{prods}} - H_{\text{reacs}}$$

In Chap. 7 the general expression for H for any phase of a multicomponent system was found to be $H = \sum N_i \bar{h}_i$, where $\bar{h}_i$ is the partial molar enthalpy. In the absence of work effects the above equation becomes

$$Q = H_P - H_R = \sum_P N_i \bar{h}_i - \sum_R N_i \bar{h}_i \equiv \sum_{P-R} N_i \bar{h}_i \qquad (10\text{-}5)$$

The symbols P and R on the summation signs represent products and reactants, respectively, and $P - R$ is the notation for products minus reactants. (This symbolism will be used throughout the chapter.) The equation is written in general terms of N_i, rather than ν_i. Some of the N_i's in the above equation may be equal to the stoichiometric coefficient for that species, but it is not necessary even for the reactants to enter in a stoichiometric ratio. In addition, for chemically inert species present the quantity ν_i has no meaning. Also, it is usually convenient to let the N_i value for the fuel equal unity, such as shown for the reactions in Eqs. (10-2) and (10-3). That is, the energy balance is typically written on the basis of 1 mol of fuel.

Equation (10-5) requires the use of partial molar enthalpy data, which commonly are not available. To overcome this difficulty, we first model reacting mixtures as ideal solutions, for which $\bar{h}_i = h_{i,\text{pure}}$ at the system T and P. For a mixture of ideal gases, as frequently found in engineering applications, this model is exact because an ideal gas behaves as if it were alone in a mixture (Gibbs-Dalton rule). Second, because enthalpy values are never absolute, we seek a convenient method for establishing a reference state for species which are formed or destroyed during a chemical reaction. A consistent accounting of the enthalpy change for a chemically reactive system is achieved by introducing the concept of the enthalpy of formation Δh_f of a substance:

> The *enthalpy of formation* Δh_f is defined as the enthalpy change that occurs when a compound is formed stoichiometrically from its *stable* elements at constant temperature and pressure.

Thus on the basis of Eq. (10-5)

$$\Delta h_f \equiv h_{\text{compound}} - \sum_i (\nu_i h_i)_{\text{stable elements}} \qquad \text{(given } T \text{ and } P\text{)}$$

where ν_i again is the stoichiometric coefficient for the given element. Basic tables of Δh_f data usually report data in a reference (dead) state of $T_0 = 25°\text{C}$ (298.15 K) and P_0 of unit pressure (1 atm or 1 bar). Table A-33 in the Appendix lists $\Delta h^o_{f,298}$ data for a number of species that are important in combustion reactions.

The superscript "o" again represents ideal gas behavior at unit pressure. It is apparent that the Δh_f^o value for all stable elements is zero. Stable elements in combustion reactions refer to species such as O_2, N_2, and H_2. Note in Table A-33 that the enthalpy of formation data for the unstable monatomic forms of these three species (O, N, H) have nonzero values.

More complete tables, such as the JANAF Thermochemical Tables [1] list Δh_f^o for hundreds of substances. These tables are similar to more conventional gas tables, but since they were designed to handle chemical reactions as well as other processes, they have certain peculiarities. The enthalpy data, listed as $(H^o - H_{298}^o)$, are obviously zero in value for every substance at the reference temperature of 25°C (298.15 K) and have negative values below the reference temperature. The entropy values, listed as S^o, are the so-called *standard state* values of an ideal gas at unit pressure (either 1 atm or 1 bar), and are zero at $T = 0$. No internal energy values are given in the JANAF tables. To evaluate the internal energy of an ideal gas, one uses the relation $u = h - pv = h - RT$. To get the entropy $S_{P,T}$ at some pressure other than unit pressure but at the same temperature, we use the relation

$$S_{P,T} = S^o - R \ln \frac{P}{1}$$

where the pressure unit for P must agree with the pressure unit chosen for evaluating S^o. The JANAF tables also list Δh_f^o data as a function of temperature for 100-degree intervals. Other data given in the JANAF tables will be referred to in later sections.

Like the sign convection for Q, the value of Δh_f^o is negative when energy is released as a compound is formed from its elements. Note also in Table A-33 in the Appendix that the data refer to a particular phase of the substance. To convert $\Delta h_{f,298}^o$ data between the gas and liquid phases, one may use the approximation

$$\Delta h_{h,298}^o \text{ (gas phase)} = \Delta h_{f,298}^o \text{ (liquid phase)} + h_{fg,298} \qquad (10\text{-}6)$$

[Note that when a substance vaporizes at 25°C, the saturation pressure is not necessarily the standard pressure of 1 atm (or 1 bar). However, the correction for the change in vapor pressure with total pressure in this case may be neglected. See Sec. 9-6.]

On the basis of the preceding discussion the base enthalpies of stable elements and compounds become

$$\text{Stable element:} \quad h(T_0, P_0) = \Delta h_{f,298}^o = 0 \qquad (10\text{-}7)$$

$$\text{Compounds:} \quad h(T_0, P_0) = \Delta h_{f,298}^o \qquad (10\text{-}8)$$

Thus for an isothermal reaction at the reference state Eq. (10-5) becomes

$$Q = \sum_{P-R} N_i \, \Delta h_{f,298,i}^o \qquad (T = 298.15 \text{ K}) \qquad (10\text{-}9)$$

and if the isothermal reaction is stoichiometrically complete,

$$Q = \sum_{P-R} \nu_i \, \Delta h^o_{f,298,i} \equiv \Delta h^o_{R,298} \qquad (T = 298.15 \text{ K}) \tag{10-10}$$

where the subscript "298" represents 298.15 K. As seen by the last equation, the heat released or absorbed during an isothermal reaction which is stoichiometrically complete is defined as the *enthalpy of reaction,* and is represented in Eq. (10-10) by $\Delta h_{R,298}$. Many engineering applications of chemical reactions involve the combustion of hydrocarbon fuel with oxygen or air. In this case the generic enthalpy of reaction is frequently called the *enthalpy of combustion* Δh_c. Both Δh_R and Δh_c are usually tabulated or calculated on a mole basis. The enthalpy of combustion on a mass basis is called the *heating value* (or the heat of combustion) of the fuel. It is listed in handbooks as a positive value, even though it represents energy released by the combustion process. When water appears as a liquid in the products of combustion, it is called the *higher heating value* (HHV). If water in the product gases is in the vapor state, the energy released is called the *lower heating value* (LHV). Heating values are tabulated in handbooks [2,3] for common fuels. (See also Table A-34 in the Appendix.)

Example 10-1. Carbon monoxide (CO) and oxygen react at 25°C according to the reaction $CO + \frac{1}{2}O_2 \rightarrow CO_2$. Determine the heat transfer during the isothermal process, in kJ/kmol CO.

Solution. On the basis of Eq. (10-10) and the data in Table A-33 we find that

$$Q = \sum_{P-R} \nu_i \, \Delta h^o_{f,298,i} = 1(-393{,}520) - 1(-110{,}530) - \frac{1}{2}(0)$$

$$= -282{,}990 \text{ kJ/kmol CO}$$

The negative sign indicates heat is released from the system, as expected. This value also is the heat of reaction $\Delta h_{R,298}$. The heating value HV of CO at 25°C would be $-\Delta h_R/M$. Hence $HV = 282{,}990/28.01 = 10{,}103$ kJ/kg CO.

In the more general case the enthalpies of the reactants and products must be evaluated at a state (T, P) different from the reference state (T_0, P_0). Symbolically, this requires that

$$h_i(T, P) = h_i(T_0, P_0) + [h_i(T, P) - h_i(T_0, P_0)]$$

where on the basis of Eq. (10-8) the first term on the right is $\Delta h^o_{f,298,i}$ and the bracketed term frequently is termed the *sensible* enthalpy change of the substance due to a temperature (and/or pressure) change. Thus the preceding equation becomes

$$h_{i,T,P} = \Delta h^o_{f,298,i} + (h_{T,P} - h^o_{298})_i \tag{10-11}$$

For an ideal-gas component h is independent of pressure so that the enthalpy relationship above is reduced to

$$h_{i,T} = \Delta h^o_{f,298,i} + (h_T - h_{298})_i \qquad \text{(ideal gas)} \qquad (10\text{-}12)$$

The term $(h_T - h_{298})$ simply represents the integral of $c_{p,o}$ between 298.15 K and T for the given species. [Recall that in the JANAF tables the enthalpy difference $(h_T - h_{298})$ is tabulated as a single value, for convenience.] If a species is in the liquid phase, the term in parentheses in Eq. (10-11) can be evaluated by modeling the substance as incompressible. In this case

$$h_{T,P} - h^o_{298} = c(T - T_0) + v_f(P - P_0)$$

where c is the specific heat and v_f is the saturated-liquid specific volume at 25°C. Thus the enthalpy of any substance in a chemical reaction is composed of two parts—the part associated with its formation from stable elements at a standard reference state (T_0, P_0) and the part associated with a change of state at constant composition.

Finally, by inserting Eq. (10-11) into Eq. (10-5), we find the steady-state energy equation, in the absence of shaft work, Δke, and Δpe, becomes

$$Q = \sum_{P-R} N_i(\Delta h^o_{f,298} + h_{T,P} - h^o_{298})_i \qquad (10\text{-}13)$$

and for ideal-gas mixtures, using Eq. (10-12),

$$Q = \sum_{P-R} N_i(\Delta h^o_{f,298} + h_T - h_{298})_i \qquad \text{(ideal gas)} \qquad (10\text{-}14)$$

The value of Q obtained by these equations, of course, may be positive or negative. For an isothermal reaction which is complete the value of Q in the above equation would equal the enthalpy of reaction $\Delta h_{R,T}$ at temperature T, similar to $\Delta h_{R,298}$ introduced in Eq. (10-10). Chemical reactions at constant T and P which release energy in the form of heat are called *exothermic* ($\Delta h_{R,T}$ is negative). Those isothermal reactions which require heat input are *endothermic* ($\Delta h_{R,T}$ is positive).

Example 10-2. Ethylene (C_2H_4) gas at 25°C is burned at 1 atm with the stoichiometric amount of air to form complete products of combustion at 1000 K. Determine the heat transfer, in kJ/kmol C_2H_4, if the inlet air is (*a*) at 25°C, and (*b*) preheated to 400 K.

Solution. The stoichiometric chemical reaction is given by

$$C_2H_4 + 3\,O_2 + 11.28\,N_2 \rightarrow 2\,CO_2 + 2\,H_2O + 11.28\,N_2$$

If the gases are modeled as ideal gases, then Eq. (10-14) is appropriate for the energy analysis.

(*a*) The enthalpy contribution for air at 25°C is zero. Hence

$$Q = \sum_{P-R} N_i(\Delta h^o_{f,298} + h_T - h_{298})_i = 2(-393{,}520 + 42{,}769 - 9364)$$

$$+ 2(-241{,}820 + 35{,}882 - 9904) + 11.28(0 + 30{,}129 - 8669) - 1(52{,}280)$$

$$= -962{,}125 \text{ kJ/kmol } C_2H_4$$

(*b*) The effect of preheating air to 400 K may be accounted for by including $(h_T - h_{298})$ terms for reactants O_2 and N_2 in the energy balance. Using the results of the preceding energy analysis, we find that

$$Q = -962{,}125 - 3(11{,}711 - 8682) - 11.28(11{,}640 - 8669)$$

$$= -1{,}004{,}725 \text{ kJ/kmol } C_2H_4$$

Preheating of the air leads roughly to a 4 percent increase in heat release.

A graphical interpretation of Eq. (10-14) is possible if we rewrite the equation for a *complete* reaction as

$$Q = \sum_{P-R} N_i\,\Delta h^o_{f,298,i} + \sum_P N_i(h_T - h_{298})_i - \sum_R N_i(h_T - h_{298})_i$$

$$= \Delta h^o_{R,298} + \Delta H_{P,S} - \Delta H_{R,S}$$

The terms $\Delta H_{P,S}$ and $\Delta H_{R,S}$ are notations for the sensible enthalpy change $\sum N_i(h_T - h_{298})_i$ of the products and reactants, respectively. The four terms in the above energy balance are shown on a $T-H$ diagram in Fig. 10-1. In this illustrated case the reactants enter at $T_1 > T_0$ (state a) and the complete products leave the reaction vessel at $T_2 > T_1$ (state e). The diagram shows clearly that the energy $\Delta H_{R,S}$ released by cooling the reactants (process a-b) and the energy released by the chemical reaction $\Delta h^o_{f,298}$ (process b-c) appears in two forms: a temperature rise of the product gases (process c-e) due to $\Delta H_{P,S}$ and a heat loss Q to the surroundings. Also, note that the larger the temperature rise of the products, the smaller the heat loss becomes, as would be expected. The enthalpy of reaction $\Delta h_{R,T_1}$ at T_1 is also shown as process a-d in Fig. 10-1.

10-2-2 Constant-Volume Combustion

When a reaction occurs in a constant-volume closed system the basic energy balance is

$$Q = \Delta U = U_p - U_R = \sum_P N_i\bar{u}_i - \sum_R N_i\bar{u}_i \tag{10-15}$$

where $\bar{u}_i$ is the partial molar internal energy for the general case. For combustion processes involving ideal-gas mixtures $\bar{u}_i$ may be replaced by the pure specific internal energy u_i. This latter quantity is related to the specific enthalpy h_i by $u_i \equiv h_i - Pv_i$. Combining this definition with Eq. (10-11), we find that

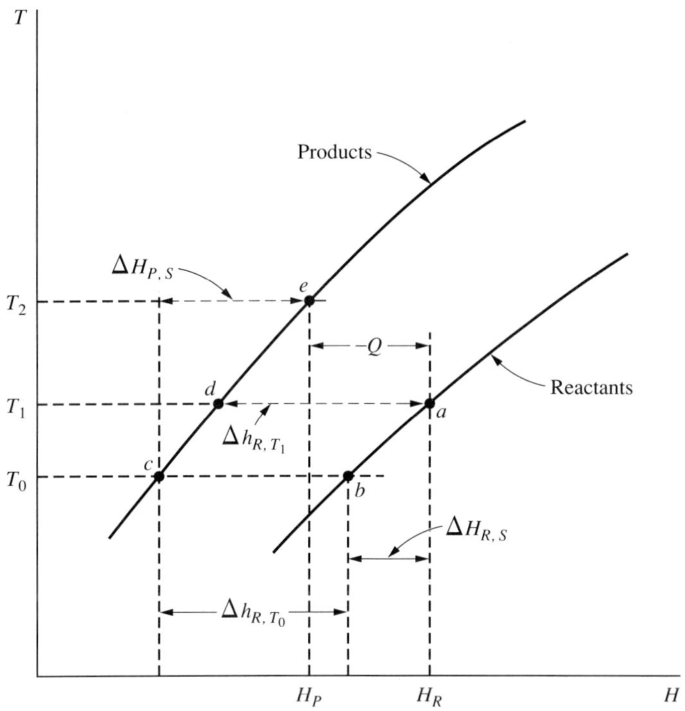

FIGURE 10-1
Illustration of energy quantities associated with a steady-state combustion process.

$$u_{i,T,P} = \Delta h^o_{f,298,i} + h_{T,P} - h^o_{298} - (Pv)_{i,T} \tag{10-16}$$

In some cases it is convenient to replace $(h_{T,P} - h^o_{298})$ by the definition $\Delta h = \Delta u + \Delta(Pv)$. This leads to a second format for $u_{i,T,P}$, namely,

$$u_{i,T,P} = [\Delta h^o_{f,298} + u_{T,P} - u^o_{298} + (Pv)_T - (Pv)^o_{298} - (Pv)_T]_i$$

$$= [\Delta h^o_{f,298} + u_{T,P} - u^o_{298} - (Pv)^o_{298}]_i \tag{10-17}$$

The substitution of Eqs. (10-16) and (10-17) into Eq. (10-15) leads to two different formats for the heat transfer to or from a closed system under constant-volume conditions. These are

$$Q = \sum_{P-R} N_i[\Delta h^o_{f,298} + h_{T,P} - h^o_{298} - (Pv)_T]_i \tag{10-18}$$

and

$$Q = \sum_{P-R} N_i[\Delta h^o_{f,298} + u_{T,P} - u^o_{298} - (Pv)^o_{298}]_i \tag{10-19}$$

Two further changes to Eqs. (10-18) and (10-19) can be made. First, the specific volume of solid and liquid species is usually negligible in comparison to that for gaseous species at a given T and P. Hence only the Pv terms for gaseous species need to be retained in the above two equations. Second, for many combustion processes the gas-phase components may be modeled as ideal gases, for which $Pv = RT$. Hence the energy balances for a constant-volume process can be rewritten in the following two forms:

$$Q = \sum_{P-R} N_i(\Delta h^o_{f,298} + h_{T,P} - h^o_{298})_i - \sum_{P-R} (N_i RT)_{\text{gases}} \tag{10-20}$$

where the last term applies only to species in the ideal-gas phase, and

$$Q = \sum_{P-R} N_i(\Delta h^o_{f,298} + u_{T,P} - u^o_{298})_i - \Delta NRT_{298} \tag{10-21}$$

where ΔN represents the moles of products minus reactants in the ideal-gas phase. The only major difference between these two equations is that one requires sensible enthalpy data $(h_{T,P} - h^o_{298})$, whereas the other requires sensible internal energy data $(u_{T,P} - u^o_{298})$.

10-2B Adiabatic Combustion Analysis

Earlier in this section Fig. 10-1 illustrated the effect of a heat loss from a steady-state reactor on the temperature rise of product combustion gases. In the limit of adiabatic operation the maximum temperature rise will occur. For a number of engineering applications it is desirable to predict this maximum temperature, which is referred to as the *adiabatic flame* (or adiabatic combustion) temperature T_{af} of the reacting mixture. Similar to the discussion in Sec. 10-2, the two cases of interest are those of steady flow and constant volume.

10-2-3 Steady-Flow Adiabatic Combustion

In order to evaluate the *theoretical* temperature limit T_{af} for a steady-flow reactor, we set $Q = 0$ in either Eq. (10-13) or (10-14) so that $H_R = H_P$. This situation is shown by the states a and f shown on a T-H diagram in Fig. 10-2. In the case of *ideal gases*

$$\sum_{P} N_i(\Delta h^o_{f,298} + h_T - h_{298})_i = \sum_{R} N_i(\Delta h^o_{f,298} + h_T - h_{298})_i \tag{10-22}$$

From knowledge of the initial temperature and composition of the reactants the right-hand side can be evaluated directly. The values of $\Delta h^o_{f,298}$ and h_{298} of the products are also known. To reach the maximum final temperature a reaction must go to completion. Thus the N_i values of the product can be found from the stoichiometry of the reaction. Therefore the only unknowns in Eq. (10-22) are the h_T values for each product gas at the unknown temperature T_{af}. The solution to the equation is then one of iteration. The theoretical adiabatic flame temperature for the combustion of common hydrocarbons with stoichiometric air typically is around 2100–2400 K. The presence of excess air can drastically lower T_{af}, depending on the percent excess used.

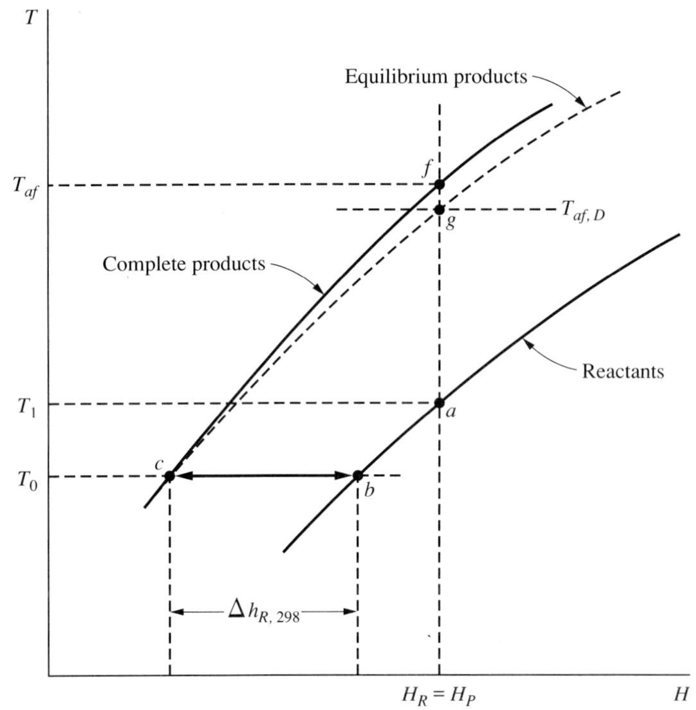

FIGURE 10-2
Illustration of the adiabatic combustion process and the effect of dissociation on T_{af}.

Example 10-3. Ethylene gas and air, both at 25°C, enter a steady-flow combustor. Determine the adiabatic combustion temperature T_{af}, in kelvins, if (*a*) stoichiometric air and (*b*) 100 percent excess air is used, and combustion is complete.

Solution. The energy analysis for adiabatic combustion involving ideal gases is given by Eq. (10-22).

(*a*) As noted in Example 10-2, the reaction for stoichiometric air is

$$C_2H_4 + 3\,O_2 + 11.28\,N_2 \rightarrow 2\,CO_2 + 2\,H_2O + 11.28\,N_2$$

The energy analysis in the form of Eq. (10-22) becomes

$$2(-393{,}520 + h_{T,CO_2} - 9364) + 2(-241{,}820 + h_{T,H_2O} - 9904) + 11.28(h_{T,N_2} - 8669) = 1(52{,}280)$$

Upon rearrangement

$$2h_{T,CO_2} + 2h_{T,H_2O} + 11.28h_{T,N_2} = 1{,}459{,}280 \text{ kJ}$$

By iteration in the gas tables, this equation is satisfied when $T \simeq 2570$ K.

(*b*) The reaction for combustion with 100 percent excess air is

$$C_2H_4 + 6\,O_2 + 22.56\,N_2 \rightarrow 2\,CO_2 + 2\,H_2O + 3\,O_2 + 22.56\,N_2$$

In this case the energy analysis is

$$2(-393{,}520 + h_{T,CO_2} - 9364) + 2(-241{,}820 + h_{T,H_2O} - 9904)$$
$$+ 3(h_{T,O_2} - 8682) + 22.56(h_{T,N_2} - 8669) = 1(52{,}280)$$

Rearrangement leads to

$$2h_{T,CO_2} + 2h_{T,H_2O} + 3h_{T,O_2} + 22.56h_{T,N_2} = 1{,}583{,}115$$

This equation is satisfied by gas table data when $T = 1605$ K. For this particular fuel the effect of 100 percent excess air is to lower the adiabatic combustion temperature from 2570 K for 0 percent excess air to 1605 K.

The adiabatic flame temperature is a theoretical *upper* limit on the product gas temperature for several reasons. First, no reactor can be perfectly insulated. Second, some of the products of combustion will *dissociate* into other chemical species as a result of the high temperature present. Typical dissociation reactions for hydrocarbon combustion might include

$$CO_2 \rightarrow CO + \tfrac{1}{2}O_2 \qquad H_2 \rightarrow 2\,H$$
$$H_2O \rightarrow H_2 + \tfrac{1}{2}O_2 \qquad O_2 \rightarrow 2\,O$$
$$H_2O \rightarrow OH + \tfrac{1}{2}H_2 \qquad N_2 \rightarrow 2\,N$$

These dissociation reactions are endothermic and consume some of the energy released by the major combustion reaction. Thus the final temperature with dissociation $T_{af,D}$ is lower than T_{af}, even when Q is essentially zero. Figure 10-2 illustrates the effect of dissociation on T_{af}. For complete combustion the temperature reaches state f, while with dissociation the final state is g. In either case $H_P = H_R$. (The quantitative difference between T_{af} and $T_{af,D}$ is overemphasized in the figure for the sake of clarity.)

10-2-4 Constant-Volume Adiabatic Combustion

The adiabatic flame temperature T_{af} for constant-volume combustion is found by a method similar to that for steady-flow processes. The value of Q is set equal to zero in either Eq. (10-20) or (10-21). In the first case the h_T data are unknown, while in the second case the u_T data for the product gases are unknown. An iteration process is used to find T_{af}. However, unlike the constant pressure assumed for steady flow, a significant pressure rise will occur at constant volume.

In general, two methods for evaluating the adiabatic combustion pressure P_{af} at constant volume are available. One requires information on the vessel volume; the other requires knowledge of the initial system pressure. Both require the assumption of an ideal gas. In the first case the expression for P_{af} for an ideal-gas mixture is

$$P_{af} = \frac{N_f R_u T_{af}}{V} \tag{10-23}$$

where T_{af} = constant-volume adiabatic flame temperature
N_f = total number of moles of gaseous products
V = bomb or vessel volume

In the second case where the initial pressure P_i and temperature T_i are known but the volume is unknown we have

$$P_{af} = P_i \frac{N_f}{N_i} \frac{T_f}{T_i} \tag{10-24}$$

where N_f and N_i represent the final and initial moles of products and reactants, respectively, that are ideal gases.

10-3 SECOND LAW, AVAILABILITY, AND IRREVERSIBILITY

Energy analysis is an important first step in the study of engineering applications of chemically reactive systems. Such studies are incomplete, however, without the additional insight provided by the second law. Further, a combined first- and second-law statement supplies information on the reversible work or availability change associated with chemical processes, as well as the irreversibility or losses. All this information is vital to the study of energy systems. The basic concepts and relationships based on the second law (and in combination with the first law) are developed below.

10-3A Second-Law Relationships

In general, a heat interaction may occur between the system and its surroundings as a reaction occurs. For this general situation the entropy generation for a steady-flow process is given by Eq. (2-43). Modified for a chemical reactor, the equation on a mole basis becomes

$$\dot{\sigma}_{\text{CV}} = \sum_P \dot{N} s - \sum_R \dot{N} s - \sum_i \frac{\dot{Q}_i}{T_i} \geq 0 \tag{10-25}$$

Similar to energy analyses, the entropy analysis of reacting systems is usually carried out on a per-mole-of-fuel basis. Division of the above equation by the molal flow rate of the fuel leads to

$$\sigma_{\text{chem}} = \sum_P N_i s_i - \sum_R N_i s_i - \sum_i \frac{Q_i}{T_i} = S_P - S_R - \sum_i \frac{Q_i}{T_i} \qquad \text{(steady flow)} \tag{10-26}$$

where Q_i is that portion of the total heat transfer per mole of fuel which crosses the section of the boundary of temperature T_i. Frequently the term $\sum Q_i/T_i$ may be difficult to evaluate at the boundary. Another approach is to modify Eq. (2-45) to account for both σ_{chem} due to the reaction and σ_Q due to heat transfer into the surroundings at T_0. This latter modification, when based on 1 mol of fuel, becomes

$$\sigma_{\text{tot}} = \sigma_{\text{chem}} + \sigma_Q = \sum_P N_i s_i - \sum_R N_i s_i - \frac{Q_0}{T_0} \qquad (10\text{-}27)$$

where Q_0 is the total heat transfer to the environment at T_0, measured relative to the system. [If heat transfer instead occurs to or from the reactor and another reservoir at $T_j \neq T_0$, the term Q_0/T_0 is replaced by Q_j/T_j. See Eq. (2-45).]

For a chemical reaction occurring in a closed system the entropy balance formulated in Chap. 2 is written as

$$S_P - S_R = \sum_i \frac{Q_i}{T_i} + \sigma_{\text{CM}}$$

When the reacting species act independently, the above equation becomes

$$\sum_P N_i s_i - \sum_R N_i s_i = \sum_i \frac{Q_i}{T_i} + \sigma_{\text{chem}} \qquad \text{(closed system)} \qquad (10\text{-}28)$$

where the N_i, Q_i, and σ_{chem} values are on a *per-mole-of-fuel* basis.

The evaluation of the adiabatic flame temperature T_{af} is an important application of the first-law analysis of combustion reactions. For this situation the entropy balances for open and closed systems represented by Eqs. (10-26) and (10-28) both reduce to

$$\sigma_{\text{chem}} = \sum_{P-R} N_i s_i \qquad (10\text{-}29)$$

Thus the entropy production of an adiabatic reaction is directly related to the entropy change for the reaction, as expected. Both terms are positive. As noted in the next section, σ_{chem} leads to a significant loss in availability during adiabatic combustion processes. The evaluation of s_i for either a steady-flow or closed system is described below.

The s_i values in the preceding equations are evaluated in a different fashion from that for h_i values in energy analyses, because the reference states are different. Compilations of data for ideal gases (such as those in the JANAF tables [1]) list absolute entropy values $s^o_{i,T}$ (by third-law convention) at various temperatures and a standard reference pressure (usually 1 atm or 1 bar). To find s_i at a different T and P, we note that

$$s_i(T, P) = s_i(T, P_{\text{ref}}) + [s_i(T, P) - s_i(T, P_{\text{ref}})]$$

or in terms of the reference entropy $s^o_{i,T}$

$$s_{i,T,P} = s^o_{i,T} + [s_i(T, P) - s_i(T, P_{\text{ref}})] \tag{10-30}$$

If the gases in the mixture can be modeled as ideal gases, then the bracketed term in the above equation is found by

$$s_i(T, P) - s_i(T, P_{\text{ref}}) = -R \ln \frac{p_i}{P_{\text{ref}}} = -R \ln \frac{y_i P}{P_{\text{ref}}}$$

where p_i is the partial pressure of the gas, and P_{ref} frequently is P_0, the dead state pressure. When all the constituents in the reaction are ideal gases, then

$$\sum_{P-R} N_i s_i = \sum_{P-R} N_i \left(s^o_{i,T} - R \ln \frac{p_i}{P_{\text{ref}}} \right) \qquad \text{(ideal gases)} \tag{10-31}$$

When one or more constituents are real gases, the value of $[s_i(T, P) - s_i(T, P_{\text{ref}})]$ can be evaluated by methods discussed in Chap. 6.

One useful model for a species in the liquid or solid phase is that of an incompressible substance. In such cases

$$s_{i,T,P} = s^o_{i,298} + c \ln \frac{T}{T_{\text{ref}}} \qquad \text{(liquid or solid phase)} \tag{10-32}$$

where c is the average specific heat between T_{ref} of 298.15 K and T, and s^o_i (liq, 298.15 K) is the absolute entropy found in the JANAF tables or similar references. The above format also applies to solid constituents. If saturated-liquid data are available, an alternative to the above equation for the liquid phase is

$$s_{i,T,P} = s^o_{i,\text{liq},298} + [s_f(T) - s_f(25°\text{C})] \qquad \text{(liquid phase)} \tag{10-33}$$

where s_f is standard notation for a *saturated-liquid* state. These models for liquids and solids, which neglect the effect of pressure on entropy, are generally quite accurate.

10-3B Availability

The concepts of maximum work and availability introduced in Chap. 3 can be readily extended to chemically reactive systems. We wish to ascertain the maximum useful work or useful mechanical power that can be produced as the result of a given chemical process while exchanging heat solely with the environment at T_0. Figure 10-3 shows reactants entering and products leaving a power plant with enthalpies H_R and H_P, entropies S_R and S_P, and stream availabilities B_R and B_P. Work W is produced and heat transfer Q_0 occurs at the system boundary to a reference reservoir at T_0. The maximum power is found by modifying Eq. (3-42) or Eq. (3-43) for a reversible process with negligible kinetic- and potential-energy terms. In either case we find that

$$\dot{W}_{\text{rev},sf} = \sum_P \dot{N}_i b_i - \sum_R \dot{N}_i b_i$$

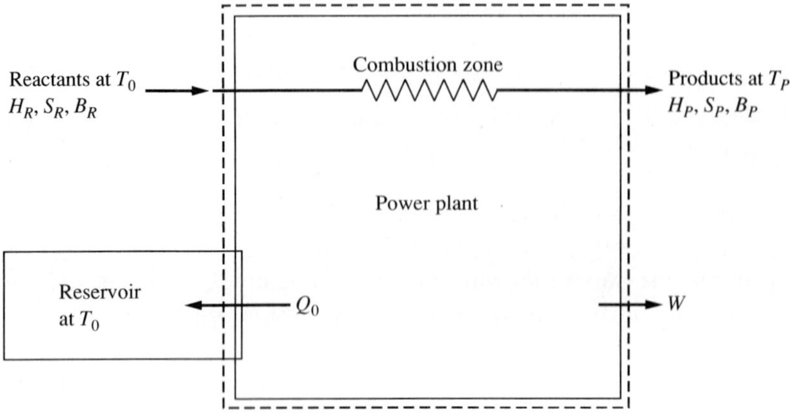

FIGURE 10-3
Schematic of a steady-flow control volume for the production of work from a combustion process with heat transfer to a reference reservoir at T_0.

where $b \equiv h - T_0 s$. Division of this equation by $\dot{N}_{\text{fuel}}$ leads to the work per mole of fuel.

$$\begin{aligned} W_{\text{rev},sf} &= \sum_{P-R} N_i b_i = \sum_{P-R} N_i (h - T_0 s)_i \\ &= H_P - H_R - T_0(S_P - S_R) \equiv B_P - B_R \end{aligned} \tag{10-34}$$

where
$$H_P - H_R = \sum_{P-R} N_i (\Delta h^o_{f,298} + h_{T,P} - h^o_{298})_i$$

$$S_P - S_R = \sum_{P-R} N_i s_{i,T,P,\text{absol}}$$

Thus the maximum work output per mole of fuel is given by the change in stream availability $(B_P - B_R)$, and this quantity will be negative in accordance with the sign convention on W in this text. In an actual power plant irreversibilities are present due to entropy generation and $W_{\text{out,act}} < W_{\text{out,rev}}$.

Equation (10-28) can also be derived by the method introduced in Chap. 3 for nonreactive processes. The steady-flow energy balance for the system in Fig. 10-3 is

$$Q_0 + W = \sum_P N_i h_i - \sum_R N_i h_i$$

and the overall entropy balance based on Eq. (10-27) is

$$\sigma = \sum_P N_i s_i - \sum_R N_i s_i - \frac{Q_0}{T_0}$$

where σ accounts for irreversibilities not only in the combustion process, but also in the power plant. Elimination of Q_0 between the two equations leads to

$$W_{\text{act},sf} = \sum_P N_i h_i - \sum_R N_i h_i + T_0 \left[\sigma - \sum_P N_i s_i + \sum_R N_i s_i \right]$$

$$= H_P - H_R - T_0(S_P - S_R) + T_0\sigma \qquad (10\text{-}35)$$

In the case of a reversible process, $T_0\sigma$ is zero and the above equation reduces to Eq. (10-34) previously derived for the maximum work.

10-3C Irreversibility

The irreversibility I of any process is still defined, as in Chap. 3, by the two relations

$$I = W_{\text{act}} - W_{\text{rev}} = T_0 \, \Delta S_{\text{tot}} \qquad (10\text{-}36)$$

where ΔS is the total entropy change of all subsystems involved in the process. For a steady-flow process $W_{\text{act}} = \Delta H - Q$ and $W_{\text{rev}} = \Delta H - T_0 \, \Delta S_{\text{reaction}}$. Hence for a reaction with heat transfer the irreversibility I_{chem} of the reaction is given by

$$I_{\text{chem}} = T_0 \, \Delta S_{\text{chem}} - Q \qquad (10\text{-}37)$$

For a combustion process where W_{act} is usually zero, $Q = H_P - H_R$. Hence the preceding equation becomes

$$I_{\text{chem}} = T_0 \sum_{P-R} N_i s_i - \sum_{P-R} N_i h_i \qquad (10\text{-}38)$$

where I_{chem} is now expressed in terms of property changes.

10-4 WORK PRODUCTION FROM CHEMICAL REACTIONS

On the basis of the preceding sections we shall develop some basic relationships for the conversion of chemical energy in work output. The general schematic in Fig. 10-3 still applies. However, the maximum work is obtained when the product gas temperature T_P is the environmental temperature T_0. Hence our initial study is directed toward processes for which the reactants enter and products leave at (T_0, P_0). The range of combustion processes of interest fall between two limits:

1. Isothermal combustion at T_0 with heat transfer with the environment at T_0 and the simultaneous production of work.
2. Adiabatic combustion to achieve a product gas at the adiabatic flame temperature T_{af}. The gases are then cooled to T_0 and the heat is used in a heat engine to produce work.

Other possibilities include case 2 but heat loss prevents attainment of T_{af} and the cooling of the product gases reaches only an intermediate temperature T_P ($T_0 < T_P < T_{af}$) rather than T_0. The two limiting cases are discussed below first with respect to physical requirements, availability changes, and irreversibilities.

10-4A Work from Isothermal Combustion

The first class will produce no work output if the reaction is allowed to proceed such that a heat loss accounts for the total energy released. That is, for steady-flow $Q = H_P - H_R$. No work is possible by means of a heat engine, because the supply and rejection temperatures are both T_0. However, theoretically a work output is possible even when $T_R = T_P = T_0$, because there is an availability change given by Eq. (10-34), namely, $(B_P - B_R)$. Unfortunately, the availability change in this case is also a measure of the irreversibility I of the process. That is, since $W_{act} = 0$,

$$I \equiv W_{act} - W_{rev} = -W_{rev} = -(B_P - B_R) \tag{10-39}$$

Thus an isothermal reaction has work capability, but that capability is totally lost if the energy exchange with the environment is solely in the form of heat.

Example 10-4. Consider the steady-flow isothermal combustion of gaseous ethylene with stoichiometric air, both initially separate and at 25°C. Assume complete combustion and that water appears as a vapor. Determine (*a*) the entropy change in kJ/K, (*b*) the stream availability change, in kilojoules, and (*c*) the irreversibility, in kJ/kmol of fuel, if $T_0 = 25°C$.

Solution

(*a*) As noted in Example 10-2, the reaction for stoichiometric air is

$$C_2H_4 + 3\,O_2 + 11.28\,N_2 \rightarrow 2\,CO_2 + 2\,H_2O + 11.28\,N_2$$

The following equations show the evaluation of the entropy values for the reactants and products. Note that the fuel and air enter separately at 1 atm.

Reactants:

$$S_{C_2H_4} = 1(219.83) = 219.83\,\text{kJ/K}$$

$$S_{O_2} = 3\left(205.03 - 8.314 \ln \frac{3}{14.28}\right) = 654.01\ \text{kJ/K}$$

$$S_{N_2} = 11.28\left(191.50 - 8.314 \ln \frac{11.28}{14.28}\right) = 2182.24\ \text{kJ/K}$$

Products:

$$S_{CO_2} = 2\left(213.685 - 8.314 \ln \frac{2}{15.28}\right) = 461.18\ \text{kJ/K}$$

$$S_{H_2O} = 2\left(188.72 - 8.314 \ln \frac{2}{15.28}\right) = 411.25\ \text{kJ/K}$$

$$S_{N_2} = 11.28\left(191.502 - 8.314 \ln \frac{11.28}{15.28}\right) = 2188.61\ \text{kJ/K}$$

For the overall reaction

$$\Delta S_{\text{adia}} = S_P - S_R = 3061.04 - 3056.08 = 4.96 \text{ kJ/K}$$

(*b*) The stream availability change, in general, is $\Delta B = \Delta H - T_0\,\Delta S$. For an isothermal steady-flow reaction at 298 K, Eq. (10-10) is used for evaluation of ΔH. Substitution of the enthalpy of formation data into this equation yields

$$\begin{aligned} Q &= H_P - H_R = \sum_{P-R} \nu_i\,\Delta h^o_{f,298,i} \\ &= 2(-393{,}520) + 2(-241{,}820) - 1(52{,}280) = -1{,}322{,}960 \text{ kJ} \end{aligned}$$

Hence for a dead state of 25°C

$$W_{\text{rev},sf} = H_P - H_R - T_0(S_P - S_R) = -1{,}322{,}960 - 298.15(4.96) = -1{,}324{,}440 \text{ kJ}$$

(*c*) Since the actual work output during isothermal combustion is zero, the irreversibility is

$$I = W_{\text{act}} - W_{\text{rev}} = 0 - (-1{,}324{,}440) = 1{,}324{,}440 \text{ kJ}$$

The isothermal combustion process is responsible for a loss of 1,324,440 kJ of work capability per kilomole of fuel.

The recovery of work output from an isothermal chemical reaction requires the use of a device called a fuel cell. A *fuel cell* is an electrochemical device in which the fuel and oxidizer do not come in direct contact, as shown in Fig. 10-4. It operates much like a battery, except that fuel and oxidizer enter and products leave continuously in steady flow. The flows are separated by two membranes

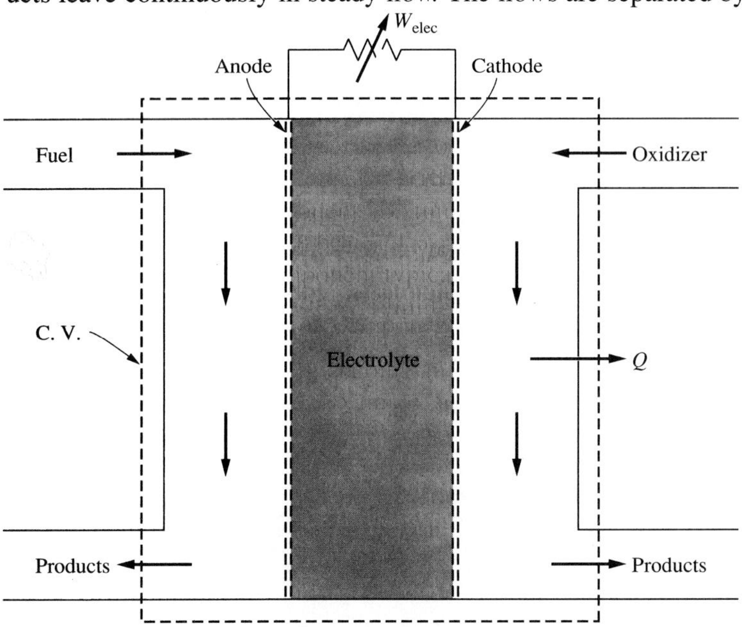

FIGURE 10-4
Schematic of an isothermal fuel cell for the production of electrical work.

which act as the anode and cathode, and the space between the electrodes is filled with a suitable electrolyte. Electrical work is produced in an external electric circuit. Heat transfer may also be necessary to keep the cell temperature constant. Such devices have played an important role in manned space travel, and earth applications are currently under active development. A more detailed analysis of fuel cells appears in the next section of this chapter.

10-4B Work from Adiabatic Combustion

The second class with an upper limit of T_{af} (or $T_{af,D}$) is important in power plant designs illustrated previously by Fig. 10-3. In a simplified model we consider the overall process within the power plant to occur in two steps. First, as shown in Fig. 10-5, combustion occurs with the product gases reaching the temperature T_{af}. In step 2 heat removed from the hot-product gases is supplied to a heat engine which produces work W and rejects heat Q_0 to the environment at T_0. In step 1 both Q and W are zero, and an energy balance leads to the adiabatic flame temperature T_{af}, as discussed earlier in the chapter. The availability change $B_P - B_R = -T_0(S_P - S_R)$, since $(H_P - H_R) = 0$. It will be negative in value, because ΔS for an adiabatic reaction must be positive in accordance with the second law. For adiabatic combustion the irreversibility is the negative of the availability change on the basis of Eq. (10-37). In summary,

$$\Delta\Psi_{\text{chem}} = B_P - B_R = -T_0(S_P - S_R) = -I_{\text{chem}} \qquad \text{(adiabatic combustion)} \tag{10-39}$$

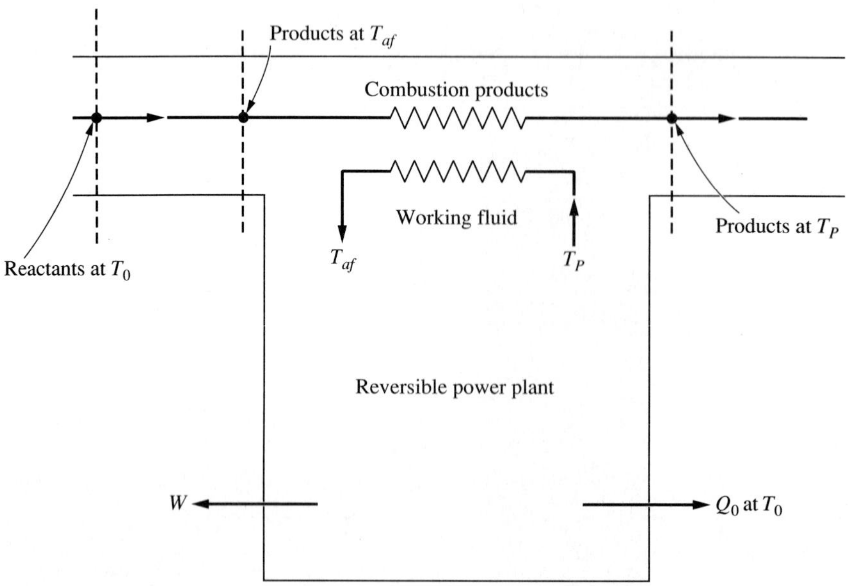

FIGURE 10-5
Schematic of reversible power plant receiving heat from product gases cooled from T_{af} to T_P.

This relation is also in agreement with Eq. (10-29) which related σ_{chem} with the entropy change for a reaction. The derivation of the above equation is straightforward. Of more significance are the numerical values of $\Delta\Psi$ or I obtained for typical hydrocarbon fuels when compared to the $\Delta\Psi$ values for isothermal combustion. The availability loss during adiabatic combustion could be 25–40 percent of the reversible work capability during isothermal combustion. Hence a large loss in work capability occurs in step 1 of Fig. 10-5 even before the hot gases reach the energy conversion portion of the device.

Example 10-5. Consider the steady-flow adiabatic combustion of gaseous ethylene with stoichiometric air, both at 25°C, and assume complete combustion. Determine (*a*) the entropy change in kJ/K, (*b*) the stream availability change, in kilojoules, and (*c*) the irreversibility, in kilojoules, per kilomole of fuel, if $T_0 = 25°\text{C}$.

Solution

(*a*) From Example 10-3 we found that the adiabatic flame temperature T_{af} was 2570 K for the gas-phase reaction

$$C_2H_4 + 3\,O_2 + 11.28\,N_2 \rightarrow 2\,CO_2 + 2\,H_2O + 11.28\,N_2$$

It is assumed that the fuel and air enter separately at 1 atm. The total entropy of the reactants at 298 K was found in Example 10-4 to be 3056.8 kJ/K. The following equations show the evaluation of the entropy values for the products at 2570 K.

Products:

$$S_{CO_2} = 2\left(324.50 - 8.314 \ln \frac{2}{15.28}\right) = 682.81 \text{ kJ/K}$$

$$S_{H_2O} = 2\left(277.78 - 8.314 \ln \frac{2}{15.28}\right) = 589.37 \text{ kJ/K}$$

$$S_{N_2} = 11.28\left(261.08 - 8.314 \ln \frac{11.28}{15.28}\right) = 2973.45 \text{ kJ/K}$$

For the overall reaction

$$\Delta S_{\text{adia}} = S_P - S_R = 4245.63 - 3056.08 = 1189.55 \text{ kJ/K}$$

The ΔS for the reaction is also a measure of the entropy production σ_{chem} per kilomole of fuel.

(*b*) The stream availability change in general is $\Delta H - T_0\,\Delta S$. For an adiabatic, steady-flow reaction $\Delta H = 0$. Hence for a dead state of 25°C

$$W_{\text{rev},sf} = -T_0(S_P - S_R) = -298.15(1189.55) = -354{,}660 \text{ kJ}$$

The adiabatic combustion process is responsible for a loss of 354,660 kJ/kmol of fuel in work capability.

(*c*) Since the actual work output during adiabatic combustion is zero, the irreversibility is

$$I = W_{\text{act}} - W_{\text{rev}} = 0 - (-354{,}660) = 354{,}660 \text{ kJ}$$

The irreversibility is roughly 27 percent of the enthalpy of reaction for the isothermal gas phase reaction at 298 K found in Example 10-4.

The inherent irreversibility of the adiabatic combustion process is coupled with the irreversibility of the energy conversion process in step 2. Unlike step 1, however, step 2 theoretically can be made reversible by the system illustrated in Fig. 10-6. Reversible heat transfer between the product gases and the working fluid of the power cycle occurs in a counterflow heat exchanger. The temperature profile in the heat exchanger is sketched on temperature-distance coordinates in Fig. 10-6(*a*). Theoretically the temperature difference between the two fluids at all cross sections in the heat exchanger approaches zero. The outlet temperature of the product gases is shown as T_P, and, in general, $T_{af} \geq T_P \geq T_0$. This means that the high-temperature side of the working fluid during heat input to the reversible power plant varies from T_{af} to T_P. To make the power plant reversible under this temperature variation, the heat removed enters a series of Carnot-type heat engines. Each engine in the series operates at a slightly lower temperature on the high-temperature side of the engine, and each rejects heat at T_0. The temperature profiles for six arbitrary heat engines in series is shown in Fig. 10-6(*b*). (In theory, the number of heat engines required would be indefinitely large.) The sum of the work outputs from the series of engines is the reversible work output of the power plant for the specified value of T_P. This summation is equal to $\Delta\Psi$ (or ΔB) for the product gases as they cool from T_{af} to T_P. Hence the reversible work produced per mole of fuel as the product gases are cooled is

$$\begin{aligned} w_{\text{rev,cooling}} &= B_P(T_P, P_0) - B_P(T_{af}, P_0) \\ &= H_P(T_P, P_0) - H_P(T_{af}, P_0) - T_0\,[S_P(T_P, P_0) - S_P(T_{af}, P_0)] \end{aligned} \tag{10-40}$$

where the calculation of ΔH_P and ΔS_P requires the use of $h(T)$ and $s^o(T)$ data from ideal-gas tables. (The evaluation of ΔH_P and ΔS_P will have to be modified if the composition changes as the gas is cooled due to dissociation effects.) The second-law efficiency η_{II} for the power plant design is found by dividing Eq. (10-40) by Eq. (10-34) applied at the restricted dead state at T_0 and P_0. Hence

$$\eta_{\text{II}} = \frac{w_{\text{rev,cooling}}}{w_{\text{rev,isoth}}} = \frac{B_P(T_P, P_0) - B_P(T_{af}, P_0)}{B_P(T_0, P_0) - B_R(T_0, P_0)} \tag{10-41}$$

Generally, the second-law efficiency would be less than that given by Eq. (10-41), because of irreversibilities within an actual heat engine, as well as irreversible heat transfer to and from the heat engine. That is, w_{out} (actual) $<$ w_{out} (reversible).

In the preceding development it was assumed that the product gases cooled to an intermediate temperature T_P. This situation fits a number of modern engineering

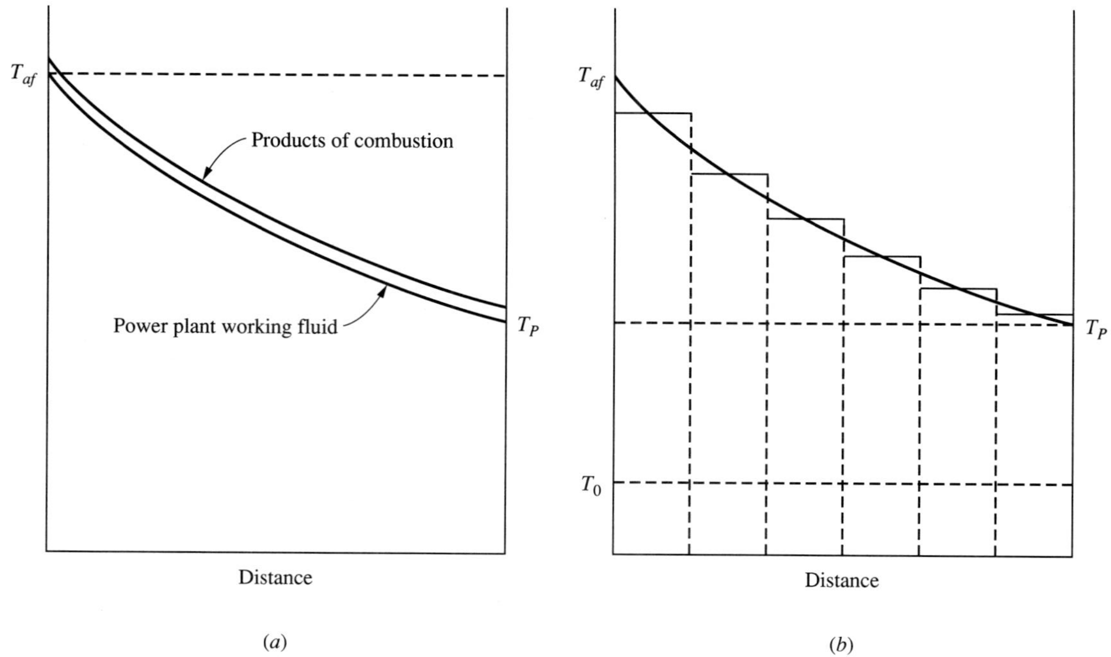

FIGURE 10-6
(*a*) Temperature profiles for a power plant receiving heat reversibly from hot-product gases; (*b*) a series of heat engines receiving heat reversibly from hot-product gases and rejecting to the environment at T_0.

applications where the products are cooled to T_P while producing work, and then are used for another purpose before being released into the environment to T_0. (A cogeneration plant for power production and industrial or municipal heating is an example.) Not unexpectedly, w_{rev} given by Eq. (10-40) monotonically increases as T_P decreases. In the limit when $T_P = T_0$ the *maximum reversible work* occurs. The *maximum second-law efficiency* also occurs when T_P is set equal to T_0 in Eq. (10-41). This limiting situation is case 2 described at the beginning of this section, where adiabatic combustion is followed by cooling to the ambient temperature T_0.

Example 10-6. Ethylene gas and stoichiometric air, both initially at 298 K, have an adiabatic combustion temperature of 2570 K. If the product gases are cooled to 298 K, determine (*a*) the entropy change, in kJ/K, (*b*) the availability change, in kilojoules, and (*c*) the maximum second-law efficiency per kilomole of fuel, if T_0 is 25°C.

Solution. From preceding examples the overall reaction is

$$C_2H_4 + 3\,O_2 + 11.28\,N_2 \rightarrow 2\,CO_2 + 2\,H_2O + 11.28\,N_2$$

(*a*) The entropy change is simply the sum of the individual changes for the products as they are cooled from 2570 to 298 K. Thus

$$\begin{aligned}\Delta S_{\text{cooling}} &= \sum N_i (s_{P,0} - s_{P,af})_i \\ &= 2(213.685 - 324.50) + 2(188.72 - 277.78) + 11.28(191.502 - 261.08) \\ &= -1184.59 \text{ kJ/K}\end{aligned}$$

Since entropy is a property, this value could also be obtained from the ΔS data in Examples 10-4 and 10-5. That is, $(S_{P,0} - S_{P,af}) = 4.96 - 1189.55 = -1184.59$ kJ/K.

(*b*) The availability change upon cooling is $\Delta H - T_0\,\Delta S$. But ΔH for the cooling process is the same as the ΔH for the isothermal combustion process in Example 10-4, namely, $-1{,}322{,}960$ kJ. Hence

$$B_{P,0} - B_{P,af} = -1{,}322{,}960 - 298.15(-1184.59) = -969{,}770 \text{ kJ}$$

This value is also found from the ΔB values for adiabatic combustion and isothermal combustion found in Examples 10-4 and 10-5. That is,

$$\Delta B_{\text{cooling}} = \Delta B_{\text{isoth}} - \Delta B_{\text{adia}} = -1{,}324{,}440 - (-354{,}660) = -969{,}780 \text{ kJ}$$

(*c*) The maximum second-law efficiency for the cooling process is found from Eq. (10-41), where $T_P = T_0$. Substitution of data yields

$$\eta_{\text{II}} = \frac{w_{\text{rev,cooling}}}{w_{\text{rev,isoth}}} = \frac{B_P(T_P, P_0) - B_P(T_{af}, P_0)}{B_P(T_0, P_0) - B_R(T_0, P_0)} = \frac{-969{,}770}{-1{,}324{,}440} = 0.732$$

Thus 73 percent of the energy released by the cooling process could theoretically be converted into useful work.

10-5 FUEL CELLS

The general characteristics of a fuel cell were introduced in the preceding section. The hydrogen-oxygen fuel cell has played an important role in manned orbiting space vehicles. Other applications under development include the electric car, power and heat for the home, and peaking power for electric utilities. As shown in Fig. 10-4, the steady-flow device produces electric power through a *controlled* chemical reaction. Oxidation of the fuel occurs at the anode, with release of electrons to the external circuit. At the cathode the oxidizer is reduced, with the consumption of electrons. Products of combustion may appear on the surface of either or both the anode and cathode, depending on the reaction involved. The circuit is completed by the movement of ions in the electrolyte. To maintain the temperature T within the cell constant, it is necessary to add or remove heat Q. An energy balance on the control volume in Fig. 10-4 shows that

$$Q + W_{\text{elec}} = \Delta H + \Delta \text{ke} + \Delta \text{pe}$$

The changes in kinetic and potential energies for fuel cells can be neglected. In addition, the maximum work occurs when a process is internally reversible. In this situation $Q = T\,\Delta S$ for an isothermal process. Therefore the *maximum* electrical work done by the fuel cell *alone* is

$$\begin{aligned} W_{\text{elec,max}} &= \Delta H - Q = \Delta H - T\,\Delta S = \Delta G_T \\ &= \sum_{P-R} N_i g_i(T, P) = \sum_{P-R} \nu_i g_i(T, P) \qquad \text{(steady flow)} \quad (10\text{-}42) \end{aligned}$$

Hence the maximum electrical work of an isothermal fuel cell alone is measured by the decrease in the Gibbs function for the chemical process. If heat transfer occurs to the environment during the process, additional work theoretically can be achieved. The total reversible work of any process which exchanges heat solely with the environment is, of course, the change in the availability function. In this case for a steady-flow process

$$W_{\text{rev,tot}} = B_P - B_R = \Delta H - T_0\,\Delta S \qquad (10\text{-}43)$$

These two results differ only in the use of T or T_0 in the calculation.

An interesting special case of Eq. (10-42) arises when the reactants and products are both in the dead state (T_0, P_0). Equation (10-42) in this case becomes

$$W_{\text{rev},sf} = \sum_{P-R} N_i(h_0 - T_0 s_0)_i = \sum_{P-R} N_i g_i(T_0, P_0)$$

But $g_i(T_0, P_0)$ is equal to the Gibbs function of formation, $\Delta g^o_{f,298}$, found in standard references when P_0 equals the standard reference pressure. Also, all the N_i's can be replaced by ν_i's for the reacting species. Hence

$$W_{\text{rev},sf} = \sum_{P-R} \nu_i\,\Delta g^o_{f,i,298} \equiv G_P(T_0, P_0) - G_R(T_0, P_0) \qquad (10\text{-}44)$$

where $G \equiv \sum N_i g_i$. If each constituent does not behave independently, then g_i must be replaced by the partial molar Gibbs function $\bar{g}_i = \mu_i$. Therefore, more generally,

$$W_{\text{rev},sf} = \sum_{P-R} \nu_i \bar{g}^o_{i,298} = \sum_{P-R} \nu_i \mu^o_{i,298}$$

It must be kept in mind that the two preceding equations apply only to the case where reactants and products are in the dead state.

The first-law efficiency η_{I} of a device is usually defined as useful energy output divided by required energy input. For a fuel cell the energy output per mole of fuel is w_{elec}, whereas the energy input is the enthalpy of reaction Δh_R released by the complete chemical reaction. The *ideal first-law efficiency* $\eta_{\text{I},i}$ of a fuel cell at temperature T is thus

$$\eta_{\text{I},i} = \frac{w_{\text{elec,max}}}{\Delta h_R} = \frac{\Delta g_R}{\Delta h_R} \qquad \text{(ideal fuel cell)} \qquad (10\text{-}45)$$

where w_{elec}, Δg_R, and Δh_R are all measured on a "per-mole-of-fuel" basis. On the basis of Eq. (10-42) the ideal first-law efficiency can also be written as

$$\eta_{\text{I},i} = 1 - \frac{T\,\Delta S}{\Delta H} = 1 - \frac{T\,\Delta s_R}{\Delta h_R} = 1 - \frac{Q_{\text{rev}}}{\Delta H} \qquad (10\text{-}46)$$

The term ΔH is always negative (exothermic), and for some fuel-cell reactions ΔS (and hence Q) is positive. In such cases $\Delta g_{R,T} > \Delta h_{R,T}$, and $\eta_{\text{I},i} > 1.0$. A first-law efficiency greater than 100 percent is of no real significance, since it depends on the arbitrary definition of the efficiency. It is not a first-law violation, however, because the energy output W_{elec} equals the sum of the energy released internally, ΔH, and the heat added, Q [see Eq. (10-42)]. Table 10-2 indicates some η_i values at 25°C, including one greater than unity, for some common reactions. Theoretical-first-law efficiencies of 80–90 percent are fairly typical for pure hydrocarbons with oxygen in a fuel cell.

Another important parameter for a fuel cell is the ideal open-circuit voltage. In Chap. 1 the reversible work of an electrochemical cell is given as $-Q_e V_i$, where Q_e is the quantity of charge C, in coulombs, passing from the cell per mole of fuel and V_i is the ideal electrostatic potential developed. The number of coulombs per mole Q_e equals the number of moles of electrons N_e produced per

TABLE 10-2
Ideal-first-law efficiencies and voltages at 25°C for some common fuel cell reactions

Reaction	Δg_R	Δh_R	$\eta_{\text{I},i}$	V_i
$C(s) + O_2 \rightarrow CO_2$	−394,410	−393,520	1.002	1.02
$H_2 + \frac{1}{2} O_2 \rightarrow H_2O$ (l)	−237,180	−285,830	0.830	1.23
$CH_4 + 2O_2 \rightarrow CO_2 + 2H_2O$ (l)	−817,980	−890,330	0.919	1.06

mole of reacting fuel multiplied by the number of coulombs per mole of electrons, $\mathcal{F}$. The quantity $\mathcal{F}$ is a universal constant called the *Faraday* constant. Its value is

$$\mathcal{F} = \frac{6.0225 \times 10^{23} \text{ electrons}}{\text{gmol electrons}} \times \frac{1.6021 \times 10^{-19} \text{ C}}{\text{electron}}$$
$$= 96{,}487 \text{ C/gmol electrons} = 96{,}487 \text{ kJ/(V} \cdot \text{kmol electrons)}$$

Therefore, using Eq. (10-42), we obtain

$$w_{\text{elec,ideal}} = -Q_e V_i = -N_e \mathcal{F} V_i = \Delta g_{R,T}$$

or written in terms of the *ideal* voltage, we have

$$V_i = \frac{-\Delta g_{R,T}}{96{,}487 N_e} \tag{10-47}$$

where $\Delta g_{R,T}$ is expressed in kilojoules per kilomole of fuel and V_i is in volts. Table 10-2 shows V_i values for three common reactions. Values around unity are typical of hydrocarbon fuels. The ideal steady-state power equals the ideal work per mole of fuel multiplied by the molar rate of fuel consumed. Hence

$$\dot{W}_{\text{ideal}} = -N_e \mathcal{F} V_i \dot{N}_{\text{fuel}} = \Delta g_{R,T} \dot{N}_{\text{fuel}} \tag{10-48}$$

On the basis of Table 10-2 the consumption of 1 gmol/min (2.1 lb/h) of methane (CH_4) ideally would produce around 13.6 kW.

The value of N_e is determined from the *half-cell* reactions which occur at the two electrodes. Fuel is oxidized at the *anode* and electrons are released to the external circuit. The oxidizer is reduced (gains electrons from the external circuit) at the *cathode*. Two specific examples are shown in Fig. 10-7. An H_2-O_2 fuel cell operating with an acidic electrolyte is shown in Fig. 10-7(*a*). For the overall reaction

$$H_2\,(g) + \tfrac{1}{2}\,O_2\,(g) \rightarrow H_2O\ (\text{liq})$$

and the half-cell reactions are

Anode: $$H_2 \rightarrow 2\,e^- + 2\,H^+$$

Cathode: $$2\,H^+ + 2\,e^- + \tfrac{1}{2}\,O_2 \rightarrow H_2O\,(\text{liq})$$

The migration of species at sites on the anode and cathode are shown in the figure. In this case N_e is 2, for the two electrons released per molecule of fuel. The situation for a hydrocarbon such as methane (CH_4) with oxygen is shown in Fig. 10-7(*b*). The overall reaction for the gaseous reactants is

$$CH_4 + 2\,O_2 \rightarrow CO_2 + 2\,H_2O\ (\text{liq})$$

and the half-cell reactions are

Anode: $$CH_4 + 2\,H_2O \rightarrow CO_2 + 8\,H^+ + 8\,e^-$$

Cathode: $$2\,O_2 + 8\,e^- + 8\,H^+ \rightarrow 4\,H_2O$$

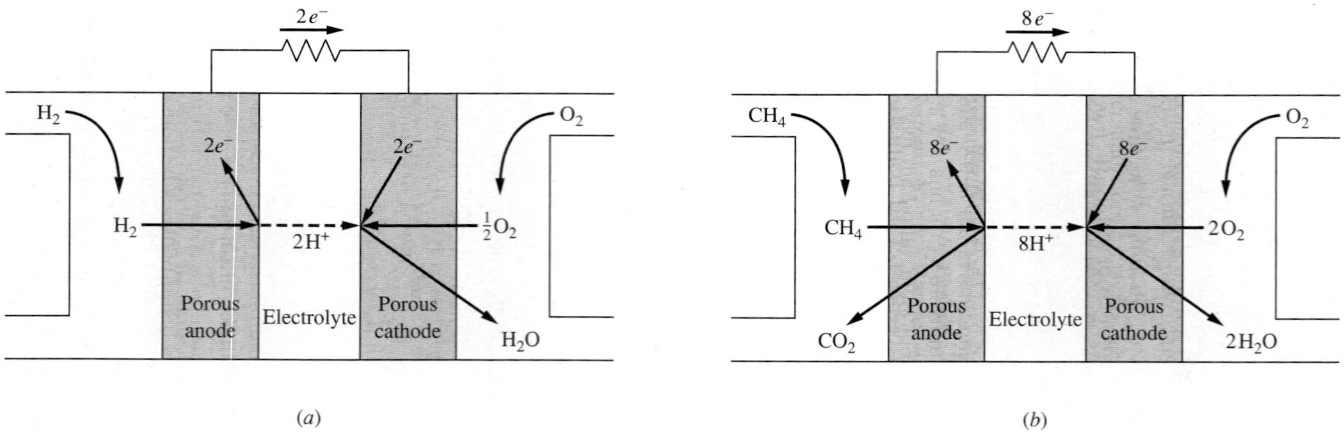

FIGURE 10-7
Schematics of (*a*) a H_2–O_2 fuel cell and (*b*) a methane-oxygen fuel cell.

Again, the migration of species for the half-cell reactions at sites on the anode and cathode are shown in Fig. 10-7(*b*).

Example 10-7. A hydrogen-oxygen fuel cell operates at 500 K. Determine (*a*) the reversible electrical work, in kJ/kmol H_2, (*b*) the ideal first-law efficiency, (*c*) the heat transfer required, in kJ/kmol H_2, and (*d*) the ideal electrical potential, in volts.

Solution. The overall reaction is $H_2 + \frac{1}{2}O_2 \rightarrow H_2O$ in the gas phase. Data from the JANAF tables are listed in Table E10-7 for hydrogen, oxygen, and water at 500 K. Δh_f^o and Δg_f^o are in kJ/kmol, s is in kJ/kmol·K.

(*a*) The reversible steady-flow work is measured by the change in the Gibbs function, namely, $\Delta H - T\Delta S$. Based on data from the table and Eq. (10-42),

$$\begin{aligned} W_{\text{rev},sf} &= \Delta H - T\,\Delta S = \sum_{P-R} \nu_i\,\Delta h^o_{f,i,T} - T \sum_{P-R} \nu_i s^o_{i,T} \\ &= -243{,}826 - 500[206.534 - 145.737 - \tfrac{1}{2}(220.693)] \\ &= -219{,}051 \text{ kJ/kmol } H_2 \end{aligned}$$

(*b*) The ideal first-law efficiency is given by Eq. (10-45).

$$\eta_{\text{I},i} = \frac{\Delta g_R}{\Delta h_R} = \frac{-219{,}051}{-243{,}826} = 0.898$$

(*c*) The heat transfer required to keep the cell isothermal is

$$Q = T\,\Delta S_R = \Delta H_R - \Delta G_R = -243{,}826 - (-219{,}051) = -24{,}775 \text{ kJ/kmol } H_2$$

Thus heat must be removed from the cell to keep it at constant temperature.

(*d*) The ideal potential of the cell is found from Eq. (10-47). From the text discussion preceding this example $N_e = 2$ for an H_2-O_2 fuel cell. Therefore

$$V_i = \frac{-\Delta g_{R,T}}{96{,}487 N_e} = \frac{-(-219{,}051)}{2(96{,}487)} = 1.135 \text{ volts}$$

TABLE E10-7

Species	Δh_f^o	Δg_f^o	s^o
H_2	0	0	145.737
O_2	0	0	220.693
H_2O	−243,826	−219,051	206.534

The maximum work output by the fuel cell discussed in the example above is measured by the change in availability, as given by $\Delta\Psi = \Delta H - T_0\,\Delta S$. From the data used above we find that $\Delta\Psi = -229{,}053$ kJ when $T_0 = 25°\text{C}$. This compares to the electrical work of −219,051 kJ. The additional work predicted by $\Delta\Psi$ is obtained from the heat transfer rejected by the cell to the environment. This additional work output (10,002 kJ) is found from the Carnot expression:

$W = Q(1 - T_0/T)$. A major benefit of fuel cells which reject heat is that the heat potentially can be used for added power production or for heating purposes elsewhere.

For a given fuel the values of both $\eta_{I,i}$ and V_i depend directly on $\Delta g_{R,T}$, and $\eta_{I,i}$ varies inversely with $\Delta h_{R,T}$. But $\Delta g_{R,T}$ and $\Delta h_{R,T}$ are both functions of temperature. Thus fuel-cell performance is affected by temperature. The effect of temperature on V_i for CH_4-O_2 and C-O_2 fuel cells is slight. However, as shown in Table 10-3, temperature has a major effect on both V_i and $\eta_{I,i}$ for the H_2-O_2 cell. Note that $\Delta h_{R,T}$ is roughly constant. But $\Delta g_{R,T}$ decreases significantly with increasing temperature, which directly affects V_i and $\eta_{I,i}$. The quantity $T\,\Delta s_{R,T}$ in the table is also important. This is a measure of the heat transfer (out, in this case) that is necessary to keep the device isothermal. The last column is a measure of the first-law efficiency of a Carnot-type heat engine which receives heat at temperature T by direct combustion of the fuel and rejects at T_0.

As noted earlier, the energy analysis for the actual fuel cell described in Fig. 10-7 is

$$W_{\text{elec}} = H_P(T, P_0) - H_R(T, P_0) - Q_{\text{act}} \tag{10-49}$$

and the reversible work from Eq. (10-43) is

$$W_{\text{rev}} = \Delta H(T, P_0) - T_0\,\Delta S(T, P_0) \tag{10-50}$$

The ratio of these two work terms can be considered as a second-law efficiency of the fuel cell operating at (T, P_0). For an isothermal cell operating at T_0,

$$\eta_{\text{II}} \equiv \frac{W_{\text{act}}}{W_{\text{rev}}} = \frac{\Delta H(T_0, P_0) - Q_{\text{act}}}{\Delta H(T_0, P_0) - T_0\,\Delta S(T_0, P_0)} \qquad \text{(fuel cell at } T_0) \tag{10-51}$$

where $\Delta H(T_0, P_0) - T_0\,\Delta S(T_0, P_0) = \Delta\Psi(T_0, P_0)$. If the fuel cell were internally reversible, $Q_{\text{act}} = T_0\,\Delta S$, $\eta_{\text{II}} = 1.0$, and $I = 0$. Fuel cells, due to their isothermal operation, are not "Carnot-limited" or "second-law-limited" devices. This is a major advantage in their favor as an energy conversion device.

The second-law efficiency for a fuel cell operating at T_0 and exchanging heat solely with the environment at T_0 was given above by Eq. (10-51). More generally,

TABLE 10-3
Variation of the fuel-cell performance with temperature for $H_2 + \frac{1}{2}O_2 \rightarrow H_2O$ in the gas phase

T (K)	Δh_R (kJ/kmol)	$T\Delta s_R$ (kJ/kmol)	Δg_R (kJ/kmol)	V_i (volts)	$\eta_{I,i}$	η_{carnot}
298	−241,820	−13,220	−228,600	1.18	0.945	0
600	−244,760	−30,710	−214,050	1.11	0.875	0.50
800	−246,440	−42,930	−203,510	1.05	0.83	0.63
1000	−247,900	−55,270	−192,630	1.00	0.78	0.70

for operation at temperature T and heat exchange solely with the environment the second-law efficiency is

$$\epsilon = \eta_{\text{II}} = \frac{W_{\text{act}}}{W_{\text{rev}}} = \frac{\Delta H_T - Q_{\text{act}}}{\Delta H_T - T_0\,\Delta S_T} \tag{10-52}$$

where the effect of pressure on ΔH_T and ΔS_T is usually negligible. The irreversibilities which reduce the terminal voltage of the cell are generally spoken of as polarization effects. These frequently are grouped into three classes—resistance or ohmic polarization, activation or chemical polarization, and concentration polarization. The magnitude of each of these effects is a function of the current density J_i, and is shown qualitatively in Fig. 10-8(a). The first of these effects is caused by the internal resistance of the cell, and it includes losses in the electrolyte as well as in the electrodes. The activation polarization arises from chemical changes occurring at the surface of the electrodes as well as absorption and desorption effects on the surface. The concentration polarization is caused by the concentration gradients set up in the electrolyte and in the gas streams in the vicinity of the electrodes.

The net result of the three polarization effects or overvoltages is a reduction in the terminal voltage at a given current density. Figure 10-8(b) presents a typical diagram for the overall cell voltage versus the current density. Current densities typically range from 0 to 1 A/cm^2. The rapid drop at low current is due to the activation or chemical polarization. In the middle range neither the activation nor concentration effects are changing appreciably, and the curve varies linearly due to the ohmic drop. Finally, at high current densities the concentration polarization drops the cell output sharply.

The large-scale fuel-cell power plant has a number of distinct advantages over conventional fossil-fuel plants. A fuel cell operates as a direct-energy converter, and not as a heat engine. Therefore it is not affected by the Carnot-efficiency limitation of the second law. Thus it has the potential for greater energy-conversion efficiencies than conventional heat-engine devices. In addition, the efficiency of a fuel cell is directly related to that of the individual cell. Hence plant size will not affect efficiency. Another major advantage of the fuel-cell power plant is its ability to respond quickly to load variations. This response can be less than a minute. Large conventional power plants have very large response times. The fuel-cell power plant has several other desirable characteristics. Sulfur oxide and nitrogen oxide emissions from the plant would be extremely small, because no direct combustion process occurs. In addition, since the device is not a heat engine, large quantities of waste heat energy do not have to be rejected to the environment. Thus large quantities of cooling water are not required, and "thermal pollution" problems are nonexistent.

Current development of the fuel cell is directed toward the production of commercial power in the range of 5–30 MW. First-generation units based on H_2 and O_2 with a phosphoric acid electrolyte operate at around 200°C. The system's efficiency (fuel to net ac power) at near full load is 35–45 percent, based on

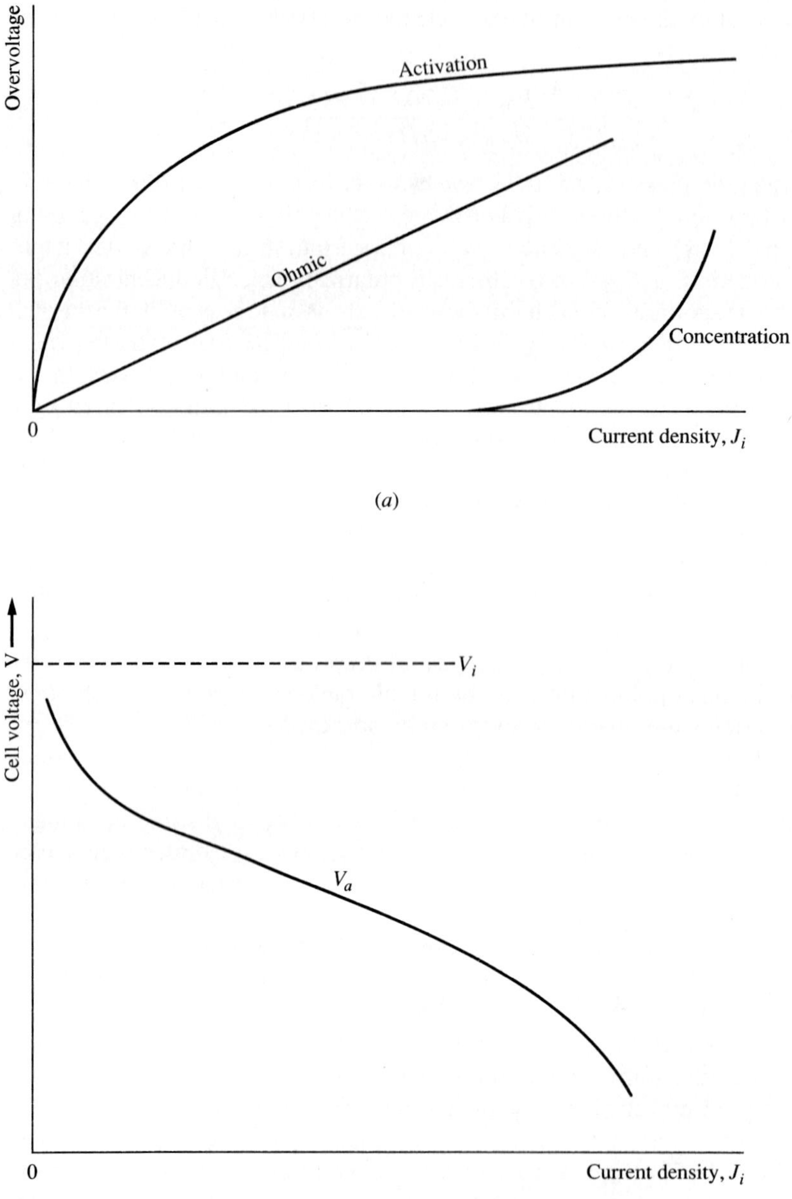

FIGURE 10-8
(*a*) Typical overvoltage curves versus current density; (*b*) typical operating curve for an actual fuel cell.

the higher heating value of the fuel. Fuel-conversion efficiencies (hydrogen to dc power) average around 58 percent. From an environmental standpoint, these fuel cells discharge around 25 percent less nitrogen oxides per megawatt than conventional generators. The fuel for second-generation fuel cells is obtained by reforming a hydrocarbon fuel into a hydrogen-rich gas. First, natural gas and steam is reacted to form hydrogen and carbon monoxide (CO). A following converter reacts the CO with steam to form more hydrogen. Both converters require a catalyst.

A more advanced fuel cell would use molten carbonate technology with operating temperatures in the 600–700°C range ($\approx$ 1200°F). This system employs natural gas or similar fuels, and lithium and potassium salts as the electrolyte. The higher temperature of operation leads to a higher efficiency than phosphoric acid cells. In addition, this type of fuel cell also eliminates the need for separate equipment for the conversion of natural gas to hydrogen, since the conversion is done inside the cell. Molten carbonate cells also can be used to produce superheated steam for additional power production or for other processes. The expected fuel-to-electricity conversion efficiency of molten carbonate cells is 54–60 percent [4].

Both phosphoric acid and molten carbonate are highly corrosive, which leads to a materials problem. To avoid this difficulty, other types of fuel cells are under active investigation. One of these is the solid oxide cell using zirconium oxide ceramic material. With an operating temperature of 1000°C (1800°F), the electrical conversion efficiency approaches 60 percent. A higher-temperature steam could also be produced as a by-product, which provides a wider range of cogeneration applications. The solid oxide units are undergoing fundamental technology developments currently.

10-6 CRITERION FOR REACTION EQUILIBRIUM

In preceding sections we assumed in first-law analyses that reactions proceed to completion. That is, one or more of the reactants is completely consumed. For many reactions, including the dissociation effects mentioned in Sec. 10-2, it is necessary to use the second law to predict the extent or degree of completion and to modify the energy analysis accordingly. Consider a general chemical reaction represented by

$$\nu_A A + \nu_B B \rightarrow \nu_E E + \nu_F F \tag{10-53}$$

where reactants A and B form products E and F according to the stoichiometric coefficients ν_A, ν_B, ν_E, and ν_F. At any instant the system is specified by the temperature, pressure, and moles of the chemical species present, that is, by T, P, N_A, N_B, N_E, and N_F. More generally, a single stoichiometric reaction may be represented by the *chemical* equation

$$0 = \sum_i \nu_i M_i$$

where the M_i's are the chemical symbols of the reacting species. The ν_i quantities are taken as *negative* values for the reactants on the left-hand side of the reaction. As any reaction proceeds, the relative magnitudes of the finite changes in composition ΔN_i of the i species are related to each other through the stoichiometric coefficients. This relationship is expressed by

$$\frac{\Delta N_A}{\nu_A} = \frac{\Delta N_B}{\nu_B} = \frac{\Delta N_E}{\nu_E} = \frac{\Delta N_F}{\nu_F} = \xi \tag{10-54}$$

where ξ (greek letter xi) is called the *extent of reaction* (or "degree of advancement") and is considered positive. ξ may be defined in physical terms by

$$\xi \equiv \frac{N_i - N_{i0}}{\nu_i} \tag{10-55}$$

where N_{i0} is some reference amount of moles; for example, the quantity present at the beginning of the reaction. It is helpful to write Eq. (10-54) also for a differential change of state. That is,

$$\frac{dN_A}{\nu_A} = \frac{dN_B}{\nu_B} = \frac{dN_E}{\nu_E} = \frac{dN_F}{\nu_F} = d\xi \tag{10-56}$$

where $d\xi$ is the differential extent of reaction. Thus, in general, $dN_i = \nu_i \, d\xi$.

Recall that the total differential of G for a system of variable composition is given by

$$dG = V\,dP - S\,dT + \sum_i \mu_i dN_i \tag{8-7}$$

In view of Eq. (10-56) the above relation can be rewritten for Eq. (10-53) as

$$dG = V dP - S dT + (\nu_E \mu_E + \nu_F \mu_F + \nu_A \mu_A + \nu_B \mu_B)\, d\xi \tag{10-57}$$

Remember that the values of ν_A and ν_B (the reactant species) are taken as negative. The negative of the quantity in parentheses is defined [5] as the affinity **A** (bold **A**). That is,

$$\mathbf{A} \equiv -\sum_i \nu_i \mu_i = \sum_{\text{reacs}} \nu_i \mu_i - \sum_{\text{prods}} \nu_i \mu_i \tag{10-58}$$

Similar to the notation used in energy analyses earlier in this chapter, when a summation over all reacting species is divided into separate summations over reactants and products, the ν_i's in the reactant summation are taken as positive, because the sign on ν has already been taken into account. (The affinity symbol **A** should not be confused with the Helmholtz function symbol A.) As a result, Eq. (10-57) becomes

$$dG = V dP - S dT - \mathbf{A}\, d\xi \tag{10-59}$$

Hence the total differential of G for a chemical reaction at constant temperature and pressure becomes

$$dG_{T,P} = -\mathbf{A}\,d\xi \qquad \text{or} \qquad (\partial G/\partial \xi)_{P,T} = -\mathbf{A}$$

Finally, on the basis of the above relation and the equilibrium criterion given by Eq. (9-4) that $dG_{T,P} \leq 0$, we can state for reactions at constant T and P that

$$\begin{aligned} \mathbf{A}\,d\xi &> 0 \qquad \text{(natural process)} \\ \mathbf{A}\,d\xi &= 0 \qquad \text{(reversible or equilibrium process)} \\ \mathbf{A}\,d\xi &< 0 \qquad \text{(unnatural process)} \end{aligned} \tag{10-60}$$

If a reaction is to proceed naturally from reactants to products [as written in Eq. (10-53)] so that $d\xi$ is positive, then the affinity $\mathbf{A}$ must also be positive for a spontaneous process in that direction. Therefore

$$\sum_i \nu_i \mu_i < 0 \qquad \text{or} \qquad \sum_{\text{reacs}} \nu_i \mu_i > \sum_{\text{prods}} \nu_i \mu_i \qquad \text{(natural process)}$$

Conversely, if in a given state the inequality expressed above is reversed in direction (such that $\mathbf{A}$ is negative), the reaction will proceed spontaneously from right to left.

In the case where the affinity is zero, or G is minimized with respect to ξ, the system is in equilibrium and no further change occurs in either direction. In this special case

$$-\mathbf{A} = \sum_{\text{prods}} \nu_i \mu_i - \sum_{\text{reacs}} \nu_i \mu_i = 0$$

or

$$\sum_i \nu_i \mu_i = 0 \qquad \text{(equilibrium criterion)} \tag{10-61}$$

Equation (10-61) is frequently called the *equation of reaction equilibrium*. It is the fundamental equation for predicting equilibrium compositions for a single reaction when T, P, and the compositions at some nonequilibrium initial state are known. It is applicable whether the species involved are gases, liquids, or solids.

The directions of chemical change outlined above in terms of the affinity function $\mathbf{A}$ are expressed graphically in Fig. 10-9. The term $\sum \nu_i \mu_i$ for either reactants or products is plotted versus the extent of reaction, ξ. The vertical distance between the reactant and the product summation for the dashed line on the left is a measure of $\mathbf{A}$, and is positive. In this case the system is not in equilibrium and will proceed toward consuming reactants and producing products ($d\xi > 0$). For the dashed line on the right the affinity is negative, and the reaction spontaneously would proceed from right to left ($d\xi < 0$). The dotted vertical line represents the case where $\mathbf{A} = 0$, and the system is in equilibrium at the specified temperature and pressure. The arrows on the reactants and products lines indicate the spontaneous direction of change. The major significance of the affinity is that it acts as the *potential for chemical change*. This is similar to T as the potential

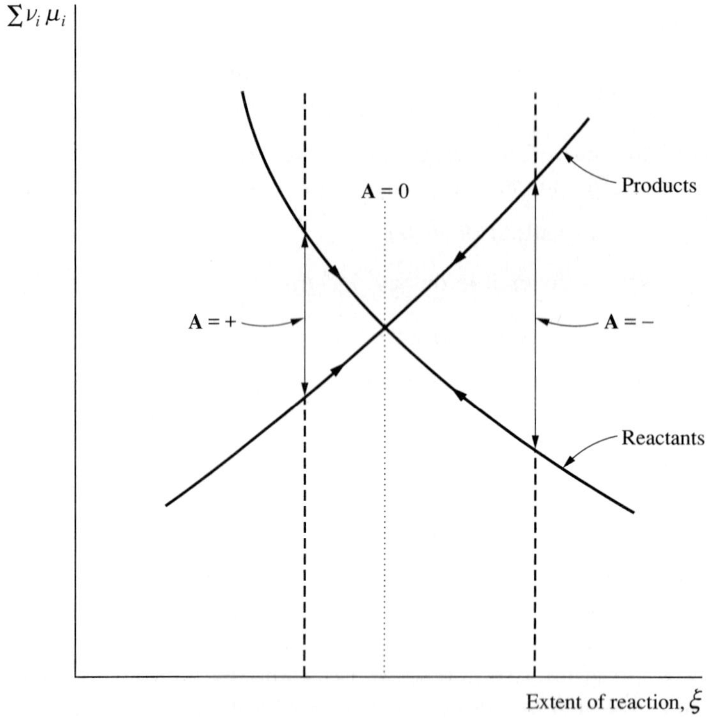

FIGURE 10-9
Variation of the affinity of a reaction with the extent of the reaction.

for heat transfer, P as the potential for mechanical work, and μ_i as the potential for mass transfer.

On the basis of Eq. (10-59) we have shown that $\mathbf{A} = -(\partial G/\partial \xi)_{P,T}$. In Sec. 8-2 it was shown that from the characteristic equations for dU, dH, dA, and dG we could develop four different expressions for the chemical potential μ_i, as given by Eq. (8-9). That is,

$$\mu_i = \left(\frac{\partial U}{\partial N_i}\right)_{S,V,N_j} = \left(\frac{\partial H}{\partial N_i}\right)_{S,P,N_j} = \left(\frac{\partial A}{\partial N_i}\right)_{T,V,N_j} = \left(\frac{\partial G}{\partial N_i}\right)_{T,P,N_j} \tag{8-9}$$

In a similar fashion, we can show that the affinity may also be expressed in terms of U, H, A, and G in the following manner. First, the characteristic equations can be transformed into a format like Eq. (10-59). Thus

$$dG = V\,dP - S\,dT - \mathbf{A}\,d\xi \tag{10-59}$$

$$dU = T\,dS - P\,dV - \mathbf{A}\,d\xi \tag{10-62a}$$

$$dH = T\,dS + V\,dP - \mathbf{A}\,d\xi \tag{10-62b}$$

$$dA = -S\,dT - V\,dP - \mathbf{A}\,d\xi \tag{10-62c}$$

From these four equations we learn that, in general,

$$\mathbf{A} = -\left(\frac{\partial U}{\partial \xi}\right)_{S,V} = -\left(\frac{\partial H}{\partial \xi}\right)_{S,P} = -\left(\frac{\partial A}{\partial \xi}\right)_{T,V} = -\left(\frac{\partial G}{\partial \xi}\right)_{T,P} \tag{10-63}$$

Moreover, because dG is an exact differential, the test for exactness on Eq. (10-59) leads to two additional relations between ξ and extensive property changes. These are

$$\left(\frac{\partial \mathbf{A}}{\partial T}\right)_{P,\xi} = \left(\frac{\partial S}{\partial \xi}\right)_{T,P} = \frac{\partial}{\partial \xi}\sum_{P-R} N_i \bar{s}_i = \sum_{P-R}\left(\frac{\partial N_i}{\partial \xi}\right)_{T,P} \bar{s}_i = \sum_i \nu_i \bar{s}_i \tag{10-64}$$

where the sum on the right can be termed the *partial molar entropy of reaction*, and in a similar fashion,

$$\left(\frac{\partial \mathbf{A}}{\partial P}\right)_{T,\xi} = -\left(\frac{\partial V}{\partial \xi}\right)_{P,T} = -\frac{\partial}{\partial \xi}\sum_{P-R} N_i \bar{v}_i = -\sum_i \nu_i \bar{v}_i \tag{10-65}$$

where the sum on the right can be termed the *partial molar volume of reaction*. These last two equations are useful, because they give the variation of the affinity with either temperature or pressure. For example, consider the situation where a positive change in the extent of reaction results in a positive volume change at constant pressure. Equation (10-65) requires that the affinity decrease with an increase in pressure on the system at the given temperature.

The preceding two equations relate the affinity to the partial molar entropies and volumes of the reacting species. It is useful also to relate the affinity to the enthalpy and the temperature of a chemically reactive system. Differentiation of Eq. (10-62b) with respect to ξ at constant temperature and pressure gives

$$\left(\frac{\partial H}{\partial \xi}\right)_{T,P} = T\left(\frac{\partial S}{\partial \xi}\right)_{T,P} - \mathbf{A} \tag{10-66}$$

The quantity $(\partial H/\partial \xi)_{T,P}$ has a special physical significance. Similar to Eq. (10-64), we can write

$$\left(\frac{\partial H}{\partial \xi}\right)_{T,P} = \frac{\partial}{\partial \xi}\sum_{P-R} N_i \bar{h}_i = \sum_{P-R}\left(\frac{\partial N_i}{\partial \xi}\right)_{T,P} \bar{h}_i = \sum_i \nu_i \bar{h}_i = \Delta h_{R,T} \tag{10-67}$$

Thus the quantity $(\partial H/\partial \xi)_{T,P}$ is the *enthalpy of reaction* $\Delta h_{R,T}$ discussed earlier in the chapter, written for a differential advancement of the reaction at constant temperature and pressure. Substitution of Eq. (10-64) into Eq. (10-66) yields

$$\left(\frac{\partial H}{\partial \xi}\right)_{T,P} = T\left(\frac{\partial \mathbf{A}}{\partial T}\right)_{P,\xi} - \mathbf{A}$$

Finally, division of the above equation by T^2 and recognition of the math identity that $(1/x)(\partial y/\partial x) - y/x^2 = (\partial/\partial x)(y/x)$ leads to (where $y = \mathbf{A}$ and $x = T$)

$$\frac{\partial}{\partial T}\left(\frac{\mathbf{A}}{T}\right)_{P,\xi} = \frac{1}{T^2}\left(\frac{\partial H}{\partial \xi}\right)_{T,P} = \frac{\Delta h_{R,T}}{T^2} \tag{10-68}$$

One of the interesting features of Eq. (10-68) is that it relates the first-law quantity $\Delta h_{R,T}$ with the second-law quantity **A**, the affinity. Subsequent integration at constant P and ξ shows the variation of $\mathbf{A}/T$ as a function of T, namely,

$$\left(\frac{\mathbf{A}_2}{T_2} - \frac{\mathbf{A}_1}{T_1}\right)_{P,\xi} = \int_{T_1}^{T_2} \frac{\Delta h_{R,T}}{T^2}\, dT \tag{10-69}$$

Knowledge of the functional relationship between $\Delta h_{R,T}$ and T would enable us to calculate the change in the affinity with temperature at constant P and ξ. In addition, we will later use Eq. (10-68) above to derive an important relationship known as the van't Hoff equation.

10-7 THE EQUILIBRIUM CONSTANT FOR GASEOUS MIXTURES

The major conclusion of the preceding section is that a reaction reaches equilibrium when the affinity **A** becomes zero for the specified temperature and pressure. For that state there is a unique equilibrium value ξ_e for the extent of reaction, and hence unique compositions for the reacting species. The method of determining the equilibrium compositions is outlined below. The starting point is the equation of reaction equilibrium ($\mathbf{A} = 0$), namely,

$$\sum_i \nu_i \mu_i = 0 \tag{10-61}$$

In addition, in Chap. 8 the chemical potential of component i in a mixture of nonideal gases at temperature T was expressed by

$$\mu_{i,T,P} = \bar{g}_i = \mu_{i,\text{ref},T} + RT \ln \frac{\bar{f}_i}{\bar{f}_{i,\text{ref}}} \tag{8-95}$$

where $\bar{f}_i$ is the fugacity of component i in the mixture. Recall that the standard reference state of a gas (indicated by the symbol superscript "o") is taken to be that of a pure ideal gas at unit pressure P_0 (usually 1 atm or 1 bar). Hence $\mu_{i,\text{ref},T} = \mu_{i,T}^o = g_{i,T}^o$, and $\bar{f}_{i,\text{ref}} = P_0$. Consequently,

$$\mu_{i,T,P} = g_{i,T}^o + RT \ln \frac{\bar{f}_i}{P_0}$$

Substitution of this relation into the equation of reaction equilibrium yields the general expression

$$\sum_i \nu_i \left(g_{i,T}^o + RT \ln \frac{\bar{f}_i}{P_0}\right) = 0$$

or

$$\sum_i \nu_i g^o_{i,T} = -RT \sum_i \nu_i \ln \frac{\bar{f}_i}{P_0} \tag{10-70}$$

The summation on the left-hand side of Eq. (10-70) is defined as ΔG^o_T. That is,

$$\sum_i \nu_i g^o_{i,T} \equiv \Delta G^o_T \tag{10-71}$$

In addition, recall that $z \ln x = \ln x^z$ and $\ln x + \ln y = \ln xy$. On this basis the summation on the right-hand side of Eq. (10-70) may be written as

$$\sum_i \nu_i \ln \frac{\bar{f}_i}{P_0} = \sum_i \left(\ln \frac{\bar{f}_i}{P_0}\right)^{\nu_i} = \ln \prod_i \left(\frac{\bar{f}_i}{P_0}\right)^{\nu_i} \equiv \ln K_{\bar{f}} \tag{10-72}$$

where the $\prod$ symbol (capital pi) mathematically represents multiplication. Hence

$$K_{\bar{f}} \equiv \prod_i \left(\frac{\bar{f}_i}{P_0}\right)^{\nu_i} \tag{10-73}$$

and Eq. (10-70) becomes

$$\ln K_{\bar{f}} = -\frac{\Delta G^o_T}{RT} \tag{10-74}$$

The quantity $K_{\bar{f}}$ is known as the *equilibrium constant K* for a reaction based on $\bar{f}_i$ values of the reacting species. ΔG^o_T is the *standard state Gibbs function change* for a reaction. Since the $g^o_{i,T}$ values must be evaluated at unit pressure (usually either 1 bar or 1 atm), the values of ΔG^o_T and $\Delta G^o_T/RT$ are solely a function of temperature.

For a large percentage of engineering applications the reactions of interest can be considered to involve ideal gases. In a development parallel to the one above for real-gas mixtures, the chemical potential for ideal-gas mixtures takes on the format

$$\mu_{i,T,\text{id}} = g^o_{i,T} + RT \ln \frac{p_i}{P_0} \tag{8-94}$$

When this equation is substituted into the equation of reaction equilibrium [Eq. (10-61)] and rearranged, we find that

$$\sum_{P-R} \nu_i g^o_{i,T} = -RT \sum_{P-R} \nu_i \ln \frac{p_i}{P_0}$$

Finally, the summation on the right is defined as $\ln K_p$; that is,

$$\ln K_p \equiv \sum_i \nu_i \ln \frac{p_i}{P_0} = \sum_i \ln \left(\frac{p_i}{P_0}\right)^{\nu_i} = \ln \prod_i \left(\frac{p_i}{P_0}\right)^{\nu_i} \tag{10-75}$$

so that
$$K_p \equiv \prod_i \left(\frac{p_i}{P_0}\right)^{\nu_i} \tag{10-76}$$

where K_p is the equilibrium constant based on partial pressures. It is apparent that $K_{\bar{f}}$ and K_p have the same numerical value for a given reaction at a given temperature. That is,

$$\ln K_{\bar{f}} = -\frac{\Delta G_T^0}{RT} \qquad \text{and} \qquad \ln K_p = -\frac{\Delta G_T^0}{RT} \tag{10-77}$$

This result is not fortuitous, but, rather, is due to the fact that the reference state of μ_i for both an ideal gas and a real gas in a mixture was chosen to be a pure ideal gas at unit pressure. However, $\ln K_{\bar{f}}$ and $\ln K_p$ are related to equilibrium compositions of the reacting species by

$$K_{\bar{f}} \equiv \prod_i \left(\frac{\bar{f}_i}{P_0}\right)^{\nu_i} \qquad \text{and} \qquad K_p \equiv \prod_i \left(\frac{p_i}{P_0}\right)^{\nu_i}$$

and, in general, $p_i \neq \bar{f}_i$. It is this inequality between p_i and $\bar{f}_i$ that leads to different predictions of the equilibrium composition for ideal-gas mixtures versus real-gas mixtures. This is particularly true for reactions at high pressure, as in ammonia synthesis.

As noted above, the numerical values of $\ln K_p$ and $\ln K_{\bar{f}}$ are both based on the same temperature function, namely, $\Delta G_T^o/RT$. For the convenience of shorthand notation we shall define the parameter K_o as

$$K_o \equiv \exp\left(\frac{-\Delta G_T^o}{RT}\right) \qquad \text{or} \qquad \ln K_o \equiv \frac{-\Delta G_T^o}{RT}$$

The term K_o might be termed the equilibrium constant based on standard state data. Due to the use of the standard state, K_o, for a *given reaction* depends *only* on the temperature. However, the *equilibrium composition* for a given reaction also depends on the initial composition and the PvT relationships governing the mixture. Based on Eq. (10-77), if a mixture fulfills the ideal-gas relationship, then $K_p = K_o$. If the mixture is gaseous, but not an ideal gas, then $K_{\bar{f}} = K_o$ and it becomes necessary to determine the fugacities. [In many texts tables of K_o data are labeled as K_p. In these cases K_p data are still based on $\exp(-\Delta G_T^o/RT)$.] Before examining the method for calculating the equilibrium composition of any gaseous mixture, we find it pertinent to review some methods for calculating ΔG_T^o and K_o.

10-8 EQUILIBRIUM CONSTANT EVALUATION

Whether the system is a real-gas mixture or an ideal-gas mixture, the equilibrium constant is calculated from the same fundamental relation

$$\ln K_{\bar{f}} = \ln K_p = -\frac{\Delta G_T^o}{RT} = \ln K_o \tag{10-77}$$

The presentation of thermochemical data in the literature permits the evaluation of ΔG_T^o or K_o by various methods. Four of these methods, based on data available from the JANAF thermochemical tables [1], are outlined below.

The first method is based on employing h and s data. In terms of more fundamental properties, $\Delta G_T^o = \Delta H_T^o - T\,\Delta S_T^o$, since $G \equiv H - TS$. For a chemical reaction

$$\Delta H_T^o = \Delta h_{R,T}^o = \sum_{P-R} \nu_i \left(\Delta h_{f,298}^o + h_T^o - h_{298}^o\right)_i$$

and

$$\Delta S_T^o = \Delta s_{R,T}^o = \sum_{P-R} \nu_i s_{T,i}^o$$

where $s_{T,i}^o$ is the absolute entropy in the sense of the third law. All the υ values are positive in this section, since we have elected to sum products minus reactants, rather than sum over all species in the reaction. The JANAF thermochemical tables is a well-known compilation of data which allows direct evaluation of ΔH_T^o and ΔS_T^o, and hence evaluation of ΔG_T^o and K.

A second and similar method is based on the presentation of Δh_f^o data as a function of temperature in the JANAF tables. The relation $\Delta h_{R,T} = \sum \nu_i\, \Delta h_{f,T,i}$ is an alternate method for evaluating the enthalpy of reaction. As a result,

$$\Delta G_T^o = \sum_{P-R} \nu_i \left(\Delta h_{f,T}^o - T s_T^o\right)_i \tag{10-78}$$

A third method is based on the appearance of $\Delta g_{f,T,i}^o$ data in the JANAF tables as a function of temperature. The use of the Gibbs function of formation Δg_f in determining ΔG_T^o follows the same rule as for Δh_f data cited above, namely,

$$\Delta g_{R,T} = \Delta G_T^o = \sum_{P-R} \nu_i \Delta g_{f,T,i}^o \tag{10-79}$$

Finally, a fourth method based on data from the third method is exploited by another column in the JANAF tables labeled "log K_P." Recall from Eq. (10-77) that $\ln K_o = -\Delta G_T^o/RT$ for any reaction. If the reaction is for the formation of a compound from its stable elements, we can write the preceding equation as

$$\ln K_{o,i} \equiv -\frac{\Delta g_{f,T,i}^o}{RT}$$

where $K_{o,i}$ is the equilibrium constant for the formation reaction of the ith species. Now, if Eq. (10-79) is divided by $-RT$, it becomes

$$-\frac{\Delta G_T^o}{RT} = \sum_{P-R} \nu_i \left(\frac{-\Delta g_{f,T,i}^o}{RT}\right)$$

Then, by using the definitions of $\ln K_o$ and $\ln K_{o,i}$, we find that

$$\ln K_o = \sum_{P-R} \nu_i \ln K_{o,i}$$

or

$$\log_{10} K_o = \sum_{P-R} \nu_i \log_{10} K_{o,i} \tag{10-80}$$

Data for $\log_{10} K_{o,i}$, listed as "Log K_P" in the JANAF tables, is available for hundreds of substances as a function of temperature. The use of logarithms to the base 10 in the tables is one of convenience. Major advantages of this last method are that Eq. (10-80) is dimensionless and that the calculation is numerically very simple. In some cases log K_P data for one or more species in a reaction of interest may not be available. Then we must evaluate the overall equilibrium constant from fundamental enthalpy and entropy data.

Table A-35 in the Appendix lists values of $\log_{10} K_o$ data for some common reactions as a function of temperature. These data are presented to illustrate trends and to simplify equilibrium composition calculations later in the chapter. It is important to note that rapid changes in K_o can occur in relatively small intervals of temperature. A change in $\log_{10} K_o$ of 3.0 changes K_o itself by a factor of 10^3. Although extensive tables of K_o data are not normally available, some care must be taken in their use. The following items need to be emphasized:

1. As noted above, the value of K_o is independent of the pressure during the actual reaction. The reason is that the standard-state Gibbs-function change ΔG_T^o is defined as the value in the standard state of unit pressure regardless of the actual conditions.
2. Equilibrium constants in this text are defined with products in the numerator of the definition and reactants in the denominator. This is fairly standard procedure, but there are authors who invert the definition. Some care must be taken in this respect when K_o values are taken from tables.
3. The value of K_o is a function of the method of writing a chemical equation for particular reactants and products. For example, consider the following three reactions at a given temperature:

$$CO + \tfrac{1}{2} O_2 \rightarrow CO_2 \tag{a}$$

$$2\,CO + O_2 \rightarrow 2\,CO_2 \tag{b}$$

$$CO_2 \rightarrow CO + \tfrac{1}{2} O_2 \tag{c}$$

The standard-state Gibbs-function change for reaction b is twice that for reaction a. Consequently, doubling the stoichiometric equation will square the value of K_o; that is, $(K_o)_b = (K_o)_a^2$. But reversal of the direction of a chemical equation reverses the sign of ΔG_T^o. Thus the K_o value for reaction c is the reciprocal of the value for reaction a. These results stem from the exponential relationship between K_o and ΔG_T^o. One can conclude from these results that tabulated values of K_o as a function of temperature are meaningless unless the chemical equation to which they refer is also cited.

The example below illustrates these four methods for evaluating the equilibrium constant.

Example 10-8. Consider the gas-phase reaction $CO + H_2O \leftrightharpoons CO_2 + H_2O$ at 1000 K. Calculate the value of $\ln K_o$ by the four methods discussed earlier. Selected data from the JANAF tables are shown in Table 10-4, expressed in kilojoules, kmoles, and kelvins.

Solution

(*a*) The first method involves use of h and s data at 1000 K and Δh_f data at 298 K. Using the relation

$$g_T^o = \Delta h_{f,298}^o + h_T^o - h_{298}^o - Ts_T^o$$

and the JANAF-based $\Delta h_{f,298}^o$ data in the Appendix for the four gases, we find that

$$\begin{aligned} CO_2\text{:}\quad g_T^o &= -393{,}520 + 33{,}405 - 1000(269.215) = -629{,}330 \\ H_2\text{:}\quad g_T^o &= 0 + 20{,}686 - 1000(166.114) = -145{,}428 \\ CO\text{:}\quad g_T^o &= -110{,}530 + 21{,}686 - 1000(234.421) = -323{,}265 \\ H_2O\text{:}\quad g_T^o &= -241{,}820 + 25{,}978 - 1000(232.597) = -448{,}439 \end{aligned}$$

Therefore

$$\begin{aligned} \Delta G_T^o &= \sum_{P-R} \nu_i g_{i,T}^o \\ &= -629{,}330 - 145{,}428 - (-323{,}265) - (-448{,}439) \\ &= -3054 \text{ kJ/kmol CO} \end{aligned}$$

and

$$\ln K_o = -\frac{\Delta G_T^o}{RT} = \frac{3054}{8.314(1000)} = 0.367$$

(*b*) The second method is based on the relation $g_T^o = \Delta h_{f,T}^o - Ts_T^o$. On the basis of data in Table 10-4, we find that

$$\begin{aligned} CO_2\text{:}\quad g_T^o &= -394{,}639 - 1000(269.215) = 663{,}854 \\ H_2\text{:}\quad g_T^o &= 0 - 1000(166.113) = -116{,}113 \\ CO\text{:}\quad g_T^o &= -112{,}010 - 1000(234.421) = -346{,}431 \\ H_2O\text{:}\quad g_T^o &= -247{,}885 - 1000(232.597) = -480{,}482 \end{aligned}$$

TABLE 10-4
Selected JANAF tabular data

Species	$h_{1000}^o - h_{298}^o$	$\Delta h_{f,1000}^o$	s_{1000}^o	$\Delta g_{f,1000}^o$	$\log K_P$
CO_2	33,405	−394,639	269.215	−395,924	20.680
H_2	20,686	0	116.113	0	0
CO	21,686	−112,010	234.421	−200,242	10.459
H_2O	25,978	−247,885	232.597	−192,631	10.062

The value of ΔG_T^o in this case again is -3054 kJ/kmol CO, even though the individual g_T^o values are all different from the preceding evaluation. Hence $\ln K_o$ again is 0.367.

(*c*) The third method directly involves $\Delta g_{f,T}^o$ data from Table 10-4,

$$\Delta G_T^o = \sum_{P-R} \nu_i \, \Delta g_{f,T}^o = -395{,}924 + 0 - (-200{,}242) - (-192{,}631)$$

$$= -3051 \text{ kJ/kmol of CO}$$

This very slightly different value of ΔG_T^o leads to the same value of 0.367 for $\ln K_o$.

(*d*) Finally, the fourth method involves use of $\log_{10} K_{o,i}$ data from the JANAF tables. In this case, for the overall reaction

$$\log_{10} K_o = \sum_{P-R} \nu_i \log_{10} K_{P,i} = 20.680 + 0 - 10.459 - 10.062 = 0.159$$

Thus $\ln K_o = 0.366$ in this final case, which essentially agrees with the preceding methods. The agreement is expected since the JANAF tables are internally consistent within the reported significant figures for the data.

In Sec. 10-6 we developed a relationship between the affinity **A** and the enthalpy of reaction, $\Delta h_{R,T}$, namely, Eq. (10-68). When this equation is written for a reaction occurring at the standard reference state, we obtain

$$\frac{\partial}{\partial T}\left(\frac{\mathbf{A}^o}{T}\right)_{P,\xi} = \frac{\Delta h_{R,T}^o}{T^2} \tag{10-81}$$

This equation leads directly to a relationship between $\ln K_o$ and $\Delta h_{R,T}$ known as the van't Hoff equation. Equation (10-58) for the affinity in this case becomes

$$-\mathbf{A}^o = \sum_{P-R} \nu_i \mu_i^o = \sum_{P-R} \nu_i g_i^o = \Delta G_T^o = -RT \ln K_o$$

Substitution of this result into Eq. (10-81) leads to the desired result,

$$\frac{d \ln K_o}{dT} = \frac{\Delta h_{R,T}^o}{RT^2}$$

or

$$\frac{d \ln K_o}{d(1/T)} = -\frac{\Delta h_{R,T}^o}{R} \tag{10-82}$$

Either one of the above forms is the *van't Hoff isobar equation*. (van't Hoff received the first Nobel prize in chemistry in 1901.) To integrate the van't Hoff equation, one needs information on the variation of $\Delta h_{R,T}^o$ with temperature. One approximation is to assume $\Delta h_{R,T}^o$ is essentially constant in the temperature range of interest. On this basis Eq. (10-82) becomes

$$\ln \frac{K_{o,2}}{K_{o,1}} = -\frac{\Delta h_{R,\text{avg}}^o}{R}\left(\frac{1}{T_2} - \frac{1}{T_1}\right) \tag{10-83}$$

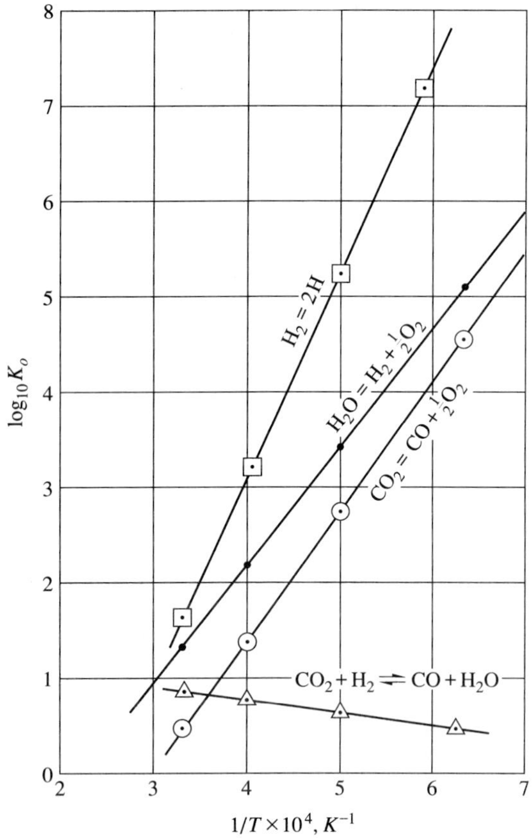

FIGURE 10-10
A plot of $\log_{10} K_o$ versus $1/T$ for several common reactions.

There are several interesting consequences of the van't Hoff equation:

1. On the basis of Eq. (10-83), $\ln K_o$ is approximately a linear function of $1/T$. Thus an appropriate way to plot K_o data is $\ln K_o$ (or $\log_{10} K_o$) versus $1/T$, as shown by Fig. 10-10 for several common reactions. The slopes of the lines on this plot are a measure of $-\Delta h^o_{R,\text{avg}}/R$. This equation also suggests that it is better to interpolate $\ln K_o$ data as a linear function of $1/T$, and not as a linear function of T.
2. The van't Hoff equation requires that the equilibrium constant decrease with increasing temperature for *all exothermic* reactions (Δh_R is negative), and increase with increasing temperatures for *all endothermic* reactions. (All the reactions in Table A-35 in the Appendix are endothermic reactions as the reaction proceeds from left to right.) This second-law requirement has a negative effect on the release of energy adiabatically from hydrocarbon combustion.

Products such as CO_2 and H_2O tend to dissociate at high temperature, and these dissociation phenomena are endothermic. The overall effect is to lower the adiabatic flame temperature from the value predicted without dissociation.

The next step is to review the methods of estimating the equilibrium composition of reacting-gas mixtures.

10-9 EQUILIBRIUM COMPOSITION EVALUATION

Methods for determining the value of the equilibrium constant K_o have been outlined in the preceding section. The equilibrium composition now can be ascertained in conjunction with either Eq. (10-73) or Eq. (10-76). That is, for a mixture of real gases

$$K_{\bar{f}} = \prod_i \left(\frac{\bar{f}_i}{P_0}\right)^{\nu_i} = \frac{\prod_P \bar{f}_i^{\nu_i}}{\prod_R \bar{f}_i^{\nu_i}} P_0^{-\Delta\nu} \qquad \text{(real-gas mixture)} \tag{10-84}$$

and for a mixture of ideal gases

$$K_p = \prod_i \left(\frac{p_i}{P_0}\right)^{\nu_i} = \frac{\prod_P p_i^{\nu_i}}{\prod_R p_i^{\nu_i}} P_0^{-\Delta\nu} \qquad \text{(ideal-gas mixture)} \tag{10-85}$$

where

$$\Delta\nu \equiv \sum_{P-R} \nu_i \tag{10-86}$$

Since P_0 is taken as unit pressure (usually 1 atm or 1 bar), the term $P_0^{-\Delta\nu}$ is unity in value. The reason for retaining the term to this point is to emphasize that $K_{\bar{f}}$ and K_p are dimensionless parameters. Consider the situation where P_0 is dropped out and forgotten (since $P_0 = 1$) in Eqs. (10-73) and (10-76). Then Eqs. (10-84) and (10-85) would have a unit of pressure to the $\Delta\nu$ power, and $\Delta\nu$ does not necessarily equal zero. This interpretation is not, however, correct. We shall drop the P_0 term in future discussions for convenience, but its symbolic presence should be noted.

The final step is to express Eqs. (10-84) and (10-85) in terms of composition variables such as mole fractions or moles for the equilibrium mixture. For convenience we shall begin with mixtures of ideal gases and then examine the approach to real-gas mixtures.

10-9-1 Ideal-Gas Mixtures

The p_i terms in Eq. (10-85) for an ideal-gas mixture can be replaced by the relation

$$p_i = y_i P = \frac{N_i}{N_T} P$$

where y_i is the gas-phase mole fraction and N_i is the moles of species i. Substitution of the above relation into Eq. (10-85) yields

$$K_p = \frac{\prod_P N_i^{\nu_i}}{\prod_R N_i^{\nu_i}} \left(\frac{P}{N_T}\right)^{\Delta\nu} \tag{10-87}$$

where $N_T = \sum N_i + N_{\text{inert}}$. (It is particularly important to include the moles of inert gas in the reacting mixture.) For a specific reaction at a specified temperature and pressure the N_i equilibrium values for the products and remaining reactants are the only unknowns. The additional equations required for the numerical solution of Eq. (10-87) are based on the conservation of atomic species. A number of standardized computer programs exist for the equilibrium analysis of simultaneous reactions involving independent K_p expressions and atomic species balances.

Example 10-9. At 500 K the equilibrium constant for the reaction $NO + \frac{1}{2}O_2 \leftrightarrows NO_2$ is 120. The initial reactants include 1 mol of NO, 10 mol of O_2, and 40 mol of N_2, and the pressure is maintained at 0.10 atm. Determine the moles of NO_2 present at equilibrium.

Solution. On the basis of Eq. (10-87) the expression for K_p is

$$K_p = \frac{n_{NO_2}}{n_{NO}(n_{O_2})^{1/2}} \left(\frac{P}{n_T}\right)^{-1/2}$$

where $n_T = n_{NO_2} + n_{NO} + n_{NO_2} + n_{\text{inert}}$. (Lowercase n is used here for the number of moles to avoid confusion with the nitrogen symbol N.) In addition, consider the overall reaction

$$1\,NO + 10\,O_2 + 40\,N_2 \rightarrow n_{NO} + n_{NO_2} + n_{O_2} + 40\,N_2$$

Two conservation equations can be written for the atomic species N and O on the basis of the above overall reaction. These are

$$\begin{aligned} \text{N bal:} \quad 1 &= n_{NO} + n_{NO_2} \\ \text{O bal:} \quad 21 &= n_{NO} + 2n_{NO_2} + 2n_{O_2} \end{aligned}$$

We have not included the N_2 species in the N balance, because we are assuming that this species acts as an inert (nonreacting) gas at this temperature. From these two conservation equations we find that in terms of NO_2 present at equilibrium

$$\begin{aligned} n_{NO} &= 1 - n_{NO_2} \\ n_{O_2} &= 10 - 0.5n_{NO_2} \\ n_T &= 51 - 0.5n_{NO_2} \end{aligned}$$

As a result, if for the moment we represent n_{NO_2} by x, the K_p expression becomes

$$K_p = \frac{x(51 - 0.5x)^{1/2}(0.1)^{-1/2}}{(1 - x)(10 - 0.5x)^{1/2}} = 120$$

The numerical solution to the above cubic equation by computer analysis is $x = n_{NO_2} = 0.943$; that is, 94.3 percent of the NO is converted to NO_2.

In the preceding example N_2 was assumed to be an inert species in the presence of oxygen at 500 K. However, two possible reactions for nitrogen must be considered. These are the direct oxidation of N_2 with O_2 and the dissociation of N_2 as given by

$$\tfrac{1}{2}N_2 + \tfrac{1}{2}O_2 \rightleftarrows NO \qquad \text{and} \qquad N_2 \rightleftarrows 2N$$

Whether to consider these reactions depends on the values of K_o for the two reactions at 500 K. From Table 35 in the Appendix the K_o values are $10^{-8.78}$ and $10^{-92.67}$, respectively. Such small K_o values imply negligible formation of either NO or N at this temperature. Thus the assumption of N_2 as nonreacting is quite appropriate. At flame temperatures around 2000–2500 K this is not true for the formation of NO, because K_o is around 10^{-1} in that temperature range. Thus NO formation from N_2 and O_2 becomes important as gas temperatures increase. This phenomenon plays an important role in air-pollution studies.

10-9-2 Real-Gas Mixtures

The equilibrium composition of real-gas mixtures is found from Eq. (10-84). In order to make the evaluation of $\bar{f}_i$ less complicated, we consider the ideal-solution model as an appropriate first approximation. On the basis of the Lewis-Randall rule

$$\bar{f}_i \approx y_i f_{i,P,T} = y_i f_i^\Delta \tag{8-120}$$

where f_i^Δ is the fugacity of a pure component at the mixture pressure and temperature. Consequently, Eq. (10-84) becomes

$$K_{\bar{f}} \approx \left[\frac{\prod_P y_i^{\nu_i}}{\prod_R y_i^{\nu_i}}\right] \times \left[\frac{\prod_P (f_i^\Delta)^{\nu_i}}{\prod_R (f_i^\Delta)^{\nu_i}}\right] = K_y K_{f^\Delta} \tag{10-88}$$

where the first bracketed term is, by definition, K_y, the equilibrium constant based on mole fractions. The second bracketed term is K_{f^Δ}, an equilibrium constant based on pure fugacities at the system temperature and pressure. However, for a pure component $f_i^\Delta/P \equiv \phi_i$, the pure fugacity coefficient. Hence

$$K_{f^\Delta} = \left[\frac{\prod_P \phi_i^{\nu_i}}{\prod_R \phi_i^{\nu_i}}\right] P^{\Delta\nu} \equiv K_\phi P^{\Delta\nu} \tag{10-89}$$

where the bracketed term is defined as K_ϕ, the equilibrium constant based on fugacity coefficients at the system pressure and temperature. Also, K_y in Eq. (10-88) can be written as

TABLE 10-5
Comparison of $K_{\bar{f}}$, K_p, and K_ϕ as a function of pressure for the gas-phase reaction:
$\frac{1}{2}N_2 + \frac{3}{2}H_2 \rightleftarrows NH_3$ at 450°C

P, atm	$K_p \times 10^3$	K_ϕ	$K_{\bar{f}} \approx K_p K_\phi \times 10^3$
10	6.59	0.988	6.51
30	6.76	0.969	6.55
50	6.90	0.953	6.57
100	7.25	0.905	6.56
300	8.84	0.750	6.63
600	12.94	0.573	7.42
1000	23.28	0.443	10.32

$$K_y = \left[\frac{\prod_P N_i^{\nu_i}}{\prod_R N_i^{\nu_i}}\right] N_T^{-\Delta\nu} \tag{10-90}$$

Thus on the basis of an ideal solution the general expression for $K_{\bar{f}}$ can be approximated by

$$K_{\bar{f}} \approx \left[\frac{\prod_P N_i^{\nu_i}}{\prod_R N_i^{\nu_i}}\right]\left(\frac{P}{N_T}\right)^{\Delta\nu}\left[\frac{\prod_P \phi_i^{\nu_i}}{\prod_R \phi_i^{\nu_i}}\right] = K_p K_\phi \tag{10-91}$$

Hence an approximate method for real-gas equilibrium involves correcting K_p for ideal-gas mixtures by the factor K_ϕ. In the limit as $P \to 0$, $K_\phi \to 1$. The values of ϕ_i can be estimated from a generalized chart, for example, at the system temperature and pressure.

As the pressure of a system increases, the accuracy of the ideal-solution model decreases. This is seen in the equilibrium data in Table 10-5 for the synthesis of ammonia. K_p data are based on experimental measurements of equilibrium compositions, whereas K_ϕ data are based on the generalized chart for ϕ. At the given temperature $K_{\bar{f}}$, and hence the product $K_p K_\phi$, should remain constant. The data of Table 10-5 indicate this is true (within 1 percent) up to 300 atm. At higher pressures the validity of the Lewis-Randall rule is considerably in error.

PROBLEMS

10-1. The combustion of methane gas (CH_4) with air leads to the following volumetric analysis of the product gases on a dry basis: 9.52 percent CO_2, 0.28 percent CO, and 3.57 percent O_2. Determine

a. the moles of air used per mole of fuel

b. the percent excess air used

10-2. The combustion of CH_4 with air leads to the following volumetric analysis of the product gases on a dry basis: 10.10 percent CO_2, 0.21 percent CO, and 2.58 percent O_2. Determine

a. the moles of air used per mole of fuel
b. the percent excess air used

10-3. Octane (C_8H_{18}) is burned with air. The volumetric analysis of the product gases on a dry basis is 10.01 percent CO_2, 0.88 percent CO, and 5.62 percent O_2. Determine

a. the moles of air used per mole of fuel
b. the percent excess air used in the test

10-4. Propane (C_3H_8) is burned in a steady-flow system with 20 percent excess air, both originally at 25°C. The products of combustion (assume complete) are cooled to 900 K. Find the amount of heat transfer, in kJ/kmol of fuel.

10-5. Butane (C_4H_{10}) is burned in a steady-flow process with 20 percent excess air, both originally at 25°C, and the complete products of combustion are cooled at 1100 K. Determine the heat transfer during the process, in kJ/kmol of gaseous fuel.

10-6. If ethane (C_2H_6) is burned in a steady-flow system with 30 percent excess air, both originally at 25°C, and the complete products of combustion are cooled to 700 K, determine the heat transfer from the product gases, in kJ/kmol of fuel.

10-7. Carbon monoxide (CO) is burned with air, both initially at 25°C. The final temperature of the product gases, assuming complete combustion, is 1100 K, and during the process 136,000 kJ of heat are removed per kilomole of CO. For the steady-flow process, determine the percent excess air used.

10-8. Acetylene (C_2H_2) is burned with air, both initially at 25°C. The final temperature of the product gases, assuming complete combustion, is 900 K. During the process 910,000 kJ of heat are removed per kilomole of C_2H_2. Determine the percent excess air used for the steady-flow process.

10-9. Ethane (C_2H_6) is burned with air, both initially at 25°C. The final temperature of the product gases, assuming complete combustion, is 1100 K, and during the process 786,000 kJ of heat are removed per kilogram mole of C_2H_6. Determine the percent excess air used during the steady-flow process.

10-10. Ethane gas (C_2H_6) is burned with 30 percent excess air, with both reactants initially at 25°C. Determine the maximum combustion temperature, in kelvins, for the steady-flow process.

10-11. Ethane gas is burned in a steady-flow process with 30 percent excess air, both of which enter separately at 25°C and 1 atm. Assume the reaction goes to completion, and the products leave as a mixture at 1 atm and the environmental temperature of 25°C. The water in the products is assumed to be all liquid. Determine

a. the entropy change for the reaction itself, in kJ/K
b. the total entropy change for the overall process, in kJ/K per kilogram mole of fuel
c. the reversible work associated with the combustion process, in kJ/kmol
d. the irreversibility of the combustion process, in kJ/kmol of fuel

10-12. Reconsider Prob. 10-11. Determine the total entropy change, in kJ/K per kilogram of fuel, if the fuel and air enter separately at 1 atm and 25°C, and all the products are gases in a mixture at 1 atm.

10-13. Reconsider Probs. 10-10 and 10-11. For the adiabatic combustion of ethane with 30 percent excess air, both initially at 1 atm and 25°C, and for $T_0 = 25°C$, determine, in kJ/kmol of fuel,

a. the availability change and the irreversibility for the adiabatic combustion process
b. the availability change when the hot-product gases are cooled to 25°C
c. the ratio of the availability change as the hot-product gases are cooled to 25°C to the availability change when both reactants and products are gases at 25°C.

10-14. Butane gas (C_4H_{10}) is burned with 20 percent excess air, with both reactants initially at 25°C. Determine the maximum combustion temperature for the steady-flow process, in kelvins.

10-15. Butane gas is burned in a steady-flow process with 20 percent excess air, both of which enter separately at 25°C and 1 atm. Assume the reaction goes to completion and the products leave as a mixture at 1 atm and the environmental temperature of 25°C. The water in the products is assumed to be all liquid. Determine

a. the entropy change for the reaction itself
b. the total entropy change for the overall process, in kJ/K per kilomole of fuel
c. the reversible work associated with the combustion process, in kJ/kmol
d. the irreversibility of the combustion process, in kJ/kmol.

10-16. Reconsider Prob. 10-15. Determine the total entropy change, in kJ/K per kmole of fuel, if the fuel and air enter separately at 1 atm and 25°C, and all the products are gases in a mixture at 1 atm.

10-17. Reconsider Probs. 10-14 and 10-15, which involve the combustion of C_4H_{10} with 20 percent excess air, both initially separate at 1 atm and 25°C. If $T_0 = 25°C$, determine, in kJ/kmol of fuel,

a. the availability change and the irreversibility for the adiabatic combustion process
b. the availability change when the hot-product gases are cooled to 25°C
c. the ratio of the availability change as the hot-product gases are cooled to 25°C to the availability change when both reactants and products are gases at 25°C

10-18. Propane gas (C_3H_8) is burned with 20 percent excess air, with both reactants initially at 25°C. Determine the maximum combustion temperature, in kelvins, for the steady-flow process.

10-19. Propane gas is burned in a steady-flow process with 20 percent excess air, both of which enter separately at 25°C and 1 atm. The reaction is complete and the products leave as a mixture at 1 atm and the environmental temperature of 25°C. The water in the products is assumed to be all liquid. Determine

a. the entropy change for the reaction itself
b. the total entropy change for the overall process, both in kJ/K per kilomole of fuel
c. the reversible work associated with the combustion process, in kJ/kmol
d. the irreversibility of the combustion process, in kJ/kmol of fuel

10-20. Reconsider Prob. 10-19. Determine the total entropy change, in kJ/K per kilomole of fuel, if the fuel and air enter separately at 1 atm and 25°C and all the products are gases in a mixture at 1 atm.

10-21. Reconsider Probs. 10-18 and 10-19. For the adiabatic combustion of C_3H_8 with 20 percent excess air, both initially separate at 1 atm and 25°C, and for $T_0 = 25°C$, determine

a. the availability change and the irreversibility for the adiabatic combustion process

b. the availability change when the hot-product gases are cooled to 25°C

c. the ratio of the availability change as the hot-product gases are cooled from the adiabatic-combustion temperature to 25°C to the availability change when both reactants and products are gases at 25°C

10-22. A fuel cell operates on the reaction of ammonia (NH_3) with hydrogen peroxide (H_2O_2) according to the overall reaction

$$2\,NH_3(g) + 3\,H_2O_2(g) \rightarrow N_2(g) + 6\,H_2O(g)$$

Each of the reactants and products is assumed to be at 298 K and 1 atm.

a. Determine the Gibbs function change for the reactions, in kJ/kmol NH_3.

b. Evaluate the ideal potential developed by the cell, in volts, if the half-cell reaction on the fuel side is: $2\,NH_3(g) \rightarrow N_2 + 6\,H^+ + 6\,e^-$.

c. Find the ideal-cell efficiency.

d. Discuss whether the cell must be cooled or heated to maintain isothermal operation.

10-23. Gaseous propane (C_3H_8) reacts with the stoichiometric amount of *oxygen* at 25°C in a fuel cell. The fuel and the pure oxidant *each* enters the cell at 1-atm pressure. Assume that CO_2 is removed from the anode at 1-atm pressure, and that liquid water is removed from the cathode. Determine

a. the maximum work that could be obtained in the steady-flow device, in kJ/kmol of fuel

b. the half-cell reactions

c. the ideal potential, in volts

d. the ideal efficiency

10-24. Consider the fuel-cell process described in Prob. 10-23. Determine the ideal efficiency if the process were carried out at 1 atm and 1000 K, rather than at 298 K. The specific heat of C_3H_8 is represented by the equation

$$c_p = -4.042 + 30.46 \times 10^{-2}T - 1.571 \times 10^{-4}T^2 + 0.0317 \times 10^{-6}T^3$$

where c_p is in kJ/kmol·K and T is in kelvins.

10-25. Gaseous acetylene (C_2H_2) reacts with the stoichiometric amount of oxygen at 25°C in a fuel cell. The fuel and the pure oxidant *each* enters the cell at 1-atm pressure. Assume that CO_2 is removed from the anode pure at 1-atm pressure, and that liquid water is removed at the cathode. Determine

a. the maximum work that could be obtained from the steady-flow device, in kJ/kmol of fuel

b. the half-cell reactions in an acidic electrolyte

c. the ideal potential of the cell, in volts

d. the ideal efficiency

10-26. Reconsider the fuel-cell process described in Prob. 10-25. Determine (a) the ideal efficiency and (b) the ideal potential, in volts, if the process were carried out at 1 atm and 800 K, rather than at 298 K. Note that water as a product will be in the gas phase in this case. The variation of the specific heat with temperature appears in Table A-3 in the Appendix for acetylene.

10-27. Gaseous methane (CH_4) reacts with the stoichiometric amount of *oxygen* at 25°C in a fuel cell. The fuel and the pure oxidant *each* enters the cell at 1 atm. Assume that CO_2 is removed from the anode at 1 atm, and that liquid water is removed from the cathode. Determine

a. the maximum work that could be obtained in the steady-flow device, in kJ/kmol of fuel
b. the half-cell reactions
c. the ideal potential, in volts
d. the ideal efficiency

10-28. Consider the fuel-cell process described in Prob. 10-27. Determine the ideal efficiency if the process were carried out at 1 atm and 800 K, rather than at 298 K. Note that water is a gas in this case. The specific-heat equation for CH_4 is found in Table A-3 in the Appendix.

10-29. Carbon monoxide and O_2 are placed in a reaction vessel at 1000 K and 4.0 bars in a one-to-one molar ratio. At that moment, before a reaction begins, determine the value of the chemical potential of CO, in kJ/kmol, assuming ideal-gas behavior.

10-30. Carbon monoxide and oxygen in a 2:1 molar ratio are maintained at 2000 K and 6.0 bars. Determine the chemical potential, in kJ/kmol, of the oxygen if it behaves as an ideal gas.

10-31. Carbon monoxide and water vapor (H_2O) are placed in a reaction vessel at 1000 K and 2.0 bars in a one-to-one molar ratio. For the given composition, determine the value of the chemical potential of the water vapor, in kJ/kmol.

10-32. At some instant a system contains 0.90 gmol of O_2 and 0.20 gmol of O in a reaction vessel at 3000 K and 0.80 bar.

a. Determine the value of the affinity **A** at that instant, in kilojoules, for the overall reaction: $O_2 \rightleftarrows 2\,O$.
b. On the basis of your answer to part (a), discuss whether the reaction will proceed toward producing more O or more O_2.
c. Determine the affinity, in kilojoules, for the given molar quantities and pressure at temperatures of 3200 and 3400 K. Estimate the temperature at which the given molar quantities would be the equilibrium values.

10-33. A system contains 0.92 gmol of H_2 and 0.16 gmol of H in a reaction vessel at 2900 K and 0.85 bar.

a. Determine the value of the affinity **A** for these conditions, in kilojoules, for the overall reaction: $H_2 \rightleftarrows 2\,H$.
b. On the basis of your answer to part (a), discuss whether the reaction will proceed toward producing more H or more H_2.
c. Determine the affinity, in kilojoules, for the given molar quantities, pressure, and temperatures of 3100 and 3400 K. Estimate the temperature at which the given molar quantities would be the equilibrium values.

10-34. Consider the gas-phase reaction at 1800 K: $CO(g) + \frac{1}{2}O_2(g) \rightleftarrows CO_2(g)$. By employing the JANAF tables, determine the equilibrium constant K_o on the basis of

a. Gibbs function of formation data
b. enthalpy of formation data at 298 K, and enthalpy and entropy data
c. $\log_{10} K_P$ of formation data

10-35. Consider the gas-phase reaction at 1600 K: $CO + H_2O \rightleftarrows CO_2 + H_2$. On the basis of the JANAF tables, determine the equilibrium constant K_o by using

a. Gibbs function of formation data
b. enthalpy of formation data at 298 K, and sensible enthalpy and entropy data
c. $\log_{10} K_P$ data for the reacting species
d. the tabulated K_o value for the overall reaction

10-36. Consider the gas-phase reaction at 2000 K: $CO(g) + \frac{1}{2}O_2(g) \rightleftarrows CO_2(g)$. On the basis of the JANAF tables, determine the equilibrium constant K_o by using

a. Gibbs function of formation data
b. enthalpy of formation data at 298 K, and enthalpy and entropy data
c. $\log_{10} K_P$ of formation data

10-37. Consider the equilibrium reaction, $\frac{1}{2}N_2 + \frac{1}{2}O_2 \rightleftarrows NO$. Calculate the equilibrium constant K_o at (a) 1200, (b) 1500, and (c) 2000 K. The enthalpy of formation of NO at 25°C is 90,290 kJ/kmol, and other appropriate data for NO are shown in Table P10-37.

TABLE P10-37

T	298	1200	1500	2000
$h^o - h^o_{298}$	0	29,120	39,729	57,859
s^o	210.758	254.816	262.703	273.128

10-38. The gas phase formation of CO_2 from CO is considered, kinetically, to be controlled by the reaction: $CO + OH \rightleftarrows CO_2 + H$.

a. Determine the equilibrium constant K_o for this reaction from log K_P of formation data from the JANAF tables at 1000 K. The log K_P data for the species OH and H are −1.218 and −8.646, respectively.
b. Discuss briefly whether or not you consider this reaction to go to completion at the given temperature from a thermodynamic viewpoint.

10-39. The reactants H_2 and F_2 are introduced into a rocket combustion chamber. In addition to the main reaction, $H_2 + F_2 \rightleftarrows 2\,HF$, one must also reflect on the possibility of either or both of the H_2 and F_2 dissociating into their monatomic species, H and F. Which, if any, of the three primary reactions would need to be considered in equilibrium calculations at 1000 K, based on the following data abstracted from the JANAF tables at 1000 K:

	F	**H**	**HF**
$\log_{10} K_P$	−0.988	−8.646	14.468

Discuss briefly in terms of the equilibrium constants for the three reactions.

10-40. Consider the equilibrium reaction $O_2 \rightleftarrows 2\,O$.

a. Determine the enthalpy of reaction at 3000 K by using enthalpy data for the individual species, in kJ/kmol O_2.
b. Estimate this value by using the van't Hoff equation and K_o data for the reaction at 2900 and 3100 K.

10-41. Calculate the enthalpy of reaction at 2500 K for the reaction $CO + \frac{1}{2}O_2 \rightleftarrows CO_2$ by using

a. the van't Hoff equation and K_o data

b. $\Delta h^o_{f,298}$ and sensible enthalpy data, in kJ/kmol

10-42. Estimate the energy required to break H_2 down into its monatomic species at 2500 K and 1 atm by using

a. the van't Hoff equation and K_o data

b. $\Delta h^o_{f,298}$ data, and by noting that the sensible enthalpy change of monatomic hydrogen is 45,770 kJ/kmol between 298 and 2500 K.

10-43. The equilibrium constant for the atmospheric gas-phase reaction $N_2 + O_2 \rightleftarrows 2\,NO$, at 2000 K is 4.0×10^{-4}. Consider an initial mixture of N_2 and O_2 in a molar ratio of (a) 10/1 and (b) 40/1. If the mixture is heated to 2000 K, find the concentration of NO in the equilibrium mixture in ppm (parts per million by volume).

10-44. Consider the gas-phase reaction $NO + \frac{1}{2}O_2 \rightleftarrows NO_2$. At 500 K the equilibrium constant K_p is 0.110. One mole of NO and 2 mol of O_2 and x mol of argon (an inert gas) are placed in a reaction chamber at 500 K and 2 atm. At the final equilibrium state there are 0.080 mol of NO_2 present. Determine the moles of argon in the reaction mixture.

10-45. Consider a reaction vessel maintained at temperature T and a pressure of 2 atm which initially contains 1 mol of NO, 2 mols of O_2, and 8 mols of an inert gas. The reaction of interest is $NO + \frac{1}{2}O_2 \rightleftarrows NO_2$. At equilibrium it is found that there are 0.70 mol of NO_2. Determine the value of K_p for the reaction at temperature T.

10-46. The equilibrium constant for the gas-phase reaction, $SO_2 + \frac{1}{2}O_2 \rightleftarrows SO_3$, at 1000 K is 1.80. The initial moles of reactants consist of (a) $1\,SO_2$, $2\,O_2$, and $8\,N_2$ at 1.1 atm, and (b) $1\,SO_2$, $2\,O_2$, and $5\,N_2$ at 0.8 atm. Determine the moles of SO_3 at equilibrium for the specified reactants, pressure, and temperature.

10-47. The equilibrium constant for the gas-phase reaction $SO_2 + \frac{1}{2}O_2 \rightleftarrows SO_3$, at 1000 K is 1.80. The initial reactants are at 2 atm and consist of 1 mol of SO_2, 2 mols of O_2, and x mol of N_2 (an inert gas). If there are 0.582 mol of SO_3 present at equilibrium at the specified pressure and temperature, determine the moles of nitrogen present.

10-48. Consider the reaction $CO_2 \rightleftarrows CO + \frac{1}{2}O_2$. The reaction occurs at 2500 K and 1 atm. Determine how much nitrogen must be added to the equilibrium mixture of ideal gases so that 80 percent of the initial pure CO_2 is present at equilibrium.

10-49. Ammonia (NH_3) is synthesized according to the equation $N_2 + 3\,H_2 \rightleftarrows 2\,NH_3$. If 1 mol of N_2 and 4 mol of H_2 are reacted at 450°C and 200 atm, determine the moles of NH_3 present at equilibrium if the substances form (a) an ideal-gas mixture and (b) an ideal solution. At 450°C the equilibrium constant for the reaction is 4.30×10^{-5}. Use generalized fugacity-coefficient data.

10-50. Consider the synthesis of NH_3 from nitrogen and hydrogen by reaction $N_2 + 3\,H_2 \rightleftarrows 2\,NH_3$. Initially, 1 mol of N_2 and 3 mols of H_2 are at 700 K. The $\log_{10} K_P$ of formation NH_3 from the JANAF tables is −2.034. Determine the moles of NH_3 formed at

a. 1 atm and the gases are ideal gases

b. 400 atm, and the gases are ideal gases

c. 400 atm, and the gases form an ideal solution

10-51. Carbon monoxide and water vapor react according to the equation $CO + H_2O \rightleftarrows CO_2 + H_2$. If 1 mol of CO and 2 mols of H_2O are reacted at 600 K and 500 atm, determine the moles of CO_2 present at equilibrium if the substances form (a) an ideal-gas mixture and (b) an ideal solution. (You will need to calculate the value of the equilibrium constant at 600 K, since its value is missing from Table A-35. To simplify the calculation of K_ϕ for part (b), use generalized fugacity-coefficient data.)

REFERENCES

1. Stull, D. R., and H. Prophet, *JANAF Thermochemical Tables*, 2nd ed., NSRDS-NBS **37,** Natl. Bureau of Stds., Washington, DC, June 1971.
2. *Marks' Standard Handbook for Mechanical Engineers*, 8th ed., McGraw-Hill, Inc., New York, 1978.
3. *Perry's Chemical Engineers' Handbook*, 6th ed., McGraw-Hill, Inc., New York, 1984.
4. Douglas, J., E. Gillis, and D. Rastler, "Fuel Cells for Urban Power," *EPRI J.*, **16**:5, September 1991.
5. De Donder, Th., *Bull. Cl. Sci., Acad. R. Belg.*, **7**(5), 197–205, 1922.

CHAPTER 11

CHEMICAL AVAILABILITY

In Chapters 3 and 4 the concept of availability was introduced in terms of thermomechanical equilibrium with the environment. The concept is extended in this chapter to include equilibrium with the chemical species in the environment. Applications include the mixing of products of combustion with the atmosphere and air-conditioning process design.

11-1 INTRODUCTION

The availability concepts introduced in Chaps. 3 and 4 applied to a *restricted dead state* associated with thermomechanical equilibrium with the environment. The mass studied was not allowed to pass into or react chemically with the environment. In this chapter we shall extend the availability or exergy analysis to include the work which theoretically can be produced as mass is permitted to pass into and mix with the atmosphere at (T_0, P_0). This mixing process will require a more detailed description of the chemical state of the environment. However, such a description will allow evaluation of availability losses in effluent streams leaving a control volume, for example. This concept will be extended further to include chemical reactions of fuel and atmospheric air in Chap. 12.

The work associated with the reversible transfer of mass between two systems initially in nonequilibrium requires a discussion of *semipermeable membranes*. Such membranes permit the passage of some substances but exclude others. There are both natural and man-made inorganic and organic membranes which act in this manner. An important commercial use of man-made membranes is in the process of reverse osmosis for the desalination of water (see Sec. 9-9). Also, nature provides such membranes in the human body. In the following use of semipermeable membranes it is assumed that a given membrane permits the passage of only one substance. Recall also from Sec. 9-2 that the reversible transfer of mass between two systems requires the chemical potential of a given species to be the same in both systems. If the chemical potential of a given substance initially is not the same in a system and its environment, for example, the substance is first removed reversibly through a semipermeable membrane into a subsystem outside the original boundary. Then a reversible work interaction is carried out on this subsystem which alters the chemical potential of the substance in the subsystem until μ_i equals that in the environment. Then the substance is transferred from the subsystem to the environment without further work required, since μ_i is now the same in the subsystem and the environment. This type of reversible mass transfer is then carried out for all the common species in the system and its environment. Thus a number of work interactions (represented by W_{mem}) may be required to transfer all the various species so that mass (chemical) equilibrium eventually is achieved between the system and its environment.

11-2 CHEMICAL AVAILABILITY

In the discussion of nonflow and flow systems which follows the semipermeable membrane previously described is essential to the development of the concept of chemical and total availability [1].

11-2-1 Closed (Nonflow) System Analysis

Consider the control mass shown in Fig. 11-1. It contains n species such that the total moles N are

$$N = N_1 + N_2 + \cdots + N_i + \cdots + N_n = \sum_{i=1}^{n} N_i$$

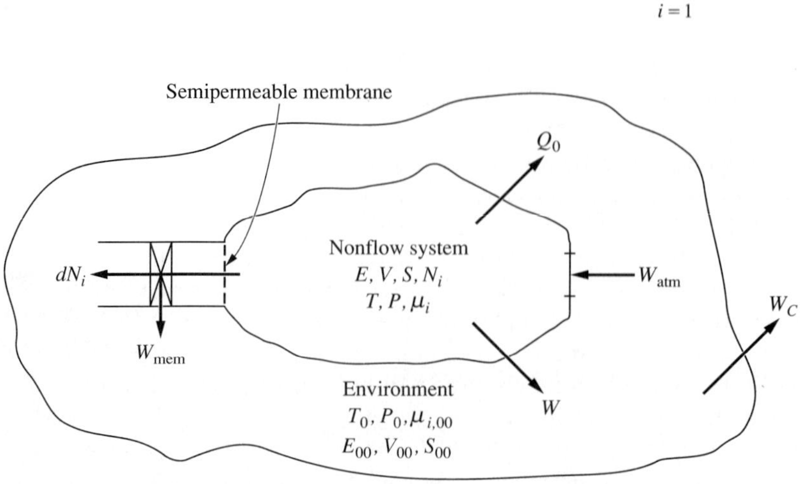

Figure 11-1
A composite system for the evaluation of the chemical availability of a nonflow system.

The system is further described by the intensive properties T, P, and the chemical potentials μ_i for the n species, as well as the extensive properties E, V, and S. The environment surrounds the system and is described by the intensive properties T_0, P_0, and $\mu_{i,00}$, where $\mu_{i,00}$ is the chemical potential of the ith species at the known state of the environment, as well as extensive properties U_{00}, V_{00}, and S_{00}. (The subscript "0" is used to identify properties of the thermomechanical dead state. The subscript "00" is now introduced to represent properties of the environment.) Hence the system and the environment are not in mutual thermal, mechanical, and chemical equilibrium. We wish to determine the maximum work output W_C for the composite system when the system is allowed to reach mutual equilibrium with the environment.

The composite system described in Fig. 11-1 is adiabatic and of constant volume, and the nonflow system is stationary. Hence the energy equation for the composite system, symbolized by a subscript "C," is $W_C = \Delta U_C$. The initial and final energies are expressed by Euler's equation found in Sec. 8-3, which states that

$$U = TS - PV + \sum_{i=1}^{n} N_i \mu_i \tag{8-23}$$

This equation, as written, applies to the initial state of the system. Euler's equation for the initial state of the environment is

$$U_{00} = T_0 S_{00} - P_0 V_{00} + \sum_{i=1}^{n} N_{i,00} \mu_{i,00}$$

where the subscript "00" again represents the environmental properties. Finally, Euler's equation for the final composite system in the environmental state is

$$U_{C,00} = T_0 S_{C,00} - P_0 V_{C,00} + \sum_{i=1}^{n} N_{i,C,00} \mu_{i,00}$$

where $S_{C,00} = S + S_{00}$, $V_{C,00} = V + V_{00}$, and $N_{i,C,00} = N_i + N_{i,00}$. The first equality holds because the maximum work occurs when the process is isentropic. The third equality holds because chemical reactions are absent. When these equations for the initial and final internal energies are substituted into the energy relation $W_C = \Delta U_C$, the result, after cancellations and rearrangement, is

$$W_C = -U + T_0 S - P_0 V + \sum_{i=1}^{n} N_i \mu_{i,00} \tag{11-1}$$

Recall that the availability function is the negative of the work output due to the adopted sign convention on work interactions. If we also replace U by E to account for possible initial kinetic and potential energy effects of the closed system, then the general formulation for the *total availability* Φ_{tot} of a mixture for a simple, compressible closed system is

$$\Phi_{\text{tot}} = E + P_0 V - T_0 S - \sum_{i=1}^{n} N_i \mu_{i,00} \tag{11-2}$$

The total closed-system availability measures the work obtainable from a system which exchanges heat solely with the environment while being brought into complete equilibrium with the environment by means of reversible processes.

There is a direct connection between the total availability Φ_{tot} derived above and the thermomechanical availability Φ_{tm} derived in Chap. 3 as simply Φ. On the basis of Chap. 3

$$\begin{aligned}\Phi_{tm} &\equiv E - U_0 + P_0(V - V_0) - T_0(S - S_0)\\ &= E + P_0V - T_0S - (U_0 + P_0V_0 - T_0S_0)\end{aligned} \tag{11-3}$$

But on the basis of Euler's equation for the restricted dead state

$$U_0 + P_0V_0 - T_0S_0 = \sum_{i=1}^{n} N_i\mu_{i,0}$$

Substitution of this relation into the preceding equation yields

$$\Phi_{tm} = E + P_0V - T_0S - \sum_{i=1}^{n} N_i\mu_{i,0} \tag{11-4}$$

If we now substitute the above equation for Φ_{tm} into Eq. (11-2) for Φ_{tot}, then

$$\Phi_{\text{tot}} = \Phi_{tm} + \sum_{i=1}^{n} N_i\,(\mu_{i,0} - \mu_{i,00}) = \Phi_{tm} + \Phi_{\text{ch}} \tag{11-5}$$

where the *nonflow chemical availability* Φ_{ch} is defined by [2]

$$\Phi_{\text{ch}} \equiv \sum_{i=1}^{n} N_i\,(\mu_{i,0} - \mu_{i,00}) \tag{11-6}$$

The quantity $\mu_{i,0}$ is the chemical potential of species i in the restricted dead state, whereas $\mu_{i,00}$ represents the value in the unrestricted or environmental dead state.

11-2-2 Control-Volume Analysis

Consider the steady-state control volume shown in Fig. 11-2. Each species of the fluid enters the control volume with properties such as (T, P, $\bar{h}_i$, $\bar{s}_i$, μ_i) and leaves at the environment state (T_0, P_0, $\bar{h}_{i,00}$, $\bar{s}_{i,00}$, $\mu_{i,00}$), where the "overbar" signifies a *partial molal* property. Any heat transfer Q_0 is exchanged solely with the environment at T_0. In order to change the composition from the inlet state to the environmental state, a semipermeable membrane and work device must be provided for each species. Only one of these for the ith species is shown in the diagram. An expression for the maximum work output is obtained by combining the first and second laws. This method is analogous to that used in Chap. 3 for the development of the thermomechanical stream availability ψ. This latter quantity will now be designated by ψ_{tm} in order to distinguish it from the total stream availability ψ_{tot} to be developed in this chapter.

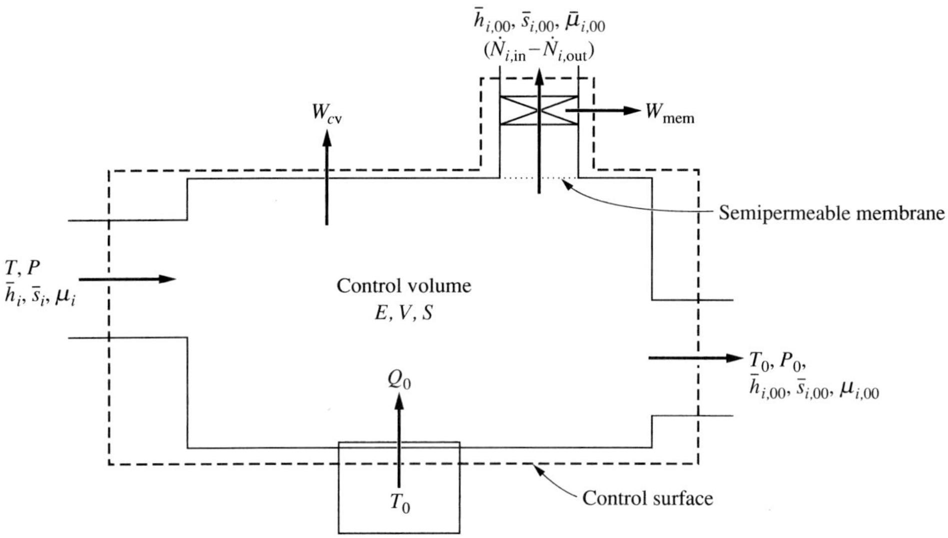

FIGURE 11-2
Schematic of a steady-flow device in which a mixture is brought into thermal, mechanical, and chemical equilibrium with the environment.

The energy equation for the control volume described in Fig. 11-2 is

$$\dot{Q}_0 + \dot{W} = \sum_i \left(\dot{N}_i \bar{h}_i\right)_{\text{out}} - \sum_i \left(\dot{N}_i \bar{h}_i\right)_{\text{in}} + \sum_i \left(\dot{N}_{i,\text{in}} - \dot{N}_{i,\text{out}}\right)\bar{h}_{i,00}$$

where the mass leaving through the membrane-work device has an environmental partial molal enthalpy value of $\bar{h}_{i,00}$. The entropy balance in terms of partial molal entropy values $\bar{s}_i$ under reversible conditions is

$$\sum_i \left(\dot{N}_i \bar{s}_i\right)_{\text{out}} - \sum_i \left(\dot{N}_i \bar{s}_i\right)_{\text{in}} + \sum_i \left(\dot{N}_{i,\text{in}} - \dot{N}_{i,\text{out}}\right)\bar{s}_{i,00} - \frac{\dot{Q}_0}{T_0} = 0$$

By eliminating $\dot{Q}_0$ between the two equations, rearranging, and recognizing that $\bar{h}_{i,00} - T_0\bar{s}_{i,00} \equiv \bar{g}_{i,00} = \mu_{i,00}$, we find that

$$\dot{W}_{\text{rev}} = \sum_i N_i\left(\bar{h}_i - T_0\bar{s}_i - \mu_{i,00}\right)_{\text{out}} - \sum_i N_i\left(\bar{h}_i - T_0\bar{s}_i - \mu_{i,00}\right)_{\text{in}}$$

However, the outlet state is the environmental dead state. Thus the first term on the right is zero; that is,

$$\left(\bar{h}_i - T_0\bar{s}_i - \mu_{i,00}\right)_{\text{out}} = \bar{h}_{i,00} - T_0\bar{s}_{i,00} - \bar{g}_{i,00} \equiv 0$$

Therefore the equation for $\dot{W}_{\text{rev}}$ reduces to

$$\dot{W}_{\text{rev},sf} = -\sum_i \dot{N}_{i,\text{in}}\left(\bar{h}_{i,\text{in}} - T_0\bar{s}_{i,\text{in}} - \mu_{i,00}\right) \qquad (11\text{-}7)$$

Division by $\dot{N}$, the total molar flow rate, gives the reversible work per mole of entering mixture. That is,

$$w_{\text{rev},sf} = -\sum_i y_i\left(\bar{h}_i - T_0\bar{s}_i - \mu_{i,00}\right) \tag{11-8}$$

where y_i is the mole fraction of the ith species in the inlet stream. Finally, as noted in Chap. 3, the reversible work in this case is the negative of the availability of the fluid entering the control volume. Therefore the desired result for the total stream availability or exergy ψ_{tot} per mole of the mixture is

$$\psi_{\text{tot}} = \sum_i y_i\left(\bar{h}_i - T_0\bar{s}_i - \mu_{i,00}\right) = h_m - T_0 s_m - \sum_{i=1}^{n} y_i \mu_{i,00} \tag{11-9}$$

where in Chap. 8 the intensive molal mixture properties h_m and s_m were defined as

$$h_m \equiv \sum_i y_i \bar{h}_i \qquad \text{and} \qquad s_m \equiv \sum_i y_i \bar{s}_i$$

and $\bar{h}_i$ and $\bar{s}_i$ again are partial molal properties. From Eq. (11-9) we find that the specific total exergy (thermomechanical plus chemical) $\psi_{\text{tot},i}$ of a pure substance "i" in steady flow is

$$\psi_{\text{tot},i} = h_i - T_0 s_i - \mu_{i,00} \tag{11-10}$$

This relation, useful in later sections, requires the evaluation of the reference chemical potential $\mu_{i,00}$ in the environmental dead state.

The final step is to relate ψ_{tot} to the thermomechanical energy ψ_{tm}. Recall from Chap. 3 that ψ_{tm} is defined as

$$\psi_{tm} = (h - h_0) - T_0(s - s_0) \tag{3-49}$$

where ke and pe effects are neglected. For a mole of mixture

$$\begin{aligned}\psi_{tm} &= \sum_i y_i\left(\bar{h}_i - \bar{h}_{i,0}\right) - \sum_i y_i T_0\left(\bar{s}_i - \bar{s}_{i,0}\right) \\ &= (h_m - T_0 s_m) - (h_{m,0} - T_0 s_{m,0}) \end{aligned} \tag{11-11}$$

where $\bar{h}_{i,0}$ and $\bar{s}_{i,0}$ are values in the *restricted* or *thermomechanical* dead state. This expression may be simplified by noting from Euler's equation and the definition of h_m that

$$u_{m,0} = T_0 s_{m,0} - P_0 v_{m,0} + \sum_i y_i \mu_{i,0} \qquad \text{and} \qquad h_{m,0} \equiv u_{m,0} + P_0 v_{m,0}$$

A combination of these two equations yields

$$h_{m,0} = T_0 s_{m,0} + \sum_i y_i \mu_{i,0}$$

Substitution of this equation into Eq. (11-11) leads to an alternate formulation for the thermomechanical stream availability for a mixture.

$$\psi_{tm} = h_m - T_0 s_m - \sum_i y_i \mu_{i,0} \tag{11-12}$$

Finally, substitution of the above equation into Eq. (11-9) for the total stream availability leads to

$$\psi_{\text{tot}} = \psi_{tm} + \sum_{i=1}^{n} y_i (\mu_{i,0} - \mu_{i,00}) = \psi_{tm} + \psi_{\text{ch}} \tag{11-13}$$

where the chemical stream availability ψ_{ch} is defined as

$$\psi_{\text{ch}} \equiv \sum_{i=1}^{n} y_i (\mu_{i,0} - \mu_{i,00}) \tag{11-14}$$

It is important to note that the chemical nonflow availability and the chemical stream availability on a molal basis are equal. In a physical context ψ_{tot} measures the availability change from an actual state to the restricted dead state, plus the change from the restricted state to the environmental dead state.

11-3 THE ENVIRONMENTAL STATE

The chemical availability of a fluid for either nonflow or steady-flow systems requires the evaluation of Eq. (11-14), namely,

$$\phi_{\text{ch}} = \psi_{\text{ch}} = \sum_{i=1}^{n} y_i (\mu_{i,0} - \mu_{i,00}) \tag{11-14}$$

The specific solution to this equation depends on the nature of the fluid. Several important cases of engineering interest are discussed below for which the fluid components appear as gases in the environment. The gas phase of the environment itself is always modeled as an ideal-gas mixture.

11-3-1 Ideal-Gas Mixtures

A number of industrial processes involve gases which can be modeled as ideal-gas mixtures. In addition, the components of the mixture are also found as gases in the atmosphere. The chemical exergy of such fluids, then, simply requires the evaluation of $(\mu_{i,0} - \mu_{i,00})$ for an ideal-gas mixture. Recall from Chap. 8 that the chemical potential of a gas in an ideal-gas mixture is given by

$$d\mu_{i,T} = RT\, d \ln p_i \qquad \text{(ideal gas)} \tag{8-92}$$

When this equation is integrated between the restricted dead state "0" and the environmental dead state "00" at constant temperature T_0, then

$$\mu_{i,0} - \mu_{i,00} = RT_0 \ln \frac{p_{i,0}}{p_{i,00}}$$

where $p_{i,0}$ is the component (partial) pressure of the ith species in the restricted state and $p_{i,00}$ is the component pressure in the environment. Note also that

$p_{i,0} = y_{i,0}P_0$ and $p_{i,00} = y_{i,00}P_0$ for an ideal-gas mixture at P_0. Substitution of these relations for $\Delta\mu$ and p_i into Eq. (11-14) leads to the stream chemical availability for an ideal-gas mixture.

$$\Psi_{\text{ch}} = RT_0 \sum_{i=1}^{n} N_i \ln \frac{p_{i,0}}{p_{i,00}} = RT_0 \sum_{i=1}^{n} N_i \ln \frac{y_{i,0}}{y_{i,00}} \qquad \text{(ideal gas)} \qquad (11\text{-}15)$$

Generally, $y_{i,0} \neq y_{i,00}$.

The thermomechanical stream availability ψ_{tm} per mole of mixture, based on Eq. (3-49) and written in terms of the ideal-gas tabular data in the Appendix, is

$$\psi_{tm} = \sum_{i=1}^{n} y_i \left[h_{i,T} - h_{i,T_0} - T_0\left(s_{i,T}^o - s_{i,T_0}^o\right) + RT_0 \ln \frac{P}{P_0} \right] \qquad (11\text{-}16)$$

The total stream availability for an *ideal-gas mixture* is the sum of the two preceding equations. The result per mole of mixture is

$$\psi_{\text{tot}} = \sum_{i=1}^{n} y_i \left[h_{i,T} - h_{i,T_0} - T_0(s_{i,T}^o - s_{i,T_0}^o) \right] + RT_0 \ln \frac{P}{P_0} + RT_0 \sum_{i=1}^{n} y_i \left(\ln \frac{y_i}{y_{i,00}} \right) \qquad (11\text{-}17)$$

The first term accounts for the system temperature $T \neq T_0$, the second term accounts for the system $P \neq P_0$, and the third term accounts for the difference between the stream and environment compositions.

All the quantities in Eq. (11-17) are known, except for the $y_{i,00}$ values of the environment. These values (except for water vapor) are based on the "standard atmosphere" described in Sec. 10-1. The mole fraction of water vapor in the atmosphere is a function of the seasons and geographic location. Hence the choice of the water vapor content in the atmosphere is arbitrary, and several values are used in the literature. It is fairly common to establish this mole fraction by arbitrarily choosing relative humidities between 60 and 100 percent at 25°C and 1-atm total pressure. Table 11-1 lists three possible sets of $y_{i,00}$ values at 25°C and 1 atm, depending on the relative humidity chosen. Note in this table that the mole fraction of water vapor is highly dependent on the relative humidity chosen, while the other $y_{i,00}$ values are much less affected. Thus one must take care when comparing exergy data from several sources, because the water vapor datum could be different [3].

Example 11-1. Methane (CH_4) combustion with 20 percent excess air at 1 atm and complete combustion leads to the following mole fractions for the product gases: CO_2, 0.0805; H_2O, 0.1610; N_2, 0.7263; and O_2, 0.0322. The environmental composition in terms of mole fractions is taken as CO_2, 0.0003; H_2O, 0.0188; N_2, 0.7662; O_2, 0.2055; and other, 0.0092. (These are the values for 60 percent relative humidity taken from Table 11-1.) Evaluate the total stream availability, in kilojoules, for the products at 1000 K, if T_0 is 298 K.

TABLE 11-1
Mole fraction $y_{i,00}$ of gases in a standard atmosphere for relative humidities of 60, 80, and 100 percent, based on data from Table 10-1

$T_0 = 25°C$ and $P_0 = 1$ atm

Substance	$\phi_{RH} = 60\%$	$\phi_{RH} = 80\%$	$\phi_{RH} = 100\%$
N_2	0.7662	0.7613	0.7564
O_2	0.2055	0.2042	0.2029
CO_2	0.0003	0.0003	0.0003
H_2O	0.0188	0.0250	0.0313
Other	0.0092	0.0092	0.0091

Solution. Equation (11-17) is applicable to this situation, where the second term is zero. Using the ideal-gas tables in the Appendix, the temperature-dependent term is

$$\begin{aligned}\psi(T) &= 0.0805[(42{,}769 - 9364) - 298(269.215 - 213.685)] \\ &\quad + 0.1610[(35{,}882 - 9904) - 298(232.597 - 188.720] \\ &\quad + 0.7263[(30{,}129 - 8669) - 298(228.057 - 191.502)] \\ &\quad + 0.0322[(31{,}389 - 8682) - 298(243.471 - 205.033)] \\ &= 11{,}471 \text{ kJ/kmol gas}\end{aligned}$$

The composition-dependent term is

$$\begin{aligned}\psi(y_i) &= 8.3145(298)\left[0.0805 \ln\frac{0.0805}{0.0003} + 0.1610 \ln\frac{0.1610}{0.0188}\right. \\ &\quad \left. + 0.7263 \ln\frac{0.7263}{0.7662} + 0.0322 \ln\frac{0.0322}{0.2055}\right] = 1730 \text{ kJ/kmol gas}\end{aligned}$$

The total stream availability is

$$\psi_{\text{tot}} = \psi(T) + \psi(y_i) = 11{,}471 + 1730 = 13{,}201 \text{ kJ/kmol gas}$$

At 1000 K the temperature-dependent term is much larger than the composition term. As the product-gas temperature approaches 298 K, the temperature term approaches zero. Hence only at temperatures slightly above 298 K are the two terms of similar magnitude.

11-3-2 Vapor and Liquid Water Streams

There are applications in steam ejector and air-conditioning design, for example, where pure water either enters or leaves the control volume. On the basis of Eq. (11-10) the total flow exergy for the water stream at (T, P) is given by

$$\psi_{\text{tot}} = h_{T,P} - T_0 s_{T,P} - \mu_{v,00} \tag{11-18}$$

where $\mu_{v,00}$ is the chemical potential of water vapor in the environment. Since the atmosphere is modeled as an ideal-gas mixture, $\mu_{v,00}$ is determined from the fundamental ideal-gas relation, $d\mu_i = RT\, d \ln p_i$. The lower limit of integration is chosen to be the pure saturation state at T_0, for which p_i is the saturation pressure p_g and $\mu = g_g$. The upper limit is the environmental mixture state with water vapor mole fraction $y_{v,00}$. Therefore at T_0,

$$\mu_{v,00} - g_{g,T_0} = RT_0 \ln \frac{y_{v,00}P_0}{p_{g,T_0}}$$

or, noting that $g_g \equiv h_g - Ts_g$, we have

$$\mu_{v,00} = h_{g,T_0} - T_0 s_{g,T_0} + RT_0 \ln \frac{y_{v,00}P_0}{p_{g,T_0}}$$

Substitution of this result into Eq. (11-18) yields, after rearrangement,

$$\psi_{\text{tot}} = (h_{T,P} - h_{g,T_0}) - T_0\,(s_{T,P} - s_{g,T_0}) - RT_0 \ln \frac{y_{v,00}P_0}{p_{g,T_0}} \qquad \text{(water stream)} \tag{11-19}$$

This equation is equally valid for a pure vapor or liquid water stream, since $g_g = g_f$ at a saturation state, or $h_g - T_0 s_g = h_f - T_0 s_f$ at T_0. All the h and s data come directly from the steam tables. As noted above for ideal-gas mixtures, the value for $y_{v,00}$ is dependent on one's choice of the water vapor concentration in a "standard" environment.

11-3-3 Incompressible Water Stream

The modeling of a liquid water stream as an incompressible fluid can be handled as an extension of Eq. (11-19). In Chap. 1 it was noted that for incompressible fluids $h_{T,P} = h_{f,T} + v_f(P - p_{\text{sat},T})$ and $s_{T,P} = s_{f,T}$. Use of these two relations, and $h_g - T_0 s_g = h_f - T_0 s_f$, in Eq. (11-19) leads to

$$\psi_{\text{tot}}\ \text{(liquid water)} = (h_{f,T} - h_{f,T_0}) + v_{f,T}\,(P - p_{\text{sat},T}) - T_0\,(s_{f,T} - s_{f,T_0}) - RT_0 \ln \frac{y_{v,00}P_0}{p_{g,T_0}} \tag{11-20}$$

where the subscripts "f" and "g" represent the saturated-liquid and saturated-vapor states, respectively. The second term on the right is usually relatively small.

Example 11-2. In Example 3-7, part 1, steam was throttled from 100 bars, 480°C to 60 bars. Determine the total flow availability for the two states, in kJ/kg, if $y_{v,00} = 0.0191$, $T_0 = 25°\text{C}$, and $P_0 = 1.01325$ bars.

Solution. The thermomechanical exergy ψ_{tm} of the initial and final states, as calculated in Example 3-7, are 1379.6 and 1314.5 kJ/kg, respectively. The difference between these values is the irreversibility i. In terms of ψ_{tm}, the total flow availability is

$$\psi_{\text{tot}} = \psi_{tm} - RT_0 \ln \frac{y_{v,00}P_0}{p_{g,T_0}} = \psi_{tm} + \psi_{\text{ch}}$$

For both states the chemical availability contribution is

$$\psi_{ch} = -\frac{8.3145(298.15)}{18.02} \ln \frac{0.0191(1.01325)}{0.03169} = 67.8 \text{ kJ/kg}$$

Therefore,

$$\psi_{tot,1} = 1379.6 + 67.8 = 1447.4 \text{ kJ/kg}$$
$$\psi_{tot,2} = 1314.5 + 67.8 = 1482.3 \text{ kJ/kg}$$

In both states the value of ψ_{tot} is roughly 5 percent greater than ψ_{tm}. The value of i is the same in either case, and the second-law effectiveness ϵ hardly changes in this case. However, in those cases where ψ_{ch} is a larger fraction of ψ_{tm}, ϵ could differ by a few percent when calculated by ψ_{tm} data versus ψ_{tot} data.

11-4 AIR-CONDITIONING PROCESSES

The concepts introduced in the preceding section are especially applicable to air-conditioning (psychrometric) processes. Atmospheric air in this case is usually assumed to be a binary mixture composed of dry air (a) and water vapor (v). The mixture behaves as an ideal-gas mixture for which $\bar{h}_i = h_{i,\text{pure}}$ and $\bar{s}_i = s^o_{i,\text{pure}} - R \ln(p_i/1)$. Thus the following rules apply:

$$P_T = \sum_i p_i = p_v + p_a \qquad \text{and} \qquad p_i = y_i P$$

$$h_m = \sum_i y_i \bar{h}_i = \sum_i y_i h_{i,T} = y_v h_v + y_a h_a$$

$$s_m = \sum y_i \bar{s}_i = \sum y_i s_{i,T,P} = y_v s_{v,T,p_v} + y_a s_{a,T,p_a}$$

One important parameter is the relative humidity ϕ_{rh}, defined by

$$\phi_{rh} \equiv \frac{p_{v,T}}{p_{g,T}} = \frac{y_v}{y_{v,g}}$$

In addition, the specific humidity (or humidity ratio) ω, a *mass ratio*, is defined by

$$\omega \equiv \frac{m_v}{m_a} = \frac{N_v M_v}{N_a M_a} = \frac{M_v p_v}{M_a p_a} = 0.622 \frac{y_v}{y_a} \tag{11-21}$$

where m_v is the mass of water vapor, m_a is the mass of dry air, $M_v = 18.02$, and $M_a = 28.97$. If the *mole fraction ratio* $\tilde{\omega}$ is defined as

$$\tilde{\omega} \equiv \frac{y_v}{y_a} \tag{11-22}$$

and since $y_v + y_a = 1$, then it follows that

$$y_v = \frac{\tilde{\omega}}{1 + \tilde{\omega}} \qquad y_a = \frac{1}{1 + \tilde{\omega}} \tag{11-23}$$

where $\tilde{\omega} = 1.608\omega$.

The total flow availability of moist air is found by starting with Eq. (11-17).

$$\psi_{\text{tot}} = \sum_{i=1}^{n} y_i\left[h_{i,T} - h_{i,T_0} - T_0\left(s^o_{i,T} - s^o_{i,T_0}\right)\right] + \sum_{i=1}^{n} y_i\left[RT_0 \ln \frac{y_i}{y_{i,00}}\right] + RT_0 \ln \frac{P}{P_0} \tag{11-17}$$

The relations for Δh and Δs in the above equation for moist air are frequently based on the assumption that the specific heats $c_{p,v}$ and $c_{p,a}$ are constant. As a result,

$$\bar{h}_v - h_{v,0} = c_{p,v}(T - T_0) \qquad \bar{h}_a - h_{a,0} = c_{p,a}(T - T_0)$$

$$\bar{s}_v - s_{v,0} = c_{p,v} \ln (T/T_0) \qquad \bar{s}_a - s_{a,0} = c_{p,a} \ln (T/T_0)$$

Inserting these relations into Eq. (11-17) and rearranging give ψ_{tot} *per mole of moist air* as

$$\psi_{\text{tot}} = (y_v c_{p,v} + y_a c_{p,a})\left(T - T_0 - T_0 \ln \frac{T}{T_0}\right) + R_u T_0 \ln \frac{P}{P_0} + R_u T_0\left(y_v \ln \frac{y_v}{y_{v,00}} + y_a \ln \frac{y_a}{y_{a,00}}\right) \tag{11-24}$$

where $c_{p,v}$ and $c_{p,a}$ are on a *mole* basis. Substitution of Eq. (11-23) into Eq. (11-24) leads to another useful format of ψ_{tot} on a *mole of mixture* basis.

$$\psi_{\text{tot}} = \frac{c_{p,a} + \tilde{\omega} c_{p,v}}{1 + \tilde{\omega}}\left(T - T_0 - T_0 \ln \frac{T}{T_0}\right) + R_u T_0 \ln \frac{P}{P_0} + R_u T_0\left(\ln \frac{1 + \tilde{\omega}_{00}}{1 + \tilde{\omega}} + \frac{\tilde{\omega}}{1 + \tilde{\omega}} \ln \frac{\tilde{\omega}}{\tilde{\omega}_{00}}\right) \tag{11-25}$$

where $\tilde{\omega}_{00}$ describes the environmental dead state. Table 11-2 lists ω_{00}, $\tilde{\omega}_{00}$, and $y_{v,00}$ for two chosen relative humidities at 25°C and 1 atm total pressure.

Psychrometric calculations are frequently carried out on a unit mass of dry air basis. The equation for ψ_{tot} on this basis is found by noting that 1 kmol of mixture corresponds to $M_a y_a$ or $M_a/(1 + \tilde{\omega})$ kilograms of dry air. Thus division of the right-hand side of Eq. (11-25) by $M_a/(1 + \tilde{\omega})$ and use of the relation $\tilde{\omega} = \omega M_a/M_v$ leads to ψ_{tot} on a *unit mass of dry air* basis.

$$\psi_{\text{tot}} = (c_{p,a} + \omega c_{p,v})\left(T - T_0 - T_0 \ln \frac{T}{T_0}\right) + (1 + \tilde{\omega}) R_a T_0 \ln \frac{P}{P_0} + R_a T_0\left[(1 + \tilde{\omega}) \ln \frac{1 + \tilde{\omega}_{00}}{1 + \tilde{\omega}} + \tilde{\omega} \ln \frac{\tilde{\omega}}{\tilde{\omega}_{00}}\right] \tag{11-26}$$

In this equation $c_{p,a}$ and $c_{p,v}$ are on a unit-mass basis and R_a is the specific-gas constant for dry air. Also, recall that $\tilde{\omega} = 1.608\omega$ and $\tilde{\omega}_{00} = 1.608\omega_{00}$. Table

TABLE 11-2
Values for ω_{00}, $\tilde{\omega}_{00}$, and $y_{v,00}$ for atmospheric air at 25°C and 1 atm

ϕ_{rh}, %	ω_{00}	$\tilde{\omega}_{00}$	$y_{v,00}$
60	0.01189	0.01912	0.01876
80	0.01596	0.02566	0.02502

11-3 lists the specific-gas constants and the average c_p values for dry air and water vapor frequently in use for room air.

11-4-1 Evaporative Cooler

A sketch of an evaporative cooling process is shown in Fig. 11-3. The device is used to humidify and cool hot dry air found in desert climates. The energy balance for the control volume on a unit of dry air is

$$h_{1,a} + \omega_1 h_{v,1} + (\omega_2 - \omega_1) h_{w,3} = h_{2,a} + \omega_2 h_{v,2}$$

where $h_a = c_{p,a}T$, $h_v = h_g$, and $h_{w,3} = h_f$ at T_3. Hence

$$c_{p,a}T_1 + \omega_1 h_{g,1} + (\omega_2 - \omega_1) h_{f,3} = c_{p,a}T_2 + \omega_2 h_{g,2} \tag{11-27}$$

where $\omega = 0.622p_v/(P - p_v)$ and $p_v = \phi_{rh}p_g$. For given inlet temperatures and relative humidity, and a desired outlet temperature T_2, the only unknown in the equation is ω_2. This permits calculating the required addition of water per unit mass of dry air.

On a second-law basis it is desirable to evaluate the irreversibility and second-law efficiency of the process. The general availability balance formulated in Chap. 3 in this case, when Q and W are zero, reduces to

$$0 = \sum_{\text{in}} \dot{m}\psi - \sum_{\text{out}} \dot{m}\psi - \dot{I}_{CV}$$

or specifically for the evaporative cooler on a unit mass of dry air basis

$$i_{\text{CV}} = \psi_{\text{tot},1} + (\omega_2 - \omega_1)\psi_{\text{tot},3} - \psi_{\text{tot},2} \tag{11-28}$$

TABLE 11-3
Values for R_i and c_p for dry air and water vapor at room conditions

Dry air	Water vapor
$R_a = 0.287$ kJ/kg·K	$R_v = 0.461$ kJ/kg·K
$c_{p,a} = 1.005$ kJ/kg·K	$c_{p,v} = 1.862$ kJ/kg·K

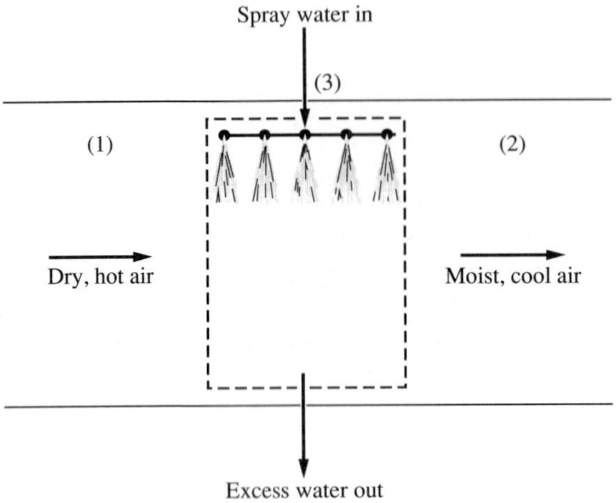

Figure 11-3
Control volume for an evaporative cooling process.

The second-law efficiency on an input/output basis can be defined as

$$\epsilon = \frac{\psi_{tot,2}}{\psi_{tot,1} + (\omega_2 - \omega_1)\,\psi_{tot,3}} \tag{11-29}$$

An example for this process is shown below.

Example 11-3. Air at 36°C, 1 atm, and 10 percent relative humidity passes through an evaporative cooler until its temperature reaches 20°C. The water is added also at 20°C. The environmental state is taken as 25°C, 1 atm, and 60 percent relative humidity. Determine (*a*) the water added in kilograms per kilogram of dry air, (*b*) the irreversibility of the process, in kJ/kg dry air, and (*c*) the second-law effectiveness.

Solution.
(*a*) The inlet ω_1 value, based on Eq. (11-21), is

$$\omega_1 = \frac{0.622(0.10)(0.05947)}{1.013 - 0.10(0.05947)} = 0.00367 \text{ kg water/kg dry air}$$

The outlet specific humidity ω_2 is found from an energy balance. Employing Eq. (11-27), we obtain

$$1.005(36) + 0.00367(2567.1) + (\omega_2 - 0.00367)(83.96) = 1.005(20) + 2538.1\omega_2$$

$$\text{or} \qquad \omega_2 = \frac{36.18 + 9.43 - 0.31 - 20.10}{2538.1 - 83.96} = 0.01027 \text{ kg water/kg dry air}$$

Therefore the amount of liquid water added is

$$\omega_2 - \omega_1 = 0.01027 - 0.00367 = 0.00660 \text{ kg water/kg dry air}$$

(*b*) Data required to find the irreversibility include $\tilde{\omega}_{00} = 0.0191$, $\tilde{\omega}_1 = 0.00590$, and $\tilde{\omega}_2 = 0.01651$. The total flow availability of the inlet moist air stream, based on Eq. (11-26), is

$$\psi_{\text{tot},1} = [1.005 + 0.00367(1.862)]\left(36 - 25 - 298 \ln \frac{309}{298}\right)$$

$$+ 0.287(298)\left(1.00590 \ln \frac{1.0191}{1.00590} + 0.00590 \ln \frac{0.00590}{0.0191}\right)$$

$$= 0.722 \text{ kJ/kg dry air}$$

By a similar calculation, $\psi_{\text{tot},2} = 0.058$ kJ/kg dry air. This value is small, since the state is very close to the environmental state. The total flow availability of the liquid stream is found from Eq. (11-20). The value of $y_{v,0}$ is 0.01876.

$$\psi_{\text{tot},3} = (83.96 - 104.89) + 1.0018(1.00 - 0.02339)(10^{-1})$$

$$- 298(0.2966 - 0.3674) - 0.461(298) \ln \frac{0.01876(1.013)}{0.03169}$$

$$= 70.4 \text{ kJ/kg water}$$

Using these data and Eq. (11-28), we find the irreversibility to be

$$i = 0.722 + (0.01027 - 0.00367)(70.4) - 0.058 = 1.129 \text{ kJ/kg dry air}$$

(*c*) Finally, the second-law efficiency for the process is

$$\epsilon = \frac{0.058}{0.722 + 0.465} = 0.05$$

These last two answers are sensitive to the choice of T_0 and $\phi_{rh,0}$ (and hence $\tilde{\omega}_{00}$) of the environmental dead state. This sensitivity occurs because all the streams have properties that are not too different from the dead state.

PROBLEMS

11-1. Propane (C_3H_8) is burned in a steady-flow process with 20 percent excess air at 1 atm, combustion is complete, and $T_0 = 298$ K. If the combustion products are at 900 K, determine the total stream availability, in kJ/kmol of C_3H_8, using data from Table 11-1 and a relative humidity of (a) 60 percent and (b) 100 percent.

11-2. Butane (C_4H_{10}) is burned in a steady-flow process with 20 percent excess air at 1 atm, combustion is complete, and $T_0 = 298$ K. If the combustion products are at 1100 K, determine the total stream availability, in kJ/kmol of C_4H_{10}, using data from Table 11-1 and a relative humidity of (a) 60 percent and (b) 100 percent.

11-3. Ethane (C_2H_6) is burned in a steady-flow process with 30 percent excess air at 1 atm, combustion is complete, and T_0 is 298 K. If the combustion products are at 700 K, determine the total stream availability, in kJ/kmol of C_2H_6, using data from Table 11-1 and a relative humidity of (a) 60 percent and (b) 100 percent.

11-4. Ethane (C_2H_6) is burned in a steady-flow process with 30 percent excess air at 1 atm, combustion is complete, and T_0 is 298 K. Determine the total stream availability of the product gases, in kJ/kmol of C_2H_6, using data from Table 11-1 for 60 percent relative humidity and a temperature of (a) 500 K and (b) 800 K.

11-5. Butane (C_4H_{10}) is burned in a steady-flow process with 20 percent excess air at 1 atm, combustion is complete, and T_0 is 298 K. Determine the total stream availability of the product gases, in kJ/kmol of C_4H_{10}, using data from Table 11-1 for 100 percent humidity and a temperature of (a) 600 K and (b) 1000 K.

11-6. A pure stream of water vapor at 3 bars enters a control volume at (a) 200 and (b) 320°C. If the relative humidity of the water in the atmosphere is 60 percent, determine the total stream availability, in kJ/kg, for $T_0 = 25°C$.

11-7. A pure stream of water vapor at 5 bars and 200°C enters a control volume. Determine the total stream availability, in kJ/kg, if the relative humidity of water in the atmosphere is (a) 60 and (b) 100 percent, and T_0 is 298 K.

11-8. A pure liquid water stream is at 25 bars and 80°C. Determine the total stream availability, in kJ/kg, by means of (a) steam table data and (b) an incompressible model if the relative humidity of the atmosphere is 60 percent, T_0 is 25°C, and P_0 is 1 atm.

11-9. A pure liquid water stream is at 50 bars and 100°C. Determine the total stream availability, in kJ/kg, by means of (a) steam table data and (b) an incompressible model if the relative humidity of the atmosphere is 60 percent, T_0 is 25°C, and P_0 is 1 atm.

11-10. Air at 34°C and 20 percent relative humidity passes through an evaporative cooler. The environmental state is 25°C, 1 atm, and 60 percent relative humidity. The final dry bulb temperature and the liquid inlet temperature are both 21°C. Determine (a) the water added in kilograms per kilogram of dry air and (b) the irreversibility of the process, in kJ/kg dry air.

11-11. An airstream at 10°C and 40 percent relative humidity is first heated to 33°C and then passed through an evaporative cooler until the temperature reaches 20°C. Heat is added from a steam coil at 200°C, and liquid water in the evaporative cooler is added at 22°C. Find (a) the water added, in kg/kg dry air, (b) the irreversibility of the heating process, in kJ/kg dry air, and (c) the irreversibility of the evaporative cooling process, in kJ/kg dry air, if T_0 is 25°C, P_0 is 1 atm, and the atmospheric relative humidity is 60 percent.

11-12. An airstream originally at 36°C, 100 kPa, and 40 percent relative humidity is cooled at constant pressure to 24°C. The heat is rejected to a sink at 20°C. Determine (a) the water condensed, in kg/kg dry air, and (b) the irreversibility of the process, in kJ/kg dry air. T_0 is 25°C, P_0 is 1 atm, and the atmospheric relative humidity is 60 percent.

REFERENCES

1. Moran, M. J., *Availability Analysis: A Guide to Efficient Energy Use,* Prentice-Hall, Englewood Cliffs, NJ, 1982.
2. Bejan, A., *Advanced Engineering Thermodynamics,* John Wiley & Sons, New York, 1988.
3. Szargut, J., D. R. Morris, and F. R. Steward, *Exergy Analysis of Thermal, Chemical, and Metallurgical Processes,* Hemisphere Publishing Corp., New York, 1988.

CHAPTER 12

CHEMICAL AVAILABILITY OF FUELS

The chemical availability introduced in Chap. 11 is a measure of the maximum work when a gas or gas mixture comes to equilibrium with the environmental composition. In this chapter the chemical availability of a substance (such as a fuel) is examined as a chemical reaction occurs and the products of the reaction reach equilibrium with the environmental composition.

12-1 INTRODUCTION

In Chaps. 10 and 11 the availability change and the irreversibility for chemical reactions of hydrocarbon fuels were studied from two viewpoints:

1. Fuel and dry air composed of O_2 and N_2 react to form products of combustion in the restricted dead state.
2. Fuel and dry air composed of O_2 and N_2 react to form products of combustion which end up in the environmental (unrestricted) dead state.

The difference between (1) and (2) is the *chemical availability* of the product gases as they proceed from the restricted to the unrestricted dead state. These evaluations were made in terms of enthalpy and entropy values of the reacting species. To extend these concepts to the most general situation, we shall consider a steady-state control volume where the fuel enters at the restricted dead state, the air (oxidant) is drawn from the environment, and the products are returned to the unrestricted dead state. Although we are focusing on combustion reactions, the procedure is generally applicable to other types of reactions.

12-2 FUEL CHEMICAL AVAILABILITY

The availability or exergy of a substance in a given state is a measure of the maximum obtainable work as the substance proceeds to the dead state while exchanging heat solely with the environment. In this section we shall develop an equation for the chemical availability, or exergy, of hydrocarbon fuels in the restricted dead state [1,2].

12-2-1 Fuel Availability

Figure 12-1 illustrates a control volume where pure fuel enters at the restricted dead state (T_0, P_0), and oxidant (O_2) enters from the environment. In general, each product is at its state in the environment, namely: $\bar{h}_{i,00}$, $\bar{s}_{i,00}$, $\bar{g}_{i,00} = \mu_{i,00}$. (The subscript "00" is used again to represent unrestricted or environmental dead-state properties, as in Chap. 11.) A semipermeable membrane is used to introduce oxygen into the control volume. The products of combustion, primarily CO_2 and H_2O, leave the control volume through semipermeable membranes in order to adjust their state following combustion to their unrestricted dead state. The maximum work output per mole of fuel is shown in Fig. 12-1 as $w_{\text{rev},f}$. In addition, heat transfer q with the environment will be necessary to assure that the products leave at T_0.

The maximum work $w_{\text{rev},f}$ is found by combining the energy and entropy statements for the process shown in Fig. 12-1. In steady state the energy equation per mole of fuel is

$$q + w_{\text{rev},f} = \Delta H = \sum_{P-R} \nu_i \bar{h}_i = \sum_{P} \nu_i \bar{h}_{i,00} - \nu_{O_2} \bar{h}_{O_2,00} - \bar{h}_{\text{fuel},0} \tag{12-1}$$

where $\bar{h}_{i,00}$ is the partial molal enthalpy in the unrestricted dead state and $\bar{h}_{\text{fuel},0}$ is the partial molal enthalpy of the fuel at the restricted dead state. For an isothermal reversible process $Q = T\,\Delta S$. In terms of the fuel oxidation process this becomes

$$q = T_0 \sum_{P-R} \nu_i \bar{s}_i = T_0 \left(\sum_{P} \nu_i \bar{s}_{i,00} - \nu_{O_2} \bar{s}_{O_2,00} - \bar{s}_{\text{fuel},0} \right) \tag{12-2}$$

Elimination of q between these two equations yields, after rearrangement,

$$w_{\text{rev},f} = \sum_{P} \nu_i (\bar{h}_{i,00} - T_0 \bar{s}_{i,00}) - \Big[\nu (\bar{h}_{00} - T_0 \bar{s}_{00}) \Big]_{O_2} - (\bar{h}_0 - T_0 \bar{s}_0)_{\text{fuel}} \tag{12-3}$$

Finally, recall that $\mu_i = \bar{g}_i = \bar{h}_i - T\bar{s}_i$. As a result, the preceding equation in terms of chemical potentials becomes

$$w_{\text{rev},f} = \sum_{P} \nu_i \mu_{i,00} - \nu_{O_2} \mu_{O_2,00} - \mu_{\text{fuel},0} \tag{12-4}$$

The first two terms on the right are measured at the environmental state, whereas the last term is evaluated at the restricted dead state.

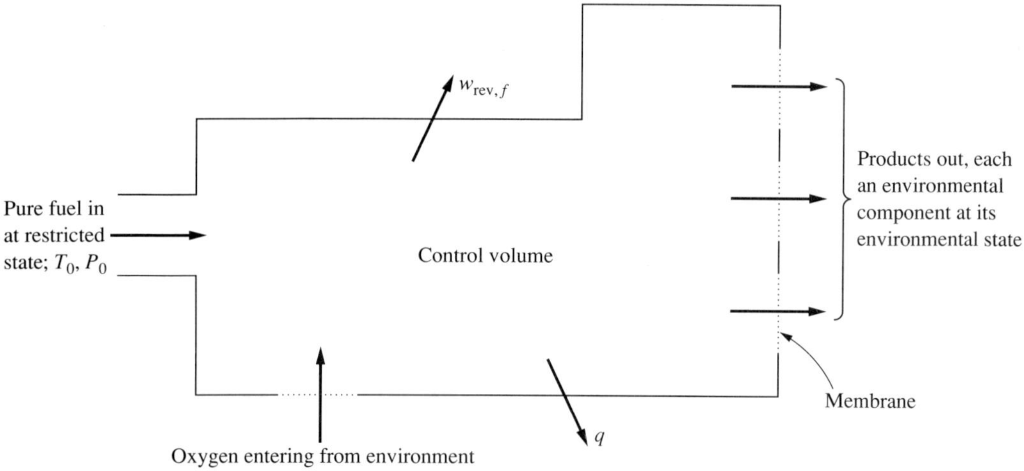

FIGURE 12-1
Control volume for the development of fuel availability. Oxygen enters and products leave through semipermeable membranes.

Similar to the discussion in Sec. 11-2, $w_{rev,f}$ measures the chemical availability of the fuel as it exists alone in the restricted dead state. Recall that the availability function ψ is the negative of the work output due to the adopted sign convention on work interactions. Therefore the *fuel chemical availability* $\psi_{ch,f}$ for a *pure* fuel in the *restricted dead state* is given by

$$\psi_{ch,f} \equiv g_{fuel,0} + \nu_{O_2}\mu_{O_2,00} - \sum_P \nu_i \mu_{i,00} \tag{12-5}$$

where $g_{fuel,0}$ replaces $\mu_{fuel,0}$ for a pure substance. Use of Eq. (12-5) requires selecting appropriate values for the reference chemical potentials.

A chemical potential can be associated with a fuel in the environmental dead state, even though it is not a compound which makes up the standard environment. Recall that the chemical stream availability ψ_{ch} is defined by Eq. (11-14) as

$$\psi_{ch} \equiv \sum_i y_i(\mu_{i,0} - \mu_{i,00})$$

For a pure fuel this becomes

$$\psi_{ch,f} = \mu_{fuel,0} - \mu_{fuel,00} = g_{fuel,0} - \mu_{fuel,00}$$

A comparison of this relation to Eq. (12-5) reveals that

$$\mu_{fuel,00} = \sum_P \nu_i \mu_{i,00} - \nu_{O_2}\mu_{O_2,00} \tag{12-6}$$

where the ν values again come from the stoichiometric combustion reaction.

12-2-2 Evaluation of the Availability of Hydrocarbon Fuels

For engineering studies of combustion reactions it is useful to evaluate and tabulate $\psi_{ch,f}$ for common hydrocarbons. On the basis of Eq. (12-5) this requires a knowledge of the values of the chemical potentials of the components of environmental air, including water vapor, as well as $g_{fuel,0}$ for the pure fuel. For ideal-gas mixtures the chemical potential of the ith component takes on the format

$$\mu_{i,T,\text{ideal}} = g_{i,T}^{o} + RT \ln \frac{p_i}{P_0} \tag{8-94}$$

For the environmental state, where $p_i = y_{i,00}P_0$, this equation becomes

$$\mu_{i,00} = g_{i,0} + RT_0 \ln y_{i,00}$$

For the complete combustion of a hydrocarbon fuel C_xH_y the only products of interest are carbon dioxide (CO_2) and water (H_2O), and the only environmental reactant is oxygen (O_2). Therefore the three $\mu_{i,0}$ equations required for Eq. (12-5) are

$$\mu_{O_2,00} = g_{O_2,0} + RT_0 \ln y_{O_2,00}$$

$$\mu_{CO_2,00} = g_{CO_2,0} + RT_0 \ln y_{CO_2,00}$$

$$\mu_{H_2O,00} = g_{H_2O,0} + RT_0 \ln y_{H_2O,00}$$

The quantity $g_{i,0} = g_{i,T_0}^{o}$, that is, the standard state Gibbs function at T_0. This latter value is readily available for numerous substances. In these equations the $y_{i,00}$ values in the atmosphere also must be known, including water vapor. As noted in Chap. 11, the choice of water vapor content in the atmosphere is arbitrary, and several values are commonly used. Table 11-1, which lists $y_{i,00}$ values at 25°C and 1 atm for two different relative humidity values, is repeated here as Table 12-1 for convenience.

TABLE 12-1
Mole fraction $y_{i,00}$ of gases in a standard atmosphere for relative humidities of 60, 80, and 100 percent, based on data from Table 10-1

T_0 = 25°C and P_0 = 1 atm

Substance	ϕ_{RH} = 60%	ϕ_{RH} = 80%	ϕ_{RH} = 100%
N_2	0.7662	0.7613	0.7564
O_2	0.2055	0.2042	0.2029
CO_2	0.0003	0.0003	0.0003
H_2O	0.0188	0.0250	0.0313
Other	0.0092	0.0092	0.0091

When the set of three equations for $\mu_{i,00}$ are substituted into Eq. (12-5), we find that

$$\psi_{\text{ch},f} = -\Delta G_{R,0} + RT_0 \ln \frac{(y_{O_2,00})^{\nu_{O_2}}}{(y_{CO_2,00})^{\nu_{CO_2}}(y_{H_2O,00})^{\nu_{H_2O}}} \tag{12-7}$$

where $\Delta G_{R,0} = \nu_{H_2O} g_{H_2O,0} + \nu_{CO_2} g_{CO_2,0} - g_{\text{fuel},0} - \nu_{O_2} g_{O_2,0}$

The quantity $\Delta G_{R,0}$ is the change in the Gibbs function per mole of fuel for the stoichiometric reaction at the state (T_0, P_0). If we can consider a hydrocarbon of the general formula C_xH_y, then the reaction with environmental O_2 to produce CO_2 and H_2O is

$$C_xH_y + \left(x + \frac{y}{4}\right) O_2 \rightarrow x\ CO_2 + \frac{y}{2}\ H_2O\ (g)$$

In this format Eq. (12-7) becomes

$$\psi_{\text{ch},f} = -\Delta G_{R,0} + RT_0 \ln \frac{(y_{O_2,00})^{x+y/4}}{(y_{CO_2,00})^{x}(y_{H_2O,00})^{y/2}} \tag{12-8}$$

where $\Delta G_{R,0} = (x) g_{CO_2,0} + \left(\frac{y}{2}\right) g_{H_2O,0} - g_{\text{fuel},0} - \left(x + \frac{y}{4}\right) g_{O_2,0}$

Equations (12-7) and (12-8) enable one to evaluate the chemical exergy of a mole of fuel C_xH_y in the restricted dead state which is transformed into the products CO_2 and H_2O in the unrestricted dead state or environmental state. (The method is easily extended to fuels of the type $C_xH_yO_z$.) All the Gibbs function data needed, $g_{i,0}$ and $g_{\text{fuel},0}$, are for the pure state at $T_0 = 25°\text{C}$ and $P_0 = 1$ atm. Hence their values are the Gibbs function of formation data, Δg_f^o, found in Table A-33 in the Appendix (and other sources). Table A-34 lists values of $\psi_{\text{ch},f}$ for some common gaseous hydrocarbon fuels at relative humidities of 60 and 100 percent. Extended lists of data are found in the literature [1,3]. An example calculation of $\psi_{\text{ch},f}$ is shown below.

Example 12-1. Determine the chemical availability of ethylene gas (C_2H_4) in the restricted dead state for a relative humidity in the environment of (*a*) 100 percent and (*b*) 60 percent. (*c*) Compare the answer to part (*a*) with the isothermal availability change found for this reaction in Example 10-4. Finally, (*d*) calculate the maximum work output when combustion products are cooled from the adiabatic flame temperature to the unrestricted dead state, and compare to the value found in Example 10-6.

Solution

(*a*) The overall reaction is

$$C_2H_4\,(g) + 3\,O_2 \rightarrow 2\,CO_2 + 2\,H_2O\,(g)$$

The $y_{i,00}$ values are found in Table 12-1. Substitution of values into either Eq. (12-7) or (12-8) leads to

$$\psi_{ch,C_2H_4} = -[2(-394{,}380) + 2(-228{,}590) - 68{,}120 - 3(0)] + 8.3145(298.15) \ln \frac{(0.2029)^3}{(0.0003)^2(0.0313)^2} = 1{,}314{,}090 + 45{,}530 = +1{,}359{,}600 \text{ kJ/kmol fuel}$$

This value of $\psi_{ch,f}$, as well as the $-\Delta G_{R,0}$ value of 1,314,090 kJ/kmol, appear in the fourth and fifth columns of Table A-34 in the Appendix, in kJ/gmol.

(*b*) For a relative humidity of 60 percent, data in Table 12-1 indicate that the last term in Eq. (12-7) has the value

$$8.3145(298.15) \ln \frac{(0.2055)^3}{(0.0003)^2(0.0188)^2} = 48{,}150 \text{ kJ}$$

In this case the fuel chemical availability would be

$$\psi_{ch,C_2H_4} = 1{,}314{,}090 + 48{,}150 = 1{,}362{,}200 \text{ kJ/kmol fuel}$$

This value, which appears in the sixth column of Table A-34, differs from that in part (*a*) by 0.2 percent.

(*c*) In Example 10-4 the availability change was found to be 1,324,440 kJ, which compares to 1,359,600 kJ in part (*a*) above. The overall chemical reaction is the same, but the models are quite different. In Example 10-4 C_2H_4 reacted with air modeled as 21.0 percent O_2 and 79.0 percent N_2. The products were modeled as 13.1 percent CO_2, 13.1 percent H_2O and 73.8 percent N_2. In the current example the entering air is modeled by the atmospheric data in Table 12-1, which includes CO_2, H_2O, and other inert gases. The product gases are in equilibrium with the environment. This difference in modeling accounts for the 2.6 percent increase in the availability change when the unrestricted dead state is part of the model. This latter model is more realistic of the actual combustion process.

(*d*) In Example 10-6 the availability change when the adiabatic combustion products at 2570 K are cooled to the restricted dead state at 298.15 K was found to be −969,780 kJ/kmol fuel. If the products are brought to the unrestricted dead state with a relative humidity of 60 percent, we must add on the correction found above in part (*b*), namely, 48,150 kJ. Therefore the maximum work output on cooling is now 1,017,930 kJ, and the second-law efficiency under this new model is

$$\epsilon = \frac{w_{max}}{\psi_{ch,f}} = \frac{1{,}017{,}930}{1{,}359{,}600} = 0.748 \qquad (74.8 \text{ percent})$$

In Example 10-6 the effectiveness was found to be 73.2 percent on the basis of the restricted dead state. In this case the environmental model shows a better theoretical performance than the thermomechanical model.

The results of Example 12-1 illustrate an important point. The main contribution to $\psi_{ch,f}$ for (C_2H_4) is the $\Delta G_{R,0}$ term, which in this case is 97 percent of the total. The data of Table A-34 confirms this for all the hydrocarbon fuels listed, when columns 4, 5, and 6 are compared. This implies that the mole fractions chosen in Table 12-1 for modeling dry atmospheric air have a very small impact when calculating $\psi_{ch,f}$ values. Secondly, part (*b*) of Example 12-1 shows that the choice of the relative humidity (and thus the water vapor content) to model the environment also has only a slight effect. This is fortunate, since it has been pointed out that there is no universal model for environmental air and its moisture content. Table A-34 also lists the lower and higher heating values for fuels. It is worth noting that these values are also close to the chemical availability of the fuels, as well as the $-\Delta G_{R,0}$ values.

When the fuel supply is a mixture of gases (such as natural gas), the chemical exergy of the fuel in the restricted state must be adjusted relative to its mole fraction $y_{f,\,\mathrm{mixt}}$ in the mixture. On the basis of Eq. (8-107) the chemical potential of the fuel $\mu_{f,\,\mathrm{mixt}}$ becomes

$$\mu_{f,\,\mathrm{mixt}} = g_{f,0} + RT_0 \ln y_{f,\,\mathrm{mixt}}$$

When this relation is used in the development of $\psi_{\mathrm{ch},f}$, then the chemical exergy of the fuel in a mixture at the restricted dead state $\psi_{\mathrm{ch},f,\mathrm{mixt}}$ becomes

$$\psi_{\mathrm{ch},f,\,\mathrm{mixt}} = \psi_{\mathrm{ch},f} + RT_0 \ln y_{f,\,\mathrm{mixt}} \tag{12-9}$$

where $\psi_{\mathrm{ch},f}$ again is for the pure fuel in the restricted dead state.

Example 12-2. A natural gas contains methane CH_4 in the amount of 70 mol percent in the restricted dead state. Find the chemical availability of the CH_4, based on the pure fuel value found in column 5 of Table A-34.

Solution. On the basis of Table A-34 and Eq. (12-9)

$$\begin{aligned}\psi_{\mathrm{ch},f,\,\mathrm{mixt}} &= \psi_{\mathrm{ch},f} + RT_0 \ln y_{f,\,\mathrm{mixt}} \\ &= 829{,}800 + 8.314(298.15) \ln 0.70 = 828{,}920 \text{ kJ/kmol}\end{aligned}$$

In this case the effect is very small (roughly a 0.1 percent correction). Even for a mole fraction of 0.10 the correction could only be 5710 kJ/kmol, which is roughly 0.7 percent of the pure fuel value. Since $\psi_{\mathrm{ch},f}$ is always a positive value, the mixture value is always less than the pure fuel value.

12-2-3 Fuel Availability for Pure Liquid Hydrocarbons

The fuel availability data in Table A-34 are for the gas phase. However, from a practical viewpoint many hydrocarbons exist in the liquid phase at 1 atm and 298.15 K. This situation occurs when the normal boiling point T_b of a substance is greater than 298.15 K. Thus some of the values for $\psi_{\mathrm{ch},f}$ in Table A-34 are for artificial or unnatural states. Methods are needed for evaluating $\psi_{\mathrm{ch},f}$ when the substance naturally occurs as a liquid in the restricted dead state.

The method developed in the preceding subsection for fuels in the gas phase is readily adaptable to the liquid phase. That is, Eq. (12-5) is general so that it applies to pure fuels in any phase. The only requirement is that the value used for $g_{\mathrm{fuel},0}$ (which equals its Gibbs functions of formation Δg_f^o at T_0) represents the phase of interest. The difficulty with the method is that Δg_f^o data for the liquid state are not as readily available as for the ideal-gas state.

As an alternative method, a relationship for $\psi_{\mathrm{ch},f}$ between the gas and liquid states is derived in terms of vapor-pressure data. The derivation starts with Eq. (11-14). Applied to a pure substance "i" the chemical availability is given by

$$\psi_{\mathrm{ch}} = \mu_0 - \mu_{i,00}$$

where $\mu_{i,00}$ again applies to substance "i" in its environmental dead state. However, for a pure substance $\mu_0 = g_0$, and the above equation becomes

$$\psi_{\mathrm{ch}} = g_0 - \mu_{i,00}$$

This equation is equally valid for pure gases or liquids. For these two phases the above equation may be written as

$$\psi_{\mathrm{ch,gas}} = g(\mathrm{gas}, T_0, P_0) - \mu_{i,00}$$

$$\psi_{\mathrm{ch,liq}} = g(\mathrm{liq}, T_0, P_0) - \mu_{i,00}$$

The value of $\mu_{i,00}$, by definition, is the same in both equations. Hence the difference in chemical availability of the pure gas and liquid phases is

$$\psi_{\mathrm{ch,gas}} - \psi_{\mathrm{ch,liq}} = g(\mathrm{gas}, T_0, P_0) - g(\mathrm{liq}, T_0, P_0)$$

By adding and subtracting the term $g(\mathrm{gas}, T_0, P^{\mathrm{sat}})$ on the right-hand side of the preceding equation, we find that

$$\begin{aligned}\psi_{\mathrm{ch,gas}} - \psi_{\mathrm{ch,liq}} = &[g(\mathrm{gas}, T_0, P_0) - g(\mathrm{gas}, T_0, p^{\mathrm{sat}})] \\ &- [g(\mathrm{liq}, T_0, P_0) - g(\mathrm{gas}, T_0, p^{\mathrm{sat}})]\end{aligned} \tag{12-10}$$

But the Gibbs function g for the pure liquid state is essentially independent of pressure (because h and s for the liquid phase are insensitive to pressure). Therefore replacement of P_0 by p^{sat} in the last bracketed term above shows that

$$[g(\mathrm{liq}, T_0, P_0) - g(\mathrm{gas}, T_0, p^{\mathrm{sat}})] = [g(\mathrm{liq}, T_0, p^{\mathrm{sat}}) - g(\mathrm{gas}, T_0, p^{\mathrm{sat}})] = 0$$

The term is zero because the Gibbs functions for vapor and liquid states in equilibrium are equal (see Sec. 6-10). The remaining bracketed term in Eq. (12-10) is evaluated by noting that since both P_0 and p^{sat} are equal to or less than 1 atm, the gas phase behaves essentially as an ideal gas. For an ideal gas $dg = RT d \ln P$. Therefore

$$g(\mathrm{gas}, T_0, P_0) - g(\mathrm{gas}, T_0, p^{\mathrm{sat}}) = RT_0 \ln \frac{P_0}{p^{\mathrm{sat}}}$$

where $P_0 = 1$ atm by choice. Substitution of this expression in Eq. (12-10) yields

$$\psi_{\mathrm{ch,gas}} - \psi_{\mathrm{ch,liq}} = -RT_0 \ln p^{\mathrm{sat}}$$

where p^{sat} must be measured in atmospheres. When the above relation is applied to a fuel in the restricted dead state, then

$$\psi_{\mathrm{ch},f,\mathrm{liq}} = \psi_{\mathrm{ch},f,\mathrm{gas}} + RT_0 \ln p^{\mathrm{sat}} \tag{12-11}$$

where the vapor pressure p^{sat} is measured at T_0. This equation is only useful when the normal boiling point is greater than $T_0 = 25°\text{C}$. The logarithmic correction term is negative because $p^{\text{sat}} < 1$ atm. The correction is relatively small. For example, when p^{sat} is 0.5, 0.1, and 0.01 atm, the correction is −1,720, −5,710, and −11,420 kJ/kmol, respectively. Since the values of $\psi_{\text{ch},f}$ for hydrocarbon fuels are typically greater than 500,000 kJ/kmol, the liquid-phase correction is usually less than 1 percent.

Example 12-3. Determine for *n*-octane the magnitude of the correction, in kJ/kmol, for the liquid phase and the percent change of $\psi_{\text{ch},f}$ when compared to the gas-phase value by means of (*a*) Δg_f^o data and (*b*) vapor-pressure data.

Solution

(*a*) The values of Δg_f^o in Table A-33 in the Appendix for gaseous and liquid *n*-octane are 17,320 and 6610 kJ/kmol, respectively. (As an alternate source, the API Research Project 44 lists these values as 16,860 and 6,940 kJ/kmol.) Use of Eq. (12-8) to evaluate $\psi_{\text{ch},f}$ for the liquid phase would show a decrease in value amounting to (17,320−6,610), or 10,710 kJ/kmol. (API 44 data require a decrease of 9,920 kJ/kmol.) Since the gas-phase value for $\psi_{\text{ch},f}$ in Table A-34 is 5,427,600 kJ/kmol for a relative humidity of 60 percent, the percent change is around −0.2 percent regardless of data source.

(*b*) An alternative method employs Eq. (12-11). The vapor pressure of *n*-octane at 25°C is 18.25 mbar, or 0.01801 atm. Therefore

$$\psi_{\text{ch,gas}} - \psi_{\text{ch,liq}} = -RT \ln p^{\text{sat}} = -8.3145(298.15) \ln 0.01801 = 9{,}960 \text{ kJ/kmol}$$

This is in substantial agreement with the values of 10,710 (or 9,920) found above.

12-2-4 Approximation Method for Fuel Availability

On the basis of previous discussions, the steady-flow fuel availability of light hydrocarbon gases and their mixtures can be evaluated accurately when the composition is known. As noted above, the exergy of pure liquid fuels can also be ascertained when vapor-pressure data are available. A question arises as to the fuel availability of heavy hydrocarbon fuels which are pure or of undetermined composition. For these latter substances data are usually lacking to calculate $\psi_{\text{ch},f}$. However, investigations indicate that for a large number of light hydrocarbons the ratio of $\psi_{\text{ch},f}$ to the lower heating value (LHV) is reasonably constant. Early work in this area was done by Szargut and Petela [4] and subsequently revised by Rodriguez [5].

More recently, Brzustowski and Brena [6] examined the ratio $\psi_{\text{ch},f}/\text{LHV}$ for homologous series of hydrocarbons. They concluded that the proportionality constant between fuel exergy and LHV is given closely by

$$\frac{\psi_{\text{ch},f}}{\text{LHV}} = 1.065 \qquad \text{(12-12)}$$

based on $\psi_{\text{ch},f}$ data at 65 percent relative humidity. More importantly, the correlation improved as the molecular weight increased and was reasonably independent of the homologous series used. Also pointed out, and shown by the data in Table A-34, fuel exergy is not very sensitive to relative humidity. [The limited data of Table A-34 support the above analysis. Excluding methane, acetylene, and ethylene (low molecular weight substances), the ratio $\psi_{\text{ch},f}/\text{LHV}$ based on Table A-34 ranges from 1.04 to 1.07, with a substantial percentage of data around 1.06.] The authors concluded that it seems likely for Eq. (12-12) "to be applicable to heavy hydrocarbons of undetermined composition." Brzustowski and Brena also show a theoretical basis for the nearly constant value for $\psi_{\text{ch},f}/\text{LHV}$.

12-2-5 Standard Chemical Availability

Although the development in this section is directed toward hydrocarbon fuels, the theory is of general applicability. The values of the fuel chemical availability $\psi_{\text{ch},f}$ listed in Table A-34 are examples of a property called the *standard chemical availability* of a substance. These are "standard" values because they are evaluated relative to a "standard" environment. In the case of Table A-34 two values of the relative humidity were chosen to show the effect of atmospheric water vapor content on $\psi_{\text{ch},f}$ values. However, availability balances for complex processes could be carried out more easily if the chemical exergy of all important solids, liquids, and gases were tabulated at a single "standard" environment. This requires that one select a number of reference substances for the natural environment, in addition to those already introduced for the development of fuel chemical availabilities.

Similar to the standard dry atmosphere and relative humidity considered earlier in this chapter, each of the additional reference substances must be assigned a "standard concentration" based on knowledge of its average concentration in the natural environment. The reference substances must be chosen in some fashion so that each important chemical element is represented. Several approaches are possible. In Ref. [3] the reference substances were selected from three obvious divisions of our natural environment: (1) gaseous components of the atmosphere, (2) ionic and nonionic substances in the oceans, and (3) solid substances found in the outer layer of the earth to a depth of a few miles. The standard chemical availabilities of other nonreference substances are then determined relative to the values assigned to the reference substances. Szargut et al. [3] list standard chemical availabilities for many inorganic and organic substances based on the method described above. The standard environment chosen includes 298.15 K, 101.325 kPa, and a relative humidity of 70 percent.

The composition of dry atmospheric air is known with considerable accuracy. Hence the standard chemical availability of each component of dry air can be evaluated easily. The stream chemical availability of the ith substance is given by Eq. (11-15), namely,

$$\psi_{\text{ch},i} = \mu_{i,0} - \mu_{i,00} = RT_0 \ln \frac{p_{i,0}}{p_{i,00}} = RT_0 \ln \frac{y_{i,0}P_0}{y_{i,00}P_{\text{atm}}}$$

For a pure substance in the restricted dead state the above equation becomes

$$\psi_{ch,i} = -RT_0 \ln \frac{y_{i,00}P_{atm}}{P_0} \tag{12-13}$$

In establishing standard chemical availabilities, it is not necessary that $P_{atm} = P_0$, but P_0 is normally set as either 1 atm or 1 bar. Thus knowledge of $y_{i,00}$ and selection of P_{atm} for the natural environment fixes the "standard" value of ψ_{ch} for a reference substance.

As an example, consider diatomic oxygen in the atmosphere. Table 12-1 lists a y_{00} value of 0.2055 for a relative humidity of 60 percent. (A standard relative humidity must be chosen.) If P_0 and P_{atm} are both taken as 1 atm for convenience, then the standard chemical availability based on Eq. (12-13) would be

$$\psi_{ch,O_2} = RT_0 \ln y_{00} = 8.3145(298.15) \ln 0.2055 = 3920 \text{ kJ/kmol}$$

In comparison, for a mole percent of 20.99 for O_2 in dry air with a relative humidity of 70 percent and an atmospheric pressure of 0.9931 bar, Ref. [3] lists ψ_{ch,O_2} as 3,970 kJ/kmol.

The content of solid carbon in the earth is not known. However, the average concentration of carbon dioxide (CO_2) in the atmosphere is known. Hence CO_2 becomes the reference compound for the element carbon. First, the standard value of ψ_{ch} for CO_2 is established by using Eq. (12-13) and data from Table 12-1 for y_{00}. This value is found to be 20,110 kJ/kmol and compares to 19,870 kJ/kmol by Ref. [3] for a slightly different choice of data. To determine the value for solid carbon itself we write the reaction

$$C(s) + O_2(g) \rightarrow CO_2(g)$$

The availability change in the restricted dead state in terms of standard availabilities is given by

$$\begin{aligned} \Delta\psi_0 &= \sum_P \nu_i \psi_{ch,i} - \sum_R \nu_i \psi_{ch,i} = \Delta H_0 - T_0 \Delta S_0 \\ &= \Delta G_0 = \sum \nu_i \Delta g_{f,i} \end{aligned} \tag{12-14}$$

where $\Delta g_{f,i}$ data are measured at T_0. Substitution of values of $\psi_{ch,i}$ and $\Delta g_{f,i}$ for the species in the chemical reaction leads to

$$20{,}110 - \psi_{ch,C(s)} - 3{,}920 = -394{,}380 - 1(0) - 1(0)$$

Upon rearrangement, the standard chemical availability for solid carbon becomes

$$\psi_{ch,C(s)} = 394{,}380 + 20{,}100 - 3{,}920 = 410{,}570 \text{ kJ/kmol}$$

The value in Ref. [3], again using slightly different standard values, is 410,260 kJ/kmol. Again, the difference in values calculated from slightly different reference data is small. This difference occurs because there is no *correct* model for the environment. Frequently it depends on the application of interest. Regardless of the arbitrary "standard" environment assumed, a slight shift in relative humidity, atmospheric pressure, or dry air composition generally seems to have little effect on availability computations based on standard chemical availabilities.

The standard value for carbon monoxide (CO) would be evaluated in a manner similar to that for solid carbon. The gas-phase reaction of pure CO in the restricted dead state with environmental oxygen to form environmental CO_2 is

$$CO + \tfrac{1}{2}O_2 \rightarrow CO_2$$

Application of Eq. (12-14) to this reaction would lead directly to the standard chemical availability of CO. Since CO may be considered a fuel, Eq. (12-7) is also applicable. That is,

$$\psi_{ch,CO} = -\left(g_{CO_2,0} - g_{CO,0} - \tfrac{1}{2}g_{O_2,0}\right) + RT_0 \ln \frac{(y_{O_2,00})^{1/2}}{y_{CO_2,00}}$$

Both methods should lead to a standard value of $\psi_{ch,i}$ for CO of approximately 275,400 kJ/kmol. If CO is in a mixture, the chemical availability is found by employing Eq. (12-9).

In the case of the element hydrogen (H_2), the standard value is built upon water vapor as the reference substance. The amount of H_2 in the air, for example, is not known, but the composition of H_2O in the natural environment may be arbitrarily chosen to model typical conditions. Then the standard reference values for O_2 and H_2O can be used to evaluate the standard value for diatomic hydrogen. In this manner a fairly extensive table of standard chemical availabilities can be generated from the known average composition of elements or compounds in the natural environment for the gas, liquid, and solid phases. Table 12-2 is a short summary of data taken from Ref. [3]. Note that the selected composition of dry air is slightly different from that in Table 10-1. The difference in the mole fraction value for CO_2 in these two tables reflects the increase in CO_2 content between 1962 and the present.

TABLE 12-2
Standard chemical availability of selected species at 298.15 K and 1.01325 bars

Chemical formula	$y_{i,00}$ in dry air	$\psi_{ch,std}$ kJ/gmol
Ar (g)	0.00933	11.69
C (s)		410.26
CO_2 (g)	0.000345	19.87
He (g)	0.000005	30.37
H_2 (g)		236.09
H_2O (g)		9.49
N_2 (g)	0.7803	0.72
O_2 (g)	0.2099	3.97

Note: Based on Ref. [3], for a relative humidity of 70 percent and an atmospheric pressure of 0.9931 bars.

12-3 AVAILABILITY ANALYSIS OF CHEMICAL PROCESSES

This section is devoted to processes illustrating the use of the chemical availability of fuels.

12-3A Gas-Fired Furnace

The gas-fired home furnace is a common example of the use of chemical energy of a fuel for heating purposes. The first-law efficiency of a furnace η_{fur} is defined as the ratio of the rate of heat delivered to the room air to the rate of energy released by the fuel. That is, $\eta_{\text{fur}} = \dot{Q}_{\text{air}}/\dot{E}_f$. Well-designed furnaces are capable of transferring 85+ percent of the lower heating value (LHV) of the fuel to the circulating air in the heating system. The rest of the energy released is lost in the stack gases. The exergy of the entering combustion air usually is negligible, because its temperature is close to that of the standard environment. The availability of the warm room air at T_H on a rate basis can be approximated by $\dot{Q}_{\text{air}}(1 - T_0/T_H)$. As a result, the second-law efficiency of a furnace ϵ_{fur} is

$$\epsilon_{\text{fur}} = \frac{\text{exergy output}}{\text{exergy input}} = \frac{\dot{Q}_{\text{air}}\left(1 - \dfrac{T_0}{T_H}\right)}{\dot{\psi}_{\text{ch},f}} \tag{12-15}$$

It was noted in the preceding section that $\psi_{\text{ch},f} \approx LHV$. As an approximation formula for the furnace effectiveness then,

$$\epsilon_{\text{fur}} \approx \frac{\eta_{\text{fur}}(\text{LHV})\left(1 - \dfrac{T_0}{T_H}\right)}{\text{LHV}} = \eta_{\text{fur}}\left(1 - \frac{T_0}{T_H}\right)$$

The use of Eq. (12-15) is examined in the example below.

Example 12-4. A furnace operates on the natural gas described in Example 12-2. If the furnace energy efficiency is 90 percent and the circulating air is heated to $T_H = 325$ K, determine (*a*) the furnace effectiveness. (*b*) Compare this to a home electric heat pump with a coefficient of performance (COP) of 3.5 if the second-law efficiency for the production and distribution of electricity, $\eta_{PP} = \dot{W}_{\text{elec}}/\psi_{\text{ch},f}$, is 36 percent and $T_0 = 5°\text{C}$.

Solution

(*a*) If we assume that methane is the only fuel component in the natural gas, Example 12-2 gives a $\psi_{\text{ch},f}$ value of 828,920 kJ/kmol, and Table A-34 indicates that LHV = 802,300 kJ/kmol. Use of Eq. (12-15) yields

$$\epsilon = \frac{\eta_{\text{fur}}(\text{LHV})\left(1 - \dfrac{T_0}{T_H}\right)}{\psi_{\text{ch},f}} = \frac{0.90(802{,}300)\left(1 - \dfrac{278}{325}\right)}{828{,}920}(100) = 12.6 \text{ percent}$$

The low effectiveness is due to the low T_H value of the heated air. Thermal energy has a very low availability when the temperature level is close to T_0. If the energy released by the fuel heats air to 300°C, the effectiveness would be 43 percent. The extremely low effectiveness of the gas home furnace is due to the large irreversibility of the combustion process.

(*b*) In Chap. 4 the effectiveness of a heat pump η_{HP} is given as

$$\epsilon_{HP} = \text{COP}\left(1 - \frac{T_0}{T_H}\right)$$

Therefore for the given data

$$\epsilon_{HP} = 3.5\left(1 - \frac{278}{325}\right) = 0.506 \quad \text{(or 50.6 percent)}$$

The overall efficiency $\epsilon_{o,HP}$ is given by

$$\epsilon_{o,HP} = \frac{\dot{Q}_{\text{air}}\left(1 - \dfrac{T_0}{T_H}\right)}{\dot{\psi}_{\text{ch},f}} = \frac{\dot{Q}_{\text{air}}\left(1 - \dfrac{T_0}{T_H}\right)}{\dot{W}_{\text{elec}}} \frac{\dot{W}_{\text{elec}}}{\dot{\psi}_{\text{ch},f}} = \epsilon_{HP}\,\epsilon_{PP}$$

where ϵ_{PP} is the effectiveness of the power plant. Substitution of data reveals that

$$\epsilon_{o,HP} = 0.506(0.36) = 0.182 \quad \text{(or 18.2 percent)}$$

This compares to 12.6 percent for the gas furnace. Thus the heat pump exhibits a higher-energy utilization. Since the heat pump is a more complex system, economics may favor the gas furnace. However, the heat pump can be used to cool, as well as to heat the home.

12-3B Steam Power Plant

In Sec. 4-2-2, a simple steam power cycle is analyzed in terms of energy and availability changes for the four basic components of the cycle: boiler-superheater, turbine, condenser, and pump. This cycle is now investigated from a slightly different perspective. As shown in Fig. 12-2, the new model involves fuel and atmospheric air entering, as well as combustion product gases leaving the control volume. The reference for stream availability will now be the environmental dead state. To find the effectiveness ϵ_{PP} for the power plant, all sections of the plant must be reexamined. In Chap. 4 the power cycle is analyzed per unit mass of steam. In this subsection the analysis is based on the rate of chemical availability $\dot{\psi}_{\text{ch},f}$ entering with the fuel in the restricted dead state. This rate of availability input must equal the rates for the three outputs of the cycle: stack gases $\dot{\psi}_{\text{gas}}$, net shaft work $\dot{W}_{\text{net}}$, and condenser cooling water $\dot{\psi}_{cw}$, plus internal losses $\dot{I}_{\text{CV}}$. That is,

$$\dot{\psi}_{\text{ch},f} = \dot{\psi}_{\text{gas}} + \dot{W}_{\text{net}} + \dot{\psi}_{cw} + \dot{I}_{\text{CV}}$$

This balance neglects external heat losses. In another format

$$1 = \frac{\dot{\psi}_{\text{gas}}}{\dot{\psi}_{\text{ch},f}} + \frac{\dot{W}_{\text{net}}}{\dot{\psi}_{\text{ch},f}} + \frac{\dot{\psi}_{cw}}{\dot{\psi}_{\text{ch},f}} + \frac{\dot{I}_{\text{CV}}}{\dot{\psi}_{\text{ch},f}} \tag{12-16}$$

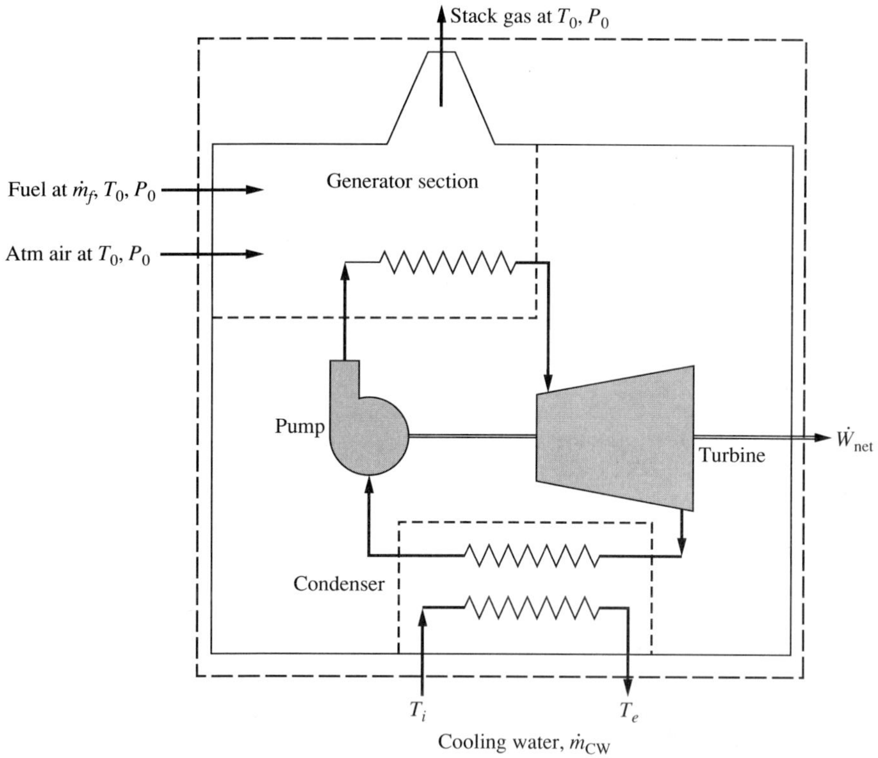

FIGURE 12-2
Overall schematic of a simple steam power plant showing the four basic components.

Each term on the right measures the fraction of the inlet rate of availability which leaves the control volume in various ways or is destroyed within the control volume. Losses within the control volume include those associated with the combustion and heat-transfer processes within the steam generator, turbine and pump irreversibilities, pipe friction, and heat losses from equipment. The following discussion is directed toward the evaluation of the first three terms of the above equation. We begin, however, with an analysis of the steam generator section.

12-3-1 Steam Generator

A schematic of a simple steam generator is shown in Fig. 12-3. The first-law generator efficiency η_{gen} for the steady-state device is

$$\eta_{gen} = \frac{\text{output}}{\text{input}} \equiv \frac{\dot{Q}_s}{\dot{E}_{fuel}} = \frac{\dot{m}_s(h_e - h_i)_s}{\dot{m}_f(\text{HHV})} \tag{12-17}$$

where $\dot{m}_s$ and $\dot{m}_f$ are the steam and fuel mass flow rates, respectively. The use of the higher heating value (HHV) implies that the water in the product gases is completely condensed. The value of η_{gen} is determined from a heat-transfer

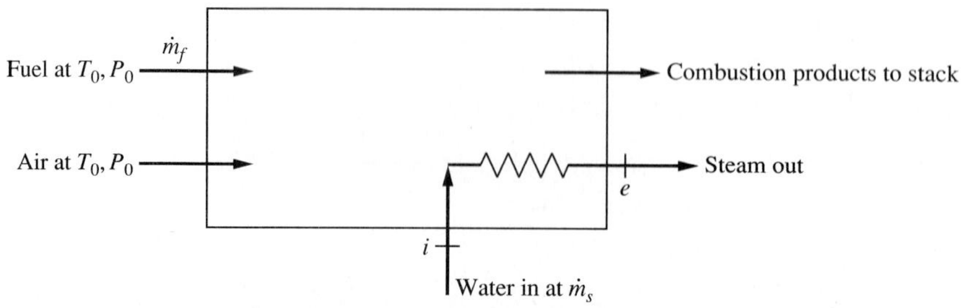

FIGURE 12-3
Schematic of a steam generator for a steam power cycle.

analysis of the generator, and hence is input information to the thermodynamic analysis. Equation (12-17) is used to determine the value of $\dot{m}_s/\dot{m}_f$. The second-law efficiency has a similar format, namely,

$$\epsilon_{\text{gen}} = \frac{\Delta\dot{\Psi}_{\text{steam}}}{\dot{\Psi}_{\text{ch},f}} = \frac{\dot{m}_s(\psi_e - \psi_i)_s}{\dot{m}_f\psi_{\text{ch},f}} = \frac{\dot{m}_s\left[(h_e - h_i)_s - T_0(s_e - s_i)_s\right]}{\dot{m}_f\psi_{\text{ch},f}} \tag{12-18}$$

Using Eq. (12-17) to replace $\dot{m}_f$ in Eq. (12-18), we find that

$$\epsilon_{\text{gen}} = \eta_{\text{gen}}(\text{HHV})\frac{\left[(h_e - h_i)_s - T_0(s_e - s_i)_s\right]}{(h_e - h_i)_s\psi_{\text{ch},f}} \tag{12-19}$$

A special case of the above general formulation is the simple phase change from saturated liquid to saturated vapor at temperature T_{sat}. In this situation $(h_e - h_i) = T_{\text{sat}}(s_e - s_i)$. If we also assume HHV $\approx \psi_{\text{ch},f}$ (see Appendix in Szargut [3] or Table A-34 for comparisons), then

$$\epsilon_{\text{gen}} = \eta_{\text{gen}}\left[1 - \frac{T_0(s_e - s_i)}{h_e - h_i}\right] = \eta_{\text{gen}}\left(1 - \frac{T_0}{T_{\text{sat}}}\right)$$

For example, at 30 bars where $T_{\text{sat}} = 507$ K and for a generator first-law efficiency of 80 percent, $\epsilon_{\text{gen}} = 33$ percent. Even when $\eta_{\text{gen}} = 100$ percent, $\epsilon = 41$ percent. The value of ϵ_{gen} is improved by raising the saturation temperature. At 100 bars where $T_{\text{sat}} = 584$ K, ϵ_{gen} reaches 39 percent when $\eta_{\text{gen}} = 80$ percent. These simple estimates indicate that ϵ of a steam generation process is relatively low.

Example 12-5. A steam generator with a generator efficiency of 82 percent uses methane gas at 25° C. For this fuel Table A-34 in the Appendix lists values of $\psi_{\text{ch},f} = 832{,}400$ kJ/kmol and HHV $= 890{,}300$ kJ/kmol for an atmospheric relative humidity of 60 percent. The generator heats water at 100 bars from 220 to 440°C. Determine the mass flow ratio of steam to fuel and the second-law efficiency for the process, for $T_0 = 25$°C.

Solution. The use of Eq. (12-17) leads to the ratio $\dot{m}_s/\dot{m}_f$. On the basis of given data and steam table data

$$\frac{\dot{m}_s}{\dot{m}_f} = \frac{\eta_{\text{gen}}(\text{HHV})}{h_e - h_i} = \frac{0.82(890{,}320)}{3213.2 - 945.9} = 322 \text{ kg steam per kmol fuel}$$

As a result, the second-law efficiency based on Eq. (12-18) is

$$\epsilon_{\text{gen}} = \frac{322\,[(3213.2 - 945.9) - 298.15(6.3805 - 2.5039)]}{829{,}320} = 0.432$$

[The generator effectiveness can also be evaluated directly from Eq. (12-19).] Thus 43 percent of the fuel availability has been transferred to the water stream, or 57 percent is lost to irreversibilities within the system. There are two sources for losses in the process. One is the irreversible heat transfer between the combustion products and the fluid in the boiler-superheater tubes. The other is the irreversibility of the combustion process. In Chap. 11 it was pointed out that the irreversibility associated with adiabatic combustion is frequently 25–35 percent of the entering fuel-air availability.

12-3-2 Stack Gas Losses

In the steam generator discussed above there are three obvious areas of availability loss. Irreversibilities exist in the combustion and in the heat-transfer processes. In addition, availability is carried out with the stack gases. The products of combustion eventually pass through a stack and are released to the environment. If the temperature of the stack gases is T_s, $T_s > T_0$, and $P_s \approx P_0$, then Eq. (11-17) for the total stream availability of an ideal-gas mixture reduces to

$$\psi_{\text{gas}} = \sum_i y_i \left[h_{i,T_s} - h_{i,T_0} - T_0(s^o_{i,T_s} - s^o_{i,T_0})\right] + \sum_i y_i \left(RT_0 \ln \frac{y_i}{y_{i,00}}\right) \qquad (12\text{-}20)$$

As noted in Example 10-1, the two terms on the right typically are of similar magnitude only when T_s is not greater than T_0 by more than a few hundred degrees Celsius. Then the fraction of the inlet fuel availability which appears in the stack gases at temperature T_s is

$$\frac{\dot{\psi}_{\text{gas}}}{\dot{\psi}_{\text{ch},f}} = \frac{\dot{N}_{\text{gas}}\psi_{\text{gas}}}{\dot{N}_{\text{fuel}}\psi_{\text{ch},f}} \qquad (12\text{-}21)$$

The ratio $\dot{N}_{\text{gas}}/\dot{N}_{\text{fuel}}$ is found from the equation for the chemical reaction between fuel and air.

12-3-3 Shaft Work

The net shaft-work output w_{net} is determined from an energy balance on the steam cycle and is expressed per unit mass of steam. For the purpose of an availability balance on the entire steam power plant the ratio of net power output to the rate of fuel availability input is needed. In addition, this fraction measures the *second-law efficiency* ϵ_{PP} of the entire power plant. Basically this quantity is given by

$$\epsilon_{PP} \equiv \frac{\dot{W}_{\text{net}}}{\dot{\psi}_{\text{ch},f}} = \frac{\dot{m}_s w_{\text{net}}}{\dot{m}_f \psi_{\text{ch},f}} \qquad (12\text{-}22)$$

The ratio $\dot{m}_s/\dot{m}_f$ is found from Eq. (12-17) and a knowledge of the generator first-law efficiency. It can also be shown that

$$\epsilon_{PP} = \frac{\dot{W}_{\text{net}}}{\dot{\psi}_{\text{ch},f}} = \frac{w_{\text{net}}}{q_{\text{in}}} \frac{\dot{m}_s q_{\text{in}}}{\dot{m}_f E_f} \frac{E_f}{\psi_{\text{ch},f}} = \eta_{th}\eta_{\text{gen}} \frac{\text{HHV}}{\psi_{\text{ch},f}} \tag{12-23}$$

where the energy E_f released by the fuel is taken as its higher-heating value, and η_{th} is the steam cycle thermal efficiency.

12-3-4 Condenser

In order to relate the availability loss in the condenser to the fuel availability, we first write an energy balance for the overall control volume in Fig. 12-2.

$$\dot{E}_{\text{fuel}} + \dot{E}_{\text{gas}} + \dot{W}_{\text{net}} + \dot{Q}_{\text{cond}} = 0$$

where heat losses from the equipment are neglected. The quantities $\dot{E}_{\text{gas}}$ and $\dot{W}_{\text{net}}$ can be expressed in terms of the first-law efficiencies of the generator η_{gen} and of the power plant η_{PP} as

$$\dot{E}_{\text{gas}} \equiv (1 - \eta_{\text{gen}})\dot{E}_{\text{fuel}} = (1 - \eta_{\text{gen}})\dot{m}_f(\text{HHV})$$
$$\dot{W}_{\text{net}} \equiv \eta_{PP}\dot{E}_{\text{fuel}} = \eta_{PP}\dot{m}_f(\text{HHV})$$

where HHV is the higher heating value of the fuel. Substituting these two relations into the energy equation, and noting that both are negative in context, we find that

$$\dot{Q}_{cw} = \dot{m}_f(\text{HHV})(\eta_{\text{gen}} - \eta_{PP}) = \dot{m}_{cw}(h_e - h_i)_{cw} \tag{12-24}$$

where $\dot{Q}_{\text{cond}}$ has been replaced by $-\dot{Q}_{cw}$ for the cooling water stream. The last term above represents the rate of energy received by the cooling water (cw) in terms of its enthalpy change. This equation relates $\dot{m}_f$ to $\dot{m}_{cw}$.

The rate of exergy received by the cooling water $\dot{\psi}_{cw}$ is, by definition,

$$\dot{\psi}_{cw} = \dot{m}_{cw}\left[(h_e - h_i) - T_0(s_e - s_i)\right]_{cw} = \dot{m}_{cw}(h_e - h_i)_{cw}\left[1 - \frac{T_0(s_e - s_i)}{(h_e - h_i)}\right]_{cw} \tag{12-25}$$

Using Eq. (12-24) to eliminate $(\dot{m}\,\Delta h)_{cw}$, Eq. (12-25) becomes

$$\dot{\psi}_{cw} = \dot{m}_f(\text{HHV})(\eta_{\text{gen}} - \eta_{PP})\left[1 - \frac{T_0(s_e - s_i)_{cw}}{(h_e - h_i)_{cw}}\right] \tag{12-26}$$

Finally, since $\dot{\psi}_{\text{ch},f} = \dot{m}_f\psi_{\text{ch},f}$, the fraction of the availability entering with the fuel that leaves in the cooling water is

$$\frac{\dot{\psi}_{cw}}{\dot{\psi}_{\text{ch},f}} = (\eta_{\text{gen}} - \eta_{PP})\frac{\text{HHV}}{\psi_{\text{ch},f}}\left[1 - \frac{T_0(s_e - s_i)_{cw}}{(h_e - h_i)_{cw}}\right] \tag{12-27}$$

The values of Δh and Δs for the cooling water can be found from a steam table from knowledge of the inlet and exit cooling water temperatures T_i and T_e.

As an alternative, the fluid may be approximated as incompressible, for which

$$h_e - h_i = c(T_e - T_i) + v_f(P_e - P_i)$$
$$s_e - s_i = c \ln(T_e/T_i)$$

The pressure correction is usually negligible when evaluating $(h_e - h_i)$. Hence Eq. (12-27) may be approximated to a reasonable accuracy by

$$\frac{\dot{\psi}_{cw}}{\dot{\psi}_{\text{ch},f}} = (\eta_{\text{gen}} - \eta_{PP})\frac{\text{HHV}}{\psi_{\text{ch},f}}\left[1 - \frac{T_0 \ln(T_e/T_i)_{cw}}{(T_e - T_i)_{cw}}\right] \tag{12-28}$$

For typical values of T_i and T_e the value of the bracketed term above is roughly 0.01–0.02. Although 40–50 percent of the fuel energy ends up in the cooling water, less than 2 percent of the availability entering the power plant is removed by the cooling water.

12-3-5 Combustion and Heat-Transfer Losses

On the basis of the results of the first two subsections the fraction of the fuel availability which appears as combustion and heat-transfer losses $\dot{I}_{\text{gen}}$ in the steam generator can be ascertained. An availability balance for the steam generator shown in Fig. 12-3 is

$$0 = \dot{\psi}_{\text{ch},f} - \dot{\psi}_{\text{gas}} - \Delta\dot{\psi}_{\text{steam}} - \dot{I}_{\text{gen}}$$

Division of the above equation by $\dot{\psi}_{\text{ch},f}$ and subsequent rearrangement leads to

$$\frac{\dot{I}_{\text{gen}}}{\dot{\psi}_{\text{ch},f}} = 1 - \frac{\dot{\psi}_{\text{gas}}}{\dot{\psi}_{\text{ch},f}} - \frac{\Delta\dot{\psi}_{\text{steam}}}{\dot{\psi}_{\text{ch},f}} \tag{12-29}$$

Knowledge of the last two terms on the right allows evaluation of the fraction of fuel availability lost to irreversibilities in the steam generator. The last term is the generator effectiveness found in subsection 12-3-1. An approximation method for evaluating the combustion and heat-transfer losses as two separate items is discussed in Sec. 12-3C that follows.

12-3-6 Other Internal Losses

The remaining internal losses include those associated with the turbine, the pump, the irreversible heat transfer in the condenser, the friction in the connecting pipes, and heat losses from all equipment. Losses for the first three items are determined from the entropy changes of the fluids in the equipment. If the devices are assumed to be adiabatic, then $i = T_0\,\Delta s$.

The example below illustrates a more complete availability analysis of a simple steam power cycle previously analyzed in Chap. 4.

Example 12-6. A simple steam power cycle generates steam at 140 bars and 560°C and condenses steam at 0.06 bar. The cooling water in the condenser experiences a temperature rise from 25 to 35°C. The adiabatic efficiency of the turbine is 85 percent and that of the pump is 70 percent. The energy source is ethylene gas in the restricted dead state at 25°C. Combustion is complete with stoichiometric air, and the stack gases leave to 440 K. The first-law efficiency of the steam generator is 82 percent, and $P = P_0$ during the combustion process. Determine (*a*) the steam generator effectiveness, (*b*) the fraction of inlet availability removed by the stack gases, (*c*) the power plant effectiveness and the fractional loss in fuel availability (*d*) in the condenser, (*e*) due to internal irreversibilities, (*f*) in the steam generator, and (*g*) in the turbine, pump, and condenser.

Solution. Data from Example 4-4 include the following:

$$q_{\text{boiler}} = 3314.4 \text{ kJ/kg} \qquad \Delta s_{\text{boiler}} = 6.0537 \text{ kJ/kg·K} \qquad w_{\text{net}} = 1216.9 \text{ kJ/kg}$$

$$q_{\text{cond}} = 41.8 \text{ kJ/kg} \qquad \dot{m}_{cw}/\dot{m}_s = 50.1$$

where $\dot{m}_{cw}$ and $\dot{m}_s$ are the cooling water and steam flow rates, respectively.

(*a*) For the steam generator the mass flow rate ratio of steam to fuel is found by Eq. (12-17).

$$\frac{\dot{m}_s}{\dot{m}_f} = \frac{\eta_{\text{gen}}(\text{HHV})}{q_{\text{boiler}}} = \frac{0.82(1{,}411{,}000)}{3314.4} = 349 \text{ kg steam per kmol fuel}$$

The effectiveness of the generator, based on Eq. (12-18), is

$$\epsilon_{\text{gen}} = \frac{349\,[3314.4 - 298.15(6.0532)]}{1{,}359{,}600} = 0.387$$

where the HHV and fuel availability (for $\phi_{RH,0} = 100\%$) are found in Table A-34.

(*b*) The availability evaluation for the stack gases at 440 K requires the mole fractions of the combustion products. To simplify the calculation, atmospheric air is taken as 21 percent oxygen and 79 percent nitrogen, rather than the standard environmental air. For stoichiometric air

$$C_2H_4 + 3\ O_2 + 11.28\ N_2 \rightarrow 2\ CO_2 + 2\ H_2O + 11.28\ N_2$$

Thus the mole fractions of CO_2, H_2O, and N_2 are 0.131, 0.131, and 0.738, respectively. Use of these values and h and s data from the ideal-gas tables in Eq. (12-20) leads to

$$\begin{aligned}\psi_{\text{gas}} &= [0.131(1055) + 0.131(884) + 0.738(844)] \\ &\quad + 8.3145(298.15)(0.7957 + 0.1873 - 0.0182) \\ &= 877 + 2392 = 3269 \text{ kJ/kmol stack gas}\end{aligned}$$

In this case at 440 K the chemical term is somewhat larger than the temperature term. The fraction of the inlet fuel availability which appears in the stack gases is

$$\frac{\dot{\psi}_{\text{gas}}}{\dot{\psi}_{\text{ch},f}} = \frac{\dot{N}_{\text{gas}}}{\dot{N}_{\text{fuel}}}\frac{\psi_{\text{gas}}}{\psi_{\text{ch},f}} = \frac{15.28}{1}\,\frac{3269}{1{,}359{,}600} = 0.037$$

Because this term is relatively small, the use of environmental air composition in the calculation would have little overall effect.

(*c*) On the basis of Eq. (12-22), and recalling that $\dot{w}_{\text{net}} = \dot{m}_s w_{\text{net}}$, we have

$$\epsilon_{PP} = \frac{\dot{w}_{\text{net}}}{\dot{\psi}_{\text{ch},f}} = \frac{349(1216.9)}{1{,}359{,}600} = 0.312$$

This value is typical of simple steam power plants. Recall from Example 4-4 that the second-law efficiency $\epsilon = 0.806$ for the steam cycle itself. The vast difference between $\epsilon_{\text{cycle}} = 0.806$ and $\epsilon_{PP} = 0.312$ reflects the large availability losses of the combustion and heat-transfer processes in the steam generator.

(*d*) The condenser evaluation is given by Eq. (12-27). First, the first-law efficiency η_{PP} is found to be

$$\eta_{PP} = \frac{\dot{m}_s w_{\text{net}}}{\dot{m}_f(\text{HHV})} = \frac{349(1216.9)}{1{,}411{,}000} = 0.301$$

which compares to a thermal efficiency $\eta_{th} = 0.367$ for the steam cycle alone. Then for the condenser cooling water temperature rise from 18 to 28°C

$$\frac{\dot{\psi}_{cw}}{\dot{\psi}_{\text{ch},f}} = (0.82 - 0.30)\frac{1{,}411{,}000}{1{,}359{,}600}\left[1 - \frac{298.15(0.1414)}{41.85}\right] = -0.004$$

This value is negative, since the temperature of the cooling water is initially below $T_0 = 25$°C. For a temperature rise from 25 to 35°C the calculated fraction would be 0.009. In either case the contribution is very small.

(*e*) At this point the availability balance based on Eq. (12-16) is

$$\frac{\dot{I}_{\text{CV}}}{\dot{\psi}_{\text{ch},f}} = 1 - 0.312 - 0.037 - (-0.004) = 0.655$$

Thus internal irreversibilities account for 65.5 percent of the fuel availability.

(*f*) The portion of the 65.5 percent losses attributed to the steam generator is found by using Eq. (12-29).

$$\frac{\dot{I}_{\text{gen}}}{\dot{\psi}_{\text{ch},f}} = 1 - 0.037 - 0.387 = 0.576$$

As a consequence, 88 percent (57.6/65.5) of the total irreversibilities is due to combustion and heat-transfer losses. The remaining losses are 65.5 − 57.6 = 7.9 percent of the fuel availability.

(*g*) The remaining losses occur in the turbine, pump, and condenser. The irreversibility in each of these adiabatic devices is $i = T_0\,\Delta s$. From Example 4-4 the i values of the turbine, pump, and condenser are 210.8, 5.78, and 91.4 kJ/kg steam, respectively. Therefore the fractional contributions relative to the fuel availability are

$$\frac{\dot{I}_t + \dot{I}_p + \dot{I}_{\text{cond}}}{\dot{\psi}_{\text{ch},f}} = \frac{349(210.8 + 5.78 + 91.4)}{1{,}359{,}600} = 0.0541 + 0.0015 + 0.0234 = 0.079$$

This value agrees with that calculated in part (f) above.

TABLE 12-3
Summary of availability outputs and losses for a simple steam power cycle

Input	
Fuel availability	100.0 units
Output	
Power developed	31.2
Cooling water	− 0.4
Stack gases	3.7
Generator losses	57.6
Turbine loss	5.41
Pump loss	0.15
Condenser loss	2.34
Total	100.0

A summary of the data for the entire plant is shown in Table 12-3. The steam generator losses overshadow all other effects. It must be kept in mind that frictional pressure drops and external heat losses have been neglected throughout the plant.

12-3C Combustion and Heat-Transfer Losses

Combustion processes frequently are affected by the simultaneous removal of energy from the combustion zone as heat transfer to another fluid. (Examples include the gas furnace and the steam generator discussed in the two preceding subsections.) The irreversibilities in the combustion and heat exchange processes are not mutually exclusive. The temperature profile in the hot-gas region affects the heat-transfer irreversibility. For this reason it is difficult to evaluate separately the two contributions to the overall loss. In an attempt to overcome this difficulty the following approximation may be made.

It has been suggested [1,7] that the effectiveness ϵ_{comb} of the combustion process alone should be measured by the parameter

$$\epsilon_{\text{comb}} = \frac{\dot{\psi}_{\text{prod},af} - \dot{\psi}_{\text{air}}}{\dot{\psi}_{\text{ch},f}} \tag{12-30}$$

where

$\dot{\psi}_{\text{prod},af}$ is the availability of the product gases at the adiabatic flame (*af*) temperature relative to the environmental dead state.
$\dot{\psi}_{\text{air}}$ is the availability of the inlet air at temperature T relative to the environmental dead state.

The value of $\dot{\psi}_{\text{prod},af}$ is found from Eq. (11-17) for a mixture of ideal gases. If the air is not preheated, then $\dot{\psi}_{\text{air}} = 0$. The fractional loss of availability in the combustion process then is

$$\frac{\dot{I}_{\text{comb}}}{\dot{\psi}_{\text{ch},f}} = 1 - \epsilon_{\text{comb}}$$

An alternative to the above calculation might be to determine the irreversibility of the adiabatic combustion reaction directly. This calculation must involve environmental air at the inlet temperature T. For an adiabatic process $I = T_0 \Delta S = T_0(S_P - S_R)$. Specification of the inlet and exit temperatures, pressures, and compositions is sufficient to determine ΔS, as noted in Sec. 11-3.

Example 12-7. Reconsider the steam generator in Example 12-6. Determine the loss in fuel availability due to the combustion process on the basis of Eq. (12-30).

Solution. The use of Eq. (12-30) requires the evaluation of the availability of the product gases at the adiabatic flame temperature. This in turn requires an evaluation of the adiabatic flame temperature when environmental air is used. Using the data from Table 12-1 based on a relative humidity of 100 percent, and ignoring the 0.0091 contribution by other gases, the combustion reaction of ethylene with stoichiometric environmental air becomes

$$C_2H_4 + 3\,O_2 + 11.18\,N_2 + 0.004\,CO_2 + 0.463\,H_2O \rightarrow 2.004\,CO_2 + 2.463\,H_2O + 11.18\,N_2$$

An energy balance on the adiabatic combustion process for both air and fuel entering at 25°C reduces to

$$2.004\,h(CO_2) + 2.463\,h(H_2O) + 11.18\,h(N_2) = 1{,}463{,}040 \text{ kJ}$$

Iteration in the gas tables leads to $T_{af} = 2506$ K. This compares to 2570 K in Example 11-3 when air is composed of only O_2 and N_2.

The total availability $\Psi_{\text{prod},af}$ of the product gases *per mole of fuel* is determined from Eq. (11-17), namely,

$$\Psi_{\text{prod},af} = \sum_i N_i \left[h_{i,T} - h_{i,T_0} - T_0(s^o_{i,T} - s^o_{i,T_0}) \right] + RT_0 \sum_i N_i \ln \frac{y_i}{y_{i,00}}$$

Based on tabular ideal-gas data at 2506 K, the product gas availability is found to be

$$\Psi_{\text{prod},af} = 179{,}790 + 180{,}150 + 604{,}410 + 38{,}370 = 1{,}002{,}720 \text{ kJ/kmol fuel}$$

Consequently, the loss of availability due to the combustion process alone is estimated to be

$$\frac{\dot{I}_{\text{comb}}}{\dot{\psi}_{\text{ch},f}} = \frac{I_{\text{comb}}}{\psi_{\text{ch},f}} = 1 - \frac{1{,}002{,}720 - 0}{1{,}359{,}600} = 0.262$$

Recall from Example 12-4 that the combined loss of the combustion process and the heat transfer to the steam is 0.576 (or 57.6 percent) of the fuel availability. Therefore the approximate loss attributed to the heat-transfer process alone is $(0.576 - 0.262) = 0.314$ (or 31.4 percent). Therefore in this particular case the combustion process loss and heat-transfer loss are roughly of the same magnitude.

12-3D Preheated Inlet Air

Preheating air is a well-known technique to increase the efficiency of combustion. Such preheating increases the adiabatic flame temperature T_{af}, and hence the value of $\dot{\psi}_{\text{prod},af}$ in Eq. (12-30). Although the term $\dot{\psi}_{\text{air}}$ obviously also increases, the net effect is an increase in the numerator of the equation. The increase in ϵ_{comb} is not linear with preheat air temperature, but falls off as the temperature increases. Curves showing the effect for methane at different percent theoretical air appear in the literature [1]. This source indicates an increase in combustion efficiency from 72 to 77.5 percent when stoichiometric air is preheated from 300 to 600 K. To account for this effect, one needs to reevaluate $\dot{\psi}_{\text{prod},af}$ for the new T_{af}, and also to evaluate $\dot{\psi}_{\text{air}}$ at the chosen preheat air temperature.

PROBLEMS

12-1. Evaluate (a) the chemical availability of propane gas (C_3H_8) in the restricted dead state for a relative humidity in the environment of 60 percent and (b) the change in availability for the isothermal combustion of C_3H_8 with stoichiometric air at 25°C and 1 atm by the method used in Sec. 10-4A.

12-2. Evaluate (a) the chemical availability of butane gas (C_4H_{10}) in the restricted dead state for a relative humidity in the environment of 60 percent and (b) the change in availability for the isothermal combustion of C_4H_{10} with stoichiometric air at 25°C and 1 atm by the method used in Sec. 10-4A.

12-3. Evaluate (a) the chemical availability of gaseous methyl alcohol [$CH_3OH(g)$] in the restricted dead state for a relative humidity in the environment of 60 percent and (b) the change in availability for the isothermal combustion of [$CH_3OH(g)$] with stoichiometric air at 25°C and 1 atm by the method used in Sec. 10-4A.

12-4. A fuel mixture is 75 percent methane (CH_4) and 25 percent ethane (C_2H_6) by volume. Employing the data of Table A-34 in the Appendix for these two fuels, find the chemical availability of the fuel, in kilojoules per mole of mixture.

12-5. A fuel mixture is 80 percent CH_4 and 20 percent C_3H_8 by volume. On the basis of data in Table A-34 for these two fuels, evaluate the chemical availability of the fuel, in kilojoules per mole of mixture.

12-6. Show that the value of the standard chemical availability of solid carbon calculated in Sec. 12-2-5 can also be calculated from Eq. (12-7), in kJ/kmol.

12-7. Calculate (a) the standard chemical availability of carbon dioxide (CO_2) by means of Eq. (12-13) and the data of Table 12-1, in kJ/kmol. Then (b) calculate the value of the standard chemical availability of carbon monoxide (CO) (1) through the use of Eq. (12-7) and (2) by use of Eq. (12-14), in kJ/kmol. Assume the relative humidity is 60 percent and the atmospheric pressure is 1 bar.

12-8. Calculate (a) the standard chemical availability of water vapor by means of Eq. (12-13) and the data of Table 12-1, in kJ/kmol. Then (b) determine the value of the standard chemical availability of diatomic hydrogen gas (1) through the use of Eq. (12-7) and (2) by use of Eq. (12-14), in kJ/kmol. Assume the relative humidity is 60 percent and the atmospheric pressure is 1 bar.

12-9. A commercial furnace operates on an 80 percent methane and 20 percent ethane by volume mixture. The furnace energy efficiency is 84 percent and the circulating air through the furnace is heated to 500 K. A competitive approach is to employ an electric heat pump with a COP (coefficient of performance) of 3.40 and a second-law efficiency for the production and distribution of electricity of 35 percent for a T_0 value of 25°C. Determine (a) the furnace effectiveness and (b) the overall second-law efficiency of the heat pump.

12-10. A simple steam power cycle has a turbine inlet state of 6.0 MPa and 540°C and an exhaust pressure of 0.008 MPa. The cooling water in the condenser experiences a temperature rise from 20 to 27°C, and T_0 is 20°C. The adiabatic efficiencies of the turbine and pump are 85 and 70 percent, respectively. The energy source is methane gas in the restricted dead state at 25°C with 25 percent excess air. Combustion is complete, and the stack gases leave at 480 K. The first-law efficiency of the steam generator is 84 percent, and $P = P_0$ during the combustion process. Determine (a) the steam generator effectiveness, (b) the fraction of the inlet fuel availability removed by the stack gases, (c) the power plant effectiveness, and the fractional loss in fuel availability (d) in the condenser cooling water, (e) due to internal irreversibilities, (f) in the steam generator, and (g) in the turbine, pump, and condenser. The atmospheric relative humidity is 100 percent.

12-11. Reconsider Prob. 12-10. The irreversibilities in the steam generator are due to the combustion process and the heat-transfer process from the hot gases to the steam. Estimate the value of $\dot{I}/\dot{\psi}_{ch,f}$ for the two processes.

12-12. The simple steam power plant in Prob. 12-10 is modified in the following manner. Steam entering the turbine at 6.0 MPa and 540°C is expanded to 0.50 MPA, reheated reversibly to 500°C, and expanded to 0.008 MPa. If all other data remain the same, determine the same quantities (a) through (g) requested in Prob. 12-10.

12-13. The simple steam power cycle in Prob. 12-10 is modified in the following way. Steam entering the turbine is expanded to 0.50 MPa, where a portion is bled to a single open heater and the remainder is expanded to 0.008 MPa. If all other data remain the same, determine the same quantities (a) through (g) requested in Prob. 12-10. In addition, (h) determine the fractional loss in fuel availability occurring in the open heater.

12-14. A simple steam power cycle has a turbine inlet state of 120 bars and 600°C and an exhaust pressure of 0.08 bar. The cooling water in the condenser experiences a temperature rise from 22 to 28°C, and T_0 is 22°C. The adiabatic efficiencies of the turbine and pump are 85 and 70 percent, respectively. The energy source is propane gas in the restricted dead state at 25°C with 20 percent excess air at 22°C. Combustion is complete, and the stack gases leave at 460 K. The first-law efficiency of the steam generator is 85 percent, the atmospheric relative humidity is 60 percent, and $P = P_0$ during the combustion process. Determine (a) the steam generator effectiveness, (b) the fraction of the inlet fuel availability removed by the stack gases, (c) the power plant effectiveness, and the fractional loss in fuel availability (d) in the condenser cooling water, (e) due to internal irreversibilities, (f) in the steam generator, and (g) in the turbine, pump, and condenser.

12-15. Reconsider Prob. 12-14. The irreversibilities in the steam generator are due to the combustion process and the heat-transfer process from the hot gases to the steam. Estimate the value of $\dot{I}/\dot{\psi}_{ch,f}$ for the two processes.

12-16. The simple steam power plant in Prob. 12-14 is modified in the following manner. Steam entering the turbine at 120 bars and 600°C is expanded to 30 bars, reheated reversibly to 540°C, and expanded to 0.08 MPa. If all other data remain the same, determine the same quantities (a) through (g) requested in Prob. 12-14.

12-17. An irreversible gas turbine power plant has a compressor with an inlet state of 1 bar and 295 K and an outlet state of 4 bars and 470 K. The turbine has an inlet state of 4 bars and 1200 K and an exhaust state of 1 bar and 900 K. The combustor is fueled with methane gas in the restricted dead state at 25°C, and the environmental state is 1 bar, 295 K, and 60 percent relative humidity. Determine (a) the moles of air supplied per mole of fuel, (b) the net work output, in kJ/kmol of fuel, (c) the power plant effectiveness, and the fraction of the inlet fuel availability lost in (d) the compressor, (e) the combustor, (f) the turbine, and (g) the turbine exhaust.

12-18. With reference to Prob. 12-17, consider that the compressor exhaust stream goes to a regenerative heat exchanger and is heated to 830 K and 1 bar. All other conditions remain the same. Determine the quantities (a) through (g) requested in Prob. 12-17. In addition, (h) determine the fraction of the inlet fuel availability lost in the regenerator.

12-19. The pressure ratio of a gas-turbine power cycle is 6.0/1, the inlet conditions to the compressor are 105 kPa and 300 K, and the exit temperature is 540 K. The turbine inlet state is 1100 K and 630 kPa, and the exhaust state is 105 kPa and 760 K. The combustor is fueled by octane gas in the restricted dead state at 25°C. The environmental state is 100 kPa, 27°C, and 60 percent relative humidity. Determine (a) the moles of air supplied per mole of fuel, (b) the net work output, in kJ/kmol of fuel, (c) the power plant effectiveness, and the fraction of the inlet fuel availability lost in (d) the compressor, (e) the combustor, (f) the turbine, and (g) the turbine exhaust.

12-20. With reference to Prob. 12-19, consider that the compressor exhaust stream goes to a regenerative heat exchanger and is heated to 730 K and 630 kPa before it enters the combustor. All other conditions remain the same. Determine the quantities (a) through (g) requested in Prob. 12-19. In addition, (h) determine the fraction of the inlet fuel availability lost in the regenerator.

REFERENCES

1. Moran, M. J., *Availability Analysis: A Guide to Efficient Energy Use*, Prentice-Hall, Englewood Cliffs, NJ, 1982.
2. Bejan, A., *Advanced Engineering Thermodynamics*, John Wiley & Sons, New York, 1988.
3. Szargut, J., D. R. Morris, and F. R. Steward, *Exergy Analysis of Thermal, Chemical, and Metallurgical Processes*, Hemisphere Publishing Corp., New York, 1988.
4. Szargut, J., and R. Petela, *Egzergia, Wyd. Nauk.,-Tech*, Warszawa, 1965 (in Polish).
5. Rodriguez, L., "Calculation of Available-energy Quantities," ACS Symposium Series No. 122, 39, American Chemical Society, 1980.
6. Brzustowski, T. A., and A. Brena, "Second Law Analyses of Energy Processes," *Trans. Can. Soc. Mech. Engr.* **10**(3), 1986.
7. Rant, Z., *Brennst. Warme-Kraft*, **13** (11), 496–500, 1961.

CHAPTER 13

A STATISTICAL VIEWPOINT OF ENTROPY

In the last half of the nineteenth century, through the work of Maxwell, Clausius, Boltzmann, and others, the concept of the entropy function was developed in terms of microscopic parameters of a macroscopic system. The modern development of the subject, begun by Planck, Einstein, and others at the beginning of the twentieth century, requires the use of quantum theory and various techniques of statistical analysis. Since a statistical analysis of the behavior of a large number of particles is involved, this microscopic approach to thermodynamics is called *statistical thermodynamics*. It is based on its own set of postulates, which are independent of those of classical or macroscopic thermodynamics.

13-1 QUANTIZATION OF ENERGY

Before 1900 two major concepts regarding the nature of matter and energy had been established. One of these was the particle nature of matter. A quantity of mass was conceived of as being composed of atoms and molecules, and each particle in theory was a separate entity and identifiable. Also, the wave nature of electromagnetic radiation was well entrenched and in common usage. In the early 1900s the concepts of energy and matter underwent a profound change. The work of Planck, Einstein, and others proposed that energy was particlelike. Light energy was considered to be absorbed or emitted in discrete quantities called *quanta*, or *photons*. In 1901 Planck derived an equation of the spectral distribution of blackbody radiation associated with electromagnetic radiation. He assumed that the energies corresponding to the various standing electromagnetic waves could take only discrete values. These values were integral multiples of $h\nu$, where ν is the frequency of the wave and h is a constant known as Planck's constant. Schrödinger in the mid-1920s suggested an equation which represents the wave nature of matter in terms of a set of standing waves. The Schrödinger wave equation is taken as a basic postulate of quantum theory. Solution of this equation for the various energy modes leads to a set of quantum, or energy, levels for each mode. The discrete or allowed energy states of particles are determined by (1) the nature of the particles and (2) constraints such as the volume of the system or the strength of the applied magnetic field.

An example of the discreteness of energy values is provided by the emission of light from substances such as sodium vapor when they are heated in a gas flame. The measured wavelength of light emitted from a sodium source is 0.5890 μm. Electrons orbiting the nucleus have been thermally excited. When an excited electron "falls" or "decays" from the excited state to the normal or "ground" state of energy, the energy difference between these states of the electron is emitted as light. The radiant energy is always emitted at the same wavelength, and it is assumed to be emitted in the form of a single photon with an energy $h\nu$. This energy difference is given by the Einstein-Planck equation

$$\epsilon_1 - \epsilon_0 = h\nu$$

where ϵ_1 = energy of excited state
ϵ_0 = ground state of energy
$h = 6.625 \times 10^{-34}$ J · s

For sodium emission at 0.5890 μm this energy-level difference is 2.10 eV.

13-2 THE ENTROPY FUNCTION

A group of particles that constitutes an isolated thermodynamic system has an overall energy U. This energy is shared by the individual particles. However, each particle is not required to have the same energy ϵ. The results of quantum mechanics indicate that a whole series of energy values must be considered. These discrete values are set by certain constraints on the system. For example, it is found that the translational (linear) kinetic energies of gas particles are inversely proportional to the volume to the two-thirds power. It is common practice to use the term "energy level" when referring to one of the allowed energy values. The series of allowed energy levels begins at some minimum value called the *ground level* (or zero level) of energy. The series then progresses through distinct steps to larger and larger energy values. Such a series of energy levels exists for each energy mode. For example, one would need to consider for a diatomic gas at low pressures at least those series of energy levels for the translational, rotational, and vibrational modes.

The term "energy level" allows one to present the concept of discrete energy values very simply. A typical diagram for energy levels is shown in Fig. 13-1. The symbols $\epsilon_0, \epsilon_1, \epsilon_2, \ldots$ represent the different energies per particle that a particle may have when occupying the ground level, the first level, and so on. By occupying a level it is meant that a particle is considered to have the energy value associated with that particular level. In Fig. 13-1 the vertical distance of

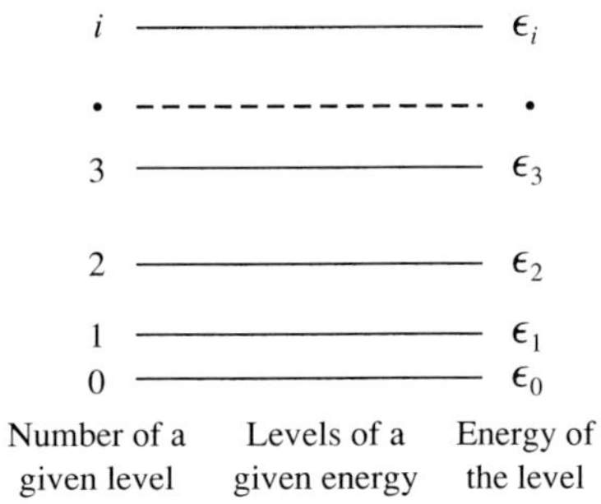

Figure 13-1
A discrete set of energy states.

any line above the baseline is proportional to the energy difference between that level and the ground level. The spacing between levels in Fig. 13-1 is shown to increase as the energy ϵ_i increases. This is in agreement with the analytical results obtained for an energy mode such as rotation. An energy mode such as vibration has equally spaced energy levels.

The microscopic picture of matter is one of an endless array of interactions or encounters between the various particles of the system. In the gas phase the particles move through the vessel undergoing encounters with other particles at the rate of roughly $10^{33}\ \mathrm{s}^{-1}/\mathrm{cm}^3$ under normal atmospheric pressure and temperature conditions. As a result of these encounters, energy is transferred between the interacting bodies. The mechanism of the energy transfer is not of concern. What is of vital interest is the fact that at each new instant of time, the microscopic representation of any isolated system has changed. The exchange of energy by molecular encounters is restrained only by the conditions that the resultant energy values associated with the various energy modes of the particles afterward satisfy the values allowed by quantum restrictions and that the total energy be conserved. Thus any system, internally in equilibrium, has a microscopic structure which leads to definite macroscopic properties such as pressure, temperature, specific volume, specific internal energy, and so on. However, a host of different descriptions on a particle basis will exist over a given time interval.

Consider an isolated system which contains n particles and has a total energy U. If the particles are considered independent of one another, then at any given time the energy of the ith molecule is given by ϵ_i. The total internal energy requirements of the system must be fulfilled by the relationship

$$U = \sum_i n_i \epsilon_i \tag{13-1}$$

where n_i represents the number of molecules whose energy is ϵ_i. The values of ϵ_i are, of course, restricted to those values allowed by quantum theory. The summation is taken over all the levels which may be expected to contain one or more particles. It has been noted already that the energy ϵ_i possessed by a particle is relative to some ground-level energy ϵ_0. For simplicity in the following examples the energy levels will be regarded as equally spaced and the value of the ground-level energy will be taken as zero. First, an isolated system will be chosen which contains four particles and a total energy of 6 units. Only one energy mode is considered, and the energy levels for this energy mode are spaced 1 unit apart, beginning at zero. The following question naturally arises: In what manner may particles be distributed among the various energy levels and still satisfy the condition that the total energy be 6 units? Obviously, only the first seven allowed energy levels must be taken into account, since higher-valued ones could not possibly be occupied. A summary of the possible distributions of particles among energy levels for the chosen example is given in Table 13-1. The nine possibilities appear in the vertical columns labeled A to I. The values in the horizontal rows give the number of particles of energy ϵ_1 for each distribution.

We shall define a *macrostate* as any possible microscopic state of an assembly of particles described in terms of the number of particles in each energy level at a given instant of time. The description of a macrostate does not require enumeration of which particles have a given energy, but only an accounting of how many particles have a given energy. In Table 13-1 the nine distributions labeled A to I will now be identified as the nine possible macrostates for a system composed of four particles and having a total energy of 6 units. Each of these macrostates satisfies the macroscopic description of the state of the system.

Further analysis of Table 13-1 leads to the concept of a microstate if we assume that the particles are distinguishable. Consider any one of the nine macrostates, for example, column A. Here three particles have a ground level of energy ϵ_0 and the remaining one is in the seventh quantum level of energy ϵ_6. Because of the random exchange of energy, the particle having 6 units of energy could be any one of the four particles in the system. Over a sufficient period of time macrostate A will occur a number of times. However, each of the four distinguishable particles will share in occupying the sixth level of energy for this particular macrostate. Thus the macrostate A really consists of four microstates, all of which are equally likely to occur over a sufficiently long period of time.

TABLE 13-1
The nine possible macrostates of four particles having a total energy of 6 units for equally spaced energy levels

Energy level	Distribution								
	A	*B*	*C*	*D*	*E*	*F*	*G*	*H*	*I*
ϵ_0	3	2	2	1	1	2			1
ϵ_1		1		2	1		3	2	
ϵ_2			1		1			2	3
ϵ_3					1	2	1		
ϵ_4			1	1					
ϵ_5		1							
ϵ_6	1								
W	4	12	12	12	24	6	4	6	4

A *microstate*, then, is a description of which particles have certain energies. Of course, we cannot specify which microstate exists at a given instant of time, or even which macrostate. Microstates do exist, however, and they are continually changing with time. The other distributions, B to I, in Table 13-1 likewise may be formed in a number of ways, depending on the placement of certain particles in various energy levels. The last horizontal row in the table, marked W, lists the number of microstates for each macrostate. We shall define the number of microstates W for a given macrostate as the *thermodynamic probability* of that macrostate. Unlike the usual mathematical probability which varies from zero to unity, the thermodynamic probability is never less than unity and is usually an extremely large number. The value of W in the case of Table 13-1 is obtained from a permutation formula

$$W = \frac{n!}{n_1!n_2!\cdots n_k!}$$

This equation gives the number of arrangements (microstates) possible for n items, where the identity of each item as residing in a particular group is recognized. However, the order of items within a group is immaterial. The symbols $n_1, n_2, \ldots$ represent the number of items in groups 1, 2, and so on. As an example, four particles are placed in three groups containing 2, 1, and 1 particles, respectively. For this situation the above equation for the possible arrangements gives

$$W = \frac{4!}{2!1!1!} = 12$$

Hence distributions B, C, and D all have 12 possible internal arrangements, or microstates. The other values of W in Table 13-1 are easily verified. It is seen from Table 13-1 that the total number of microstates is 84. Since all the microstates are assumed to occur with equal likelihood, over a long period of time each microstate would be formed $\frac{1}{84}$ of the time.

For a system of n particles and total energy U there is a finite fixed number of macrostates for each set of allowed energy levels. Since the number of macrostates is fixed in each case, the number of microstates is also set. The total number of microstates for a given thermodynamic state is represented by W_{tot}. This simply is the sum of the W values for each possible macrostate. The number of microstates W_{tot} thus is a unique and identifiable number (at least in theory) for every equilibrium state. If W_{tot} has a certain fixed value for each equilibrium state, it partially fulfills the concept of a property in at least one respect. It has one and only one value for each equilibrium state. If a system is carried through a cyclic process, the value of W_{tot} will necessarily be the same for the initial and final states of the cycle.

Although the value of W_{tot} is fixed for a given equilibrium state, it is not a useful measure of a thermodynamic property. If any property is to be tabulated and correlated with other properties, its value should vary in direct proportion to its mass; that is, it should be extensive in nature. The thermodynamic probability W is not an extensive variable. Probability theory states that the total arrangements of two independent systems taken together are the product of the arrangements of each individual group. In terms of thermodynamic probability, if system A can be arranged in three ways and system B in two ways, there are six possible arrangements that might occur when both systems are considered simultaneously. For two thermodynamic systems considered simultaneously $W_{AB} = W_A W_B$. Hence the property W is multiplicative in nature, not additive.

Although W itself is multiplicative, there exists a mathematical function of W such that one of its basic characteristics is its additivity. In fact, the only function of W which has this feature is a logarithmic function. We have already seen that for a composite system the value of W is

$$W_{AB} = W_A W_B$$

By taking the logarithm of each side, we find that

$$\log W_{AB} = \log W_A + \log W_B$$

Therefore a new thermodynamic property can be defined in terms of the thermodynamic probability of the system. This property is called the *statistical* entropy of the system, and is designated again by the symbol S. In general, S is related to W_{tot} by the definition,

$$S \equiv k \ln W_{\text{tot}} \qquad (13\text{-}2)$$

where k is a proportionality constant. [It is necessary to choose the logarithm to the base e in order to make the numerical values of the statistical entropy function agree with values calculated from macroscopic measurements.] This equation is known as the *Boltzmann-Planck equation*. Its general validity has been well established. The constant k, which is called the *Boltzmann constant* is given by

$$k = \frac{R_u}{N_A}$$

where R_u, again, is the universal gas constant, and N_A is Avogadro's number. In the SI system of units the value of k is 1.380×10^{-23} J/K · molecule.

The approach used in Secs. 13-2 and 13-3 to develop a statistical entropy function is based on the Boltzmann definition of entropy. It is applicable to isolated systems where each microstate is equally likely. A more general approach developed by Gibbs is in the form

$$S = -k \sum_i p_i \ln p_i \qquad (13\text{-}3)$$

where the p_i are the probabilities of individual quantum states which may not be equally likely in occurrence. If all quantum states are equally likely, then the number of microscopic states is W_{tot}, or $p_i = 1/W_{\text{tot}}$. In this case Eq. (13-3) becomes

$$\begin{aligned} S &= -k \sum_{i=1}^{W_{\text{tot}}} \frac{1}{W_{\text{tot}}} \ln \frac{1}{W_{\text{tot}}} = -k\, W_{\text{tot}} \left(\frac{1}{W_{\text{tot}}} \ln \frac{1}{W_{\text{tot}}} \right) \\ &= k \ln W_{\text{tot}} \end{aligned}$$

Hence Eq. (13-3) reduces to the Boltzmann format of Eq. (13-2) in this restricted case.

13-3 ENTROPY AND THE MOST PROBABLE MICROSTATE

The entropy function, as defined by the Boltzmann-Planck equation, is given by the relation $S = k \ln W_{\text{tot}}$. In practice, however, it is much more convenient to evaluate S by using a modified form of the Boltzmann-Planck equation. The validity of this modification can be justified by examining one further characteristic of systems on the microscopic level. Recall the example presented in Sec. 13-2 which illustrated the concepts of macrostates and microstates. The system considered had four distinguishable particles and a total energy of 6 units, which resulted in nine macrostates (see Table 13-1). With each macrostate there is associated a certain number of microstates. All microstates are tacitly assumed to be equally likely in occurrence in a random molecular system. Macrostate E has 24 possible microstates out of the 84 total. This is 28 percent of the total number of arrangements. The macrostate which has the largest number of microstates is defined as the *most probable macrostate*. Hence macrostate E is the most probable macrostate for the given system. Associated with every thermodynamic system is a macrostate which can be called the most probable one.

The most probable macrostate defines the energy distribution which would be observed more frequently than any other distribution. For thermodynamic systems where n is very large it turns out that this distribution is overwhelmingly more probable than any other. Therefore it is the characteristics of the most probable macrostate that determine the equilibrium thermodynamic properties. The most probable distribution is, in a sense, the equilibrium distribution. Although there are innumerable macrostates possible for a given thermodynamic state, experience dictates that all of them can be ignored, except the most probable one, in computing thermodynamic properties.

This situation is made clear when one studies two subsystems, each of which has an extremely large number of particles and the total shared energy is large. When n is very large, it should be realized that W for a given subsystem is an extremely rapidly increasing function of its energy. In this case since the total energy is shared, as subsystem X acquires an increasing amount of energy, the energy of Y decreases. Therefore W_X increases extremely rapidly while W_Y decreases extremely rapidly as X acquires an increasing share of the total energy. Let us represent the thermodynamic probability of the composite system by W_C. Since $W_C = W_X W_Y$, the product of W_X and W_Y exhibits an extremely sharp maximum at some value of U_X. Thus a plot of the probability of various microscopic descriptions versus U_X must also show the same sharp maximum. This state of maximum W_C is, of course, the most probable (*mp*) split in energy, and we might symbolize the energy of subsystem X in this state by $U_{X,mp}$. The general characteristics of a plot of p_i (the probability of a certain split in the shared energy) versus U_X is shown in Fig. 13-2. The region ΔU in the figure where p_i has a significant value is such that $\Delta U \ll U_{X,mp}$. We know that when two large

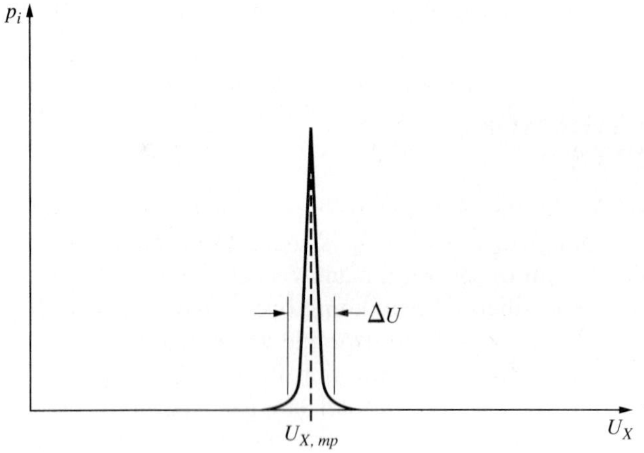

Figure 13-2
Probability of occurrence of subsystem X when its energy is U_X.

systems are in thermal equilibrium, they do not exhibit a widely varying energy for either one in a macroscopic sense. The probability that U_X will be appreciably different from $U_{X,mp}$ is negligible. On the microscopic level we see that wide variations are possible but highly improbable. Hence our instruments never detect any fluctuations. The most probable arrangement and those only slightly different from it in their arrangement predominate over all other possibilities.

This preceding discussion was based on two systems which shared a common quantity of energy. The same general conclusions can be drawn for a single isolated system of energy U. Although the energy is fixed, there is a huge number of macrostates that might exist. However, for any large system of particles, only a small fraction of the possible macrostates account for a large fraction of the total number of microstates. In addition, the characteristics of the energy distributions of this small number of macrostates are not very different from the most probable distribution. That is, the probability of occurrence of a macrostate falls off extremely fast for very small changes in the n_i values associated with the most probable macrostate. For example, consider a system of distinguishable particles. Small changes δn_i are made at various energy levels, starting with the system in the most probable macrostate. The value of W_{mp} is altered to a new value $(W_{mp} + \delta W)$. A mathematical analysis of this situation leads to the following approximation:

$$\frac{W_{mp} + \delta W}{W_{mp}} \approx \exp\left[-\frac{1}{2}\sum_i \frac{(\delta n_i)^2}{n_i}\right]$$

Note that whether the δn_i values are positive or negative, the new macrostate always has fewer arrangements than the most probable one. Hence the equation is consistent with the definition of the most probable macrostate. In most systems n_i will usually be large, such as 10^{15} to 10^{20}. Therefore even when δn_i is a fraction of a percent, the bracketed term is extremely large. This is due to the fact that the δn_i terms are all squared. Consequently, small changes in δn_i lead to a drastic reduction in the number of microstates when starting from the most probable macrostate. Obviously, only those macrostates which are nearly like the most probable one in their distribution of particles need to be considered. For this reason it is reasonable to base macroscopic properties solely on the particle distribution of the most probable macrostate.

If the most probable macrostate is a sufficient model, how is the concept of the entropy function affected, since it is defined on the basis of W_{tot} and not of W_{mp}? Actually, since W_{mp} has a fixed value for any equilibrium state, it could serve as the basis for the definition of the entropy function, as well as W_{tot}. However, it is not necessary to modify our definition of S as given by the Boltzmann-Planck equation. As a result of the mathematics used when dealing with systems of a large number of particles, it turns out that for all practical purposes $\ln W_{mp} = \ln W_{\text{tot}}$. For evaluating the entropy, then, we use the most probable macrostate to find W, rather than account for all macrostates. The reasonableness of this approximation is illustrated by the following example. It is necessary to limit the example to a study of relatively small systems, but the extrapolation to larger systems is quite obvious.

Table 13-2 lists the arrangement of distinguishable particles among three equally spaced energy levels for systems $x, y,$ and z. The specific macrostate shown in each case has been chosen to be the most probable one. The only difference among the systems is that the values of n and U have been varied. (Note that none of the systems is large enough to be considered macroscopic in size.) In each case the average energy per particle is exactly the same. For these three systems we wish to compare the values of W_{mp} to W_{tot} and $\ln W_{mp}$ to $\ln W_{\text{tot}}$. The results are summarized as shown. Note that the values of W_{mp} and W_{tot} are extremely large, even for these relatively small systems. As the number of particles increases, the ratio W_{mp}/W_{tot} becomes very small. More importantly, as the number of particles increases, it appears that $\log W_{mp}$ approaches $\log W_{\text{tot}}$. (*Note:* Logarithms to the base 10 have been used in the table for convenience. The ratio of logarithms would be the same regardless of the base used.) If the example were extrapolated to a system of 10^{23} particles, the ratio of logarithms would be unity for all practical purposes. Although we have used an example involving a system of distinguishable particles with a limited size in all cases, the conclusion reached is applicable to real systems as well. That is, $\ln W_{mp} = \ln W_{\text{tot}}$ to any desired accuracy if the system is large. Hence the evaluation of the entropy function by use of the Boltzmann-Planck equation can be made in terms of a modified equation,

$$S = k \ln W_{mp} \tag{13-4}$$

In evaluating the entropy function, we do not need to bother with the details of the possible macrostates that could possibly exist for a particular equilibrium state. Attention is focused simply on the most probable macrostate.

In summary, for macroscopic thermodynamic systems, the most probable macrostate and those nearly identical to it predominate. Therefore the macroscopic properties of a system can be based entirely on the microscopic characteristics of

TABLE 13-2
Comparison of W_{mp} versus W_{tot}

n	W_{mp}	W_{tot}	W_{mp}/W_{tot}	$\log_{10} W_{mp}$	$\log_{10} W_{tot}$	Log ratio
62	1.4×10^{14}	4.25×10^{16}	3.3×10^{-3}	14.154	16.623	0.851
310	5.95×10^{75}	1.13×10^{80}	5.25×10^{-5}	75.783	80.053	0.948
1550	4.4×10^{386}	5.7×10^{395}	7.7×10^{-10}	386.642	395.757	0.978

the most probable macrostate. Calculation of properties based on this approach will agree with experimental measurements. (Exceptions to this rule must be permitted, since in a few cases fluctuations from the mean are significant.) It is of particular interest that the entropy function can be correctly evaluated in terms of the Boltzmann-Planck equation by employing the logarithm of the thermodynamic probability associated with the most probable macrostate.

13-4 MICROSCOPIC IMPLICATIONS OF WORK AND HEAT

It is important that we examine interactions on a microscopic basis, especially with regard to internally reversible processes. We shall find that the effects of heat and work interactions on changes at the microscopic level are quite different.

In an earlier section it was noted that the discrete values of energy allowed for a given energy mode are set by the value of one of the generalized coordinates (displacements) of the system and by the nature of the substance. As an example, consider the translational kinetic energies of gas particles maintained in a box of volume V. The values of the ϵ_i's are found to be proportional to $1/V^{2/3}$. That is, the volume of the system determines the spacing of the energy levels for this particular form of energy. Other generalized displacements will determine the magnitude of the energy-level spacing for other energy modes. Consequently, the energy-level spacing for a certain energy mode will change only if the corresponding generalized displacement is altered. Let us consider a simple compressible system containing a monatomic gas. The volume is slowly reduced by a compression process. As the volume diminishes, the energy levels become further apart. For each energy level $\epsilon_i \rightarrow \epsilon_i + d\epsilon_i$, where $d\epsilon_i$ is positive for a compression process. Each level contains n_i particles. Therefore an increment of energy $n_i\, d\epsilon_i$ must be supplied at each level of energy. If the only energy added to all levels is due to boundary work; that is, if the process is adiabatic and internally reversible, then

$$\delta W_{\text{int rev}} = \sum_i n_i\, d\epsilon_i \qquad (13\text{-}5)$$

Internal reversibility is required, since any dissipative effects present (such as friction) would negate the statement that the generalized work interaction exactly equaled the change in the energy of the closed system. The above equation is not valid for nonquasistatic types of work interactions. For example, it has been pointed out that paddle-wheel effects, the flow of electric current through a resistor, and work done against frictional forces do not necessarily involve a change in a generalized displacement. Consequently, these effects do not contribute to the change in the ϵ_i values of the particles. It is also important to note that the n_i values do not change if the only interaction is the internally reversible change of a generalized displacement. The addition of energy to a closed system due to internally reversible work effects is not to be associated with a change in the particle distribution. The energy of each level changes with a constant population of particles for a given macrostate. Figure 13-3(*a*) illustrates a hypothetical change in volume of a system so that the energy spacing is doubled. Note that the n_i values are constant.

On a microscopic basis we have seen that $U = \sum_i n_i \epsilon_i$. Therefore for a differential change of state

$$dU = \sum_i n_i \, d\epsilon_i + \sum_i \epsilon_i \, dn_i \tag{13-6}$$

According to this equation, any change in the energy U must be associated with two effects: (1) the population of each energy level is held fixed while the energy values themselves change and (2) the values of the energy levels are held fixed while the distribution of particles among the levels changes. In addition, the conservation of energy principle for closed systems states that

$$dU = \delta Q + \delta W \tag{1-34}$$

For an internally reversible change of state $W = \sum_i n_i \, d\epsilon_i$. Therefore it follows from the above equations that the macroscopic heat effects during an internally reversible process bring about a microscopic change in the state of the system given by

$$\delta Q_{\text{int rev}} = \sum_i \epsilon_i \, dn_i \tag{13-7}$$

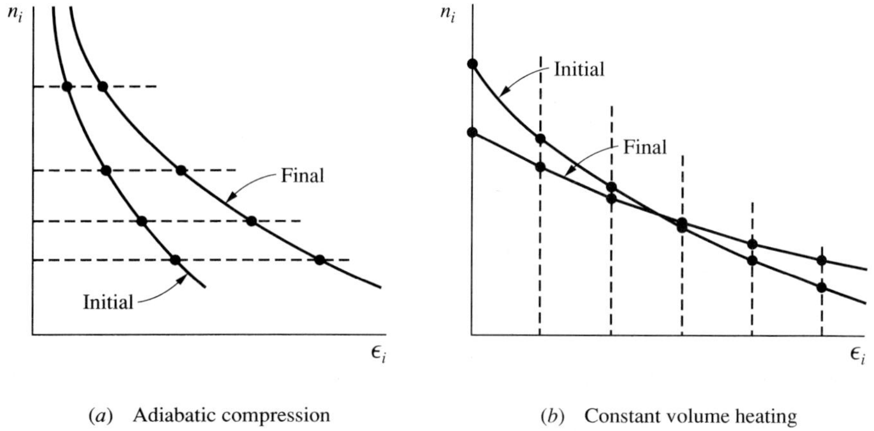

Figure 13-3
Shifting of ϵ_i or n_i values by work or heat effects during an internally reversible process.

Thus the population of each level changes by dn_i when a differential heating effect occurs. Since $\sum_i dn_i = 0$ for a closed system, some of the levels become more highly populated, whereas other levels decrease in their population during heat transfer. Figure 13-3(*b*) shows the hypothetical case of heat addition in the absence of work interactions.

13-5 ENTROPY CHANGES IN TERMS OF MACROSCOPIC VARIABLES

The entropy function S has been defined in terms of microscopic concepts. Now, it is desirable to relate the entropy function to macroscopic phenomena. For an infinitesimal process the change in the entropy of a closed system is found by differentiating Eq. (13-2). Because $\ln W_{\text{tot}} = \ln W_{mp}$ for a large system of particles, we may write

$$dS = k\, d(\ln W_{mp}) \tag{13-8}$$

That is, the entropy change is determined by the change in the number of arrangements of the most probable distribution as the equilibrium state is altered. Our goal is to relate this change in $\ln W_{mp}$ to macroscopically measurable quantities.

We have already noted in preceding examples of isolated systems that the thermodynamic probability W for any macrostate depends solely on the n_i values associated with each energy level. Since the logarithm of W would also depend solely on the n_i values for that macrostate, we may denote functionally that

$$\ln W = f(n_1, n_2, n_3, \ldots, n_i, \ldots)$$

If a differential change of state occurred, the change in $\ln W$ for any macrostate would be given by the total differential,

$$d \ln W = \frac{\partial \ln W}{\partial n_1} dn_1 + \frac{\partial \ln W}{\partial n_2} dn_2 + \cdots = \sum_i \frac{\partial \ln W}{\partial n_i}$$

This equation is valid for any macrostate, including the most probable macrostate. Thus we may write specifically that

$$d \ln W_{mp} = \sum_i \frac{\partial \ln W_{mp}}{\partial n_i} dn_i \tag{13-9}$$

The change in the entropy function for an infinitesimal change of state between two equilibrium states then becomes, if we substitute Eq. (13-9) into Eq. (13-8),

$$dS = k \sum_i \frac{\partial \ln W_{mp}}{\partial n_i} dn_i \tag{13-10}$$

The evaluation of the quantity $\partial \ln W_{mp}/\partial n_i$ is fairly straightforward, although there is a considerable amount of unavoidable mathematics. Since the most probable macrostate is, by definition, the one with the largest number of arrangements, we may approach the problem as one requiring the maximization of a function through the techniques of differential calculus. One additional complication must be considered, however. In addition to the general equation relating the dependent variable, $\ln W$, to the independent variables, the n_i's, there are two more equations relating the n_i's which must be satisfied during the maximization process. These are

$$\sum_i n_i = n \tag{13-11}$$

and

$$\sum_i \epsilon_i n_i = U \tag{13-1}$$

The problem we have, then, is one of maximization in the presence of constraining equations. The most general method of solution is one developed by Lagrange.

If the constraining equations cited above are differentiated, holding the ϵ_i value constant during the differentiation, we find that

$$dn_1 + dn_2 + \cdots + dn_i = \sum_i dn_i = 0 \tag{13-12}$$

and

$$\epsilon_1\, dn_1 + \epsilon_2\, dn_2 + \cdots + \epsilon_i\, dn_i = \sum_i \epsilon_i\, dn_i = 0 \tag{13-13}$$

Maximization of the entropy requires that Eq. (13-8) be written as

$$\sum_i \frac{\partial \ln W_{mp}}{\partial n_i}\, dn_i = 0 \tag{13-14}$$

The method of Lagrange's undetermined multipliers requires first that Eqs. (13-12) and (13-13) each be multiplied by different constants (chosen to be A and B in this case). Then the resulting equations are subtracted from Eq. (13-14). The result is

$$\sum_i \left(\frac{\partial \ln W_{mp}}{\partial n_i} - A - B\epsilon_i \right) dn_i = 0 \tag{13-15}$$

Since the variations dn_i on each energy level are arbitrary and independent, Eq. (13-15) is only satisfied if the coefficients of dn_i equal zero. (A more detailed reasoning is found in books on statistical thermodynamics.) For any energy level where dn_i particles are shifted when the system changes from one equilibrium state to another, maximization of the statistical entropy function requires that

$$\frac{\partial \ln W_{mp}}{\partial n_i} = A + B\epsilon_i \tag{13-16}$$

The two quantities A and B represent two constants, as yet undetermined. This relationship is the one we sought in order to evaluate dS through the use of Eq. (13-10). If we combine Eqs. (13-10) and (13-16), then

$$\begin{aligned} dS &= k \sum_i \frac{\partial \ln W_{mp}}{\partial n_i}\, dn_i = k \sum_i (A + B\epsilon_i)\, dn_i \\ &= kA \sum_i dn_i + kB \sum_i \epsilon_i\, dn_i \end{aligned} \tag{13-17}$$

The terms A and B have been brought outside the summation signs, since they are constants. The first summation on the right-hand side of Eq. (13-17) is zero on the basis of Eq. (13-12). Consequently, Eq. (13-17) reduces to the form

$$dS = kB \sum_i \epsilon_i \, dn_i \tag{13-18}$$

Equation (13-18) represents the change in entropy due to a differential change of state between two equilibrium states, expressed in terms of the shifting of particles among the energy levels of the most probable macrostate. The equation is completely general in this form. The change of state is not restricted to any particular type of process and may involve both heat and work interactions.

Equation (13-18) contains two quantities that must be replaced by their counterparts in macroscopic thermodynamics, namely, kB and $\sum_i \epsilon_i \, dn_i$. The first of these is found by considering two simple compressible control masses X and Y, which are brought into thermal contact through a rigid wall. The composite system $X + Y$ is isolated from its environment. Thus the volume of each system remains constant, and no energy transfer as work occurs. If the two systems are not initially in mutual equilibrium, a heat interaction will occur between systems X and Y until a new equilibrium state is reached.

During the process the change in entropy for each simple compressible system can be expressed as $S = S(U, V)$. For a differential change of state

$$dS = \left(\frac{\partial S}{\partial U}\right)_V dU + \left(\frac{\partial S}{\partial V}\right)_U dV$$

In our particular example the volume is held fixed on each system, and therefore the second term on the right is zero for each system. Consequently, the total entropy change for the composite system C may be written as

$$dS_C = dS_X + dS_Y = \left(\frac{\partial S_X}{\partial U_X}\right)_{V_X} dU_X + \left(\frac{\partial S_Y}{\partial U_Y}\right)_{V_Y} dU_Y \tag{13-19}$$

As the composite system approaches equilibrium, $dS_C = 0$. The variation of the composite entropy with energy U_X is shown in Fig. 13-4. In addition, the total energy of the composite system is a constant given by $U_C = U_X + U_Y$. Thus $dU_X = -dU_Y$. These two facts in combination with Eq. (13-14) indicate that at the state of equilibrium

$$\left(\frac{\partial S_X}{\partial U_X}\right)_{V_X} = \left(\frac{\partial S_Y}{\partial U_Y}\right)_{V_Y} \tag{13-20}$$

The derivatives expressed in this equation are properties of the systems X and Y. However, the only property of each system that necessarily must be the same at the state of thermal equilibrium is the temperature. Therefore it is to be expected that the partial derivative $(\partial S/\partial U)_V$ is related in some way to the temperature of the system. The dimensions of the partial derivative give a clue to this relationship. The dimensions of S are energy/(absolute temperature). When divided by the

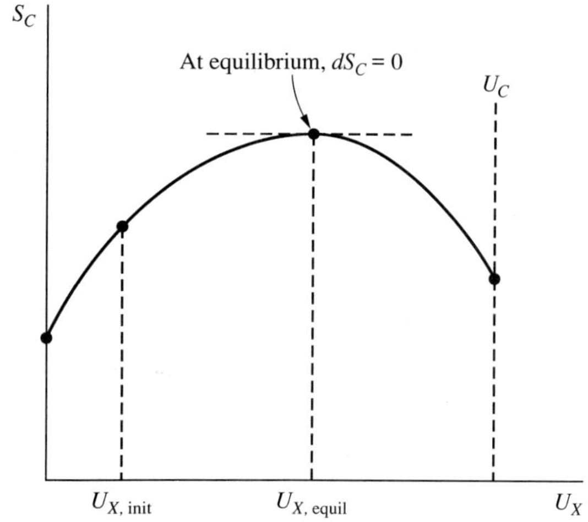

Figure 13-4
Variation of composite entropy with energy shared by systems X and Y.

dimension of U, which is energy, we see that the dimension of the partial derivative is the reciprocal of the absolute temperature, T^{-1}. Hence it is appropriate to define a thermodynamic temperature which is the reciprocal of the derivative found in Eq. (13-20). Consequently,

$$T \equiv \frac{1}{(\partial S/\partial U)_V} \qquad \text{or} \qquad \left(\frac{\partial S}{\partial U}\right)_V = \frac{1}{T} \tag{13-21}$$

It can be shown that this thermodynamic temperature is the same as that given by the ideal-gas temperature scale.

As a final step, the partial derivative in Eq. (13-21) can be related to the constant B. Recall that, in general, the change in U is given by $dU = \sum_i \epsilon_i \, dn_i + \sum_i n_i \, d\epsilon_i$. In our example involving systems X and Y the volume of each is constant. Hence the term $\sum_i n_i \, d\epsilon_i$ is zero in the expression of dU, since a change in ϵ depends on a change in V. As a result, we find that $dU_V = \sum_i \epsilon_i \, dn_i$. If we substitute this result into Eq. (13-18) for dS, then for this specific example

$$dS = kB \, dU \qquad \text{(at constant volume)}$$

or

$$\left(\frac{\partial S}{\partial U}\right)_V = kB \tag{13-22}$$

If Eqs. (13-21) and (13-22) are now equated, we find that

$$kB = \frac{1}{T} \qquad \text{or} \qquad B = \frac{1}{kT} \tag{13-23}$$

Thus the constant B which appeared originally as one of the constants in Eq. (13-16) is inversely proportional to the thermodynamic temperature or ideal-gas temperature, in degrees kelvin. Although derived for a special process, Eq. (13-23) is of general validity. (Another approach is to define $1/kB$ as the *statistical* definition of temperature. Then one can show that this temperature is identical with the classical definition of temperature.)

Direct substitution of the above relations into Eq. (13-18) leads to the expression

$$dS = \frac{1}{T}\sum_i \epsilon_i \, dn_i \tag{13-24}$$

All that remains now is the replacement of the summation term by an equivalent macroscopic quantity. In Sec. 13-4 it was demonstrated that a heat interaction affects the number of particles associated with each energy level if the process is internally reversible. That is,

$$\delta Q_{\text{int rev}} = \sum_i \epsilon_i \, dn_i \tag{13-7}$$

If this relation is substituted into Eq. (13-22), we obtain the familiar result that

$$dS = \left(\frac{\delta Q}{T}\right)_{\text{int rev}} \tag{13-25}$$

Thus statistical thermodynamics leads to the classical expression for the entropy change as a function of δQ and T during internally reversible processes.

13-6 THE SECOND AND THIRD LAWS REVISITED

On the basis of microscopic concepts entropy has been defined in terms of the number of microscopic arrangements, the thermodynamic probability W of an isolated system of given energy U and volume V. There is a certain value of W_{tot} for each equilibrium state of a macroscopic system, and therefore a unique value of S associated with each equilibrium state. As any system undergoes a change of state, it generally is expected that W_{tot} will change, and thus that the system will have a new value for its entropy. There is also the possibility, however, that W_{tot} may not change in some cases. Hence the chance that the entropy will not be altered by some changes of state should not be ruled out. This would simply mean that the total number of arrangements did not change from one equilibrium state to another. Two examples are now presented which illustrate the behavior of W_{tot} when a constraint is removed on an isolated system.

First, consider the free expansion of a gas from one-half an insulated, rigid vessel into the other half of the vessel, which is initially evacuated. If the inside of the entire vessel walls is chosen as the boundary of the system, then the system is an isolated one, since both Q and W are zero and no mass crosses the boundaries. All changes of state occur within the vessel, and the environment is unaffected. On the basis of the conservation of energy principle the energy of the gas does not change during the expansion process (when the internal partition is removed).

However, the values of the permitted energy levels for the particles do change. As noted earlier, for a simple compressible substance the energy levels for the translational kinetic energy of gases are inversely proportional to the volume to the two-thirds power. When gas fills the entire vessel, the energy spacings $\Delta\epsilon$ are somewhat smaller than before. Thus we have the situation of a system of n particles with fixed energy U, where the particles now redistribute themselves over allowed energy levels which are more closely spaced. It should be apparent that in this situation the value of W_{tot} must increase. That is, there are many more ways to distribute a given amount of energy among a given number of particles if the energy spacings decrease. For example, recall the data from Table 13-1, where four distinguishable particles having a total energy of 6 units were distributed over energy levels 1 unit apart. Nine macrostates resulted. If the energy levels become $\frac{1}{2}$ unit apart (due to a volume change), 30 macrostates would be possible. The important result is that the change in the total number of arrangements, and hence the entropy change, during a free expansion is always *positive*.

As a final example, consider an isolated system consisting of two subsystems, X and Y. Initially, they are isolated from each other by means of a rigid, thermally insulating partition. The partition acts as a constraint on the energy of each subsystem. If the total energy of the composite isolated system is $U_C = U_X + U_Y$, then the energy of X must remain constant, and likewise for subsystem Y. Associated with each subsystem are thermodynamic probabilities W_X and W_Y, and for the composite isolated system $W_C = W_X W_Y$. The two subsystems are now brought into thermal contact through the rigid partition. Since the volume of each subsystem is fixed, no boundary work is possible. However, an energy exchange in the form of heat is permitted. Now the total energy U_C is shared by the two subsystems. On a microscopic basis there is no real limitation to the manner in which the energy is shared between X and Y. As the energy of each subsystem varies, the value for W for each subsystem changes.

As an illustrative example, let $\epsilon_0 = 0$ in both subsystems, with $\Delta\epsilon = 1$. Since the volumes of X and Y are constant, the energy-level values do not change for gases contained within the subsystems. Subsystem X will contain four distinguishable particles, subsystem Y will contain three distinguishable particles, and the total energy shared by the two subsystems will be 8 units. Table 13-3 summarizes the various possible microscopic descriptions. No matter what the original split in the total energy had been, the final value of 3003 for W_{tot} would have been much larger than the initial value once the constraint between the two subsystems was removed. Apparently the removal of a thermal constraint so that an isolated system changes to a new equilibrium state always leads to an increase in W_{tot}, and hence to an increase in the entropy of the isolated system.

(The preceding example simply illustrates the point that the removal of a thermal constraint brings about an increase in the thermodynamic probability of an isolated system. However, even though the total energy may now be divided between the two subsystems in nine different ways, we would not expect to find the energy of either subsystem to vary significantly at equilibrium if the subsystems are relatively large. That is, at macroscopic equilibrium all the thermodynamic properties, including each subsystem's energy, should be constant on the basis

TABLE 13-3
Microstates for two systems in thermal contact

U_X	U_Y	W_X	W_Y	W_C	P_I
0	8	1	45	45	0.015
1	7	4	36	144	0.048
2	6	10	28	280	0.093
3	5	20	21	420	0.140
4	4	35	15	525	0.175
5	3	56	10	560	0.186
6	2	84	6	504	0.168
7	1	120	3	360	0.120
8	0	165	1	165	0.055
				3003	1.000

of experimental measurements. Therefore many of the possible splits in energy never occur, even though they are theoretically possible.)

The preceding discussion can be summarized in the following way. If a constraint of some type is removed from an isolated system initially in equilibrium, the number of internal arrangements W_{tot} will either increase or remain the same. That is, $W_{\text{tot,final}} \geq W_{\text{tot,initial}}$. Although in all our examples W_{tot} increased, the possibility that the thermodynamic probability of an isolated system will not change must be included. Experience simply dictates that the number of arrangements never decreases. It is this characteristic that makes the second law a *directional* or *limiting* law. Thus on the basis of the Boltzmann-Planck equation the entropy of an isolated system never decreases. Hence the microscopic interpretation of the behavior of the entropy function for an isolated system is in agreement with classical thermodynamics.

In Chap. 7 the third law was stated in terms of the entropy change of an internally reversible, isothermal change of state at absolute zero temperature. Because this change in entropy is zero, a convention has been established in macroscopic thermodynamics to set the entropy of pure crystalline substances as zero at 0 K. As a substance approaches 0 K upon cooling, the total energy of the system decreases. From a microscopic viewpoint the particles now have energies associated only with the ground level and a few energy levels above it. Upon further cooling every system approaches the lowest possible energy U_0 of its ground state. Every particle would have the ground-level energy ϵ_0. Since Boltzmann statistics for an isolated system assumes equally probable states, there is only one macrostate for the system at absolute zero temperature. Since $W_{\text{tot}} = 1$, the Boltzmann equation yields $S = 0$. (This analysis also assumes that the ground level is not degenerate; that is, there are not several microscopic ground-level states of energy ϵ_0.) Thus the Boltzmann equation for entropy establishes a value at 0 K which is the same as that chosen conventionally in classical thermodynamics.

APPENDIX A

SUPPLEMENTARY TABLES AND FIGURES

TABLE A-1
Physical constants

Physical constants	
Avogadro's number	$N_A = 6.023 \times 10^{26}$ atoms/kgmol
Universal gas constant	$R_u = 0.08205$ L · atm/(gmol · K)
	$= 8.314$ kJ/(kgmol · K)
	$= 0.08314$ bar · m^3/(kgmol · K)
	$= 8.314$ kPa · m^3/(kgmol · K)
Planck's constant	$h = 6.626 \times 10^{-34}$ J · s/molecule
Boltzmann's constant	$k = 1.380 \times 10^{-23}$ J/(K · molecule)
Speed of light	$c = 2.988 \times 10^{10}$ cm/s
Standard gravity	$g = 9.80665$ m/s^2

TABLE A-2

Molar mass, critical constants, gas-phase specific heats, and acentric factor at 25°C and 1 atm for some common substances

T_c, K; P_c, bars; v_c, m^3/kmol; c_v and c_p, kJ/(kg·K)

Substance	Molar mass	T_c	P_c	v_c	Z_c	c_v	c_p	ω
Acetylene (C_2H_2)	26.04	308.3	61.4	0.113	0.270	1.37	1.69	0.190
Air (equivalent)	28.97	133	37.7	0.0829	0.284	0.718	1.005	
Ammonia (NH_3)	17.04	405.5	113.5	0.0723	0.244	1.66	2.15	0.250
Benzene (C_6H_6)	78.11	562.2	48.9	0.259	0.271	0.67	0.775	0.212
n-Butane (C_4H_{10})	58.12	425.2	38.0	0.255	0.274	1.56	1.71	0.199
Carbon dioxide (CO_2)	44.02	304.2	73.8	0.0941	0.274	0.657	0.846	0.224
Carbon monoxide (CO)	28.01	132.9	35.0	0.0928	0.294	0.744	1.04	0.049
Refrigerant-134a ($C_2F_4H_2$)	102.03	374.3	40.6	0.201	0.262	0.76	0.85	0.326
Ethane (C_2H_6)	30.07	305.4	48.8	0.148	0.285	1.48	1.75	0.099
Ethanol (C_2H_5OH)	46.07	513.9	61.4	0.167	0.240	1.26	1.43	0.644
Ethylene (C_2H_4)	28.05	282.3	50.4	0.130	0.280	1.23	1.53	0.089
Helium (He)	4.003	5.2	2.3	0.0579	0.302	3.12	5.19	−0.365
Hydrogen (H_2)	2.016	33.2	13.0	0.0648	0.306	10.2	14.3	−0.218
Methane (CH_4)	16.04	190.6	46.0	0.0991	0.288	1.70	2.22	0.008
Methanol (CH_3OH)	32.04	512.6	80.9	0.118	0.224	1.14	1.37	0.556
Nitrogen (N_2)	28.01	126.2	33.9	0.0897	0.290	0.743	1.04	0.039
Nitric oxide (NO)	30.01	180	64.8	0.0578	0.250	0.82	1.10	0.588
Oxygen (O_2)	32.00	154.6	50.4	0.0734	0.288	0.658	0.918	0.025
Propane (C_3H_8)	44.09	369.8	42.5	0.203	0.281	1.48	1.67	0.153
Sulfur dioxide (SO_2)	64.06	431	78.8	0.122	0.269	0.471	0.601	0.256
Water (H_2O)	18.02	647.3	220.9	0.0558	0.233	1.40	1.86	0.344

Sources: Principally from D. Ambrose, "Vapor-Liquid Critical Properties," National Physical Laboratory Report 107, 1980, and R. C. Reid et al., *The Properties of Gases and Liquids*, 4th ed., McGraw-Hill, Inc., New York, 1987.

TABLE A-3
Specific heats of some common liquids and solids c_p, kJ/(kg·°C)

A. Liquids

Substance	State	c_p	Substance	State	c_p
Water	1 atm, 273 K	4.217	Benzene	1 atm, 15°C	1.80
	1 atm, 280 K	4.198		1 atm, 65°C	1.92
	1 atm, 300 K	4.179	Glycerin	1 atm, 10°C	2.32
	1 atm, 320 K	4.180		1 atm, 50°C	2.58
	1 atm, 340 K	4.188	Mercury	1 atm, 10°C	0.138
	1 atm, 360 K	4.203		1 atm, 315°C	0.134
	1 atm, 373 K	4.218	Sodium	1 atm, 95°C	1.38
Ammonia	sat., −20°C	4.52		1 atm, 540°C	1.26
	sat., 50°C	5.10	Propane	1 atm, 0°C	2.41
Refrigerant-12	sat., −40°C	0.883	Bismuth	1 atm, 425°C	0.144
	sat., −20°C	0.908		1 atm, 760°C	0.164
	sat., 50°C	1.02	Ethyl alcohol	1 atm, 25°C	2.43

B. Solids

Substance	Temp.	c_p	Substance	Temp.	c_p
Ice	200 K	1.56	Silver	20°C	0.233
	220 K	1.71		200°C	0.243
	240 K	1.86	Lead	−173°C	0.118
	260 K	2.01		−50°C	0.126
	270 K	2.08		27°C	0.129
	273 K	2.11		100°C	0.131
Aluminum	200 K	0.797		200°C	0.136
	250 K	0.859	Copper	−173°C	0.254
	300 K	0.902		−100°C	0.342
	350 K	0.929		−50°C	0.367
	400 K	0.949		0°C	0.381
	450 K	0.973		27°C	0.386
	500 K	0.997		100°C	0.393
Iron	293 K	0.448		200°C	0.403

TABLE A-4
Ideal-gas specific-heat data for selected gases, kJ/(kg·K)

1. Zero-pressure specific heats for six common gases, where $k = c_p/c_v$

Temp., K	c_p	c_v	k	c_p	c_v	k	c_p	c_v	k	Temp., K
	Air			Carbon dioxide (CO_2)			Carbon monoxide (CO)			
250	1.003	0.716	1.401	0.791	0.602	1.314	1.039	0.743	1.400	250
300	1.005	0.718	1.400	0.846	0.657	1.288	1.040	0.744	1.399	300
350	1.008	0.721	1.398	0.895	0.706	1.268	1.043	0.746	1.398	350
400	1.013	0.726	1.395	0.939	0.750	1.252	1.047	0.751	1.395	400
450	1.020	0.733	1.391	0.978	0.790	1.239	1.054	0.757	1.392	450
500	1.029	0.742	1.387	1.014	0.825	1.229	1.063	0.767	1.387	500
550	1.040	0.753	1.381	1.046	0.857	1.220	1.075	0.778	1.382	550
600	1.051	0.764	1.376	1.075	0.886	1.213	1.087	0.790	1.376	600
650	1.063	0.776	1.370	1.102	0.913	1.207	1.100	0.803	1.370	650
700	1.075	0.788	1.364	1.126	0.937	1.202	1.113	0.816	1.364	700
750	1.087	0.800	1.359	1.148	0.959	1.197	1.126	0.829	1.358	750
800	1.099	0.812	1.354	1.169	0.980	1.193	1.139	0.842	1.353	800
900	1.121	0.834	1.344	1.204	1.015	1.186	1.163	0.866	1.343	900
1000	1.142	0.855	1.336	1.234	1.045	1.181	1.185	0.888	1.335	1000
	Hydrogen (H_2)			Nitrogen (N_2)			Oxygen (O_2)			
250	14.051	9.927	1.416	1.039	0.742	1.400	0.913	0.653	1.398	250
300	14.307	10.183	1.405	1.039	0.743	1.400	0.918	0.658	1.395	300
350	14.427	10.302	1.400	1.041	0.744	1.399	0.928	0.668	1.389	350
400	14.476	10.352	1.398	1.044	0.747	1.397	0.941	0.681	1.382	400
450	14.501	10.377	1.398	1.049	0.752	1.395	0.956	0.696	1.373	450
500	14.513	10.389	1.397	1.056	0.759	1.391	0.972	0.712	1.365	500
550	14.530	10.405	1.396	1.065	0.768	1.387	0.988	0.728	1.358	550
600	14.546	10.422	1.396	1.075	0.778	1.382	1.003	0.743	1.350	600
650	14.571	10.447	1.395	1.086	0.789	1.376	1.017	0.758	1.343	650
700	14.604	10.480	1.394	1.098	0.801	1.371	1.031	0.771	1.337	700
750	14.645	10.521	1.392	1.110	0.813	1.365	1.043	0.783	1.332	750
800	14.695	10.570	1.390	1.121	0.825	1.360	1.054	0.794	1.327	800
900	14.822	10.698	1.385	1.145	0.849	1.349	1.074	0.814	1.319	900
1000	14.983	10.859	1.380	1.167	0.870	1.341	1.090	0.830	1.313	1000

Source: Data adapted from *Tables of Thermal Properties of Gases*, NBS Circular 564, 1955.

TABLE A-4 (*Continued*)

2. Specific-heat data for monatomic gases

Over a wide range of temperatures at low pressures, the specific heats c_v and c_p of all monatomic gases are essentially independent of temperature and pressure. In addition, on a molar basis all monatomic gases have the same value for either c_v or c_p in a given set of units. One set of values is

$$c_v = 12.5 \text{ kJ/(kgmol} \cdot \text{K)} \qquad \text{and} \qquad c_p = 20.8 \text{ kJ/(kgmol} \cdot \text{K)}$$

Source: Data adapted from *Tables of Thermal Properties of Gases*, NBS Circular 564, 1955.

3. Constant-pressure specific-heat equations for various gases at zero pressure (SI units)

$$\frac{c_p}{R_u} = a + bT + cT^2 + dT^3 + eT^4$$

where T is in kelvins, equation valid from 300 to 1000 K

Gas	a	$b \times 10^3$	$c \times 10^6$	$d \times 10^9$	$e \times 10^{12}$
CO	3.710	−1.619	3.692	−2.032	0.240
CO_2	2.401	8.735	−6.607	2.002	
H_2	3.057	2.677	−5.180	5.521	−1.812
H_2O	4.070	−1.108	4.152	−2.964	0.807
O_2	3.626	−1.878	7.056	−6.764	2.156
N_2	3.675	−1.208	2.324	−0.632	−0.226
Air (dry)	3.653	−1.334	3.291	−1.910	0.275
NH_3	3.591	0.494	8.345	−8.383	2.730
NO	4.046	−3.418	7.982	−6.114	1.592
NO_2	3.459	2.065	6.687	−9.556	3.620
SO_2	3.267	5.324	0.684	−5.281	2.559
SO_3	2.578	14.556	−9.176	−0.792	1.971
CH_4	3.826	−3.979	24.558	−22.733	6.963
C_2H_2	1.410	19.057	−24.501	16.391	−4.135
C_2H_4	1.426	11.383	7.989	−16.254	6.749

Source: Adapted from the data in NASA SP-273, Government Printing Office, Washington, 1971.

TABLE A-5

Ideal-gas properties of air

T, K; h, kJ/kg; u, kJ/kg; s^o, kJ/(kg·K); P^o = 1 atm

T	h	p_r	u	v_r	s^o	T	h	p_r	u	v_r	s^o
200	199.97	0.3363	142.56	1707.	1.29559	460	462.02	6.245	329.97	211.4	2.13407
210	209.97	0.3987	149.69	1512.	1.34444	470	472.24	6.742	337.32	200.1	2.15604
220	219.97	0.4690	156.82	1346.	1.39105	480	482.49	7.268	344.70	189.5	2.17760
230	230.02	0.5477	164.00	1205.	1.43557	490	492.74	7.824	352.08	179.7	2.19876
240	240.02	0.6355	171.13	1084.	1.47824	500	503.02	8.411	359.49	170.6	2.21952
250	250.05	0.7329	178.28	979.	1.51917	510	513.32	9.031	366.92	162.1	2.23993
260	260.09	0.8405	185.45	887.8	1.55848	520	523.63	9.684	374.36	154.1	2.25997
270	270.11	0.9590	192.60	808.0	1.59634	530	533.98	10.37	381.84	146.7	2.27967
280	280.13	1.0889	199.75	738.0	1.63279	540	544.35	11.10	389.34	139.7	2.29906
285	285.14	1.1584	203.33	706.1	1.65055	550	554.74	11.86	396.86	133.1	2.31809
290	290.16	1.2311	206.91	676.1	1.66802	560	565.17	12.66	404.42	127.0	2.33685
295	295.17	1.3068	210.49	647.9	1.68515	570	575.59	13.50	411.97	121.2	2.35531
300	300.19	1.3860	214.07	621.2	1.70203	580	586.04	14.38	419.55	115.7	2.37348
305	305.22	1.4686	217.67	596.0	1.71865	590	596.52	15.31	427.15	110.6	2.39140
310	310.24	1.5546	221.25	572.3	1.73498	600	607.02	16.28	434.78	105.8	2.40902
315	315.27	1.6442	224.85	549.8	1.75106	610	617.53	17.30	442.42	101.2	2.42644
320	320.29	1.7375	228.42	528.6	1.76690	620	628.07	18.36	450.09	96.92	2.44356
325	325.31	1.8345	232.02	508.4	1.78249	630	638.63	19.84	457.78	92.84	2.46048
330	330.34	1.9352	235.61	489.4	1.79783	640	649.22	20.64	465.50	88.99	2.47716
340	340.42	2.149	242.82	454.1	1.82790	650	659.84	21.86	473.25	85.34	2.49364
350	350.49	2.379	250.02	422.2	1.85708	660	670.47	23.13	481.01	81.89	2.50985
360	360.58	2.626	257.24	393.4	1.88543	670	681.14	24.46	488.81	78.61	2.52589
370	370.67	2.892	264.46	367.2	1.91313	680	691.82	25.85	496.62	75.50	2.54175
380	380.77	3.176	271.69	343.4	1.94001	690	702.52	27.29	504.45	72.56	2.55731
390	390.88	3.481	278.93	321.5	1.96633	700	713.27	28.80	512.33	69.76	2.57277
400	400.98	3.806	286.16	301.6	1.99194	710	724.04	30.38	520.23	67.07	2.58810
410	411.12	4.153	293.43	283.3	2.01699	720	734.82	32.02	528.14	64.53	2.60319
420	421.26	4.522	300.69	266.6	2.04142	730	745.62	33.72	536.07	62.13	2.61803
430	431.43	4.915	307.99	251.1	2.06533	740	756.44	35.50	544.02	59.82	2.63280
440	441.61	5.332	315.30	236.8	2.08870	750	767.29	37.35	551.99	57.63	2.64737
450	451.80	5.775	322.62	223.6	2.11161	760	778.18	39.27	560.01	55.54	2.66176

T	h	p_r	u	v_r	s^o	T	h	p_r	u	v_r	s^o
780	800.03	43.35	576.12	51.64	2.69013	1360	1467.49	399.1	1077.10	9.780	3.32724
800	821.95	47.75	592.30	48.08	2.71787	1380	1491.44	424.2	1095.26	9.337	3.34474
820	843.98	52.59	608.59	44.84	2.74504	1400	1515.42	450.5	1113.52	8.919	3.36200
840	866.08	57.60	624.95	41.85	2.77170	1420	1539.44	478.0	1131.77	8.526	3.37901
860	888.27	63.09	641.40	39.12	2.79783	1440	1563.51	506.9	1150.13	8.153	3.39586
880	910.56	68.98	657.95	36.61	2.82344	1460	1587.63	537.1	1168.49	7.801	3.41247
900	932.93	75.29	674.58	34.31	2.84856	1480	1611.79	568.8	1186.95	7.468	3.42892
920	955.38	82.05	691.28	32.18	2.87324	1500	1635.97	601.9	1205.41	7.152	3.44516
940	977.92	89.28	708.08	30.22	2.89748	1520	1660.23	636.5	1223.87	6.854	3.46120
960	1000.55	97.00	725.02	28.40	2.92128	1540	1684.51	672.8	1242.43	6.569	3.47712
980	1023.25	105.2	741.98	26.73	2.94468	1560	1708.82	710.5	1260.99	6.301	3.49276
1000	1046.04	114.0	758.94	25.17	2.96770	1580	1733.17	750.0	1279.65	6.046	3.50829
1020	1068.89	123.4	776.10	23.72	2.99034	1600	1757.57	791.2	1298.30	5.804	3.52364
1040	1091.85	133.3	793.36	22.39	3.01260	1620	1782.00	834.1	1316.96	5.574	3.53879
1060	1114.86	143.9	810.62	21.14	3.03449	1640	1806.46	878.9	1335.72	5.355	3.55381
1080	1137.89	155.2	827.88	19.98	3.05608	1660	1830.96	925.6	1354.48	5.147	3.56867
1100	1161.07	167.1	845.33	18.896	3.07732	1680	1855.50	974.2	1373.24	4.949	3.58335
1120	1184.28	179.7	862.79	17.886	3.09825	1700	1880.1	1025	1392.7	4.761	3.5979
1140	1207.57	193.1	880.35	16.946	3.11883	1750	1941.6	1161	1439.8	4.328	3.6336
1160	1230.92	207.2	897.91	16.064	3.13916	1800	2003.3	1310	1487.2	3.944	3.6684
1180	1254.34	222.2	915.57	15.241	3.15916	1850	2065.3	1475	1534.9	3.601	3.7023
1200	1277.79	238.0	933.33	14.470	3.17888	1900	2127.4	1655	1582.6	3.295	3.7354
1220	1301.31	254.7	951.09	13.747	3.19834	1950	2189.7	1852	1630.6	3.022	3.7677
1240	1324.93	272.3	968.95	13.069	3.21751	2000	2252.1	2068	1678.7	2.776	3.7994
1260	1348.55	290.8	986.90	12.435	3.23638	2050	2314.6	2303	1726.8	2.555	3.8303
1280	1372.24	310.4	1004.76	11.835	3.25510	2100	2377.4	2559	1775.3	2.356	3.8605
1300	1395.97	330.9	1022.82	11.275	3.27345	2150	2440.3	2837	1823.8	2.175	3.8901
1320	1419.76	352.5	1040.88	10.747	3.29160	2200	2503.2	3138	1872.4	2.012	3.9191
1340	1443.60	375.3	1058.94	10.247	3.30959	2250	2566.4	3464	1921.3	1.864	3.9474

Source: Adapted from J. H. Keenan and J. Kaye, *Gas Tables*, Wiley, New York, 1945. (Original tables were revised in 1983.) Reproduced by permission of John Wiley & Sons, Inc.

TABLE A-6

Ideal-gas enthalpy, internal energy, and absolute entropy of diatomic nitrogen (N_2)

$\Delta h_f = 0$ kJ/kmol

T, K; h and u, kJ/kmol; s, kJ/(kmol·K); P^o = 1 atm

T	h	u	s^o	T	h	u	s^o
0	0	0	0	600	17,563	12,574	212.066
220	6,391	4,562	182.639	610	17,864	12,792	212.564
230	6,683	4,770	183.938	620	18,166	13,011	213.055
240	6,975	4,979	185.180	630	18,468	13,230	213.541
250	7,266	5,188	186.370	640	18,772	13,450	214.018
260	7,558	5,396	187.514	650	19,075	13,671	214.489
270	7,849	5,604	188.614	660	19,380	13,892	214.954
280	8,141	5,813	189.673	670	19,685	14,114	215.413
290	8,432	6,021	190.695	680	19,991	14,337	215.866
298	8,669	6,190	191.502	690	20,297	14,560	216.314
300	8,723	6,229	191.682	700	20,604	14,784	216.756
310	9,014	6,437	192.638	710	20,912	15,008	217.192
320	9,306	6,645	193.562	720	21,220	15,234	217.624
330	9,597	6,853	194.459	730	21,529	15,460	218.059
340	9,888	7,061	195.328	740	21,839	15,686	218.472
350	10,180	7,270	196.173	750	22,149	15,913	218.889
360	10,471	7,478	196.995	760	22,460	16,141	219.301
370	10,763	7,687	197.794	770	22,772	16,370	219.709
380	11,055	7,895	198.572	780	23,085	16,599	220.113
390	11,347	8,104	199.331	790	23,398	16,830	220.512
400	11,640	8,314	200.071	800	23,714	17,061	220.907
410	11,932	8,523	200.794	810	24,027	17,292	221.298
420	12,225	8,733	201.499	820	24,342	17,524	221.684
430	12,518	8,943	202.189	830	24,658	17,757	222.067
440	12,811	9,153	202.863	840	24,974	17,990	222.447
450	13,105	9,363	203.523	850	25,292	18,224	222.822
460	13,399	9,574	204.170	860	25,610	18,459	223.194
470	13,693	9,786	204.803	870	25,928	18,695	223.562
480	13,988	9,997	205.424	880	26,248	18,931	223.927
490	14,285	10,210	206.033	890	26,568	19,168	224.288
500	14,581	10,423	206.630	900	26,890	19,407	224.647
510	14,876	10,635	207.216	910	27,210	19,644	225.002
520	15,172	10,848	207.792	920	27,532	19,883	225.353
530	15,469	11,062	208.358	930	27,854	20,122	225.701
540	15,766	11,277	208.914	940	28,178	20,362	226.047
550	16,064	11,492	209.461	950	28,501	20,603	226.389
560	16,363	11,707	209.999	960	28,826	20,844	226.728
570	16,662	11,923	210.528	970	29,151	21,086	227.064
580	16,962	12,139	211.049	980	29,476	21,328	227.398
590	17,262	12,356	211.562	990	29,803	21,571	227.728

TABLE A-6 *(Continued)*

T	h	u	s^o	T	h	u	s^o
1000	30,129	21,815	228.057	1760	56,227	41,594	247.396
1020	30,784	22,304	228.706	1780	56,938	42,139	247.798
1040	31,442	22,795	229.344	1800	57,651	42,685	248.195
1060	32,101	23,288	229.973	1820	58,363	43,231	248.589
1080	32,762	23,782	230.591	1840	59,075	43,777	248.979
1100	33,426	24,280	231.199	1860	59,790	44,324	249.365
1120	34,092	24,780	231.799	1880	60,504	44,873	249.748
1140	34,760	25,282	232.391	1900	61,220	45,423	250.128
1160	35,430	25,786	232.973	1920	61,936	45,973	250.502
1180	36,104	26,291	233.549	1940	62,654	46,524	250.874
1200	36,777	26,799	234.115	1960	63,381	47,075	251.242
1220	37,452	27,308	234.673	1980	64,090	47,627	251.607
1240	38,129	27,819	235.223	2000	64,810	48,181	251.969
1260	38,807	28,331	235.766	2050	66,612	49,567	252.858
1280	39,488	28,845	236.302	2100	68,417	50,957	253.726
1300	40,170	29,361	236.831	2150	70,226	52,351	254.578
1320	40,853	29,878	237.353	2200	72,040	53,749	255.412
1340	41,539	30,398	237.867	2250	73,856	55,149	256.227
1360	42,227	30,919	238.376	2300	75,676	56,553	257.027
1380	42,915	31,441	238.878	2350	77,496	57,958	257.810
1400	43,605	31,964	239.375	2400	79,320	59,366	258.580
1420	44,295	32,489	239.865	2450	81,149	60,779	259.332
1440	44,988	33,014	240.350	2500	82,981	62,195	260.073
1460	45,682	33,543	240.827	2550	84,814	63,613	260.799
1480	46,377	34,071	241.301	2600	86,650	65,033	261.512
1500	47,073	34,601	241.768	2650	88,488	66,455	262.213
1520	47,771	35,133	242.228	2700	90,328	67,880	262.902
1540	48,470	35,665	242.685	2750	92,171	69,306	263.577
1560	49,168	36,197	243.137	2800	94,014	70,734	264.241
1580	49,869	36,732	243.585	2850	95,859	72,163	264.895
1600	50,571	37,268	244.028	2900	97,705	73,593	265.538
1620	51,275	37,806	244.464	2950	99,556	75,028	266.170
1640	51,980	38,344	244.896	3000	101,407	76,464	266.793
1660	52,686	38,884	245.324	3050	103,260	77,902	267.404
1680	53,393	39,424	245.747	3100	105,115	79,341	268.007
1700	54,099	39,965	246.166	3150	106,972	80,782	268.601
1720	54,807	40,507	246.580	3200	108,830	82,224	269.186
1740	55,516	41,049	246.990	3250	110,690	83,668	269.763

Source: Based on data from the JANAF Thermochemical Tables, NSRDS-NBS-37, 1971.

TABLE A-7

Ideal-gas enthalpy, internal energy, and absolute entropy of diatomic oxygen (O_2)

$\Delta h_f = 0$ kJ/kmol

T, K; h and u, kJ/kmol; s, kJ/(kmol·K); P^o = 1 atm

T	h	u	s^o	T	h	u	s^o
0	0	0	0	600	17,929	12,940	226.346
220	6,404	4,575	196.171	610	18,250	13,178	226.877
230	6,694	4,782	197.461	620	18,572	13,417	227.400
240	6,984	4,989	198.696	630	18,895	13,657	227.918
250	7,275	5,197	199.885	640	19,219	13,898	228.429
260	7,566	5,405	201.027	650	19,544	14,140	228.932
270	7,858	5,613	202.128	660	19,870	14,383	229.430
280	8,150	5,822	203.191	670	20,197	14,626	229.920
290	8,443	6,032	204.218	680	20,524	14,871	230.405
298	8,682	6,203	205.033	690	20,854	15,116	230.885
300	8,736	6,242	205.213	700	21,184	15,364	231.358
310	9,030	6,453	206.177	710	21,514	15,611	231.827
320	9,325	6,664	207.112	720	21,845	15,859	232.291
330	9,620	6,877	208.020	730	22,177	16,107	232.748
340	9,916	7,090	208.904	740	22,510	16,357	233.201
350	10,213	7,303	209.765	750	22,844	16,607	233.649
360	10,511	7,518	210.604	760	23,178	16,859	234.091
370	10,809	7,733	211.423	770	23,513	17,111	234.528
380	11,109	7,949	212.222	780	23,850	17,364	234.960
390	11,409	8,166	213.002	790	24,186	17,618	235.387
400	11,711	8,384	213.765	800	24,523	17,872	235.810
410	12,012	8,603	214.510	810	24,861	18,126	236.230
420	12,314	8,822	215.241	820	25,199	18,382	236.644
430	12,618	9,043	215.955	830	25,537	18,637	237.055
440	12,923	9,264	216.656	840	25,877	18,893	237.462
450	13,228	9,487	217.342	850	26,218	19,150	237.864
460	13,535	9,710	218.016	860	26,559	19,408	238.264
470	13,842	9,935	218.676	870	26,899	19,666	238.660
480	14,151	10,160	219.326	880	27,242	19,925	239.051
490	14,460	10,386	219.963	890	27,584	20,185	239.439
500	14,770	10,614	220.589	900	27,928	20,445	239.823
510	15,082	10,842	221.206	910	28,272	20,706	240.203
520	15,395	11,071	221.812	920	28,616	20,967	240.580
530	15,708	11,301	222.409	930	28,960	21,228	240.953
540	16,022	11,533	222.997	940	29,306	21,491	241.323
550	16,338	11,765	223.576	950	29,652	21,754	241.689
560	16,654	11,998	224.146	960	29,999	22,017	242.052
570	16,971	12,232	224.708	970	30,345	22,280	242.411
580	17,290	12,467	225.262	980	30,692	22,544	242.768
590	17,609	12,703	225.808	990	31,041	22,809	243.120

TABLE A-7 (*Continued*)

T	h	u	s^o	T	h	u	s^o
1000	31,389	23,075	243.471	1760	58,880	44,247	263.861
1020	32,088	23,607	244.164	1780	59,624	44,825	264.283
1040	32,789	24,142	244.844	1800	60,371	45,405	264.701
1060	33,490	24,677	245.513	1820	61,118	45,986	265.113
1080	34,194	25,214	246.171	1840	61,866	46,568	265.521
1100	34,899	25,753	246.818	1860	62,616	47,151	265.925
1120	35,606	26,294	247.454	1880	63,365	47,734	266.326
1140	36,314	26,836	248.081	1900	64,116	48,319	266.722
1160	37,023	27,379	248.698	1920	64,868	48,904	267.115
1180	37,734	27,923	249.307	1940	65,620	49,490	267.505
1200	38,447	28,469	249.906	1960	66,374	50,078	267.891
1220	39,162	29,018	250.497	1980	67,127	50,665	268.275
1240	39,877	29,568	251.079	2000	67,881	51,253	268.655
1260	40,594	30,118	251.653	2050	69,772	52,727	269.588
1280	41,312	30,670	252.219	2100	71,668	54,208	270.504
1300	42,033	31,224	252.776	2150	73,573	55,697	271.399
1320	42,753	31,778	253.325	2200	75,484	57,192	272.278
1340	43,475	32,334	253.868	2250	77,397	58,690	273.136
1360	44,198	32,891	254.404	2300	79,316	60,193	273.981
1380	44,923	33,449	254.932	2350	81,243	61,704	274.809
1400	45,648	34,008	255.454	2400	83,174	63,219	275.625
1420	46,374	34,567	255.968	2450	85,112	64,742	276.424
1440	47,102	35,129	256.475	2500	87,057	66,271	277.207
1460	47,831	35,692	256.978	2550	89,004	67,802	277.979
1480	48,561	36,256	257.474	2600	90,956	69,339	278.738
1500	49,292	36,821	257.965	2650	92,916	70,883	279.485
1520	50,024	37,387	258.450	2700	94,881	72,433	280.219
1540	50,756	37,952	258.928	2750	96,852	73,987	280.942
1560	51,490	38,520	259.402	2800	98,826	75,546	281.654
1580	52,224	39,088	259.870	2850	100,808	77,112	282.357
1600	52,961	39,658	260.333	2900	102,793	78,682	283.048
1620	53,696	40,227	260.791	2950	104,785	80,258	283.728
1640	54,434	40,799	261.242	3000	106,780	81,837	284.399
1660	55,172	41,370	261.690	3050	108,778	83,419	285.060
1680	55,912	41,944	262.132	3100	110,784	85,009	285.713
1700	56,652	42,517	262.571	3150	112,795	86,601	286.355
1720	57,394	43,093	263.005	3200	114,809	88,203	286.989
1740	58,136	43,669	263.435	3250	116,827	89,804	287.614

Source: Based on data from the JANAF Thermochemical Tables, NSRDS-NBS-37, 1971.

TABLE A-8

Ideal-gas enthalpy, internal energy, and absolute entropy of carbon monoxide (CO)

$\Delta h_f = -110{,}530$ kJ/kmol

***T*, K; *h* and *u*, kJ/kmol; *s*, kJ/(kmol·K); P° = 1 atm**

T	h	u	s^o	T	h	u	s^o
0	0	0	0	600	17,611	12,622	218.204
220	6,391	4,562	188.683	610	17,915	12,843	218.708
230	6,683	4,771	189.980	620	18,221	13,066	219.205
240	6,975	4,979	191.221	630	18,527	13,289	219.695
250	7,266	5,188	192.411	640	18,833	13,512	220.179
260	7,558	5,396	193.554	650	19,141	13,736	220.656
270	7,849	5,604	194.654	660	19,449	13,962	221.127
280	8,140	5,812	195.173	670	19,758	14,187	221.592
290	8,432	6,020	196.735	680	20,068	14,414	222.052
298	8,669	6,190	197.543	690	20,378	14,641	222.505
300	8,723	6,229	197.723	700	20,690	14,870	222.953
310	9,014	6,437	198.678	710	21,002	15,099	223.396
320	9,306	6,645	199.603	720	21,315	15,328	223.833
330	9,597	6,854	200.500	730	21,628	15,558	224.265
340	9,889	7,062	201.371	740	21,943	15,789	224.692
350	10,181	7,271	202.217	750	22,258	16,022	225.115
360	10,473	7,480	203.040	760	22,573	16,255	225.533
370	10,765	7,689	203.842	770	22,890	16,488	225.947
380	11,058	7,899	204.622	780	23,208	16,723	226.357
390	11,351	8,108	205.383	790	23,526	16,957	226.762
400	11,644	8,319	206.125	800	23,844	17,193	227.162
410	11,938	8,529	206.850	810	24,164	17,429	227.559
420	12,232	8,740	207.549	820	24,483	17,665	227.952
430	12,526	8,951	208.252	830	24,803	17,902	228.339
440	12,821	9,163	208.929	840	25,124	18,140	228.724
450	13,116	9,375	209.593	850	25,446	18,379	229.106
460	13,412	9,587	210.243	860	25,768	18,617	229.482
470	13,708	9,800	210.880	870	26,091	18,858	229.856
480	14,005	10,014	211.504	880	26,415	19,099	230.227
490	14,302	10,228	212.117	890	26,740	19,341	230.593
500	14,600	10,443	212.719	900	27,066	19,583	230.957
510	14,898	10,658	213.310	910	27,392	19,826	231.317
520	15,197	10,874	213.890	920	27,719	20,070	231.674
530	15,497	11,090	214.460	930	28,046	20,314	232.028
540	15,797	11,307	215.020	940	28,375	20,559	232.379
550	16,097	11,524	215.572	950	28,703	20,805	232.727
560	16,399	11,743	216.115	960	29,033	21,051	233.072
570	16,701	11,961	216.649	970	29,362	21,298	233.413
580	17,003	12,181	217.175	980	29,693	21,545	233.752
590	17,307	12,401	217.693	990	30,024	21,793	234.088

TABLE A-8 *(Continued)*

T	h	u	s^o	T	h	u	s^o
1000	30,355	22,041	234.421	1760	56,756	42,123	253.991
1020	31,020	22,540	235.079	1780	57,473	42,673	254.398
1040	31,688	23,041	235.728	1800	58,191	43,225	254.797
1060	32,357	23,544	236.364	1820	58,910	43,778	255.194
1080	33,029	24,049	236.992	1840	59,629	44,331	255.587
1100	33,702	24,557	237.609	1860	60,351	44,886	255.976
1120	34,377	25,065	238.217	1880	61,072	45,441	256.361
1140	35,054	25,575	238.817	1900	61,794	45,997	256.743
1160	35,733	26,088	239.407	1920	62,516	46,552	257.122
1180	36,406	26,602	239.989	1940	63,238	47,108	257.497
1200	37,095	27,118	240.663	1960	63,961	47,665	257.868
1220	37,780	27,637	241.128	1980	64,684	48,221	258.236
1240	38,466	28,426	241.686	2000	65,408	48,780	258.600
1260	39,154	28,678	242.236	2050	67,224	50,179	259.494
1280	39,884	29,201	242.780	2100	69,044	51,584	260.370
1300	40,534	29,725	243.316	2150	70,864	52,988	261.226
1320	41,266	30,251	243.844	2200	72,688	54,396	262.065
1340	41,919	30,778	244.366	2250	74,516	55,809	262.887
1360	42,613	31,306	244.880	2300	76,345	57,222	263.692
1380	43,309	31,836	245.388	2350	78,178	58,640	264.480
1400	44,007	32,367	245,889	2400	80,015	60,060	265.253
1420	44,707	32,900	246,385	2450	81,852	61,482	266.012
1440	45,408	33,434	246.876	2500	83,692	62,906	266.755
1460	46,110	33,971	247.360	2550	85,537	64,335	267.485
1480	46,813	34,508	247.839	2600	87,383	65,766	268.202
1500	47,517	35,046	248.312	2650	89,230	67,197	268.905
1520	48,222	35,584	248.778	2700	91,077	68,628	269.596
1540	48,928	36,124	249.240	2750	92,930	70,066	270.285
1560	49,635	36,665	249.695	2800	94,784	71,504	270.943
1580	50,344	37,207	250.147	2850	96,639	72,945	271.602
1600	51,053	37,750	250.592	2900	98,495	74,383	272.249
1620	51,763	38,293	251.033	2950	100,352	75,825	272.884
1640	52,472	38,837	251.470	3000	102,210	77,267	273.508
1660	53,184	39,382	251.901	3050	104.073	78,715	274.123
1680	53,895	39,927	252.329	3100	105,939	80,164	274.730
1700	54,609	40,474	252.751	3150	107,802	81,612	275.326
1720	55,323	41,023	253.169	3200	109,667	83,061	275.914
1740	56,039	41,572	253.582	3250	111,534	84,513	276.494

Source: Based on data from the JANAF Thermochemical Tables, NSRDS-NBS-37, 1971.

TABLE A-9
Ideal-gas enthalpy, internal energy, and absolute entropy of carbon dioxide (CO_2)
$\Delta h_f = -393{,}520$ kJ/kmol
T, K; h and u, kJ/kmol; s, kJ/(kmol·K); P^o = 1 atm

T	h	u	s^o	T	h	u	s^o
0	0	0	0	600	22,280	17,291	243.199
220	6,601	4,772	202.966	610	22,754	17,683	243.983
230	6,938	5,026	204.464	620	23,231	18,076	244.758
240	7,280	5,285	205.920	630	23,709	18,471	245.524
250	7,627	5,548	207.337	640	24,190	18,869	246.282
260	7,979	5,817	208.717	650	24,674	19,270	247.032
270	8,335	6,091	210.062	660	25,160	19,672	247.773
280	8,697	6,369	211.376	670	25,648	20,078	248.507
290	9,063	6,651	212.660	680	26,138	20,484	249.233
298	9,364	6,885	213.685	690	26,631	20,894	249.952
300	9,431	6,939	213.915	700	27,125	21,305	250.663
310	9,807	7,230	215.146	710	27,622	21,719	251.368
320	10,186	7,526	216.351	720	28,121	22,134	252.065
330	10,570	7,826	217.534	730	28,622	22,552	252.755
340	10,959	8,131	218.694	740	29,124	22,972	253.439
350	11,351	8,439	219.831	750	29,629	23,393	254.117
360	11,748	8,752	220.948	760	30,135	23,817	254.787
370	12,148	9,068	222.044	770	30,644	24,242	255.452
380	12,552	9,392	223.122	780	31,154	24,669	256.110
390	12,960	9,718	224.182	790	31,665	25,097	256.762
400	13,372	10,046	225.225	800	32,179	25,527	257.408
410	13,787	10,378	226.250	810	32,694	25,959	258.048
420	14,206	10,714	227.258	820	33,212	26,394	258.682
430	14,628	11,053	228.252	830	33,730	26,829	259.311
440	15,054	11,393	229.230	840	34,251	27,267	259.934
450	15,483	11,742	230.194	850	34,773	27,706	260.551
460	15,916	12,091	231.144	860	35,296	28,125	261.164
470	16,351	12,444	232.080	870	35,821	28,588	261.770
480	16,791	12,800	233.004	880	36,347	29,031	262.371
490	17,232	13,158	233.916	890	36,876	29,476	262.968
500	17,678	13,521	234.814	900	37,405	29,922	263.559
510	18,126	13,885	235.700	910	37,935	30,369	264.146
520	18,576	14,253	236.575	920	38,467	30,818	264.728
530	19,029	14,622	237.439	930	39,000	31,268	265.304
540	19,485	14,996	238.292	940	39,535	31,719	265.877
550	19,945	15,372	239.135	950	40,070	32,171	266.444
560	20,407	15,751	239.962	960	40,607	32,625	267.007
570	20,870	16,131	240.789	970	41,145	33,081	267.566
580	21,337	16,515	241.602	980	41,685	33,537	268.119
590	21,807	16,902	242.405	990	42,226	33,995	268.670

TABLE A-9 ***(Continued)***

T	h	u	s^o	T	h	u	s^o
1000	42,769	34,455	269.215	1760	86,420	71,787	301.543
1020	43,859	35,378	270.293	1780	87,612	72,812	302.271
1040	44,953	36,306	271.354	1800	88,806	73,840	302.884
1060	46,051	37,238	272.400	1820	90,000	74,868	303.544
1080	47,153	38,174	273.430	1840	91,196	75,897	304.198
1100	48,258	39,112	274.445	1860	92,394	76,929	304.845
1120	49,369	40,057	275.444	1880	93,593	77,962	305.487
1140	50,484	41,006	276.430	1900	94,793	78,996	306.122
1160	51,602	41,957	277.403	1920	95,995	80,031	306.751
1180	52,724	42,913	278.362	1940	97,197	81,067	307.374
1200	53,848	43,871	279.307	1960	98,401	82,105	307.992
1220	54,977	44,834	280.238	1980	99,606	83,144	308.604
1240	56,108	45,799	281.158	2000	100,804	84,185	309.210
1260	57,244	46,768	282.066	2050	103,835	86,791	310.701
1280	58,381	47,739	282.962	2100	106,864	89,404	312.160
1300	59,522	48,713	283.847	2150	109,898	92,023	313.589
1320	60,666	49,691	284.722	2200	112,939	94,648	314.988
1340	61,813	50,672	285.586	2250	115,984	97,277	316.356
1360	62,963	51,656	286.439	2300	119,035	99,912	317.695
1380	64,116	52,643	287.283	2350	122,091	102,552	319.011
1400	65,271	53,631	288.106	2400	125,152	105,197	320.302
1420	66,427	54,621	288.934	2450	128,219	107,849	321.566
1440	67,586	55,614	289.743	2500	131,290	110,504	322.808
1460	68,748	56,609	290.542	2550	134,368	113,166	324.026
1480	69,911	57,606	291.333	2600	137,449	115,832	325.222
1500	71,078	58,606	292.114	2650	140,533	118,500	326.396
1520	72,246	59,609	292.888	2700	143,620	121,172	327.549
1540	73,417	60,613	292.654	2750	146,713	123,849	328.684
1560	74,590	61,620	294.411	2800	149,808	126,528	329.800
1580	76,767	62,630	295.161	2850	152,908	129,212	330.896
1600	76,944	63,741	295.901	2900	156,009	131,898	331.975
1620	78,123	64,653	296.632	2950	159,117	134,589	333.037
1640	79,303	65,668	297.356	3000	162,226	137,283	334.084
1660	80,486	66,592	298.072	3050	165,341	139,982	335.114
1680	81,670	67,702	298.781	3100	168,456	142,681	336.126
1700	82,856	68,721	299.482	3150	171,576	145,385	337.124
1720	84,043	69,742	300.177	3200	174,695	148,089	338.109
1740	85,231	70,764	300.863	3250	177,822	150,801	339.069

Source: Based on data from the JANAF Thermochemical Tables, NSRDS-NBS-37, 1971.

TABLE A-10

Ideal-gas enthalpy, internal energy, and absolute entropy of water (H_2O)
$\Delta h_f = -241{,}820$ kJ/kmol
T, K; h and u, kJ/kmol; s, kJ/(kmol·K); P° = 1 atm

T	h	u	s^o	T	h	u	s^o
0	0	0	0	600	20,402	15,413	212.920
220	7,295	5,466	178.576	610	20,765	15,693	213.529
230	7,628	5,715	180.054	620	21,130	15,975	214.122
240	7,961	5,965	181.471	630	21,495	16,257	214.707
250	8,294	6,215	182.831	640	21,862	16,541	215.285
260	8,627	6,466	184.139	650	22,230	16,826	215.856
270	8,961	6,716	185.399	660	22,600	17,112	216.419
280	9,296	6,968	186.616	670	22,970	17,399	216.976
290	9,631	7,219	187.791	680	23,342	17,688	217.527
298	9,904	7,425	188.720	690	23,714	17,978	218.071
300	9,966	7,472	188.928	700	24,088	18,268	218.610
310	10,302	7,725	190.030	710	24,464	18,561	219.142
320	10,639	7,978	191.098	720	24,840	18,854	219.668
330	10,976	8,232	192.136	730	25,218	19,148	220.189
340	11,314	8,487	193.144	740	25,597	19,444	220.707
350	11,652	8,742	194.125	750	25,977	19,741	221.215
360	11,992	8,998	195.081	760	26,358	20,039	221.720
370	12,331	9,255	196.012	770	26,741	20,339	222.221
380	12,672	9,513	196.920	780	27,125	20,639	222.717
390	13,014	9,771	197.807	790	27,510	20,941	223.207
400	13,356	10,030	198.673	800	27,896	21,245	223.693
410	13,699	10,290	199.521	810	28,284	21,549	224.174
420	14,043	10,551	200.350	820	28,672	21,855	224.651
430	14,388	10,813	201.160	830	29,062	22,162	225.123
440	14,734	11,075	201.955	840	29,454	22,470	225.592
450	15,080	11,339	202.734	850	29,846	22,779	226.057
460	15,428	11,603	203.497	860	30,240	23,090	226.517
470	15,777	11,869	204.247	870	30,635	23,402	226.973
480	16,126	12,135	204.982	880	31,032	23,715	227.426
490	16,477	12,403	205.705	890	31,429	24,029	227.875
500	16,828	12,671	206.413	900	31,828	24,345	228.321
510	17,181	12,940	207.112	910	32,228	24,662	228.763
520	17,534	13,211	207.799	920	32,629	24,980	229.202
530	17,889	13,482	208.475	930	33,032	25,300	229.637
540	18,245	13,755	209.139	940	33,436	25,621	230.070
550	18,601	14,028	209.795	950	33,841	25,943	230.499
560	18,959	14,303	210.440	960	34,247	26,265	230.924
570	19,318	14,579	211.075	970	34,653	26,588	231.347
580	19,678	14,856	211.702	980	35,061	26,913	231.767
590	20,039	15,134	212.320	990	35,472	27,240	232.184

TABLE A-10 *(Continued)*

T	h	u	s^o	T	h	u	s^o
1000	35,882	27,568	232.597	1760	70,535	55,902	258.151
1020	36,709	28,228	233.415	1780	71,523	56,723	258.708
1040	37,542	28,895	234.223	1800	72,513	57,547	259.262
1060	38,380	29,567	235.020	1820	73,507	58,375	259.811
1080	39,223	30,243	235.806	1840	74,506	59,207	260.357
1100	40,071	30,925	236.584	1860	75,506	60,042	260.898
1120	40,923	31,611	237.352	1880	76,511	60,880	261.436
1140	41,780	32,301	238.110	1900	77,517	61,720	261.969
1160	42,642	32,997	238.859	1920	78,527	62,564	262.497
1180	43,509	33,698	239.600	1940	79,540	63,411	263.022
1200	44,380	34,403	240.333	1960	80,555	64,259	263.542
1220	45,256	35,112	241.057	1980	81,573	65,111	264.059
1240	46,137	35,827	241.773	2000	82,593	65,965	264.571
1260	47,022	36,546	242.482	2050	85,156	68,111	265.838
1280	47,912	37,270	243.183	2100	87,735	70,275	267.081
1300	48,807	38,000	243.877	2150	90,330	72,454	268.301
1320	49,707	38,732	244.564	2200	92,940	74,649	269.500
1340	50,612	39,470	245.243	2250	95,562	76,855	270.679
1360	51,521	40,213	245.915	2300	98,199	79,076	271.839
1380	52,434	40,960	246.582	2350	100,846	81,308	272.978
1400	53,351	41,711	247.241	2400	103,508	83,553	274.098
1420	54,273	42,466	247.895	2450	106,183	85,811	275.201
1440	55,198	43,226	248.543	2500	108,868	88,082	276.286
1460	56,128	43,989	249.185	2550	111,565	90,364	277.354
1480	57,062	44,756	249.820	2600	114,273	92,656	278.407
1500	57,999	45,528	250.450	2650	116,991	94,958	279.441
1520	58,942	46,304	251.074	2700	119,717	97,269	280.462
1540	59,888	47,084	251.693	2750	122,453	99,588	281.464
1560	60,838	47,868	252.305	2800	125,198	101,917	282.453
1580	61,792	48,655	252.912	2850	127,952	104,256	283.429
1600	62,748	49,445	253.513	2900	130,717	106,605	284.390
1620	63,709	50,240	254.111	2950	133,486	108,959	285.338
1640	64,675	51,039	254.703	3000	136,264	111,321	286.273
1660	65,643	51,841	255.290	3050	139,051	113,692	287.194
1680	66,614	52,646	255.873	3100	141,846	116,072	288.102
1700	67,589	53,455	256.450	3150	144,648	118,458	288.999
1720	68,567	54,267	257.022	3200	147,457	120,851	289.884
1740	69,550	55,083	257.589	3250	150,272	123,250	290.756

Source: Based on data from the JANAF Thermochemical Tables, NSRDS-NBS-37, 1971.

TABLE A-11

Ideal-gas enthalpy, internal energy, and absolute entropy of diatomic hydrogen (H_2)

$\Delta h_f = 0$ kJ/kmol

T, K; h and u, kJ/kmol; s, kJ/(kmol·K); P° = 1 atm

T	h	u	s^o	T	h	u	s^o
0	0	0	0	1440	42,808	30,835	177.410
260	7,370	5,209	126.636	1480	44,091	31,786	178.291
270	7,657	5,412	127.719	1520	45,384	32,746	179.153
280	7,945	5,617	128.765	1560	46,683	33,713	179.995
290	8,233	5,822	129.775	1600	47,990	34,687	180.820
298	8,468	5,989	130.574	1640	49,303	35,668	181.632
300	8,522	6,027	130.754	1680	50,662	36,654	182.428
320	9,100	6,440	132.621	1720	51,947	37,646	183.208
340	9,680	6,853	134.378	1760	53,279	38,645	183.973
360	10,262	7,268	136.039	1800	54,618	39,652	184.724
380	10,843	7,684	137.612	1840	55,962	40,663	185.463
400	11,426	8,100	139.106	1880	57,311	41,680	186.190
420	12,010	8,518	140.529	1920	58,668	42,705	186.904
440	12,594	8,936	141.888	1960	60,031	43,735	187.607
460	13,179	9,355	143.187	2000	61,400	44,771	188.297
480	13,764	9,773	144.432	2050	63,119	46,074	189.148
500	14,350	10,193	145.628	2100	64,847	47,386	189.979
520	14,935	10,611	146.775	2150	66,584	48,708	190.796
560	16,107	11,451	148.945	2200	68,328	50,037	191.598
600	17,280	12,291	150.968	2250	70,080	51,373	192.385
640	18,453	13,133	152.863	2300	71,839	52,716	193.159
680	19,630	13,976	154.645	2350	73,608	54,069	193.921
720	20,807	14,821	156.328	2400	75,383	55,429	194.669
760	21,988	15,669	157.923	2450	77,168	56,798	195.403
800	23,171	16,520	159.440	2500	78,960	58,175	196.125
840	24,359	17,375	160.891	2550	80,755	59,554	196.837
880	25,551	18,235	162.277	2600	82,558	60,941	197.539
920	26,747	19,098	163.607	2650	84,386	62,335	198.229
960	27,948	19,966	164.884	2700	86,186	63,737	198.907
1000	29,154	20,839	166.114	2750	88,008	65,144	199.575
1040	30,364	21,717	167.300	2800	89,838	66,558	200.234
1080	31,580	22,601	168.449	2850	91,671	67,976	200.885
1120	32,802	23,490	169.560	2900	93,512	69,401	201.527
1160	34,028	24,384	170.636	2950	95,358	70,831	202.157
1200	35,262	25,284	171.682	3000	97,211	72,268	202.778
1240	36,502	26,192	172.698	3050	99,065	73,707	203.391
1280	37,749	27,106	173.687	3100	100,926	75,152	203.995
1320	39,002	28,027	174.652	3150	102,793	76,604	204.592
1360	40,263	28,955	175.593	3200	104,667	78,061	205.181
1400	41,530	29,889	176.510	3250	106,545	79,523	205.765

TABLE A-11 *(Continued)*

Ideal-gas enthalpy, internal energy, and absolute entropy of monatomic oxygen (O)
$\Delta h_f = 249{,}190$ kJ/kmol

T	h	u	s^o	T	h	u	s^o
0	0	0	0	2400	50,894	30,940	204.932
298	6,852	4,373	160.944	2450	51,936	31,566	205.362
300	6,892	4,398	161.079	2500	52,979	32,193	205.783
500	11,197	7,040	172.088	2550	54,021	32,820	206.196
1000	21,713	13,398	186.678	2600	55,064	33,447	206.601
1500	32,150	19,679	195.143	2650	56,108	34,075	206.999
1600	34,234	20,931	196.488	2700	57,152	34,703	207.389
1700	36,317	22,183	197.751	2750	58,196	35,332	207.772
1800	38,400	23,434	198.941	2800	59,241	35,961	208.148
1900	40,482	24,685	200.067	2850	60,286	36,590	208.518
2000	42,564	25,935	201.135	2900	61,332	37,220	208.882
2050	43,605	26,560	201.649	2950	62,378	37,851	209.240
2100	44,646	27,186	202.151	3000	63,425	38,482	209.592
2150	45,687	27,811	202.641	3100	65,520	39,746	210.279
2200	46,728	28,436	203.119	3200	67,619	41,013	210.945
2250	47,769	29,062	203.588	3300	69,720	42,283	211.592
2300	48,811	29,688	204.045	3400	71,824	43,556	212.220
2350	49,852	30,314	204.493	3500	73,932	44,832	212.831

Ideal-gas enthalpy, internal energy, and absolute entropy of hydroxyl (OH)
$\Delta h_f = 39{,}040$ kJ/kmol

T	h	u	s^o	T	h	u	s^o
0	0	0	0	2400	77,015	57,061	248.628
298	9,188	6,709	183.594	2450	78,801	58,431	249.364
300	9,244	6,749	183.779	2500	80,592	59,806	250.088
500	15,181	11,024	198.955	2550	82,388	61,186	250.799
1000	30,123	21,809	219.624	2600	84,189	62,572	251.499
1500	46,046	33,575	232.506	2650	85,995	63,962	252.187
1600	49,358	36,055	234.642	2700	87,806	65,358	252.864
1700	52,706	38,571	236.672	2750	89,622	66,757	253.530
1800	56,089	41,123	238.606	2800	91,442	68,162	254.186
1900	59,505	43,708	240.453	2850	93,266	69,570	254.832
2000	62,952	46,323	242.221	2900	95,095	70,983	255.468
2050	64,687	47,642	243.077	2950	96,927	72,400	256.094
2100	66,428	48,968	243.917	3000	98,763	73,820	256.712
2150	68,177	50,301	244.740	3100	102,447	76,673	257.919
2200	69,932	51,641	245.547	3200	106,145	79,539	259.093
2250	71,694	52,987	246.338	3300	109,855	82,418	260.235
2300	73,462	54,339	247.116	3400	113,578	85,309	261.347
2350	75,236	55,697	247.879	3500	117,312	88,212	262.429

Source: Based on data from the JANAF Thermochemical Tables, NSRDS-NBS-37, 1971.

TABLE A-12

Properties of saturated water: Temperature table

v, cm³/g; u, kJ/kg; h, kJ/kg; s, kJ/(kg·K); 1 bar = 0.1 MPa

		Specific volume		Internal energy		Enthalpy			Entropy	
Temp., °C T	Press., bars P	Sat. liquid v_f	Sat. vapor v_g	Sat. liquid u_f	Sat. vapor u_g	Sat. liquid h_f	Evap. h_{fg}	Sat. vapor h_g	Sat. liquid s_f	Sat. vapor s_g
0	0.00611	1.0002	206278	−0.03	2375.4	−0.02	2501.4	2501.3	0.0001	9.1565
4	0.00813	1.0001	157232	16.77	2380.9	16.78	2491.9	2508.7	0.0610	9.0514
5	0.00872	1.0001	147120	20.97	2382.3	20.98	2489.6	2510.6	0.0761	9.0257
6	0.00935	1.0001	137734	25.19	2383.6	25.20	2487.2	2512.4	0.0912	9.0003
8	0.01072	1.0002	120917	33.59	2386.4	33.60	2482.5	2516.1	0.1212	8.9501
10	0.01228	1.0004	106379	42.00	2389.2	42.01	2477.7	2519.8	0.1510	8.9008
11	0.01312	1.0004	99857	46.20	2390.5	46.20	2475.4	2521.6	0.1658	8.8765
12	0.01402	1.0005	93784	50.41	2391.9	50.41	2473.0	2523.4	0.1806	8.8524
13	0.01497	1.0007	88124	54.60	2393.3	54.60	2470.7	2525.3	0.1953	8.8285
14	0.01598	1.0008	82848	58.79	2394.7	58.80	2468.3	2527.1	0.2099	8.8048
15	0.01705	1.0009	77926	62.99	2396.1	62.99	2465.9	2528.9	0.2245	8.7814
16	0.01818	1.0011	73333	67.18	2397.4	67.19	2463.6	2530.8	0.2390	8.7582
17	0.01938	1.0012	69044	71.38	2398.8	71.38	2461.2	2532.6	0.2535	8.7351
18	0.02064	1.0014	65038	75.57	2400.2	75.58	2458.8	2534.4	0.2679	8.7123
19	0.02198	1.0016	61293	79.76	2401.6	79.77	2456.5	2536.2	0.2823	8.6897
20	0.02339	1.0018	57791	83.95	2402.9	83.96	2454.1	2538.1	0.2966	8.6672
21	0.02487	1.0020	54514	88.14	2404.3	88.14	2451.8	2539.9	0.3109	8.6450
22	0.02645	1.0022	51447	92.32	2405.7	92.33	2449.4	2541.7	0.3251	8.6229
23	0.02810	1.0024	48574	96.51	2407.0	96.52	2447.0	2543.5	0.3393	8.6011
24	0.02985	1.0027	45883	100.70	2408.4	100.70	2444.7	2545.4	0.3534	8.5794
25	0.03169	1.0029	43360	104.88	2409.8	104.89	2442.3	2547.2	0.3674	8.5580
26	0.03363	1.0032	40994	109.06	2411.1	109.07	2439.9	2549.0	0.3814	8.5367
27	0.03567	1.0035	38774	113.25	2412.5	113.25	2437.6	2550.8	0.3954	8.5156
28	0.03782	1.0037	36690	117.42	2413.9	117.43	2435.2	2552.6	0.4093	8.4946
29	0.04008	1.0040	34733	121.60	2415.2	121.61	2432.8	2554.5	0.4231	8.4739
30	0.04246	1.0043	32894	125.78	2416.6	125.79	2430.5	2556.3	0.4369	8.4533
31	0.04496	1.0046	31165	129.96	2418.0	129.97	2428.1	2558.1	0.4507	8.4329
32	0.04759	1.0050	29540	134.14	2419.3	134.15	2425.7	2559.9	0.4644	8.4127
33	0.05034	1.0053	28011	138.32	2420.7	138.33	2423.4	2561.7	0.4781	8.3927
34	0.05324	1.0056	26571	142.50	2422.0	142.50	2421.0	2563.5	0.4917	8.3728
35	0.05628	1.0060	25216	146.67	2423.4	146.68	2418.6	2565.3	0.5053	8.3531
36	0.05947	1.0063	23940	150.85	2424.7	150.86	2416.2	2567.1	0.5188	8.3336
38	0.06632	1.0071	21602	159.20	2427.4	159.21	2411.5	2570.7	0.5458	8.2950
40	0.07384	1.0078	19523	167.56	2430.1	167.57	2406.7	2574.3	0.5725	8.2570
45	0.09593	1.0099	15258	188.44	2436.8	188.45	2394.8	2583.2	0.6387	8.1648

TABLE A-12 *(Continued)*

		Specific volume		Internal energy		Enthalpy			Entropy	
Temp., °C T	Press., bars P	Sat. liquid v_f	Sat. vapor v_g	Sat. liquid u_f	Sat. vapor u_g	Sat. liquid h_f	Evap. h_{fg}	Sat. vapor h_g	Sat. liquid s_f	Sat. vapor s_g
50	.1235	1.0121	12032	209.32	2443.5	209.33	2382.7	2592.1	.7038	8.0763
55	.1576	1.0146	9568	230.21	2450.1	230.23	2370.7	2600.9	.7679	7.9913
60	.1994	1.0172	7671	251.11	2456.6	251.13	2358.5	2609.6	.8312	7.9096
65	.2503	1.0199	6197	272.02	2463.1	272.06	2346.2	2618.3	.8935	7.8310
70	.3119	1.0228	5042	292.95	2469.6	292.98	2333.8	2626.8	.9549	7.7553
75	.3858	1.0259	4131	313.90	2475.9	313.93	2321.4	2635.3	1.0155	7.6824
80	.4739	1.0291	3407	334.86	2482.2	334.91	2308.8	2643.7	1.0753	7.6122
85	.5783	1.0325	2828	355.84	2488.4	355.90	2296.0	2651.9	1.1343	7.5445
90	.7014	1.0360	2361	376.85	2494.5	376.92	2283.2	2660.1	1.1925	7.4791
95	.8455	1.0397	1982	397.88	2500.6	397.96	2270.2	2668.1	1.2500	7.4159
100	1.0133	1.0435	1673.	418.94	2506.5	419.04	2257.0	2676.1	1.3069	7.3549
110	1.433	1.0516	1210.	461.14	2518.1	461.30	2230.2	2691.5	1.4185	7.2387
120	1.985	1.0603	891.9	503.50	2529.3	503.71	2202.6	2706.3	1.5276	7.1296
130	2.701	1.0697	668.5	546.02	2539.9	546.31	2174.2	2720.5	1.6344	7.0269
140	3.613	1.0797	508.9	588.74	2550.0	589.13	2144.7	2733.9	1.7391	6.9299
150	4.758	1.0905	392.8	631.68	2559.5	632.20	2114.3	2746.5	1.8418	6.8379
160	6.178	1.1020	307.1	674.86	2568.4	675.55	2082.6	2758.1	1.9427	6.7502
170	7.917	1.1143	242.8	718.33	2576.5	719.21	2049.5	2768.7	2.0419	6.6663
180	10.02	1.1274	194.1	762.09	2583.7	763.22	2015.0	2778.2	2.1396	6.5857
190	12.54	1.1414	156.5	806.19	2590.0	807.62	1978.8	2786.4	2.2359	6.5079
200	15.54	1.1565	127.4	850.65	2595.3	852.45	1940.7	2793.2	2.3309	6.4323
210	19.06	1.1726	104.4	895.53	2599.5	897.76	1900.7	2798.5	2.4248	6.3585
220	23.18	1.1900	86.19	940.87	2602.4	943.62	1858.5	2802.1	2.5178	6.2861
230	27.95	1.2088	71.58	986.74	2603.9	990.12	1813.8	2804.0	2.6099	6.2146
240	33.44	1.2291	59.76	1033.2	2604.0	1037.3	1766.5	2803.8	2.7015	6.1437
250	39.73	1.2512	50.13	1080.4	2602.4	1085.4	1716.2	2801.5	2.7927	6.0730
260	46.88	1.2755	42.21	1128.4	2599.0	1134.4	1662.5	2796.6	2.8838	6.0019
270	54.99	1.3023	35.64	1177.4	2593.7	1184.5	1605.2	2789.7	2.9751	5.9301
280	64.12	1.3321	30.17	1227.5	2586.1	1236.0	1543.6	2779.6	3.0668	5.8571
290	74.36	1.3656	25.57	1278.9	2576.0	1289.1	1477.1	2766.2	3.1594	5.7821
300	85.81	1.4036	21.67	1332.0	2563.0	1344.0	1404.9	2749.0	3.2534	5.7045
320	112.7	1.4988	15.49	1444.6	2525.5	1461.5	1238.6	2700.1	3.4480	5.5362
340	145.9	1.6379	10.80	1570.3	2464.6	1594.2	1027.9	2622.0	3.6594	5.3357
360	186.5	1.8925	6.945	1725.2	2351.5	1760.5	720.5	2481.0	3.9147	5.0526
374.14	220.9	3.155	3.155	2029.6	2029.6	2099.3	0	2099.3	4.4298	4.4298

Source: J. H. Keenan, F. G. Keyes, P. G. Hill, and J. G. Moore, *Steam Tables*, Wiley, New York, 1969. Reprinted by permission of John Wiley & Sons, Inc.

TABLE A-13

Properties of saturated water: Pressure table

v, cm³/g; u, kJ/kg; h, kJ/kg; s, kJ/(kg·K); 1 bar = 0.1 MPa

		Specific volume		Internal energy		Enthalpy			Entropy	
Press., bars P	Temp., °C T	Sat. liquid v_f	Sat. vapor v_g	Sat. liquid u_f	Sat. vapor u_g	Sat. liquid h_f	Evap. h_{fg}	Sat. vapor h_g	Sat. liquid s_f	Sat. vapor s_g
.040	28.96	1.0040	34800	121.45	2415.2	121.46	2432.9	2554.4	.4226	8.4746
.060	36.16	1.0064	23739	151.53	2425.0	151.53	2415.9	2567.4	.5210	8.3304
.080	41.51	1.0084	18103	173.87	2432.2	173.88	2403.1	2577.0	.5926	8.2287
0.10	45.81	1.0102	14674	191.82	2437.9	191.83	2392.8	2584.7	.6493	8.1502
0.20	60.06	1.0172	7649	251.38	2456.7	251.40	2358.3	2609.7	.8320	7.9085
0.30	69.10	1.0223	5229.	289.20	2468.4	289.23	2336.1	2625.3	.9439	7.7686
0.40	75.87	1.0265	3993.	317.53	2477.0	317.58	2319.2	2636.8	1.0259	7.6700
0.50	81.33	1.0300	3240.	340.44	2483.9	340.49	2305.4	2645.9	1.0910	7.5939
0.60	85.94	1.0331	2732.	359.79	2489.6	359.86	2293.6	2653.5	1.1453	7.5320
0.70	89.95	1.0360	2365.	376.63	2494.5	376.70	2283.3	2660.0	1.1919	7.4797
0.80	93.50	1.0380	2087.	391.58	2498.8	391.66	2274.1	2665.8	1.2329	7.4346
0.90	96.71	1.0410	1869.	405.06	2502.6	405.15	2265.7	2670.9	1.2695	7.3949
1.00	99.63	1.0432	1694.	417.36	2506.1	417.46	2258.0	2675.5	1.3026	7.3594
1.50	111.4	1.0528	1159.	466.94	2519.7	467.11	2226.5	2693.6	1.4336	7.2233
2.00	120.2	1.0605	885.7	504.49	2529.5	504.70	2201.9	2706.7	1.5301	7.1271
2.50	127.4	1.0672	718.7	535.10	2537.2	535.37	2181.5	2716.9	1.6072	7.0527
3.00	133.6	1.0732	605.8	561.15	2543.6	561.47	2163.8	2725.3	1.6718	6.9919
3.50	138.9	1.0786	524.3	583.95	2546.9	584.33	2148.1	2732.4	1.7275	6.9405
4.00	143.6	1.0836	462.5	604.31	2553.6	604.74	2133.8	2738.6	1.7766	6.8959
4.50	147.9	1.0882	414.0	622.25	2557.6	623.25	2120.7	2743.9	1.8207	6.8565
5.00	151.9	1.0926	374.9	639.68	2561.2	640.23	2108.5	2748.7	1.8607	6.8212
6.00	158.9	1.1006	315.7	669.90	2567.4	670.56	2086.3	2756.8	1.9312	6.7600
7.00	165.0	1.1080	272.9	696.44	2572.5	697.22	2066.3	2763.5	1.9922	6.7080
8.00	170.4	1.1148	240.4	720.22	2576.8	721.11	2048.0	2769.1	2.0462	6.6628
9.00	175.4	1.1212	215.0	741.83	2580.5	742.83	2031.1	2773.9	20946	6.6226
10.0	179.9	1.1273	194.4	761.68	2583.6	762.81	2015.3	2778.1	2.1387	6.5863
15.0	198.3	1.1539	131.8	843.16	2594.5	844.84	1947.3	2792.2	2.3150	6.4448
20.0	212.4	1.1767	99.63	906.44	2600.3	908.79	1890.7	2799.5	2.4474	6.3409
25.0	224.0	1.1973	79.98	959.11	2603.1	962.11	1841.0	2803.1	2.5547	6.2575
30.0	233.9	1.2165	66.68	1004.8	2604.1	1008.4	1795.7	2804.2	2.6457	6.1869
35.0	242.6	1.2347	57.07	1045.4	2603.7	1049.8	1753.7	2803.4	2.7253	6.1253
40.0	250.4	1.2522	49.78	1082.3	2602.3	1087.3	1714.1	2801.4	2.7964	6.0701
45.0	257.5	1.2692	44.06	1116.2	2600.1	1121.9	1676.4	2798.3	2.8610	6.0199
50.0	264.0	1.2859	39.44	1147.8	2597.1	1154.2	1640.1	2794.3	2.9202	5.9734
60.0	275.6	1.3187	32.44	1205.4	2589.7	1213.4	1571.0	2784.3	3.0267	5.8892
70.0	285.9	1.3513	27.37	1257.6	2580.5	1267.0	1505.1	2772.1	3.1211	5.8133
80.0	295.1	1.3842	23.52	1305.6	2569.8	1316.6	1441.3	2758.0	3.2068	5.7432
90.0	303.4	1.4178	20.48	1350.5	2557.8	1363.3	1378.9	2742.1	3.2858	5.6772
100.	311.1	1.4524	18.03	1393.0	2544.4	1407.6	1317.1	2724.7	3.3596	5.6141
110.	318.2	1.4886	15.99	1433.7	2529.8	1450.1	1255.5	2705.6	3.4295	5.5527
120.	324.8	1.5267	14.26	1473.0	2513.7	1491.3	1193.6	2684.9	3.4962	5.4924
130.	330.9	1.5671	12.78	1511.1	2496.1	1531.5	1130.7	2662.2	3.5606	5.4323
140.	336.8	1.6107	11.49	1548.6	2476.8	1571.1	1066.5	2637.6	3.6232	5.3717
150.	342.2	1.6581	10.34	1585.6	2455.5	1610.5	1000.0	2610.5	3.6848	5.3098
160.	347.4	1.7107	9.306	1622.7	2431.7	1650.1	930.6	2580.6	3.7461	5.2455
170.	352.4	1.7702	8.364	1660.2	2405.0	1690.3	856.9	2547.2	3.8079	5.1777
180.	357.1	1.8397	7.489	1698.9	2374.3	1732.0	777.1	2509.1	3.8715	5.1044
190.	361.5	1.9243	6.657	1739.9	2338.1	1776.5	688.0	2464.5	3.9388	5.0228
200.	365.8	2.036	5.834	1785.6	2293.0	1826.3	583.4	2409.7	4.0139	4.9269
220.9	374.1	3.155	3.155	2029.6	2029.6	2099.3	0	2099.3	4.4298	4.4298

Source: J. H. Keenan, F. G. Keyes, P. G. Hill, and J. G. Moore, *Steam Tables*, Wiley, New York, 1969. Reprinted by permission of John Wiley & Sons, Inc.

TABLE A-14

Properties of water: Superheated-vapor table

***v*, cm^3/g; *u*, kJ/kg; *h*, kJ/kg; *s*, kJ/(kg·K)**

Temp., °C	v	u	h	s	v	u	h	s
	0.06 bar (0.006 MPa) (T_{sat} = 36.16°C)				**0.35 bar (0.035 MPa) (T_{sat} = 72.69°C)**			
Sat.	23739	2425.0	2567.4	8.3304	4526.	2473.0	2631.4	7.7158
80	27132	2487.3	2650.1	8.5804	4625.	2483.7	2645.6	7.7564
120	30219	2544.7	2726.0	8.7840	5163.	2542.4	2723.1	7.9644
160	33302	2602.7	2802.5	8.9693	5696.	2601.2	2800.6	8.1519
200	36383	2661.4	2879.7	9.1398	6228.	2660.4	2878.4	8.3237
240	39462	2721.0	2957.8	9.2982	6758.	2720.3	2956.8	8.4828
280	42540	2781.5	3036.8	9.4464	7287.	2780.9	3036.0	8.6314
320	45618	2843.0	3116.7	9.5859	7815.	2842.5	3116.1	8.7712
360	48696	2905.5	3197.7	9.7180	8344.	2905.1	3197.1	8.9034
400	51774	2969.0	3279.6	9.8435	8872.	2968.6	3279.2	9.0291
440	54851	3033.5	3362.6	9.9633	9400.	3033.2	3362.2	9.1490
500	59467	3132.3	3489.1	10.134	10192.	3132.1	3488.8	9.3194
	0.70 bar (0.07 MPa) (T_{sat} = 89.95°C)				**1.0 bar (0.10 MPa) (T_{sat} = 99.63°C)**			
Sat.	2365.	2494.5	2660.0	7.4797	1694.	2506.1	2675.5	7.3594
100	2434.	2509.7	2680.0	7.5341	1696.	2506.7	2676.2	7.3614
120	2571.	2539.7	2719.6	7.6375	1793.	2537.3	2716.6	7.4668
160	2841.	2599.4	2798.2	7.8279	1984.	2597.8	2796.2	7.6597
200	3108.	2659.1	2876.7	8.0012	2172.	2658.1	2875.3	7.8343
240	3374.	2719.3	2955.5	8.1611	2359.	2718.5	2954.5	7.9949
280	3640.	2780.2	3035.0	8.3162	2546.	2779.6	3034.2	8.1445
320	3905.	2842.0	3115.3	8.4504	2732.	2841.5	3114.6	8.2849
360	4170.	2904.6	3196.5	8.5828	2917.	2904.2	3195.9	8.4175
400	4434.	2968.2	3278.6	8.7086	3103.	2967.9	3278.2	8.5435
440	4698.	3032.9	3361.8	8.8286	3288.	3032.6	3361.4	8.6636
500	5095.	3131.8	3488.5	8.9991	3565.	3131.6	3488.1	8.8342
	1.5 bars (0.15 MPa) (T_{sat} = 111.37°C)				**3.0 bars (0.30 MPa) (T_{sat} = 133.55°C)**			
Sat.	1159	2519.7	2693.6	7.2233	606.	2543.6	2725.3	6.9919
120	1188.	2533.3	2711.4	7.2693				
160	1317.	2595.2	2792.8	7.4665	651.	2587.1	2782.3	7.1276
200	1444.	2656.2	2872.9	7.6433	716.	2650.7	2865.5	7.3115
240	1570.	2717.2	2952.7	7.8052	781.	2713.1	2947.3	7.4774
280	1695.	2778.6	3032.8	7.9555	844.	2775.4	3028.6	7.6299
320	1819.	2840.6	3113.5	8.0964	907.	2838.1	3110.1	7.7722
360	1943.	2903.5	3195.0	8.2293	969.	2901.4	3192.2	7.9061
400	2067.	2967.3	3277.4	8.3555	1032.	2965.6	3275.0	8.0330
440	2191.	3032.1	3360.7	8.4757	1094.	3030.6	3358.7	8.1538
500	2376.	3131.2	3487.6	8.6466	1187.	3130.0	3486.0	8.3251
600	2685.	3301.7	3704.3	8.9101	1341.	3300.8	3703.2	8.5892

TABLE A-14 (*Continued*)

Temp., °C	v	u	h	s	v	u	h	s
	5.0 bars (0.50 MPa) (T_{sat} = 151.86°C)				**7.0 bars (0.70 MPa) (T_{sat} = 164.97°C)**			
Sat.	374.9	2561.2	2748.7	6.8213	272.9	2572.5	2763.5	6.7080
180	404.5	2609.7	2812.0	6.9656	284.7	2599.8	2799.1	6.7880
200	424.9	2642.9	2855.4	7.0592	299.9	2634.8	2844.8	6.8865
240	464.6	2707.6	2939.9	7.2307	329.2	2701.8	2932.2	7.0641
280	503.4	2771.2	3022.9	7.3865	357.4	2766.9	3017.1	7.2233
320	541.6	2834.7	3105.6	7.5308	385.2	2831.3	3100.9	7.3697
360	579.6	2898.7	3188.4	7.6660	412.6	2895.8	3184.7	7.5063
400	617.3	2963.2	3271.9	7.7938	439.7	2960.9	3268.7	7.6350
440	654.8	3028.6	3356.0	7.9152	466.7	3026.6	3353.3	7.7571
500	710.9	3128.4	3483.9	8.0873	507.0	3126.8	3481.7	7.9299
600	804.1	3299.6	3701.7	8.3522	573.8	3298.5	3700.2	8.1956
700	896.9	3477.5	3925.9	8.5952	640.3	3476.6	3924.8	8.4391
	10.0 bars (1.0 MPa) (T_{sat} = 179.91°C)				**15.0 bars (1.5 MPa) (T_{sat} = 198.32°C)**			
Sat.	194.4	2583.6	2778.1	6.5865	131.8	2594.5	2792.2	6.4448
200	206.0	2621.9	2827.9	6.6940	132.5	2598.1	2796.8	6.4546
240	227.5	2692.9	2920.4	6.8817	148.3	2676.9	2899.3	6.6628
280	248.0	2760.2	3008.2	7.0465	162.7	2748.6	2992.7	6.8381
320	267.8	2826.1	3093.9	7.1962	176.5	2817.1	3081.9	6.9938
360	287.3	2891.6	3178.9	7.3349	189.9	2884.4	3169.2	7.1363
400	306.6	2957.3	3263.9	7.4651	203.0	2951.3	3255.8	7.2690
440	325.7	3023.6	3349.3	7.5883	216.0	3018.5	3342.5	7.3940
500	354.1	3124.4	3478.5	7.7622	235.2	3120.3	3473.1	7.5698
540	372.9	3192.6	3565.6	7.8720	247.8	3189.1	3560.9	7.6805
600	401.1	3296.8	3697.9	8.0290	266.8	3293.9	3694.0	7.8385
640	419.8	3367.4	3787.2	8.1290	279.3	3364.8	3783.8	7.9391
	20.0 bars (2.0 MPa) (T_{sat} = 212.42°C)				**30.0 bars (3.0 MPa) (T_{sat} = 233.90°C)**			
Sat.	99.6	2600.3	2799.5	6.3409	66.7	2604.1	2804.2	6.1869
240	108.5	2659.6	2876.5	6.4952	68.2	2619.7	2824.3	6.2265
280	120.0	2736.4	2976.4	6.6828	77.1	2709.9	2941.3	6.4462
320	130.8	2807.9	3069.5	6.8452	85.0	2788.4	3043.4	6.6245
360	141.1	2877.0	3159.3	6.9917	92.3	2861.7	3138.7	6.7801
400	151.2	2945.2	3247.6	7.1271	99.4	2932.8	3230.9	6.9212
440	161.1	3013.4	3335.5	7.2540	106.2	3002.9	3321.5	7.0520
500	175.7	3116.2	3467.6	7.4317	116.2	3108.0	3456.5	7.2338
540	185.3	3185.6	3556.1	7.5434	122.7	3178.4	3546.6	7.3474
600	199.6	3290.9	3690.1	7.7024	132.4	3285.0	3682.3	7.5085
640	209.1	3362.2	3780.4	7.8035	138.8	3357.0	3773.5	7.6106
700	223.2	3470.9	3917.4	7.9487	148.4	3466.5	3911.7	7.7571

TABLE A-14 (*Continued*)

Temp., °C	v	u	h	s	v	u	h	s
	40 bars (4.0 MPa) (T_{sat} = 250.40°C)				**60 bars (6.0 MPa) (T_{sat} = 275.64°C)**			
Sat.	49.78	2602.3	2801.4	6.0701	32.44	2589.7	2784.3	5.8892
280	55.46	2680.0	2901.8	6.2568	33.17	2605.2	2804.2	5.9252
320	61.99	2767.4	3015.4	6.4553	38.76	2720.0	2952.6	6.1846
360	67.88	2845.7	3117.2	6.6215	43.31	2811.2	3071.1	6.3782
400	73.41	2919.9	3213.6	6.7690	47.39	2892.9	3177.2	6.5408
440	78.72	2992.2	3307.1	6.9041	51.22	2970.0	3277.3	6.6853
500	86.43	3099.5	3445.3	7.0901	56.65	3082.2	3422.2	6.8803
540	91.45	3171.1	3536.9	7.2056	60.15	3156.1	3517.0	6.9999
600	98.85	3279.1	3674.4	7.3688	65.25	3266.9	3658.4	7.1677
640	103.7	3351.8	3766.6	7.4720	68.59	3341.0	3752.6	7.2731
700	111.0	3462.1	3905.9	7.6198	73.52	3453.1	3894.1	7.4234
740	115.7	3536.6	3999.6	7.7141	76.77	3528.3	3989.2	7.5190
	80 bars (8.0 MPa) (T_{sat} = 295.06°C)				**100 bars (10.0 MPa) (T_{sat} = 311.06°C)**			
Sat.	23.52	2569.8	2758.0	5.7432	18.03	2544.4	2724.7	5.6141
320	26.82	2662.7	2877.2	5.9489	19.25	2588.8	2781.3	5.7103
360	30.89	2772.7	3019.8	6.1819	23.31	2729.1	2962.1	6.0060
400	34.32	2863.8	3138.3	6.3634	26.41	2832.4	3096.5	6.2120
440	37.42	2946.7	3246.1	6.5190	29.11	2922.1	3213.2	6.3805
480	40.34	3025.7	3348.4	6.6586	31.60	3005.4	3321.4	6.5282
520	43.13	3102.7	3447.7	6.7871	33.94	3085.6	3425.1	6.6622
560	45.82	3178.7	3545.3	6.9072	36.19	3164.1	3526.0	6.7864
600	48.45	3254.4	3642.0	7.0206	38.37	3241.7	3625.3	6.9029
640	51.02	3330.1	3738.3	7.1283	40.48	3318.9	3723.7	7.0131
700	54.81	3443.9	3882.4	7.2812	43.58	3434.7	3870.5	7.1687
740	57.29	3520.4	3978.7	7.3782	45.60	3512.1	3968.1	7.2670
	120 bars (12.0 MPa) (T_{sat} = 324.75°C)				**140 bars (14.0 MPa) (T_{sat} = 336.75°C)**			
Sat.	14.26	2513.7	2684.9	5.4924	11.49	2476.8	2637.6	5.3717
360	18.11	2678.4	2895.7	5.8361	14.22	2617.4	2816.5	5.6602
400	21.08	2798.3	3051.3	6.0747	17.22	2760.9	3001.9	5.9448
440	23.55	2896.1	3178.7	6.2586	19.54	2868.6	3142.2	6.1474
480	25.76	2984.4	3293.5	6.4154	21.57	2962.5	3264.5	6.3143
520	27.81	3068.0	3401.8	6.5555	23.43	3049.8	3377.8	6.4610
560	29.77	3149.0	3506.2	6.6840	25.17	3133.6	3486.0	6.5941
600	31.64	3228.7	3608.3	6.8037	26.83	3215.4	3591.1	6.7172
640	33.45	3307.5	3709.0	6.9164	28.43	3296.0	3694.1	6.8326
700	36.10	3425.2	3858.4	7.0749	30.75	3415.7	3846.2	6.9939
740	37.81	3503.7	3957.4	7.1746	32.25	3495.2	3946.7	7.0952

TABLE A-14 (*Continued*)

Temp., °C	v	u	h	s	v	u	h	s
	160 bars (16.0 MPa) (T_{sat} = 347.44°C)				**180 bars (18.0 MPa) (T_{sat} = 357.06°C)**			
Sat.	9.31	2431.7	2580.6	5.2455	7.49	2374.3	2509.1	5.1044
360	11.05	2539.0	2715.8	5.4614	8.09	2418.9	2564.5	5.1922
400	14.26	2719.4	2947.6	5.8175	11.90	2672.8	2887.0	5.6887
440	16.52	2839.4	3103.7	6.0429	14.14	2808.2	3062.8	5.9428
480	18.42	2939.7	3234.4	6.2215	15.96	2915.9	3203.2	6.1345
520	20.13	3031.1	3353.3	6.3752	17.57	3011.8	3378.0	6.2960
560	21.72	3117.8	3465.4	6.5132	19.04	3101.7	3444.4	6.4392
600	23.23	3201.8	3573.5	6.6399	20.42	3188.0	3555.6	6.5696
640	24.67	3284.2	3678.9	6.7580	21.74	3272.3	3663.6	6.6905
700	26.74	3406.0	3833.9	6.9224	23.62	3396.3	3821.5	6.8580
740	28.08	3486.7	3935.9	7.0251	24.83	3478.0	3925.0	6.9623
	200 bars (20.0 MPa) (T_{sat} = 365.81°C)				**240 bars (24.0 MPa)**			
Sat.	5.83	2293.0	2409.7	4.9269				
400	9.94	2619.3	2818.1	5.5540	6.73	2477.8	2639.4	5.2393
440	12.22	2774.9	3019.4	5.8450	9.29	2700.6	2923.4	5.6506
480	13.99	2891.2	3170.8	6.0518	11.00	2838.3	3102.3	5.8950
520	15.51	2992.0	3302.2	6.2218	12.41	2950.5	3248.5	6.0842
560	16.89	3085.2	3423.0	6.3705	13.66	3051.1	3379.0	6.2448
600	18.18	3174.0	3537.6	6.5048	14.81	3145.2	3500.7	6.3875
640	19.40	3260.2	3648.1	6.6286	15.88	3235.5	3616.7	6.5174
700	21.13	3386.4	3809.0	6.7993	17.39	3366.4	3783.8	6.6947
740	22.24	3469.3	3914.1	6.9052	18.35	3451.7	3892.1	6.8038
800	23.85	3592.7	4069.7	7.0544	19.74	3578.0	4051.6	6.9567
	280 bars (28.0 MPa)				**320 bars (32.0 MPa)**			
400	3.83	2223.5	2330.7	4.7494	2.36	1980.4	2055.9	4.3239
440	7.12	2613.2	2812.6	5.4494	5.44	2509.0	2683.0	5.2327
480	8.85	2780.8	3028.5	5.7446	7.22	2718.1	3949.2	5.5968
520	10.20	3906.8	3192.3	5.9566	8.53	2860.7	3133.7	5.8357
560	11.36	3015.7	3333.7	6.1307	9.63	2979.0	3287.2	6.0246
600	12.41	3115.6	3463.0	6.2823	10.61	3085.3	3424.6	6.1858
640	13.38	3210.3	3584.8	6.4187	11.50	3184.5	3552.5	6.3290
700	14.73	3346.1	3758.4	6.6029	12.73	3325.4	3732.8	6.5203
740	15.58	3433.9	3870.0	6.7153	13.50	3415.9	3847.8	6.6361
800	16.80	3563.1	4033.4	6.8720	14.60	3548.0	4015.1	6.7966
900	18.73	3774.3	4298.8	7.1084	16.33	3762.7	4285.1	7.0372

Source: J. H. Keenan, F. G. Keyes, P. G. Hill, and J. G. Moore, *Steam Tables*, Wiley, New York, 1969. Reprinted by permission of John Wiley & Sons, Inc.

TABLE A-15
Properties of water: Compressed-liquid table
v, cm³/g; u, kJ/kg; h, kJ/kg; s, kJ/(kg·K)

Temp., °C	v	u	h	s	v	u	h	s
	25 bars (2.5 MPa) (223.99°C)				**50 bars (5.0 MPa) (263.99°C)**			
20	1.0006	83.80	86.30	.2961	.9995	83.65	88.65	.2956
40	1.0067	167.25	169.77	.5715	1.0056	166.95	171.97	.5705
80	1.0280	334.29	336.86	1.0737	1.0268	333.72	338.85	1.0720
100	1.0423	418.24	420.85	1.3050	1.0410	417.52	422.72	1.3030
140	1.0784	587.82	590.52	1.7369	1.0768	586.76	592.15	1.7343
180	1.1261	761.16	763.97	2.1375	1.1240	759.63	765.25	2.1341
200	1.1555	849.9	852.8	2.3294	1.1530	848.1	848.1	2.3255
220	1.1898	940.7	943.7	2.5174	1.1866	938.4	944.4	2.5128
Sat.	1.1973	959.1	962.1	2.5546	1.2859	1147.8	1154.2	2.9202
	75 bars (7.5 MPa) (290.59°C)				**100 bars (10 MPa) (311.06°C)**			
20	.9984	83.50	90.99	.2950	.9972	83.36	93.33	.2945
40	1.0045	166.64	174.18	.5696	1.0034	166.35	176.38	.5686
80	1.0256	333.15	340.84	1.0704	1.0245	332.59	342.83	1.0688
100	1.0397	416.81	424.62	1.3011	1.0385	416.12	426.50	1.2992
140	1.0752	585.72	593.78	1.7317	1.0737	584.68	595.42	1.7292
180	1.1219	758.13	766.55	2.1308	1.1199	756.65	767.84	2.1275
220	1.1835	936.2	945.1	2.5083	1.1805	934.1	945.9	2.5039
260	1.2696	1124.4	1134.0	2.8763	1.2645	1121.1	1133.7	2.8699
Sat.	1.3677	1282.0	1292.2	3.1649	1.4524	1393.0	1407.6	3.3596
	150 bars (15 MPa) (342.24°C)				**200 bars (20 MPa) (365.81°C)**			
20	.9950	83.06	97.99	.2934	.9928	82.77	102.62	.2923
40	1.0013	165.76	180.78	.5666	.9992	165.17	185.16	.5646
80	1.0222	331.48	346.81	1.0656	1.0199	330.40	350.80	1.0624
100	1.0361	414.74	430.28	1.2955	1.0337	413.39	434.06	1.2917
140	1.0707	582.66	598.72	1.7242	1.0678	580.69	602.04	1.7193
180	1.1159	753.76	770.50	2.1210	1.1120	750.95	773.20	2.1147
220	1.1748	929.9	947.5	2.4953	1.1693	925.9	949.3	2.4870
260	1.2550	1114.6	1133.4	2.8576	1.2462	1108.6	1133.5	2.8459
300	1.3770	1316.6	1337.3	3.2260	1.3596	1306.1	1333.3	3.2071
Sat.	1.6581	1585.6	1610.5	3.6848	2.036	1785.6	1826.3	4.0139
	250 bars (25 MPa)				**300 bars (30 MPa)**			
20	.9907	82.47	107.24	.2911	.9886	82.17	111.84	.2899
40	.9971	164.60	189.52	.5626	.9951	164.04	193.89	.5607
100	1.0313	412.08	437.85	1.2881	1.0290	410.78	441.66	1.2844
200	1.1344	834.5	862.8	2.2961	1.1302	831.4	865.3	2.2893
300	1.3442	1296.6	1330.2	3.1900	1.3304	1287.9	1327.8	3.1741

Source: J. H. Keenan, F. G. Keyes, P. G. Hill, and J. G. Moore, *Steam Tables*, Wiley, New York, 1969. Reprinted by permission of John Wiley & Sons, Inc.

TABLE A-16

Properties of saturated refrigerant-134a (CF_4H_2): Temperature table

v, m^3/kg; u, kJ/kg; h, kJ/kg; s, kJ/(kg·K)

		Specific volume		Internal energy		Enthalpy			Entropy	
Temp., °C	Press., bars	Sat. liquid $v_f \times 10^3$	Sat. vapor v_g	Sat. liquid u_f	Sat. vapor u_g	Sat. liquid h_f	Evap. h_{fg}	Sat. vapor h_g	Sat. liquid s_f	Sat. vapor s_g
−40	0.5164	0.7055	0.3569	−0.04	204.45	0.00	222.88	222.88	0.0000	0.9560
−36	0.6332	0.7113	0.2947	4.68	206.73	4.73	220.67	225.40	0.0201	0.9506
−32	0.7704	0.7172	0.2451	9.47	209.01	9.52	218.37	227.90	0.0401	0.9456
−28	0.9305	0.7233	0.2052	14.31	211.29	14.37	216.01	230.38	0.0600	0.9411
−26	1.0199	0.7265	0.1882	16.75	212.43	16.82	214.80	231.62	0.0699	0.9390
−24	1.1160	0.7296	0.1728	19.21	213.57	19.29	213.57	232.85	0.0798	0.9370
−22	1.2192	0.7328	0.1590	21.68	214.70	21.77	212.32	234.08	0.0897	0.9351
−20	1.3299	0.7361	0.1464	24.17	215.84	24.26	211.05	235.31	0.0996	0.9332
−18	1.4483	0.7395	0.1350	26.67	216.97	26.77	209.76	236.53	0.1094	0.9315
−16	1.5748	0.7428	0.1247	29.18	218.10	29.30	208.45	237.74	0.1192	0.9298
−12	1.8540	0.7498	0.1068	34.25	220.36	34.39	205.77	240.15	0.1388	0.9267
−8	2.1704	0.7569	0.0919	39.38	222.60	39.54	203.00	242.54	0.1583	0.9239
−4	2.5274	0.7644	0.0794	44.56	224.84	44.75	200.15	244.90	0.1777	0.9213
0	2.9282	0.7721	0.0689	49.79	227.06	50.02	197.21	247.23	0.1970	0.9190
4	3.3765	0.7801	0.0600	55.08	229.27	55.35	194.19	249.53	0.2162	0.9169
8	3.8756	0.7884	0.0525	60.43	231.46	60.73	191.07	251.80	0.2354	0.9150
12	4.4294	0.7971	0.0460	65.83	233.63	66.18	187.85	254.03	0.2545	0.9132
16	5.0416	0.8062	0.0405	71.29	235.78	71.69	184.52	256.22	0.2735	0.9116
20	5.7160	0.8157	0.0358	76.80	237.91	77.26	181.09	258.36	0.2924	0.9102
24	6.4566	0.8257	0.0317	82.37	240.01	82.90	177.55	260.45	0.3113	0.9089

TABLE A-16 *(Continued)*

Temp., °C	Press., bars	Specific volume		Internal energy		Enthalpy			Entropy	
		Sat. liquid $v_f \times 10^3$	Sat. vapor v_g	Sat. liquid u_f	Sat. vapor u_g	Sat. liquid h_f	Evap. h_{fg}	Sat. vapor h_g	Sat. liquid s_f	Sat. vapor s_g
26	6.8530	0.8309	0.0298	85.18	241.05	85.75	175.73	261.48	0.3208	0.9082
28	7.2675	0.8362	0.0281	88.00	242.08	88.61	173.89	262.50	0.3302	0.9076
30	7.7006	0.8417	0.0265	90.84	243.10	91.49	172.00	263.50	0.3396	0.9070
32	8.1528	0.8473	0.0250	93.70	244.12	94.39	170.09	264.48	0.3490	0.9064
34	8.6247	0.8530	0.0236	96.58	245.12	97.31	168.14	265.45	0.3584	0.9058
36	9.1168	0.8590	0.0223	99.47	246.11	100.25	166.15	266.40	0.3678	0.9053
38	9.6298	0.8651	0.0210	102.38	247.09	103.21	164.12	267.33	0.3772	0.9047
40	10.164	0.8714	0.0199	105.30	248.06	106.19	162.05	268.24	0.3866	0.9041
42	10.720	0.8780	0.0188	108.25	249.02	109.19	159.94	269.14	0.3960	0.9035
44	11.299	0.8847	0.0177	111.22	249.96	112.22	157.79	270.01	0.4054	0.9030
48	12.526	0.8989	0.0159	117.22	251.79	118.35	153.33	271.68	0.4243	0.9017
52	13.851	0.9142	0.0142	123.31	253.55	124.58	148.66	273.24	0.4432	0.9004
56	15.278	0.9308	0.0127	129.51	255.23	130.93	143.75	274.68	0.4622	0.8990
60	16.813	0.9488	0.0114	135.82	256.81	137.42	138.57	275.99	0.4814	0.8973
70	21.162	1.0027	0.0086	152.22	260.15	154.34	124.08	278.43	0.5302	0.8918
80	26.324	1.0766	0.0064	169.88	262.14	172.71	106.41	279.12	0.5814	0.8827
90	32.435	1.1949	0.0046	189.82	261.34	193.69	82.63	276.32	0.6380	0.8655
100	39.742	1.5443	0.0027	218.60	248.49	224.74	34.40	259.13	0.7196	0.8117

Source: Produced from a computer program provided by R. S. Basu, Allied Signal Corporation. *Literature Source:* D. P. Wilson and R. S. Basu, "Thermodynamic Properties of a New Stratospherically Safe Working Fluid—Refrigerant-134a," *ASHRAE Trans.*, **94**(Pt 2); 2095–2118, 1988.

TABLE A-17

Properties of saturated refrigerant-134a (CF_4H_2): Pressure table

v, m^3/kg; u, kJ/kg; h, kJ/kg; s, kJ/(kg·K)

		Specific volume		Internal energy		Enthalpy			Entropy	
Press., bars	Temp., °C	Sat. liquid $v_f \times 10^3$	Sat. vapor v_g	Sat. liquid u_f	Sat. vapor u_g	Sat. liquid h_f	Evap. h_{fg}	Sat. vapor h_g	Sat. liquid s_f	Sat. vapor s_g
0.6	−37.07	0.7097	0.3100	3.41	206.12	3.46	221.27	224.72	0.0147	0.9520
0.8	−31.21	0.7184	0.2366	10.41	209.46	10.47	217.92	228.39	0.0440	0.9447
1.0	−26.43	0.7258	0.1917	16.22	212.18	16.29	215.06	231.35	0.0678	0.9395
1.2	−22.36	0.7323	0.1614	21.23	214.50	21.32	212.54	233.86	0.0879	0.9354
1.4	−18.80	0.7381	0.1395	25.66	216.52	25.77	210.27	236.04	0.1055	0.9322
1.6	−15.62	0.7435	0.1229	29.66	218.32	29.78	208.19	237.97	0.1211	0.9295
1.8	−12.73	0.7485	0.1098	33.31	219.94	33.45	206.26	239.71	0.1352	0.9273
2.0	−10.09	0.7532	0.0993	36.69	221.43	36.84	204.46	241.30	0.1481	0.9253
2.4	−5.37	0.7618	0.0834	42.77	224.07	42.95	201.14	244.09	0.1710	0.9222
2.8	−1.23	0.7697	0.0719	48.18	226.38	48.39	198.13	246.52	0.1911	0.9197
3.2	2.48	0.7770	0.0632	53.06	228.43	53.31	195.35	248.66	0.2089	0.9177
3.6	5.84	0.7839	0.0564	57.54	230.28	57.82	192.76	250.58	0.2251	0.9160

TABLE A-17 *(Continued)*

Press., bars	Temp., °C	Specific volume: Sat. liquid $v_f \times 10^3$	Specific volume: Sat. vapor v_g	Internal energy: Sat. liquid u_f	Internal energy: Sat. vapor u_g	Enthalpy: Sat. liquid h_f	Enthalpy: Evap. h_{fg}	Enthalpy: Sat. vapor h_g	Entropy: Sat. liquid s_f	Entropy: Sat. vapor s_g
4.0	8.93	0.7904	0.0509	61.69	231.97	62.00	190.32	252.32	0.2399	0.9145
5.0	15.74	0.8056	0.0409	70.93	235.64	71.33	184.74	256.07	0.2723	0.9117
6.0	21.58	0.8196	0.0341	78.99	238.74	79.48	179.71	259.19	0.2999	0.9097
7.0	26.72	0.8328	0.0292	86.19	241.42	86.78	175.07	261.85	0.3242	0.9080
8.0	31.33	0.8454	0.0255	92.75	243.78	93.42	170.73	264.15	0.3459	0.9066
9.0	35.53	0.8576	0.0226	98.79	245.88	99.56	166.62	266.18	0.3656	0.9054
10.0	39.39	0.8695	0.0202	104.42	247.77	105.29	162.68	267.97	0.3838	0.9043
12.0	46.32	0.8928	0.0166	114.69	251.03	115.76	155.23	270.99	0.4164	0.9023
14.0	52.43	0.9159	0.0140	123.98	253.74	125.26	148.14	273.40	0.4453	0.9003
16.0	57.92	0.9392	0.0121	132.52	256.00	134.02	141.31	275.33	0.4714	0.8982
18.0	62.91	0.9631	0.0105	140.49	257.88	142.22	134.60	276.83	0.4954	0.8959
20.0	67.49	0.9878	0.0093	148.02	259.41	149.99	127.95	277.94	0.5178	0.8934
25.0	77.59	1.0562	0.0069	165.48	261.84	168.12	111.06	279.17	0.5687	0.8854
30.0	86.22	1.1416	0.0053	181.88	262.16	185.30	92.71	278.01	0.6156	0.8735

Source: Produced from a computer program provided by R. S. Basu, Allied Signal Corporation. *Literature Source:* D. P. Wilson and R. S. Basu, "Thermodynamic Properties of a New Stratospherically Safe Working Fluid—Refrigerant-134a," *ASHRAE Trans.*, **94**(Pt. 2): 2095–2118, 1988.

TABLE A-18

Properties of superheated refrigerant-134a (CF_4H_2)

T, °C; *v*, m^3/kg; *u*, kJ/kg; *h*, kJ/kg; *s*, kJ/(kg·K)

T	*v*	*u*	*h*	*s*	*v*	*u*	*h*	*s*
	0.6 bar (0.060 MPa) (T_{sat} = − 37.07°C)				**1.0 bar (0.10 MPa) (T_{sat} = − 26.43°C)**			
Sat.	0.31003	206.12	224.72	0.9520	0.19170	212.18	231.35	0.9395
−20	0.33536	217.86	237.98	1.0062	0.19770	216.77	236.54	0.9602
−10	0.34992	224.97	245.96	1.0371	0.20686	224.01	244.70	0.9918
0	0.36433	232.24	254.10	1.0675	0.21587	231.41	252.99	1.0227
10	0.37861	239.69	262.41	1.0973	0.22473	238.96	261.43	1.0531
20	0.39279	247.32	270.89	1.1267	0.23349	246.67	270.02	1.0829
30	0.40688	255.12	279.53	1.1557	0.24216	254.54	278.76	1.1122
40	0.42091	263.10	288.35	1.1844	0.25076	262.58	287.66	1.1411
50	0.43487	271.25	297.34	1.2126	0.25930	270.79	296.72	1.1696
60	0.44879	279.58	306.51	1.2405	0.26779	279.16	305.94	1.1977
70	0.46266	288.08	315.84	1.2681	0.27623	287.70	315.32	1.2254
80	0.47650	296.75	325.34	1.2954	0.28464	296.40	324.87	1.2528
90	0.49031	305.58	335.00	1.3224	0.29302	305.27	334.57	1.2799
	1.4 bars (0.14 MPa) (T_{sat} = − 18.80°C)				**1.8 bars (0.18 MPa) (T_{sat} = − 12.73°C)**			
Sat.	0.13945	216.52	236.04	0.9322	0.10983	219.94	239.71	0.9273
−10	0.14549	223.03	243.40	0.9606	0.11135	222.02	242.06	0.9362
0	0.15219	230.55	251.86	0.9922	0.11678	229.67	250.69	0.9684
10	0.15875	238.21	260.43	1.0230	0.12207	237.44	259.41	0.9998
20	0.16520	246.01	269.13	1.0532	0.12723	245.33	268.23	1.0304
30	0.17155	253.96	277.97	1.0828	0.13230	253.36	277.17	1.0604
40	0.17783	262.06	286.96	1.1120	0.13730	261.53	286.24	1.0898
50	0.18404	270.32	296.09	1.1407	0.14222	269.85	295.45	1.1187
60	0.19020	278.74	305.37	1.1690	0.14710	278.31	304.79	1.1472
70	0.19633	287.32	314.80	1.1969	0.15193	286.93	314.28	1.1753
80	0.20241	296.06	324.39	1.2244	0.15672	295.71	323.92	1.2030
90	0.20846	304.95	334.14	1.2516	0.16148	304.63	333.70	1.2303
	2.0 bars (0.20 MPa) (T_{sat} = − 10.09°C)				**2.4 bars (0.24 MPa) (T_{sat} = − 5.37°C)**			
Sat.	0.09933	221.43	241.30	0.9253	0.08343	224.07	244.09	0.9222
−10	0.09938	221.50	241.38	0.9256				
0	0.10438	229.23	250.10	0.9582	0.08574	228.31	248.89	0.9399
10	0.10922	237.05	258.89	0.9898	0.08993	236.26	257.84	0.9721
20	0.11394	244.99	267.78	1.0206	0.09399	244.30	266.85	1.0034
30	0.11856	253.06	276.77	1.0508	0.09794	252.45	275.95	1.0339
40	0.12311	261.26	285.88	1.0804	0.10181	260.72	285.16	1.0637
50	0.12758	269.61	295.12	1.1094	0.10562	269.12	294.47	1.0930
60	0.13201	278.10	304.50	1.1380	0.10937	277.67	303.91	1.1218
70	0.13639	286.74	314.02	1.1661	0.11307	286.35	313.49	1.1501
80	0.14073	295.53	323.68	1.1939	0.11674	295.18	323.19	1.1780
90	0.14504	304.47	333.48	1.2212	0.12037	304.15	333.04	1.2055

TABLE A-18 (*Continued*)

T	*v*	*u*	*h*	*s*	*v*	*u*	*h*	*s*
	2.8 bars (0.28 MPa) (T_{sat} = − 1.23°C)				**3.2 bars (0.32 MPa) (T_{sat} = 2.48°C)**			
Sat.	0.07193	226.38	246.52	0.9197	0.06322	228.43	248.66	0.9177
0	0.07240	227.37	247.64	0.9238				
10	0.07613	235.44	256.76	0.9566	0.06576	234.61	255.65	0.9427
20	0.07972	243.59	265.91	0.9883	0.06901	242.87	264.95	0.9749
30	0.08320	251.83	275.12	1.0192	0.07214	251.19	274.28	1.0062
40	0.08660	260.17	284.42	1.0494	0.07518	259.61	283.67	1.0367
50	0.08992	268.64	293.81	1.0789	0.07815	268.14	293.15	1.0665
60	0.09319	277.23	303.32	1.1079	0.08106	276.79	302.72	1.0957
70	0.09641	285.96	312.95	1.1364	0.08392	285.56	312.41	1.1243
80	0.09960	294.82	322.71	1.1644	0.08674	294.46	322.22	1.1525
90	0.10275	303.83	332.60	1.1920	0.08953	303.50	332.15	1.1802
100	0.10587	312.98	342.62	1.2193	0.09229	312.68	342.21	1.2076
	4.0 bars (0.40 MPa) (T_{sat} = 8.93°C)				**5.0 bars (0.50 MPa) (T_{sat} = 15.74°C)**			
Sat.	0.05089	231.97	252.32	0.9145	0.04086	235.64	256.07	0.9117
10	0.05119	232.87	253.35	0.9182				
20	0.05397	241.37	262.96	0.9515	0.04188	239.40	260.34	0.9264
30	0.05662	249.89	272.54	0.9837	0.04416	248.20	270.28	0.9597
40	0.05917	258.47	282.14	1.0148	0.04633	256.99	280.16	0.9918
50	0.06164	267.13	291.79	1.0452	0.04842	265.83	290.04	1.0229
60	0.06405	275.89	301.51	1.0748	0.05043	274.73	299.95	1.0531
70	0.06641	284.75	311.32	1.1038	0.05240	283.72	309.92	1.0825
80	0.06873	293.73	321.23	1.1322	0.05432	292.80	319.96	1.1114
90	0.07102	302.84	331.25	1.1602	0.05620	302.00	330.10	1.1397
100	0.07327	312.07	341.38	1.1878	0.05805	311.31	340.33	1.1675
110	0.07550	321.44	351.64	1.2149	0.05988	320.74	350.68	1.1949
	6.0 bars (0.60 MPa) (T_{sat} = 21.58°C)				**7.0 bars (0.70 MPa) (T_{sat} = 26.72°C)**			
Sat.	0.03408	238.74	259.19	0.9097	0.02918	241.42	261.85	0.9080
30	0.03581	246.41	267.89	0.9388	0.02979	244.51	265.37	0.9197
40	0.03774	255.45	278.09	0.9719	0.03157	253.83	275.93	0.9539
50	0.03958	264.48	288.23	1.0037	0.03324	263.08	286.35	0.9867
60	0.04134	273.54	298.35	1.0346	0.03482	272.31	296.69	1.0182
70	0.04304	282.66	308.48	1.0645	0.03634	281.57	307.01	1.0487
80	0.04469	291.86	318.67	1.0938	0.03781	290.88	317.35	1.0784
90	0.04631	301.14	328.93	1.1225	0.03924	300.27	327.74	1.1074
100	0.04790	310.53	339.27	1.1505	0.04064	309.74	338.19	1.1358
110	0.04946	320.03	349.70	1.1781	0.04201	319.31	348.71	1.1637
120	0.05099	329.64	360.24	1.2053	0.04335	328.98	359.33	1.1910
130	0.05251	339.38	370.88	1.2320	0.04468	338.76	370.04	1.2179

TABLE A-18 (*Continued*)

T	*v*	*u*	*h*	*s*	*v*	*u*	*h*	*s*
	8.0 bars (0.80 MPa) (T_{sat} = 31.33°C)				**9.0 bars (0.90 MPa) (T_{sat} = 35.53°C)**			
Sat.	0.02547	243.78	264.15	0.9066	0.02255	245.88	266.18	0.9054
40	0.02691	252.13	273.66	0.9374	0.02325	250.32	271.25	0.9217
50	0.02846	261.62	284.39	0.9711	0.02472	260.09	282.34	0.9566
60	0.02992	271.04	294.98	1.0034	0.02609	269.72	293.21	0.9897
70	0.03131	280.45	305.50	1.0345	0.02738	279.30	303.94	1.0214
80	0.03264	289.89	316.00	1.0647	0.02861	288.87	314.62	1.0521
90	0.03393	299.37	326.52	1.0940	0.02980	298.46	325.28	1.0819
100	0.03519	308.93	337.08	1.1227	0.03095	308.11	335.96	1.1109
110	0.03642	318.57	347.71	1.1508	0.03207	317.82	346.68	1.1392
120	0.03762	328.31	358.40	1.1784	0.03316	327.62	357.47	1.1670
130	0.03881	338.14	369.19	1.2055	0.03423	337.52	368.33	1.1943
140	0.03997	348.09	380.07	1.2321	0.03529	347.51	379.27	1.2211
	10.0 bars (1.00 MPa) (T_{sat} = 39.39°C)				**12.0 bars (1.20 MPa) (T_{sat} = 46.32°C)**			
Sat.	0.02020	247.77	267.97	0.9043	0.01663	251.03	270.99	0.9023
40	0.02029	248.39	268.68	0.9066				
50	0.02171	258.48	280.19	0.9428	0.01712	254.98	275.52	0.9164
60	0.02301	268.35	291.36	0.9768	0.01835	265.42	287.44	0.9527
70	0.02423	278.11	302.34	1.0093	0.01947	275.59	298.96	0.9868
80	0.02538	287.82	313.20	1.0405	0.02051	285.62	310.24	1.0192
90	0.02649	297.53	324.01	1.0707	0.02150	295.59	321.39	1.0503
100	0.02755	307.27	334.82	1.1000	0.02244	305.54	332.47	1.0804
110	0.02858	317.06	345.65	1.1286	0.02335	315.50	343.52	1.1096
120	0.02959	326.93	356.52	1.1567	0.02423	325.51	354.58	1.1381
130	0.03058	336.88	367.46	1.1841	0.02508	335.58	365.68	1.1660
140	0.03154	346.92	378.46	1.2111	0.02592	345.73	376.83	1.1933
	14.0 bars (1.40 MPa) (T_{sat} = 52.43°C)				**16.0 bars (1.60 MPa) (T_{sat} = 57.92°C)**			
Sat.	0.01405	253.74	273.40	0.9003	0.01208	256.00	275.33	0.8982
60	0.01495	262.17	283.10	0.9297	0.01233	258.48	278.20	0.9069
70	0.01603	272.87	295.31	0.9658	0.01340	269.89	291.33	0.9457
80	0.01701	283.29	307.10	0.9997	0.01435	280.78	303.74	0.9813
90	0.01792	293.55	318.63	1.0319	0.01521	291.39	315.72	1.0148
100	0.01878	303.73	330.02	1.0628	0.01601	301.84	327.46	1.0467
110	0.01960	313.88	341.32	1.0927	0.01677	312.20	339.04	1.0773
120	0.02039	324.05	352.59	1.1218	0.01750	322.53	350.53	1.1069
130	0.02115	334.25	363.86	1.1501	0.01820	332.87	361.99	1.1357
140	0.02189	344.50	375.15	1.1777	0.01887	343.24	373.44	1.1638
150	0.02262	354.82	386.49	1.2048	0.01953	353.66	384.91	1.1912
160	0.02333	365.22	397.89	1.2315	0.02017	364.15	396.43	1.2181

Source: Produced from a computer program provided by R. S. Basu, Allied Signal Corporation. *Literature Source:* D. P. Wilson and R. S. Basu, "Thermodynamic Properties of a New Stratospherically Safe Working Fluid—Refrigerant-134a," *ASHRAE Trans.*, **94**(Pt 2): 2095–2118, 1988.

TABLE A-19

Properties of saturated nitrogen (N_2): Temperature and pressure tables v, L/kg or cm^3/g; u and h, kJ/kg; s, kJ/(kg·K); 1 bar = 0.1 MPa

		Specific volume		Internal energy		Enthalpy			Entropy	
Temp., K T	Press., bar(s) P	Sat. liquid v_f	Sat. vapor v_g	Sat. liquid u_f	Sat. vapor u_g	Sat. liquid h_f	Evap. h_{fg}	Sat. vapor h_g	Sat. liquid s_f	Sat. vapor s_g
63.15	0.125	1.152	1477.	−50.8	145.6	−50.8	214.9	164.1	2.423	5.826
70.0	0.386	1.191	526.	−37.1	150.2	−37.0	207.5	170.5	2.631	5.595
80.0	1.369	1.257	164.	−16.4	156.4	−16.2	195.1	178.9	2.906	5.345
90.0	3.600	1.340	66.3	4.5	161.1	5.0	180.0	185.0	3.152	5.152
100.0	7.775	1.448	31.3	25.7	163.4	26.8	160.9	187.7	3.376	4.985
110.0	14.67	1.606	16.0	48.3	162.2	50.7	134.9	185.6	3.594	4.820
120.0	25.15	1.908	8.03	76.9	154.1	81.7	92.6	174.3	3.847	4.619
126.25	33.96	3.388	3.39	123.2	123.2	134.7	0	134.7	4.257	4.257
77.24	1.00	1.238	219.1	−22.0	154.9	−21.9	198.7	176.8	2.835	5.407
83.63	2.00	1.285	115.3	−8.9	158.3	−8.6	190.0	181.4	2.998	5.270
91.25	4.00	1.352	60.0	7.1	161.5	7.6	177.9	185.5	3.180	5.130
96.41	6.00	1.406	40.46	18.0	162.9	18.8	168.4	187.2	3.297	5.044
100.41	8.00	1.454	30.37	26.4	163.4	27.6	160.1	187.7	3.384	4.978
103.76	10.00	1.500	24.16	33.9	163.4	35.4	152.2	187.6	3.457	4.924
110.38	15.00	1.614	15.57	49.3	162.0	51.7	133.7	185.4	3.603	4.814
115.56	20.00	1.741	11.04	63.0	159.1	66.5	114.7	181.2	3.726	4.718
119.88	25.00	1.902	8.11	76.5	154.3	81.3	93.3	174.6	3.844	4.622
123.61	30.00	2.152	5.87	91.6	146.3	98.1	65.8	163.9	3.974	4.506
125.63	33.00	2.497	4.550	104.7	137.1	112.9	39.2	152.1	4.088	4.400

Source: Data adapted from A. A. Vasserman, Ya. Z. Kazavchinskii, and V. A. Rabinovich, *Thermophysical Properties of Air and Air Components*, Izdatel'stvo Nauka, Moscow, 1966.

TABLE A-20

Properties of nitrogen (N_2): Superheated-vapor table
v, L/kg or cm^3/g; u and h, kJ/kg; s, kJ/(kg·K)

Temp., K	v	u	h	s	v	u	h	s
	1 bar (0.1 MPa) (T_{sat} = 77.2 K)				**5 bars (0.5 MPa) (T_{sat} = 94.0 K)**			
100	290.8	172.8	201.9	5.691	52.9	167.7	194.1	5.162
150	442.6	210.4	254.7	6.120	86.4	208.1	251.3	5.627
200	592.3	247.8	307.0	6.421	117.5	246.4	305.1	5.937
250	741.4	285.0	359.1	6.654	147.8	284.0	357.9	6.172
300	890.2	322.2	411.2	6.844	177.9	321.4	410.3	6.364
350	1039.	359.4	463.3	7.005	207.9	358.8	462.7	6.525
400	1187.	396.7	515.4	7.144	237.7	396.2	515.0	6.665
	10 bars (1.0 MPa) (T_{sat} = 108.8 K)				**20 bars (2.0 MPa) (T_{sat} = 115.6 K)**			
150	41.9	205.0	246.9	5.401	19.6	198.4	237.5	5.150
200	58.1	244.6	302.7	5.722	28.5	240.9	297.8	5.498
250	73.6	282.7	356.3	5.962	36.6	280.1	353.2	5.746
300	88.9	320.3	409.2	6.155	44.4	318.3	407.1	5.942
350	104.0	357.9	461.9	6.317	52.1	356.2	460.4	6.107
400	119.0	395.5	514.5	6.457	59.7	394.1	513.5	6.248
450	134.0	433.2	567.2	6.582	67.3	432.0	566.5	6.373
	50 bars (5.0 MPa)				**100 bars (10.0 MPa)**			
150	5.92	171.5	201.1	4.692	2.39	123.5	147.4	4.218
200	10.7	229.3	282.9	5.169	5.02	208.8	259.0	4.870
250	14.4	272.2	344.1	5.443	7.12	259.3	330.4	5.190
300	17.8	312.4	401.1	5.651	8.95	302.7	392.2	5.416
350	21.0	351.4	456.3	5.821	10.7	343.6	450.3	5.596
400	24.1	390.1	510.7	5.967	12.3	383.6	506.8	5.746
450	27.2	428.7	564.8	6.094	13.9	423.1	562.4	5.877
	150 bars (15.0 MPa)				**200 bars (20.0 MPa)**			
150	1.95	94.3	134.9	4.064	1.78	96.1	131.6	3.980
200	3.37	190.8	241.3	4.680	2.69	177.2	231.0	4.554
250	4.80	247.1	319.2	5.030	3.73	236.6	311.2	4.913
300	6.09	293.5	384.8	5.269	4.70	285.3	379.3	5.162
350	7.28	336.3	445.5	5.456	5.62	329.5	441.8	5.355
400	8.42	377.5	503.7	5.612	6.49	371.8	501.5	5.515
450	9.52	418.0	560.7	5.746	7.33	413.0	559.6	5.651

Source: Data adapted from A. A. Vasserman, Ya. Z. Kazavchinskii, and V. A. Rabinovich, *Thermophysical Properties of Air and Air Components*, Izdatel'stvo Nauka, Moscow, 1966.

TABLE A-21

Properties of saturated carbon dioxide (CO_2): Temperature table

v, m³/kg; u and h, kJ/kg; s, kJ/(kg·K)

Temp., K T	Press., bars P	Specific volume: Sat. liquid $v_f \times 10^3$	Specific volume: Sat. vapor $v_g \times 10^3$	Internal energy: Sat. liquid u_f	Internal energy: Sat. vapor u_g	Enthalpy: Sat. liquid h_f	Enthalpy: Evap. h_{fg}	Enthalpy: Sat. vapor h_g	Entropy: Sat. liquid s_f	Entropy: Sat. vapor s_g
216.58	5.180	0.8486	71.216	−0.44	307.33	0.00	344.22	344.22	0.0000	1.5892
220	5.996	0.8575	62.368	5.76	309.59	6.27	340.72	346.99	0.0279	1.5766
230	8.935	0.8857	42.781	23.93	312.97	24.72	326.47	351.19	0.1070	1.5265
240	12.830	0.9179	29.966	42.83	314.47	44.01	308.91	352.92	0.1863	1.4735
250	17.856	0.9555	21.403	62.62	315.00	64.33	288.89	353.22	0.2670	1.4226
260	24.194	1.0005	15.509	83.31	314.38	85.73	266.17	351.90	0.3492	1.3729
270	32.034	1.0566	11.316	105.12	312.17	108.50	239.92	348.42	0.4331	1.3217
280	41.595	1.1309	8.222	128.57	307.66	133.27	208.59	341.86	0.5203	1.2654
282	43.737	1.1492	7.697	133.86	306.35	138.56	201.45	340.01	0.5383	1.2527
284	45.960	1.1690	7.198	138.64	304.87	144.01	193.93	334.95	0.5567	1.2395
286	48.269	1.1907	6.721	143.88	303.17	149.63	185.98	335.61	0.5758	1.2259
288	50.665	1.2147	6.263	149.31	301.19	155.47	177.46	332.93	0.5951	1.2113
290	53.152	1.2416	5.823	154.93	298.91	161.53	168.33	329.86	0.6151	1.1956
292	55.734	1.2720	5.396	160.83	296.24	167.92	158.40	326.31	0.6360	1.1784
294	58.415	1.3072	4.979	167.96	293.95	175.60	147.44	323.04	0.6580	1.1595
296	61.198	1.3491	4.568	173.72	289.17	181.98	135.15	317.13	0.6817	1.1382
298	64.090	1.4011	4.155	181.07	284.35	190.05	120.93	310.98	0.7073	1.1132
300	67.095	1.4699	3.726	189.44	278.00	199.30	103.70	303.00	0.7371	1.0827
302	70.220	1.5746	3.250	199.96	268.86	211.02	80.66	291.68	0.7744	1.0414
304.21	73.825	2.1459	2.146	234.63	234.63	250.47	0.00	250.47	0.9023	0.9023

Source: Calculated values based on equations from *International Thermodynamic Tables of the Fluid State Carbon Dioxide*, International Union of Pure and Applied Chemistry, 1973.

TABLE A-22

Properties of carbon dioxide (CO_2): Superheated-vapor table

v, m^3/kg; u and h, kJ/kg; s, kJ/(kg·K)

Temp., K	v	u	h	s	v	u	h	s
	1 bar (0.1 MPa)				**5 bars (0.5 MPa)**			
300	0.5640	367.37	423.77	2.2065	0.1105	364.79	420.04	1.8939
350	0.6592	401.59	467.51	2.3395	0.1303	399.68	464.83	2.0300
400	0.7541	438.13	513.54	2.4615	0.1497	436.62	511.47	2.1536
450	0.8491	476.67	561.58	2.5751	0.1690	475.46	559.96	2.2683
500	0.9439	517.08	611.47	2.6814	0.1882	516.08	610.18	2.3751
	10 bars (1.0 MPa)				**20 bars (2.0 MPa)**			
300	0.05381	361.39	415.20	1.7514	0.02536	354.05	404.77	1.5955
350	0.06414	397.23	461.37	1.8921	0.03106	392.14	454.26	1.7464
400	0.06819	440.67	508.86	2.0177	0.03640	430.77	503.57	1.8771
450	0.08403	473.89	557.92	2.1341	0.04154	470.75	553.83	1.9961
500	0.09377	515.68	609.45	2.2418	0.04656	512.17	605.29	2.1057
	30 bars (3.0 MPa)				**40 bars (4.0 MPa)**			
300	0.01577	345.78	393.09	1.4903	0.01086	336.04	379.48	1.4013
350	0.02002	386.77	446.83	1.6542	0.01447	381.13	439.01	1.5835
400	0.02381	426.75	498.18	1.7905	0.01750	422.68	492.68	1.7257
450	0.02738	467.57	549.71	1.9125	0.02029	464.40	545.56	1.8512
500	0.03081	509.64	602.07	2.0239	0.02295	507.02	598.82	1.9643
	50 bars (5.0 MPa)				**100 bars (10.0 MPa)**			
300	0.00779	323.72	362.64	1.3145	0.00125	165.51	177.98	0.6514
350	0.01114	375.11	430.81	1.5237	0.00437	337.45	381.14	1.2818
400	0.01372	418.49	487.09	1.6733	0.00618	395.79	457.62	1.4858
450	0.01605	461.15	541.40	1.8019	0.00759	444.43	520.34	1.6344
500	0.01824	504.39	595.59	1.9171	0.00884	491.14	579.55	1.7603

Source: Calculated values based on equations from *International Thermodynamic Tables of the Fluid State Carbon Dioxide*, International Union of Pure and Applied Chemistry, 1973.

TABLE A-23

Constants for the Benedict-Webb-Rubin, Redlich-Kwong, and van der Waals equations of state

1. Benedict-Webb-Rubin; Units are bar(s), m^3/kgmol, and K

Constants	n-Butane (C_4H_{10})	Carbon dioxide (CO_2)	Carbon monoxide (CO)	Methane (CH_4)	Nitrogen (N_2)
a	1.9068	0.1386	0.0371	0.0500	0.0254
A_0	10.216	2.7730	1.3587	1.8791	1.0673
b	0.039998	0.007210	0.002632	0.003380	0.002328
B_0	0.12436	0.04991	0.05454	0.04260	0.04074
c	3.205×10^5	1.511×10^4	1.054×10^3	2.578×10^3	7.379×10^2
C_0	1.006×10^6	1.404×10^5	8.673×10^3	2.286×10^4	8.164×10^3
α	1.101×10^{-3}	8.470×10^{-5}	1.350×10^{-4}	1.244×10^{-4}	1.272×10^{-4}
γ	0.0340	0.00539	0.0060	0.0060	0.0053

Source: H. W. Cooper and J. C. Goldfrank, *Hydrocarbon Processing*, **46** (12): 141 (1967).

2. Redlich-Kwong; a is in bar(m^3/kgmol)^2(K)$^{0.5}$, b is in m^3/kgmol

Substance	a	b	Substance	a	b
Carbon dioxide (CO_2)	64.64	0.02969	Oxygen (O_2)	17.38	0.02199
Carbon monoxide (CO)	17.26	0.02743	Propane (C_3H_8)	183.070	0.06269
Methane (CH_4)	32.19	0.02969	Refrigerant-134a	197.0	0.06634
Nitrogen (N_2)	15.59	0.02681	Water (H_2O)	142.64	0.02110

Source: Computed from critical data.

3. van der Waals; a is in bar(m^3/kgmol)^2, b is in m^3/kgmol

Substance	a	b	Substance	a	b
Acetylene (C_2H_2)	4.410	0.0510	Ethylene (C_2H_4)	4.563	0.0574
Air (equivalent)	1.358	0.0364	Helium (He)	0.0341	0.0234
Ammonia (NH_3)	4.233	0.0373	Hydrogen (H_2)	0.247	0.0265
Benzene (C_6H_6)	18.63	0.1181	Methane (CH_4)	2.285	0.0427
n-Butane (C_4H_{10})	13.80	0.1196	Nitrogen (N_2)	1.361	0.0385
Carbon dioxide (CO_2)	3.643	0.0427	Oxygen (O_2)	1.369	0.0315
Carbon monoxide (CO)	1.463	0.0394	Propane (C_3H_8)	9.315	0.0900
Refrigerant-134a (CF_4H_2)	10.05	0.0957	Sulfur dioxide (SO_2)	6.837	0.0304
Ethane (C_2H_6)	5.575	0.0650	Water (H_2O)	5.507	0.0304

Source: Computed from critical data.

TABLE A-24
Lee-Kesler values for $Z^{(0)}$ (T_r, P_r)

T_r	P_r = 0.010	0.050	0.100	0.200	0.400	0.600	0.800
0.30	0.0029	0.0145	0.0290	0.0579	0.1158	0.1737	0.2315
0.35	0.0026	0.0130	0.0261	0.0522	0.1043	0.1564	0.2084
0.40	0.0024	0.0119	0.0239	0.0477	0.0953	0.1429	0.1904
0.45	0.0022	0.0110	0.0221	0.0442	0.0882	0.1322	0.1762
0.50	0.0021	0.0103	0.0207	0.0413	0.0825	0.1236	0.1647
0.55	0.9804	0.0098	0.0195	0.0390	0.0778	0.1166	0.1553
0.60	0.9849	0.0093	0.0186	0.0371	0.0741	0.1109	0.1476
0.65	0.9881	0.9377	0.0178	0.0356	0.0710	0.1063	0.1415
0.70	0.9904	0.9504	0.8958	0.0344	0.0687	0.1027	0.1366
0.75	0.9922	0.9598	0.9165	0.0336	0.0670	0.1001	0.1330
0.80	0.9935	0.9669	0.9319	0.8539	0.0661	0.0985	0.1307
0.85	0.9946	0.9725	0.9436	0.8810	0.0661	0.0983	0.1301
0.90	0.9954	0.9768	0.9528	0.9015	0.7800	0.1006	0.1321
0.93	0.9959	0.9790	0.9573	0.9115	0.8059	0.6635	0.1359
0.95	0.9961	0.9803	0.9600	0.9174	0.8206	0.6967	0.1410
0.97	0.9963	0.9815	0.9625	0.9227	0.8338	0.7240	0.5580
0.98	0.9965	0.9821	0.9637	0.9253	0.8398	0.7360	0.5887
0.99	0.9966	0.9826	0.9648	0.9277	0.8455	0.7471	0.6138
1.00	0.9967	0.9832	0.9659	0.9300	0.8509	0.7574	0.6353
1.01	0.9968	0.9837	0.9669	0.9322	0.8561	0.7671	0.6542
1.02	0.9969	0.9842	0.9679	0.9343	0.8610	0.7761	0.6710
1.05	0.9971	0.9855	0.9707	0.9401	0.8743	0.8002	0.7130
1.10	0.9975	0.9874	0.9747	0.9485	0.8930	0.8323	0.7649
1.15	0.9978	0.9891	0.9780	0.9554	0.9081	0.8576	0.8032
1.20	0.9981	0.9904	0.9808	0.9611	0.9205	0.8779	0.8330
1.30	0.9985	0.9926	0.9852	0.9702	0.9396	0.9083	0.8764
1.40	0.9988	0.9942	0.9884	0.9768	0.9534	0.9298	0.9062
1.50	0.9991	0.9954	0.9909	0.9818	0.9636	0.9456	0.9278
1.60	0.9993	0.9964	0.9928	0.9856	0.9714	0.9575	0.9439
1.70	0.9994	0.9971	0.9943	0.9886	0.9775	0.9667	0.9563
1.80	0.9995	0.9977	0.9955	0.9910	0.9823	0.9739	0.9659
1.90	0.9996	0.9982	0.9964	0.9929	0.9861	0.9796	0.9735
2.00	0.9997	0.9986	0.9972	0.9944	0.9892	0.9842	0.9796
2.20	0.9998	0.9992	0.9983	0.9967	0.9937	0.9910	0.9886
2.40	0.9999	0.9996	0.9991	0.9983	0.9969	0.9957	0.9948
2.60	1.0000	0.9998	0.9997	0.9994	0.9991	0.9990	0.9990
2.80	1.0000	1.0000	1.0001	1.0002	1.0007	1.0013	1.0021
3.00	1.0000	1.0002	1.0004	1.0008	1.0018	1.0030	1.0043
3.50	1.0001	1.0004	1.0008	1.0017	1.0035	1.0055	1.0075
4.00	1.0001	1.0005	1.0010	1.0021	1.0043	1.0066	1.0090

TABLE A-24 (*Continued*)

T_r	P_r 1.000	1.200	1.500	2.000	3.000	5.000	7.000	10.000
0.30	0.2892	0.3470	0.4335	0.5775	0.8648	1.4366	2.0048	2.8507
0.35	0.2604	0.3123	0.3901	0.5195	0.7775	1.2902	1.7987	2.5539
0.40	0.2379	0.2853	0.3563	0.4744	0.7095	1.1758	1.6373	2.3211
0.45	0.2200	0.2638	0.3294	0.4384	0.6551	1.0841	1.5077	2.1338
0.50	0.2056	0.2465	0.3077	0.4092	0.6110	1.0094	1.4017	1.9801
0.55	0.1939	0.2323	0.2899	0.3853	0.5747	0.9475	1.3137	1.8520
0.60	0.1842	0.2207	0.2753	0.3657	0.5446	0.8959	1.2398	1.7440
0.65	0.1765	0.2113	0.2634	0.3495	0.5197	0.8526	1.1773	1.6519
0.70	0.1703	0.2038	0.2538	0.3364	0.4991	0.8161	1.1241	1.5729
0.75	0.1656	0.1981	0.2464	0.3260	0.4823	0.7854	1.0787	1.5047
0.80	0.1626	0.1942	0.2411	0.3182	0.4690	0.7598	1.0400	1.4456
0.85	0.1614	0.1924	0.2382	0.3132	0.4591	0.7388	1.0071	1.3943
0.90	0.1630	0.1935	0.2383	0.3114	0.4527	0.7220	0.9793	1.3496
0.93	0.1664	0.1963	0.2405	0.3122	0.4507	0.7138	0.9648	1.3257
0.95	0.1705	0.1998	0.2432	0.3138	0.4501	0.7092	0.9561	1.3108
0.97	0.1779	0.2055	0.2474	0.3164	0.4504	0.7052	0.9480	1.2968
0.98	0.1844	0.2097	0.2503	0.3182	0.4508	0.7035	0.9442	1.2901
0.99	0.1959	0.2154	0.2538	0.3204	0.4514	0.7018	0.9406	1.2835
1.00	0.2901	0.2237	0.2583	0.3229	0.4522	0.7004	0.9372	1.2772
1.01	0.4648	0.2370	0.2640	0.3260	0.4533	0.6991	0.9339	1.2710
1.02	0.5146	0.2629	0.2715	0.3297	0.4547	0.6980	0.9307	1.2650
1.05	0.6026	0.4437	0.3131	0.3452	0.4604	0.6956	0.9222	1.2481
1.10	0.6880	0.5984	0.4580	0.3953	0.4770	0.6950	0.9110	1.2232
1.15	0.7443	0.6803	0.5798	0.4760	0.5042	0.6987	0.9033	1.2021
1.20	0.7858	0.7363	0.6605	0.5605	0.5425	0.7069	0.8990	1.1844
1.30	0.8438	0.8111	0.7624	0.6908	0.6344	0.7358	0.8998	1.1580
1.40	0.8827	0.8595	0.8256	0.7753	0.7202	0.7761	0.9112	1.1419
1.50	0.9103	0.8933	0.8689	0.8328	0.7887	0.8200	0.9297	1.1339
1.60	0.9308	0.9180	0.9000	0.8738	0.8410	0.8617	0.9518	1.1320
1.70	0.9463	0.9367	0.9234	0.9043	0.8809	0.8984	0.9745	1.1343
1.80	0.9583	0.9511	0.9413	0.9275	0.9118	0.9297	0.9961	1.1391
1.90	0.9678	0.9624	0.9552	0.9456	0.9359	0.9557	1.0157	1.1452
2.00	0.9754	0.9715	0.9664	0.9599	0.9550	0.9772	1.0328	1.1516
2.20	0.9865	0.9847	0.9826	0.9806	0.9827	1.0094	1.0600	1.1635
2.40	0.9941	0.9936	0.9935	0.9945	1.0011	1.0313	1.0793	1.1728
2.60	0.9993	0.9998	1.0010	1.0040	1.0137	1.0463	1.0926	1.1792
2.80	1.0031	1.0042	1.0063	1.0106	1.0223	1.0565	1.1016	1.1830
3.00	1.0057	1.0074	1.0101	1.0153	1.0284	1.0635	1.1075	1.1848
3.50	1.0097	1.0120	0.0156	1.0221	1.0368	1.0723	1.1138	1.1834
4.00	1.0115	1.0140	1.0179	1.0249	1.0401	1.0747	1.1136	1.1773

Source: From Lee, B. I., and M. G. Kesler, "A Generalized Thermodynamic Correlation Based on Three-Parameter Corresponding States," *AIChE Journal,* **21**(3): 510–527, 1975. Reproduced by permission of the American Institute of Chemical Engineers.

TABLE A-25
Lee-Kesler values for $Z^{(1)}(T_r, P_r)$

T_r	P_r = 0.010	0.050	0.100	0.200	0.400	0.600	0.800
0.30	−0.0008	−0.0040	−0.0081	−0.0161	−0.0323	−0.0484	−0.0645
0.35	−0.0009	−0.0046	−0.0093	−0.0185	−0.0370	−0.0554	−0.0738
0.40	−0.0010	−0.0048	−0.0095	−0.0190	−0.0380	−0.0570	−0.0758
0.45	−0.0009	−0.0047	−0.0094	−0.0187	−0.0374	−0.0560	−0.0745
0.50	−0.0009	−0.0045	−0.0090	−0.0181	−0.0360	−0.0539	−0.0716
0.55	−0.0314	−0.0043	−0.0086	−0.0172	−0.0343	−0.0513	−0.0682
0.60	−0.0205	−0.0041	−0.0082	−0.0164	−0.0326	−0.0487	−0.0646
0.65	−0.0137	−0.0772	−0.0078	−0.0156	−0.0309	−0.0461	−0.0611
0.70	−0.0093	−0.0507	−0.1161	−0.0148	−0.0294	−0.0438	−0.0579
0.75	−0.0064	−0.0339	−0.0744	−0.0143	−0.0282	−0.0417	−0.0550
0.80	−0.0044	−0.0228	−0.0487	−0.1160	−0.0272	−0.0401	−0.0526
0.85	−0.0029	−0.0152	−0.0319	−0.0715	−0.0268	−0.0391	−0.0509
0.90	−0.0019	−0.0099	−0.0205	−0.0442	−0.1118	−0.0396	−0.0503
0.93	−0.0015	−0.0075	−0.0154	−0.0326	−0.0763	−0.1662	−0.0514
0.95	−0.0012	−0.0062	−0.0126	−0.0262	−0.0589	−0.1110	−0.0540
0.97	−0.0010	−0.0050	−0.0101	−0.0208	−0.0450	−0.0770	−0.1647
0.98	−0.0009	−0.0044	−0.0090	−0.0184	−0.0390	−0.0641	−0.1100
0.99	−0.0008	−0.0039	−0.0079	−0.0161	−0.0335	−0.0531	−0.0796
1.00	−0.0007	−0.0034	−0.0069	−0.0140	−0.0285	−0.0435	−0.0588
1.01	−0.0006	−0.0030	−0.0060	−0.0120	−0.0240	−0.0351	−0.0429
1.02	−0.0005	−0.0026	−0.0051	−0.0102	−0.0198	−0.0277	−0.0303
1.05	−0.0003	−0.0015	−0.0029	−0.0054	−0.0092	−0.0097	−0.0032
1.10	−0.0000	0.0000	0.0001	0.0007	0.0038	0.0106	0.0236
1.15	0.0002	0.0011	0.0023	0.0052	0.0127	0.0237	0.0396
1.20	0.0004	0.0019	0.0039	0.0084	0.0190	0.0326	0.0499
1.30	0.0006	0.0030	0.0061	0.0125	0.0267	0.0429	0.0612
1.40	0.0007	0.0036	0.0072	0.0147	0.0306	0.0477	0.0661
1.50	0.0008	0.0039	0.0078	0.0158	0.0323	0.0497	0.0677
1.60	0.0008	0.0040	0.0080	0.0162	0.0330	0.0501	0.0677
1.70	0.0008	0.0040	0.0081	0.0163	0.0329	0.0497	0.0667
1.80	0.0008	0.0040	0.0081	0.0162	0.0325	0.0488	0.0652
1.90	0.0008	0.0040	0.0079	0.0159	0.0318	0.0477	0.0635
2.00	0.0008	0.0039	0.0078	0.0155	0.0310	0.0464	0.0617
2.20	0.0007	0.0037	0.0074	0.0147	0.0293	0.0437	0.0579
2.40	0.0007	0.0035	0.0070	0.0139	0.0276	0.0411	0.0544
2.60	0.0007	0.0033	0.0066	0.0131	0.0260	0.0387	0.0512
2.80	0.0006	0.0031	0.0062	0.0124	0.0245	0.0365	0.0483
3.00	0.0006	0.0029	0.0059	0.0117	0.0232	0.0345	0.0456
3.50	0.0005	0.0026	0.0052	0.0103	0.0204	0.0303	0.0401
4.00	0.0005	0.0023	0.0046	0.0091	0.0182	0.0270	0.0357

TABLE A-25 *(Continued)*

	P_r							
T_r	**1.000**	**1.200**	**1.500**	**2.000**	**3.000**	**5.000**	**7.000**	**10.000**
0.30	−0.0806	−0.0966	−0.1207	−0.1608	−0.2407	−0.3996	−0.5572	−0.7915
0.35	−0.0921	−0.1105	−0.1379	−0.1834	−0.2738	−0.4523	−0.6279	−0.8863
0.40	−0.0946	−0.1134	−0.1414	−0.1879	−0.2799	−0.4603	−0.6365	−0.8936
0.45	−0.0929	−0.1113	−0.1387	−0.1840	−0.2734	−0.4475	−0.6162	−0.8606
0.50	−0.0893	−0.1069	−0.1330	−0.1762	−0.2611	−0.4253	−0.5831	−0.8099
0.55	−0.0849	−0.1015	−0.1263	−0.1669	−0.2465	−0.3991	−0.5446	−0.7521
0.60	−0.0803	−0.0960	−0.1192	−0.1572	−0.2312	−0.3718	−0.5047	−0.6928
0.65	−0.0759	−0.0906	−0.1122	−0.1476	−0.2160	−0.3447	−0.4653	−0.6346
0.70	−0.0718	−0.0855	−0.1057	−0.1385	−0.2013	−0.3184	−0.4270	−0.5785
0.75	−0.0681	−0.0808	−0.0996	−0.1298	−0.1872	−0.2929	−0.3901	−0.5250
0.80	−0.0648	−0.0767	−0.0940	−0.1217	−0.1736	−0.2682	−0.3545	−0.4740
0.85	−0.0622	−0.0731	−0.0888	−0.1138	−0.1602	−0.2439	−0.3201	−0.4254
0.90	−0.0604	−0.0701	−0.0840	−0.1059	−0.1463	−0.2195	−0.2862	−0.3788
0.93	−0.0602	−0.0687	−0.0810	−0.1007	−0.1374	−0.2045	−0.2661	−0.3516
0.95	−0.0607	−0.0678	−0.0788	−0.0967	−0.1310	−0.1943	−0.2526	−0.3339
0.97	−0.0623	−0.0669	−0.0759	−0.0921	−0.1240	−0.1837	−0.2391	−0.3163
0.98	−0.0641	−0.0661	−0.0740	−0.0893	−0.1202	−0.1783	−0.2322	−0.3075
0.99	−0.0680	−0.0646	−0.0715	−0.0861	−0.1162	−0.1728	−0.2254	−0.2989
1.00	−0.0879	−0.0609	−0.0678	−0.0824	−0.1118	−0.1672	−0.2185	−0.2902
1.01	−0.0223	−0.0473	−0.0621	−0.0778	−0.1072	−0.1615	−0.2116	−0.2816
1.02	−0.0062	0.0227	−0.0524	−0.0722	−0.1021	−0.1556	−0.2047	−0.2731
1.05	0.0220	0.1059	0.0451	−0.0432	−0.0838	−0.1370	−0.1835	−0.2476
1.10	0.0476	0.0897	0.1630	0.0698	−0.0373	−0.1021	−0.1469	−0.2056
1.15	0.0625	0.0943	0.1548	0.1667	0.0332	0.0611	−0.1084	−0.1642
1.20	0.0719	0.0991	0.1477	0.1990	0.1095	−0.0141	−0.0678	−0.1231
1.30	0.0819	0.1048	0.1420	0.1991	0.2079	0.0875	0.0176	−0.0423
1.40	0.0857	0.1063	0.1383	0.1894	0.2397	0.1737	0.1008	0.0350
1.50	0.0864	0.1055	0.1345	0.1806	0.2433	0.2309	0.1717	0.1058
1.60	0.0855	0.1035	0.1303	0.1729	0.2381	0.2631	0.2255	0.1673
1.70	0.0838	0.1008	0.1259	0.1658	0.2305	0.2788	0.2628	0.2179
1.80	0.0816	0.0978	0.1216	0.1593	0.2224	0.2846	0.2871	0.2576
1.90	0.0792	0.0947	0.1173	0.1532	0.2144	0.2848	0.3017	0.2876
2.00	0.0767	0.0916	0.1133	0.1476	0.2069	0.2819	0.3097	0.3096
2.20	0.0719	0.0857	0.1057	0.1374	0.1932	0.2720	0.3135	0.3355
2.40	0.0675	0.0803	0.0989	0.1285	0.1812	0.2602	0.3089	0.3459
2.60	0.0634	0.0754	0.0929	0.1207	0.1706	0.2484	0.3009	0.3475
2.80	0.0598	0.0711	0.0876	0.1138	0.1613	0.2372	0.2915	0.3443
3.00	0.0565	0.0672	0.0828	0.1076	0.1529	0.2268	0.2817	0.3385
3.50	0.0497	0.0591	0.0728	0.0949	0.1356	0.2042	0.2584	0.3194
4.00	0.0443	0.0527	0.0651	0.0849	0.1219	0.1857	0.2378	0.2994

Source: From Lee, B. I., and M. G. Kesler, "A Generalized Thermodynamic Correlation Based on Three-Parameter Corresponding States," *AIChE Journal,* **21**(3): 510–527, 1978. Reproduced by permission of the American Institute of Chemical Engineers.

TABLE A-26
Lee-Kesler residual enthalpy values for $[(h^* - h)/RT_c]^{(0)}$

	P_r						
T_r	**0.010**	**0.050**	**0.100**	**0.200**	**0.400**	**0.600**	**0.800**
0.30	6.045	6.043	6.040	6.034	6.022	6.011	5.999
0.35	5.906	5.904	5.901	5.895	5.882	5.870	5.858
0.40	5.763	5.761	5.757	5.751	5.738	5.726	5.713
0.45	5.615	5.612	5.609	5.603	5.590	5.577	5.564
0.50	5.465	5.463	4.459	5.453	5.440	5.427	5.414
0.55	0.032	5.312	5.309	5.303	5.290	5.278	5.265
0.60	0.027	5.162	5.159	5.153	5.141	5.129	5.116
0.65	0.023	0.118	5.008	5.002	4.991	4.980	4.968
0.70	0.020	0.101	0.213	4.848	4.838	4.828	4.818
0.75	0.017	0.088	0.183	4.687	4.679	4.672	4.664
0.80	0.015	0.078	0.160	0.345	4.507	4.504	4.499
0.85	0.014	0.069	0.141	0.300	4.309	4.313	4.316
0.90	0.012	0.062	0.126	0.264	0.596	4.074	4.094
0.93	0.011	0.058	0.118	0.246	0.545	0.960	3.920
0.95	0.011	0.056	0.113	0.235	0.516	0.885	3.763
0.97	0.011	0.054	0.109	0.225	0.490	0.824	1.356
0.98	0.010	0.053	0.107	0.221	0.478	0.797	1.273
0.99	0.010	0.052	0.105	0.216	0.466	0.773	1.206
1.00	0.010	0.051	0.103	0.212	0.455	0.750	1.151
1.01	0.010	0.050	0.101	0.208	0.445	0.728	0.102
1.02	0.010	0.049	0.099	0.203	0.434	0.708	1.060
1.05	0.009	0.046	0.094	0.192	0.407	0.654	0.955
1.10	0.008	0.042	0.086	0.175	0.367	0.581	0.827
1.15	0.008	0.039	0.079	0.160	0.334	0.523	0.732
1.20	0.007	0.036	0.073	0.148	0.305	0.474	0.657
1.30	0.006	0.031	0.063	0.127	0.259	0.399	0.545
1.40	0.005	0.027	0.055	0.110	0.224	0.341	0.463
1.50	0.005	0.024	0.048	0.097	0.196	0.297	0.400
1.60	0.004	0.021	0.043	0.086	0.173	0.261	0.350
1.70	0.004	0.019	0.038	0.076	0.153	0.231	0.309
1.80	0.003	0.017	0.034	0.068	0.137	0.206	0.275
1.90	0.003	0.015	0.031	0.062	0.123	0.185	0.246
2.00	0.003	0.014	0.028	0.056	0.111	0.167	0.222
2.20	0.002	0.012	0.023	0.046	0.092	0.137	0.182
2.40	0.002	0.010	0.019	0.038	0.076	0.114	0.150
2.60	0.002	0.008	0.016	0.032	0.064	0.095	0.125
2.80	0.001	0.007	0.014	0.027	0.054	0.080	0.105
3.00	0.001	0.006	0.011	0.023	0.045	0.067	0.088
3.50	0.001	0.004	0.007	0.015	0.029	0.043	0.056
4.00	0.000	0.002	0.005	0.009	0.017	0.026	0.033

TABLE A-26 *(Continued)*

	P_r							
T_r	**1.000**	**1.200**	**1.500**	**2.000**	**3.000**	**5.000**	**7.000**	**10.000**
0.30	5.987	5.975	5.957	5.927	5.868	5.748	5.628	5.446
0.35	5.845	5.833	5.814	5.783	5.721	5.595	5.469	5.278
0.40	5.700	5.687	5.668	5.636	5.572	5.442	5.311	5.113
0.45	5.551	5.538	5.519	5.486	5.421	5.288	5.154	4.950
0.50	5.401	5.388	5.369	5.336	5.270	5.135	4.999	4.791
0.55	5.252	5.239	5.220	5.187	5.121	4.986	4.849	4.638
0.60	5.104	5.091	5.073	5.041	4.976	4.842	4.704	4.492
0.65	4.956	4.945	4.927	4.896	4.833	4.702	4.565	4.353
0.70	4.808	4.797	4.781	4.752	4.693	4.566	4.432	4.221
0.75	4.655	4.646	4.632	4.607	4.554	4.434	4.303	4.095
0.80	4.494	4.488	4.478	4.459	4.413	4.303	4.178	3.974
0.85	4.316	4.316	4.312	4.302	4.269	4.173	4.056	3.857
0.90	4.108	4.118	4.127	4.132	4.119	4.043	3.935	3.744
0.93	3.953	3.976	4.000	4.020	4.024	3.963	3.863	3.678
0.95	3.825	3.865	3.904	3.940	3.958	3.910	3.815	3.634
0.97	3.658	3.732	3.796	3.853	3.890	3.856	3.767	3.591
0.98	3.544	3.652	3.736	3.806	3.854	3.829	3.743	3.569
0.99	3.376	3.558	3.670	3.758	3.818	3.801	3.719	3.548
1.00	2.584	3.441	3.598	3.706	3.782	3.774	3.695	3.526
1.01	1.796	3.283	3.516	3.652	3.744	3.746	3.671	3.505
1.02	1.627	3.039	3.442	3.595	3.705	3.718	3.647	3.484
1.05	1.359	2.034	3.030	3.398	3.583	3.632	3.575	3.420
1.10	1.120	1.487	2.203	2.965	3.353	3.484	3.453	3.315
1.15	0.968	1.239	1.719	2.479	3.091	3.329	3.329	3.211
1.20	0.857	1.076	1.443	2.079	2.807	3.166	3.202	3.107
1.30	0.698	0.860	1.116	1.560	2.274	2.825	2.942	2.899
1.40	0.588	0.716	0.915	1.253	1.857	2.486	2.679	2.692
1.50	0.505	0.611	0.774	1.046	1.549	2.175	2.421	2.486
1.60	0.440	0.531	0.667	0.894	1.318	1.904	2.177	2.285
1.70	0.387	0.466	0.583	0.777	1.139	1.672	1.953	2.091
1.80	0.344	0.413	0.515	0.683	0.996	1.476	1.751	1.908
1.90	0.307	0.368	0.458	0.606	0.880	1.309	1.571	1.736
2.00	0.276	0.330	0.411	0.541	0.782	1.167	1.411	1.577
2.20	0.226	0.269	0.334	0.437	0.629	0.937	1.143	1.295
2.40	0.187	0.222	0.275	0.359	0.513	0.761	0.929	1.058
2.60	0.155	0.185	0.228	0.297	0.422	0.621	0.756	0.858
2.80	0.130	0.154	0.190	0.246	0.348	0.508	0.614	0.689
3.00	0.109	0.129	0.159	0.205	0.288	0.415	0.495	0.545
3.50	0.069	0.081	0.099	0.127	0.174	0.239	0.270	0.264
4.00	0.041	0.048	0.058	0.072	0.095	0.116	0.110	0.061

Source: From Lee, B. I., and M. G. Kesler, "A Generalized Thermodynamic Correlation Based on Three-Parameter Corresponding States," *AIChE Journal*, **21**(3): 510–527, 1975. Reproduced by permission of the American Institute of Chemical Engineers.

TABLE A-27
Lee-Kesler residual enthalpy values for $[(h^* - h)/RT_c]^{(1)}$

	P_r						
T_r	0.010	0.050	0.100	0.200	0.400	0.600	0.800
0.30	11.098	11.096	11.095	11.091	11.083	11.076	11.069
0.35	10.656	10.655	10.654	10.653	10.650	10.646	10.643
0.40	10.121	10.121	10.121	10.120	10.121	10.121	10.121
0.45	9.515	9.515	9.516	9.517	9.519	9.521	9.523
0.50	8.868	8.869	8.870	8.872	8.876	8.880	8.884
0.55	0.080	8.211	8.212	8.215	8.221	8.226	8.232
0.60	0.059	7.568	7.570	7.573	7.579	7.585	7.591
0.65	0.045	0.247	6.949	6.952	6.959	6.966	6.973
0.70	0.034	0.185	0.415	6.360	6.367	6.373	6.381
0.75	0.027	0.142	0.306	5.796	5.802	5.809	5.816
0.80	0.021	0.110	0.234	0.542	5.266	5.271	5.278
0.85	0.017	0.087	0.182	0.401	4.753	4.754	4.758
0.90	0.014	0.070	0.144	0.308	0.751	4.254	4.248
0.93	0.012	0.061	0.126	0.265	0.612	1.236	3.942
0.95	0.011	0.056	0.115	0.241	0.542	0.994	3.737
0.97	0.010	0.052	0.105	0.219	0.483	0.837	1.616
0.98	0.010	0.050	0.101	0.209	0.457	0.776	1.324
0.99	0.009	0.048	0.097	0.200	0.433	0.722	1.154
1.00	0.009	0.046	0.093	0.191	0.410	0.675	1.034
1.01	0.009	0.044	0.089	0.183	0.389	0.632	0.940
1.02	0.008	0.042	0.085	0.175	0.370	0.594	0.863
1.05	0.007	0.037	0.075	0.153	0.318	0.498	0.691
1.10	0.006	0.030	0.061	0.123	0.251	0.381	0.507
1.15	0.005	0.025	0.050	0.099	0.199	0.296	0.385
1.20	0.004	0.020	0.040	0.080	0.158	0.232	0.297
1.30	0.003	0.013	0.026	0.052	0.100	0.142	0.177
1.40	0.002	0.008	0.016	0.032	0.060	0.083	0.100
1.50	0.001	0.005	0.009	0.018	0.032	0.042	0.048
1.60	0.000	0.002	0.004	0.007	0.012	0.013	0.011
1.70	0.000	0.000	0.000	−0.000	−0.003	−0.009	−0.017
1.80	−0.000	−0.001	−0.003	−0.006	−0.015	−0.025	−0.037
1.90	−0.001	−0.003	−0.005	−0.011	−0.023	−0.037	−0.053
2.00	−0.001	−0.003	−0.007	−0.015	−0.030	−0.047	−0.065
2.20	−0.001	−0.005	−0.010	−0.020	−0.040	−0.062	−0.083
2.40	−0.001	−0.006	−0.012	−0.023	−0.047	−0.071	−0.095
2.60	−0.001	−0.006	−0.013	−0.026	−0.052	−0.078	−0.104
2.80	−0.001	−0.007	−0.014	−0.028	−0.055	−0.082	−0.110
3.00	−0.001	−0.007	−0.014	−0.029	−0.058	−0.086	−0.114
3.50	−0.002	−0.008	−0.016	−0.031	−0.062	−0.092	−0.122
4.00	−0.002	−0.008	−0.016	−0.032	−0.064	−0.096	−0.127

TABLE A-27 (*Continued*)

	P_r							
T_r	**1.000**	**1.200**	**1.500**	**2.000**	**3.000**	**5.000**	**7.000**	**10.000**
0.30	11.062	11.055	11.044	11.027	10.992	10.935	10.872	10.781
0.35	10.640	10.637	10.632	10.624	10.609	10.581	10.554	10.529
0.40	10.121	10.121	10.121	10.122	10.123	10.128	10.135	10.150
0.45	9.525	9.527	9.531	9.537	9.549	9.576	9.611	9.663
0.50	8.888	8.892	8.899	8.909	8.932	8.978	9.030	9.111
0.55	8.238	8.243	8.252	8.267	8.298	8.360	8.425	8.531
0.60	7.596	7.603	7.614	7.632	7.669	7.745	7.824	7.950
0.65	6.980	6.987	6.997	7.017	7.059	7.147	7.239	7.381
0.70	6.388	6.395	6.407	6.429	6.475	6.574	6.677	6.837
0.75	5.824	5.832	5.845	5.868	5.918	6.027	6.142	6.318
0.80	5.285	5.293	5.306	5.330	5.385	5.506	5.632	5.824
0.85	4.763	4.771	4.784	4.810	4.872	5.008	5.149	5.358
0.90	4.249	4.255	4.268	4.298	4.371	4.530	4.688	4.916
0.93	3.934	3.937	3.951	3.987	4.073	4.251	4.422	4.662
0.95	3.712	3.713	3.730	3.773	3.873	4.068	4.248	4.497
0.97	3.470	3.467	3.492	3.551	3.670	3.885	4.077	4.336
0.98	3.332	3.327	3.363	3.434	3.568	3.795	3.992	4.257
0.99	3.164	3.164	3.223	3.313	3.464	3.705	3.909	4.178
1.00	2.471	2.952	3.065	3.186	3.358	3.615	3.825	4.100
1.01	1.375	2.595	2.880	3.051	3.251	3.525	3.742	4.023
1.02	1.180	1.723	2.650	2.906	3.142	3.435	3.661	3.947
1.05	0.877	0.878	1.496	2.381	2.800	3.167	3.418	3.722
1.10	0.617	0.673	0.617	1.261	2.167	2.720	3.023	3.362
1.15	0.459	0.503	0.487	0.604	1.497	2.275	2.641	3.019
1.20	0.349	0.381	0.381	0.361	0.934	1.840	2.273	2.692
1.30	0.203	0.218	0.218	0.178	0.300	1.066	1.592	2.086
1.40	0.111	0.115	0.108	0.070	0.044	0.504	1.012	1.547
1.50	0.049	0.046	0.032	−0.008	−0.078	0.142	0.556	1.080
1.60	0.005	−0.004	−0.023	−0.065	−0.151	−0.082	0.217	0.689
1.70	−0.027	−0.040	−0.063	−0.109	−0.202	−0.223	−0.028	0.369
1.80	−0.051	−0.067	−0.094	−0.143	−0.241	−0.317	−0.203	0.112
1.90	−0.070	−0.088	−0.117	−0.169	−0.271	−0.381	−0.330	−0.092
2.00	−0.085	−0.105	−0.136	−0.190	−0.295	−0.428	−0.424	−0.255
2.20	−0.106	−0.128	−0.163	−0.221	−0.331	−0.493	−0.551	−0.489
2.40	−0.120	−0.144	−0.181	−0.242	−0.356	−0.535	−0.631	−0.645
2.60	−0.130	−0.156	−0.194	−0.257	−0.376	−0.567	−0.687	−0.754
2.80	−0.137	−0.164	−0.204	−0.269	−0.391	−0.591	−0.729	−0.836
3.00	−0.142	−0.170	−0.211	−0.278	−0.403	−0.611	−0.763	−0.899
3.50	−0.152	−0.181	−0.224	−0.294	−0.425	−0.650	−0.827	−1.015
4.00	−0.158	−0.188	−0.233	−0.306	−0.442	−0.680	−0.874	−1.097

Source: From Lee, B. I., and M. G. Kesler, "A Generalized Thermodynamic Correlation Based on Three-Parameter Corresponding States," *AIChE Journal*, **21**(3): 510–527, 1975. Reproduced by permission of the American Institute of Chemical Engineers.

TABLE A-28
Lee-Kesler residual entropy values for $[(s^* - s)/R]^{(0)}$

	P_r						
T_r	**0.010**	**0.050**	**0.100**	**0.200**	**0.400**	**0.600**	**0.800**
0.30	11.614	10.008	9.319	8.635	7.961	7.574	7.304
0.35	11.185	9.579	8.890	8.205	7.529	7.140	6.869
0.40	10.802	9.196	8.506	7.821	7.144	6.755	6.483
0.45	10.453	8.847	8.157	7.472	6.794	6.404	6.132
0.50	10.137	8.531	7.841	7.156	6.479	6.089	5.816
0.55	0.038	8.245	7.555	6.870	6.193	5.803	5.531
0.60	0.029	7.983	7.294	6.610	5.933	5.544	5.273
0.65	0.023	0.122	7.052	6.368	5.694	5.306	5.036
0.70	0.018	0.096	0.206	6.140	5.467	5.082	4.814
0.75	0.015	0.078	0.164	5.917	5.248	4.866	4.600
0.80	0.013	0.064	0.134	0.294	5.026	4.649	4.388
0.85	0.011	0.054	0.111	0.239	4.785	4.418	4.166
0.90	0.009	0.046	0.094	0.199	0.463	4.145	3.912
0.93	0.008	0.042	0.085	0.179	0.408	0.750	3.723
0.95	0.008	0.039	0.080	0.168	0.377	0.671	3.556
0.97	0.007	0.037	0.075	0.157	0.350	0.607	1.056
0.98	0.007	0.036	0.073	0.153	0.337	0.580	0.971
0.99	0.007	0.035	0.071	0.148	0.326	0.555	0.903
1.00	0.007	0.034	0.069	0.144	0.315	0.532	0.847
1.01	0.007	0.033	0.067	0.139	0.304	0.510	0.799
1.02	0.006	0.032	0.065	0.135	0.294	0.491	0.757
1.05	0.006	0.030	0.060	0.124	0.267	0.439	0.656
1.10	0.005	0.026	0.053	0.108	0.230	0.371	0.537
1.15	0.005	0.023	0.047	0.096	0.201	0.319	0.452
1.20	0.004	0.021	0.042	0.085	0.177	0.277	0.389
1.30	0.003	0.017	0.033	0.068	0.140	0.217	0.298
1.40	0.003	0.014	0.027	0.056	0.114	0.174	0.237
1.50	0.002	0.011	0.023	0.046	0.094	0.143	0.194
1.60	0.002	0.010	0.019	0.039	0.079	0.120	0.162
1.70	0.002	0.008	0.017	0.033	0.067	0.102	0.137
1.80	0.001	0.007	0.014	0.029	0.058	0.088	0.117
1.90	0.001	0.006	0.013	0.025	0.051	0.076	0.102
2.00	0.001	0.006	0.011	0.022	0.044	0.067	0.089
2.20	0.001	0.004	0.009	0.018	0.035	0.053	0.070
2.40	0.001	0.004	0.007	0.014	0.028	0.042	0.056
2.60	0.001	0.003	0.006	0.012	0.023	0.035	0.046
2.80	0.000	0.002	0.005	0.010	0.020	0.029	0.039
3.00	0.000	0.002	0.004	0.008	0.017	0.025	0.033
3.50	0.000	0.001	0.003	0.006	0.012	0.017	0.023
4.00	0.000	0.001	0.002	0.004	0.009	0.013	0.017

TABLE A-28 (*Continued*)

	P_r							
T_r	**1.000**	**1.200**	**1.500**	**2.000**	**3.000**	**5.000**	**7.000**	**10.000**
0.30	7.099	6.935	6.740	6.497	6.182	5.847	5.683	5.578
0.35	6.663	6.497	6.299	6.052	5.728	5.376	5.194	5.060
0.40	6.275	6.109	5.909	5.660	5.330	4.967	4.772	4.619
0.45	5.924	5.757	5.557	5.306	4.974	4.603	4.401	4.234
0.50	5.608	5.441	5.240	4.989	4.656	4.282	4.074	3.899
0.55	5.324	5.157	4.956	4.706	4.373	3.998	3.788	3.607
0.60	5.066	4.900	4.700	4.451	4.120	3.747	3.537	3.353
0.65	4.830	4.665	4.467	4.220	3.892	3.523	3.315	3.131
0.70	4.610	4.446	4.250	4.007	3.684	3.322	3.117	2.935
0.75	4.399	4.238	4.045	3.807	3.491	3.138	2.939	2.761
0.80	4.191	4.034	3.846	3.615	3.310	2.970	2.777	2.605
0.85	3.976	3.825	3.646	3.425	3.135	2.812	2.629	2.463
0.90	3.738	3.599	3.434	3.231	2.964	2.663	2.491	2.334
0.93	3.569	3.444	3.295	3.108	2.860	2.577	2.412	2.262
0.95	3.433	3.326	3.193	3.023	2.790	2.520	2.362	2.215
0.97	3.259	3.188	3.081	2.932	2.719	2.463	2.312	2.170
0.98	3.142	3.106	3.019	2.884	2.682	2.436	2.287	2.148
0.99	2.972	3.010	2.953	2.835	2.646	2.408	2.263	2.126
1.00	2.178	2.893	2.879	2.784	2.609	2.380	2.239	2.105
1.01	1.391	2.736	2.798	2.730	2.571	2.352	2.215	2.083
1.02	1.225	2.495	2.706	2.673	2.533	2.325	2.191	2.062
1.05	0.965	1.523	2.328	2.483	2.415	2.242	2.121	2.001
1.10	0.742	1.012	1.557	2.081	2.202	2.104	2.007	1.903
1.15	0.607	0.790	1.126	1.649	1.968	1.966	1.897	1.810
1.20	0.512	0.651	0.890	1.308	1.727	1.827	1.789	1.722
1.30	0.385	0.478	0.628	0.891	1.299	1.554	1.581	1.556
1.40	0.303	0.372	0.478	0.663	0.990	1.303	1.386	1.402
1.50	0.246	0.299	0.381	0.520	0.777	1.088	1.208	1.260
1.60	0.204	0.247	0.312	0.421	0.628	0.913	1.050	1.130
1.70	0.172	0.208	0.261	0.350	0.519	0.773	0.915	1.013
1.80	0.147	0.177	0.222	0.296	0.438	0.661	0.799	0.908
1.90	0.127	0.153	0.191	0.255	0.375	0.570	0.702	0.815
2.00	0.111	0.134	0.167	0.221	0.325	0.497	0.620	0.733
2.20	0.087	0.105	0.130	0.172	0.251	0.388	0.492	0.599
2.40	0.070	0.084	0.104	0.138	0.201	0.311	0.399	0.496
2.60	0.058	0.069	0.086	0.113	0.164	0.255	0.329	0.416
2.80	0.048	0.058	0.072	0.094	0.137	0.213	0.277	0.353
3.00	0.041	0.049	0.061	0.080	0.116	0.181	0.236	0.303
3.50	0.029	0.034	0.042	0.056	0.081	0.126	0.166	0.216
4.00	0.021	0.025	0.031	0.041	0.059	0.093	0.123	0.162

Source: From Lee, B. I., and M. G. Kesler, "A Generalized Thermodynamic Correlation Based on Three-Parameter Corresponding States," *AIChE Journal*, **21**(3): 510–527, 1975. Reproduced by permission of the American Institute of Chemical Engineers.

TABLE A-29
Lee-Kesler residual entropy values for $[(s^* - s)/R]^{(1)}$

	P_r						
T_r	**0.010**	**0.050**	**0.100**	**0.200**	**0.400**	**0.600**	**0.800**
0.30	16.782	16.774	16.764	16.744	16.705	16.665	16.626
0.35	15.413	15.408	15.401	15.387	15.359	15.333	15.305
0.40	13.990	13.986	13.981	13.972	13.953	13.934	13.915
0.45	12.564	12.561	12.558	12.551	12.537	12.523	12.509
0.50	11.202	11.200	11.197	11.192	11.182	11.172	11.162
0.55	0.115	9.948	9.946	9.942	9.935	9.928	9.921
0.60	0.078	8.828	8.826	8.823	8.817	8.811	8.806
0.65	0.055	0.309	7.832	7.829	7.824	7.819	7.815
0.70	0.040	0.216	0.491	6.951	6.945	6.941	6.937
0.75	0.029	0.156	0.340	6.173	6.167	6.162	6.158
0.80	0.022	0.116	0.246	0.578	5.475	5.468	5.462
0.85	0.017	0.088	0.183	0.408	4.853	4.841	4.832
0.90	0.013	0.068	0.140	0.301	0.744	4.269	4.249
0.93	0.011	0.058	0.120	0.254	0.593	1.219	3.914
0.95	0.010	0.053	0.109	0.228	0.517	0.961	3.697
0.97	0.010	0.048	0.099	0.206	0.456	0.797	1.570
0.98	0.009	0.046	0.094	0.196	0.429	0.734	1.270
0.99	0.009	0.044	0.090	0.186	0.405	0.680	1.098
1.00	0.008	0.042	0.086	0.177	0.382	0.632	0.977
1.01	0.008	0.040	0.082	0.169	0.361	0.590	0.883
1.02	0.008	0.039	0.078	0.161	0.342	0.552	0.807
1.05	0.007	0.034	0.069	0.140	0.292	0.460	0.642
1.10	0.005	0.028	0.055	0.112	0.229	0.350	0.470
1.15	0.005	0.023	0.045	0.091	0.183	0.275	0.361
1.20	0.004	0.019	0.037	0.075	0.149	0.220	0.286
1.30	0.003	0.013	0.026	0.052	0.102	0.148	0.190
1.40	0.002	0.010	0.019	0.037	0.072	0.104	0.133
1.50	0.001	0.007	0.014	0.027	0.053	0.076	0.097
1.60	0.001	0.005	0.011	0.021	0.040	0.057	0.073
1.70	0.001	0.004	0.008	0.016	0.031	0.044	0.056
1.80	0.001	0.003	0.006	0.013	0.024	0.035	0.044
1.90	0.001	0.003	0.005	0.010	0.019	0.028	0.036
2.00	0.000	0.002	0.004	0.008	0.016	0.023	0.029
2.20	0.000	0.001	0.003	0.006	0.011	0.016	0.021
2.40	0.000	0.001	0.002	0.004	0.008	0.012	0.015
2.60	0.000	0.001	0.002	0.003	0.006	0.009	0.012
2.80	0.000	0.001	0.001	0.003	0.005	0.008	0.010
3.00	0.000	0.001	0.001	0.002	0.004	0.006	0.008
3.50	0.000	0.000	0.001	0.001	0.003	0.004	0.006
4.00	0.000	0.000	0.001	0.001	0.002	0.003	0.005

TABLE A-29 *(Continued)*

	P_r							
T_r	**1.000**	**1.200**	**1.500**	**2.000**	**3.000**	**5.000**	**7.000**	**10.000**
0.30	16.586	16.547	16.488	16.390	16.195	15.837	15.468	14.925
0.35	15.278	15.251	15.211	15.144	15.011	14.751	14.496	14.153
0.40	13.896	13.877	13.849	13.803	13.714	13.541	13.376	13.144
0.45	12.496	12.482	12.462	12.430	12.367	12.248	12.145	11.999
0.50	11.153	11.143	11.129	11.107	11.063	10.985	10.920	10.836
0.55	9.914	9.907	9.897	9.882	9.853	9.806	9.769	9.732
0.60	8.799	8.794	8.787	8.777	8.760	8.736	8.723	8.720
0.65	7.810	7.807	7.801	7.794	7.784	7.779	7.785	7.811
0.70	6.933	6.930	6.926	6.922	6.919	6.929	6.952	7.002
0.75	6.155	6.152	6.149	6.147	6.149	6.174	6.213	6.285
0.80	5.458	5.455	5.453	5.452	5.461	5.501	5.555	5.648
0.85	4.826	4.822	4.820	4.822	4.839	4.898	4.969	5.082
0.90	4.238	4.232	4.230	4.236	4.267	4.351	4.442	4.578
0.93	3.894	3.885	3.884	3.896	3.941	4.046	4.151	4.300
0.95	3.658	3.647	3.648	3.669	3.728	3.851	3.966	4.125
0.97	3.406	3.391	3.401	3.437	3.517	3.661	3.788	3.957
0.98	3.264	3.247	3.268	3.318	3.412	3.569	3.701	3.875
0.99	3.093	3.082	3.126	3.195	3.306	3.477	3.616	3.796
1.00	2.399	2.868	2.967	3.067	3.200	3.387	3.532	3.717
1.01	1.306	2.513	2.784	2.933	3.094	3.297	3.450	3.640
1.02	1.113	1.655	2.557	2.790	2.986	3.209	3.369	3.565
1.05	0.820	0.831	1.443	2.283	2.655	2.949	3.134	3.348
1.10	0.577	0.640	0.618	1.241	2.067	2.534	2.767	3.013
1.15	0.437	0.489	0.502	0.654	1.471	2.138	2.428	2.708
1.20	0.343	0.385	0.412	0.447	0.991	1.767	2.115	2.430
1.30	0.226	0.254	0.282	0.300	0.481	1.147	1.569	1.944
1.40	0.158	0.178	0.200	0.220	0.290	0.730	1.138	1.544
1.50	0.115	0.130	0.147	0.166	0.206	0.479	0.823	1.222
1.60	0.086	0.098	0.112	0.129	0.159	0.334	0.604	0.969
1.70	0.067	0.076	0.087	0.102	0.127	0.248	0.456	0.775
1.80	0.053	0.060	0.070	0.083	0.105	0.195	0.355	0.628
1.90	0.043	0.049	0.057	0.069	0.089	0.160	0.286	0.518
2.00	0.035	0.040	0.048	0.058	0.077	0.136	0.238	0.434
2.20	0.025	0.029	0.035	0.043	0.060	0.105	0.178	0.322
2.40	0.019	0.022	0.027	0.034	0.048	0.086	0.143	0.254
2.60	0.015	0.018	0.021	0.028	0.041	0.074	0.120	0.210
2.80	0.012	0.014	0.018	0.023	0.035	0.065	0.104	0.188
3.00	0.010	0.012	0.015	0.020	0.031	0.058	0.093	0.158
3.50	0.007	0.009	0.011	0.015	0.024	0.046	0.073	0.122
4.00	0.006	0.007	0.009	0.012	0.020	0.038	0.060	0.100

Source: From Lee, B. I., and M. G. Kesler, "A Generalized Thermodynamic Correlation Based on Three-Parameter Corresponding States," *AIChE Journal,* **21**(3): 510–527, 1975. Reproduced by permission of the American Institute of Chemical Engineers.

TABLE A-30
Pitzer generalized saturation data

			Vaporization		Vapor		Liquid	
T_r	$-(\log P_r)^{(0)}$	$-\left(\frac{\partial \log P_r}{\partial \omega}\right)$	$\frac{\Delta s_v^{(0)}}{R}$	$\frac{\Delta s_v^{(1)}}{R}$	$Z^{(0)}$	$Z^{(1)}$	$Z^{(0)}$	$Z^{(1)}$
1.00	0.000	0.000	0.00	0.00	0.291	−0.080	0.291	−0.080
0.99	0.025	0.021	1.29	1.42	0.43	−0.030	0.202	−0.090
0.98	0.050	0.042	1.70	1.97	0.47	0.000	0.179	−0.093
0.97	0.076	0.064	2.01	2.38	0.51	+0.020	0.162	−0.095
0.96	0.102	0.086	2.28	2.71	0.54	0.035	0.148	−0.095
0.95	0.129	0.109	2.52	3.00	0.565	0.045	0.136	−0.095
0.94	0.156	0.133	2.74	3.28	0.59	0.055	0.125	−0.094
0.92	0.212	0.180	3.14	3.80	0.63	0.075	0.108	−0.092
0.90	0.270	0.230	3.50	4.29	0.67	0.095	0.0925	−0.087
0.88	0.330	0.285	3.82	4.73	0.70	0.110	0.0790	−0.080
0.86	0.391	0.345	4.12	5.18	0.73	0.125	0.0680	−0.075
0.84	0.455	0.405	4.42	5.64	0.756	0.135	0.0585	−0.068
0.82	0.522	0.475	4.72	6.09	0.781	0.140	0.0498	−0.062
0.80	0.592	0.545	5.02	6.54	0.804	0.144	0.0422	−0.057
0.78	0.665	0.620	5.32	7.00	0.826	0.144	0.0360	−0.053
0.76	0.742	0.705	5.64	7.50	0.846	0.142	0.0300	−0.048
0.74	0.823	0.800	5.96	8.05	0.864	0.137	0.0250	−0.043
0.72	0.909	0.895	6.29	8.56	0.881	0.131	0.0210	−0.037
0.70	1.000	1.00	6.64	9.11	0.897	0.122	0.0172	−0.032
0.68	1.096	1.12	6.99	9.72	0.911	0.113	0.0138	−0.027
0.66	1.198	1.25	7.36	10.3	0.922	0.104	0.0111	−0.022
0.64	1.308	1.39	7.73	11.0	0.932	0.097	0.0088	−0.018
0.62	1.426	1.54	8.11	11.7	0.940	0.090	0.0068	−0.015
0.60	1.552	1.70	8.52	12.4	0.947	0.083	0.0052	−0.012
0.58	1.688	1.88	8.93	13.2	0.953	0.077	0.0039	−0.009
0.56	1.834	2.08	9.38	14.0	0.959	0.070	0.0028	−0.007

Based on data appearing in Lewis, G. N., and M. Randall, *Thermodynamics*, 2nd ed., McGraw-Hill, Inc., New York, 1961. *Source:* K. S. Pitzer et al., *J. Am. Chem. Soc.*, **77**:3439, 1955.

TABLE A-31
Lee-Kesler fugacity coefficient values for $(\log_{10} f/P)^{(0)}$

T_r	P_r 0.010	0.050	0.100	0.200	0.400	0.600	0.800
0.30	−3.708	−4.402	−4.696	−4.985	−5.261	−5.412	−5.512
0.35	−2.471	−3.166	−3.461	−3.751	−4.029	−4.183	−4.285
0.40	−1.566	−2.261	−2.557	−2.848	−3.128	−3.283	−3.387
0.45	−0.879	−1.575	−1.871	−2.162	−2.444	−2.601	−2.707
0.50	−0.344	−1.040	−1.336	−1.628	−1.912	−2.070	−2.177
0.55	−0.008	−0.614	−0.911	−1.204	−1.488	−1.647	−1.755
0.60	−0.007	−0.269	−0.566	−0.859	−1.144	−1.304	−1.413
0.65	−0.005	−0.026	−0.283	−0.576	−0.862	−1.023	−1.132
0.70	−0.004	−0.021	−0.043	−0.341	−0.627	−0.789	−0.899
0.75	−0.003	−0.017	−0.035	−0.144	−0.430	−0.592	−0.703
0.80	−0.003	−0.014	−0.029	−0.059	−0.264	−0.426	−0.537
0.85	−0.002	−0.012	−0.024	−0.049	−0.123	−0.285	−0.396
0.90	−0.002	−0.010	−0.020	−0.041	−0.086	−0.166	−0.276
0.93	−0.002	−0.009	−0.018	−0.037	−0.077	−0.122	−0.214
0.95	−0.002	−0.008	−0.017	−0.035	−0.072	−0.113	−0.176
0.97	−0.002	−0.008	−0.016	−0.033	−0.067	−0.105	−0.148
0.98	−0.002	−0.008	−0.016	−0.032	−0.065	−0.101	−0.142
0.99	−0.001	−0.007	−0.015	−0.031	−0.063	−0.098	−0.137
1.00	−0.001	−0.007	−0.015	−0.030	−0.061	−0.095	−0.132
1.01	−0.001	−0.007	−0.014	−0.029	−0.059	−0.091	−0.127
1.02	−0.001	−0.007	−0.014	−0.028	−0.057	−0.088	−0.122
1.05	−0.001	−0.006	−0.013	−0.025	−0.052	−0.080	−0.110
1.10	−0.001	−0.005	−0.011	−0.022	−0.045	−0.069	−0.093
1.15	−0.001	−0.005	−0.009	−0.019	−0.039	−0.059	−0.080
1.20	−0.001	−0.004	−0.008	−0.017	−0.034	−0.051	−0.069
1.30	−0.001	−0.003	−0.006	−0.013	−0.026	−0.039	−0.052
1.40	−0.001	−0.003	−0.005	−0.010	−0.020	−0.030	−0.040
1.50	−0.000	−0.002	−0.004	−0.008	−0.016	−0.024	−0.032
1.60	−0.000	−0.002	−0.003	−0.006	−0.012	−0.019	−0.025
1.70	−0.000	−0.001	−0.002	−0.005	−0.010	−0.015	−0.020
1.80	−0.000	−0.001	−0.002	−0.004	−0.008	−0.012	−0.015
1.90	−0.000	−0.001	−0.002	−0.003	−0.006	−0.009	−0.012
2.00	−0.000	−0.001	−0.001	−0.002	−0.005	−0.007	−0.009
2.20	−0.000	−0.000	−0.001	−0.001	−0.003	−0.004	−0.005
2.40	−0.000	−0.000	−0.000	−0.001	−0.001	−0.002	−0.003
2.60	−0.000	−0.000	−0.000	−0.000	−0.000	−0.001	−0.001
2.80	0.000	0.000	0.000	0.000	0.000	0.000	0.001
3.00	0.000	0.000	0.000	0.000	0.001	0.001	0.002
3.50	0.000	0.000	0.000	0.001	0.001	0.002	0.003
4.00	0.000	0.000	0.000	0.001	0.002	0.003	0.004

TABLE A-31 (*Continued*)

	P_r							
T_r	1.000	1.200	1.500	2.000	3.000	5.000	7.000	10.000
0.30	−5.584	−5.638	−5.697	−5.759	−5.810	−5.782	−5.679	−5.461
0.35	−4.359	−4.416	−4.479	−4.547	−4.611	−4.608	−4.530	−4.352
0.40	−3.463	−3.522	−3.588	−3.661	−3.735	−3.752	−3.694	−3.545
0.45	−2.785	−2.845	−2.913	−2.990	−3.071	−3.104	−3.063	−2.938
0.50	−2.256	−2.317	−2.387	−2.468	−2.555	−2.601	−2.572	−2.468
0.55	−1.835	−1.897	−1.969	−2.052	−2.145	−2.201	−2.183	−2.096
0.60	−1.494	−1.557	−1.630	−1.715	−1.812	−1.878	−1.869	−1.795
0.65	−1.214	−1.278	−1.352	−1.439	−1.539	−1.612	−1.611	−1.549
0.70	−0.981	−1.045	−1.120	−1.208	−1.312	−1.391	−1.396	−1.344
0.75	−0.785	−0.850	−0.925	−1.015	−1.121	−1.204	−1.215	−1.172
0.80	−0.619	−0.685	−0.760	−0.851	−0.958	−1.046	−1.062	−1.026
0.85	−0.479	−0.544	−0.620	−0.711	−0.819	−0.911	−0.930	−0.901
0.90	−0.359	−0.424	−0.500	−0.591	−0.700	−0.794	−0.817	−0.793
0.93	−0.296	−0.361	−0.437	−0.527	−0.637	−0.732	−0.756	−0.735
0.95	−0.258	−0.322	−0.398	−0.488	−0.598	−0.693	−0.719	−0.699
0.97	−0.223	−0.287	−0.362	−0.452	−0.561	−0.657	−0.683	−0.665
0.98	−0.206	−0.270	−0.344	−0.434	−0.543	−0.639	−0.666	−0.649
0.99	−0.191	−0.254	−0.328	−0.417	−0.526	−0.622	−0.649	−0.633
1.00	−0.176	−0.238	−0.312	−0.401	−0.509	−0.605	−0.633	−0.617
1.01	−0.168	−0.224	−0.297	−0.385	−0.493	−0.589	−0.617	−0.602
1.02	−0.161	−0.210	−0.282	−0.370	−0.477	−0.573	−0.601	−0.588
1.05	−0.143	−0.180	−0.242	−0.327	−0.433	−0.529	−0.557	−0.546
1.10	−0.120	−0.148	−0.193	−0.267	−0.368	−0.462	−0.491	−0.482
1.15	−0.102	−0.125	−0.160	−0.220	−0.312	−0.403	−0.433	−0.426
1.20	−0.088	−0.106	−0.135	−0.184	−0.266	−0.352	−0.382	−0.377
1.30	−0.066	−0.080	−0.100	−0.134	−0.195	−0.269	−0.296	−0.293
1.40	−0.051	−0.061	−0.076	−0.101	−0.146	−0.205	−0.229	−0.226
1.50	−0.039	−0.047	−0.059	−0.077	−0.111	−0.157	−0.176	−0.173
1.60	−0.031	−0.037	−0.046	−0.060	−0.085	−0.120	−0.135	−0.129
1.70	−0.024	−0.029	−0.036	−0.046	−0.065	−0.092	−0.102	−0.094
1.80	−0.019	−0.023	−0.028	−0.036	−0.050	−0.069	−0.075	−0.066
1.90	−0.015	−0.018	−0.022	−0.028	−0.038	−0.052	−0.054	−0.043
2.00	−0.012	−0.014	−0.017	−0.021	−0.029	−0.037	−0.037	−0.024
2.20	−0.007	−0.008	−0.009	−0.012	−0.015	−0.017	−0.012	0.004
2.40	−0.003	−0.004	−0.004	−0.005	−0.006	−0.003	0.005	0.024
2.60	−0.001	−0.001	−0.001	−0.001	0.001	0.007	0.017	0.037
2.80	0.001	0.001	0.002	0.003	0.005	0.014	0.025	0.046
3.00	0.002	0.003	0.003	0.005	0.009	0.018	0.031	0.053
3.50	0.004	0.005	0.006	0.008	0.013	0.025	0.038	0.061
4.00	0.005	0.006	0.007	0.010	0.016	0.028	0.041	0.064

Source: From Lee, B. I., and M. G. Kesler, "A Generalized Thermodynamic Correlation Based on Three-Parameter Corresponding States," *AIChE Journal*, **21**(3): 510–527, 1975. Reproduced by permission of the American Institute of Chemical Engineers.

TABLE A-32
Lee-Kesler fugacity coefficient values for $(\log_{10} f/P)^{(1)}$

T_r	P_r 0.010	0.050	0.100	0.200	0.400	0.600	0.800
0.30	−8.778	−8.779	−8.781	−8.785	−8.790	−8.797	−8.804
0.35	−6.528	−6.530	−6.532	−6.536	−6.544	−6.551	−6.559
0.40	−4.912	−4.914	−4.916	−4.919	−4.929	−4.937	−4.945
0.45	−3.726	−3.728	−3.730	−3.734	−3.742	−3.750	−3.758
0.50	−2.838	−2.839	−2.841	−2.845	−2.853	−2.861	−2.869
0.55	−0.013	−2.163	−2.165	−2.169	−2.177	−2.184	−2.192
0.60	−0.009	−1.644	−1.646	−1.650	−1.657	−1.664	−1.671
0.65	−0.006	−0.031	−1.242	−1.245	−1.252	−1.258	−1.265
0.70	−0.004	−0.021	−0.044	−0.927	−0.934	−0.940	−0.946
0.75	−0.003	−0.014	−0.030	−0.675	−0.682	−0.688	−0.694
0.80	−0.002	−0.010	−0.020	−0.043	−0.481	−0.487	−0.493
0.85	−0.001	−0.006	−0.013	−0.028	−0.321	−0.327	−0.332
0.90	−0.001	−0.004	−0.009	−0.018	−0.039	−0.199	−0.204
0.93	−0.001	−0.003	−0.007	−0.013	−0.029	−0.048	−0.141
0.95	−0.001	−0.003	−0.005	−0.011	−0.023	−0.037	−0.103
0.97	−0.000	−0.002	−0.004	−0.009	−0.018	−0.029	−0.042
0.98	−0.000	−0.002	−0.004	−0.008	−0.016	−0.025	−0.035
0.99	−0.000	−0.002	−0.003	−0.007	−0.014	−0.021	−0.030
1.00	−0.000	−0.001	−0.003	−0.006	−0.012	−0.018	−0.025
1.01	−0.000	−0.001	−0.003	−0.005	−0.010	−0.016	−0.021
1.02	−0.000	−0.001	−0.002	−0.004	−0.009	−0.013	−0.017
1.05	−0.000	−0.001	−0.001	−0.002	−0.005	−0.006	−0.007
1.10	−0.000	−0.000	0.000	0.000	0.001	0.002	0.004
1.15	0.000	0.000	0.001	0.002	0.005	0.008	0.011
1.20	0.000	0.001	0.002	0.003	0.007	0.012	0.017
1.30	0.000	0.001	0.003	0.005	0.011	0.017	0.023
1.40	0.000	0.002	0.003	0.006	0.013	0.020	0.027
1.50	0.000	0.002	0.003	0.007	0.014	0.021	0.028
1.60	0.000	0.002	0.003	0.007	0.014	0.021	0.029
1.70	0.000	0.002	0.004	0.007	0.014	0.021	0.029
1.80	0.000	0.002	0.003	0.007	0.014	0.021	0.028
1.90	0.000	0.002	0.003	0.007	0.014	0.021	0.028
2.00	0.000	0.002	0.003	0.007	0.013	0.020	0.027
2.20	0.000	0.002	0.003	0.006	0.013	0.019	0.025
2.40	0.000	0.002	0.003	0.006	0.012	0.018	0.024
2.60	0.000	0.001	0.003	0.006	0.011	0.017	0.023
2.80	0.000	0.001	0.003	0.005	0.011	0.016	0.021
3.00	0.000	0.001	0.003	0.005	0.010	0.015	0.020
3.50	0.000	0.001	0.002	0.004	0.009	0.013	0.018
4.00	0.000	0.001	0.002	0.004	0.008	0.012	0.016

TABLE A-32 *(Continued)*

T_r	P_r 1.000	1.200	1.500	2.000	3.000	5.000	7.000	10.000
0.30	−8.811	−8.818	−8.828	−8.845	−8.880	−8.953	−9.022	−9.126
0.35	−6.567	−6.575	−6.587	−6.606	−6.645	−6.723	−6.800	−6.919
0.40	−4.954	−4.962	−4.974	−4.995	−5.035	−5.115	−5.195	−5.312
0.45	−3.766	−3.774	−3.786	−3.806	−3.845	−3.923	−4.001	−4.114
0.50	−2.877	−2.884	−2.896	−2.915	−2.953	−3.027	−3.101	−3.208
0.55	−2.199	−2.207	−2.218	−2.236	−2.273	−2.342	−2.410	−2.510
0.60	−1.677	−1.684	−1.695	−1.712	−1.747	−1.812	−1.875	−1.967
0.65	−1.271	−1.278	−1.287	−1.304	−1.336	−1.397	−1.456	−1.539
0.70	−0.952	−0.958	−0.967	−0.983	−1.013	−1.070	−1.124	−1.201
0.75	−0.700	−0.705	−0.714	−0.728	−0.756	−0.809	−0.858	−0.929
0.80	−0.499	−0.504	−0.512	−0.526	−0.551	−0.600	−0.645	−0.709
0.85	−0.338	−0.343	−0.351	−0.364	−0.388	−0.432	−0.473	−0.530
0.90	−0.210	−0.215	−0.222	−0.234	−0.256	−0.296	−0.333	−0.384
0.93	−0.146	−0.151	−0.158	−0.170	−0.190	−0.228	−0.262	−0.310
0.95	−0.108	−0.114	−0.121	−0.132	−0.151	−0.187	−0.220	−0.265
0.97	−0.075	−0.080	−0.087	−0.097	−0.116	−0.149	−0.180	−0.223
0.98	−0.059	−0.064	−0.071	−0.081	−0.099	−0.132	−0.162	−0.203
0.99	−0.044	−0.050	−0.056	−0.066	−0.084	−0.115	−0.144	−0.184
1.00	−0.031	−0.036	−0.042	−0.052	−0.068	−0.099	−0.127	−0.166
1.01	−0.024	−0.024	−0.030	−0.038	−0.054	−0.084	−0.111	−0.149
1.02	−0.019	−0.015	−0.018	−0.026	−0.041	−0.069	−0.095	−0.132
1.05	−0.007	−0.002	0.008	0.007	−0.005	−0.029	−0.052	−0.085
1.10	0.007	0.012	0.025	0.041	0.042	0.026	0.008	−0.019
1.15	0.016	0.022	0.034	0.056	0.074	0.069	0.057	0.036
1.20	0.023	0.029	0.041	0.064	0.093	0.102	0.096	0.081
1.30	0.030	0.038	0.049	0.071	0.109	0.142	0.150	0.148
1.40	0.034	0.041	0.053	0.074	0.112	0.161	0.181	0.191
1.50	0.036	0.043	0.055	0.074	0.112	0.167	0.197	0.218
1.60	0.036	0.043	0.055	0.074	0.110	0.167	0.204	0.234
1.70	0.036	0.043	0.054	0.072	0.107	0.165	0.205	0.242
1.80	0.035	0.042	0.053	0.070	0.104	0.161	0.203	0.246
1.90	0.034	0.041	0.052	0.068	0.101	0.157	0.200	0.246
2.00	0.034	0.040	0.050	0.066	0.097	0.152	0.196	0.244
2.20	0.032	0.038	0.047	0.062	0.091	0.143	0.186	0.236
2.40	0.030	0.036	0.044	0.058	0.086	0.134	0.176	0.227
2.60	0.028	0.034	0.042	0.055	0.080	0.127	0.167	0.217
2.80	0.027	0.032	0.039	0.052	0.076	0.120	0.158	0.208
3.00	0.025	0.030	0.037	0.049	0.072	0.114	0.151	0.199
3.50	0.022	0.026	0.033	0.043	0.063	0.101	0.134	0.179
4.00	0.020	0.023	0.029	0.038	0.057	0.090	0.121	0.163

Source: From Lee, B. I., and M. G. Kesler, "A Generalized Thermodynamic Correlation Based on Three-Parameter Corresponding States," *AIChE Journal,* **21**(3): 510–527, 1975. Reproduced by permission of the American Institute of Chemical Engineers.

TABLE A-33

Values of the enthalpy of formation, Gibbs function of formation, absolute entropy, and enthalpy of vaporization at 25°C and 1 atm

Δh_f^o, Δg_f^o, and h_{fg} in kJ/kgmol; s^o in kJ/(kgmol·K)

Substance	Formula	Δh_f^0	Δg_f^0	s^0	h_{fg}
Carbon	$C(s)$	0	0	5.74	
Hydrogen	$H_2(g)$	0	0	130.57	
Nitrogen	$N_2(g)$	0	0	191.50	
Oxygen	$O_2(g)$	0	0	205.04	
Carbon monoxide	$CO(g)$	−110,530	−137,150	197.56	
Carbon dioxide	$CO_2(g)$	−393,520	−394,380	213.67	
Water	$H_2O(g)$	−241,820	−228,590	188.72	
Water	$H_2O(l)$	−285,830	−237,180	69.95	44,010
Hydrogen peroxide	$H_2O_2(g)$	−136,310	−105,600	232.63	61,090
Ammonia	$NH_3(g)$	−46,190	−16,590	192.33	
Oxygen	$O(g)$	249,170	231,770	160.95	
Hydrogen	$H(g)$	218,000	203,290	114.61	
Nitrogen	$N(g)$	472,680	455,510	153.19	
Hydroxyl	$OH(g)$	39,040	34,280	183.75	
Methane	$CH_4(g)$	−74,850	−50,790	186.16	
Acetylene (Ethyne)	$C_2H_2(g)$	226,730	209,170	200.85	
Ethylene (Ethene)	$C_2H_4(g)$	52,280	68,120	219.83	
Ethane	$C_2H_6(g)$	−84,680	−32,890	229.49	
Propylene (Propene)	$C_3H_6(g)$	20,410	62,720	266.94	
Propane	$C_3H_8(g)$	−103,850	−23,490	269.91	15,060
n-Butane	$C_4H_{10}(g)$	−126,150	−15,710	310.03	21,060
n-Pentane	$C_5H_{12}(g)$	−146,440	−8,200	348.40	31,410
n-Octane	$C_8H_{18}(g)$	−208,450	17,320	463.67	41,460
n-Octane	$C_8H_{18}(l)$	−249,910	6,610	360.79	
Benzene	$C_6H_6(g)$	82,930	129,660	269.20	33,830
Methyl alcohol	$CH_3OH(g)$	−200,890	−162,140	239.70	37,900
Methyl alcohol	$CH_3OH(l)$	−238,810	−166,290	126.80	
Ethyl alcohol	$C_2H_5OH(g)$	−235,310	−168,570	282.59	42,340
Ethyl alcohol	$C_2H_5OH(l)$	−277,690	−174,890	160.70	
Mercury	$Hg(l)$	0	0	77.24	
Mecuric oxide	$HgO(c)$	−90,210	−58,400	70.45	
Manganese	$Mn(c)$	0	0	31.8	
Manganese dioxide	$MnO_2(c)$	−520,030	−465,180	53.14	
Manganese trioxide	$Mn_2O_3(c)$	−958,970	−881,150	110.5	
Lead	$Pb(c)$	0	0	64.81	
Lead oxide	$PbO_2(c)$	−277,400	−217,360	68.6	
Lead sulfate	$PbSO_4(c)$	−919,940	−813,200	148.57	
Zinc	$Zn(c)$	0	0	41.63	
Zinc oxide	$ZnO(c)$	−348,280	−318,320	43.64	
Sulfuric acid	$H_2SO_4(l)$	−813,990	−690,100	156.90	
Sulfuric acid	(aq, m = 1)	−909,270	−744,630	20.1	
Silver oxide	$Ag_2O(c)$	−31,050	−11,200	121.7	

Sources: From the JANAF Thermochemical Tables, Dow Chemical Co., 1971; *Selected Values of Chemical Thermodynamic Properties*, NBS Technical Note 270-3, 1968; and *API Research Project 44*, Carnegie Press, 1953.

TABLE A-34

The lower heating value (LHV), higher heating value (HHV), Gibbs function of reaction ($\Delta G_{R,0}$), and chemical availability ($\psi_{ch,f}$) of various pure fuels, in kJ/gmol, in the restricted dead state of T_0 = 25°C, P_0 = 1 atm, $\phi_{RH,0}$ = 100% and 60%

Fuel	LHV	HHV	$-\Delta G_{R,0}$	$\psi_{ch,f,100}$	$\psi_{ch,f,60}$
Hydrogen, H_2(g)	241.8	285.8	228.65	235.2	237.6
Carbon monoxide, CO(g)	283.0	283.0	257.3	275.4	275.4
Carbon, C(s)	393.5	393.5	394.4	410.5	410.2
Methane, CH_4(g)	802.3	890.3	800.8	829.8	832.4
Acetylene, C_2H_2(g)	1255.6	1299.6	1226.6	1264.8	1266.2
Ethylene, C_2H_4(g)	1323.0	1411.0	1314.1	1359.6	1362.2
Ethane, C_2H_6(g)	1427.8	1559.9	1441.7	1493.2	1497.1
Propylene, C_3H_6(g)	1926.4	2058.5	1931.7	1999.0	2002.9
Propane, C_3H_8(g)	2044.0	2220.0	2074.1	2148.0	2153.2
Butane, C_4H_{10}(g)	2657.0	2877.1	2704.9	2801.2	2807.7
n-Pentane, C_5H_{12}(g)	3272.1	3536.1	3335.4	3454.1	3461.9
n-Octane, C_8H_8(g)	5116.1	5512.2	5229.9	5415.9	5427.6
Benzene, C_6H_6(g)	3169.5	3301.5	3181.9	3296.6	3300.6
Methyl alcohol, CH_3OH(g)	676.3	764.3	689.5	720.5	723.0
Ethyl alcohol, C_2H_5OH(g)	1277.2	1409.2	1306.0	1359.4	1363.3

Property values based on data from Table A-33.

TABLE A-35
Logarithms to the base 10 of the equilibrium constant K_0

$$K_p = \frac{(p_E)^{\nu_E}(p_F)^{\nu_F}}{(p_A)^{\nu_A}(p_B)^{\nu_B}} \text{ for the reaction } \nu_A A + \nu_B B \leftrightharpoons \nu_E E + \nu_F F$$

Numbered reactions:

(1) $H_2 \leftrightharpoons 2H$
(2) $O_2 \leftrightharpoons 2O$
(3) $N_2 \leftrightharpoons 2N$
(4) $\frac{1}{2}O_2 + \frac{1}{2}N_2 \leftrightharpoons NO$
(5) $H_2O \leftrightharpoons H_2 + \frac{1}{2}O_2$
(6) $H_2O \leftrightharpoons OH + \frac{1}{2}H_2$
(7) $CO_2 \leftrightharpoons CO + \frac{1}{2}O_2$
(8) $CO_2 + H_2 \leftrightharpoons CO + H_2O$

Temp., K	log K_0 values for reactions numbered above							
	(1)	(2)	(3)	(4)	(5)	(6)	(7)	(8)
298	−71.224	−81.208	−159.600	−15.171	−40.048	−46.054	−45.066	−5.018
500	−40.316	−45.880	−92.672	−8.783	−22.886	−26.130	−25.025	−2.139
1000	−17.292	−19.614	−43.056	−4.062	−10.062	−11.280	−10.221	−0.159
1200	−13.414	−15.208	−34.754	−3.275	−7.899	−8.789	−7.764	+0.135
1400	−10.630	−12.054	−28.812	−2.712	−6.347	−7.003	−6.014	+0.333
1600	−8.532	−9.684	−24.350	−2.290	−5.180	−5.662	−4.706	+0.474
1700	−7.666	−8.706	−22.512	−2.116	−4.699	−5.109	−4.169	+0.530
1800	−6.896	−7.836	−20.874	−1.962	−4.270	−4.617	−3.693	+0.577
1900	−6.204	−7.058	−19.410	−1.823	−3.886	−4.177	−3.267	+0.619
2000	−5.580	−6.356	−18.092	−1.699	−3.540	−3.780	−2.884	+0.656
2100	−5.016	−5.720	−16.898	−1.586	−3.227	−3.422	−2.539	+0.688
2200	−4.502	−5.142	−15.810	−1.484	−2.942	−3.095	−2.226	+0.716
2300	−4.032	−4.614	−14.818	−1.391	−2.682	−2.798	−1.940	+0.742
2400	−3.600	−4.130	−13.908	−1.305	−2.443	−2.525	−1.679	+0.764
2500	−3.202	−3.684	−13.070	−1.227	−2.224	−2.274	−1.440	+0.784
2600	−2.836	−3.272	−12.298	−1.154	−2.021	−2.042	−1.219	+0.802
2700	−2.494	−2.892	−11.580	−1.087	−1.833	−1.828	−1.015	+0.818
2800	−2.178	−2.536	−10.914	−1.025	−1.658	−1.628	−0.825	+0.833
2900	−1.882	−2.206	−10.294	−0.967	−1.495	−1.442	−0.649	+0.846
3000	−1.606	−1.898	−9.716	−0.913	−1.343	−1.269	−0.485	+0.858
3100	−1.348	−1.610	−9.174	−0.863	−1.201	−1.107	−0.332	+0.869
3200	−1.106	−1.340	−8.664	−0.815	−1.067	−0.955	−0.189	+0.878
3300	−0.878	−1.086	−8.186	−0.771	−0.942	−0.813	−0.054	+0.888
3400	−0.664	−0.846	−7.736	−0.729	−0.824	−0.679	+0.071	+0.895
3500	−0.462	−0.620	−7.312	−0.690	−0.712	−0.552	+0.190	+0.902

Source: Based on data from the JANAF Thermochemical Tables, NSRDS-NBS-37, 1971, and revisions published in *Journal of Physical and Chemical Data* through 1982.

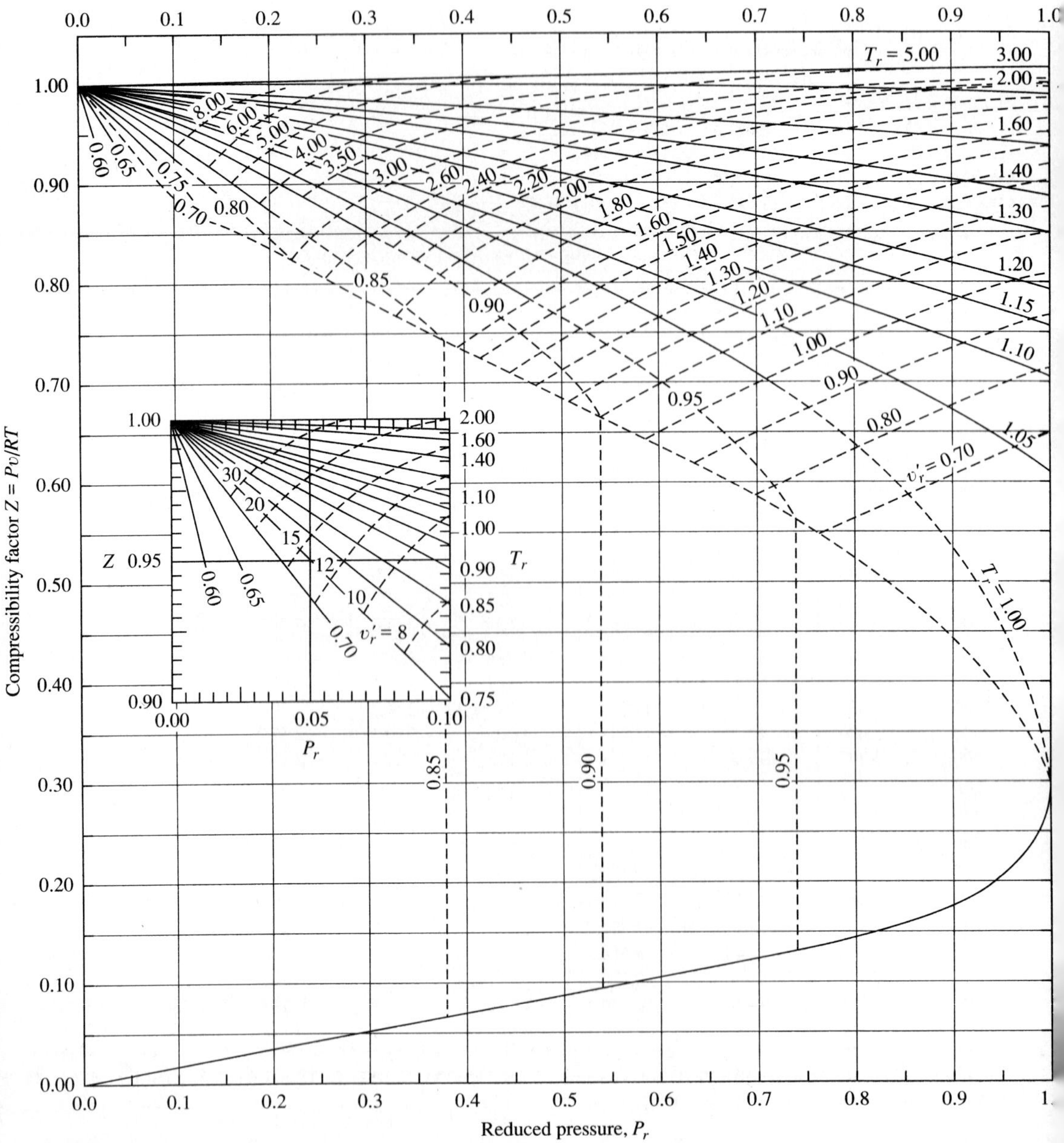

FIGURE A-36
Generalized compressibility chart, $P_r \leq 1$. [Modified by Peter E. Liley, *Chemical Engineering*, **94**, (1987). *Original source:* E. F. Obert, *Concepts of Thermodynamics*, McGraw-Hill, Inc., (1960)].

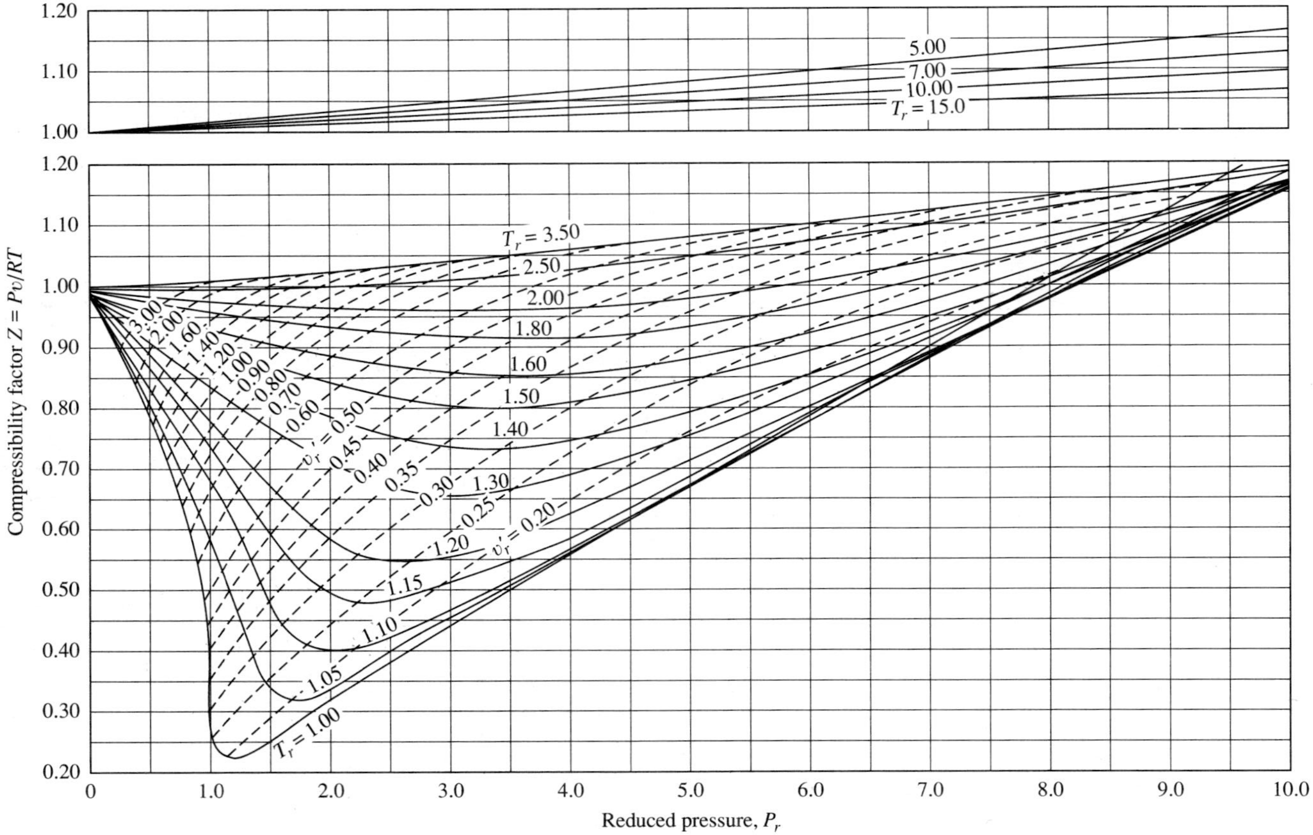

FIGURE A-37
Generalized compressibility chart, $P_r \leq 10$. (*Source:* E. F. Obert, *Concepts of Thermodynamics,* McGraw-Hill, Inc, 1960.

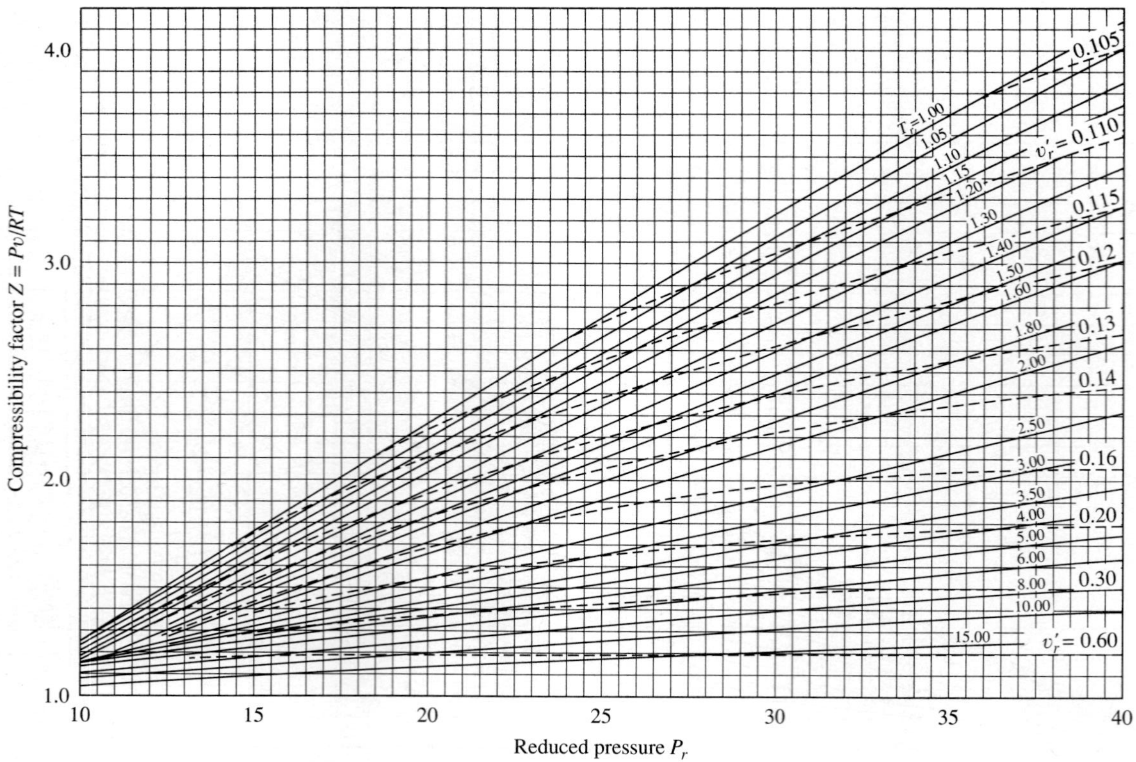

FIGURE A-38
Generalized compressibility chart, $10 \leq P_r \leq 40$. (*Source:* E. F. Obert, *Concepts of Thermodynamics*, McGraw-Hill, Inc., 1960.)

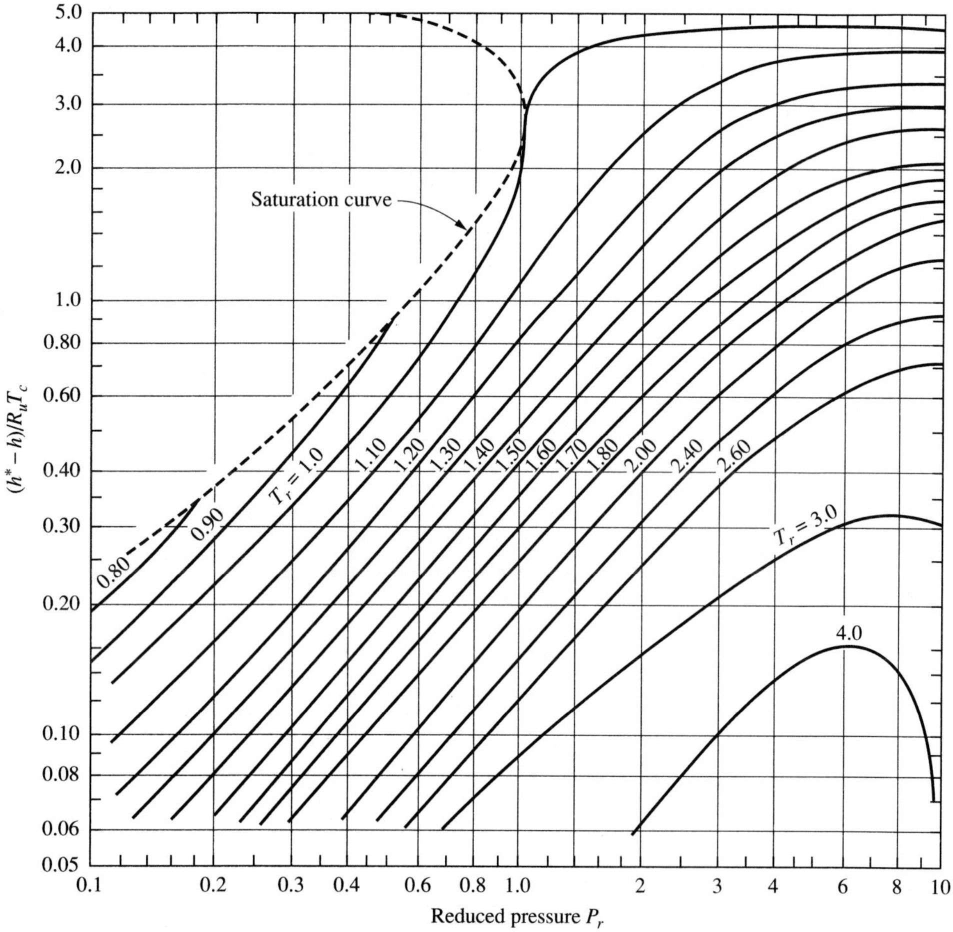

FIGURE A-39

Generalized residual enthalpy chart. (*Source:* K. Wark, *Thermodynamics,* 5th ed., McGraw-Hill, INc., New York, 1988. Based on data from A. L. Lydersen, R. A. Greenkorn, and O. A. Hougen, "Engineering Experiment Station Report No. 4," University of Wisconsin, 1955.). $h*$ = enthalpy of an ideal gas, h = enthalpy of an actual gas.

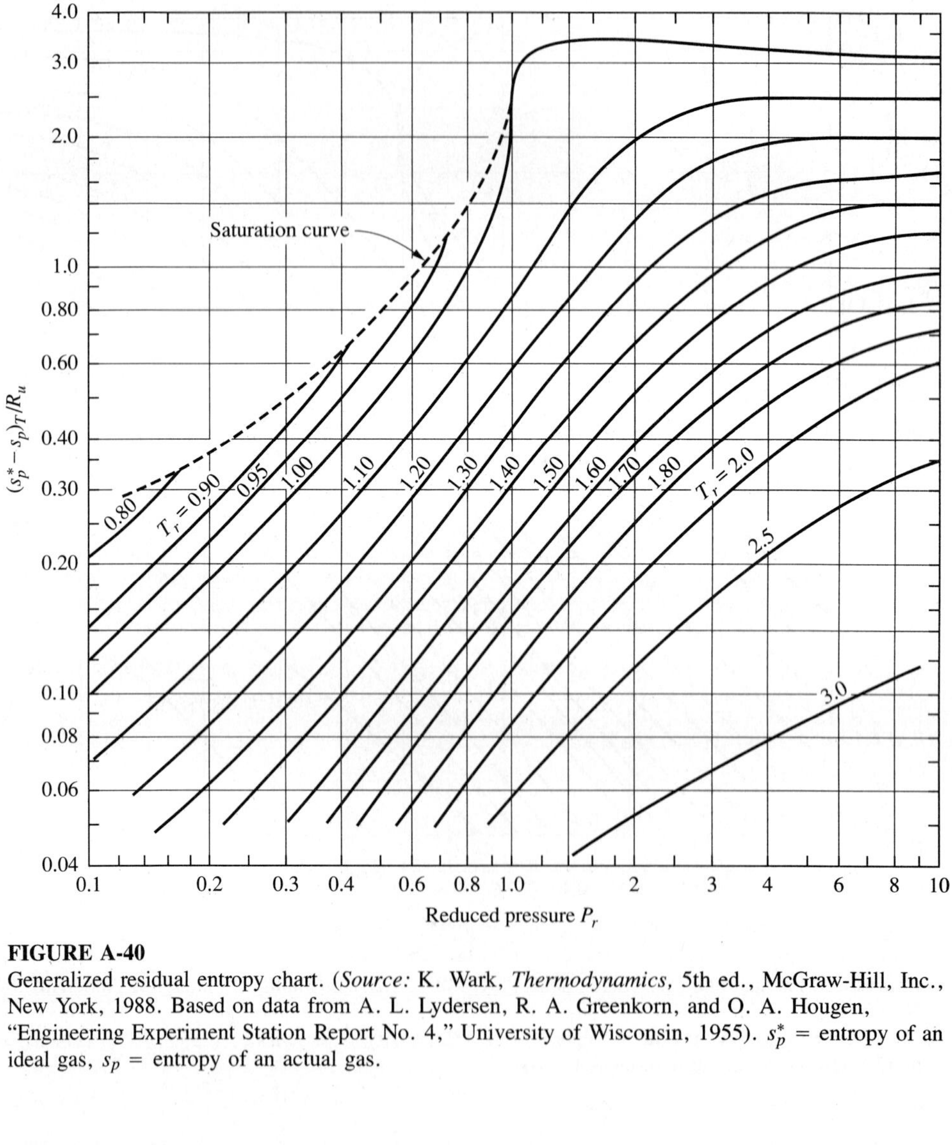

FIGURE A-40

Generalized residual entropy chart. (*Source:* K. Wark, *Thermodynamics,* 5th ed., McGraw-Hill, Inc., New York, 1988. Based on data from A. L. Lydersen, R. A. Greenkorn, and O. A. Hougen, "Engineering Experiment Station Report No. 4," University of Wisconsin, 1955). s_p^* = entropy of an ideal gas, s_p = entropy of an actual gas.

APPENDIX B

SYMBOLS

A	Area; Helmholtz function, $A = U - TS$
$\mathbf{A}$	Affinity
a	Specific Helmholtz function, $a = u - Ts$
B	Virial coefficient; Darrieus function
b	Specific Darrieus function, $b = h + ke + pe - T_0 s$; virial coefficient
C	Number of components (in Gibbs' phase rule); a constant
COP	Coefficient of performance
CV	Control volume
c	Virial coefficient
c_v	Specific heat at constant volume $(\partial u/\partial T)_v$
c_p	Specific heat at constant pressure $(\partial h/\partial T)_p$
d	An infinitesimal increase in a state function
E	Stored energy
$\mathbf{E}$	Electric field strength
e	Specific stored energy
F	Force; variance or degrees of freedom (in phase rule)
F_k	Measurable generalized force
$F_{k,\mathrm{eq}}$	Equilibrium generalized force
f	Fugacity
f_i	Fugacity of component i in a mixture
G	Gibbs function, $G = H - TS$
ΔG	Gibbs-function change for a unit reaction

ΔG_T^o	Standard-state Gibbs-function change for a unit reaction
g	Local acceleration of gravity; specific Gibbs function, $g = h - Ts$
H	Enthalpy, $H = U + PV$
HHV	Higher heating value
$\mathbf{H}$	Magnetic field strength
Δh_R	Enthalpy of reaction
Δh_f	Enthalpy of formation
Δh_c	Enthalpy of combustion
h	Specific enthalpy, $h = u + Pv$; Planck's constant
h^*	$h^* = h + ke + pe$
I	Irreversibility
I_Q	Irreversibility due to heat transfer
i	Specific irreversibility
K_T	Isothermal coefficient of compressibility
$K_{\bar{f}}$	Equilibrium constant based on component fugacities (real gas)
K_o	Equilibrium constant based on ΔG_T^o
K_y	Equilibrium constant based on mole fractions
K_ϕ	Equilibrium constant based on fugacity coefficients
K_p	Equilibrium constant based on component pressures (ideal gas)
k	Specific-heat ratio c_p/c_v; Boltzmann's constant
k_s	Spring constant
ke	Specific kinetic energy
L	Length
LLV	Lower heating value
M	Molar mass, or molecular weight
$\mathbf{M}$	Magnetization per unit volume
m	Mass of a substance
$\dot{m}$	Mass flow rate
N	Number of moles
N_A	Avogadro's number
N_e	Moles of electrons
n	Number of particles; polytropic constant
P	Pressure; number of phases (in the phase rule)
$\mathbf{P}$	Polarization
P_r	Reduced pressure
p_r	Relative pressure in air table
p_i	Component pressure
pe	Specific potential energy

Q	Heat interaction
Q_e	Electric charge
q	Heat interaction per unit mass
q_f	Frictional heating
R	Specific-gas constant, R_u/M
R_u	Universal gas constant
r	Distance between masses; radius of curvature
r_p	Pressure ratio
S	Entropy
s	Specific entropy; distance
T	Temperature
T_{af}	Adiabatic flame temperature
T_b	Normal boiling point
T_R	Temperature of a thermal reservoir
T_r	Reduced temperature
t	Time
U	Internal energy
u	Specific internal energy
V	Volume; velocity; electrostatic potential
V_i	Ideal voltage
v	Specific volume
v_r	Relative volume in air table
W	Work interaction; thermodynamic probability
w	Work interaction per unit mass
X	Mass fraction
X_k	Generalized displacement
x	Quality; mole fraction; cartesian coordinate
y	Mole fraction in the vapor phase; cartesian coordinate
$\bar{y}_i$	Partial molar property
Z	Compressibility factor, $Z = Pv/(RT)$
Z_c	Critical compressibility factor, $Z_c = P_c v_c/RT_c$
z	Height; cartesian coordinate

B-1 GREEK SYMBOLS AND SPECIAL NOTATION

β	Volumetric expansion coefficient
γ	Surface tension
Δ	A finite increase in a property

δ	An infinitesimal increase in a path function
ϵ	Energy of a particle; second-law effectiveness; strain
ϵ_F	Fermi energy of a particle
ϵ_0	Permittivity
η	Efficiency
η_{II}	Second-law efficiency
$\mathcal{F}$	Faraday constant
Θ_{D}	Debye characteristic temperature
Θ_{E}	Einstein characteristic temperature
μ_{JT}	Joule-Thomson coefficient
μ_i	Chemical potential of the ith component
ν	Fundamental frequency of an oscillator; stoichiometric coefficient
$\Delta\nu$	Change in the stoichiometric coefficients for a unit reaction
ξ	Extent of reaction
Π	Osmotic pressure
ρ	Density
σ	Entropy production; stress
σ_Q	Entropy production associated with heat transfer
σ_m	Specific entropy production
τ	Time; torque
Φ	Extensive closed-system availability (exergy)
Φ_Q	Availability transfer due to heat transfer
$\Phi_{Q,R}$	Availability transfer to or from a heat reservoir at T_R
$\Phi_{Q,i}$	Availability transfer associated with Q_i at T_i
$\Phi_{Q,j}$	Availability transfer to or from a heat reservoir at T_j
Φ_{ch}	Closed-system chemical availability
Φ_{tm}	Closed-system thermomechanical availability
Φ_{tot}	Sum of thermomechanical and chemical availability
ϕ	Closed-system availability per unit mass; relative humidity
$\overline{\phi}_i$	Fugacity coefficient of ith component
ψ	Stream availability per unit mass
ψ_{ch}	Chemical stream availability per unit mass
$\psi_{\mathrm{ch},f}$	Chemical stream availability of a fuel
ψ_{tm}	Thermomechanical stream availability per unit mass
ψ_{tot}	Sum of thermomechanical and chemical stream availability
ω	Acentric factor; specific humidity, humidity ratio
$\sum_i x_i$	The sum $x_i + x_2 + \cdots + x_n$
$\Pi_i x_i$	The product $x_1 x_2 x_3 \cdots x_n$

$\equiv$	Identity symbol
$\oint$	Closed-line integral

B-2 SUBSCRIPTS

a	End state of actual process
CV	Control volume
cw	Cooling water
c	Critical state
e	Exit state value
f	Final state value; saturated-liquid value
fg	Change in value between saturated-liquid and saturated-vapor phases
g	Saturated-vapor value
H	High temperature (as in T_H and Q_H)
HE	Heat engine
HP	Heat pump
i	Inlet state value
L	Low temperature (as in T_L and Q_L)
m	Mixture value
R	Reservoir value
r	Reduced state value; relative state value
s	End state of isentropic process
tp	Triple point
v	Vapor state
x	Mole fraction; quality, fraction vapor in liquid-vapor mixture
y	Mole fraction in vapor phase
0	Thermochemical (restricted) dead state
00	Environmental (unrestricted) dead state

INDEX